Endosymbiotic Organelle Acquisition

Steven D. Schwartzbach
Peter G. Kroth • Miroslav Oborník
Editors

Endosymbiotic Organelle Acquisition

Solutions to the Problem of Protein Localization and Membrane Passage

Springer

Editors
Steven D. Schwartzbach
Department of Biological Sciences
University of Memphis
Memphis, TN, USA

Peter G. Kroth
Department of Biology
Universität Konstanz
Konstanz, Germany

Miroslav Oborník
Institute of Parasitology
Biology Centre
Czech Academy of Sciences
České Budějovice, Czech Republic

ISBN 978-3-031-57444-3 ISBN 978-3-031-57446-7 (eBook)
https://doi.org/10.1007/978-3-031-57446-7

This Springer imprint is published by the registered company Springer Nature Switzerland AG
The registered company address is: Gewerbestrasse 11, 6330 Cham, Switzerland

This volume is dedicated to Elaine Schwartzbach, beloved wife, mother, grandmother and a friend to all. Her memory is a blessing.

Preface

The acquisition of mitochondria and chloroplasts, the respective sites of aerobic respiration and photosynthesis, was a pivotal event in the evolution of eukaryotic organisms. It is generally accepted that mitochondria and chloroplasts evolved through an endosymbiotic relationship between a primitive phagocytic host cell and a bacterium. While the principle combination of host cell and endosymbiont for the origin of mitochondria, e.g., whether the endosymbiont was anaerobic or aerobic, is still under debate, the origin of the first plastids involved a eukaryotic host cells and a cyanobacterium. In both cases, the host cells benefitted from the metabolic activities of the endosymbiont, while the endosymbiont benefitted from the nutrients and stable environment provided by the host. The initial endosymbiont evolved into a subcellular organelle through the massive transfer of genes from the endosymbiont genome to the host genome. Accordingly, gene transfer to the host nucleus resulted in reduced organelle genomes. Conversely, the loss of host genes for duplicated functions resulted in the host being unable to survive independently of the organelle and requiring the organelle to maintain biochemical functionality in the host.

Proteins encoded by endosymbiont genes transferred to the host nucleus were translated in the cytoplasm. Thus, the evolution of a mechanism for transporting the proteins into the endosymbiont-derived organelle was essential for the success of the endosymbiotic partnership. Mitochondria and primary chloroplasts are surrounded by two membranes and specific transport systems evolved for direct post-translational cytoplasm to organelle transport. In the case of chloroplasts, additional transport pathways evolved for targeting proteins further to the thylakoid.

Complex plastids that evolved by secondary or higher endosymbioses are enclosed by two to four membranes. It is thought that complex plastids arose through an endosymbiotic association between a eukaryotic host and a eukaryotic photosynthetic endosymbiont. Genes needed to be transferred from both the endosymbiont nucleus and the organelle genome to the host nucleus. For some complex plastids, proteins are co-translationally transported across the ER membrane, and vesicular transport is used to deliver proteins to the outermost plastid membrane. In other complex plastids, the outer plastid membrane is continuous with the ER so that translocation across the ER represents protein translocation across the outermost plastid membrane. There are in fact some extant organisms where loss of the endosymbiont nucleus is not complete but rather a reduced endosymbiont genome, the nucleomorph, is found within the periplastidic space; the region between the outer two and inner two plastid membranes. For organisms possessing a nucleomorph, a more complex targeting transport system was required. Nuclear-encoded proteins needed to be targeted and transported from the cytoplasm to the plastid and from the cytoplasm to the periplastidic space, while nucleomorph-encoded proteins needed to be targeted and transported solely from the periplastidic space to the plastid. Homologues of some of the proteins found in the translocation complex of simple chloroplasts, in the ER translocon, and in the ERAD system for retrotranslocation of proteins out of the ER to the cytoplasm are found in the membranes surrounding complex plastids. Proteins required for membrane localization of bacterial proteins have also been identified in thylakoids. Modification of host protein targeting membrane translocation systems was clearly part of the evolutionary process that produced the protein import membrane translocation machinery currently found in mitochondria, simple and complex plastids and mitochondria-related organelles.

To reconstruct the events leading to the evolution of endosymbionts into mitochondria, primary chloroplasts, and complex plastids thus require the integration of diverse aspects of host and endosymbiont cell biology that include: phylogenetics, mechanisms for genome reduction and gene transfer between genomes, phagocytosis, vacuole protein targeting, protein translocation into the ER, the ERAD system utilized for retrotranslocation of proteins from the ER, vesicular transport between endomembrane compartments, protein insertion into the bacterial membrane, mechanisms of protein targeting to mitochondria, primary chloroplasts, and complex chloroplasts. Reviews of our current knowledge regarding these topics are plentiful but scattered throughout the bacterial, parasite, plant, and animal literature. This volume will provide in one place both our current understanding of the protein targeting translocation processes, gene transfer processes, and genome reduction processes in the host and in the endosymbiont, which were likely utilized during the evolution of an endosymbiont into an organelle as well as the resultant evolutionary outcomes of these modifications used by mitochondria, mitochondria-related organelles, and primary and complex chloroplasts for protein targeting and translocation.

Memphis, TN, USA — Steven D. Schwartzbach
Konstanz, Germany — Peter G. Kroth
České Budějovice, Czech Republic — Miroslav Oborník

Contents

Contributors

Sébastien Besteiro LPHI, Univ Montpellier, CNRS, Montpellier, France

Debashish Bhattacharya Department of Biochemistry and Microbiology, Rutgers University, New Brunswick, NJ, USA

Rutgers, The State University of New Jersey, Department of Biochemistry and Microbiology, New Brunswick, NJ, USA

Fabien Burki Department of Organismal Biology, Program in Systematic Biology, Uppsala University, Uppsala, Sweden

Victoria Calatrava The Carnegie Institution for Science, Department of Plant Biology, Stanford, CA, USA

Gregor Christa Institute for Zoology and Didactics of Biology, University of Wuppertal, Wuppertal, Germany

Jenny Melo Clavijo Institute for Zoology and Didactics of Biology, University of Wuppertal, Wuppertal, Germany

Ross E. Dalbey Department of Chemistry and Biochemistry, The Ohio State University, Columbus, OH, USA

Richard G. Dorrell Sorbonne Université, Paris, France

Kai Dünser Faculty of Biology, Chair of Molecular Plant Physiology (MoPP), University of Freiburg, Freiburg, Germany

Center for Integrative Biological Signalling Studies (CIBSS), University of Freiburg, Freiburg, Germany

Institute of Plant Biotechnology and Cell Biology, University of Natural Resources and Life Sciences, Wien, Austria

Dion G. Durnford Department of Biology, University of New Brunswick, Fredericton, NB, Canada

Zoltán Füssy Faculty of Science, BIOCEV, Charles University, Prague, Czech Republic

Faculty of Science, University of South Bohemia, České Budějovice, Czech Republic

Scripps Institution of Oceanography, University of California San Diego, La Jolla, CA, United States of America

Sven B. Gould Institute for Molecular Evolution, HHU Düsseldorf, Düsseldorf, Germany

Arthur R. Grossman The Carnegie Institution for Science, Department of Plant Biology, Stanford, CA, USA

Ansgar Gruber Institute of Parasitology, Biology Centre, Czech Academy of Sciences, České Budějovice, Czech Republic

Faculty of Science, University of South Bohemia, České Budějovice, Czech Republic

Vladimír Hampl Department of Parasitology, Faculty of Science, Charles University, BIOCEV, Vestec, Czech Republic

Elisabeth Hehenberger Institute of Parasitology, Biology Centre, CAS, České Budějovice, Czech Republic

Takeo Horiguchi Department of Biological Sciences, Hokkaido University, Sapporo, Japan

Anna Karnkowska Institute of Evolutionary Biology, University of Warsaw, Warsaw, Poland

Philip Kauffman Department of Chemistry and Biochemistry, The Ohio State University, Columbus, OH, USA

Sharbani Kaushik Department of Chemistry and Biochemistry, The Ohio State University, Columbus, OH, USA

Department of Chemical Engineering at Maharashtra Institute of Technology, World Peace University, Pune, Maharashtra, India

Felix Kessler Laboratoire de Physiologie Végétale, Université de Neuchâtel, Neuchâtel, Switzerland

Jürgen Kleine-Vehn Faculty of Biology, Chair of Molecular Plant Physiology (MoPP), University of Freiburg, Freiburg, Germany

Center for Integrative Biological Signalling Studies (CIBSS), University of Freiburg, Freiburg, Germany

Martin Kolisko Institute of Parasitology, Biology Centre, Czech Academy of Sciences, České Budějovice, Czech Republic

Anežka Konupková Faculty of Science, BIOCEV, Charles University, Prague, Czech Republic

Peter G. Kroth Department of Biology, Universität Konstanz, Konstanz, Germany

Andreas Kuhn Institute of Biology, University of Hohenheim, Stuttgart, Germany

William H. Lewis Department of Biochemistry, University of Cambridge, Cambridge, UK

Duckhyun Lhee Department of Biological Sciences, Sungkyunkwan University, Suwon, South Korea

Kacper Maciszewski Institute of Parasitology, Biology Centre, Czech Academy of Sciences, České Budějovice, Czech Republic

Anna M. G. Novák Vanclová Institute Jacques Monod, Paris, France

Miroslav Oborník Institute of Parasitology, Biology Centre, Czech Academy of Sciences, České Budějovice, Czech Republic

Faculty of Science, University of South Bohemia, České Budějovice, Czech Republic

Eléa A. Renaud LPHI, Univ Montpellier, CNRS, Montpellier, France

Andrew J. Roger Institute for Comparative Genomics, Department of Biochemistry and Molecular Biology, Dalhousie University, Halifax, NS, Canada

Ann-Kathrin Rößling Faculty of Biology, Chair of Molecular Plant Physiology (MoPP), University of Freiburg, Freiburg, Germany

Syrian G. Sanchez LPHI, Univ Montpellier, CNRS, Montpellier, France

Steven D. Schwartzbach Department of Biological Sciences, University of Memphis, Memphis, TN, USA

Venkatasalam Shanmugabalaji Laboratoire de Physiologie Végétale, Université de Neuchâtel, Neuchâtel, Switzerland

Timothy G. Stephens Rutgers, The State University of New Jersey, Department of Biochemistry and Microbiology, New Brunswick, NJ, USA

Lucia Tomečková Faculty of Science, BIOCEV, Charles University, Prague, Czech Republic

Ross F. Waller Department of Biochemistry, University of Cambridge, Cambridge, UK

Norico Yamada Department of Biology, University of Konstanz, Konstanz, Germany

Hwan Su Yoon Department of Biological Sciences, Sungkyunkwan University, Suwon, South Korea

Kristína Záhonová Faculty of Science, BIOCEV, Charles University, Prague, Czech Republic

Institute of Parasitology, Biology Centre, Czech Academy of Sciences, České Budějovice, Czech Republic

Life Science Research Centre, Department of Biology and Ecology, Faculty of Science, University of Ostrava, Ostrava, Czech Republic

Division of Infectious Diseases, Department of Medicine, Faculty of Medicine and Dentistry, University of Alberta, Edmonton, AB, Canada

Julie A. Z. Zedler Synthetic Biology of Photosynthetic Organisms, Matthias Schleiden Institute for Genetics, Bioinformatics and Molecular Botany, Friedrich Schiller University Jena, Jena, Germany

Part I
Endosymbiotic Organelle Evolution

The Evolutionary Origin of Primary Plastids

Duckhyun Lhee, Debashish Bhattacharya, and Hwan Su Yoon

Abstract This chapter delves into primary endosymbiosis, which gave rise to algal and plant photosynthetic organelles. Primary plastids originated from the integration of a cyanobacterial endosymbiont into a host cell. The process of primary endosymbiosis raises fundamental questions about when, where, who, what, why, and how this intricate relationship evolved. By focusing on the only two known instances of primary plastid origin—those found in the ancestor of Archaeplastida and photosynthetic *Paulinella*—this chapter aims to provide the reader with a comprehensive understanding of the evolution of primary plastids. We explore the similarities and differences between these two cases, shedding light on the pathways that led to the establishment of photosynthetic capacity in eukaryotes.

Keywords Primary plastids · Origin of plastids · Endosymbiosis · Organellogenesis · Photosynthesis · Gene transfer · *Paulinella* · Archaeplastida · Cyanobacterium · Chlamydiales

1 Introduction

Primary plastids are photosynthetic organelles that have evolved through primary endosymbiosis, whereby a permanent cyanobacterial endosymbiont is transformed into an organelle [1]. There are two known cases of primary plastid origin. One putatively occurred in the common ancestor of Archaeplastida (i.e., green algae/land plants, red algae, and glaucophytes). Although multiple origins of plastids in

D. Lhee · H. S. Yoon (✉)
Department of Biological Sciences, Sungkyunkwan University, Suwon, South Korea
e-mail: hsyoon2011@skku.edu

D. Bhattacharya
Department of Biochemistry and Microbiology, Rutgers University,
New Brunswick, NJ, USA
e-mail: dbhattac@rutgers.edu

S. D. Schwartzbach et al. (eds.), *Endosymbiotic Organelle Acquisition*,
https://doi.org/10.1007/978-3-031-57446-7_1

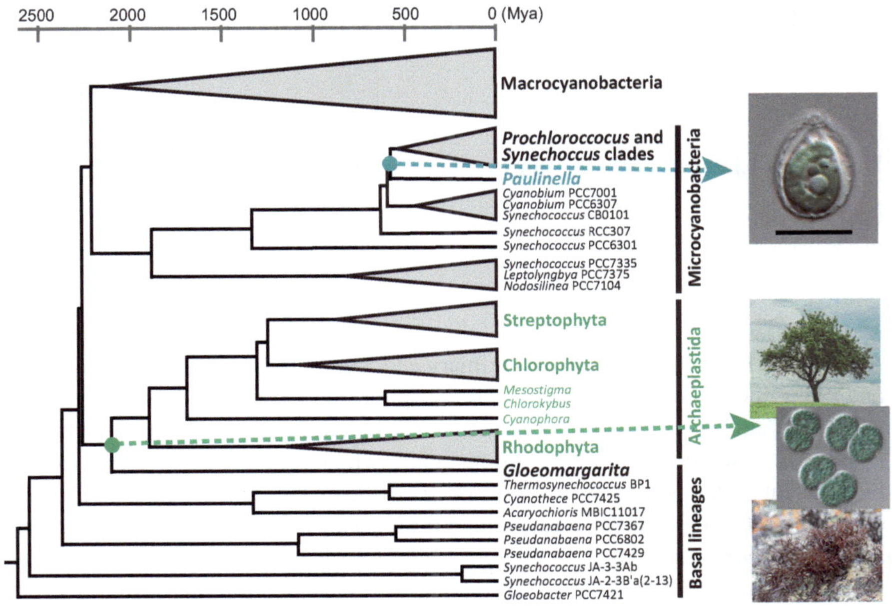

Fig. 1 Independent primary plastid endosymbiosis in Archaeplastida and photosynthetic *Paulinella*. The circles on the branches indicate the cyanobacterial ancestors of plastids. (Modified from Sánchez-Baracaldo et al. [6])

the Archaeplastida have been proposed [2] and will be discussed below, we will assume the broadly accepted single primary endosymbiosis hypothesis to simplify the initial discussion. The other case of primary plastid origin occurred in the ancestor of photosynthetic *Paulinella* species, the filose, testate amoebae within Cercozoa, Rhizaria [3, 4] (Fig. 1). Primary plastid endosymbiosis in Archaeplastida occurred 1.5–2.1 billion years ago (bya) [5, 6], and this organelle was transferred multiple times thereafter into other eukaryotic lineages through secondary and additional rounds (complex) of endosymbiosis. These complex endosymbioses resulted in the emergence of a diverse array of photosynthetic eukaryotes, including diatoms and dinoflagellates that are significant primary producers in many ecosystems [7, 8]. The union of a prokaryotic endosymbiont with a host cell, as in the case of mitochondria and plastids, is one of the most transformative forces in the evolution of eukaryotes. Therefore, understanding the mechanisms that underlie primary endosymbiosis is important. In addition, studying the evolution of primary plastids can provide useful information for engineering synthetic or genetically manipulated photosynthetic organelles for applied uses. However, based on studies of primary plastids that evolved more than 1.5 billion years ago, it is difficult to investigate early events in the formation of primary plastids.

In this regard, photosynthetic *Paulinella* provides a unique model for studying the early stages of primary plastids, because this event occurred relatively recently in this lineage, 90–600 million years ago (mya) [6, 9, 10]. The current data strongly support the scenario that a heterotrophic ancestor of photosynthetic *Paulinella* engulfed a cyanobacterial cell that evolved into the photosynthetic organelle termed the chromatophore [4, 11, 12]. After plastid acquisition, photosynthetic *Paulinella* diverged into different species; currently, *P. chromatophora*, *P. micropora*, and *P. longichromatophora* [11, 13, 14]. In this chapter, both the terms 'chromatophore' and 'plastid' will refer to the photosynthetic organelle in *Paulinella*.

If plastids are derived from cyanobacteria, how do they differ from cyanobacterial endosymbionts found in many organisms [15]? Several criteria have been commonly used to distinguish between organelles and endosymbionts [16, 17]. These include genetic integration (gene transfer from the endosymbiont to the host nucleus and nuclear-encoded proteins targeted to the endosymbiont), cellular integration (synchronization of cell cycles and vertical organelle inheritance), and metabolic integration (interdependence of metabolism). However, it is difficult to clearly distinguish between endosymbionts and organelles [17, 18], as in cases where host--encoded proteins are targeted to endosymbionts [19–22]. This has led to discussion that plastids and mitochondria may be regarded as bacteria [23], or that sexual symbiotic integration should be considered [24]. The distinction between endosymbionts and organelles is an additional point of discussion in this chapter, which can be addressed with Archaeplastida primary plastids as well as the plastid in *Paulinella* [4, 25, 26]. We will provide a comprehensive description of the evolution of primary plastids by comparing the known in Archaeplastida and photosynthetic *Paulinella*.

2 Sources of Primary Plastids

The endosymbiotic theory proposes that plastids and mitochondria were derived from prokaryotes that became organelles in eukaryotic (or proto-eukaryotic, in the case of mitochondria) cells [27]. Influenced by A. Schimper's notes in 1883, C. Mereschkowsky first advocated this theory in 1905, which was later popularized and more strongly established by the work of Lynn Margulis in 1967 [28–31]. For a history of the endosymbiotic theory, readers are referred to the following articles [32–37]. In the case of plastid formation, a great deal of evidence has accumulated in support of the endosymbiotic theory, but there exist alternative hypotheses (e.g., host eukaryote or noncyanobacterial bacterial origin). In the following sections of this review, we will summarize the evidence for these theories.

2.1 Evidence Supporting a Cyanobacterial Origin of Plastids

Plastids in Archaeplastida

The endosymbiotic theory is supported by several features (i.e., cell substance, structure, functions, phylogeny) that are shared by free-living cyanobacteria and plastids in Archaeplastida (Table 1). Both these lineages carry out oxygenic photosynthesis using two photosystems (PSI and PSII). In contrast, photosynthetic

Table 1 Similarities of cyanobacteria and plastids of Archaeplastida and chromatophore of *Paulinella*

Item	Cyanobacteria	Plastid of Archaeplastida	Chromatophore of *Paulinella*
Energy source	Oxygenic photosynthesis	Oxygenic photosynthesis	Oxygenic photosynthesis
Chlorophyll *a*	Present	Present	Present
Chlorophyll *b*	Present in *Prochlorococcus*	Present in green algae/ plants	Unknown
Phycobilisomes	Present except in *Prochlorococcus*	Present in red algae and glaucophytes	Genes are present
Envelope membrane	Outer and inner membranes	Outer and inner membranes	Outer and inner membranes
Beta barrel protein	Present in the outer membrane	Present in the outer membrane	Not identified
Peptidoglycan	Thick and conspicuous	Conspicuous in glaucophytes. Thin in mosses and ferns	Present
Thylakoid membrane	Present except in *Gloeobacter*	Present	Present
Glycolipids	MGDG, DGDG	MGDG, DGDG	MGDG and DGDG were found in the whole lysate
Reproduction	Binary fission	Binary fission	Binary fission
Protein synthesis	Prokaryotic	Prokaryotic	Prokaryotic
DNA	Present	Present	Present
Ribosomal RNA	Prokaryotic	Prokaryotic	Prokaryotic

Modified from Sato [37]

bacteria other than cyanobacteria contain a single photosystem [38]. Cyanobacteria and plastids also share components of the photosynthetic machinery such as chlorophyll *a,* chlorophyll *b*, phycobilisomes, and thylakoid membranes containing galactolipids. Galactolipids are thought to be a common trait of oxygenic photosynthetic organisms [36, 39]. In addition, the plastid membranes of glaucophytes and mosses contain peptidoglycan, which forms a layer outside the plasma membrane in most bacteria [40, 41]. Beta-barrel proteins, such as porins and Omp85 are present in the outer membrane of plastids, mitochondria, and Gram-negative bacteria but absent in the plasma membranes of eukaryotes [42, 43]. These features suggest that the two limiting membranes of primary plastids are of Gram-negative cyanobacterial origin, although they have some eukaryotic features. Another similarity between cyanobacteria and plastids is that they both undergo binary fission during cell division [44, 45].

In 1962, DNA was discovered in the plastid of *Chlamydomonas reinhardtii* [46] and the first plastid genome was completed in 1986 [47]. The development of DNA sequencing technology and phylogenetic analysis provided convincing evidence for the cyanobacterial origin of plastids. Plastid-encoded rRNA genes [48], ribosomal proteins [49, 50], and photosynthetic components [51] demonstrate the cyanobacterial origin in phylogenetic analyses. During primary endosymbiosis, endosymbiont genes were thought to be transferred into the host nuclear genome [52]. Approximately 6–20% of the genes in the nuclear genome of algae and plants are of cyanobacterial origin [53–58]. For instance, in experimentally determined plastid proteomes, 12% (*Cyanophora paradoxa*) to 24% (*Arabidopsis thaliana*) of the protein complement are derived from nuclear-encoded genes of cyanobacterial origin [58]. In the GreenCut2 resource, when considering conserved nucleus-encoded proteins in plants and green algae that are absent in nonphotosynthetic organisms, 32% of these proteins have homologs in cyanobacteria [59]. These phylogenetic studies provide strong evidence for the cyanobacterial origin of plastids. Similarities are also found in the order of genes as well as gene content. For example, comparison of cyanobacteria and plastids of primary and secondary origins, ribosomal proteins, ATPase subunits, and RNA polymerase subunits show conservation of gene order [60, 61].

Chromatophore of *Paulinella*

Most of the components shared by plastids and cyanobacteria are also present in the chromatophore of *Paulinella* (Table 1). *P. chromatophora* has been shown to assimilate radiocarbon photosynthetically, using $H^{14}CO_3^-$ labeling experiments [62]. All genes essential for photosystems I and II are present in the chromatophore and nuclear genomes of photosynthetic *Paulinella*. These results indicate that photosynthetic *Paulinella* performs oxygenic photosynthesis. The fluorescence of chlorophyll *a* was measured, indicating its presence [63]. Although the presence of phycobilisomes is not experimentally confirmed, a complete set of antenna protein genes (except for phycoerythrin) are encoded in the chromatophore genome and show expression [64]. Photosynthetic *Paulinella* also has a thylakoid membrane in the chromatophore [65], and galactolipids are detected in lipids extracted from *P. micropora* cell lysates [66]. The chromatophore has two limiting membranes. Transmission electron microscopy reveals a peptidoglycan layer in the chromatophore membrane [65–67]. The chromatophores also divide by binary fission [65, 68, 69].

DNA was found in the central part of the chromatophore, emitting an intense bluish fluorescence after DAPI staining [68]. Using a total of 406 conserved genes shared by the chromatophore genome and cyanobacteria, the phylogeny shows that the chromatophore of *Paulinella* is sister to *Prochlorococcus*–*Synechococcus* strains [70]. The number of genes of cyanobacterial origin in the nuclear genome is estimated to be less than 0.001% (32 in *P. chromatophora* and 58 in *P. micropora*), which is very few compared to Archaeplastida species [71–74]. However, it should be noted that the number of genes encoded in the chromatophore genome (~900) is much larger than the plastids of Archaeplastida (80–250 genes). One of the EGT--derived genes is high light-inducible (HLI) protein, which is required for cell viability under excess light in cyanobacteria [64]. The expansion of the HLI gene family and its light regulation likely reflects the important role of these genes in the light/

oxidative stress experienced by *Paulinella* during primary endosymbiosis [75]. In addition, the alignment of chromatophore genome regions with homologous sites in cyanobacteria shows conservation of gene order and orientation [9, 76].

2.2 Noncyanobacterial Contributions to Plastid Endosymbiosis

Plastids in Archaeplastida

Primary endosymbiosis relies on more than the genetic contributions made by the cyanobacterial endosymbiont to ensure success. About 40% of the plastid proteome shared by red algae and green plants originated from the eukaryote host and various lineages of noncyanobacterial prokaryotes [77]. Some important plastid pathways, such as amino acid biosynthesis and the Calvin–Benson cycle include eukaryotic and noncyanobacterial genes [78, 79]. In GreenCut2, of the nucleus-encoded proteins conserved in plants and green algae but absent in nonphotosynthetic organisms, 30% are eukaryote-specific [59]. Most plastid envelope metabolite transporter proteins in *Arabidopsis* are of host origin [80–82]. One of these is the family of Nucleotide Sugar Translocators (NSTs) that are the primary transporters of fixed carbon in red and green algae. This transporter family is originally derived from a sugar-phosphate transporter in the host endomembrane system and is now targeted to the inner membrane of the plastid [83].

Among experimentally determined plastid proteomes, 7% (*Cyanophora paradoxa*) to 15% (*Arabidopsis thaliana*) are of noncyanobacterial horizontal gene transfer (HGT) origin [58]. Interestingly, 30/40 of the HGT-derived genes in the *Cyanophora paradoxa* plastid proteome appear to originate from the common ancestor of Archaeplastida. There are even noncyanobacterial HGTs encoded in the plastid genome. In red algae, plastid-encoded RuBisCO has a proteobacterial origin [84] and the menaquinone biosynthesis genes have a chlamydial origin in the extremophilic lineage [85, 86].

Chromatophore of *Paulinella*

Unlike the plastids of Archaeplastida, the outer membrane is interpreted to be host--derived [65–67], whereas the inner membrane is derived from the cyanobacterial plasma membrane. In addition, porin seems to be missing in the outer membrane of the chromatophore and is not encoded in this genome [76].

In the chromatophore proteome, ~32% of the proteome comprises genes of noncyanobacterial origin [87]. Using the chromatophore transit peptide sequence, chromatophore-targeted proteins were predicted in silico. Of the 433 chromatophore import candidate proteins, only 17 are of α-cyanobacterial origin and were most likely acquired by EGT. Another 26 import candidates are of bacterial origin and were most likely acquired by HGT. Thus, host or uncertain origin accounts for the remaining 390 import candidates. Although the number of noncyanobacterial HGTs in photosynthetic *Paulinella* is small compared to Archaeplastida, many of the HGT-derived bacterial genes fill gaps in critical chromatophore pathways/processes, suggesting an important role for HGT in this primary endosymbiosis.

3 Hypotheses About Plastid Formation

3.1 Autogenous Formation of Plastid

Under the autogenous hypothesis, plastids and mitochondria are derived from invaginations of the plasma membrane of eukaryotes [88, 89]. After the double membrane structure was formed, organelles that contain a complete genome differentiated into mitochondria and plastids. This hypothesis competed with the endosymbiotic theory in the 1970s. The discovery of DNA in the plastid and phylogenetic analysis excluded the autogenous formation of the plastid in the 1980s [37].

3.2 Host-Directed Plastid Formation

Sato suggested that glycolipid biosynthesis could have preceded plastid formation as an adaptation to phosphate limitation, and it could have supported the formation of membranes made of glycolipids, which then facilitated the incorporation of the photosynthetic machinery from cyanobacteria [90]. This hypothesis is based on the study that many enzymes involved in the synthesis of plastid peptidoglycan and glycolipids are not of cyanobacterial origin [91]. In addition, the lipid biosynthetic pathways in cyanobacteria and plastids are unrelated in terms of biochemistry and phylogeny [92], and many lipid biosynthetic enzymes in plastids were acquired prior to the acquisition of ribosomes [51]. This hypothesis argues that a cyanobacterium is not the ancestor of plastids but rather, it was a gene donor. However, if a massive number of cyanobacterial genes were transferred one by one, the problem of low selective advantage during the intermediate stages of this process would need to be addressed [37].

3.3 Acquisition of an Exogenous Endosymbiont

The endosymbiotic theory proposes that an entire cyanobacterial cell was incorporated within the cytosol of the host cell and became an endosymbiont, which evolved into a photosynthetic organelle [29]. After the endosymbiotic theory was proposed, various hypotheses were proposed to explain the process of primary plastid endosymbiosis in more detail.

Endosymbiosis as a Rapid Transition

In this model, cyanobacteria become endosymbionts within a heterotrophic host cell, initiating long-term changes in both cells [93]. One of the key features of the early success of endosymbiosis must have been the integration of the metabolism of the two cells [80]. Therefore, a system of metabolite transporters was critical to supply the endosymbiont with essential nutrients and to allow the host to benefit directly from photosynthetic primary production. In Archaeplastida, most plastid transporters are of host origin [80], therefore, these transporters should have been transferred to endosymbionts in the early stage of endosymbiosis.

Several scenarios have been proposed to explain protein transport system evolution in primary plastids [94]. The discovery of a protein transported to the plastid via the secretory pathway [95], combined with the complicated structure of the TOC/TIC machinery, led to the proposal that plastid-targeted proteins were initially imported via the endomembrane system [96]. In another scenario, after the loss of the phagosomal membrane in the early evolution of primary plastids, a TOC/TIC-based import system would have evolved, bypassing the endomembrane system [94].

For the genome of the endosymbiont, the intracellular environment leads to the loss of genes for many metabolic functions that are no longer needed. Over time, endosymbiont genes could be transferred to the host nucleus, expressed on the cytosolic ribosome, and then transported to the organelle. This would lead to genetic integration of endosymbiont and host. As the host cell continued to target proteins into the endosymbiont and regulated metabolic fluxes between the cytoplasm and the endosymbiont, it gained control over the growth and division of the endosymbiont.

Shopping Bag Model

The host may form a series of transient endosymbioses before a stable relationship is finally achieved [97]. This shopping bag model allows the protein targeting system to be established before the final endosymbiont [93]. In this model, a heterotroph begins to transiently retain cyanobacteria. To benefit from retaining cyanobacteria instead of digesting them, the heterotroph would need transporters that could extract photosynthates from the transient endosymbiont. This could be achieved by targeting the transporter proteins to the membranes of transient symbionts. The existence of a protein-targeting system allows for an increasing number of proteins to be targeted to transient endosymbionts, thus promoting the integration process. Over time, the retention period could increase as the host developed a regulatory system to control the endosymbiont. This transient endosymbiosis stage is possibly mixotrophic, and HGT could occur from a variety of prey. Finally, some endosymbionts are retained and become an organelle. Therefore, cyanobacterial genes in the host nucleus are not necessarily all derived from the same cyanobacterium. As evidence, phylogenetic analysis of plastid genes often indicates different origins [51]. Furthermore, the acquisition time for each plastid gene varies, indicating that multiple gene transfers may have occurred. However, the deep internal branches of single gene trees are difficult to reconstruct because they are often sensitive to the methods used, in particular when inferring phylogenies from anciently diverged sequences [98–100].

Ménage à Trois Hypothesis (MATH)

This hypothesis proposes that Chlamydiales played an active role in the establishment of primary plastids in Archaeplastida [101, 102]. Phylogenomic analyses using algal and plant genome data found chlamydial genes in the nuclear genome [101, 103, 104]. In this scenario, Chlamydiales sheltered the cyanobacterial plastid ancestor from host defenses and provided carbohydrate transporters and protein effectors that allowed the storage of exported carbohydrates in host glycogen pools [105, 106]. Initially, the photosynthetic process would not have been adequately regulated by the host. Therefore, the storage system would have acted as a buffer between the unsynchronized carbon supply and demand during primary endosymbiosis. When Deschamps et al. reconstructed the storage polysaccharide metabolism of the common ancestor of Archaeplastida, there was a complete set of eukaryotic glycogen metabolism enzymes with only three prokaryotic genes [107]. To store photosynthate from the endosymbiont, chlamydial effectors and transporters related to glycogen metabolism could have been essential. Therefore, intracellular chlamydia has been proposed as partners in a tripartite symbiosis. This hypothesis was questioned because it was not strongly supported by phylogenetic analysis [100]. However, more recent work has demonstrated the major contribution of Chlamydiales to menaquinone (vitamin K_2) metabolism in Cyanidiophyceae red algae and to algal isoprenoid biosynthesis [86]. These data, combined with the finding that one-half of the genes involved in tryptophan synthesis in Archaeplastida are of chlamydial origin [108], support the MATH and provide impetus for future research to explore this hypothesis [109].

3.4 *Plastid Formation Model for* Paulinella

In the case of *Paulinella,* it has been proposed that the heterotrophic ancestor engulfed the cyanobacterium, and during the initial mixotrophic phase, the endosymbiont evolved into a fully functioning photosynthetic organelle [4, 11, 12] (Fig. 2). In the heterotrophic phase, HGT could occur, supported by the finding that cyanobacterial genes are present in the nuclear genome of heterotrophic *Paulinella* [12]. The mixotrophic cells initially delayed digestion of cyanobacteria, and during this stage, HGTs of noncyanobacterial origin may have played an important role because the number of HGTs (98–171) is greater than the number of EGTs (32–58), and these foreign genes fill gaps in critical chromatophore pathways/processes [73,

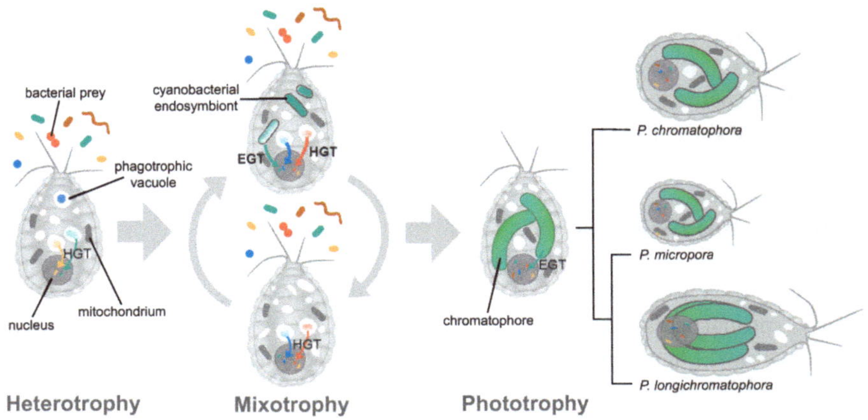

Fig. 2 Proposed model for the evolution of photosynthetic *Paulinella*. Heterotrophic ancestor (left) engulfed the cyanobacteria, and through the mixotrophic state (middle), the cyanobacterial endosymbiont became a photosynthetic organelle. After acquiring plastids, photosynthetic *Paulinella* diverged into different species (right)

74]. The host may have begun to target proteins to the endosymbionts. The proposed mixotrophic stage in *Paulinella* is similar to the shopping bag model. Interestingly, some nuclear genes of cyanobacterial origin clearly clustered with a specific clade of beta-cyanobacteria rather than with the alpha-cyanobacterial ancestor of the chromatophore [73]. This could be evidence for multiple endosymbioses of different cyanobacteria, or it could simply be HGT from lysed cyanobacterial prey.

Can other models of plastid formation in Archaeplastida be applied to *Paulinella*? It is difficult to support the hypothesis of host-directed plastid formation in *Paulinella* because the chromatophore genome encodes all the enzymes involved in the synthesis of lipids from the chromatophore membrane [66]. In addition, it is difficult to argue that chlamydia played an important role in the primary endosymbiosis of *Paulinella* because there are 10 chlamydial HGTs and none of them are annotated as transporters or genes related to the carbon storage system [73]. The number of proteobacterial HGTs (50) is greater than from chlamydia (10). There is stronger evidence for plastid acquisition via a cyanobacterial endosymbiont in *Paulinella* than in Archaeplastida. However, it is not clear whether there were multiple transient endosymbioses or a single event that evolved into the chromatophore.

Because the primary endosymbiosis in *Paulinella* occurred relatively recently when compared to Archaeplastida, studies in *Paulinella* have focused on finding factors (retrotransposition, trans-splicing, and the thioredoxin system) that facilitate chromatophore formation [75, 110, 111]. Calatrava et al. revealed that EGT-derived HLI genes, involved in photoprotection, were expanded and acquired light--responsive regulatory elements that function in the host through retrotransposition [75]. The authors hypothesized that during the early stages of host-endosymbiont metabolic integration, the lack of coordination between the metabolisms of two organisms led to high and sustained levels of reactive oxygen species (ROS) production, which induced retrotransposon-dependent gene duplications. This resulted in the modification of the ancestral promoter to allow the subfunctionalization of different HLI genes, allowing the organism to cope more effectively with different features of light stress. Trans-splicing has been suggested as another factor that contributed to the establishment of chromatophore [112]. Due to the presence of splice leader sequences only in photosynthetic *Paulinella* and the high rate of splice leader addition on EGT/HGT-derived gene transcripts, the authors hypothesized that splice leader trans-splicing has played an important role in the evolution of the plastid in *Paulinella*. Because the thioredoxin (TRX) system is linked to photosynthetic electron transport and the fine-tuning of metabolic pathways to fluctuating light levels in plastids of Archaeplastida [113], the TRX system was investigated in the chromatophore of *Paulinella*.The chromatophore, protein candidates involved in the TRX system were derived from the existing genetic toolkit in diverse organisms [110], implying the importance of thiol-based redox regulation in the primary endosymbiosis. However, these hypotheses need more solid evidence in future studies. For more details about the *Paulinella* primary endosymbiosis, see Chap. 12 in this collection, which addresses the metabolic integration of endosymbionts.

3.5 Endosymbiotic Ratchet

It is difficult to ascertain how primary plastid endosymbiosis influenced subsequent speciation using data from the anciently derived Archaeplastida. The *Paulinella* lineage offers a much better model because of the recentness of this endosymbiotic event and the availability of genome data from the known host cell and endosymbiont lineages. With this in mind, Bhattacharya et al. explored the short-term consequences of primary plastid origin in photosynthetic *Paulinella* from the perspective of population genetics and proposed the endosymbiotic ratchet hypothesis to explain the available data [114] (Fig. 3). They propose that post-endosymbiosis, the growth rate of novel photosynthetic lineages will be slowed by the poor integration of the nascent organelle in host cell biology, as evident in extant *Paulinella* lineages that divide every ca. 5–7 days in replete culture conditions. This low growth rate will lead to a stark reduction in effective population size and isolation of local populations. The resulting genetic bottleneck will weaken the strength of selection and allow genome growth in survivors vis-à-vis the proliferation of selfish genetic elements and gene family expansion. These predictions are borne out by the available data with the heterotrophic ancestors of photosynthetic *Paulinella* having compact genomes (likely <50 Mbp in size), whereas the photosynthetic lineages have massive genomes ranging in size from 707 Mbp in *P. micropora* KR01 to ca. 9.6 Gbp in *P. chromatophora* CCAC 0185. This genome expansion would, in some lineages, feed genetic innovation, which could drive organelle domestication and the origin of gene regulation pathways. This idea is supported by the HLI gene family

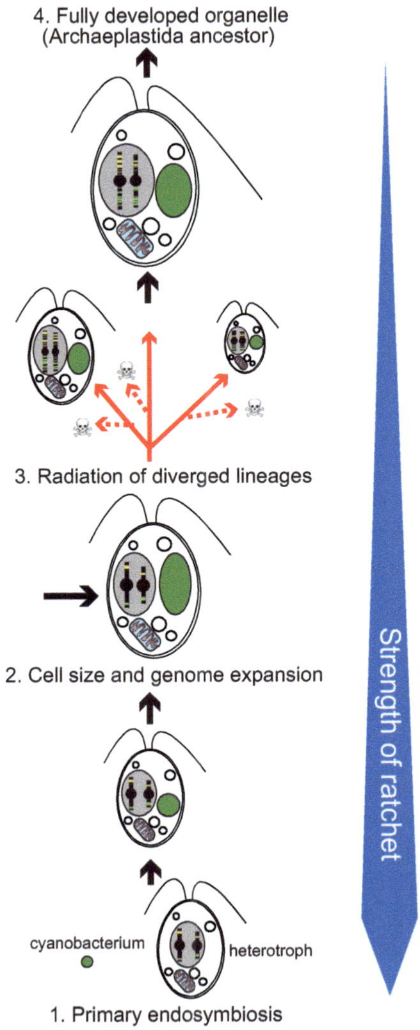

Fig. 3 The endosymbiotic ratchet hypothesis posits an explanation for the pre- and post-endosymbiotic evolutionary trajectory of heterotrophs that become photosynthetic via primary plastid endosymbiosis. The ratchet proceeds as follows: (1) a heterotrophic cell captures and retains a cyanobacterium as an endosymbiont. (2) This event precipitates a reduction in effective population size, which leads to cell size and genome expansion due to the proliferation of transposable elements/repeats/gene families, which is coincident with endosymbiont genome reduction, EGT to the host nuclear genome, HGTs, inefficient photosynthesis, inefficient plastid protein import, and slow growth rate. (3) Ecological diversification occurs in different habitats with genome and cell size reflecting effective population size. The dashed lines and the skull and crossbones indicate presumptive losses due to extinction. (4) The ubiquitous, surviving algal lineages contain fully developed organelles with highly reduced genomes and effective population size recovers due to high photosynthetic efficiency leading to rapid growth rates. Cell at this phase also have efficient plastid protein import systems and plastid gene regulation allows full integration with host cell biology, as in extant Archaeplastida

evolution described above [75]. In summary, the endosymbiotic ratchet hypothesis posits that isolated populations exploited divergent strategies for survival post-primary endosymbiosis and assorted themselves in non-overlapping niches to minimize competition during the early phase of organelle integration. These forces led to rapid speciation and divergence among the radiating lines of descent, which is apparent in both the *Paulinella* and Archaeplastida lineages. This hypothesis provides a general framework for interpreting post-endosymbiosis lineage evolution that is driven by disruptive selection and demographic and population shifts.

4 Conditions at the Time when Primary Plastid Endosymbiosis Occurred

4.1 Archaeplastida Host and Cyanobacterial Endosymbiont

Because primary endosymbiosis in the Archaeplastida ancestor occurred 1.5–2.1 bya, it is challenging to determine the putative cyanobacterial donor lineage of primary plastids [115]. In a recent study, it was suggested that the primary plastid of Archaeplastida traces its origin to *Gloeomargarita* ~ 2.1 bya and that the common ancestor of Archaeplastida appeared ~1.9 bya in the Paleoproterozoic [6, 50, 116]. Ancestral trait reconstruction reveals that primary plastids diverged from cyanobacteria with small cell diameters and probably inhabited low-salinity environments characteristic of fresh water [6, 117]. However, it is unknown whether the extant cyanobacteria are representative of their ancestors. Given the immense timescales involved, it should be kept in mind that the genomes of endosymbionts were already chimeras and the descendants of free-living relatives continued to experience HGT with other prokaryotes [27].

The common ancestor of Archaeplastida is presumed to be a cell wall-less, phagotrophic unicell. It was suggested that cyanobacteria may have entered the cell through phagotrophy. However, most species belonging to Archaeplastida lack phagotrophy except for the early diverging green algae *Cymbomonas* [118] and *Rhodelphis* [119]. *Rhodelphis* are flagellate predators with nonphotosynthetic plastids and are sister to red algae. Because information about the common ancestor of Archaeplastida is limited, studying organisms that have symbiotic relationships with cyanobacteria can potentially provide novel insights into this area of inquiry [120]. If cyanobacteria enter the cell by phagocytosis, they must avoid digestion. In ciliates, *Paramecium caudatum*, a bacterium has been found that can escape from food vacuoles [121]. Cyanobacteria can also enter the symbiosome (specialized compartment in a host cell that houses an endosymbiont in a symbiotic relationship) through phagocytosis [122]. Because successful pathogens have evolved multiple strategies to prevent or escape phagocytosis [123], it was suggested that living cyanobacteria were driven into endosymbiosis with the help of a chlamydial pathogen [102]. Entering the cell without phagocytosis can be another possible scenario.

Cyanobacteria could gradually fold into the cytoplasm of a host by nonphagocytotic invagination of the plasma membrane. Diatoms or the *Geosiphon* fungus have cyanobacterial endosymbionts, although they appear to lack phagocytosis [15]. Because some symbiotic cyanobacteria associate extracellularly with diatoms and dinoflagellates [124], a prolonged ecto-symbiotic association may have occurred prior to endosymbiotic uptake of cyanobacteria [120].

4.2 **Paulinella** *Host and Cyanobacterial Endosymbiont*

In the case of the chromatophore of *Paulinella*, its putative donor is a diverged member of the marine clade that includes *Prochlorococcus* and *Synechococcus* [6, 9, 125]. Currently, three photosynthetic *Paulinella* species have been described, two from freshwater and one from marine environments [13, 14]. There have also been reports of brackish water strains, which may be new species [114, 126]. Nine heterotrophic sister species are all marine [127–130]; two of them were sampled from seawater and seven from benthic sediments of beach sand. Based on these results, the ancestor of photosynthetic *Paulinella* was a heterotrophic marine species. However, it is not clear whether the common ancestor of photosynthetic *Paulinella* was a planktonic or benthic species. It seems that primary endosymbiosis occurred in the marine environment and after acquiring the photosynthetic organelle, *Paulinella* diverged into species adapted to different salinity environments.

In the case of *Paulinella,* the entry of cyanobacteria by phagocytosis is more strongly supported. Digestive vacuoles were not found in photosynthetic *Paulinella* [65], whereas transmission electron microscopy observation of heterotrophic *Paulinella ovalis* frequently revealed cyanobacteria within food vacuoles [127]. In addition, single-cell genome analysis of heterotrophic *Paulinella ovalis*-like cells showed an intron-containing nuclear gene of α-cyanobacterial origin, suggesting that phagotrophy may be related to HGT from prey bacteria in *Paulinella* [12]. The outer membrane of the chromatophore has been interpreted as being of host origin [65, 67] and maybe the remnant of the host phagosome membrane.

4.3 Possible Initial Benefits of Primary Plastids

Carbon Products

Since Mereschkowsky's initial endosymbiotic theory [29], the production of carbohydrates by the cyanobacterial endosymbiont was thought to be a crucial key for the establishment of plastids [107].

Provision of Oxygen

An alternative suggestion is that the initial advantage of plastids may have been their metabolic end-product, O_2 [131]. At the time and place where the primary plastid endosymbiosis occurred, there may have been a lack of freely available oxygen, because an oxygen minimum zone that tended toward euxinia (anoxic and sulfidic) existed under moderately oxic surface waters during most of the Proterozoic [132]. In this situation, the host with plastids would benefit better from mitochondrial respiration using released oxygen from plastids. It has been shown in ciliates that oxygen produced by photosynthetic algae can power host niche expansion into anaerobic habitats, granting access to new food sources and refuge from aerobic predators and competitors [133, 134].

Nitrogen Fixation

Because many modern cyanobacterial symbioses usually provide nitrogen [135], it has been hypothesized that primary endosymbionts also arose from nitrogen-fixing cyanobacteria [57]. This hypothesis is supported by studies suggesting that plastid ancestors may have the phenotypic characteristics of filamentous phenotype and heterocyst differentiation [55]. However, in the analysis including the recently discovered *Gloeomargarita* clade, *Gloeomargarita*-like cyanobacteria were the closest modern relatives of Archaeplastida plastids [116]. Because *Gloeomargarita lithophora* lacks the ability to fix nitrogen, fixation of N_2 was unlikely to have played a role during the primary endosymbiosis in Archaeplastida [116].

Initial Benefits of Plastids in *Paulinella*

The oxygen product of photosynthesis as an advantage is ruled out by the fact that photosynthetic *Paulinella* species have been found in aerobic habitats [14]. It is unlikely that nitrogen fixation was the advantage because the marine clades of *Prochlorococcus* and *Synechococcus* do not fix nitrogen [136]. The advantage could be the production of carbohydrates by the cyanobacterial endosymbiont. However, unlike the case in Archaeplastida, there were various photosynthetic eukaryotes that already colonized marine environments at the time of the *Paulinella* primary endosymbiosis [116]. What was the advantage of primary endosymbiosis in *Paulinella* to overcome massive genetic innovation and not be left behind in the competition of other photosynthetic organisms? The fact that photosynthetic *Paulinella* survives and grows only at low light intensities [64] also shows that the advantage of gaining the photosynthetic ability only works at limited light intensities.

5 How Many Primary Plastid Endosymbioses Have Occurred, Giving Rise to Photosynthetic Organelles?

There are only two known cases in which a cyanobacterial endosymbiont evolved into a photosynthetic organelle. However, instances of secondary and complex endosymbioses are more common. Why is primary plastid endosymbiosis rare? The 'chassis and engine' model suggests that not only is an 'engine (cyanobacterial endosymbiont)' important, but also 'chassis (host cell)' including ROS detoxification, regulatory mechanism, and photosynthate storage is crucial for primary plastid endosymbiosis to occur [137]. In secondary and complex endosymbioses, the chassis already formed in the primary endosymbiosis process could facilitate the integration of the plastid into the metabolic systems of the new host eukaryote. Therefore, it is difficult to overcome the complex evolutionary forces associated with primary plastid endosymbiosis, which is why they are rare. So far, we have mentioned that there are two independent cases of primary plastid endosymbiosis in Archaeplastida and *Paulinella,* but there could be more than a single endosymbiosis during the evolution of primary plastids [138] (Fig. 4).

5.1 Single Endosymbiosis

Many components shared by the three lineages (glaucophyte, red algae, and green lineage) were presented as supporting evidence that primary plastid endosymbiosis occurred only once in the common ancestor of Archaeplastida. For example, the plastid metabolite transporter family is derived from the common ancestor of Archaeplastida [80, 83]. The chimeric origins of the translocon component [61, 139] and Calvin–Benson cycle enzymes [140, 141] are conserved in the three lineages. In addition, the monophyly of the Archaeplastida lineages was revealed by phylogenetic analysis using concatenated plastid genes [50, 116, 142], and the order

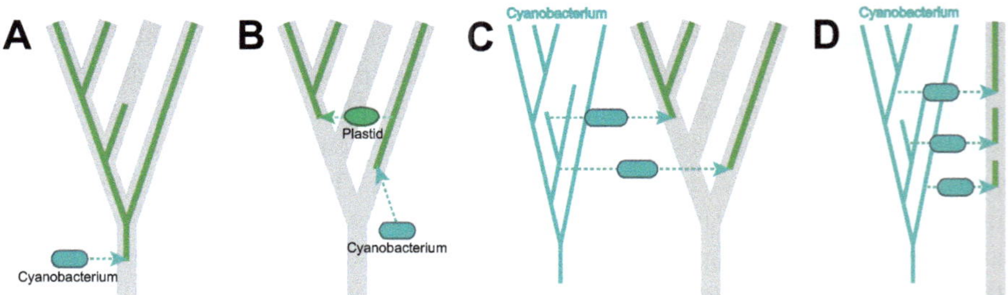

Fig. 4 Possible biological scenarios for the origin of plastids, modified from Mackiewicz & Gagat [138]. (**a**) Single endosymbiosis and secondary loss. (**b**) Serial endosymbiosis. (**c**) Multiple endosymbioses from closely related cyanobacteria (now extinct). (**d**) Multiple endosymbioses in the same host at different times

was conserved in the gene cluster of Archaeplastida lineages and cyanobacteria [61]. Unlike the phylogenetic analysis of plastid genes, the monophyly of three lineages of Archaeplastida has not been clearly recovered using nuclear genes [119, 143, 144]. However, recent studies have shown the monophyly of three lineages of Archaeplastida, including *Rhodelphis* and Picozoa using nuclear data [145–148]. Irisarri et al. demonstrated that taxon sampling, selection of alignment positions, and substitution models for phylogeny are critical to recovering the monophyly of Archaeplastida. However, this evidence does not necessarily support a single primary endosymbiosis event, which will be discussed in Sect. 5.4 [149].

5.2 Single Endosymbiosis and Secondary Loss

Recently, Schön et al. argued that Picozoa is the first Archaeplastida lineage without a plastid [150]. This finding could be explained either by a complete loss of plastids in Picozoa or by multiple independent endosymbioses. Complete plastid loss in Picozoa follows the scenario of single endosymbiosis and secondary loss (Fig. 4a).

5.3 Serial Endosymbiosis

One way to explain the multiple independent endosymbioses is serial endosymbiosis (Fig. 4b). In this scenario, primary endosymbiosis occurs in the ancestor of the glaucophyte and green lineages, and then the ancestor of red algae and *Rhodelphis* obtains the primary plastid of the green lineage, which is a secondary endosymbiosis [151]. This scenario can explain the difference in the phylogenetic relationship between plastid genes and nuclear genes. Nuclear phylogenomic studies showed a sister-group relationship between the glaucophyte and the green lineages [119, 147, 150], while the plastid phylogenies showed a closer relationship between the green lineage and red algae [116, 142]. If the plastids of red algae originated from this scenario, they should be bound by more than two membranes like other secondary plastids. However, two membranes have also been found in some complex plastids, which may be due to the loss of a membrane [152].

5.4 Multiple Endosymbioses from Closely Related Cyanobacteria (Now Extinct)

It is possible that primary endosymbiosis occurred multiple times from closely related cyanobacteria (now extinct) (Fig. 4c); for example, once in the ancestor of the glaucophyte and green lineages, and once in the ancestor of red algae and *Rhodelphis*. Because independent endosymbioses involved closely related cyanobacteria (now extinct), monophyly of Archaeplastida could be achieved from plastid genes of extant taxa. Other evidence commonly cited to support a single plastid origin may also support this scenario. Similarities in plastid gene content are most consistent with the hypothesis of convergent evolution [149]. The conservation of chimeric origins can be explained by (yet unknown) biochemical advantages of using a particular copy of the gene (e.g., the host isoform) in any plastid [97]. Even the influx of plastid DNA into nuclear genomes could lead to artifacts in the phylogeny of nuclear genes [2].

Additionally, as in the shopping bag model, multiple endosymbioses can occur in the same host at different times, and the resulting organelle is a chimera of products from different predecessors (Fig. 4D). However, it would be difficult to find evidence for this model with the current data if the endosymbiosis came from a closely related cyanobacterium, rendering this type of alternative hypothesis essentially untestable.

5.5 *Case of Photosynthetic* Paulinella

In the case of photosynthetic *Paulinella,* phylogenetic analysis of chromatophore--encoded genes supports the monophyly of three photosynthetic *Paulinella* species that diverged from alpha cyanobacteria, i.e., the *Prochlorococcus* and *Synechococcus* clades [14]. Furthermore, the 18S rRNA phylogeny also revealed the monophyly of photosynthetic *Paulinella* species. The chromatophore genome structure of photosynthetic *Paulinella* species was highly constrained [70]. The small number of EGTs (32–58) in photosynthetic *Paulinella* also supports a single primary endosymbiosis [73, 74]. However, these data cannot exclude the possibility of multiple transient endosymbioses from closely related cyanobacteria, although it is highly unlikely.

6 Experimental Tests of Primary Plastid Endosymbiosis

Knowledge about the ancient Archaeplastida endosymbiosis can be gleaned using experimental methods with other model systems [120]. For example, the ciliate genus *Paramecium* has several endosymbionts. An algae-free strain of *Paramecium* can be artificially created, and a symbiotic relationship re-established. Therefore, this organism is now becoming a model to understand the mechanisms for establishing primary and secondary endosymbioses [153]. For photosynthetic endosymbionts, most of the experiments were carried out with *Chlorella*-like species. However, it is possible to establish cyanobacteria as endosymbionts [154]. This model organism can be used to understand the selective pressures behind the primary plastid endosymbiosis. It is possible to study different sets of conditions that promote mutualism and interdependence. In another experiment, in which heterotrophic *Dictyostelium* cells were cultured with two types of photosynthetic bacteria as food sources, unique autotrophic organisms were formed via bacterial endosymbiosis [155]. Other organisms that have a transient symbiotic relationship with cyanobacteria may be a new candidate for studying primary plastid endosymbiosis.

Genetic engineering can be used to test primary endosymbiosis. *Synechococcus elongatus* PCC 7942 engineered with invasin (bacterial virulence factors that encode mammalian cell invasion) and listeriolysin-O (a membrane-degrading enzyme that allows bacteria to escape the digestive vacuoles of their hosts) showed escape from lysosomal digestion [156]. In addition, the knockout of glycogen synthase genes in *Synechococcus elongatus* PCC 7942 results in 'metabolic overflow' mutants that leak photosynthate [157]. The use of photosynthate from endosymbionts has been suggested as an important step in maintaining the endosymbiotic relationship. This hypothesis can be tested using engineered cyanobacteria and a model host such as *Paramecium.*

An artificial photosynthetic endosymbiosis was created using engineered cyanobacteria and budding yeast to test different hypotheses [158]. Yeast mutants deficient in oxidative phosphorylation-mediated ATP synthesis under defined photosynthetic selection conditions were used as host strains. The engineered cyanobacterial strains were then introduced into yeast cells through cell fusion. Under optimal conditions, the engineered yeast/cyanobacteria chimera propagated for at least 15–20 generations. In addition, Cournoyer et al. tested the role of ADP/ATP translocases in primary endosymbiosis. A cyanobacterial mutant expressing the recombinant ADP/ATP translocase gene (*ntt1*) was introduced into the yeast mutant and this chimera had a faster growth rate and better stability. To test whether parasitic genetic factors helped the endosymbiotic relationship, cyanobacterial mutants were generated that recombinantly expressed various SNARE-like proteins from pathogenic intracellular bacteria. Chimeras with these mutants resulted in faster growth rates and better stability. These platforms could facilitate laboratory-based evolutionary experiments of primary endosymbiosis.

Bottom-up constructed artificial cells may provide an opportunity to study the metabolic mechanism during primary endosymbiosis. An artificial cell containing cyanobacteria capable of light harvesting and carbon dioxide fixation has been shown to produce glucose molecules by converting light energy into chemical energy [159]. In addition, these glucose molecules were used to build a cascade cycle of NADH/NAD+ within the artificial cells. In another experiment, a switchable photosynthetic organelle was artificially designed [160]. In this artificial organelle, an ATP synthase and two photoconverters (plant-derived photosystem II and bacteria-derived proteorhodopsin) enabled ATP synthesis. These artificial photosynthetic systems could be used to understand the evolution of primary plastid endosymbiosis.

7 Divergence After Primary Plastid Endosymbiosis

After primary endosymbiosis in the common ancestor of Archaeplastida, the eukaryotic photoautotrophs diverged into different lineages. Among the three lineages in Archaeplastida (Glaucophyta, Rhodophyta, Chloroplastida = Viridiplantae), the order of lineage differentiation has been a subject of debate [138]. Recently, nuclear phylogenies including *Rhodelphis* and Picozoa supported the early divergence of the Rhodophyta [119, 147, 150]. These three photosynthetic lineages have many differences among them, excluding the shared feature of a 2-membrane bound plastid [161–164] (Table 2). These differences allowed them to adapt to different environments. To use light in different environments, each of the three plant lineages modified photosynthetic light-harvesting antennae inherited from the cyanobacteria [165]. Chloroplastida contain chlorophylls-*a* and -*b*, whereas Rhodophyta and Glaucophyta contain chlorophyll-*a* and phycobilisomes. Another difference is the carbohydrate storage system. Interestingly, starch is found in the cytoplasm of Rhodophyta and Glaucophyta but is localized in the chloroplast stroma of Chloroplastida [166]. The pathway is thought to have been cytosolic in the common ancestor of the three Archaeplastida lineages. It has been suggested that starch was selectively returned to the chloroplasts of Chloroplastida because these organisms experienced higher levels of chlorophyll synthesis and higher light intensities than their sister taxa.

Photosynthetic *Paulinella* diverged into three species, showing differences in habitat, cell size, and number of oral scales [14]. An interesting point is the difference in the size of the genome, which is approximately 700 Mb for *P. micropora*, but is predicted to exceed 10 Gb for *P. chromarophora* [73, 74]. Considering that heterotrophic *Paulinella* is predicted to have a genome size of about 11 Mb, a massive change in genome size occurred as a result of primary endosymbiosis. To explain the accelerated evolution in the post-primary endosymbiosis, Bhattacharya et al. proposed the endosymbiotic ratchet hypothesis [114]. Another difference between photosynthetic *Paulinella* species is that double-stranded DNA virus regions were found in the genome of *P. micropora* [167]. However, double-stranded DNA viral regions observed in *P. micropora* are absent or poorly represented in the genome of *P. chromatophora,* implying recent invasion by large dsDNA viruses in *P. micropora* [74].

8 Comparison of Archaeplastida and Photosynthetic *Paulinella*

What can we learn from comparing the two independent primary plastid endosymbiosis in Archaeplastida and photosynthetic *Paulinella*? Is there a model that can explain both? There are many differences between the two, such as the origin of the host and the cyanobacterial endosymbiont, and the time and environment in which

Table 2 Major trait distribution among photosynthetic Archaeplastida lineages [114]

Lineage	Plastid	Mitochondria	Morphology	Species	Storage	Nuclear genome
Chloroplastida (Viridiplantae)	Two membranes, chlorophylls-*a*, -*b*, variable genome size, often highly reduced, stacked thylakoids often in grana	Flat cristae	Ubiquitous, unicellular flagellate, colonial, filamentous, seaweeds, plants, parenchyma	450,000–500,000	Starch in plastid	Highly variable, 13 Mbp in *Ostreococcus* to >1 Gbp in plants
Rhodophyta	Two membranes, chlorophyll-*a*, phycobilisomes, large genomes, ungrouped thylakoids	Flat cristae	Mostly marine, lack flagella and centrioles, unicellular, colonial, filamentous, seaweeds, pseudoparenchyma, pit connections between cells	>7000	Floridean starch in the cytosol	Massive (25%) gene inventory loss in the ancestor, 16–100 Mbp in size
Glaucophyta	Two membranes, chlorophyll-*a*, phycobilisomes, peptidoglycan, concentric unstacked thylakoids, moderate-sized genomes	Flat cristae	Freshwater, unicellular, colonial	14–26	Starch in cytosol	100 Mbp in *Cyanophora paradoxa*

Table 3 Comparison of traits in Archaeplastida and photosynthetic *Paulinella*

Characteristic	Archaeplastida	Photosynthetic *Paulinella*
Host eukaryote	Common ancestor of Archaeplastida	Rhizaria heterotrophic *Paulinella*
Origin of plastid	*Gloeomargarita*-like cyanobacteria	Marine *Prochlorococcus* and *Synechococcus* clades (α-cyanobacteria)
Lineage age estimation	1.5–2.1 bya	90–600 mya
Phagotrophic capacity	Lost in Rhodophyta, Glaucophyta, and most Chloroplastida, but preserved in some prasinophytes, *Rhodelphis*, Picozoa	Lost
Plastid genome size	80–200 kbp	~1000 kbp
Number of plastid genes	80–250	~900
Cyanobacterial proteins in plastid proteome	70–390	17
Noncyanobacterial prokaryotic proteins in plastid proteome	40–240	26
Host-derived and proteins of uncertain origin in plastid proteome	320–900	390
Cyanobacterial genes in the nucleus	274–4500	32–58
Retrograde signaling	Present	Unknown
Redox regulation system	Present	Putative thioredoxin system has been proposed
Beta barrel outer membrane pores	Present	Not identified
Import system of nucleus-encoded proteins into the plastid	Translocons at the outer and inner plastid membranes (TOC/TIC complex)	No TOC/TIC complex, model of chromatophore transit peptide--mediated protein trafficking via the Golgi has been proposed
Solute transporters in plastid inner membrane	100–150	25
Storage system	Alpha-glucan accumulator (e.g., starch)	Beta-glucan accumulator (unpublished)

Modified from Ponce-Toledo et al. [185]

the primary plastid endosymbiosis occurred. Despite these differences, if there is a similarity between the two, it may provide clues about the elements essential for photosynthetic primary endosymbiosis. In fact, most of what is known about photosynthetic organelles is based on studies in model species such as *Arabidopsis thaliana* and *Chlamydomonas reinhardtii*. Because many genetic tools have not yet been developed in *Paulinella*, it is difficult to study them at the same level as in plants. However, other aspects of plastid evolution can be found in the study of *Paulinella*. In this section, different mechanisms for the proper functioning of plastids will be compared between Archaeplastida and photosynthetic *Paulinella* (Table 3).

8.1 Regulatory Mechanism of Plastid Metabolism

Archaeplastida have several regulatory mechanisms to maintain metabolic integration between the nucleus and the plastid. These mechanisms include transcriptional regulation, post-transcriptional regulation, anterograde signaling (from the nucleus to the organelle), retrograde signaling (from the organelle to the nucleus), and redox regulation [113, 168–170].

One way to gain control over the endosymbiont is to encode key endosymbiont genes in the host nuclear genome [111]. Control is gained by losing key genes from the chromatophore genome (either by outright gene loss or by transfer to the nuclear genome) that occupy central parts of the pathways. This leaves only the nuclear--encoded versions, which confer control of functions in the chromatophore to the host. In photosynthetic *Paulinella*, this process was observed to have occurred with the nucleotide precursor, nucleotide synthesis, stringent response pathways, and DNA replication complex. This process likely gave the host control of these pathways in both compartments in a small number of evolutionary steps.

In Archaeplastida species, plastid metabolism is controlled not only by regulation of transcription of nuclear-encoded plastid-targeted genes but also by direct modulation of transcription of plastid-encoded genes. This is achieved, at least in part, by a nuclear-encoded plastid-targeted RNA polymerase and nuclear-encoded sigma factors that modulate the promoter selectivity of the plastid-encoded RNA polymerase [169]. However, only two sigma factors were found in the chromatophore genome and there was no evidence for a nuclear-encoded chromatophore--targeted RNA polymerase [171]. Differential expression analysis under light/dark conditions showed that *P. chromatophora* lacks light-induced transcriptional regulation of chromatophore genes and most EGT-derived nuclear genes [64]. However, Zhang et al. used only a small number of chromatophore genes for expression analysis, so more in-depth research is needed.

Among the chromatophore-targeted proteins, many are annotated as transcription factors [87] and octotrico peptide repeat (OPR) proteins [172]. Based on the similarity of their domain architecture to known organelle-targeted expression regulators, OPRs likely serve as nuclear factors that modulate chromatophore gene expression by direct binding to specific target RNAs.

The correlation between chromatophore ROS production and the nuclear HLI gene activation raises the possibility of retrograde signaling in photosynthetic *Paulinella* because ROS is known to be involved in retrograde signaling in plants [170]. However, more work is needed to assess the existence of retrograde signaling in the *Paulinella* lineage.

In plastids, the thioredoxin (TRX) system is associated with photosynthetic electron transport and the fine-tuning of metabolic pathways to fluctuating light levels [113]. In photosynthetic *Paulinella*, the TRX system has evolved independently, with parts derived from the existing genetic toolkit in diverse organisms [110]. These results suggest that plastid regulatory mechanisms have evolved in *Paulinella*, although they differ from those in plants. However, the regulatory mechanisms proposed by the observations in *Paulinella* remain to be tested experimentally.

8.2 Establishment of New Transport Routes for Nucleus-Encoded Proteins

In Archaeplastida, most plastid proteins are synthesized in the cytosol and then imported via the TOC/TIC complex, which recognizes N-terminal targeting sequences [173]. In photosynthetic *Paulinella*, a translocon similar to TOC/TIC appears to be missing. However, nucleus-encoded chromatophore-targeted proteins with conserved N-terminal sequence extension have been found [87]. The small, imported protein PsaE was detected in the Golgi [174], and a model of chromatophore transit peptide-mediated protein trafficking through the Golgi has been proposed [175]. The details of protein translocation will be discussed in another chapter.

8.3 Metabolite Transporters of the Plastid Envelope

The transport of metabolites across the cyanobacterial membrane may have been important for the establishment of endosymbiosis. In Archaeplastida, most plastid transporters are of host origin, suggesting that plastid establishment was a host-driven process [81]. In the case of photosynthetic *Paulinella*, the chromatophore genome probably lost 70 solute transporters during primary endosymbiosis, and the remaining transport systems do not appear to be suitable for establishing metabolic connectivity [172]. However, there is no evidence for the insertion of nucleus-encoded transporters (or any other multispanning transmembrane proteins) into the inner membrane (or thylakoids) of the chromatophore. Instead, single transmembrane helices and amphipathic helices have been identified in the short chromatophore-targeted orphan proteins that are expected to be involved in modulating membrane permeability. Therefore, the mechanism of metabolic connectivity between the endosymbiont and the host in *Paulinella* is different from that in Archaeplastida.

8.4 Light Acclimation

Excess light absorbed by photosynthetic organisms can generate ROS that can cause cellular damage. Therefore, photosynthetic organisms have evolved direct and indirect mechanisms to sense and respond to excess light [176]. These mechanisms include plastid avoidance movement, nonphotochemical quenching, ROS detoxification, and changes in nuclear gene expression responses via retrograde signaling pathways.

Photosynthetic *Paulinella* survive and grow only at low light intensities [64], although biophysical measurements of photosystem II show no obvious evidence of impairment [63]. This may be explained by the finding that many photoprotective

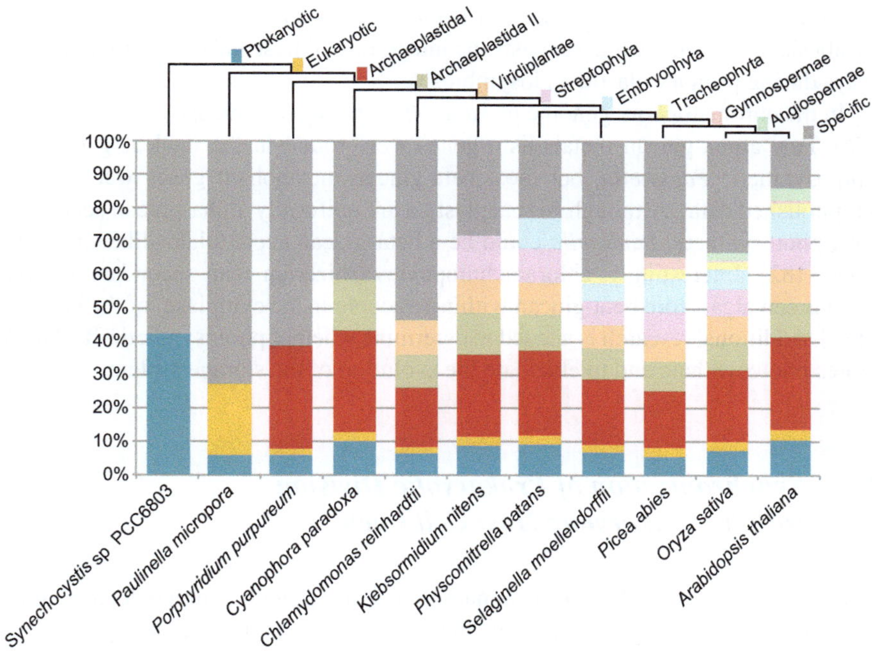

Fig. 5 Percentage of OGFs sorted by relative age (tree drawn at the top) for each species in diurnally rhythmic genes [74]

genes have been lost or transferred to the host genome, and light-based transcriptional regulation may have been lost or significantly reduced in the chromatophore genome [64, 74]. Although the host acquired light-responsive regulatory elements for the EGT-derived HLI genes [75], this was likely insufficient to adapt to higher light intensities.

Adaptation to the diurnal light–dark cycle is also important for photosynthetic organisms. Ferrari et al. showed that despite large phylogenetic distances, diurnally rhythmic genes in Archaeplastida species are conserved across lineages [177]. Furthermore, many diurnal rhythmic genes were inherited from the common ancestor of Archaeplastida. However, when comparing *P. micropora* KR01 with Archaeplastida species, diurnally rhythmic genes show a different pattern even though they share a large fraction of genes of eukaryotic origin (Fig. 5). These results suggest that *Paulinella* has an essentially novel set of diurnally regulated genes that respond to light stress and the light–dark cycle [74].

8.5 Storage System

In photosynthetic organisms, disruption of carbohydrate storage affects cellular energy balance and results in photochemical damage [178–180]. The major storage polysaccharide in Archaeplastida is starch, which is synthesized by a mosaic of enzymes, some of which are derived from cyanobacteria [181]. Therefore, starch synthesis in Archaeplastida suggests the integration of host and endosymbiont pathways during primary plastid endosymbiosis.

In the chromatophore genome of *Paulinella,* glycogen synthase genes are absent [76]. Few alpha-glucan metabolism genes were found in the nuclear genome of photosynthetic *Paulinella*, but most beta-glucan metabolism genes were present (unpublished data). Although Archaeplastida are uniformly alpha-glucan accumulators, photosynthetic *Paulinella* could be a beta-glucan accumulator-like stramenopiles [182]. Interestingly, chlorarachniophytes (rhizarian with secondary plastids from green alga) store beta-glucan within a vacuole in the cytoplasm of the host cell [183]. Additional research is needed to determine whether photosynthetic *Paulinella* stores photosynthate and to elucidate the evolution of the storage system.

8.6 Synchronization of Prokaryotic Division with the Eukaryotic Host Cell Cycle

Plastid division is coordinated by a macromolecular protein complex located at the division site. In the division complex, the stromal components are largely of cyanobacterial origin, whereas the cytosolic components are strictly of eukaryotic origin [184]. These proteins are encoded in the nuclear genome and targeted to the plastid.

Although the chromatophore and host cell division are coupled in *Paulinella,* the molecular machinery of chromatophore division is poorly understood. On the chromatophore genome, most of the cell-division genes commonly found in α-cyanobacteria have been identified [76]. However, it is unknown whether these genes are involved in chromatophore division and whether other nuclear genes may also play a role.

9 Concluding Remarks

Despite the enormous impact of primary endosymbiosis on ecosystems, much remains to be learned about the evolution of primary plastids. In this chapter, we highlight similarities and differences between the independent events in Archaeplastida and photosynthetic *Paulinella* and describe how they inform the field of organellogenesis. Among the similarities and differences, the ones we should pay attention to are those that have evolved to perform the same function despite having different evolutionary trajectories. For example, by examining how *Paulinella* evolved to respond to light stress or diurnal rhythms and comparing it to plant systems, we can learn about the evolution of the primary plastid. In particular, we can infer ancestral traits and the forces driving genome evolution in photosynthetic *Paulinella* by studying their heterotrophic sister species. Here, only the two primary plastids were compared, but by comparing obligate vs. facultative endosymbionts, primary vs. complex plastids, and photosynthetic vs. non-photosynthetic endosymbionts, it will be possible to identify features essential for the success of primary plastid endosymbiosis. The evolution of the primary plastid can be revealed through in-depth studies of various endosymbiotic organisms, including *Paulinella,* and experimental tests of primary endosymbiosis using the tools of genetic engineering.

Acknowledgments This study was supported by grants funded by the National Research Foundation of Korea (NRF- 2022R1A2B5B03002312, 2022R1A5A1031361, 2021R1C1C 2010445).

References

1. Gould SB, Waller RF, McFadden GI (2008) Plastid evolution. Annu Rev Plant Biol 59(1):491–517. https://doi.org/10.1146/annurev.arplant.59.032607.092915
2. Stiller JW (2007) Plastid endosymbiosis, genome evolution and the origin of green plants. Trends Plant Sci 12(9):391–396. https://doi.org/10.1016/j.tplants.2007.08.002
3. Bhattacharya D, Helmchen T, Melkonian M (1995) Molecular evolutionary analyses of nuclear-encoded small subunit ribosomal RNA identify an independent rhizopod lineage containing the Euglyphina and the Chlorarachniophyta. J Eukaryot Microbiol 42(1):65–69. https://doi.org/10.1111/j.1550-7408.1995.tb01541.x
4. Nowack ECM (2014) *Paulinella chromatophora*—rethinking the transition from endosymbiont to organelle. Acta Soc Bot Pol 83(4):387–397. https://doi.org/10.5586/asbp.2014.049
5. Yoon HS, Hackett JD, Ciniglia C et al (2004) A molecular timeline for the origin of photosynthetic eukaryotes. Mol Biol Evol 21(5):809–818. https://doi.org/10.1093/molbev/msh075
6. Sánchez-Baracaldo P, Raven JA, Pisani D et al (2017) Early photosynthetic eukaryotes inhabited low-salinity habitats. Proc Natl Acad Sci U S A 114(37):E7737–E7E45. https://doi.org/10.1073/pnas.1620089114
7. Keeling PJ (2010) The endosymbiotic origin, diversification and fate of plastids. Philos Trans R Soc Lond B Biol Sci 365(1541):729–748. https://doi.org/10.1098/rstb.2009.0103
8. Not F, Siano R, Kooistra WHCF et al (2012) Chapter one - Diversity and ecology of eukaryotic marine phytoplankton. In: Piganeau G (ed) Advances in botanical research. Academic Press, pp 1–53
9. Yoon HS, Reyes-Prieto A, Melkonian M et al (2006) Minimal plastid genome evolution in the *Paulinella* endosymbiont. Curr Biol 16(17):R670–R6R2. https://doi.org/10.1016/j.cub.2006.08.018
10. Delaye L, Valadez-Cano C, Perez-Zamorano B (2016) How really ancient is *Paulinella chromatophora*? PLoS Curr 8:ecurrents.tol.e68a099364bb1a1e129a17b4e06b0c6b. https://doi.org/10.1371/currents.tol.e68a099364bb1a1e129a17b4e06b0c6b
11. Lauterborn R (1895) Protozoenstudien II. *Paulinella chromatophora* nov. gen., nov. spec., ein beschalter Rhizopode des Süßwassers mit blaugrünen chromatophorenartigen Einschlüssen. Z Wiss Zool 59:537–544
12. Bhattacharya D, Price DC, Yoon HS et al (2012) Single cell genome analysis supports a link between phagotrophy and primary plastid endosymbiosis. Sci Rep 2:356. https://doi.org/10.1038/srep00356
13. Kim S, Park MG (2016) *Paulinella longichromatophora* sp. nov., a new marine photosynthetic testate amoeba containing a chromatophore. Protist 167(1):1–12. https://doi.org/10.1016/j.protis.2015.11.003
14. Lhee D, Yang EC, Kim JI et al (2017) Diversity of the photosynthetic *Paulinella* species, with the description of *Paulinella micropora* sp. nov. and the chromatophore genome

sequence for strain KR01. Protist 168(2):155–170. https://doi.org/10.1016/j.protis.2017.01.003
15. Thompson AW, Zehr JP (2013) Cellular interactions: lessons from the nitrogen-fixing cyanobacteria. J Phycol 49(6):1024–1035. https://doi.org/10.1111/jpy.12117
16. Cavalier-Smith T, Lee JJ (1985) Protozoa as hosts for endosymbioses and the conversion of symbionts into organelles. J Protozool 32(3):376–379. https://doi.org/10.1111/j.1550-7408.1985.tb04031.x
17. Keeling PJ, Archibald JM (2008) Organelle evolution: what's in a name? Curr Biol 18(8):R345–R3R7. https://doi.org/10.1016/j.cub.2008.02.065
18. Bodył A, Mackiewicz P, Stiller JW (2007) The intracellular cyanobacteria of *Paulinella chromatophora*: endosymbionts or organelles? Trends Microbiol 15(7):295–296. https://doi.org/10.1016/j.tim.2007.05.002
19. Van de Velde W, Zehirov G, Szatmari A et al (2010) Plant peptides govern terminal differentiation of bacteria in symbiosis. Science 327(5969):1122–1126. https://doi.org/10.1126/science.1184057
20. McCutcheon JP, Keeling PJ (2014) Endosymbiosis: protein targeting further erodes the organelle/symbiont distinction. Curr Biol 24(14):R654–R6R5. https://doi.org/10.1016/j.cub.2014.05.073
21. Nakabachi A, Ishida K, Hongoh Y et al (2014) Aphid gene of bacterial origin encodes a protein transported to an obligate endosymbiont. Curr Biol 24(14):R640–R6R1. https://doi.org/10.1016/j.cub.2014.06.038
22. Morales J, Kokkori S, Weidauer D et al (2016) Development of a toolbox to dissect host-endosymbiont interactions and protein trafficking in the trypanosomatid *Angomonas deanei*. BMC Ecol Evol 16(1):247. https://doi.org/10.1186/s12862-016-0820-z
23. Oborník M (2019) In the beginning was the word: how terminology drives our understanding of endosymbiotic organelles. Microbial Cell (Graz, Austria) 6(2):134–141. https://doi.org/10.15698/mic2019.02.669
24. Gruber A (2019) What's in a name? How organelles of endosymbiotic origin can be distinguished from endosymbionts. Microbial Cell (Graz, Austria) 6(2):123–133. https://doi.org/10.15698/mic2019.02.668
25. Theissen U, Martin W (2006) The difference between organelles and endosymbionts. Curr Biol 16(24):R1016–R10R7. https://doi.org/10.1016/j.cub.2006.11.020
26. Bhattacharya D, Archibald JM (2006) Response to Theissen and Martin. Curr Biol 16(24):R1017–R10R8. https://doi.org/10.1016/j.cub.2006.11.021
27. Zimorski V, Ku C, Martin WF et al (2014) Endosymbiotic theory for organelle origins. Curr Opin Microbiol 22:38–48. https://doi.org/10.1016/j.mib.2014.09.008
28. Schimper AFW (1883) Über die Entwicklung der Chlorophyllkörper und Farbkörper. Botanische Zeitung 41(7):105–162
29. Mereschkowsky K (1905) Über Natur und Ursprung der Chromatophoren im Pflanzenreiche. Biologisches Centralblatt 25(18):593–604
30. Margulis L (1970) Origin of eukaryotic cells : evidence and research implications for a theory of the origin and evolution of microbial, plant, and animal cells on the Precambrian earth. Yale University Press, New Haven
31. Martin W, Kowallik K (1999) Annotated English translation of Mereschkowsky's 1905 paper 'Über Natur und Ursprung der Chromatophoren imPflanzenreiche'. Eur J Phycol 34(3):287–295. https://doi.org/10.1080/09670269910001736342
32. Sapp J (2004) The dynamics of symbiosis: an historical overview. Can J Bot 82(8):1046–1056. https://doi.org/10.1139/b04-055
33. Archibald JM (2015) Endosymbiosis and eukaryotic cell evolution. Curr Biol 25(19):R911–RR21. https://doi.org/10.1016/j.cub.2015.07.055
34. O'Malley MA (2015) Endosymbiosis and its implications for evolutionary theory. Proc Natl Acad Sci U S A 112(33):10270–10277. https://doi.org/10.1073/pnas.1421389112
35. Martin WF (2017) Physiology, anaerobes, and the origin of mitosing cells 50 years on. J Theor Biol 434:2–10. https://doi.org/10.1016/j.jtbi.2017.01.004

36. Sato N (2017) Revisiting the theoretical basis of the endosymbiotic origin of plastids in the original context of Lynn Margulis on the origin of mitosing, eukaryotic cells. J Theor Biol 434:104–113. https://doi.org/10.1016/j.jtbi.2017.08.028
37. Sato N (2019) Endosymbiotic theories of organelles revisited. In retrospects and prospects. Springer, Singapore
38. Hohmann-Marriott MF, Blankenship RE (2011) Evolution of photosynthesis. Annu Rev Plant Biol 62(1):515–548. https://doi.org/10.1146/annurev-arplant-042110-103811
39. Zepke HD, Heinz E, Radunz A et al (1978) Combination and positional distribution of fatty acids in lipids from blue-green algae. Arch Microbiol 119(2):157–162. https://doi.org/10.1007/BF00964267
40. Takano H, Takechi K (2010) Plastid peptidoglycan. Biochimica et Biophysica Acta (BBA) – General Subjects 1800(2):144–151. https://doi.org/10.1016/j.bbagen.2009.07.020
41. Hirano T, Tanidokoro K, Shimizu Y et al (2016) Moss chloroplasts are surrounded by a peptidoglycan wall containing D-amino acids. Plant Cell 28(7):1521–1532. https://doi.org/10.1105/tpc.16.00104
42. Fischer K, Weber A, Brink S et al (1994) Porins from plants. Molecular cloning and functional characterization of two new members of the porin family. J Biol Chem 269(41):25754–25760. https://doi.org/10.1016/S0021-9258(18)47312-7
43. Walther DM, Rapaport D, Tommassen J (2009) Biogenesis of beta-barrel membrane proteins in bacteria and eukaryotes: evolutionary conservation and divergence. Cell Mol Life Sci 66(17):2789–2804. https://doi.org/10.1007/s00018-009-0029-z
44. Possingham JV, Lawrence ME (1983) Controls to plastid division. In: Bourne GH, Danielli JF, Jeon KW (eds) International review of cytology. Academic Press, pp 1–56
45. Miyagishima S-y (2005) Origin and evolution of the chloroplast division machinery. J Plant Res 118(5):295–306. https://doi.org/10.1007/s10265-005-0226-2
46. Ris H, Plaut W (1962) Ultrastructure of DNA-containing areas in the chloroplast of *Chlamydomonas*. J Cell Biol 13(3):383–391. https://doi.org/10.1083/jcb.13.3.383
47. Shinozaki K, Ohme M, Tanaka M et al (1986) The complete nucleotide sequence of the tobacco chloroplast genome: its gene organization and expression. EMBO J 5(9):2043–2049. https://doi.org/10.1002/j.1460-2075.1986.tb04464.x
48. Giovannoni SJ, Turner S, Olsen GJ et al (1988) Evolutionary relationships among cyanobacteria and green chloroplasts. J Bacteriol 170(8):3584–3592. https://doi.org/10.1128/jb.170.8.3584-3592.1988
49. Ohta N, Sato N, Nozaki H et al (1997) Analysis of the cluster of ribosomal protein genes in the plastid genome of a unicellular red alga *Cyanidioschyzon merolae*: Translocation of the str cluster as an early event in the Rhodophyte-Chromophyte lineage of plastid evolution. J Mol Evol 45(6):688–695. https://doi.org/10.1007/PL00006273
50. Moore KR, Magnabosco C, Momper L et al (2019) An expanded ribosomal phylogeny of cyanobacteria supports a deep placement of plastids. Front Microbiol 10:1612. https://doi.org/10.3389/fmicb.2019.01612
51. Sato N (2020) Complex origins of chloroplast membranes with photosynthetic machineries: multiple transfers of genes from divergent organisms at different times or a single endosymbiotic event? J Plant Res 133(1):15–33. https://doi.org/10.1007/s10265-019-01157-z
52. Weeden NF (1981) Genetic and biochemical implications of the endosymbiotic origin of the chloroplast. J Mol Evol 17(3):133–139. https://doi.org/10.1007/bf01733906
53. Martin W, Rujan T, Richly E et al (2002) Evolutionary analysis of *Arabidopsis*, cyanobacterial, and chloroplast genomes reveals plastid phylogeny and thousands of cyanobacterial genes in the nucleus. Proc Natl Acad Sci U S A 99(19):12246–12251. https://doi.org/10.1073/pnas.182432999
54. Sato N, Ishikawa M, Fujiwara M et al (2005) Mass identification of chloroplast proteins of endosymbiont origin by phylogenetic profiling based on organism-optimized homologous protein groups. Genome Inform 16(2):56–68
55. Deusch O, Landan G, Roettger M et al (2008) Genes of cyanobacterial origin in plant nuclear genomes point to a heterocyst-forming plastid ancestor. Mol Biol Evol 25(4):748–761. https://doi.org/10.1093/molbev/msn022

56. Moustafa A, Bhattacharya D (2008) PhyloSort: a user-friendly phylogenetic sorting tool and its application to estimating the cyanobacterial contribution to the nuclear genome of *Chlamydomonas*. BMC Evol Biol 8(1):6. https://doi.org/10.1186/1471-2148-8-6
57. Dagan T, Roettger M, Stucken K et al (2013) Genomes of Stigonematalean cyanobacteria (subsection V) and the evolution of oxygenic photosynthesis from prokaryotes to plastids. Genome Biol Evol 5(1):31–44. https://doi.org/10.1093/gbe/evs117
58. Qiu H, Price DC, Weber APM et al (2013) Assessing the bacterial contribution to the plastid proteome. Trends Plant Sci 18(12):680–687. https://doi.org/10.1016/j.tplants.2013.09.007
59. Karpowicz SJ, Prochnik SE, Grossman AR et al (2011) The GreenCut2 resource, a phylogenomically derived inventory of proteins specific to the plant lineage. J Biol Chem 286(24):21427–21439. https://doi.org/10.1074/jbc.M111.233734
60. Stoebe B, Kowallik KV (1999) Gene-cluster analysis in chloroplast genomics. Trends Genet 15(9):344–347. https://doi.org/10.1016/S0168-9525(99)01815-6
61. Löffelhardt W (2014) The single primary endosymbiotic event. In: Löffelhardt W (ed) Endosymbiosis. Springer Vienna, Vienna, pp 39–52
62. Kies LKBP (1979) Function of cyanelles in the thecamoeba *Paulinella chromatophora*. Naturwissenschaften 66(11):578–579
63. Gabr A, Zournas A, Stephens TG et al (2022) Evidence for a robust photosystem II in the photosynthetic amoeba *Paulinella*. New Phytol 234(3):934–945. https://doi.org/10.1111/nph.18052
64. Zhang R, Nowack ECM, Price DC et al (2017) Impact of light intensity and quality on chromatophore and nuclear gene expression in *Paulinella chromatophora*, an amoeba with nascent photosynthetic organelles. Plant J 90(2):221–234. https://doi.org/10.1111/tpj.13488
65. Kies L (1974) Elektronenmikroskopische Untersuchungen an *Paulinella chromatophora* Lauterborn, einer Thekamöbe mit blau-grünen Endosymbionten (Cyanellen). Protoplasma 80(1):69–89. https://doi.org/10.1007/bf01666352
66. Sato N, Yoshitomi T, Mori-Moriyama N (2020) Characterization and biosynthesis of lipids in *Paulinella micropora* MYN1: evidence for efficient integration of chromatophores into cellular lipid metabolism. Plant Cell Physiol 61(5):869–881. https://doi.org/10.1093/pcp/pcaa011
67. Nomura M, Ishida K-i (2016) Fine-structural observations on siliceous scale production and shell assembly in the testate amoeba *Paulinella chromatophora*. Protist 167(4):303–318. https://doi.org/10.1016/j.protis.2016.05.002
68. Lukavský J, Cepak V (1992) DAPI fluorescent staining of DNA material in cyanelles of the Rhizopod *Paulinella chromatophora* LAUTERB. Archiv für Protistenkunde 142(3):207–212. https://doi.org/10.1016/S0003-9365(11)80086-2
69. Nomura M, Nakayama T, Ishida K-i (2014) Detailed process of shell construction in the photosynthetic testate amoeba *Paulinella chromatophora* (Euglyphid, Rhizaria). J Eukaryot Microbiol 61(3):317–321. https://doi.org/10.1111/jeu.12102
70. Lhee D, Ha J-S, Kim S et al (2019) Evolutionary dynamics of the chromatophore genome in three photosynthetic *Paulinella* species. Sci Rep 9(1):2560. https://doi.org/10.1038/s41598-019-38621-8
71. Nakayama T, Ishida K-i (2009) Another acquisition of a primary photosynthetic organelle is underway in *Paulinella chromatophora*. Curr Biol 19(7):R284–R2R5. https://doi.org/10.1016/j.cub.2009.02.043
72. Reyes-Prieto A, Yoon HS, Moustafa A et al (2010) Differential gene retention in plastids of common recent origin. Mol Biol Evol 27(7):1530–1537. https://doi.org/10.1093/molbev/msq032
73. Nowack ECM, Price DC, Bhattacharya D et al (2016) Gene transfers from diverse bacteria compensate for reductive genome evolution in the chromatophore of *Paulinella chromatophora*. Proc Natl Acad Sci U S A 113(43):12214–12219. https://doi.org/10.1073/pnas.1608016113
74. Lhee D, Lee J, Ettahi K et al (2020) Amoeba genome reveals dominant host contribution to plastid endosymbiosis. Mol Biol Evol 38(2):344–357. https://doi.org/10.1093/molbev/msaa206

75. Calatrava V, Stephens TG, Gabr A et al (2022) Retrotransposition facilitated the establishment of a primary plastid in the thecate amoeba *Paulinella*. Proc Natl Acad Sci U S A 119(23):e2121241119. https://doi.org/10.1073/pnas.2121241119
76. Nowack EC, Melkonian M, Glockner G (2008) Chromatophore genome sequence of *Paulinella* sheds light on acquisition of photosynthesis by eukaryotes. Curr Biol 18(6):410–418. https://doi.org/10.1016/j.cub.2008.02.051
77. Suzuki K, Miyagishima S-y (2010) Eukaryotic and eubacterial contributions to the establishment of plastid proteome estimated by large-scale phylogenetic analyses. Mol Biol Evol 27(3):581–590. https://doi.org/10.1093/molbev/msp273
78. Reyes-Prieto A, Bhattacharya D (2007) Phylogeny of Calvin cycle enzymes supports Plantae monophyly. Mol Phylogenet Evol 45(1):384–391. https://doi.org/10.1016/j.ympev.2007.02.026
79. Reyes-Prieto A, Moustafa A (2012) Plastid-localized amino acid biosynthetic pathways of Plantae are predominantly composed of non-cyanobacterial enzymes. Sci Rep 2(1):955. https://doi.org/10.1038/srep00955
80. Tyra HM, Linka M, Weber APM et al (2007) Host origin of plastid solute transporters in the first photosynthetic eukaryotes. Genome Biol 8(10):R212. https://doi.org/10.1186/gb-2007-8-10-r212
81. Facchinelli F, Weber A (2011) The metabolite transporters of the plastid envelope: an update. Front Plant Sci 2:50. https://doi.org/10.3389/fpls.2011.00050
82. Karkar S, Facchinelli F, Price DC et al (2015) Metabolic connectivity as a driver of host and endosymbiont integration. Proc Natl Acad Sci U S A 112(33):10208–10215. https://doi.org/10.1073/pnas.1421375112
83. Weber AP, Linka M, Bhattacharya D (2006) Single, ancient origin of a plastid metabolite translocator family in Plantae from an endomembrane-derived ancestor. Eukaryot Cell 5(3):609–612. https://doi.org/10.1128/ec.5.3.609-612.2006
84. Delwiche CF, Palmer JD (1996) Rampant horizontal transfer and duplication of rubisco genes in eubacteria and plastids. Mol Biol Evol 13(6):873–882. https://doi.org/10.1093/oxfordjournals.molbev.a025647
85. Gross J, Meurer J, Bhattacharya D (2008) Evidence of a chimeric genome in the cyanobacterial ancestor of plastids. BMC Evol Biol 8(1):117. https://doi.org/10.1186/1471-2148-8-117
86. Cenci U, Qiu H, Pillonel T et al (2018) Host-pathogen biotic interactions shaped vitamin K metabolism in Archaeplastida. Sci Rep 8(1):15243. https://doi.org/10.1038/s41598-018-33663-w
87. Singer A, Poschmann G, Muhlich C et al (2017) Massive protein import into the early-evolutionary-stage photosynthetic organelle of the amoeba *Paulinella chromatophora*. Curr Biol 27(18):2763–73 e5. https://doi.org/10.1016/j.cub.2017.08.010
88. Uzzell T, Spolsky C (1974) Mitochondria and plastids as endosymbionts: a revival of special creation? The similarities between cellular organelles and prokaryotes are probably primitive features retained independently from a common ancestor. Am Sci 62(3):334–343
89. Cavalier-Smith T (1975) The origin of nuclei and of eukaryotic cells. Nature 256(5517):463–468. https://doi.org/10.1038/256463a0
90. Sato N (2021) Are cyanobacteria an ancestor of chloroplasts or just one of the gene donors for plants and algae? Genes 12(6):823. https://doi.org/10.3390/genes12060823
91. Sato N, Takano H (2017) Diverse origins of enzymes involved in the biosynthesis of chloroplast peptidoglycan. J Plant Res 130(4):635–645. https://doi.org/10.1007/s10265-017-0935-3
92. Sato N, Awai K (2017) "Prokaryotic pathway" is not prokaryotic: noncyanobacterial origin of the chloroplast lipid biosynthetic pathway revealed by comprehensive phylogenomic analysis. Genome Biol Evol 9(11):3162–3178. https://doi.org/10.1093/gbe/evx238
93. Keeling PJ (2013) The number, speed, and impact of plastid endosymbioses in eukaryotic evolution. Annu Rev Plant Biol 64(1):583–607. https://doi.org/10.1146/annurev-arplant-050312-120144
94. Bodył A, Mackiewicz P, Stiller JW (2009) Early steps in plastid evolution: current ideas and controversies. Bioessays 31(11):1219–1232. https://doi.org/10.1002/bies.200900073

95. Villarejo A, Burén S, Larsson S et al (2005) Evidence for a protein transported through the secretory pathway en route to the higher plant chloroplast. Nat Cell Biol 7(12):1224–1231. https://doi.org/10.1038/ncb1330
96. Bhattacharya D, Archibald JM, Weber AP et al (2007) How do endosymbionts become organelles? Understanding early events in plastid evolution. Bioessays 29(12):1239–1246. https://doi.org/10.1002/bies.20671
97. Larkum AWD, Lockhart PJ, Howe CJ (2007) Shopping for plastids. Trends Plant Sci 12(5):189–195. https://doi.org/10.1016/j.tplants.2007.03.011
98. Foster PG, Cox CJ, Embley TM (2009) The primary divisions of life: a phylogenomic approach employing composition-heterogeneous methods. Philos Trans R Soc Lond B Biol Sci 364(1527):2197–2207. https://doi.org/10.1098/rstb.2009.0034
99. Philippe H, Roure B (2011) Difficult phylogenetic questions: more data, maybe; better methods, certainly. BMC Biol 9(1):91. https://doi.org/10.1186/1741-7007-9-91
100. Domman D, Horn M, Embley TM et al (2015) Plastid establishment did not require a chlamydial partner. Nat Commun 6(1):6421. https://doi.org/10.1038/ncomms7421
101. Huang J, Gogarten JP (2007) Did an ancient chlamydial endosymbiosis facilitate the establishment of primary plastids? Genome Biol 8(6):R99. https://doi.org/10.1186/gb-2007-8-6-r99
102. Ball SG, Colleoni C, Kadouche D et al (2015) Toward an understanding of the function of Chlamydiales in plastid endosymbiosis. Biochim Biophys Acta 1847(6):495–504. https://doi.org/10.1016/j.bbabio.2015.02.007
103. Becker B, Hoef-Emden K, Melkonian M (2008) Chlamydial genes shed light on the evolution of photoautotrophic eukaryotes. BMC Evol Biol 8(1):203. https://doi.org/10.1186/1471-2148-8-203
104. Moustafa A, Reyes-Prieto A, Bhattacharya D (2008) Chlamydiae has contributed at least 55 genes to Plantae with predominantly plastid functions. PloS One 3(5):e2205. https://doi.org/10.1371/journal.pone.0002205
105. Ball SG, Subtil A, Bhattacharya D et al (2013) Metabolic effectors secreted by bacterial pathogens: essential facilitators of plastid endosymbiosis? Plant Cell 25(1):7–21. https://doi.org/10.1105/tpc.112.101329
106. Ball SG (2014) Evolution of storage polysaccharide metabolism in Archaeplastida opens an unexpected window on the molecular mechanisms that drove plastid endosymbiosis. In: Löffelhardt W (ed) Endosymbiosis. Springer Vienna, Vienna, pp 111–134
107. Deschamps P, Colleoni C, Nakamura Y et al (2008) Metabolic symbiosis and the birth of the plant kingdom. Mol Biol Evol 25(3):536–548. https://doi.org/10.1093/molbev/msm280
108. Cenci U, Bhattacharya D, Weber APM et al (2017) Biotic host–pathogen interactions as major drivers of plastid endosymbiosis. Trends Plant Sci 22(4):316–328. https://doi.org/10.1016/j.tplants.2016.12.007
109. Ball SG, Greub G (2015) Blurred pictures from the crime scene: the growing case for a function of Chlamydiales in plastid endosymbiosis. Microbes Infect 17(11):723–726. https://doi.org/10.1016/j.micinf.2015.09.007
110. Lhee D, Bhattacharya D, Yoon HS (2021) Independent evolution of the thioredoxin system in photosynthetic *Paulinella* species. Curr Biol 31(7):R328–R3r9. https://doi.org/10.1016/j.cub.2021.02.029
111. Gabr A, Stephens TG, Bhattacharya D (2022) Loss of key endosymbiont genes may facilitate early host control of the chromatophore in *Paulinella*. iScience 25(9):104974. https://doi.org/10.1016/j.isci.2022.104974
112. Gabr A, Stephens TG, Bhattacharya D (2022) Hypothesis: trans-splicing generates evolutionary novelty in the photosynthetic amoeba *Paulinella*. J Phycol 58(3):392–405. https://doi.org/10.1111/jpy.13247
113. Cejudo FJ, González MC, Pérez-Ruiz JM (2021) Redox regulation of chloroplast metabolism. Plant Physiol 186(1):9–21. https://doi.org/10.1093/plphys/kiaa062
114. Bhattacharya D, Etten JV, Benites LF et al (2022) Endosymbiotic ratchet accelerates divergence after organelle origin. Bioessays n/a(n/a):2200165. https://doi.org/10.1002/bies.202200165

115. Ku C, Roettger M, Zimorski V et al (2014) Plastid origin: who, when and why? Acta Soc Bot Pol 83(4):281–289. https://doi.org/10.5586/asbp.2014.045
116. Ponce-Toledo RI, Deschamps P, López-García P et al (2017) An early-branching freshwater cyanobacterium at the origin of plastids. Curr Biol 27(3):386–391. https://doi.org/10.1016/j.cub.2016.11.056
117. Sánchez-Baracaldo P (2015) Origin of marine planktonic cyanobacteria. Sci Rep 5(1):17418. https://doi.org/10.1038/srep17418
118. Maruyama S, Kim E (2013) A modern descendant of early green algal phagotrophs. Curr Biol 23(12):1081–1084. https://doi.org/10.1016/j.cub.2013.04.063
119. Gawryluk RMR, Tikhonenkov DV, Hehenberger E et al (2019) Non-photosynthetic predators are sister to red algae. Nature 572(7768):240–243. https://doi.org/10.1038/s41586-019-1398-6
120. Gavelis GS, Gile GH (2018) How did cyanobacteria first embark on the path to becoming plastids?: lessons from protist symbioses. FEMS Microbiol Lett 365(19):fny209. https://doi.org/10.1093/femsle/fny209
121. Siegmund L, Burmester A, Fischer MS et al (2013) A model for endosymbiosis: interaction between *Tetrahymena pyriformis* and *Escherichia coli*. Eur J Protistol 49(4):552–563. https://doi.org/10.1016/j.ejop.2013.04.007
122. Kodama Y, Sumita H (2022) The ciliate *Paramecium bursaria* allows budding of symbiotic *Chlorella variabilis* cells singly from the digestive vacuole membrane into the cytoplasm during algal reinfection. Protoplasma 259(1):117–125. https://doi.org/10.1007/s00709-021-01645-x
123. Uribe-Querol E, Rosales C (2017) Control of phagocytosis by microbial pathogens. Front Immunol 8:1368. https://doi.org/10.3389/fimmu.2017.01368
124. Gómez F, Furuya K, Takeda S (2005) Distribution of the cyanobacterium *Richelia intracellularis* as an epiphyte of the diatom *Chaetoceros compressus* in the western Pacific Ocean. J Plankton Res 27(4):323–330. https://doi.org/10.1093/plankt/fbi007
125. Marin B, Nowack ECM, Melkonian M (2005) A Plastid in the making: evidence for a second primary endosymbiosis. Protist 156(4):425–432. https://doi.org/10.1016/j.protis.2005.09.001
126. Pankow VH (1982) *Paulinella chromatophora* Lauterb., eine bisher nur im Süßwasser nachgewiesene Thekamöbe, in den Boddengewässern des Darß and des Zingst (südliche Ostsee). Archiv für Protistenkunde 126(3):261–263. https://doi.org/10.1016/s0003-9365(82)80036-5
127. Johnson PW, Hargraves PE, Sieburth JM (1988) Ultrastructure and ecology of *Calycomonas ovalis* Wulff, 1919, (Chrysophyceae) and its redescription as a testate rhizopod, *Paulinella ovalis* N. Comb. (Filosea: Euglyphina). J Protozool 35(4):618–626. https://doi.org/10.1111/j.1550-7408.1988.tb04160.x
128. VØRs N (1993) Marine heterotrophic amoebae, flagellates and heliozoa from Belize (Central America) and Tenerife (Canary Islands), with descriptions of new species, *Luffisphaera bulbochaete* n. sp., *L. longihastis* n. sp., *L. turriformis* n. sp. and *Paulinella intermedia* n. sp. J Eukaryot Microbiol 40(3):272–287. https://doi.org/10.1111/j.1550-7408.1993.tb04917.x
129. Hannah F, Rogerson A, Anderson OR (1996) A description of *Paulinella indentata* n. sp. (Filosea: Euglyphina) from subtidal coastal benthic sediments. J Eukaryot Microbiol 43(1):1–4. https://doi.org/10.1111/j.1550-7408.1996.tb02464.x
130. Nicholls KH (2009) Six new marine species of the genus *Paulinella* (Rhizopoda: Filosea, or Rhizaria: Cercozoa). J Mar Biol Assoc U K 89(07):1415. https://doi.org/10.1017/s0025315409000514
131. Martin W, Müller M (1998) The hydrogen hypothesis for the first eukaryote. Nature 392(6671):37–41. https://doi.org/10.1038/32096
132. Johnston DT, Wolfe-Simon F, Pearson A et al (2009) Anoxygenic photosynthesis modulated Proterozoic oxygen and sustained Earth's middle age. Proc Natl Acad Sci U S A 106(40):16925–16929. https://doi.org/10.1073/pnas.0909248106

133. Finlay BJ, Maberly SC, Esteban GF (1996) Spectacular abundance of ciliates in anoxic pond water: contribution of symbiont photosynthesis to host respiratory oxygen requirements. FEMS Microbiol Ecol 20(4):229–235. https://doi.org/10.1111/j.1574-6941.1996.tb00321.x
134. Esteban GF, Finlay BJ, Clarke KJ (2009) Sequestered organelles sustain aerobic microbial life in anoxic environments. Environ Microbiol 11(2):544–550. https://doi.org/10.1111/j.1462-2920.2008.01797.x
135. Bustos-Díaz ED, Barona-Gómez F, Cibrián-Jaramillo A (2019) Chapter 2: Cyanobacteria in nitrogen-fixing symbioses. In: Mishra AK, Tiwari DN, Rai AN (eds) Cyanobacteria. Academic Press, pp 29–42
136. Latysheva N, Junker VL, Palmer WJ et al (2012) The evolution of nitrogen fixation in cyanobacteria. Bioinformatics 28(5):603–606. https://doi.org/10.1093/bioinformatics/bts008
137. Stephens TG, Gabr A, Calatrava V et al (2021) Why is primary endosymbiosis so rare? New Phytol 231(5):1693–1699. https://doi.org/10.1111/nph.17478
138. Mackiewicz P, Gagat P (2014) Monophyly of Archaeplastida supergroup and relationships among its lineages in the light of phylogenetic and phylogenomic studies. Are we close to a consensus? Acta Soc Bot Pol 83(4):263
139. McFadden GI, van Dooren GG (2004) Evolution: Red algal genome affirms a common origin of all plastids. Curr Biol 14(13):R514–R5R6. https://doi.org/10.1016/j.cub.2004.06.041
140. Matsuzaki M, Misumi O, Shin-i T et al (2004) Genome sequence of the ultrasmall unicellular red alga *Cyanidioschyzon merolae* 10D. Nature 428(6983):653–657. https://doi.org/10.1038/nature02398
141. Price DC, Chan CX, Yoon HS et al (2012) *Cyanophora paradoxa* genome elucidates origin of photosynthesis in algae and plants. Science 335(6070):843–847. https://doi.org/10.1126/science.1213561
142. Shih PM, Wu D, Latifi A et al (2013) Improving the coverage of the cyanobacterial phylum using diversity-driven genome sequencing. Proc Natl Acad Sci U S A 110(3):1053–1058. https://doi.org/10.1073/pnas.1217107110
143. Burki F, Kaplan M, Tikhonenkov DV et al (2016) Untangling the early diversification of eukaryotes: a phylogenomic study of the evolutionary origins of Centrohelida, Haptophyta and Cryptista. Proc R Soc Lond B Biol Sci 283(1823):20152802. https://doi.org/10.1098/rspb.2015.2802
144. Strassert JFH, Jamy M, Mylnikov AP et al (2019) New phylogenomic analysis of the enigmatic phylum telonemia further resolves the eukaryote tree of life. Mol Biol Evol 36(4):757–765. https://doi.org/10.1093/molbev/msz012
145. Lax G, Eglit Y, Eme L et al (2018) Hemimastigophora is a novel supra-kingdom-level lineage of eukaryotes. Nature 564(7736):410–414. https://doi.org/10.1038/s41586-018-0708-8
146. Strassert JFH, Irisarri I, Williams TA et al (2021) A molecular timescale for eukaryote evolution with implications for the origin of red algal-derived plastids. Nat Commun 12(1):1879. https://doi.org/10.1038/s41467-021-22044-z
147. Irisarri I, Strassert JFH, Burki F (2022) Phylogenomic insights into the origin of primary plastids. Syst Biol 71(1):105–120. https://doi.org/10.1093/sysbio/syab036
148. Yazaki E, Yabuki A, Imaizumi A et al (2022) The closest lineage of Archaeplastida is revealed by phylogenomics analyses that include *Microheliella maris*. Open Biol 12(4):210376. https://doi.org/10.1098/rsob.210376
149. Stiller JW, Reel DC, Johnson JC (2003) A single origin of plastids revisited: convergent evolution in organellar genome content. J Phycol 39(1):95–105. https://doi.org/10.1046/j.1529-8817.2003.02070.x
150. Schön ME, Zlatogursky VV, Singh RP et al (2021) Single cell genomics reveals plastid-lacking Picozoa are close relatives of red algae. Nat Commun 12(1):6651. https://doi.org/10.1038/s41467-021-26918-0
151. Kim E, Maruyama S (2014) A contemplation on the secondary origin of green algal and plant plastids. Acta Soc Bot Pol 83(4):331–336. https://doi.org/10.5586/asbp.2014.040
152. Schnepf E, Elbrächter M (1988) Cryptophycean-like double membrane-bound chloroplast in the dinoflagellate, *Dinophysis Ehrenb.*: evolutionary, phylogenetic and toxicological implications. Plant Biol 101:196–203

153. Fujishima M, Kodama Y (2022) Mechanisms for establishing primary and secondary endosymbiosis in *Paramecium*. J Eukaryot Microbiol 69(5):e12901. https://doi.org/10.1111/jeu.12901
154. Ohkawa H, Hashimoto N, Furukawa S et al (2011) Forced symbiosis between *Synechocystis* spp. PCC 6803 and apo-symbiotic *Paramecium bursaria* as an experimental model for evolutionary emergence of primitive photosynthetic eukaryotes. Plant Signal Behav 6(6):773–776. https://doi.org/10.4161/psb.6.6.15239
155. Maeda Y, Abe T (2022) Experimental induction of heterotrophic to autotrophic conversion, realized by the enforced primary endosymbiosis of photosynthetic bacteria onto eukaryotic amoebae. Nat Sci 14(9):364–385
156. Agapakis CM, Niederholtmeyer H, Noche RR et al (2011) Towards a synthetic chloroplast. PloS One 6(4):e18877. https://doi.org/10.1371/journal.pone.0018877
157. Gründel M, Scheunemann R, Lockau W et al (2012) Impaired glycogen synthesis causes metabolic overflow reactions and affects stress responses in the cyanobacterium *Synechocystis* sp. PCC 6803. Microbiology 158(12):3032–3043. https://doi.org/10.1099/mic.0.062950-0
158. Cournoyer JE, Altman SD, Gao Y-l et al (2022) Engineering artificial photosynthetic lifeforms through endosymbiosis. Nat Commun 13(1):2254. https://doi.org/10.1038/s41467-022-29961-7
159. Yang B, Li S, Mu W et al (2022) Light-harvesting artificial cells containing cyanobacteria for CO_2 fixation and further metabolism mimicking. Small 19(13):e2201305. https://doi.org/10.1002/smll.202201305
160. Lee KY, Park S-J, Lee KA et al (2018) Photosynthetic artificial organelles sustain and control ATP-dependent reactions in a protocellular system. Nat Biotechnol 36(6):530–535. https://doi.org/10.1038/nbt.4140
161. Yoon HS, Müller KM, Sheath RG et al (2006) Defining the major lineages of red algae (Rhodophyta). J Phycol 42(2):482–492. https://doi.org/10.1111/j.1529-8817.2006.00210.x
162. Leliaert F, Smith DR, Moreau H et al (2012) Phylogeny and molecular evolution of the green algae. Crit Rev Plant Sci 31(1):1–46. https://doi.org/10.1080/07352689.2011.615705
163. Jackson CJ, Clayden SL, Reyes-Prieto A (2015) The Glaucophyta: the blue-green plants in a nutshell. Acta Soc Bot Pol 84:149–165. https://doi.org/10.5586/asbp.2015.020
164. Brawley SH, Blouin NA, Ficko-Blean E et al (2017) Insights into the red algae and eukaryotic evolution from the genome of *Porphyra umbilicalis* (Bangiophyceae, Rhodophyta). Proc Natl Acad Sci U S A 114(31):E6361–E6E70. https://doi.org/10.1073/pnas.1703088114
165. Bag P (2021) Light harvesting in fluctuating environments: evolution and function of antenna proteins across photosynthetic lineage. Plants 10(6):1184. https://doi.org/10.3390/plants10061184
166. Deschamps P, Haferkamp I, d'Hulst C et al (2008) The relocation of starch metabolism to chloroplasts: when, why and how. Trends Plant Sci 13(11):574–582. https://doi.org/10.1016/j.tplants.2008.08.009
167. Matsuo M, Katahata A, Tachikawa M et al (2019) Large DNA virus promoted the endosymbiotic evolution to make a photosynthetic eukaryote. bioRxiv:809541. https://doi.org/10.1101/809541
168. Stern DS, Higgs DC, Yang J (1997) Transcription and translation in chloroplasts. Trends Plant Sci 2(8):308–315. https://doi.org/10.1016/S1360-1385(97)89953-0
169. Puthiyaveetil S, McKenzie SD, Kayanja GE et al (2021) Transcription initiation as a control point in plastid gene expression. Biochim Biophys Acta Gene Regul Mech 1864(3):194689. https://doi.org/10.1016/j.bbagrm.2021.194689
170. Chi W, Sun X, Zhang L (2013) Intracellular signaling from plastid to nucleus. Annu Rev Plant Biol 64:559–582. https://doi.org/10.1146/annurev-arplant-050312-120147
171. Nowack ECM, Vogel H, Groth M et al (2011) Endosymbiotic gene transfer and transcriptional regulation of transferred genes in *Paulinella chromatophora*. Mol Biol Evol 28(1):407–422. https://doi.org/10.1093/molbev/msq209

172. Oberleitner L, Poschmann G, Macorano L et al (2020) The puzzle of metabolite exchange and identification of putative octotrico peptide repeat expression regulators in the nascent photosynthetic organelles of *Paulinella chromatophora*. Front Microbiol 11:607182. https://doi.org/10.3389/fmicb.2020.607182
173. Chen K, Chen X, Schnell DJ (2000) Mechanism of protein import across the chloroplast envelope. Biochem Soc Trans 28(4):485–491
174. Nowack ECM, Grossman AR (2012) Trafficking of protein into the recently established photosynthetic organelles of *Paulinella chromatophora*. Proc Natl Acad Sci U S A 109(14):5340–5345. https://doi.org/10.1073/pnas.1118800109
175. Oberleitner L, Perrar A, Macorano L et al (2022) A bipartite chromatophore transit peptide and N-terminal protein processing in the *Paulinella chromatophore*. Plant Physiol 189(1):152–164. https://doi.org/10.1093/plphys/kiac012
176. Li Z, Wakao S, Fischer BB et al (2009) Sensing and responding to excess light. Annu Rev Plant Biol 60(1):239–260. https://doi.org/10.1146/annurev.arplant.58.032806.103844
177. Ferrari C, Proost S, Janowski M et al (2019) Kingdom-wide comparison reveals the evolution of diurnal gene expression in Archaeplastida. Nat Commun 10(1):737. https://doi.org/10.1038/s41467-019-08703-2
178. Ragel P, Streb S, Feil R et al (2013) Loss of starch granule initiation has a deleterious effect on the growth of arabidopsis plants due to an accumulation of ADP-glucose. Plant Physiol 163(1):75–85. https://doi.org/10.1104/pp.113.223420
179. Krishnan A, Kumaraswamy GK, Vinyard DJ et al (2015) Metabolic and photosynthetic consequences of blocking starch biosynthesis in the green alga *Chlamydomonas reinhardtii* sta6 mutant. Plant J 81(6):947–960. https://doi.org/10.1111/tpj.12783
180. Huang W, Haferkamp I, Lepetit B et al (2018) Reduced vacuolar β-1,3-glucan synthesis affects carbohydrate metabolism as well as plastid homeostasis and structure in *Phaeodactylum tricornutum*. Proc Natl Acad Sci U S A 115(18):4791–4796. https://doi.org/10.1073/pnas.1719274115
181. Ball S, Colleoni C, Cenci U et al (2011) The evolution of glycogen and starch metabolism in eukaryotes gives molecular clues to understand the establishment of plastid endosymbiosis. J Exp Bot 62(6):1775–1801. https://doi.org/10.1093/jxb/erq411
182. Ball S, Colleoni C, Arias MC (2015) The transition from glycogen to starch metabolism in cyanobacteria and eukaryotes. In: Nakamura Y (ed) Starch: metabolism and structure. Springer Japan, Tokyo, pp 93–158
183. McFadden GI, Gilson PR, Sims IM (1997) Preliminary characterization of carbohydrate stores from chlorarachniophytes (Division: Chlorarachniophyta). Phycol Res 45(3):145–151. https://doi.org/10.1111/j.1440-1835.1997.tb00087.x
184. Miyagishima SY, Nakanishi H, Kabeya Y (2011) Structure, regulation, and evolution of the plastid division machinery. Int Rev Cell Mol Biol 291:115–153. https://doi.org/10.1016/b978-0-12-386035-4.00004-5
185. Ponce-Toledo RI, López-García P, Moreira D (2019) Horizontal and endosymbiotic gene transfer in early plastid evolution. New Phytol 224(2):618–624. https://doi.org/10.1111/nph.15965

Complex Plastids Across the Eukaryotes: An Overview of Inherited and Convergently Evolved Characters

Anna M. G. Novák Vanclová and Richard G. Dorrell

Abstract All plastids are ultimately descended from a cyanobacterial ancestor integrated by a eukaryotic cell through primary endosymbiosis. However, most of the diversity of plastid-bearing organisms is the result of endosymbiotic exchange between eukaryotes that have occurred independently and repeatedly in various unrelated lineages. During each of these plastid re-establishments, gene transfer to the host nucleus, formation of protein and metabolite import into the plastid, spatial integration of the new organelle into the host cell and membrane architecture, and rewiring of metabolic pathways around it must take place. Remarkably, highly similar solutions using similar or identical tools and building blocks are often employed in unrelated host-organelle systems. In this chapter, we attempt to summarize and compare various characters across the currently known plastid-type diversity and discuss the trends in their evolution—conservation, reduction, diversification, or convergence—and the possible driving or constraining mechanisms behind them.

Keywords Serial endosymbiosis · Plastid replacement · Endosymbiont integration · Convergent evolution · Pre-adaptation · Character evolution

1 Introduction

The acquisition of photosynthesis, via the endosymbiotic uptake of chloroplasts or "plastids", has had dramatic impacts on the functional biology of eukaryotic life. Photosynthetic eukaryotes, i.e. plants and eukaryotic algae, are evolutionarily diverse and are broadly distributed across the eukaryotic tree of life (Fig. 1). However, their plastids can all be ultimately traced to a singular or possibly two

A. M. G. Novák Vanclová (✉)
Institute Jacques Monod, Paris, France

R. G. Dorrell (✉)
Sorbonne Université, Paris, France

S. D. Schwartzbach et al. (eds.), *Endosymbiotic Organelle Acquisition*,
https://doi.org/10.1007/978-3-031-57446-7_2

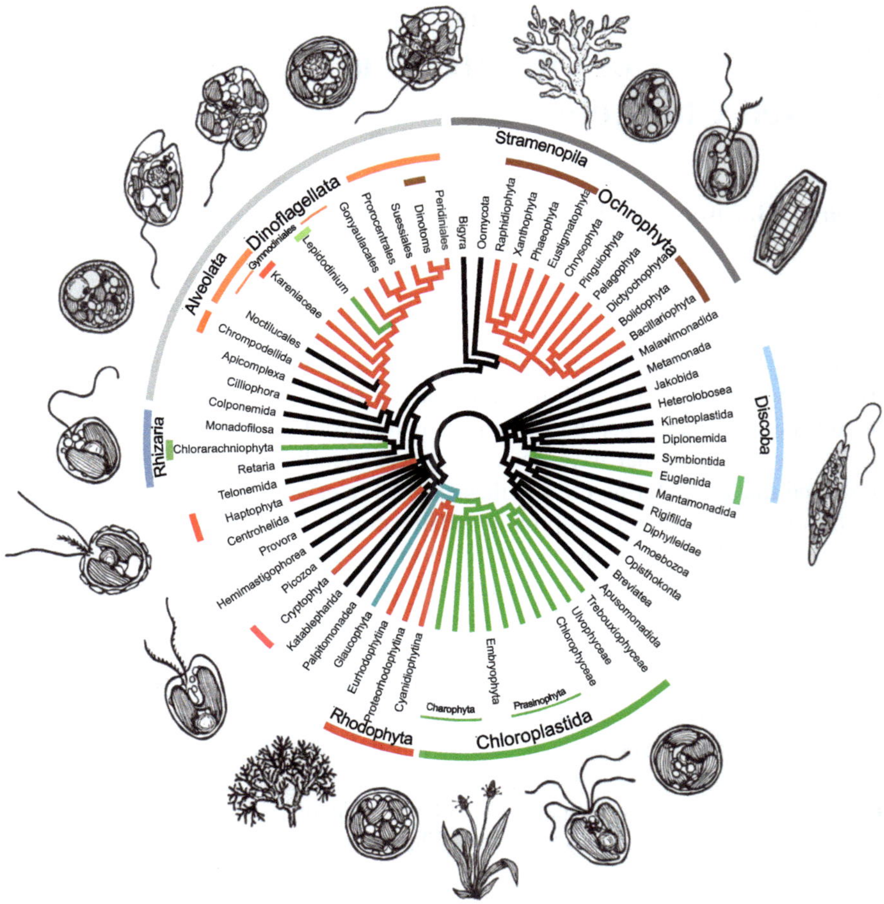

Fig. 1 Overview of the currently accepted eukaryotic tree of life with focus on plastid-bearing lineages discussed in this chapter. Lineages harbouring green- or red-derived plastids are denoted by the respective branch colour. Larger taxa comprised of or containing these photosynthetic lineages are delineated by arcs of different illustrative colours, and examples of their representatives are drawn along their side. The topology of the unrooted pan-eukaryotic tree is synthesized from recent publications focusing on new deep eukaryote relationships and new lineages as well as the inner taxonomy of particular groups [30, 48, 105, 126, 138, 155, 193, 206, 219, 220]

symbiogenetic events [201] between photosynthetic cyanobacterium and a suspected common ancestor of Archaeplastida. This supergroup is regrouped as follows: land plants, green algae, red algae, rhodelphids (secondarily non-photosynthetic relatives of red algae; [73]), picozoa (non-photosynthetic protozoa with no apparent endosymbiotic history; [193]), and glaucophytes. This occurred between 1.5 and 2 billion years ago (bya) [169, 205, 242], i.e. very early in the history of eukaryotes that arose between 2 and 2.5 bya [14, 169, 205]. The freshwater cyanobacterium *Gloeomargarita* was identified as the closest known relative of the plastid ancestor [176]. The deep relationships between the three groups of Archaeplastida are not yet resolved.

Glaucophytes are unicellular freshwater and terrestrial algae and the least diversified and studied group of Archaeplastida. Their plastids (also known as cyanelles) contain chlorophyll *a*, are blue-green in colour, and retain several ancestral features reminiscent of their cyanobacterial predecessor, such as a peptidoglycan envelope and bulky light-harvesting structures called phycobilisomes [115]. Rhodophyta and Chloroplastida, on the other hand, are large and diverse groups that contain unicellular, colonial, and multicellular representatives, both microscopic and macroscopic. Rhodophytes are characterized by their lack of flagella and centrioles and very limited motility, almost exclusively marine habitat, and presence of phycobilins and chlorophyll *a*. They are termed red algae, but their colour varies from dark red to light green, depending on the ratio between chlorophyll and other pigments.

Rhodophyta are grouped into two major lineages. Cyanidiophyceae are extremophilic single-celled algae native to acidic and hot environments with highly reduced nuclear genomes (e.g. <5000 nucleus-encoded genes, compared to >10,000 for most eukaryotic algal groups), but substantial differences and limited synapomorphies between the two major orders Cyanidiales and Galdieriales [38, 180]. Rhodophytina are by contrast mesophilic, and in certain cases (the Eurohodophytina, including Bangiophyceae and Florideophyceae) multi-cellular algae, although they also contain the largely unicellular Proteorhodophytina (Composogonophytes, Stylonematophytes, Porphyridiophytes, Rhodellophytes; [155]). Rhodophytina have likewise undergone substantial reductive evolution in their nuclear genomes (with typically between 5000 and 10,000 genes in their nuclear genomes; [181]), although the mitochondrial and plastid genes of several orders within the Proteorhodophytina have undergone dramatic expansion of their non-coding DNA, resulting in some of the largest organelle genomes documented [155, 224].

Chloroplastida (also termed Viridiplantae) are typified by the presence of chlorophylls *a* and *b*. They are found in marine, freshwater, and terrestrial habitats and range from unicellular actively swimming algae with light-sensing eyespots and multiple flagella to coccal immobile cells (some of which form photobionts of lichens) and filamentous multicellular algae to land plants. They comprise of two major groups, chlorophytes (including core chlorophytes: Chlorophyceae, Ulvophyceae, Trebouxiophyceae, and Chlorodendrophyceae, and a paraphylum of various, mostly unicellular and marine, lineages termed prasinophytes) and streptophytes (including land plants and charophytes, a paraphylum of flagellated or filamentous unicellular and truly multicellular algae) [138].

Despite the ecological and evolutionary importance of the Archaeplastida, endosymbiotic events between a cyanobacterium and a eukaryote leading to the establishment of fully integrated organelles appear to be quite a rare evolutionary phenomenon. Only one case of primary endosymbiosis resulting in a plastid-like photosynthetic organelle outside Archaeplastida has been reported to this day: the chromatophore of the rhizarian thecate amoeba *Paulinella*.

Nonetheless, plastids are no longer confined to the Archaeplastida supergroup (Fig. 1). This is due to the repeated acquisition of eukaryotic algae containing plastids by other branches of the tree of life, via "secondary" or higher (tertiary, quaternary) endosymbioses. These predominantly eukaryote-to-eukaryote endosymbiotic events, which account for the majority of algae in the modern ocean, can be grouped together under the umbrella term "complex plastids".

Historically, major groups of complex plastid-bearing algae were correctly identified as unrelated to plants. A major subset of these algae (cryptomonads, haptophytes, stramenopiles) were grouped together under the name Chromista, underlined by the presence of distinctive chloroplast features found in most of their members such as chlorophyll *c* (found in all Chromistan algae except eustigmatophytes) and membrane connections linking the plastid and nuclear-ER (except for in synurophytes and raphidophytes; [4]). Considering both their structural similarities and eventually their encoded genomes and gene sequences, the chromist plastids were determined to be the result of a secondary endosymbiotic transfer from a rhodophyte-derived, plastid-bearing donor and considered their key synapomorphy.

After ciliates were identified as closely related to dinoflagellates (with which they share subcortical alveoles), the "chromalveolate" hypothesis was formulated [32–34]. This scenario proposed that the common ancestor of "chromists" and non-photosynthetic alveolates harboured plastids that were secondarily lost in ciliates. This hypothesis implicitly considers complete plastid loss to occur more readily than independent plastid acquisition. Documented cases of complete plastid loss have subsequently been documented from clades known to ancestrally possess plastids (*Cryptosporidium* in the apicomplexans [250], *Hematodinium* in the dinoflagellates [79], and potentially SAGH1/2 lineages in the chrysophytes [194]), and it is possible that ancestral presence and differential loss are indeed the most parsimonious explanation for the discontinuous presence of plastids specific chromalveolate lineages. Nonetheless, an ever-growing number of phylogenetically distinct non-photosynthetic "chromalveolates" have since been identified, thanks to rapidly improving availability of genomic data and new phylogenetic algorithms and models that, for example, significantly mitigated the long branch attraction artefact in phylogenetic trees [47]. This gradually made the chromalveolate hypothesis less and less parsimonious as its reconciliation with new data required growing number of independent complete plastid losses.

At the same time, the traditionally "chromist" lineages began to split in modern phylogenomic analyses; for example, haptophytes and cryptophytes, long considered to be sisters forming the group Hacrobia, were shown as only distantly related. Moreover, cryptophytes consistently exhibit strong phylogenetic affiliation to Archaeplastida, often resolving as either their sister or inner lineage [2, 12, 28, 30]. If genuine, the position of cryptophytes inside Archaeplastida would conclusively disprove the chromalveolate hypothesis, as a secondary red plastid would have to "travel back in time" to have been acquired by a common ancestor of cryptophytes and Archaeplastida, including the primary red algae, and all other "chromalveolate" algal groups.

At the current stage, the chromalvolate scenario remains possible but highly non-parsimonous and has been mostly abandoned in favour of the alternative "serial hypothesis". This integrates the diverse relationships between the formerly "chromist" nuclear lineages with the apparent and phylogenetically supported monophyly of their plastids [117] by postulating a series of endosymbiotic transfers from one eukaryotic group to another. This greatly reduces the number of proposed secondary plastid losses, limiting them to lineages in which either a plastid residue (e.g. apicomplexans and non-photosynthetic cryptomonads [214], diatoms [111], and chrysophytes [50], all of which possess leucoplasts) or genetic evidence of its presence has been detected (debatably in the non-photosynthetic stramenopile *Actinophrys* [9], which lacks evident plastids but may possess plastid-derived tRNA-processing enzymes). At the same time, it significantly increases the number of endosymbiotic transfers associated with complex plastids from simply secondary to tertiary, quaternary, and potentially even quinary. However, the exact identities of the plastid donors and recipients in most of these endosymbiotic transfers, as well as the order and timing of these events, remain highly debated [5, 25, 30, 53, 59, 177, 190, 202, 205]. It is also conceivable that plastids of some of the lineages were acquired independently from different rhodophyte donors and resolve as monophyletic due to the vast temporal distance and absence of closer relatives among extant rhodophytes. However, no molecular or phylogenetic evidence for multiple origins of "chromist" plastids was presented to this day.

At this point, it is important to clarify the terminology that will be used throughout this chapter (see Glossary box), as some of the terms are frequently used in the literature to describe multiple phenomena. The word "serial" in this context should not be used to describe plastids but rather the evolutionary process that led to their establishment, as in "serial endosymbiosis" or "serial acquisition", referring to the evolutionary history of the organelle that underwent multiple eukaryote-to-eukaryote transfers. This makes it independently tied to the "serial hypothesis" itself as this is the source of the term and as no cases of serial endosymbiosis have been proposed outside the former "chromists" to this date: all complex plastids originating in Chloroplastida (as opposed to Rhodophyta) are secondary, were gained independently, and are traceable to their respective donor lineages. Another evolutionary phenomenon in reference to which the term "serial" is often used is the "serial replacement" of plastids that is interestingly confined to the Dinoflagellata (and possibly Chrompodellida) lineage. This is unfortunate since the same term is used to refer to different evolutionary phenomena: both the exchange of multiple eukaryotic hosts by the plastid and the exchange of multiple plastids in a single host, which often—but not always—coincides. For clear distinction, this text will use the term "plastid replacement" for the latter and reserve the term "serial endosymbiosis" or "serial plastid acquisition" for the cases specifically relating to tertiary or higher endosymbiotic transfers between organisms with complex red plastids.

Glossary

Complex plastid plastid originated from other than primary endosymbiosis, typically bound by more than two membranes.
EGT endosymbiotic gene transfer; transfer of genes from endosymbiont/organelle genome to the host nucleus.
Chromists a polyphylum of organisms with complex red plastids of a shared origin and originally considered to have been vertically inherited; includes cryptophytes, haptophytes, ochrophytes, and plastid-bearing alveolates (within the myzozoans: dinoflagellates, and apicomplexans).
Kleptoplasty possession of a non-permanent plastid-like endosymbiont (kleptoplastid) acquired through incomplete digestion of algal prey.
LGT lateral gene transfer; transfer of genes between otherwise unrelated lineages, as opposed to vertical inheritance.
Monoplastidic bottleneck a hypothesis postulating that a period of monoplastidy (presence of a single plastid per cell) is necessary for complete plastid integration through division synchronization.
Plastid replacement plastid acquisition in a lineage that previously possessed a plastid of independent origin that was secondarily lost or reduced.
Primary endosymbiosis endosymbiotic relationship between a eukaryote and a prokaryote. Primary endosymbioses have given rise to two major classes of organelles: plastids, or "chloroplasts", and mitochondria.
Primary plastid plastid originated from a primary endosymbiosis between a eukaryte and a cyanobacterium, bound by two membranes, possessed by plants, green and red algae, and glaucophytes within the Archaeplastida. A primary, plastid-like organelle termed "chromatophore" acquired through a separate endosymbiotic event has been characterized in the euglyphyid *Paulinella.*
Secondary endosymbiosis endosymbiotic relationship between eukaryotes, one of which possesses a primary plastid, resulting in a secondary plastid.
Serial endosymbiosis process of subsequent transfers of plastids across multiple eukaryotic host lineages, involving tertiary and/or higher endosymbiotic events.
Shopping bag model a hypothesis postulating that the acquisition of a stable plastid can be preceeded by a number of transient symbiotic relationships with other algae that leave genetic traces in the host genome.
Schnepf theorem a proposed rule postulating that a biological membrane separates a plasmatic and non-plasmatic compartment.
Tertiary (quaternary, quinary) endosymbiosis endosymbiotic relationship between eukaryotes, one of which possesses a secondary (or higher) plastid, resulting in a tertiary (or higher) plastid.

As mentioned above, currently known complex plastids of green algal origin have been acquired from members of Chloroplastida by the ancestors of their current hosts through independent secondary endosymbioses. Lineages with these green secondary plastids include the following:

1.1 Euglenophyta

Euglenophytes are a phototrophic subclade of Euglenida (Euglenozoa, Discoba). They are characterized by a proteinaceous pellicle underlying the cell surface that often lends them great flexibility and the capability of a unique slime-like motion (metaboly), as well as usually two anterior flagella with thick paraxonemal rods emerging from a deep paraflagellar pocket that can support swimming as well as flagellar gliding [135]. Members of this group are generally free-living and nutritionally diverse and plastic, with osmotrophic and phagotrophic members, as well as representatives able to thrive in both aerobic and anaerobic conditions, thanks to their facultatively anaerobic mitochondria [77]. Euglenophytes diverged from the rest of euglenids after the acquisition of plastids closely related to those of the genus *Pyramimonas*, a member of prasinophytes (a paraphylum of early-diverging chlorophyte lineages; [138]). This event is dated to approximately 650–530 million years ago (mya) [103, 222] and believed to have taken place in marine environment since both the plastid donor lineage and the earliest branching euglenophytes are marine, even though most of their current diversity is found within freshwater habitats.

1.2 Chlorarachniophyta

Chlorarachniophytes belong among Cercozoa (Rhizaria), a group comprising of morphologically diverse amoebae and amoeboflagellates forming pseudopodia, filopodia, or reticulopodia, some of which possess siliceous scales that can form thecae or lattices [36]. The placement of the Rhizaria (which also includes the primary, plastid-like organelle-containing *Paulinella*) within the SAR group also containing stramenopiles (e.g., diatoms) and alveolates (dinoflagellates) was a key early argument against the chromalveolate hypothesis [29]. Nonetheless, chloroarachniophytes additionally contain substantial numbers of genes of inferred red origin (c.f. the "red carpet" hypothesis), calling into question the possibility of a historical evolutionary interaction with red algae or their endosymbiotic derivatives [6, 177]. Chlorarachniophytes usually produce three different cell types: the amoeboid stage consisting of single amoebae forming pseudopodia or nets of multiple cells connected by reticulopodia that feeds by phagocytosis in addition to photosynthesizing; ovoid swimming zoospores with one flagellum; and thick-walled cysts [116]. The origin of their plastids has been traced to the core chlorophyte group Bryopsidales and dated to approximately 580 mya [103].

1.3 Lepidodinium, *MGD and TGD*

Lepidodinium is a genus in the dinoflagellate order Gymnodiniales that generally possess the typical dinokont cell plan with two heterodynamic flagella in roughly perpendicular grooves (cingulum and sulcus) but lack theca. There are currently two described species (*L. chlorophorum* and *L. viride*). *Lepidodinium* underwent a plastid replacement having lost the peridinin-containing plastid ancestral to Gymnodiniales [63] and gaining a green one that was originally proposed to be of prasinophyte origin but later determined to be most closely related to the core chlorophyte group Pedinophyta and up to 550 million years old [63, 147]. The presence of nucleomorphs in *Lepidodinium* has been inferred (e.g. by the presence of plastid-associated ribosomes) but not conclusively shown [63, 110]. It is of note, however, that molecular clock analyses of any dinoflagellates may be significantly overestimated due to their atypical genetic features [169, 205]. MGD and TGD are two recently discovered and not yet formally described strains of dinoflagellates that also host pedinophyte-like secondary plastids including clearly defined nucleomorphs (in this case traceable to the genus *Pedinomonas*). These organelles, however, probably do not share origin with the one of *Lepidodinium* due to their early-branching position within dinoflagellates and presumably not with one another either [191] and their plastid acquisitions were not yet dated.

On the other hand, the story of the red complex plastids is older, richer in terms of the number of concerned groups of algae, but much less clear in regard to the order and number of endosymbiotic plastid transfers and to the identities of the donor group in these events. The absence of evidence for multiple transfers from rhodophytes, the relatively more "ancestral" structure of the cryptophyte plastid (retention of a nucleomorph and phycobilisomes, *see below*), and the close genetic relationship between cryptophyte and haptophyte plastid genomes, and haptophyte and ochrophyte plastid proteomes [53, 173, 182] are in favour of the group Cryptophyta being the only truly secondary plastid-bearing algae with other red complex plastids having been "passaged" through them in a series of independent endosymbiotic events. This might be further supported by the recent identification of the unique enzyme agarase in *Goniomonas*, the closest non-photosynthetic lineage of cryptophytes [37]. Rhodophytes are typically surrounded by a durable polysaccharide layer and agarase might have represented a crucial prerequisite for their engulfment and acquisition of red plastids and a gateway for their subsequent spread across the tree of life [196]. However, determining the identity of the plastid donors in these transfers, including the original red plastid source among Rhodophyta, has proven to be much more problematic than in the case of the green plastid lineage. This is only consistent with the proposed timing of these events of up to over 1.5 bya, an estimate overlapping with the time window proposed for primary endosymbiosis at the base of Archaeplastida, suggesting that the first complex plastid might have emerged soon after the establishment of primary plastids [169, 205].

Following is a brief summary of algal groups with complex red plastids.

1.4 Cryptophyta

Cryptophytes are a photosynthetic group of cryptomonads. They are small ubiquitous organisms with two flagella and typical cell asymmetry and swimming motion [94]. One genus, the marine benthic bacterivore *Goniomonas*, likely represents the ancestral aplastidic state of this lineage [37]. Along with katablepharids, palpitomonads, and some lineages of "heliozoans" with whom they share some characters such as ejectosomes, they form the group Cryptista whose phylogenetic position remains somewhat controversial as they place in the vicinity of and frequently even inside Archaeplastida [30]. However, it has also been proposed that this phylogenetic attraction could simply be an artefact of the presumably heavy endosymbiotic transfer from the rhodophyte donor of the cryptophyte plastid [35]. The more parsimonious scenario requiring no cryptic secondary plastid losses that will be considered in this chapter is that they represent a clade sister to Archaeplastida and never possessed primary plastids (or, equally, place inside a polyphyletic Archaeplastida clade but not with any lineage that has independently acquired primary plastids). Cryptophyte plastids are the only red complex plastids known to contain a remnant of the symbiont nucleus, termed a "nucleomorph", whose phylogenetic signal identifies Rhodophyta as their source but (apart of placing them as sister to Proteorhodophytina and Eurhodophytina, to the exclusion of Cyanidiales) does not allow for further tracing to a specific donor lineage [100, 106, 243].

1.5 Ochrophyta

Ochrophytes are an extremely diverse group spanning from picoplankton to the largest macroalgae on Earth. They are part of the kingdom Stramenopiles (also termed Heterokonta) with which they share some of the ancestral characters present in only some of their protist lineages, namely two flagella one of which bears the eponymous straw-like hairs. The best known and ecologically most significant ochrophytes are diatoms (Bacillariophyceae), whose cells are covered in siliceous frustules and move by gliding (only sexual stages of some subgroups are flagellated), and Phaeophytes that include macroscopic and truly multicellular seaweeds such as kelps [51]. Members of the group Xanthophyta range from macroscopic filamentous algae to unicellular flagellates, and the less known or "visible" but more typical morphology exhibiting protist groups include Eustigmatophyta, Chrysophyceae, Raphidiophyta, Pelagophyta, Pinguiophyta, Bolidophyta, Dictyochophyta, and Picophagea [7]. Plastids of these organisms are believed to have originated in a single endosymbiotic event that was already a part of the serial endosymbiotic trend, possibly from a cryptophyte donor which would make ochrophyte plastids tertiary [53, 204, 205], as there is no evidence for cryptic plastid or secondary plastid loss in their close relatives, the heterotrophic and often parasitic oomycetes [53, 203].

1.6 *Haptophyta*

Haptophyta are a group of mostly marine and mostly phototropic or mixotrophic algae with two, usually equal flagella and a third flagellum-like appendage, termed the haptonema, unique to these organisms [62]. Some haptophytes synthesize and don calcium carbonate-based scales or "coccoliths" of various shapes and degrees of complexity and represent major players in the oceanic carbon and calcium cycles [221]. Together with Centrohelida, a group of predatory axopodia-forming and scale or spike-bearing spherical amoebae, they form the clade Haptista which was originally suspected to be closely related to Cryptista (forming the group Hacrobia) but whose phylogenetic position remains uncertain (they are often placed sister to either TSAR (supergroup comprised of Telonemida, Stramenopila, Alveolata, and Rhizaria) or to Archaeplastida + Cryptista) [30]. Rappephytes, pavlovophytes, and prymnesiophytes, previously considered separate lineages sister to haptophytes, were recently reported as true members of this group containing plastids of the same origin [114]. Plastids of haptophytes are most likely the result of one of the older serial endosymbiotic transitions and might have originated directly from cryptophytes (which would make them tertiary), based on phylogenetic affinities of their plastid genomes; or from ochrophytes, specifically pelagophytes or dictyochophytes (which would make them quarternary), based on phylogenetic signals associated with their nucleus-encoded and plastid-targeted proteomes [17, 53, 204, 205]. One possible scenario that could explain the contradicting evidence, i.e. the cryptophyte-like plastid genome and pelagophyte-/dictyochophyte-like plastid proteome, is a hybrid origin of haptophyte plastids through cryptic endosymbiosis and plastid replacement [53].

1.7 *Chromerids*

Chromerids is an informal term for the photosynthetic members of Colpodellida, a taxon related to Sporozoa (obligatory parasitic apicomplexans) with whom it forms the recently proposed phylum Apicomonada [35], and Dinoflagellata. Although chrompodellids are relatively abundant and widespread, chromerids comprise of two relatively recently described and obscure lineages of marine algae associated with corals—*Chromera* and *Vitrella*. Both organisms form non-motile, thick-walled spherical vegetative cells and flagellated zoospores that resemble other chrompodellids [69, 108, 226]. Their plastids were originally suspected to share evolutionary history with both the highly reduced apicoplasts of apicomplexan parasites and the peridinin-containing plastids of dinoflagellates and represent a "missing link" between these two major alveolate groups [106, 233]. Indeed, plastid genome phylogenies clearly place chromerid and apicomplexan plastids as monophyletic, suggesting a common origin. In some plastid phylogenies, the *Vitrella* plastid genome further resolves as a sister-group to dinoflagellates [57], and both *Chromera* and

Vitrella plastid possess nucleus-encoded synapomorphies common with perdinin dinoflagellates (form II, nucleus-encoded RuBisCo, and 3' terminal transcript poly(U) tail addition), which would suggest a common ancestry of all myzozoan plastids [106, 153, 229]. However, the characters of chromerid plastids are suspiciously distinct from dinoflagellates in terms of pigmentation and are relatively similar to those of some ochrophytes (specifically, similar accessory pigmentation and lack of chlorophyll c in *Vitrella* and the eustigmatophytes [164]). In some phylogenetic analyses based on plastid genes, they may have originated by serial endosymbiosis from ochrophytes [195], which would make them quaternary and one of the few red complex plastids with somewhat traceable origin [60, 106].

1.8 Dinoflagellata

Dinoflagellates are a very abundant and diversified group of alveolates. Although they are ancestrally photosynthetic, roughly half of their diversity have lost photosynthetic capability and rely on predation or parasitism [31]. Their most notable shared characteristics include the presence of subcortical alveoles that typically hold cellulose plates forming a rigid theca around the cell and a unique organization of their chromatin which is heavily expanded in size, permanently condensed, and bound to histone-like proteins of bacterial origin instead of otherwise pan-eukaryotic and strictly conserved histones [189]. The ancestral form of dinoflagellate plastid conserved in most photosynthetic lineages contains the apocarotenoid peridinin, has a highly reduced and fragmented genome, utilizes the same 3′ poly(U) tail and nucleus-encoded RuBisCo found in chromerids (see above), and its evolutionary history is unclear; they may possibly be derived from haptophytes or ochrophytes which means they are at least quaternary and they may at least partially share an origin with the plastids of chromerids [56, 60, 106].

This is, however, not the only type of dinoflagellate plastid. In addition to the multiple obligately heterotrophic dinoflagellate groups, many dinoflagellate lineages that have lost (or extremely reduced) the original peridinin-containing plastids have acquired a replacement in the form of a kleptoplastid (transient plastid "stolen" from algal or other kleptoplastidic prey) or stable integrated plastid or endosymbiont. The latter include the genus *Lepidodinium* discussed above, and a subset of another Gymnodiniales family, Kareniaceae with fucoxanthin-containing plastids derived from haptophytes (i.e. being at least quarternary) up to 200–250 mya [169, 205, 218]. Finally, the "dinotoms" are dinoflagellates within the family Peridiniales that possess complex diatom symbionts (acquired multiple times, traceable to particular genera, and not of a singular origin across all representatives; [128, 236, 237]). The dinotom endosymbionts, acquired only about 50 mya retain a high degree of genetic independence, still possessing their own nucleus, cytoplasm, and mitochondria and not importing host-encoded protein products [101, 205].

The last group with complex plastids of red origin is Apicomplexa. However, due to their typically parasitic or commensal (e.g., *Nephromyces* [188]) lifestyle and varying degree of plastid reduction (up to the complete loss of genome and presence of only several essential enzymatic reactions preventing its complete redundancy), they are not relevant for discussion of the inherited or convergent nature of their plastid characters and are omitted from this chapter. Non-photosynthetic chrompodellids, cryptomonads, and chrysophytes will also be considered out of scope of this review. The overall picture of the relationship between these groups and the serial plastid transfers between them with their hypothetical relative order and timescale and basic characters of their plastids are summarized in Fig. 2.

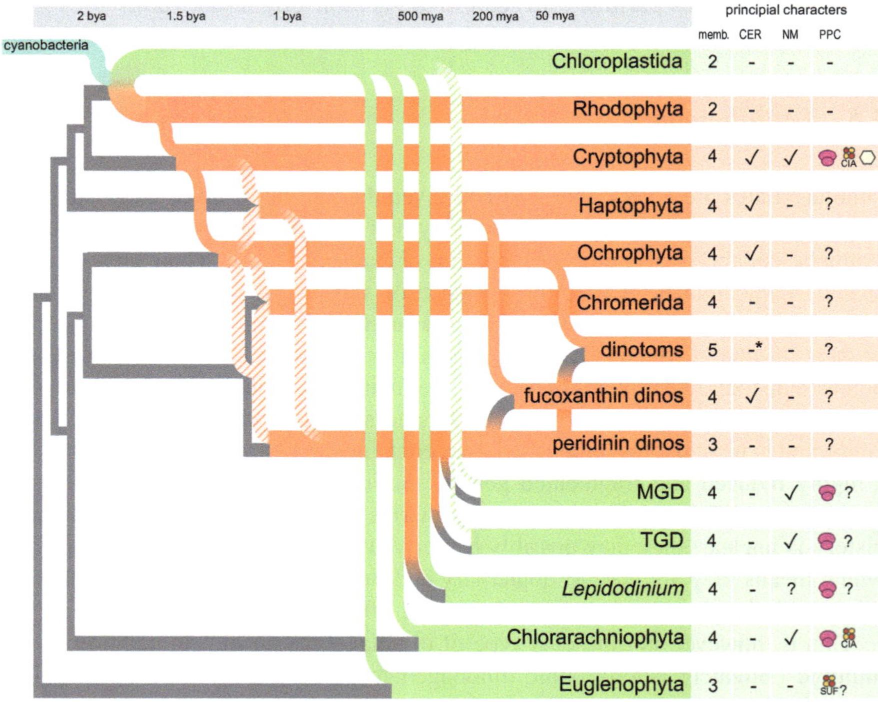

Fig. 2 Left: A summary of the proposed gains, transfers, and replacements of known complex plastids. The grey lines represent branches of an unrooted subset of a pan-eukaryotic tree showing the vertical evolutionary backbone, while the thick, coloured lines covering and cross-linking the grey branches represent plastid history. Timings given reflect the predicted origin of each plastid lineage, rather than the evolutionary divergences of the host lineages. Semi-transparent and/or redundant lines pertain to as yet unresolved relationships and events with unclear dating, based on different hypothetical scenarios for plastid acquisition as discussed in the main text. Right: the most important evolutionary characters related to plastid integration: number of envelope membranes; discontinuity or continuity with ER (i.e. presence of CER); presence or absence of nucleomorph and other characters in the periplastidial compartment. Checkmarks indicate the presence of a character, minus signs its absence, and question marks denote an unclear status. For the periplastidial characters: purple icons define ribosomes; red and yellow icons specify iron-sulphur cluster assembly pathways; and hexagonal icons refer to storage polysaccharides. The asterisk denotes the special "dual" case in dinotoms where the CER is in fact present inside the endosymbiont (as per the setup in diatoms) but not as a connection to the host ER

In the next sections, various morphological and molecular characters of plastids will be discussed and compared between these lineages, exploring which of the common characters can be explained by vertical or horizontal inheritance and by convergent evolution reflecting similar evolutionary pressures on otherwise unrelated systems, and how does this pertain to the possible scenarios of the serial hypothesis.

2 Envelope Membranes

The higher number of envelope membranes was one of the first noticed defining characters of complex plastids that identified them as the results of secondary endosymbiotic transfer [74]. Interestingly, the envelope membrane number seems to consistently settle on four (including unusual cases of Aurearenophyceae and Synchromophycea within the ochrophytes in which the fourth outermost membrane is shared between multiple plastids; [97, 109]). Complex plastids with five (one case), three (two cases), or even two membranes (a single genus within ochrophytes, *Chrysoparadoxa*; [232]) are much less frequent, and it does not grow with the increasing degree of endosymbiotic transfer.

Every additional membrane represents a barrier that needs to be crossed by any imported molecule and especially host nucleus-encoded proteins via specifically evolved import machinery—and it is possible that the number of envelope membranes saturates at four as the successful establishment of a novel (or duplicated) import system is consistently less probable than the loss of the extra membrane(s). In particular, the loss of the endosymbiont cER may be made more likely due to redundancy next to its host analogue, and ER protein import through the Sec61 machinery is typically cotranslational. For example, the expression products of endosymbiont-derived genes that had been relocated to the host nuclear genome would be imported through the host ER during translation, but pre-translated proteins would not be able to interact (via ribosomal docking) with a remnant symbiont ER protein import machinery contained within the plastid.

The five membranes of the dinotom plastids/endosymbionts, which include a symbiont ER but are not connected to the host ER, may specifically limit the integration of dinotom plastids with their host. On the one hand, the retention of a symbiont ER machinery would necessitate the translation of plastid-targeted proteins within the symbiont cytoplasm, thus selecting against endosymbiotic gene relocation to the host nucleus, as described above. Moreover, the absence of a direct connection between the host ER and the symbiont might impede the import of host-derived metabolites and proteins into the symbiont cytoplasm. In regard to the relatively young age of dinotoms, it remains a question of whether this arrangement is an evolutionarily stable state or represents an early stage of complex plastid integration with the host machinery. Of note, dinotoms contain many kleptoplastidic and some fully heterotrophic species, alongside those in which the symbionts is stably passed from parent to daughter generations [235].

An alternative way to compare the complex plastid envelope membranes is to look at whether the outermost membrane forms the chloroplast endoplasmic reticulum (CER), a spatial and functional continuum with host ER, how are the host nucleus-encoded plastid proteins imported across the additional membrane(s), and which parts of the endomembrane system do these membranes correspond to in the particular arrangement. In plastids with a CER, i.e. almost all four membrane–bound complex red plastids (chromerids and apicomplexans being the exceptions), the organelle is effectively located inside the ER lumen. The passage of the nascent protein across the outermost membrane is cotranslational and functionally indistinguishable from that of secretory pathway proteins. In order to proceed further into the organelle, a dedicated system derived by paralogous duplication of ER-associated degradation (ERAD) pathway termed SELMA (symbiont-specific ERAD-like machinery) is used in all concerned groups in a process spatially analogous to the excretion of damaged proteins destined for degradation in proteasomes in cytosol. Therefore, the intermembrane space between the two outermost and the two innermost plastid membranes is analogous to cytoplasm (which then holds the double-membrane bound plastid itself), in accordance with the Schnepf theorem stating that a biological membrane typically separates a plasmatic and non-plasmatic phase [151, 192].

In case of the two lineages with three membrane-bound plastids, euglenophytes and peridinin dinoflagellates, the situation is very different from the typical "chromists" but remarkably evolutionarily convergent. In both, the plastid protein import starts by cotranslational entry to ER lumen but continues as vesicular transport through Golgi (not directly from the ER as is the case in four membrane–bound plastids of chlorarachniophytes; [158, 198]). The outer plastid membrane in these organisms is characterized by its fusion with Golgi vesicles and therefore analogous to that of an endosome or a vacuole, making the outermost intermembrane space a non-plasmatic space *sensu* the Schnepf theorem but breaking this rule by having two such spaces border one another.

Finally, the four membrane–bound plastids without a CER are less legible and may be more diversified in their evolutionary history. The one in chromerids may simply represent a special case of the typical "chromist" arrangement as they also use SELMA which clearly points to the outermost membranes being functionally analogous to ER and the intermembrane space between them representing a non-plasmatic phase similar to ER lumen even if the compartments are not physically connected and only communicate via vesicles. This unusual status of chromerid plastids may be related to the way their heterotrophic ancestors (belonging to Myzozoa, same as colpodellids, dinoflagellates, and apicomplexans) may have ingested their prey [16], which has been proposed to be by the myzocytotic puncturing of membrane of the prey and sucking out the chloroplasts which might therefore face different mechanical stresses, although this hypothesis (and its impact on the possibility for genes to relocate from the symbiont nucleus to the host) requires further experimental investigation.

The four membrane–bound plastids of chlorarachniophytes and *Lepidodinium* lack SELMA and therefore evidence for the direct relationship between their additional membranes and the host ER. The import pathways associated with these plastids remain poorly defined but may possibly involve vesicles—and these have indeed been observed in the periplastidial compartment (PPC) of *Lepidodinium*, although their origin and significance remain unclear [63].

3 Plastid Protein Targeting

The innermost two membranes of all complex plastids are presumably homologous to those of primary plastids and by extension to the ones of free-living cyanobacteria. As a result, the machinery for protein import across these two membranes is widely conserved, consisting of outer and inner chloroplast translocase complexes (TOC and TIC). However, there is a noticeable trend toward reduction of these complexes, especially in case of TOC [19, 89, 144]. Only the central channel subunit (homologous of Toc75) has been found in most complex algae. Moreover, many of the complex algal Toc75 homologues are highly divergent, and it is not clear which are true homologues as opposed to more distantly related members of the Omp85 superfamily to which Toc75 belongs [24]. The TIC of complex algae generally retains more subunits. In cryptophytes and chlorarachniophytes, both being secondary algae that retain nucleomorphs, an almost full complex is present, while at least a core set of Tic110, Tic20, and Tic22 is kept in most other groups with the notable exception of euglenophytes whose TIC possibly consists of multiple copies of a single subunit, Tic21 [162, 246]. This pattern begs the question of whether the higher conservation of TIC could be tied to the presence of nucleomorphs as both the cryptophyte and chloroarachniophyte nucleomorph genomes retain genes for TIC subunits absent from other complex algae [76, 93]. More sequence data for the relatively recently described MGD and TGD may help clarify this possibility.

It is possible that the overall reduction of both complexes makes the translocases less finely tuned or specific. This, however, may no longer pose a problem in a complex plastid as opposed to a primary one whose TOC is in immediate contact with cytoplasm containing a wide spectrum of soluble proteins among which the plastid-targeted ones need to be detected and distinguished from those destined to other compartments. In complex plastids with additional membranes, i.e., additional selective steps in the protein import pathway, the selective pressure on the specificity of TOC/TIC may be relaxed and allow their simplification.

It is also possible that the TOC/TIC complexes of complex algae are reduced in their original protein composition but supplemented by novel subunits or interaction partners specific to complex algae or their individual lineages. This process of eukaryotic innovation of the originally prokaryotic system is already well understood in the establishment of primary plastids [45]. To truly explore this possibility, in silico reconstructions are not sufficient and new data based on proteomics and molecular characterization of the translocase complexes is necessary.

As briefly discussed in the section on envelope membranes, the SELMA machinery was found in all red complex plastids except peridinin dinoflagellates [173], despite its presence in fucoxanthin dinoflagellates. The pre-SELMA part of the protein import, however, is highly similar in all complex plastids, always employing cotranslational import to ER or CER based on the recognition of N-terminal signal peptide (SP) that is usually not distinguishable from the canonical signal peptides destining proteins to secretory pathway. These systems are derived from the respective host endomembrane systems rather than inherited with plastids and therefore not homologous.

After the imported protein passes the first membrane in its path, its signal peptide is cleaved off revealing the next immediate part of the targeting presequence, chloroplast transit peptide (cTP; termed as such to avoid confusion with mitochondrial transit peptide which can be similar, especially in complex algae where the additional plastid membranes alleviate the selective pressure on cTP/mTP distinctiveness). Notably, proteins with dual localization (e.g. aminoacyl-tRNA synthetases) can possess a presequence of intermediate hydrophobicity that can be recognized as both SP and mTP [53, 75]. Transit peptides are not conserved in primary structure and even their length can vary greatly [99, 170, 240], which may relate to the fact that they function inside the ER lumen, and are not (as in primary plastid) under selection for differentiation from mitochondrial targeting sequences. Nonetheless, chloroplast-targeting sequences in secondary plastids are recognized by downstream parts of the import machinery based on their overall chemical properties and, in case of systems employing SELMA, several conserved positions surrounding the SP/cTP boundary. The most widely conserved part of this motif is the phenylalanine on +1 position which already frequently occurs in rhodophyte cTPs but seems to be under stricter selection in complex algae, especially ochrophytes [66, 68, 81, 89, 118, 170]. Notably, it is also conserved in cTPs of peridinin dinoflagellates suggesting that, assuming there is no cryptic SELMA missed by previous studies [173], it is not strictly tied to this machinery and probably also plays a role in recognition by TOC/TIC or other, not yet described translocase or factor. On the other hand, it is less conserved in cTPs on fucoxanthin dinoflagellates where the most common residue on both +1 and +2 positions is arginine [161, 170]. Several positions preceding the SP/cTP cleavage site are also conserved as “AXA” in cryptophytes and “ASA” and ochrophytes and chromerids (although not as strictly in the latter) [82], whereas fucoxanthin dinoflagellate cTPs are instead characterized by a single somewhat conserved glycine residue precede by a relatively conserved signal peptide with a dicystein motif [161]. Neither the +1 phenylalanine nor any other sequence conservation was identified in green secondary plastids, although significant overall enrichment in proline seems to be characteristic for euglenophyte cTPs [162] (Fig. 3).

While plastid-targeted proteins of most complex algae possess only this bipartite N-terminal presequence (not counting the thylakoid lumen-targeting or internal signals that direct further intraplastidial sorting), in euglenophytes and peridinin dinoflagellates, the only two groups with three-membrane bound plastids, roughly half of them possesses a second hydrophobic domain following the cTP. This feature is

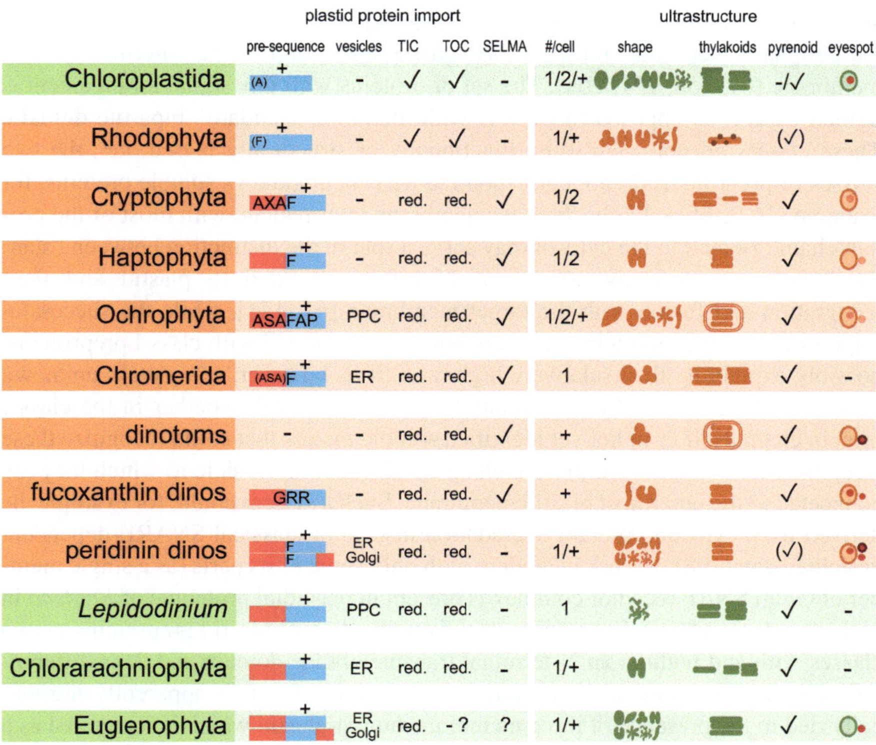

Fig. 3 Summary of the discussed features of plastidial protein import machineries and plastid ultrastructure. Targeting pre-sequences are shown including their multiple parts (red: signal peptide and second ("stop-transfer") hydrophobic domain, blue: transit peptide), conserved amino acid motifs (parentheses denote less strict conservation), and net charge (+) if applicable. Where applicable, both classes of pre-sequences are shown in parallel. In the category of vesicle transport, protein-carrying vesicles passing directly from the ER to the plastid are distinguished from those that passing through the Golgi and those only observed in the PPC. The number of plastids per cell is given as 1, 2, or more; in many groups more or all options occur in different subgroups or representatives. The listed plastid shapes are not exhaustive and include the most typical ones described in the literature to demonstrate the morphological diversity (e.g. compare haptophytes and euglenophyte, groups with similar numbers of described species), as is the case of shown thylakoid structures: stacks of one, two, three, or more (grana), presence of phycobilisomes (small darker circles) or girdle lamella (double-line circle). Eyespots are imaged in regard to their localization (inside or outside plastid), number of distinct types occurring in various representative of the given group, possession of their own membrane (border), and are rendered in lighter colour to denote they are not universally present throughout the given group. Checkmarks indicate the presence of a character, minus signs its absence, question marks its unclear status, and checkmarks in parentheses denote the character is commonly present only in a subset of the given algal group

also termed "stop-transfer signal" (STS) as it serves as a membrane anchor during the obligatory vesicular transport through ER and Golgi to the outermost plastid membrane [61, 158, 171, 208]. The set of proteins with this signal arrangement is termed class I, as opposed to class II with the more "standard" bipartite domain. These classes do not seem to be functionally or structurally biased, i.e., the two classes of preproteins do not correspond to transmembrane vs soluble proteins, for example [162]. More likely, the transmembrane intermediate with most of the protein chain oriented to the cytosol may serve a role in the distinction between "standard" secretory pathway vesicles and vesicles destined to plastid and their recognition and fusion with the target membrane [61, 171], while the successful import of class II preproteins may rely on "piggybacking" with class I preproteins, possibly explaining their relatively high ratio. It is, however, completely unknown how and where would this additional "address" be coded, whether in the class I protein chain itself or in not yet identified soluble factors that could recognize these free, most likely misfolded protein chains, and direct the vesicle to which they are connected to the outermost plastid membrane. Such a system could, for example, be derived from cytosolic protein degradation instead of classical SNARE-dependent vesicle fusion [198] or work in tandem with only some of its parts (as a single member of Golgi SNAP receptor complex is present in plastidial proteome of *Euglena* in high abundance [162]). In any case, the diversification of plastid preproteins to two classes with and without an N-terminal transmembrane domain and their differential vesicular transport (in transmembrane vs soluble form) is apparently intrinsically tied to the presence of three membrane bound-plastid which is integrated as a de facto downstream compartment of the secretory pathway following Golgi. This represents a remarkable case of convergent evolution in two algal groups whose neither hosts nor plastids are related [141].

4 Plastid Morphology and Ultrastructure

Both primary and complex plastids come in all shapes and sizes, in some groups they are more diversified, in others, relatively uniform. Among the most common plastid shape (especially in monoplastidic or diplastidic unicellular algae) is a H- or U-shaped arrangement of either a single bilobed plastid or two plastids connected by a narrow junction. This is the case of many green algae including the closest relatives/potential donors of the euglenophyte and green dinoflagellate plastids, *Pyramimonas* and *Pedinomonas*. The secondary plastids derived from them, however, do not exhibit the same morphology—*Lepidodinium* possesses a single reticular plastid, distinct in shape from the likely phylogenetically related plastids of MGD and TGD [191, 230], and euglenophytes evolved a wide variety of plastid morphologies from simple ovoid, through stellate and reticulate to elongated and ribbon-shaped and only some lineages retain a single plastid while the majority possesses much higher and usually not stable amount [136]. The morphology of neither the rhodophyte predecessor of complex plastids nor the bryopsidales-like source of chlorarachniophyte plastids is known, but interestingly, both resulting secondary algal groups typically possess one bilobed (H-shaped) plastid as well [92, 94, 116, 215]. Haptophytes most commonly have two plastids (one in the case of the Pavlovales), also in an H-shaped arrangement [62] while the number and shape of plastids are more diversified in ochrophytes [7]. The H-shaped arrangement, however, occurs as a transitional stage of division of ovoid plastids, e.g., those of the model diatom *Phaeodactylum* [213]. In some polyplastidic ochrophytes, the outermost membrane may be shared between multiple plastid units creating a de facto transitional stage [97, 109]. Most plastids further derived from these groups are less varied: chromerids possess a single ovoid, cone-shaped, or moderately lobed plastid [69]; fucoxanthin dinoflagellates have multiple plastids of elongated U-shaped or ribbon-like plastids; and dinotom endosymbionts are amorphous and lobed as they lack the siliceous frustule of their free-living pennate diatom ancestors [227]. Plastid morphology and numbers in peridinin dinoflagellates vary greatly, consistent with the vast diversity of this group [189, 227]. Interestingly, in the complex kleptoplastidic relationship involving a cryptophyte, a ciliate (*Mesodinium*), and a dinoflagellate (*Dinophysis*), the kleptoplastid shape changes with each transfer to a new host [71, 86, 122].

The pattern of occurrence of monoplastidic, biplastidic, and polyplastidic representatives may support the notion of the "monoplastidic bottleneck". This theory states that following a stage during which multiple endosymbionts co-occur in the same host cell to allow for the lysis of some of them, which may constitute the major source of endosymbiont-to-host gene transfer [133, 140], a period of strict monoplastidy may be essential for new organelle establishment as it enables the initial synchronization of the host and its nascent organelle division [46]. Biplastidy could then occur as a transitional stage during the division of monoplastidic cells and become fixated following a simple shift in division, representing a first stage of host and organelle division desynchronization. Polyplastidy would therefore generally occur in later-branching taxa while their sister lineages or paraphyla at their bases are often monoplastidic or biplastidic. This is the case of polyplastidic streptophytes within the generally monoplastidic Chloroplastida, and generally polyplastidic ochrophytes versus their early-branching monoplastidic lineages such as bolidophytes, pinguiophytes, and synchromophytes [84]. This scenario is less true within euglenophytes, where the few monoplastidic genera are more scattered among polyplastidic representatives with only *Eutreptiella* being both monoplastidic and early-branching [46, 225], and indeed in the apicomplexa, where *Chromera* and *Vitrella* are polyplastidic while core apicomplexans are monoplastidic; however, we note that the early evolution of both these plastid lineages remains poorly understood [142]. Additionally, the concept of the monoplastic bottleneck might only apply to organisms reproducing somatically, as sexually reproducing species such as *Chlamydomonas* that perform uniparental plastid inheritance lose one of the parent plastid types at each sexual reproduction [1]. The number of plastids per cell may also be a directly proportional function of cell size as it exhibits this pattern in at least some groups, e.g. in diatoms [197]. This could be related to the diffusion constraints in providing the larger cell with photosynthetic assimilates and possibly also plastid portability in cells with larger and more developed cytoskeleton.

Unlike plant plastids, both primary and complex algal plastids contain pyrenoids. These are often centrally localized or close to the nucleomorph in the case of cryptophytes and chlorarachniophytes. They are sometimes penetrated by thylakoid lamellae (e.g., in cryptophytes, haptophytes, and euglenophytes) and are typically spatially associated with polysaccharides whose synthesis and storage are almost universally extraplastidial [200, 209]. The exceptions are Chloroplastida with intraplastidial starch and cryptophytes whose floridean starch is synthesized and deposited in the PPC. The presence of starch in the cryptophyte PPC likely reflects its origins, as the cytoplasmic compartment of the rhodophyte plastid donor where the synthesis and storage of this compound is extraplastidial but plastid-adjacent. This arrangement is not observed in other complex red plastids.

Red and green primary plastids are quite distinct in their thylakoid ultrastructures. While the thylakoids of Chloroplastida take the form of often highly stacked lamellae or grana, the thylakoids of rhodophytes are single and their structure is distinctly characterized by the presence of knob-like phycobilisomes [94]. Thylakoids of complex plastids, on the other hand, are rarely single (this is only the case of chlorarachniophytes and some cryptophytes) and do not form grana, instead they are generally in the form of longer lamellae most often stacked in threes. Cryptophytes, a group most directly derived from the rhodophyte complex plastid ancestor, possess phycobiliproteins, but these do not from typical phycobilisomes, instead they are a part of a second light-harvesting complex located in the thylakoid lumen [94]. A very typical structure almost universally present in ochrophyte plastids (and conserved in dinotoms) is the girdle lamella, a triple thylakoid layer enclosing the rest of the plastid stroma [51, 199]. Notably, the possibly ochrophyte-derived plastids of chromerids do not contain girdle lamellae consistent with the fact that their proposed donor lineage, eustigmatophytes, is one of the few groups that lack this structure [4]. A similar enclosing layer, in this case formed by a single thylakoid, can also be found in some rhodophytes, but this structure exhibits no evolutionary continuity to the ochrophyte girdle lamella as it is not conserved in cryptophytes or any other complex algae and the two therefore likely evolved independently [199].

5 Eyespots

Intraplastidial eyespots (stigma) containing carotenoids and rhodopsins are a typical feature of flagellated and actively swimming unicellular green algae, as opposed to aflagellated rhodophytes with only very limited motility [146, 174]. However, some kind of light-sensing organelle has evolved in at least some representatives of most complex algal groups, sometimes multiple times. Surprisingly, most of the green secondary algae did not inherit eyespots from their plastid donors: neither chlorarachniophytes nor any of the green dinoflagellates possess any eyespots, while the distinctive eyespots of euglenophytes have clearly evolved independently as they are plastid-independent, associated with flagella and not membrane-bound [27, 104, 167, 230]. The eyespot present in some cryptophytes consists of intraplastidial carotenoid-containing globules similar to the eyespots of green algae and represents another independent evolution of such a structure. Intraplastidial eyespots are present in some ochrophytes and some peridinin dinoflagellates, but other representatives can also possess multiple types of extraplastidial carotenoid globules that are free-floating (found in both groups) or vacuole-associated (dinoflagellates; [96, 127]). In Pavlovales (Haptophyta), extraplastidial lipid globule-based light-sensing structures associated with the flagella are present, and both extra- and intraplastidial analogues can be found in fucoxanthin dinoflagellates [40, 62, 189]. All the aforementioned structures seem to be independently evolved de novo, given their cellular localizations and phylogenetic relationships between their hosts, suggesting their invention is quite straightforward and the selective pressure on their fixation is high. We note that algal eyespots are often associated with flagella, whose movement they regulate with plastids themselves. This could partially explain the specific absence in chlorarachniophytes, amoebae moving via pseudopodia, and chromerids whose main vegetative stage is non-flagellated. Finally, the only case of eyespots of somewhat traceable symbiogenetic origin can be found in dinotoms in the form of independent organelles bound by three membranes that strongly suggests they are extremely reduced ancestral peridinin-containing plastids [88, 96]. Unfortunately, proteins associated with the function and biogenesis of algal eyespots remain mostly uncharacterized, preventing their phylogenetic tracing.

6 Nucleomorphs and Periplastidial Compartments

Nucleomorphs are residual forms of nuclei of the plastid donors residing in the periplastidial compartment between the innermost two and outermost two membranes. Their genomes are greatly reduced in size (several hundred kbp) and code mostly genes required for their own genetic housekeeping, and only encode a small number of proteins destined for import to the plastid itself. They are found in chlorarachniophytes and cryptophytes where they have been sequenced and studied in multiple representatives [76, 131, 210, 216, 217], but also the two newly discovered dinoflagellates with pedinophyte-like plastids: MGD and TGD [191]. In *Lepidodinium*, ribosomes and membrane-bound structures were observed in the PPC [63, 85, 231], but no genetic evidence of a nucleomorph was presented as of today. The genetic features of nucleomorphs—size, organization on three chromosomes, GC content, expression pattern, and even protein contents (and their functional distribution)—are remarkably similar in cryptophytes and chlorarachniophytes, having evolved independently but seemingly under very similar selective pressure and/or mechanistic constraints in a case of remarkable evolutionary convergence. The number of introns in nucleomorph genomes, on the other hand,

Table 1 Comparison of the nucleomorph genome features of three cryptophytes and four chlorarachniophytes; data regarding nucleomorphs of MGD and TGD are not available to date

	Size (kb)	GC%	Protein-coding genes	CP-targeted genes	Introns
Guillardia theta	551	26.4	487	30	17
Hemiselmis andersenii	571	25.2	472	30	0
Cryptomonas paramecium	486	26.1	466	18	2
Bigelowiella natans	373	28.5	288	17	865
Lotharella vacuolata	432	24.7	294	17	1028
Lotharella oceanica	612	33.0	338	17	1021
Amorphochlora amoebiformis	374	30.0	295	17	793

differs very significantly between these groups and to some degree also between the individual representatives ([152, 210, 217]; summarized in Table 1).

Even if they do not keep the remnant symbiont nucleus, the PPCs of various groups possibly still perform some metabolic functions that may need to be spatially separated from the plastid stroma. In both cryptophytes and chlorarachniophytes, the CIA pathway for iron-sulphur cluster assembly resides in the PPC in addition to the analogous SUF pathway in plastid stroma, most likely serving nucleomorph-encoded iron-sulphur proteins in situ. Interestingly, multiple pathways for likely spatially separated iron-sulphur cluster assemblies were also reported in euglenophytes, in this case the standard stromal SUF and an additional copy of this pathway of prokaryote origin that was proposed to be localized in one of the two PPC-like intermembrane spaces [162]. In diatoms, C4 metabolism is localized in PPC which probably enables carbon and nitrogen exchange with mitochondria via aspartate/glutamate shuttling [10, 179, 234, 244].

Dinotoms represent their own case and cannot really be compared to plastids with nucleomorphs and/or distinct periplastidial compartments as their analogous structures are not at all reduced and they contain a full-sized diatom nucleus, ER, and mitochondria in what is virtually a full-fledged cytoplasm of the endosymbiont [87].

7 Plastid Genomes and Transcript Processing

Plastid genomes, ultimately derived from their cyanobacterial predecessors, are a unique marker for understanding plastid evolutionary history as they undergo almost exclusively reductive evolution regarding their coding capacity. Among notable exceptions and unique cases of lateral gene transfer from bacteria to the plastid genome are a ribosomal protein (rpl36) conserved in cryptophytes and haptophytes [182] and an operon for scytonemin biosynthesis in some eustigmatophytes [245]. The typical number of genes coded in the plastid genome is notably lower in green primary algae than their red counterparts (90–130 vs. 200–250), and so is the total genome size (100–200 kbp vs. 200–250 kbp). One of the differences in gene content is the status of the two genes for RuBisCO subunits: in the green plastid lineage (including plants), the small subunit (RbcS) is nucleus-encoded and plastid-imported, while in red plastids with the canonical version of this enzyme, both subunits are plastid-encoded [211]. With additional plastid transfers, the plastid genome undergoes losses to its gene content. All secondary green plastid genomes retain 80–85% of the genes found in primary green plastids. The same degree of reduction is found in cryptophytes relative to primary red plastids, which would be consistent with a secondary origin of this lineage. Moreover, in green secondary plastids, a similar percentage of this subset represents a shared core gene inventory—27 tRNAs and approximately 50 protein-coding genes [110, 149, 187, 223, 225]—demonstrating that the plastid gene loss/retention has been under similar evolutionary pressure in different secondary hosts. The same metric is not easily applicable to genomes of more derived plastids as the exact degree of endosymbiosis is not safely established for most of them, and there are often only one or two cases in each category, but it may be following a trend of gradual gene loss with each endosymbiotic transition. Around 75% of the original rhodophyte gene content is retained in genomes of the likely tertiary ochrophyte plastids, around 60% in the possibly tertiary or quaternary plastids of haptophytes, 50% in the possibly quaternary plastids of chromerids, and 45% in the quaternary or quinary, haptophyte-derived plastids of fucoxanthin dinoflagellates (the latter number may be underestimated due to incompleteness of available datasets). Genomes of the possibly quaternary plastids of peridinin dinoflagellates represent a special case due to their extreme reduction (only about 15 genes of strictly photosynthesis-related function, 7% of the original rhodophyte gene content) and a unique organization on multiple and usually single gene-coding minicircles. It is also of note that plastid genomes of fucoxanthin dinoflagellates exhibit significant differential gene loss suggesting that dinoflagellates may be generally prone to elevated frequency of plastid genome reduction [54, 70, 184]. A detailed "roadmap" of the reductive evolution of plastid genomes in green secondary algae was recently published in [223].

The gradual loss of plastid genes presumably associated with successive endosymbioses, however, does not necessarily reflect on the total genome size as non-coding regions and introns may expand as is the case in fucoxanthin dinoflagellates with a large amount of pseudogenes and especially euglenophytes where plastid introns, otherwise present in moderate number in most plastid-bearing lineages, diversified (adding unique type III introns and matryoshka doll-like "twintrons" to the inventory) and proliferated massively with well over 100 plastid introns being a common occurrence in these organisms [175, 225]. Fucoxanthin dinoflagellate plastid introns are not out of the ordinary in their number (two reported in plastid genome of *Karlodinium micrum*; [70]), but they are of interest due to their evolutionary context as the ancestors of neither the haptophyte nor the dinoflagellate party of these chimeric organisms typically possess any typical plastidial introns. However, the latter case is unique as the individual genes are coded by separate minicircles, and this arrangement could be an extreme case of intron expansion, i.e., to the degree of physical separation of the genes [65, 184].

The typical plastid genome organization is a single circular molecule and this was seemingly indeed conserved in the vast majority of plastids regardless of the degree of their derivativeness, the only clear exceptions being the aforementioned peridinin dinoflagellates with highly streamlined plastid genomes fragmented to many small (~3 kbp) circular molecules [249], fucoxanthin dinoflagellates which have at least some of their genes encoded on minicircles in addition to the main molecule (which may be circular or linear; [184]), and *Chromera velia* within the chromerids whose plastid genomes are linear and lacks apparent inverted repeats containing ribosomal DNA genes but has a terminal inverted repeat containing the *psbA* gene [107, 163]. This supposed structural uniformity, however, may reflect bias borne from the universal expectation on organellar genomes to be circular like their prokaryotic predecessors who undergo markedly simpler handling during their replication and transcription, while in reality the shape and size of the organellar genomes may be much more heterogeneous and fluid [4, 13], and the data that would allow us to identify notable commonalities or differences between complex plastids in regard to this phenomenon is currently lacking (Fig. 4).

In addition to intron splicing, editing of transcripts takes place in some plastids. Limited substitutional editing (C/U, occurring in around 0.1% of residues, 1% in extreme cases) has been described in various plants [150, 166], but the process seems to be largely absent in most other primary and complex algae [55]. This is not the case for dinoflagellates and their chromerid relatives. Substitutional editing of plastid transcripts is widespread across dinoflagellates, affecting over 5% of residues in some species, and all twelve potential editing events have been detected across this lineage [56, 121]. Another unique type of plastid transcript modification is 3′ polyuridinylylation (or poly(U) tail addition) which seems to be even more ancestral as it is shared between all dinoflagellate groups and with chromerids as well [106]. In chromerids, it predominantly targets mRNA of photosynthesis-related genes and may play a role in their expression, e.g. the discrimination of pseudogene transcripts [52, 184] and the coordination of other editing and transcript processing events [42, 160]. Both of these processes also occur in fucoxanthin dinoflagellates, a group with haptophyte plastid replacement [54, 55, 102, 121, 184]. Haptophytes do not edit nor polyuridinylate their plastid transcripts, so this mechanism was inherited from the peridinin-containing predecessor and secondarily introduced into the novel plastids, possibly as a part of the process of gaining control over the new organelle enabled by the pre-remediation of substitutional mutations in its genome. This may be related to the observed increased evolutionary rate in plastid-encoded genes of fucoxanthin dinoflagellates [114], whose quality control may have become more relaxed due to the availability of post-transcriptional repair via editing (although not all reported cases of editing in this group are corrective [121, 184]). Remarkably, substitutional editing was recently described also in dinoflagellate lineages with green secondary plastids, *Lepidodinium*, MGD, and TGD [149], most likely representing another case of repurposing of the editing tools retained from peridinin-containing (or even primarily heterotrophic) ancestor rather than a newly and convergently evolved system, especially since some degree of transcript editing was also reported for the mitochondrial and nuclear genomes in some of these

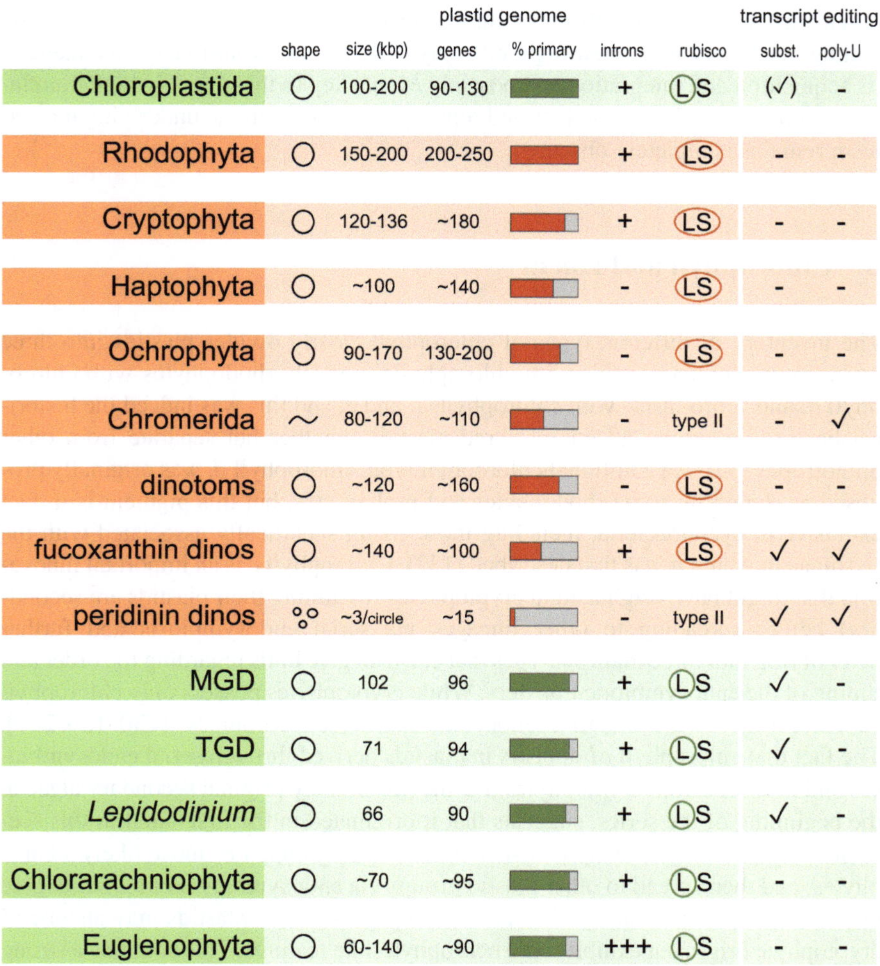

Fig. 4 Summary of the discussed characters of plastid genomes and their molecular genetics. The conservative form of plastid genome is represented as a circular, linear, or minicircle diagram. The number of genes includes genes for both proteins and RNAs and is also represented as a bar plot of the approximate ratio of genes kept in comparison to an "average" green or red algal genome. Exact numbers of introns are not given but instead divided to three categories: no or next to no introns (–), fair number of introns (+), and elevated numbers of introns (+++). The letters L and S represent the large and small subunits of the canonical RuBisCO and their encirclement denotes which subunits are plastid-encoded. Note that some algal groups possess an alternative, type II RuBisCO which will be discussed further in the text. Checkmarks indicate the presence of the character, minus signs its absence, and checkmarks in parentheses denote the character is commonly present only in a subset of the given algal group

organisms [139]. As such, this may suggest that the secondary introduction of "old" editing machinery into a new organelle may in fact be an evolutionary advantage in its acquisition and integration and possibly even a key to the uniquely high dinoflagellate propensity for plastid loss and replacement, although the underlying mechanism remains completely obscured.

8 Chlorophyll and Haem

The inventory of different types of chlorophyll clearly divides plastids into three main groups: green plastids with chlorophylls *a* and *b*, rhodophytes with chlorophyll *a*, and "chromists" with chlorophylls *a* and *c*, and this was indeed the historical base for grouping all complex red plastids together but separate from other phototrophs [32]. An additional chlorophyll type, chlorophyll *d*, was originally proposed as shared by some rhodophytes and prokaryotes, but this pigment is in fact exclusive to cyanobacteria, including those living specifically associated with the red algae in which it was first described [132]. Chlorophyll *c* is an important innovation that might have originated in cryptophytes (assuming their plastids are secondary) before spreading to other lineages via serial endosymbiosis and further diversifying into three main subtypes that can also give hints regarding the order and timing of the endosymbiotic transfers. While cryptophytes possess only chlorophyll *c2*, chlorophyll *c1* is found throughout almost all other "chromists" [130, 156, 247]. The fact that chlorophyll *c1* appears in plastids derived through serial endosymbioses and is absent from cryptophytes, i.e. the presumably original secondary algae at the beginning of the series, suggests that it originated in the first—and in this scenario only—tertiary endosymbiosis-derived algal group, i.e. most likely ochrophytes, and then spread to other plastid groups via endosymbioses of higher degree or possibly a cryptic endosymbiosis in haptophytes whose plastids may also be of cryptophyte origin but exhibit some ochrophyte-like features. This could be a strong argument for a single plastid transfer from cryptophytes (e.g., to ochrophytes) if we were to assume very low odds of two independent "inventions" of chlorophyll *c1* or endosymbiosis-unrelated horizontal gain of its biosynthetic pathway. Unfortunately, the identity of the enzymes directly involved in its synthesis (or conversion from other type) is not currently known which prevents the potentially simple solution of this conundrum by comparative genomics and phylogenetics. The last type, chlorophyll *c3*, is seemingly a unique invention of haptophytes that subsequently spread to fucoxanthin dinoflagellates with their haptophyte-derived plastids [238, 248].

Meanwhile, chlorophyll *c* was completely secondarily lost in a subset of ochrophytes, namely eustigmatophytes and some xanthophytes [72, 156]. This lack of the otherwise typical and universally "chromist" pigment is shared with chromerids providing character-level support for the proposed origin of chromerid plastids in one of these groups (eustigmatophytes coincidentally also lack the girdle lamella, also quite typical of ochrophytes and notably absent from chromerid plastids [90]).

Chlorophyll *c* is no longer synthesized in *Lepidodinium,* and the identifiable enzymes of the chlorophyll synthesis pathway (those leading to protochlorophyllide and chlorophyll *a*) are of green algal origin, which could be interpreted as the result of secondary loss of photosynthesis along with the synthesis of the necessary light-harvesting pigments preceding the plastid replacement [148].

The backbone of the tetrapyrrole pathway (up to protoporphyrin) and its extension to haem (and potentially bilins) are essential and present in all free-living eukaryotes. What differs is its subcellular localization and the several initial steps leading to 5-aminolevulinate (ALA). This precursor can be produced either from glycine in a one-step process termed Shemin or C4 pathway or from glutamate in a three-step C5 pathway. The former is typical for animals, fungi, and other heterotrophs and connected to a mitochondrial tetrapyrrole and haem synthesis, whereas the latter is typical for plants and most algae where not only the chlorophyll branch but also the whole tetrapyrrole biosynthesis takes place in plastid and the mitochondrial version was lost due to redundancy. This is however not the case in several lineages with complex plastids. For a start, in euglenophytes and chlorarachniophytes, both pathways are conserved, although it is not clear whether the non-plastidial versions are fully mitochondrion-localized or whether they partially take place in cytosol [39, 78]. These pathways may have specialized towards different types of haem and/or bilins and their parallel retention may be adaptive rather than coincidental. It is also possible that the presence of a plastid-to-mitochondrial haem transporter is the main driver behind the loss of the non-plastidial pathway, and only lineages capable of haem export could afford to undergo the otherwise advantageous differential loss.

The plastidial pathway is present in peridinin dinoflagellates—including at least some of the free-living secondarily heterotrophic representatives, some of which are bioluminescent and possibly use a modified version of this pathway for the synthesis of luciferin—and in non-photosynthetic chrysophytes [50]. In cryptophytes, ferrochelatase converting protoporphyrin to protohaem is the only enzyme present left from the mitochondrial pathway, as the so far purely plastidial tetrapyrrole pathway seemingly bifurcates after protoporphyrin which is then modified independently to the final haem (or bilin) molecule in each organelle. Two versions of the plastidial C5-starting pathway are kept in dinotoms—one in the endosymbiont and one in a putative remnant of the peridinin plastid based on in silico prediction—reflecting the very limited metabolite exchange between the endosymbiont and the host that seemingly does not extend to the export of the essential haem [39, 88].

Last but not the least, the most unique arrangement of the tetrapyrrole pathway can be found in chromerids and apicomplexans. These algae are the only ones to lack the C5 pathway enzymes. The apicomplexan tetrapyrrole pathway starts by the C4 reaction in the mitochondrion. The remainder of the common pathway and the branch leading to chlorophyll localise to the plastid, while the haem part of the pathway localises to the mitochondrion [39, 185]. The chromerid pathway is similar, but with the terminal steps of the haem branch of the pathway (protophyrinogen oxidase and ferrochelatase) putatively localized to the plastid [123].

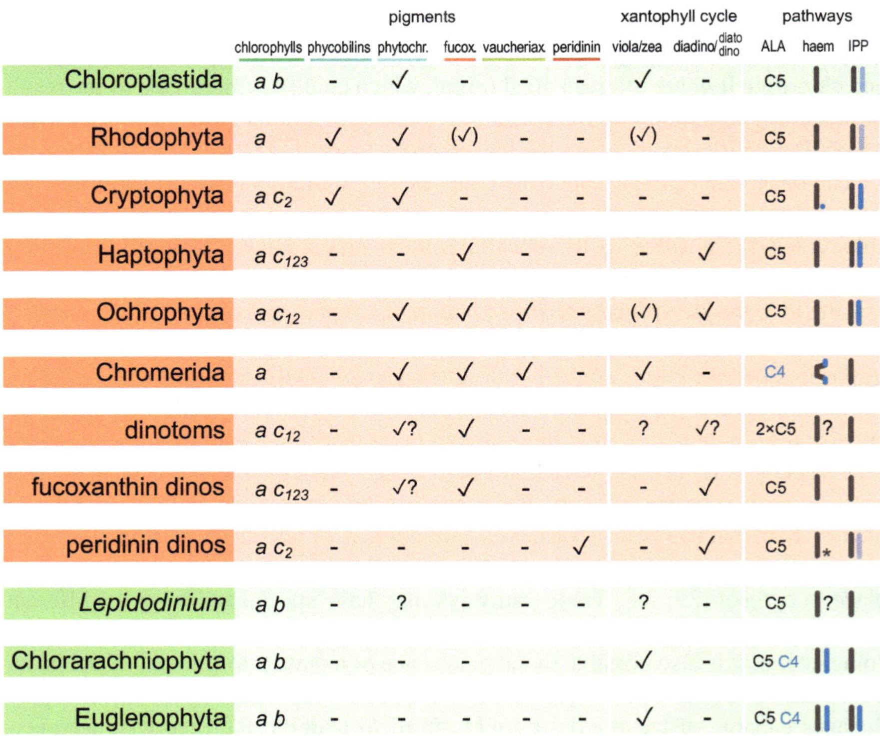

Fig. 5 Summary of the major photosynthetic and photoprotective pigments in the discussed algae, as well as features of the metabolic pathways involved in their syntheses. Checkmarks indicate presence of the character, minus signs its absence, question marks its unclear status, and checkmarks in parentheses denote the character is commonly present only in a subset of the given algal group. Grey lines and text represent pathways localized in plastids, while blue lines and text represent that their non-plastidial alternative is present in the cell as well (the lighter colour means this is the case of only some representatives). The alternating line of the chromerid haem pathway refers to the spatial uniqueness of this pathway in this group (see text) and the asterisk denotes the presumed modification of this pathway in bioluminescent dinoflagellates

The bilin-derived light-sensing pigments, phytochromes, that are generally present in primary algae and plants were conserved throughout the red plastid lineage but not in green secondary plastids [41] (Fig. 5).

9 Isoprenoids and Carotenoids

The second type of molecules bound to light-harvesting proteins are carotenoids (mostly carotenes and xanthophylls) whose function is photoprotective as they dissipate the excess energy as heat and prevent oxidative damage by ROS and inhibition of photosynthesis in a process termed non-photochemical quenching (NPQ).

The main mechanism for this is the light-induced de-epoxidation of a xanthophyll molecule associated with dissipation of excitation energy and its conversion back to the epoxidized state under low light [134, 154]. This system is present throughout primary and complex algae, but the types of involved xanthophylls vary. The "canonical" version of this cycle present in most plants and green and red primary algae involves de-epoxidation of violaxanthin to zeaxanthin, but there are variations utilizing the same enzymes for de-epoxidation of antheraxanthin (which could be considered a partially de-epoxidized violaxanthin) to zeaxanthin (in some rhodophytes) or lutein epoxide to lutein (in some plants) [43, 80, 134].

This system is also present in secondary green algae, some ochrophytes (namely chrysophytes, xanthophytes, and eustigmatophytes), and chromerids, with possible slight variations in the main xanthophylls used. Other ochrophytes (pelagophytes, dictyochophytes, diatoms, and bolidophytes) have evolved an innovation of this system utilizing unique carotenoids, diadinoxanthin (epoxidized) and diatoxanthin (de-epoxidized), and specific (though not very phylogenetically distant) enzymes, diadinoxanthin de-epoxidase and diatoxanthin epoxidase. This system is also shared with haptophytes (and fucoxanthin dinoflagellates with haptophyte-derived plastids), further substantiating the similarity of their plastid protein inventory to that of ochrophytes [4, 53] and partially with peridinin dinoflagellates that also use diadinoxanthin but de-epoxidize it to dinoflagellate-specific dinoxanthin. Some of the enzymes involved in xanthophyll compound synthesis and cycles in algae with secondary red chloroplasts are of probable red algal origin [44], while others may be the products of horizontally transfered genes (e.g. from prasinophytes; [67]), and others still may be the result of gene duplications unique to this lineage.

Algae contain many other carotenoids not directly involved in the xanthophyll cycle but instead bound to light-harvesting antennae and dissipating excess energy as heat caused via structural changes in their chains. The inventory of these accessory carotenoids is very wide and can be highly specific even at low taxonomic level and change depending on current conditions. The most universal type of carotene seems to be β-carotene, whereas α-carotene that is present, which is otherwise common (but not ubiquitous) in plants and primary algae, is surprisingly missing from almost all complex algae [212]. Some types of carotenoids are worth discussing for their uniqueness or distribution that reflects the serial plastid evolution. These include peridinin, an apocarotenoid unique to peridinin dinoflagellates not present in any other lineage, fucoxanthin present in a subset of rhodophytes, some ochrophytes (most notably kelps and diatoms), haptophytes, and dinoflagellates with haptophyte-derived plastids and diatom endosymbionts, and vaucheriaxanthin present only in chrysophytes, eustigmatophytes, and chromerids [172, 212].

Unlike all other complex red plastid groups, cryptophytes lack any form of xanthophyll cycle and β-carotene and utilize an alternative strategy of NPQ by direct antennae protonation [112]. If this pathway were ancestral to the secondary red plastid, its absence from cryptophytes might reflect a secondary loss after their endosymbiotic acquisition. Alternatively, the distribution of the xanthophyll cycle and different pigments specific to individual lineages (e.g. diadino-/diatoxanthin) might reflect the order of serial endosymbioses between cryptophytes and other complex red plastid groups. The violaxanthin-based cycle might have been horizontally acquired (e.g. via the acquisition of specific prasinophyte-derived genes) in the common ochrophyte ancestor, potentially following the tertiary acquisition of a cryptophyte-derived plastid. Violaxanthin- and vaucheriaxanthin-type xanthophyll pathways could have likewise been transferred by tertiary or quaternary endosymbiosis from an ochrophyte related to eustigmatophytes to the common ancestor of chromerids. Finally, diatoxanthin- and fucoxanthin-type xanthophylls pathway could have been endosymbiotically transferred from a pelagophyte relative to the common haptophyte ancestor and subsequently into multiple dinoflagellate plastid lineages [53, 60, 67, 241]. This scenario would be congruent with the proposed status of ochrophytes as the only tertiary algae in the endosymbiotic series.

Carotenoid main chains are composed of isoprene units whose double bonds are the key to their biophysical properties. Other important isoprenoid compounds include various phytohormones, tocopherols, tocotrienols, phylloquinone, and phytol that is added as a side chain to tetrapyrrole ring in chlorophyll synthesis. The main isoprenoid precursor is isopentenyl pyrophosphate (IPP) that can be produced by two alternative pathways. The mevalonate pathway is localized in cytosol and uses acetyl-CoA as the input molecule. On the other hand, the non-mevalonate or MEP/DOXP (for methyl-D-erythritol 4-phosphate/deoxyxylulose phosphate) pathway using phosphoenolpyruvate, originated in cyanobacteria and was inherited with plastids [168]. Both pathways are kept and function in their respective cellular compartments in plants, but only the plastidial MEP/DOXP pathway is retained and serves as the sole source of IPP in green algae and in some rhodophytes [113, 157]. A singular MEP/DOXP pathway is also a hallmark of many complex plastid bearing lineages including chlorarachniophytes, chromerids, and most dinoflagellates (including lineages that underwent plastid replacement). The dependence of the cell on a plastid-associated MEP/DOXP pathway likely represents a major evolutionary driver behind plastid retention in secondarily non-photosynthetic members of these groups, including the apicomplexan malaria parasite *Plasmodium* where this pathway was experimentally demonstrated to be the last linchpin preventing plastid loss [239], and serves as an important pharmaceutical target (fosmidomycin).

Exceptionally, the parasitic dinoflagellate *Hematodinium* still operates the mevalonate pathway in the cytosol and was able to lose plastids completely [79]. The genes encoding this pathway in *Hematodinium* resolve phylogenetically with primarily non-photosynthetic alveolates and oomycetes, suggesting they were vertically retained from the common dinoflagellate ancestor. With this in mind, the probable multiple loss of the mevalonate pathway from dinoflagellates, chromerids and apicomplexans, despite its widespread retention in other complex plastid-containing groups, merits further attention. It remains to be determined if the dinoflagellate, chromerids, and apicomplexan mevalonate pathway is more functionally dispensible than that of other algal groups. In euglenophytes, it has been biochemically observed that the plastidial (MEP/DOXP) and non-plastidial (mevalonate) pools of IPP created by the two alternative pathways are chemically isolated, the plastid-produced IPP supplementing carotenoid but not phytol synthesis [119]. If other complex plastid-containing groups use a similar organization, it might explain the widespread retention of the mevalonate pathway across the tree of life, except in dinoflagellates, chromerids, and apicomplexans where its dispensability may paradoxically select for the retention of plastids even in non-photosynthetic groups.

10 Light Harvesting

Light-harvesting complex (LHC) proteins are an extended protein superfamily originating in the cyanobacterial ancestors of plastids. These proteins are embedded in thylakoid membranes, typically by one transmembrane helix, and closely associated with the two photosystems. Their overall arrangement and functionality are relatively conserved but their evolution in different algal groups is underpinned by differential expansions of their specialized subfamilies [49].

The *sensu stricto* LHCs include chlorophyll-binding protein subfamilies (specialized for specific chlorophyll combinations, e.g., the uniquely "chromist" CAC for chlorophyll *a*,*c*-binding) some of which also bind carotenes and fucoxanthins, as well as clades Li818/LHCx and LHCz whose function is more photoprotective [11, 23]. Li818/LHCx family proteins are significantly expanded in ochrophytes with several subunits present exclusively in this lineage [49], but are further present in euglenophytes, chlorarachniophytes, and haptophytes, and furthermore have green algal homologues which are likely the origin of this subfamily [64]. Members of LHCz subfamily can be found in cryptophytes, ochrophytes, haptophytes, and chlorarachniophytes [159], suggesting they might have originated and diversified in the red complex plastid lineage and been acquired via lateral transfer [177]. In chlorarachniophytes, a unique clade of LHCs, LHCy is also present, and given the notable reduction in their photosystem I, especially the absence of subunits putatively involved with LHC organizing, this novel subfamily has been proposed to function specifically in connection to this derived system [95, 159]. Cryptophyte LHC structures reflect the presence of phycobiliproteins that are, as opposed to their stroma-facing chlorophyll-containing antennae, localized in the thylakoid lumen. While in rhodophytes they form bulky phycobilisomes on the thylakoid membrane, their cryptophyte counterparts are only loosely associated with the lumen-facing chains of some CAC proteins (specifically those associated with photosystem I), mostly soluble and not detectable as distinct physical structures [21, 22, 64]. Euglenophyte LHCs are not very divergent but expanded by gene duplication and most notably synthesized in the form of large polyproteins that are cut to mature size only after import to plastid [98]. This may have facilitated the gain of control over plastid protein expression or help maintain a desirable ratio between different subunits [125]. The situation in peridinin dinoflagellates is remarkably similar as their LHCs underwent extensive paralogization and diversification and are also synthesized in the form of polyproteins [18, 91].

Another specific subfamily of LHCs, red-shifted violaxanthin and chlorophyll *a*-binding proteins (rVCP), was recently discovered in chromerids and subsequently identified in several ochrophytes including eustigmatophytes [15, 22, 124]. Further members of the LHC superfamily include various "LHC-like" protein subfamilies that are typically functionally connected to stress responses (e.g. ELIP, SEP, HLIP, OHP). At least some of these are conserved throughout red complex plastids and euglenophytes but seemingly absent from peridinin dinoflagellates and chlorarachniophytes [64]. An additional subfamily termed RedCAP (Red lineage CAB-like proteins), whose members are nucleus-encoded and whose exact localization and functional specialization is unclear, is almost universally conserved in the red plastid lineage as a whole, seemingly having originated in rhodophytes [64, 207]. Finally, a unique peridinin-chlorophyll *a*-protein (PCP) that does not belong to the LHC superfamily is present in peridinin dinoflagellates and is most remarkable for its status as the only membrane-extrinsic light-harvesting antenna [22, 64, 165].

11 Carbon and Nitrogen Metabolism

Glycolysis is present and well studied in plant plastids, and its presence was also noted in some lineages of complex algae, especially several groups of ochrophytes [58, 129]. In other groups, only a subset of glycolytic enzymes are localized to plastids, and the nature of this pathway is not clear. Many glycolytic enzymes intersect with the Calvin cycle or oxidative pentose phosphate pathway, and they may not in fact form a single functional unit with more downstream enzymes which could represent more of a metabolic crossroads. In ochrophytes, hexokinase is absent, and phosphofructokinase uses up pyrophosphate instead of ATP [120], and the pathway does not seem to be coregulated as a single unit. Enolase and phospho-glycerate mutase, which are largely absent from primary plastid–containing lineages including red algae and many plants, both exhibit strong transcriptional coregulation in diatoms with enzymes of Calvin cycle and Krebs cycle, suggesting important roles in ochrophyte primary carbon metabolism [3, 58]. More significant discontinuity with two initial and two terminal enzymes putatively missing was reported in euglenophytes [162].

The central plastidial enzyme RuBisCO is present throughout complex algae including some secondarily non-photosynthetic members [246], but surprisingly, it is not homologous in all: peridinin dinoflagellates and chromerids share an alternative form of this enzyme, the type II RuBisCO of prokaryotic origin that was acquired via lateral gene transfer by the common ancestor of these two groups [106, 153]. This may point to a common endosymbiotic ancestry of both plastids, or at least a common ancestor already possessing a plastid supported by the type II RuBisCO. Of note, none of the dinoflagellates with replacement plastids are known to use the nuclear form II RuBisCO, with all studied serially acquired dinoflagellate plastid genomes encoding one or both canonical form I RuBisCO subunits. It remains to be determined what evolutionary mechanistics link the acquisition and retention of a nucleus-encoded RuBisCO isoform to the unique and extremely reduced nature of the peridinin dinoflagellate plastid genome [249].

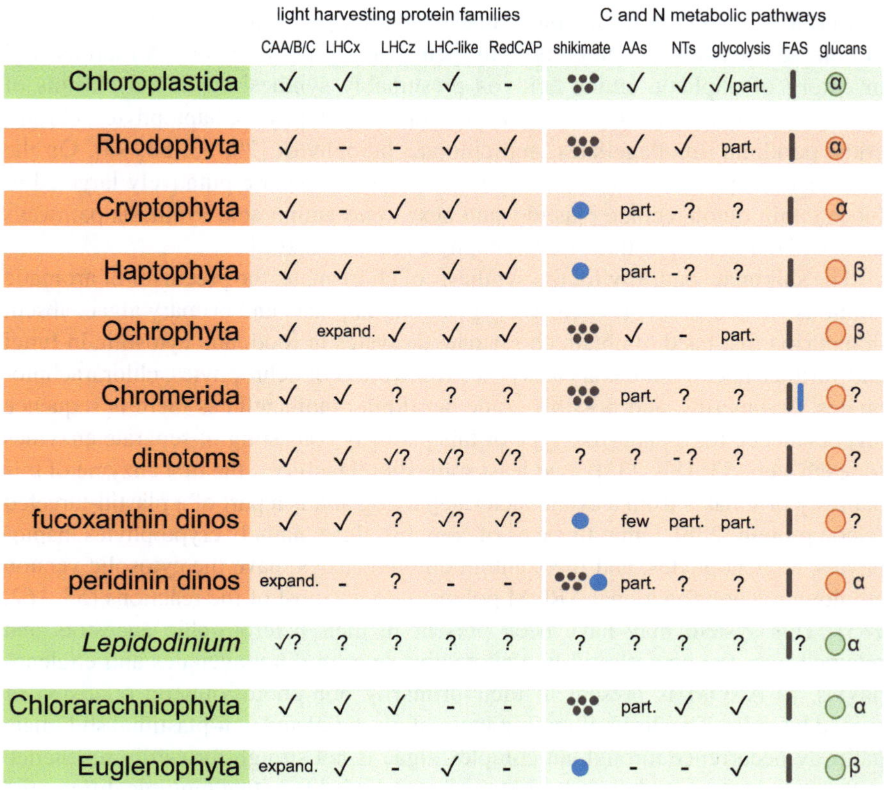

Fig. 6 Summary of the discussed features of plastidial light harvesting and nitrogen and carbon metabolism. Shikimate pathways can employ either the polyfunctional AROM polypeptide (large circle) or multiple single-step enzymes (smaller circles) and take place in the plastid (grey) or cytosol (blue). The latter colour-coding is also used for the fatty acid synthesis pathway represented as a simple line. Checkmarks indicate the presence of the full or almost full pathway or inventory of reactions, with partial or significantly reduced pathways denoted in the text, minus signs denote complete or almost complete absence, and question marks denote unclear status. The significantly expanded protein families are marked as such instead of a simple checkmark. Glucans are distinguished as alpha (e.g. starch) or beta (e.g. paramylon) and the site of their synthesis or storage (intraplastidial, periplastidial, plastid-associated, extraplastidial) is represented by the relative positioning of the Greek letter to the circle representing the plastid

While primary plastids synthesize their own nucleotides in situ, this is not the case for most complex plastids based on the available data regarding the putative localization of purine and pyrimidine metabolism enzymes [113]. Only chlorarachniophytes presumably have plastidial nucleotide synthesis pathways [95]. Other groups such as ochrophytes and euglenophytes rely on import of nucleotides synthesized in cytosol and some of the dedicated membrane transporters were indeed identified in transcriptomic or proteomic data of these organisms [8, 162] (Fig. 6).

Plastidial pathways for the biosynthesis of amino acids are consistently reduced in complex plastids in comparison to their primary predecessors. An almost full inventory of aliphatic amino acids is presumably synthesized in the plastids of ochrophytes and an incomplete set is present in cryptophytes, haptophytes, chromerids, peridinin dinoflagellates, and chlorarachniophytes [20, 53, 83, 95]. On the other hand, very few amino acid–synthesizing enzymes are putatively targeted to fucoxanthin dinoflagellate plastids, and next to no amino acid synthesis pathways are localized to the plastids of euglenophytes [83, 161, 162].

The shikimate pathway for the synthesis of chorismate, the precursor of aromatic amino acids and folate, is exclusively plastidial in plants and primary algae, absent in metazoa that need to obtain chorismate derivates in food, and cytosolic in fungi and various unicellular eukaryotes [143, 183, 186]. For ochrophytes, chlorarachniophytes, chromerids, and possibly some peridinin dinoflagellates there is sequence level evidence for a plant-like type of this pathway consisting of separate enzymes for each step [53, 183, 233]. In at least some dinoflagellates, the first enzyme of this pathway originates from a cyanobacterial paralog and is a part of a plastid-targeted fusion protein [228]. Other lineages of complex algae, namely cryptophytes, haptophytes, euglenophytes, and fucoxanthin dinoflagellates, have the cytosolic version employing a polyfunctional AROM polypeptide for most of the reactions [53, 162, 183]. This system may have been present in their heterotrophic ancestors and retained over the new plastid-like alternative (e.g. in dinoflagellates and euglenophytes, as AROM is present in their primarily non-photosynthetic relatives) or gained laterally. The discontinuous pattern of plastidial and non-plastidial shikimate pathway occurrence throughout complex algae is not straightforwardly reconciled with any proposed model of plastid transfer, and it is likely that multiple differential losses and/or lateral gene transfers have occurred.

12 Conclusions

Complex plastids occur in many independent lineages and as such they are quite diversified in a number of their molecular and structural characters. In others, however, they exhibit a surprising degree of conservation. In the case of red plastids acquired through serial endosymbioses, this can be attributed to either homology as all these organelles are most likely ultimately derived from the combined impacts of a singular ancestor and post-endosymbiotic evolutionary convergences.

The relationships between different "chromist" plastids are by no means resolved, and the molecular taxonomic methods employed to answer this evolutionary question have their limits. For instance, most red plastid endosymbioses seem to have been very ancient events which, combined with the often accelerated reductive evolution of the respective plastid genomes and diversification of their proteomes, hinders more accurate tracing of their evolutionary path. Additionally, there are the general phylogenetic problems such as distinction between EGT and posterior LGT (including the controversial question of how probable the latter even is and whether it really occurs as frequently as the current paradigm assumes [137, 145]) and phylogenetic artefacts (such as long branch attraction) that have been significantly reduced but not eliminated. In this regard, a careful analysis and comparison of non-genetic characters (structure, biochemistry, functionality, and finer architecture of molecular machineries, etc.) may represent a somewhat overlooked alternative source of evolutionary hints to complement and possibly help interpret molecular taxonomic data. In comparison to sequences, character evolution can be easier to read as some state changes are strictly unidirectional (e.g., membrane loss) or easily identifiable as rare evolutionary phenomena (e.g., presence of a unique biochemical pathway).

For example, while the molecular taxonomic evidence for the origin of chromerid plastids in eustigmatophytes/chrysophytes [195] could have been affected by long branch attraction or limited sampling, the characters shared by these groups (loss of chlorophyll *c*, violaxanthin-type xantophyll cycle and red-shifted violaxanthin-binding proteins, vaucheriaxanthin, absence of girdle lamella) provide additional support for this relationship. The presence of a nucleomorph in plastids of cryptophytes could be used as an argument for their secondary endosymbiotic origin, at the base of other red complex plastids (those of ochrophytes, haptophytes, chromerids, and dinoflagellates) that may have lost their nucleomorphs during a tertiary or higher endosymbiotic acquisition. The feasibility of this transfer remains to be determined, considering both nuclear gene data (e.g., if certain genes associated with the nucleus-encoded proteomes of red complex plastids evolved within cryptomonads, following the divergence of sister groups such as *Goniomonas* [37]) and molecular clock data (i.e., if the stem lineages of cryptomonads and other red complex plastids coexisted, allowing a tertiary endosymbiotic acquisition [205]).

On the other hand, other than strictly genetic characters can also be at odds with the phylogenetic signal from plastid genomes, suggesting a more complex evolutionary history than simple inheritance of the plastid from a donor group to which the plastid genome is most closely related. This is, for example, the case of haptophytes whose plastids appear closely related to those of cryptophytes based on plastid phylogenomics (and the shared rare case of prokaryote-to-organelle LGT), but also exhibit a number of characters absent from cryptophytes and instead shared with ochrophytes (chlorophyll *c1*, diadinoxanthin/diatoxanthin type of xantophyll cycle, β-glucan as a main storage compound, plastocyanin present in some representatives). Additionally, their predicted plastid proteome exhibits mixed phylogenetic signal pointing towards pelagophytes or dictyochophytes as major sources of plastid-related nucleus-encoded proteins [53]. In this case, the comparison problematizes the supposed straightforward evolutionary history through tertiary endosymbiosis from a single source and supports the notion of a chimeric plastid, possibly supplemented by nucleus-encoded genes recycled from a previous (or even partially concurrent) cryptic endosymbiotic or kleptoplastidic relationship [133].

In a most extreme scenario, the presence of a shared nucleus-encoded bacterial-type RuBisCO and plastidial poly(U) tail addition pathway, both rare evolutionary events, unifies the plastids of chromerids and peridinin dinoflagellates, suggesting a common deep evolutionary origin [106]. The presence of further phylogenetic signals and other structural features (e.g., xanthophyll pigment types) that respectively link chromerids to eustigmatophytes/chrysophytes, and dinoflagellates to haptophytes, deserves deeper inspection. One possible scenario for this seeming phylogenetic discrepancy is a hitherto unidentified serial endosymbiosis deep in dinoflagellate or chromerid evolution, where a common ancestor of one lineage, already possessing a plastid with nuclear RuBisCo and poly(U) tail addition, acquired a large number of additional plastid-related functions from an additional source [16]. That said, peridinin dinoflagellate and chromerid plastids show monophyly, albeit with long branches, in the limited number of plastid genome trees in which they have both been included, and the total composition and phylogenetic origins of the plastid proteomes of both groups remain poorly understood [57, 68, 106].

In conclusion, in order to thoroughly examine (and hopefully eventually understand) the convoluted complex plastid evolution, it is essential to consider these fascinating organelles as a whole and not to reduce them solely to their genome sequences. Plastids, especially once they start to migrate from host to host, might represent intrinsically chimeric entities, combining not only their host and donor but possibly also other plastids and their hosts through traces of their previous relationships, lateral gene transfer, and possibly other, not yet understood evolutionary mechanisms. Significant genetic exchange between red and green lineages has been repeatedly reported [26, 59, 162, 177, 178], and this mosaicity should be studied on larger scale and with higher resolution. It is also important to give thought to the microevolutionary forces that may drive significant diversification of plastid features even between species or strains [161].

Acknowledgements AMGNV and RGD acknowledge funding from an ANR JCJC ("PanArctica", grant number ANR-21-CE02-0014) and ERC Starting Grant ("ChloroMosaic", grant number 101039760) awarded to RGD.

References

1. Adams GMW (1978) Chloroplast gene transmission in Chlamydomonas reinhardtii: a random choice model. Plasmid 1:522–535. https://doi.org/10.1016/0147-619X(78)90009-4
2. Adl SM, Bass D, Lane CE, Lukeš J, Schoch CL, Smirnov A et al (2019) Revisions to the classification, nomenclature, and diversity of eukaryotes. J Eukaryot Microbiol 66:4–119. https://doi.org/10.1111/jeu.12691
3. Ait-Mohamed O, Novák Vanclová AMG, Joli N, Liang Y, Zhao X, Genovesio A et al (2020) PhaeoNet: a holistic RNAseq-based portrait of transcriptional coordination in the model diatom phaeodactylum tricornutum. Front Plant Sci 11. https://doi.org/10.3389/fpls.2020.590949
4. Andersen RA (2004) Biology and systematics of heterokont and haptophyte algae. Am J Bot 91:1508–1522. https://doi.org/10.3732/ajb.91.10.1508
5. Archibald JM (2015) Genomic perspectives on the birth and spread of plastids. Proc Natl Acad Sci U S A 112:1421374112. https://doi.org/10.1073/pnas.1421374112
6. Archibald JM, Rogers MB, Toop M, Ishida KI, Keeling PJ (2003) Lateral gene transfer and the evolution of plastid-targeted proteins in the secondary plastid-containing alga

Bigelowiella natans. Proc Natl Acad Sci U S A 100:7678–7683. https://doi.org/10.1073/PNAS.1230951100/SUPPL_FILE/0951DATASET1.TXT

7. Archibald JM, Simpson AGB (2017) Handbook of the protists. https://doi.org/10.1007/978-3-319-32669-6
8. Ast M, Gruber A, Schmitz-Esser S, Neuhaus HE, Kroth PG, Horn M et al (2009) Diatom plastids depend on nucleotide import from the cytosol. Proc Natl Acad Sci U S A 106:3621–3626. https://doi.org/10.1073/PNAS.0808862106/SUPPL_FILE/0808862106SI.PDF
9. Azuma T, Pánek T, Tice AK, Kayama M, Kobayashi M, Miyashita H et al (2022) An enigmatic stramenopile sheds light on early evolution in ochrophyta plastid organellogenesis. Mol Biol Evol 39. https://doi.org/10.1093/MOLBEV/MSAC065
10. Bailleul B, Berne N, Murik O, Petroutsos D, Prihoda J, Tanaka A et al (2015) Energetic coupling between plastids and mitochondria drives CO2 assimilation in diatoms. Nature 524:366–369. https://doi.org/10.1038/nature14599
11. Bailleul B, Rogato A, de Martino A, Coesel S, Cardol P, Bowler C et al (2010) An atypical member of the light-harvesting complex stress-related protein family modulates diatom responses to light. Proc Natl Acad Sci U S A 107:18214–18219. https://doi.org/10.1073/PNAS.1007703107/SUPPL_FILE/PNAS.201007703SI.PDF
12. Baurain D, Brinkmann H, Petersen J, Rodríguez-Ezpeleta N, Stechmann A, Demoulin V et al (2010) Phylogenomic evidence for separate acquisition of plastids in cryptophytes, haptophytes, and stramenopiles. Mol Biol Evol 27:1698–1709. https://doi.org/10.1093/molbev/msq059
13. Bendich AJ (2004) Circular chloroplast chromosomes: the grand illusion. Plant Cell 16:1661–1666. https://doi.org/10.1105/tpc.160771
14. Betts HC, Puttick MN, Clark JW, Williams TA, Donoghue PCJ, Pisani D (2018) Integrated genomic and fossil evidence illuminates life's early evolution and eukaryote origin. Nature Ecology & Evolution 2(10):1556–1562. https://doi.org/10.1038/s41559-018-0644-x
15. Bína D, Durchan M, Kuznetsova V, Vácha F, Litvín R, Polívka T (2019) Energy transfer dynamics in a red-shifted violaxanthin-chlorophyll a light-harvesting complex. Biochim Biophys Acta Bioenerg 1860:111–120. https://doi.org/10.1016/j.bbabio.2018.11.006
16. Bodył A, Moszczyński K (2006) Did the peridinin plastid evolve through tertiary endosymbiosis? A hypothesis. Eur J Phycol 41:435–448. https://doi.org/10.1080/09670260600961080
17. Bodył A, Stiller JW, Mackiewicz P (2009) Chromalveolate plastids: direct descent or multiple endosymbioses? Trends Ecol Evol 24:119–121; author reply 121-2. https://doi.org/10.1016/j.tree.2008.11.003
18. Boldt L, Yellowlees D, Leggat W (2012) Hyperdiversity of genes encoding integral light-harvesting proteins in the dinoflagellate Symbiodinium sp. PLoS One 7:1–13. https://doi.org/10.1371/journal.pone.0047456
19. Bolte K, Bullmann L, Hempel F, Bozarth A, Zauner S, Maier U-G (2009) Protein targeting into secondary plastids. J Eukaryot Microbiol 56:9–15. https://doi.org/10.1111/j.1550-7408.2008.00370.x
20. Bromke, M. A. (2013). Amino acid biosynthesis pathways in diatoms. Metabolites 3, 294–311. doi: https://doi.org/10.3390/METABO3020294.
21. Broughton MJ, Howe CJ, Hiller RG (2006) Distinctive organization of genes for light-harvesting proteins in the cryptophyte alga Rhodomonas. Gene 369:72–79. https://doi.org/10.1016/j.gene.2005.10.026
22. Büchel C (2020) Light harvesting complexes in chlorophyll c-containing algae. Biochim Biophys Acta Bioenerg 1861. https://doi.org/10.1016/j.bbabio.2019.05.003
23. Buck JM, Sherman J, Bártulos CR, Serif M, Halder M, Henkel J et al (2019) Lhcx proteins provide photoprotection via thermal dissipation of absorbed light in the diatom Phaeodactylum tricornutum. Nat Commun 10(1):1–12. https://doi.org/10.1038/s41467-019-12043-6
24. Bullmann L, Haarmann R, Mirus O, Bredemeier R, Hempel F, Maier UG et al (2010) Filling the gap, evolutionarily conserved Omp85 in plastids of chromalveolates. J Biol Chem 285:6848–6856. https://doi.org/10.1074/jbc.M109.074807

25. Burki F (2017) The convoluted evolution of eukaryotes with complex plastids, 1st edn. Elsevier Ltd. https://doi.org/10.1016/bs.abr.2017.06.001
26. Burki F, Flegontov P, Oborník M, Cihlář J, Pain A, Lukeš J et al (2012a) Re-evaluating the green versus red signal in eukaryotes with secondary plastid of red algal origin. Genome Biol Evol 4:626–635. https://doi.org/10.1093/gbe/evs049
27. Burki F, Kaplan M, Tikhonenkov DV, Zlatogursky V, Minh BQ, Radaykina LV et al (2016) Untangling the early diversification of eukaryotes: a phylogenomic study of the evolutionary origins of Centrohelida, Haptophyta and Cryptista. Proc R Soc B Biol Sci 283:20152802. https://doi.org/10.1098/rspb.2015.2802
28. Burki F, Okamoto N, Pombert J-F, Keeling PJ (2012b) The evolutionary history of haptophytes and cryptophytes: phylogenomic evidence for separate origins. Proc Biol Sci 279:2246–2254. https://doi.org/10.1098/rspb.2011.2301
29. Burki F, Pawlowski J (2006) Monophyly of rhizaria and multigene phylogeny of unicellular bikonts. Mol Biol Evol 23:1922–1930. https://doi.org/10.1093/MOLBEV/MSL055
30. Burki F, Roger AJ, Brown MW, Simpson AGB (2020) The new tree of eukaryotes. Trends Ecol Evol 35:43–55. https://doi.org/10.1016/J.TREE.2019.08.008
31. Carradec Q, Pelletier E, da Silva C, Alberti A, Seeleuthner Y, Blanc-Mathieu R et al (2018) A global ocean atlas of eukaryotic genes. Nat Commun 9(1):1–13. https://doi.org/10.1038/s41467-017-02342-1
32. Cavalier-Smith T (1981) Eukaryote kingdoms: seven or nine? Biosystems 14:461–481. https://doi.org/10.1016/0303-2647(81)90050-2
33. Cavalier-Smith T (1999) Principles of protein and lipid targeting in secondary symbiogenesis: euglenoid, dinoflagellate, and sporozoan plastid origins and the eukaryote family tree. J Eukaryot Microbiol 46:347–366. https://doi.org/10.1111/j.1550-7408.1999.tb04614.x
34. Cavalier-Smith T (2002) Chloroplast evolution: secondary symbiogenesis and multiple losses. Curr Biol 12, 62–64. Available at: http://www.sciencedirect.com/science/article/pii/S0960982201006753. Accessed 28 June 2014
35. Cavalier-Smith T (2017) Kingdom Chromista and its eight phyla: a new synthesis emphasising periplastid protein targeting, cytoskeletal and periplastid evolution, and ancient divergences. Protoplasma 255(1):297–357. https://doi.org/10.1007/S00709-017-1147-3
36. Cavalier-Smith T, Chao EE, Lewis R (2018) Multigene phylogeny and cell evolution of chromist infrakingdom Rhizaria: contrasting cell organisation of sister phyla Cercozoa and Retaria. Protoplasma 255(5):1517–1574. https://doi.org/10.1007/S00709-018-1241-1
37. Cenci U, Sibbald SJ, Curtis BA, Kamikawa R, Eme L, Moog D et al (2018) Nuclear genome sequence of the plastid-lacking cryptomonad Goniomonas avonlea provides insights into the evolution of secondary plastids. BMC Biol 16:137. https://doi.org/10.1186/s12915-018-0593-5
38. Cho CH, Park SI, Huang TY, Lee Y, Ciniglia C, Yadavalli HC et al (2023) Genome-wide signatures of adaptation to extreme environments in red algae. Nat Commun 14(1):1–16. https://doi.org/10.1038/s41467-022-35566-x
39. Cihlář J, Füssy Z, Horák A, Oborník M (2016) Evolution of the tetrapyrrole biosynthetic pathway in secondary algae: conservation, redundancy and replacement. PLoS One 11:e0166338. https://doi.org/10.1371/journal.pone.0166338
40. Craveiro SC, Moestrup Ø, Daugbjerg N, Calado AJ (2010) Ultrastructure and large subunit rDNA-based phylogeny of sphaerodinium cracoviense, an unusual freshwater dinoflagellate with a novel type of eyespot. J Eukaryot Microbiol 57:568–585. https://doi.org/10.1111/j.1550-7408.2010.00512.x
41. Dammeyer T, Frankenberg-Dinkel N (2008) Function and distribution of bilin biosynthesis enzymes in photosynthetic organisms. Photochem Photobiol Sci 7:1121–1130. https://doi.org/10.1039/B807209B
42. Dang Y, Green BR (2009) Substitutional editing of Heterocapsa triquetra chloroplast transcripts and a folding model for its divergent chloroplast 16S rRNA. Gene 442:73–80. https://doi.org/10.1016/J.GENE.2009.04.006

43. Dautermann O, Lohr M (2017) A functional zeaxanthin epoxidase from red algae shedding light on the evolution of light-harvesting carotenoids and the xanthophyll cycle in photosynthetic eukaryotes. Plant J 92:879–891. https://doi.org/10.1111/tpj.13725
44. Dautermann O, Lyska D, Andersen-Ranberg J, Becker M, Fröhlich-Nowoisky J, Gartmann H et al (2020) An algal enzyme required for biosynthesis of the most abundant marine carotenoids. Sci Adv 6. https://doi.org/10.1126/SCIADV.AAW9183/SUPPL_FILE/AAW9183_SM.PDF
45. Day PM, Theg SM (2018) Evolution of protein transport to the chloroplast envelope membranes. Photosynth Res 138:315–326. https://doi.org/10.1007/s11120-018-0540-x
46. de Vries J, Gould SB (2018) The monoplastidic bottleneck in algae and plant evolution. J Cell Sci 131:1–13. https://doi.org/10.1242/jcs.203414
47. Delmont TO, Gaia M, Hinsinger DD, Frémont P, Vanni C, Fernandez-Guerra A et al (2022) Functional repertoire convergence of distantly related eukaryotic plankton lineages abundant in the sunlit ocean. Cell Genom 2:100123. https://doi.org/10.1016/J.XGEN.2022.100123
48. Derelle R, López-García P, Timpano H, Moreira D (2016) A phylogenomic framework to study the diversity and evolution of stramenopiles (=heterokonts). Mol Biol Evol 33:2890–2898. https://doi.org/10.1093/MOLBEV/MSW168
49. Dittami SM, Michel G, Collén J, Boyen C, Tonon T (2010) Chlorophyll-binding proteins revisited – a multigenic family of light-harvesting and stress proteins from a brown algal perspective. BMC Evol Biol 10:365. https://doi.org/10.1186/1471-2148-10-365
50. Dorrell RG, Azuma T, Nomura M, de Kerdrel GA, Paoli L, Yang S et al (2019) Principles of plastid reductive evolution illuminated by nonphotosynthetic chrysophytes. Proc Natl Acad Sci U S A 116:6914–6923. https://doi.org/10.1073/pnas.1819976116
51. Dorrell RG, Bowler C (2017) Secondary plastids of stramenopiles, 1st edn. Elsevier Ltd. https://doi.org/10.1016/bs.abr.2017.06.003
52. Dorrell RG, Drew J, Nisbet RER, Howe CJ (2014) Evolution of chloroplast transcript processing in plasmodium and its chromerid algal relatives. PLoS Genet 10:e1004008. https://doi.org/10.1371/JOURNAL.PGEN.1004008
53. Dorrell RG, Gile G, McCallum G, Méheust R, Bapteste EP, Klinger CM et al (2017a) Chimeric origins of ochrophytes and haptophytes revealed through an ancient plastid proteome. Elife 6. https://doi.org/10.7554/eLife.23717.001
54. Dorrell RG, Hinksman GA, Howe CJ (2016) Diversity of transcripts and transcript processing forms in plastids of the dinoflagellate alga Karenia mikimotoi. Plant Mol Biol 90:233–247. https://doi.org/10.1007/s11103-015-0408-9
55. Dorrell RG, Howe CJ (2012) Functional remodeling of RNA processing in replacement chloroplasts by pathways retained from their predecessors. Proc Natl Acad Sci U S A 109:18879–18884. https://doi.org/10.1073/pnas.1212270109
56. Dorrell RG, Howe CJ (2015) Integration of plastids with their hosts: Lessons learned from dinoflagellates. Proc Natl Acad Sci U S A 112:10247–10254. https://doi.org/10.1073/pnas.1421380112
57. Dorrell RG, Klinger CM, Newby RJ, Butterfield ER, Richardson E, Dacks JB et al (2017b) Progressive and biased divergent evolution underpins the origin and diversification of peridinin dinoflagellate plastids. Mol Biol Evol 34:361–379. https://doi.org/10.1093/MOLBEV/MSW235
58. Dorrell RG, Liang Y, Gueguen N, Nonoyama T, Zhang Y, Croteau D et al (2022) A plastidial glycolytic-gluconeogenic switch of mitochondrial origin enables diatom adaptations to high latitudes. bioRxiv:2022.09.08.507166. https://doi.org/10.1101/2022.09.08.507166
59. Dorrell RG, Smith AG (2011) Do red and green make brown?: perspectives on plastid acquisitions within chromalveolates. Eukaryot Cell 10:856–868. https://doi.org/10.1128/EC.00326-10
60. Dorrell RG, Villain A, Perez-Lamarque B, de Kerdrel GA, McCallum G, Watson AK et al (2021) Phylogenomic fingerprinting of tempo and functions of horizontal gene transfer within ochrophytes. Proc Natl Acad Sci U S A 118. https://doi.org/10.1073/PNAS.2009974118

61. Durnford DG, Gray MW (2006) Analysis of Euglena gracilis plastid-targeted proteins reveals different classes of transit sequences. Eukaryot Cell 5:2079–2091. https://doi.org/10.1128/EC.00222-06
62. Eikrem W, Medlin LK, Henderiks J, Rokitta S, Rost B, Probert I et al (2017) Haptophyta. In: Handbook of the protists: second edition, pp 893–953. https://doi.org/10.1007/978-3-319-28149-0_38/COVER
63. Elbrächter M, Schnepf E (1996) Gymnodinium chlorophorum, a new, green, bloom-forming dinoflagellate (Gymnodiniales, Dinophyceae) with a vestigial prasinophyte endosymbiont. Phycologia 35:381–393. https://doi.org/10.2216/i0031-8884-35-5-381.1
64. Engelken J, Funk C, Adamska I (2012) The extended light-harvesting complex (lhc) protein superfamily: classification and evolutionary dynamics. pp 265–284. https://doi.org/10.1007/978-94-007-1533-2_11
65. Espelund M, Minge MA, Gabrielsen TM, Nederbragt AJ, Shalchian-Tabrizi K, Otis C et al (2012) Genome fragmentation is not confined to the peridinin plastid in Dinoflagellates. PLoS One 7. https://doi.org/10.1371/journal.pone.0038809
66. Felsner G, Sommer MS, Maier UG (2010) The physical and functional borders of transit peptide-like sequences in secondary endosymbionts. BMC Plant Biol 10:223. https://doi.org/10.1186/1471-2229-10-223
67. Frommolt R, Werner S, Paulsen H, Goss R, Wilhelm C, Zauner S et al (2008) Ancient recruitment by chromists of green algal genes encoding enzymes for carotenoid biosynthesis. Mol Biol Evol 25:2653–2667. https://doi.org/10.1093/molbev/msn206
68. Füssy Z, Faitová T, Oborník M (2019) Subcellular compartments interplay for carbon and nitrogen allocation in Chromera velia and Vitrella brassicaformis. Genome Biol Evol 11:1765–1779. https://doi.org/10.1093/gbe/evz123
69. Füssy Z, Oborník M (2017) Chromerids and their plastids. Adv Bot Res 84:187–218. https://doi.org/10.1016/bs.abr.2017.07.001
70. Gabrielsen TM, Minge MA, Espelund M, Tooming-Klunderud A, Patil V, Nederbragt AJ et al (2011) Genome evolution of a tertiary dinoflagellate plastid. PLoS One 6:e19132. https://doi.org/10.1371/journal.pone.0019132
71. Gantt E, Edwards MR, Provasoli L (1971) Chloroplast structure of the cryptophyceae: evidence for phycobiliproteins within intrathylakoidal spaces. J Cell Biol 48:280–290. https://doi.org/10.1083/JCB.48.2.280
72. Gardian Z, Tichý J, Vácha F (2011) Structure of PSI, PSII and antennae complexes from yellow-green alga Xanthonema debile. Photosynth Res 108:25–32. https://doi.org/10.1007/S11120-011-9647-Z/FIGURES/7
73. Gawryluk RMR, del Campo J, Okamoto N, Strassert JFH, Lukeš J, Richards TA et al (2016) Morphological identification and single-cell genomics of marine diplonemids. Curr Biol 26:3053–3059. https://doi.org/10.1016/J.CUB.2016.09.013
74. Gibbs SP (1981) The chloroplasts of some algal groups may have evolved from endosymbiotic eukaryotic algae. Ann N Y Acad Sci 361:193–208
75. Gile GH, Moog D, Slamovits CH, Maier U-G, Archibald JM (2015) Dual organellar targeting of aminoacyl-tRNA synthetases in diatoms and cryptophytes. Genome Biol Evol 7:1728–1742. https://doi.org/10.1093/gbe/evv095
76. Gilson PR, Su V, Slamovits CH, Reith ME, Keeling PJ, McFadden GI (2006) Complete nucleotide sequence of the chlorarachniophyte nucleomorph: nature's smallest nucleus. Proc Natl Acad Sci U S A 103:9566–9571. https://doi.org/10.1073/pnas.0600707103
77. Ginger ML, Fritz-Laylin LK, Fulton C, Cande WZ, Dawson SC (2010) Intermediary metabolism in protists: a sequence-based view of facultative anaerobic metabolism in evolutionarily diverse eukaryotes. Protist 161:642. https://doi.org/10.1016/J.PROTIS.2010.09.001
78. Gomez-Silva B, Timko MP, Schiff JA (1985) Chlorophyll biosynthesis from glutamate or 5-aminolevulinate in intact Euglena chloroplasts. Planta 165:12–22. https://doi.org/10.1007/BF00392206

79. Gornik SG, Febrimarsa, Cassin AM, MacRae JI, Ramaprasad A, Rchiad Z et al (2015) Endosymbiosis undone by stepwise elimination of the plastid in a parasitic dinoflagellate. Proc Natl Acad Sci U S A 112:5767–5772. https://doi.org/10.1073/pnas.1423400112
80. Goss R, Jakob T (2010) Regulation and function of xanthophyll cycle-dependent photoprotection in algae. Photosynth Res 106:103–122. https://doi.org/10.1007/s11120-010-9536-x
81. Gould SB, Sommer MS, Hadfi K, Zauner S, Kroth PG, Maier U-G (2006) Protein targeting into the complex plastid of cryptophytes. J Mol Evol 62(6):674–681. https://doi.org/10.1007/S00239-005-0099-Y
82. Gruber A, Rocap G, Kroth PG, Armbrust EV, Mock T (2015) Plastid proteome prediction for diatoms and other algae with secondary plastids of the red lineage. Plant J 81:519–528. https://doi.org/10.1111/tpj.12734
83. Guedes R, Prosdocimi F, Fernandes G, Moura L, Ribeiro H, Ortega J (2011) Amino acids biosynthesis and nitrogen assimilation pathways: a great genomic deletion during eukaryotes evolution. BMC Genomics 12:S2. https://doi.org/10.1186/1471-2164-12-S4-S2
84. Guillou L, Moon-Van Der Staay SY, Claustre H, Partensky F, Vaulot D (1999) Diversity and abundance of Bolidophyceae (Heterokonta) in two oceanic regions. Appl Environ Microbiol 65:4528–4536. https://doi.org/10.1128/aem.65.10.4528-4536.1999
85. Hansen G, Botes L, de Salas M (2007) Ultrastructure and large subunit rDNA sequences of Lepidodinium viride reveal a close relationship to Lepidodinium chlorophorum comb. nov. (= Gymnodinium chlorophorum). Phycological Res 55:25–41. https://doi.org/10.1111/j.1440-1835.2006.00442.x
86. Hansen PJ, Nielsen LT, Johnson M, Berge T, Flynn KJ (2013) Acquired phototrophy in Mesodinium and Dinophysis – a review of cellular organization, prey selectivity, nutrient uptake and bioenergetics. Harmful Algae 28:126–139. https://doi.org/10.1016/J.HAL.2013.06.004
87. Hehenberger E, Burki F, Kolisko M, Keeling PJ (2016) Functional relationship between a dinoflagellate host and its diatom endosymbiont. Mol Biol Evol 33:2376–2390. https://doi.org/10.1093/MOLBEV/MSW109
88. Hehenberger E, Imanian B, Burki F, Keeling PJ (2014) Evidence for the retention of two evolutionary distinct plastids in dinoflagellates with diatom endosymbionts. Genome Biol Evol 6:2321–2334. https://doi.org/10.1093/gbe/evu182
89. Hempel F, Felsner G, Maier UG (2010) New mechanistic insights into pre-protein transport across the second outermost plastid membrane of diatoms. Mol Microbiol 76:793–801. https://doi.org/10.1111/j.1365-2958.2010.07142.x
90. Hibberd DJ, Leedale GF (1972) Observations on the cytology and ultrastructure of the new algal class, eustigmatophyceae. Ann Bot 36:49–71. https://doi.org/10.1093/OXFORDJOURNALS.AOB.A084577
91. Hiller RG, Wrench PM, Sharples FP (1995) The light-harvesting chlorophyll a-c-binding protein of dinoflagellates: a putative polyprotein. FEBS Lett 363:175–178. https://doi.org/10.1016/0014-5793(95)00297-m
92. Hirakawa Y (2017) Chlorarachniophytes with complex secondary plastids of green algal origin, 1st edn. Elsevier Ltd. https://doi.org/10.1016/bs.abr.2017.06.006
93. Hirakawa Y, Burki F, Keeling PJ (2012) Genome-based reconstruction of the protein import machinery in the secondary plastid of a chlorarachniophyte alga. Eukaryot Cell 11:324–333. https://doi.org/10.1128/EC.05264-11
94. Hoef-Emden K, Archibald JM (2017) Cryptophyta (cryptomonads). In: Handbook of the protists: second edition, pp 851–891. https://doi.org/10.1007/978-3-319-28149-0_35/COVER
95. Hopkins JF, Spencer DF, Laboissiere S, Neilson JAD, Eveleigh RJM, Durnford DG et al (2012) Proteomics reveals plastid- and periplastid-targeted proteins in the chlorarachniophyte alga Bigelowiella natans. Genome Biol Evol 4:1391–1406. https://doi.org/10.1093/gbe/evs115
96. Horiguchi T, Kawai H, Kubota M, Takahashi T, Watanabe M (1999) Phototactic responses of four marine dinoflagellates with different types of eyespot and chloroplast. Phycological Res 47:101–107. https://doi.org/10.1046/j.1440-1835.1999.47220158.x

97. Horn S, Ehlers K, Fritzsch G, Gil-Rodríguez MC, Wilhelm C, Schnetter R (2007) Synchroma grande spec. nov. (Synchromophyceae class. nov., Heterokontophyta): an amoeboid marine alga with unique plastid complexes. Protist 158:277–293. https://doi.org/10.1016/J.PROTIS.2007.02.004
98. Houlné G, Schantz R (1993) Expression of polyproteins in euglena. CRC Crit Rev Plant Sci 12:1–17. https://doi.org/10.1080/07352689309382354
99. Huesgen PF, Alami M, Lange PF, Foster LJ, Schröder WP, Overall CM et al (2013) Proteomic amino-termini profiling reveals targeting information for protein import into complex plastids. PLoS One 8:e74483. https://doi.org/10.1371/JOURNAL.PONE.0074483
100. Iida K, Takishita K, Ohshima K, Inagaki Y (2007) Assessing the monophyly of chlorophyll-c containing plastids by multi-gene phylogenies under the unlinked model conditions. Mol Phylogenet Evol 45:227–238. https://doi.org/10.1016/J.YMPEV.2007.05.003
101. Imanian B, Keeling PJ (2007) The dinoflagellates Durinskia baltica and Kryptoperidinium foliaceum retain functionally overlapping mitochondria from two evolutionarily distinct lineages. BMC Evol Biol 7:1–11. https://doi.org/10.1186/1471-2148-7-172
102. Jackson CJ, Gornik SG, Waller RF (2013) A tertiary plastid gains RNA editing in its new host. Mol Biol Evol 30:788–792. https://doi.org/10.1093/molbev/mss270
103. Jackson C, Knoll AH, Chan CX, Verbruggen H (2018) Plastid phylogenomics with broad taxon sampling further elucidates the distinct evolutionary origins and timing of secondary green plastids. Sci Rep 8:1523. https://doi.org/10.1038/s41598-017-18805-w
104. James TW, Crescitelli F, Loew ER, McFarland WN (1992) The eyespot of euglena gracilis: a microspectrophotometric study. Vision Res 32:1583–1591. https://doi.org/10.1016/0042-6989(92)90151-8
105. Janouskovec J, Gavelis GS, Burki F, Dinh D, Bachvaroff TR, Gornik SG et al (2017) Major transitions in dinoflagellate evolution unveiled by phylotranscriptomics. Proc Natl Acad Sci U S A 114:E171–E180. https://doi.org/10.1073/pnas.1614842114
106. Janouškovec J, Horák A, Oborník M, Lukeš J, Keeling PJ (2010) A common red algal origin of the apicomplexan, dinoflagellate, and heterokont plastids. Proc Natl Acad Sci U S A 107:10949–10954. https://doi.org/10.1073/PNAS.1003335107
107. Janouškovec J, Sobotka R, Lai DH, Flegontov P, Koník P, Komenda J et al (2013) Split photosystem protein, linear-mapping topology, and growth of structural complexity in the plastid genome of Chromera velia. Mol Biol Evol 30:2447–2462. https://doi.org/10.1093/MOLBEV/MST144
108. Janouškovec J, Tikhonenkov DV, Burki F, Howe AT, Kolísko M, Mylnikov AP et al (2015) Factors mediating plastid dependency and the origins of parasitism in apicomplexans and their close relatives. Proc Natl Acad Sci U S A 112:10200–10207. https://doi.org/10.1073/pnas.1423790112
109. Kai A, Yoshii Y, Nakayama T, Inouye I (2008) Aurearenophyceae classis nova, a new class of heterokontophyta based on a new marine unicellular alga Aurearena cruciata gen. et sp. nov. inhabiting sandy beaches. Protist 159:435–457. https://doi.org/10.1016/J.PROTIS.2007.12.003
110. Kamikawa R, Tanifuji G, Kawachi M, Miyashita H, Hashimoto T, Inagaki Y (2015a) Plastid genome-based phylogeny pinpointed the origin of the green-colored plastid in the dinoflagellate Lepidodinium chlorophorum. *Genome Biol Evol* 7:1133–1140. https://doi.org/10.1093/GBE/EVV060
111. Kamikawa R, Yubuki N, Yoshida M, Taira M, Nakamura N, Ishida KI et al (2015b) Multiple losses of photosynthesis in Nitzschia (Bacillariophyceae). Phycological Res 63:19–28. https://doi.org/10.1111/PRE.12072
112. Kaňa R, Kotabová E, Sobotka R, Prášil O (2012) Non-photochemical quenching in cryptophyte alga Rhodomonas salina is located in chlorophyll a/c antennae. PLoS One 7. https://doi.org/10.1371/journal.pone.0029700
113. Kanehisa M, Goto S (2000) KEGG: Kyoto encyclopedia of genes and genomes. Nucleic Acids Res 28:27–30. https://doi.org/10.1093/nar/28.1.27

114. Kawachi M, Nakayama T, Kayama M, Nomura M, Miyashita H, Bojo O et al (2021) Rappemonads are haptophyte phytoplankton. Curr Biol 31:2395–2403.e4. https://doi.org/10.1016/j.cub.2021.03.012
115. Keeling PJ (2004) Diversity and evolutionary history of plastids and their hosts. Am J Bot 91:1481–1493. https://doi.org/10.3732/AJB.91.10.1481
116. Keeling PJ (2016) Chlorarachniophytes. In: Handbook of the protists, pp 1–17. https://doi.org/10.1007/978-3-319-32669-6_34-1
117. Khan H, Parks N, Kozera C, Curtis BA, Parsons BJ, Bowman S et al (2007) Plastid genome sequence of the cryptophyte alga Rhodomonas salina CCMP1319: lateral transfer of putative DNA replication machinery and a test of chromist plastid phylogeny. Mol Biol Evol 24:1832–1842. https://doi.org/10.1093/MOLBEV/MSM101
118. Kilian O, Kroth PG (2005) Identification and characterization of a new conserved motif within the presequence of proteins targeted into complex diatom plastids. Plant J 41:175–183. https://doi.org/10.1111/j.1365-313X.2004.02294.x
119. Kim D, Filtz MR, Proteau PJ (2004) The methylerythritol phosphate pathway contributes to carotenoid but not phytol biosynthesis in Euglena gracilis. J Nat Prod 67:1067–1069. https://doi.org/10.1021/np049892x
120. Kim J, Fabris M, Baart G, Kim MK, Goossens A, Vyverman W et al (2016) Flux balance analysis of primary metabolism in the diatom *Phaeodactylum tricornutum*. Plant J 85:161–176. https://doi.org/10.1111/tpj.13081
121. Klinger CM, Paoli L, Newby RJ, Wang MYW, Carroll HD, Leblond JD et al (2018) Plastid transcript editing across dinoflagellate lineages shows lineage-specific application but conserved trends. Genome Biol Evol 10:1019–1038. https://doi.org/10.1093/gbe/evy057
122. Koike K, Sekiguchi H, Kobiyama A, Takishita K, Kawachi M, Koike K et al (2005) A novel type of kleptoplastidy in Dinophysis (Dinophyceae): Presence of haptophyte-type plastid in Dinophysis mitra. Protist 156:225–237. https://doi.org/10.1016/j.protis.2005.04.002
123. Koreny L, Sobotka R, Janouskovec J, Keeling PJ, Oborník M (2011) Tetrapyrrole synthesis of photosynthetic chromerids is likely homologous to the unusual pathway of apicomplexan parasites. Plant Cell 23:3454–3462. https://doi.org/10.1105/tpc.111.089102
124. Kotabová E, Jarešová J, Kaňa R, Sobotka R, Bína D, Prášil O (2014) Novel type of red-shifted chlorophyll a antenna complex from Chromera velia. I. Physiological relevance and functional connection to photosystems. Biochim Biophys Acta Bioenerg 1837:734–743. https://doi.org/10.1016/j.bbabio.2014.01.012
125. Koziol AG, Durnford DG (2008) Euglena light-harvesting complexes are encoded by multifarious polyprotein mRNAs that evolve in concert. Mol Biol Evol 25:92–100. https://doi.org/10.1093/molbev/msm232
126. Krabberød AK, Orr RJS, Bråte J, Kristensen T, Bjørklund KR, Shalchian-Tabrizi K (2017) Single cell transcriptomics, mega-phylogeny, and the genetic basis of morphological innovations in Rhizaria. Mol Biol Evol 34:1557–1573. https://doi.org/10.1093/MOLBEV/MSX075
127. Kreimer G (1999) Reflective properties of different eyespot types in dinoflagellates. Available at: http://www.urbanfischer.de/journals/protist
128. Kretschmann J, Žerdoner Čalasan A, Gottschling M (2018) Molecular phylogenetics of dinophytes harboring diatoms as endosymbionts (Kryptoperidiniaceae, Peridiniales), with evolutionary interpretations and a focus on the identity of Durinskia oculata from Prague. Mol Phylogenet Evol 118:392–402. https://doi.org/10.1016/J.YMPEV.2017.10.011
129. Kroth PG, Chiovitti A, Gruber A, Martin-Jezequel V, Mock T, Parker MS et al (2008) A model for carbohydrate metabolism in the diatom Phaeodactylum tricornutum deduced from comparative whole genome analysis. PLoS One 3:e1426. https://doi.org/10.1371/journal.pone.0001426
130. Kuczynska P, Jemiola-Rzeminska M, Strzalka K (2015) Photosynthetic pigments in diatoms. Mar Drugs 13:5847–5881. https://doi.org/10.3390/md13095847
131. Lane CE, Khan H, MacKinnon M, Fong A, Theophilou S, Archibald JM (2006) Insight into the diversity and evolution of the cryptomonad nucleomorph genome. Mol Biol Evol 23:856–865. https://doi.org/10.1093/molbev/msj066

132. Larkum AWD, Kühl M (2005) Chlorophyll d: the puzzle resolved. Trends Plant Sci 10:355–357. https://doi.org/10.1016/j.tplants.2005.06.005
133. Larkum AWD, Lockhart PJ, Howe CJ (2007) Shopping for plastids. Trends Plant Sci 12:189–195. https://doi.org/10.1016/j.tplants.2007.03.011
134. Latowski D, Kuczyńska P, Strzałka K (2011) Xanthophyll cycle – a mechanism protecting plants against oxidative stress. Redox Rep 16:78–90. https://doi.org/10.1179/174329211X13020951739938
135. Leander BS, Lax G, Karnkowska A, Simpson AGB (2017) Euglenida. In: Handbook of the protists. Springer International Publishing, Cham, pp 1–42. https://doi.org/10.1007/978-3-319-32669-6_13-1
136. Leedale GF (1967) Euglenoid flagellates. Prentice-Hall
137. Leger MM, Eme L, Stairs CW, Roger AJ (2018) Demystifying eukaryote lateral gene transfer (Response to Martin 2017 DOI: 10.1002/bies.201700115). BioEssays 40. https://doi.org/10.1002/BIES.201700242
138. Leliaert F, Smith DR, Moreau H, Herron MD, Verbruggen H, Delwiche CF et al (2012) Phylogeny and molecular evolution of the green algae. CRC Crit Rev Plant Sci 31:1–46. https://doi.org/10.1080/07352689.2011.615705
139. Liew YJ, Li Y, Baumgarten S, Voolstra CR, Aranda M (2017) Condition-specific RNA editing in the coral symbiont Symbiodinium microadriaticum. https://doi.org/10.1371/journal.pgen.1006619
140. Lister DL, Bateman JM, Purton S, Howe CJ (2003) DNA transfer from chloroplast to nucleus is much rarer in Chlamydomonas than in tobacco. Gene 316:33–38. https://doi.org/10.1016/S0378-1119(03)00754-6
141. Lukeš J, Leander BS, Keeling PJ (2009) Cascades of convergent evolution: The corresponding evolutionary histories of euglenozoans and dinoflagellates. Proc Natl Acad Sci U S A 106:9963–9970. https://doi.org/10.1073/pnas.0901004106
142. Lukes J, Leander BS, Keeling PJ (2009) Cascades of convergent evolution: the corresponding evolutionary histories of euglenozoans and dinoflagellates. Proc Natl Acad Sci U S A 106(Suppl 1):9963–9970. https://doi.org/10.1073/PNAS.0901004106
143. Maeda H, Dudareva N (2012) The Shikimate pathway and aromatic amino acid biosynthesis in plants. Annu Rev Plant Biol 63:73–105. https://doi.org/10.1146/annurev-arplant-042811-105439
144. Maier UG, Zauner S, Hempel F (2015) Protein import into complex plastids: cellular organization of higher complexity. Eur J Cell Biol. https://doi.org/10.1016/j.ejcb.2015.05.008
145. Martin WF (2017) Too much eukaryote LGT. BioEssays 39:1700115. https://doi.org/10.1002/BIES.201700115
146. Maschmann S, Ruban K, Wientapper J, Walter WJ (2020) Phototaxis of the unicellular red alga cyanidioschyzon merolae is mediated by novel actin-driven tentacles. Int J Mol Sci 21:1–12. https://doi.org/10.3390/ijms21176209
147. Matsumoto T, Shinozaki F, Chikuni T, Yabuki A, Takishita K, Kawachi M et al (2011) Green-colored plastids in the dinoflagellate genus Lepidodinium are of core chlorophyte origin. Protist 162:268–276. https://doi.org/10.1016/J.PROTIS.2010.07.001
148. Matsuo E, Inagaki Y (2018) Patterns in evolutionary origins of heme, chlorophyll a and isopentenyl diphosphate biosynthetic pathways suggest non-photosynthetic periods prior to plastid replacements in dinoflagellates. PeerJ 6:e5345. https://doi.org/10.7717/peerj.5345
149. Matsuo E, Morita K, Nakayama T, Yazaki E, Sarai C, Takahashi K et al (2022) Comparative plastid genomics of green-colored dinoflagellates unveils parallel genome compaction and RNA editing. https://doi.org/10.3389/fpls.2022.918543
150. Monde RA, Schuster G, Stern DB (2000) Processing and degradation of chloroplast mRNA. Biochimie 82:573–582. https://doi.org/10.1016/S0300-9084(00)00606-4
151. Moog D, Maier UG (2017) Cellular compartmentation follows rules: the Schnepf theorem, its consequences and exceptions. BioEssays 39:1700030. https://doi.org/10.1002/bies.201700030

152. Mori S, Sumiya N, Matsunaga S (2022) Nucleomorph: a fascinating remnant of endosymbiosis. Cytologia (Tokyo) 87:203–208. https://doi.org/10.1508/cytologia.87.203
153. Morse D, Salois P, Markovic P, Hastings J (1995) A nuclear-encoded form II RuBisCO in dinoflagellates. Science 268(5217):1622–1624. https://doi.org/10.1126/science.7777861
154. Müller P, Li XP, Niyogi KK (2001) Non-photochemical quenching. a response to excess light energy. Plant Physiol 125:1558–1566. https://doi.org/10.1104/PP.125.4.1558
155. Muñoz-Gómez SA, Mejía-Franco FG, Durnin K, Colp M, Grisdale CJ, Archibald JM et al (2017) The new red algal subphylum proteorhodophytina comprises the largest and most divergent plastid genomes known. Curr Biol 27:1677–1684.e4. https://doi.org/10.1016/J.CUB.2017.04.054
156. Myśliwa-Kurdziel B, Latowski D, Strzałka K (2019) Chlorophylls c—occurrence, synthesis, properties, photosynthetic and evolutionary significance. In: Advances in botanical research. Academic, pp 91–119. https://doi.org/10.1016/bs.abr.2019.04.002
157. Narang PK, Dey J, Mahapatra SR, Roy R, Kushwaha GS, Misra N et al (2021) Genome-based identification and comparative analysis of enzymes for carotenoid biosynthesis in microalgae. World J Microbiol Biotechnol 38(1):1–22. https://doi.org/10.1007/S11274-021-03188-Y
158. Nassoury N, Cappadocia M, Morse D (2003) Plastid ultrastructure defines the protein import pathway in dinoflagellates. J Cell Sci 116:2867–2874. https://doi.org/10.1242/jcs.00517
159. Neilson JAD, Rangsrikitphoti P, Durnford DG (2017) Evolution and regulation of Bigelowiella natans light-harvesting antenna system. J Plant Physiol 217:68–76. https://doi.org/10.1016/J.JPLPH.2017.05.019
160. Nelson MJ, Dang Y, Filek E, Zhang Z, Yu VWC, Ishida KI et al (2007) Identification and transcription of transfer RNA genes in dinoflagellate plastid minicircles. Gene 392:291–298. https://doi.org/10.1016/J.GENE.2007.01.018
161. Novák Vanclová AMG, Nef C, Vancl A, Liu F, Füssy Z, Bowler C et al (2022) Divergent and diversified proteome content across a serially acquired plastid lineage. *bioRxiv*:2022.11.30.518497. https://doi.org/10.1101/2022.11.30.518497
162. Novák Vanclová AMG, Zoltner M, Kelly S, Soukal P, Záhonová K, Füssy Z et al (2020) Metabolic quirks and the colourful history of the Euglena gracilis secondary plastid. New Phytologist 225:1578–1592. https://doi.org/10.1111/nph.16237
163. Oborník M, Lukeš J (2015) The organellar genomes of chromera and vitrella, the phototrophic relatives of apicomplexan parasites. Annu Rev Microbiol 69:129–144. https://doi.org/10.1146/annurev-micro-091014-104449
164. Oborník M, Modrý D, Lukeš M, Černotíková-Stříbrná E, Cihlář J, Tesařová M et al (2012) Morphology, ultrastructure and life cycle of Vitrella brassicaformis n. sp., n. gen., a novel chromerid from the great barrier reef. Protist 163:306–323. https://doi.org/10.1016/J.PROTIS.2011.09.001
165. Ogata T, Kodama M, Nomura S, Kobayashi M, Nozawa T, Katoh T et al (1994) A novel peridinin—chlorophyll a protein (PCP) from the marine dinoflagellate Alexandrium cohorticula: A high pigment content and plural spectral forms of peridinin and chlorophyll a. FEBS Lett 356:367–371. https://doi.org/10.1016/0014-5793(94)01314-4
166. Oldenkott B, Yamaguchi K, Tsuji-Tsukinoki S, Knie N, Knoop V (2014) Chloroplast RNA editing going extreme: more than 3400 events of C-to-U editing in the chloroplast transcriptome of the lycophyte Selaginella uncinata. RNA 20:1499–1506. https://doi.org/10.1261/rna.045575.114
167. Osafune T, Schiff JA (1980) Stigma and flagellar swelling in relation to light and carotenoids in Euglena gracilis var. bacillaris. J Ultrasruct Res 73:336–349. https://doi.org/10.1016/S0022-5320(80)90093-3
168. Paniagua-Michel J, Olmos-Soto J, Ruiz MA (2012) Pathways of carotenoid biosynthesis in bacteria and microalgae. Methods Mol Biol 892:1–12. https://doi.org/10.1007/978-1-61779-879-5_1/COVER

169. Parfrey LW, Lahr DJG, Knoll AH, Katz LA (2011) Estimating the timing of early eukaryotic diversification with multigene molecular clocks. Proc Natl Acad Sci 108:13624–13629. https://doi.org/10.1073/pnas.1110633108
170. Patron NJ, Waller RF (2007) Transit peptide diversity and divergence: a global analysis of plastid targeting signals. Bioessays 29:1048–1058. https://doi.org/10.1002/bies.20638
171. Patron NJ, Waller RF, Archibald JM, Keeling PJ (2005) Complex protein targeting to dinoflagellate plastids. J Mol Biol 348:1015–1024. https://doi.org/10.1016/j.jmb.2005.03.030
172. Peng J, Yuan JP, Wu CF, Wang JH (2011) Fucoxanthin, a marine carotenoid present in brown seaweeds and diatoms: metabolism and bioactivities relevant to human health. Mar Drugs 9:1806–1828. https://doi.org/10.3390/MD9101806
173. Petersen J, Ludewig AK, Michael V, Bunk B, Jarek M, Baurain D et al (2014) Chromera velia, endosymbioses and the rhodoplex hypothesis—plastid evolution in cryptophytes, alveolates, stramenopiles, and haptophytes (CASH lineages). Genome Biol Evol 6:666–684. https://doi.org/10.1093/GBE/EVU043
174. Pickett-Heaps J, West J, Wilson S, McBride D (2001) Time-lapse videomicroscopy of cell (spore) movement in red algae. Eur J Phycol 36:9–22. https://doi.org/10.1080/09670260110001735148
175. Pombert JF, James ER, Janouškovec J, Keeling PJ (2012) Evidence for transitional stages in the evolution of euglenid group II introns and twintrons in the Monomorphina aenigmatica plastid genome. PLoS One 7. https://doi.org/10.1371/journal.pone.0053433
176. Ponce-Toledo RI, Deschamps P, López-García P, Zivanovic Y, Benzerara K, Moreira D (2017) An early-branching freshwater cyanobacterium at the origin of plastids. Curr Biol 27:386–391. https://doi.org/10.1016/J.CUB.2016.11.056
177. Ponce-Toledo RI, López-García P, Moreira D (2019) Horizontal and endosymbiotic gene transfer in early plastid evolution. New Phytol 224:618–624. https://doi.org/10.1111/nph.15965
178. Ponce-Toledo RI, Moreira D, López-García P, Deschamps P, Ruiz-Trillo I (2018) Secondary plastids of euglenids and chlorarachniophytes function with a mix of genes of red and green algal ancestry. Mol Biol Evol. https://doi.org/10.1093/molbev/msy121
179. Prihoda J, Tanaka A, de Paula WBM, Allen JF, Tirichine L, Bowler C (2012) Chloroplast-mitochondria cross-talk in diatoms. J Exp Bot 63:1543–1557. https://doi.org/10.1093/jxb/err441
180. Qiu H, Price DC, Weber APM, Reeb V, Chan Yang E, Lee JM et al (2013) Adaptation through horizontal gene transfer in the cryptoendolithic red alga Galdieria phlegrea. Curr Biol 23:R865–R866. https://doi.org/10.1016/J.CUB.2013.08.046
181. Qiu H, Price DC, Yang EC, Yoon HS, Bhattacharya D (2015) Evidence of ancient genome reduction in red algae (Rhodophyta). J Phycol 51:624–636. https://doi.org/10.1111/JPY.12294
182. Rice DW, Palmer JD (2006) An exceptional horizontal gene transfer in plastids: gene replacement by a distant bacterial paralog and evidence that haptophyte and cryptophyte plastids are sisters. BMC Biol 4:31. https://doi.org/10.1186/1741-7007-4-31
183. Richards TA, Dacks JB, Campbell SA, Blanchard JL, Foster PG, McLeod R et al (2006) Evolutionary origins of the eukaryotic shikimate pathway: gene fusions, horizontal gene transfer, and endosymbiotic replacements. Eukaryot Cell 5:1517–1531. https://doi.org/10.1128/EC.00106-06
184. Richardson E, Dorrell RG, Howe CJ (2014) Genome-wide transcript profiling reveals the coevolution of plastid gene sequences and transcript processing pathways in the fucoxanthin dinoflagellate Karlodinium veneficum. Mol Biol Evol 31:2376–2386. https://doi.org/10.1093/MOLBEV/MSU189
185. Richtová J, Sheiner L, Gruber A, Yang SM, Kořený L, Striepen B et al (2021) Using diatom and apicomplexan models to study the heme pathway of Chromera velia. Int J Mol Sci 22. https://doi.org/10.3390/ijms22126495
186. Rippert P, Puyaubert J, Grisollet D, Derrier L, Matringe M (2009) Tyrosine and phenylalanine are synthesized within the plastids in arabidopsis. Plant Physiol 149:1251–1260. https://doi.org/10.1104/pp.108.130070

187. Rogers MB, Gilson PR, Su V, McFadden GI, Keeling PJ (2007) The complete chloroplast genome of the chlorarachniophyte bigelowiella natans: evidence for independent origins of chlorarachniophyte and euglenid secondary endosymbionts. Mol Biol Evol 24:54–62. https://doi.org/10.1093/molbev/msl129
188. Saffo MB, McCoy AM, Rieken C, Slamovits CH (2010) Nephromyces, a beneficial apicomplexan symbiont in marine animals. Proc Natl Acad Sci U S A 107:16190–16195. https://doi.org/10.1073/PNAS.1002335107/SUPPL_FILE/PNAS.201002335SI.PDF
189. Saldarriaga JF, "Max" Taylor FJR (2017) Dinoflagellata. In: Handbook of the protists: second edition, pp 625–678. https://doi.org/10.1007/978-3-319-28149-0_22/COVER
190. Sanchez-Puerta MV, Delwiche CF (2020) A hypothesis for plastid evolution in chromalveolates. J Phycol 44:1097–1107. https://doi.org/10.1111/J.1529-8817.2008.00559.X
191. Sarai C, Tanifuji G, Nakayama T, Kamikawa R, Takahashi K, Yazaki E et al (2020) Dinoflagellates with relic endosymbiont nuclei as models for elucidating organellogenesis. Proc Natl Acad Sci U S A 117:5364–5375. https://doi.org/10.1073/pnas.1911884117
192. Schnepf E, Elbrächter M (1988) Cryptophycean-like double membrane-bound chloroplast in the dinoflagellate, Dinophysis Ehrenb.: evolutionary, phylogenetic and toxicological implications. Botanica Acta 101:196–203. https://doi.org/10.1111/j.1438-8677.1988.tb00033.x
193. Schön ME, Zlatogursky VV, Singh RP, Poirier C, Wilken S, Mathur V et al (2021) Single cell genomics reveals plastid-lacking Picozoa are close relatives of red algae. Nat Commun 12(1):1–10. https://doi.org/10.1038/s41467-021-26918-0
194. Seeleuthner Y, Mondy S, Lombard V, Carradec Q, Pelletier E, Wessner M et al (2018) Single-cell genomics of multiple uncultured stramenopiles reveals underestimated functional diversity across oceans. Nat Commun 9(1):1–10. https://doi.org/10.1038/s41467-017-02235-3
195. Ševčíková T, Horák A, Klimeš V, Zbránková V, Demir-Hilton E, Sudek S et al (2015) Updating algal evolutionary relationships through plastid genome sequencing: did alveolate plastids emerge through endosymbiosis of an ochrophyte? https://doi.org/10.1038/srep10134
196. Sibbald SJ, Archibald JM (2020) Genomic insights into plastid evolution. Genome Biol Evol 12:978–990. https://doi.org/10.1093/GBE/EVAA096
197. Silkin V, Fedorov A, Flynn KJ, Paramonov L, Pautova L (2021) Protoplasmic streaming of chloroplasts enables rapid photoacclimation in large diatoms. J Plankton Res 43:831–845. https://doi.org/10.1093/PLANKT/FBAB071
198. Sláviková S, Vacula R, Fang Z, Ehara T, Osafune T, Schwartzbach SD (2005) Homologous and heterologous reconstitution of Golgi to chloroplast transport and protein import into the complex chloroplasts of Euglena. J Cell Sci 118:1651–1661. https://doi.org/10.1242/jcs.02277
199. Solymosi K (2012) Plastid structure, diversification and interconversions I. Algae. Curr Chem Biol 6, 167–186(20). Available at: https://www.ingentaconnect.com/content/ben/ccb/2012/00000006/00000003/art00002. Accessed 8 July 2019
200. Solymosi K (2013) Plastid structure, diversification and interconversions I. Algae. Curr Chem Biol 6:167–186. https://doi.org/10.2174/2212796811206030002
201. Stephens TG, Gabr A, Calatrava V, Grossman AR, Bhattacharya D (2021) Why is primary endosymbiosis so rare? New Phytol 231:1693–1699. https://doi.org/10.1111/NPH.17478
202. Stiller JW (2014) Toward an empirical framework for interpreting plastid evolution. J Phycol. https://doi.org/10.1111/jpy.12178
203. Stiller JW, Huang J, Ding Q, Tian J, Goodwillie C (2009) Are algal genes in nonphotosynthetic protists evidence of historical plastid endosymbioses? BMC Genomics 10:484. https://doi.org/10.1186/1471-2164-10-484
204. Stiller JW, Schreiber J, Yue J, Guo H, Ding Q, Huang J (2014) The evolution of photosynthesis in chromist algae through serial endosymbioses. Nat Commun 5:5764. https://doi.org/10.1038/ncomms6764
205. Strassert JFH, Irisarri I, Williams TA, Burki F (2021) A molecular timescale for eukaryote evolution with implications for the origin of red algal-derived plastids. Nat Commun 12. https://doi.org/10.1038/s41467-021-22044-z

206. Strassert JFH, Jamy M, Mylnikov AP, Tikhonenkov DV, Burki F (2019) New phylogenomic analysis of the enigmatic phylum telonemia further resolves the eukaryote tree of life. Mol Biol Evol 36:757–765. https://doi.org/10.1093/molbev/msz012
207. Sturm S, Engelken J, Gruber A, Vugrinec S, Kroth PG, Adamska I et al (2013) A novel type of light-harvesting antenna protein of red algal origin in algae with secondary plastids. BMC Evol Biol 13. https://doi.org/10.1186/1471-2148-13-159
208. Sulli C, Fang ZW, Muchhal U, Schwartzbach SD (1999) Topology of Euglena chloroplast protein precursors within endoplasmic reticulum to Golgi to chloroplast transport vesicles. J Biol Chem 274:457–463. https://doi.org/10.1074/jbc.274.1.457
209. Suzuki E, Suzuki R (2013) Variation of storage polysaccharides in phototrophic microorganisms. J Appl Glycosci 60(1):21–27. https://doi.org/10.5458/jag.jag.jag-2012_016
210. Suzuki S, Shirato S, Hirakawa Y, Ishida KI (2015) Nucleomorph genome sequences of two chlorarachniophytes, Amorphochlora amoebiformis and Lotharella vacuolata. Genome Biol Evol 7:1533–1545. https://doi.org/10.1093/gbe/evv096
211. Tabita FR, Hanson TE, Li H, Satagopan S, Singh J, Chan S (2007) Function, structure, and evolution of the RubisCO-like proteins and their RubisCO homologs. Microbiol Mol Biol Rev 71:576–599. https://doi.org/10.1128/MMBR.00015-07
212. Takaichi S (2011) Carotenoids in algae: distributions, biosyntheses and functions. Mar Drugs 9:1101–1118. https://doi.org/10.3390/md9061101
213. Tanaka A, de Martino A, Amato A, Montsant A, Mathieu B, Rostaing P et al (2015) Ultrastructure and membrane traffic during cell division in the marine pennate diatom Phaeodactylum tricornutum. Protist 166:506–521. https://doi.org/10.1016/J.PROTIS.2015.07.005
214. Tanifuji G, Kamikawa R, Moore CE, Mills T, Onodera NT, Kashiyama Y et al (2020) Comparative plastid genomics of cryptomonas species reveals fine-scale genomic responses to loss of photosynthesis. Genome Biol Evol 12:3926–3937. https://doi.org/10.1093/GBE/EVAA001
215. Tanifuji G, Onodera NT (2017) Cryptomonads: a model organism sheds light on the evolutionary history of genome reorganization in secondary endosymbioses, 1st edn. Elsevier Ltd. https://doi.org/10.1016/bs.abr.2017.06.005
216. Tanifuji G, Onodera NT, Brown MW, Curtis BA, Roger AJ, Ka-Shu Wong G et al (2014) Nucleomorph and plastid genome sequences of the chlorarachniophyte Lotharella oceanica: convergent reductive evolution and frequent recombination in nucleomorph-bearing algae. BMC Genomics 15:374. https://doi.org/10.1186/1471-2164-15-374
217. Tanifuji G, Onodera NT, Wheeler TJ, Dlutek M, Donaher N, Archibald JM (2011) Complete nucleomorph genome sequence of the nonphotosynthetic alga Cryptomonas paramecium reveals a core nucleomorph gene set. Genome Biol Evol 3:44–54. https://doi.org/10.1093/gbe/evq082
218. Tengs T, Dahlberg OJ, Shalchian-Tabrizi K, Klaveness D, Rudi K, Delwiche CF et al (2000) Phylogenetic analyses indicate flint the 19'hexanoyloxy-fucoxanthin- containing dinoflagellates have tertiary plastids of haptophyte origin. Mol Biol Evol 17:718–729. https://doi.org/10.1093/oxfordjournals.molbev.a026350
219. Tikhonenkov DV, Mikhailov KV, Gawryluk RMR, Belyaev AO, Mathur V, Karpov SA et al (2022) Microbial predators form a new supergroup of eukaryotes. Nature 612(7941):714–719. https://doi.org/10.1038/s41586-022-05511-5
220. Tikhonenkov DV, Strassert JFH, Janouškovec J, Mylnikov AP, Aleoshin VV, Burki F et al (2020) Predatory colponemids are the sister group to all other alveolates. Mol Phylogenet Evol 149:106839. https://doi.org/10.1016/J.YMPEV.2020.106839
221. Tsuji Y, Yoshida M (2017) Biology of haptophytes: complicated cellular processes driving the global carbon cycle, 1st edn. Elsevier Ltd. https://doi.org/10.1016/bs.abr.2017.07.002
222. Turmel M, Gagnon MC, O'Kelly CJ, Otis C, Lemieux C (2009) The chloroplast genomes of the green algae Pyramimonas, Monomastix, and Pycnococcus shed new light on the evolutionary history of prasinophytes and the origin of the secondary chloroplasts of euglenids. Mol Biol Evol 26:631–648. https://doi.org/10.1093/molbev/msn285

223. Uthanumallian K, Iha C, Repetti SI, Chan CX, Bhattacharya D, Duchene S et al (2021) Tightly constrained genome reduction and relaxation of purifying selection during secondary plastid endosymbiosis. Mol Biol Evol 39. https://doi.org/10.1093/molbev/msab295
224. Van Beveren F, Eme L, López-García P, Ciobanu M, Moreira D (2022) Independent size expansions and intron proliferation in red algal plastid and mitochondrial genomes. Genome Biol Evol 14. https://doi.org/10.1093/GBE/EVAC037
225. Vanclová AMG, Hadariová L, Hrdá Š, Hampl V (2017) Secondary plastids of euglenophytes. In: Hirakawa Y (ed) Advances in botanical research. Academic, pp 321–358. https://doi.org/10.1016/bs.abr.2017.06.008
226. Votýpka J, Modrý D, Oborník M, Šlapeta J, Lukeš J (2017) Apicomplexa. In: Handbook of the protists: second edition, pp 567–624. https://doi.org/10.1007/978-3-319-28149-0_20/COVER
227. Waller RF, Kořený L (2017) Plastid complexity in dinoflagellates: a picture of gains, losses, replacements and revisions. Adv Bot Res 84:105–143. https://doi.org/10.1016/bs.abr.2017.06.004
228. Waller RF, Slamovits CH, Keeling PJ (2006) Lateral gene transfer of a multigene region from cyanobacteria to dinoflagellates resulting in a novel plastid-targeted fusion protein. Mol Biol Evol 23:1437–1443. https://doi.org/10.1093/molbev/msl008
229. Wang Y, Morse D (2006) Rampant polyuridylylation of plastid gene transcripts in the dinoflagellate Lingulodinium. Nucleic Acids Res 34:613–619. https://doi.org/10.1093/nar/gkj438
230. Watanabe MM, Suda S, Inouya I, Sawaguchi T, Chihara M (1990) Lepidodinium viride gen. et sp. nov. (Gymnodiniales, Dinophyta), a green dinoflagellate with a chlorophyll a- and b-containing endosymbiont. J Phycol 26:741–751. https://doi.org/10.1111/J.0022-3646.1990.00741.X
231. Watanabe MM, Takeda Y, Sasa T, Inouye I, Suda S, Sawaguchi T et al (1987) A green dinoflagellate with chlorophylls a and b: morphology, fine structure of the chloroplast and chlorophyll composition. J Phycol 23:382–389. https://doi.org/10.1111/J.1529-8817.1987.TB04148.X
232. Wetherbee R, Jackson CJ, Repetti SI, Clementson LA, Costa JF, van de Meene A et al (2018) The golden paradox – a new heterokont lineage with chloroplasts surrounded by two membranes. J Phycol:jpy.12822. https://doi.org/10.1111/jpy.12822
233. Woo YH, Ansari H, Otto TD, Klinger CM, Kolisko M, Michálek J et al (2015) Chromerid genomes reveal the evolutionary path from photosynthetic algae to obligate intracellular parasites. Elife 4. https://doi.org/10.7554/eLife.06974
234. Wu S, Gu W, Jia S, Wang L, Wang L, Liu X et al (2021) Proteomic and biochemical responses to different concentrations of CO2 suggest the existence of multiple carbon metabolism strategies in Phaeodactylum tricornutum. Biotechnol Biofuels 14. https://doi.org/10.1186/s13068-021-02088-5
235. Yamada N, Bolton JJ, Trobajo R, Mann DG, Dąbek P, Witkowski A et al (2019) Discovery of a kleptoplastic 'dinotom' dinoflagellate and the unique nuclear dynamics of converting kleptoplastids to permanent plastids. Sci Rep 9:1–13. https://doi.org/10.1038/s41598-019-46852-y
236. Yamada N, Sakai H, Onuma R, Kroth PG, Horiguchi T (2020) Five non-motile dinotom dinoflagellates of the genus dinothrix. Front Plant Sci 11. https://doi.org/10.3389/fpls.2020.591050
237. Yamada N, Sym SD, Horiguchi T (2017) Identification of highly divergent diatom-derived chloroplasts in dinoflagellates, including a description of Durinskia kwazulunatalensis sp. nov. (Peridiniales, Dinophyceae). Mol Biol Evol 34:1335–1351. https://doi.org/10.1093/molbev/msx054
238. Yamada N, Tanaka A, Horiguchi T (2015) Pigment compositions are linked to the habitat types in dinoflagellates. J Plant Res 128:923–932. https://doi.org/10.1007/s10265-015-0745-4
239. Yeh E, DeRisi JL (2011) Chemical rescue of malaria parasites lacking an apicoplast defines organelle function in blood-stage plasmodium falciparum. PLoS Biol 9. https://doi.org/10.1371/journal.pbio.1001138

240. Yokoyama A, Takahashi F, Kataoka H, Hara Y, Nozaki H (2011) Evolutionary analyses of the nuclear-encoded photosynthetic gene psbo from tertiary plastid-containing algae in dinophyta. J Phycol 47:407–414. https://doi.org/10.1111/j.1529-8817.2011.00961.x
241. Yoon HS, Hackett JD, Bhattacharya D (2002a) A single origin of the peridinin- and fucoxanthin-containing plastids in dinoflagellates through tertiary endosymbiosis. Proc Natl Acad Sci U S A 99:11724–11729. https://doi.org/10.1073/pnas.172234799
242. Yoon HS, Hackett JD, Ciniglia C, Pinto G, Bhattacharya D (2004) A molecular timeline for the origin of photosynthetic eukaryotes. Mol Biol Evol 21:809–818. https://doi.org/10.1093/molbev/msh075
243. Yoon HS, Hackett JD, Pinto G, Bhattacharya D (2002b) The Single, Ancient Origin of Chromist Plastids. J Phycol 38:40–40. https://doi.org/10.1046/J.1529-8817.38.S1.8.X
244. Yu G, Nakajima K, Gruber A, Rio Bartulos C, Schober AF, Lepetit B et al (2022) Mitochondrial phosphoenolpyruvate carboxylase contributes to carbon fixation in the diatom Phaeodactylum tricornutum at low inorganic carbon concentrations. New Phytologist 235:1379–1393. https://doi.org/10.1111/nph.18268
245. Yurchenko T, Ševčíková T, Strnad H, Butenko A, Eliáš M (2016) The plastid genome of some eustigmatophyte algae harbours a bacteria-derived six-gene cluster for biosynthesis of a novel secondary metabolite. Open Biol 6. https://doi.org/10.1098/rsob.160249
246. Záhonová K, Füssy Z, Birčák E, Novák Vanclová AMG, Klimeš V, Vesteg M et al (2018) Peculiar features of the plastids of the colourless alga Euglena longa and photosynthetic euglenophytes unveiled by transcriptome analyses. Sci Rep 8:17012. https://doi.org/10.1038/s41598-018-35389-1
247. Zapata M, Fraga S, Rodríguez F, Garrido JL (2012) Pigment-based chloroplast types in dinoflagellates. Mar Ecol Prog Ser 465:33–52. https://doi.org/10.3354/meps09879
248. Zapata M, Jeffrey SW, Wright SW, Rodríguez F, Garrido JL, Clementson L (2004) Photosynthetic pigments in 37 species (65 strains) of Haptophyta: implications for oceanography and chemotaxonomy. Mar Ecol Prog Ser 270. https://doi.org/10.3354/meps270083
249. Zhang Z, Green BR, Cavalier-Smith T (1999) Single gene circles in dinoflagellate chloroplast genomes. Nature 400:155–159. https://doi.org/10.1038/22099
250. Zhu G, Marchewka MJ, Keithly JS (2000) Cryptosporidium parvum appears to lack a plastid genome. Microbiology (Reading) 146(Pt 2):315–321. https://doi.org/10.1099/00221287-146-2-315

The Evolutionary Origin of Mitochondria and Mitochondrion-Related Organelles

Vladimír Hampl and Andrew J. Roger

Abstract Mitochondria and related organelles—hydrogenosomes, mitosomes and other mitochondrion-related organelles (MROs)—are essential compartments in almost all eukaryotic cells. Although mitochondria are renowned for their role in high-efficiency ATP synthesis via oxidative phosphorylation, they have a wide spectrum of other functions within eukaryotic cells including processes such as cofactor biosynthesis, lipid and amino acid metabolism to name a few. In addition, their functions markedly differ between major eukaryote lineages and even between cell types or life stages of the same organism. From an evolutionary standpoint, the origin of mitochondria was clearly an important step in one of the most significant evolutionary transitions in the history of life—the origin of modern eukaryotic cells. Yet, both the timing of mitochondrial symbiosis in eukaryogenesis and the role of mitochondria in this evolutionary transition remain shrouded in mystery and continue to inspire vigorous debate. In this chapter, we will summarise and discuss the various forms of evidence relating to mitochondrial origins, the integration of this symbiotic bacterium into the eukaryotic cellular lineage and the diversification of its properties in the eukaryote tree of life.

Keywords Mitochondria · Endosymbiosis · Alphaproteobacteria · Evolution · Protein import · Division · Eukaryogenesis · Phylogeny · Mitochondrial cristae · Mitochondrial genome · Adaptations to anaerobiosis

V. Hampl (✉)
Department of Parasitology, Faculty of Science, Charles University, BIOCEV, Vestec, Czech Republic
e-mail: vlada@natur.cuni.cz

A. J. Roger
Institute for Comparative Genomics, Department of Biochemistry and Molecular Biology, Dalhousie University, Halifax, NS, Canada
e-mail: Andrew.Roger@dal.ca

S. D. Schwartzbach et al. (eds.), *Endosymbiotic Organelle Acquisition*,
https://doi.org/10.1007/978-3-031-57446-7_3

1 The Early History of Research on Mitochondria

The mitochondrion is a key functional organelle of most eukaryotic cells with notable exceptions of highly specialised cell types (mammalian erythrocytes) and particular groups of anaerobic protists (oxymonads). Mitochondria come in a variety of structural and functional types, which complicates generalisations of their roles in the cells. The best-studied mitochondria in human and yeast (*Saccharomyces cerevisiae*) cells house pathways of carbon metabolism (e.g. the Krebs cycle, fatty acid metabolism and one-carbon metabolism), nitrogen metabolism (the urea cycle), synthesis of amino acids, lipids including cardiolipin and cofactors (ubiquinol, haeme, FeS clusters) and others [1]. German pathologist Richard Altmann was likely the first person to observe mitochondria naming them bioblasts and outlined an idea that was to be largely confirmed in the future that mitochondria are elementary units of life with metabolic and genetic autonomy [2]. The ideas on the autonomy of mitochondria were further elaborated by Paul Portier and Ivan Wallin, who championed the hypothesis that mitochondria were intracellular bacteria (see in [3–5]). Both even attempted to cultivate these organelles, which failed for the reasons we now well understand. Nevertheless, the partial autonomy of mitochondria was supported by demonstrations that they were capable of protein synthesis and that they contained DNA [6–8]. The latter evidence was used by Lynn Margulis to bolster her arguments for their bacterial origin as a part of her serial endosymbiosis theory (SET) [9, 10]. Once it was clarified that the mitochondrial genome and ribosomal RNA were specifically related to those of eubacteria [11, 12] and had distinct origins from genes of nucleocytoplasmic provenance, the endosymbiotic origin of mitochondria gained widespread acceptance.

2 The Prokaryotic Ancestry of Mitochondria

As the hypothesis of a bacterial endosymbiotic origin for mitochondria started to gain traction in the 1970s, it remained unclear what type of bacteria were their closest relatives. Margulis' original proposal implicated an 'aerobic bacterium' as the endosymbiont in an anaerobic heterotrophic 'amitotic' amoeba [9]. Carl Woese, on the other hand, suggested the symbiosis could have taken place pre-oxygenation of the planet and so the endosymbiont could have been an anaerobic photosynthetic bacterium that later evolved into an aerobically respiring organelle within eukaryotes [13]. However, by far the most influential early ideas about the nature of the endosymbiont were advanced in 1975 by John and Whatley who proposed that the bacterium *Paracoccus denitrificans* had specific similarities with mitochondria with respect to aerobic respiratory biochemistry and its electron transport chain [14]. With the subsequent sequencing of proteins and ribosomal RNAs, a phylogenetic picture of the mitochondrial symbiosis began to emerge. Comparisons of oligonucleotide catalogues of small subunit (SSU) ribosomal RNAs strongly suggested a phylogenetic origin of the mitochondrial SSU rRNA from within the bacteria completely distinct from cytosolic SSU rRNAs of eukaryotes [15]. Analyses of cytochrome c [16] found an even more specific phylogenetic affinity between mitochondrial homologues and those of purple non-sulphur bacteria (e.g. species of *Rhodobacter, Rhodomicrobium, Rhodospirillum,* etc.) and *P. denitrificans.* In the mid-1980s, full-length ribosomal RNA sequencing confirmed and extended these earlier observations, clearly showing that mitochondria were sisters to a clade of purple non-sulphur bacteria and relatives [17] that was defined as the alpha subdivision of the *Proteobacteria* by Woese and colleagues. As more data accumulated from mitochondrial and nuclear protein-coding genes of mitochondrial origin and genomes of diverse bacteria were sequenced, phylogenetic analyses appeared to settle on a specific relationship between mitochondria and a subgroup of obligate intracellular *Alphaproteobacteria*, the *Rickettsiales* [18–20].

In the past two decades, with the explosive growth in the microbial genome and transcriptome sequencing and the widespread adoption of phylogenomic analysis of multiple protein genes, several alternative hypotheses for the mitochondrial sister group have been advanced. For example, once genome sequences became available for members of the hyper-abundant planktonic alphaproteobacterial order *Pelagibacterales* (also known as SAR11), several analyses suggested they could be the immediate sister group of mitochondria [21, 22]. Subsequent analyses [23–25] suggested this could be an artefact owing to the long branches of mitochondria and *Pelagibacterales* coupled with a shared compositional bias (both mitochondrial and pelagibacterial genomes are A + T rich and therefore enriched in the amino acids with A + T-rich codons such as F, Y, M, I, N and K versus G, A, R and P that have G + C-rich codons). In fact, similar arguments have been made (see [26] and references therein) questioning the phylogenetic association of long-branched A + T-rich *Rickettsiales* and *Holosporales* with mitochondria frequently found in some recent phylogenomic analyses (see, for example [27]).

The flood of metagenomic data from diverse environments across the planet has led to the discovery of a vast number of previously unknown, diverse novel lineages within bacteria. The opportunity this afforded to re-evaluate the question of mitochondrial origins was first seized upon by Martijn and colleagues [28]. In their phylogenomic analyses of 24 concatenated mitochondrial genome-encoded proteins, they included orthologs from cultivated proteobacteria as well as from substantial numbers of novel alphaproteobacterial taxa represented in metagenomic assembled genomes (MAG) assembled from the TARA oceans data set. They also used several strategies to ameliorate amino acid composition bias including filtering out the most compositionally heterogeneous sites and recoding the amino acid sequence data to fewer states and conducting phylogenetic estimation under biologically realistic site-heterogeneous mixture models. Once the most biased sites were removed, they found the mitochondrial clade forms an independent sister group to the *Alphaproteobacteria* to the exclusion of the *Magnetococcales* and a novel clade comprised of oceanic proteobacterial MAGs they called MarineProteo1. A similar result was also recovered using a compositionally site-filtered data set of

29 nucleus-encoded mitochondrial proteins, albeit with lower statistical support [28]. However, this work was criticized by Fan and colleagues [29] who suggested that site removal methods remove true phylogenetic signal and that the '*Alphaproteobacteria*-sister' position represents a long-branch attraction artefact whereby the outgroup is attracting the long-branch leading to mitochondria. In their analyses, they sample the most G + C-rich and shortest-branched rickettsial and mitochondrial lineages to ameliorate composition bias. Under these conditions, they recover mitochondria as sisters to *Rickettsiales* from within a large subgroup of the *Alphaproteobacteria* clade they call 'Alpha-II' group. Martijn and colleagues [30], have, in turn, challenged these latter analyses on a number of grounds. They documented mistaken paralog inclusion and ortholog absence in some of Fan and colleagues' protein data sets and they provide arguments, and conduct analyses to show that the more G + C-rich ricksettsial and mitochondrial sequences selected by Fan and colleagues are convergently secondarily enriched in G + C from A + T-rich ancestors and consequently have a similar amino acid composition bias stemming from this G + C-richness. Once the convergently G + C-rich sites were removed, the *Alphaproteobacteria*-sister topology was recovered, albeit with weak support in their analyses.

Muñoz-Gómez and colleagues [31] recently revisited this question and attempted to improve on previous analyses by (i) inclusion of additional alphaproteobacterial MAGs from sedimentary lacustrine environments, microbialites and microbial mats; (ii) expansion of the phylogenomic data set to include a total of 108 proteins of mitochondrial origins and their alphaproteobacterial orthologs; and (iii) application of a new model of protein sequence evolution—named GFmix—that allows for both site heterogeneity and shifts in the overall amino acid composition of proteins along branches in a phylogenetic tree. Muñoz-Gómez and colleagues' analyses of this expanded data set using a site-heterogeneous profile mixture model recover the *Alphaproteobacteria*-sister topology with full support even without the removal of compositionally heterogeneous sites. Removal of compositionally heterogeneous sites did not strongly affect this position for mitochondria but did disrupt the suspected composition-based artefactual clustering of *Rickettsiales*, *Holosporales* and *Pelagibacteriales* in the tree. Similarly, the use of the GFmix model on the complete data set also strongly favoured the *Alphaproteobacteria*-sister arrangement in a tree where the above three A + T-rich alphaproteobacterial groups were separated. Furthermore, statistical topology tests using GFmix excluded the *Rickettsiales*-sister and other alternative positions for mitochondria. Interestingly, a search for the Mic60 protein involved in cristae formation across the full diversity of alphaproteobacterial genomes represented in these analyses revealed that this protein appears to be a synapomorphy uniting *Alphaproteobacteria* and mitochondria to the exclusion of *Magnetococcales* and the MarineProteo1 clade. The former two groups appear to ancestrally share this protein, whereas the latter two groups and outgroup proteobacteria lack it. Overall, this suggests that mitochondria evolved from a lineage that was the immediate sister group of extant *Alphaproteobacteria*.

3 Genome Evolution and Endosymbiotic Gene Transfer

Alphaproteobacteria represent a large and diverse prokaryotic group with complete genome sizes and coding capacities recorded in the NCBI genome database (as of 20th February 2023) spanning from 11.5 MB (*Bradyrhizobium barranii*, Bioproject PRJNA608878) and 17,443 coding sequences (CDS; Caulobacteraceae bacterium, PRJNA495371) down to 106 kB and 39 CDS in highly reduced intracellular symbiont of cicadas (*Candidatus* Hodgkinia cicadicola, Bioproject PRJNA385841). The available estimate of the genome coding capacity of the *Alphaproteobacteria* common ancestor (3000–5000 genes) [32] is based on two decades-old data sets and should be updated. In any case, the coding capacities of mitochondrial genomes are orders of magnitude lower, spanning from 0 in MROs to 100 in the jakobid *Andalucia godoyi* [33], underscoring the substantial gene loss from the mitochondrial genome since the first establishment of a mitochondrial symbiont. Many losses resulted from the redundancy of genes in the intracellular environment, whereas others occurred after the evolution of the translocation systems capable of importing proteins from the cytosol (see below), which allowed many genes to be relocated to the host nucleus, a process known as endosymbiotic gene transfer. As a consequence of this process, we find nucleus-encoded pan-eukaryotic genes for proteins that function in mitochondria and have a phylogenetic affiliation with *Alphaproteobacteria* [32]. Unfortunately, a more precise determination of the contribution from mitochondrial endosymbiont is hindered by weak phylogenetic signal due to the long time since the transfer events and a fast rate of evolution of some of the relevant proteins.

To clarify the rates and patterns of gene transfer from the mitochondrial genome to the nucleus, Janouškovec et al. [34] estimated the ancestral gene contents of mtDNA at various nodes of a time-calibrated deep eukaryotic phylogeny. Based on the recovered patterns, they suggested that the *'rate of gene transfer to the nucleus changed profoundly in time'*. Specifically, the number of genes encoded on mtDNA over time appeared to fit an exponential decay model rather than a linear decay model, with most of the losses taking place early in eukaryote evolution and in a massively parallel fashion in separate major lineages. These authors also found that the gene loss process appeared to reach an asymptote in various lineages (i.e. the loss process seems to have all but halted) and that this asymptote differed in major lineages (e.g. jakobids had a higher asymptotic gene content than stramenopiles, which were higher than unicellular opisthokonts). Interestingly, they also detected a later 'secondary' phase of exponential loss for several lineages, such as animals, euglenozoans and myzozoans. The animal secondary reduction phase, for example, occurred in the late Proterozoic, with the resulting asymptotic gene content about half as much as their unicellular opisthokont sisters. The authors suggested that exponential decreases in gene content of organellar DNA over time they observed indicated a rapid early transfer rate, followed by deceleration over time as the asymptotic gene content was approached. However, an alternative interpretation is simply that transferable or non-essential genes in the ancestral mitochondrial genome were randomly lost or transferred over time with a constant mean rate over time as expected for a stochastic process (a Poisson process). The remainder of mitochondrial genes in each lineage could not be transferred to the nucleus (untransferable) and could not be lost (i.e. essential). If we assume this Poisson process as a null transfer/loss model, then the reason the gene content over time curves are exponentially decreasing is not because of a slowing transfer rate over time, but rather that, at the beginning of the gene loss process, there are many genes on the mtDNA that can be transferred, and at the end, there are only a few such genes left plus a set of 'untransferable/essential' genes. The equation describing this null model is a variant of the well-known radioactive decay formula, except that a term is added on to account for the number of untransferable/essential genes: $N_{\text{genes left}} = N_0 e^{-\lambda t} + N_u$, where N_0 is the original number of transferable genes on mtDNA in a particular descendant lineage of last eukaryotic common ancestor (LECA), λ is the constant transfer/loss rate per unit time in the lineage, t is the time elapsed on the lineage and N_u is the number of untransferable/essential genes. In this framework, the numbers and types of untransferable and essential mtDNA genes could be properties of the eukaryotic lineage that episodically change because of changing biochemical needs of the host, changes in the 'barriers' to gene transfers (discussed above and in [34]) or features such as G + C-content, pKa and environmental conditions (see [35, 36] and discussed below). Such episodic changes would shift some of the untransferable genes into the transferable/losable category, allowing secondary phases of exponential genome reduction to occur, as suggested by Janouškovec and colleagues [36]. Using the foregoing as a null model, a variety of more complex models could be compared for their relative fit to real data. The latter models could account for various possible phenomena such as lineage-specific gene transfer/loss rates, multiple different lineage-specific proportions of untransferable genes and different transfer rates among different transferable genes. Such analyses have the potential to improve our understanding of the mechanisms and dynamics of organellar gene content over eukaryotic evolution.

It should be mentioned that the processes underpinning EGT, insertion of mtDNA fragments into nuclear genomes resulting in nuclear copies of mitochondrial DNA (NUMTs), are ongoing even in already highly reduced mitochondrial genomes. These mutations occur with relatively high frequency ($2*10^{-5}$ per generation in baker's yeast, $1*10^{-4}$ in humans) [37, 38]. On the other hand, the majority of NUMTs are non-functional and evolutionarily neutral, and are degraded by mutations over time; only in unfortunate cases do they cause genetic disorders [39].

The above discussion raises the question of why most aerobic (or facultatively aerobic) eukaryotes have retained mtDNA-encoded genes and their mitochondrial chromosomes at all—that is why there are untransferable genes (N_u)? The 'hydrophobicity hypothesis' holds that some proteins—such as those that are particularly large, hydrophobic and with many transmembrane domains—if expressed in the nucleus, would be difficult to transport into mitochondria and there would be a high risk of mistargeting to the endoplasmic reticulum [40]. The Co-location for Redox Regulation (CoRR) hypothesis [41] provides another explanation based on the regulatory role of the genes encoding electron transport chain (ETC) components on the mitochondrial genome. CoRR holds that the expression of critical ETC genes needs to be tightly controlled in an organelle-specific manner which may rapidly affect the expression of components of the electron transport chain (ETC) in individual organelles to ensure optimal stoichiometry and biochemical response to local subcellular conditions. Giannakis and colleagues [36] investigated various possible features of organellar genes and their encoded proteins to determine the best predictors of gene retention in organellar genomes. Consistent with earlier analyses [42, 43], they found clear evidence that hydrophobicity and binding energy centrality (i.e. a measure related to how central a protein is in a complex) positively predict organellar retention. These findings are consistent with both the hydrophobicity hypothesis for obvious reasons and with the CoRR hypothesis because the energy centrality feature the redox regulatory role. They also identified several additional predictors including G + C-content and carboxyl pK_a (the pK_a of the set of all carboxyl groups in the protein), although these appeared to be less important. Soon after, García Pascual, Nordbotten and Johnston [35] identified the environmental setting of an organism as another factor influencing organellar gene retention. Organisms living in environments with high-amplitude regular oscillations in conditions (e.g. tidal or diurnal light–dark cycles) were more likely to retain more organellar genes compared to those in more stable or 'noisy' environments. Clearly, many factors are now known to influence organellar gene retention and many other potential factors remain to be tested (e.g. effective population size, amino acid substitution rate, etc.). In general, the presence of the second half of the electron transport chain (complexes III, IV and F_0F_1ATPsynthase) correlates most strongly with the presence of the mitochondrial genome. Among organisms possessing these complexes, alveolates are the most reduced with the *Chromera velia* mt genome encoding for only two protein-coding genes (cox1 and cox3) [44] and *Amoebophrya ceratii*, a parasitic dinoflagellate that putatively lost its mt genome completely, although this result is disputed [45, 46].

The sizes and architectures of mitochondrial genomes vary even more drastically across eukaryotic diversity than gene content and are apparently more subject to evolutionary tinkering [47, 48]. The NCBI genome database (as of 20th February 2023) comprises almost 17,000 mitochondrial genomes of which more than 14,000 are from animals. Their haploid sizes span from 6 kB in Apicomplexa [49], over 16 kB in humans [12] and approximately 70 kB in choanoflagellates [50] and ciliates [51] to more than 11 MB in flowering plants with high content of repetitive elements [52]. The architecture of the mitochondrial genomes varies from a classical circular molecule as in the case of a human to linear or branched forms and multipartite organisations in which the genome is organised into multiple linear or circular chromosomes, extensively reviewed in [48, 53]. It is possible that various replication models, such as strand displacement (in animals), rolling circle (in fungi) or recombination-dependent (in plants), may explain the differences in substitution rates, non-coding DNA content and architecture [54]. Well known are extreme examples of mitochondrial genomes of kinetoplastids [55] and diplonemids [56] organised in multiple classes of circles either loose or catenated structures but always heavily multiplied resulting in the DNA content reaching 250 Mb in *Diplonema papilatum*, which exceeds the nuclear DNA content of the same organism [57].

4 Integration of Endosymbiont into the Cell—Evolution of Machineries for Import and Export of Proteins

The origin of a protein import system that translocates proteins into mitochondria was a key innovation that permitted the mitochondrial endosymbiont to become more integrated into the eukaryotic cell. Extant aerobic mitochondria contain typically more than a thousand proteins [58], which is less than a free-living alphaproteobacterium, but 1–2 orders of magnitude more than what extant mitochondrial genomes encode [48]. Therefore, hundreds of proteins synthesized on cytosolic ribosomes must be translocated into the mitochondrion. Unsurprisingly, the very core of the mitochondrial protein import systems is derived from the endosymbiont (Fig. 1). This includes, on the outer membrane, a β-barrel porin Tom40 acting as the main import gate and its insertase Sam50, on the inner membrane, the complex of proteins PAM acting as import motor and Oxa1 inserting proteins from the mitochondrial matrix into the inner membrane. In the mitochondrial matrix, the molecular chaperones chaperonin-60 (cpn60) and cpn10 are responsible for the folding of proteins after their import [59]. Aside from the core subunits of the above complexes, the majority of proteins involved in the import process are of eukaryotic origin; that is, they originated from proteins in the host cell [59–62].

Despite common basic principles, the composition of protein translocation systems varies across lineages exhibiting convergent, reductive and potentially also constructive neutral evolution leading sometimes to baroque unnecessarily complex systems (a comprehensive review of the composition, function and evolution of import complexes can be found in [59–62]). Most transported proteins are recognised by the presence of N-terminal pre-sequences, and these are imported across both mitochondrial membranes all the way into the mitochondrial matrix. Other classes of imported proteins comprise hydrophobic mitochondrial carrier family proteins (MCF) targeted to the inner membrane, β-barrel porins and alpha helix-containing tail-anchored proteins targeted to the outer membrane, and finally cysteine motif-containing proteins targeted to the intermembrane space [62]. Typically, two types of receptors associated with the Tom40 channel and exposed to the cytosol are present. The smaller of the two, represented by Tom20 in yeast and *Arabidopsis thaliana*, ATOM46 in *T. brucei* and TvTom36 in *T. vaginalis*, is responsible for binding the pre-sequence-containing proteins, while the larger, represented by Tom70 in yeast, OM64 in *A. thaliana*, ATOM69 in *T. brucei* and likely also *Tv*Tom46 in *T. vaginalis*, recognises hydrophobic MCF family proteins. Interestingly, these receptors in the listed organisms do not seem to be specifically related and some occupy mutually inverted topological conformations (N-terminus in/out). The

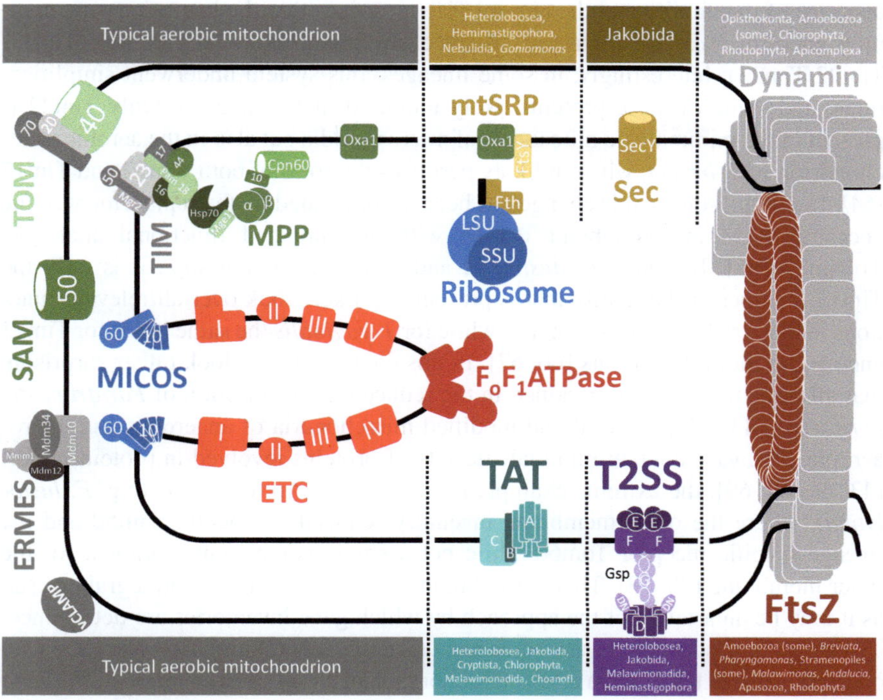

Fig. 1 Schematic representation of functional systems discussed in the text with the information on their taxonomic distribution. The left smaller half of the scheme illustrates the situation in typical aerobic mitochondrion. *Saccharomyces cerevisiae* nomenclature is used for protein labelling. The middle-right part provides information on non-canonical membrane transport complexes reported from various eukaryotic taxa. The rightmost part illustrates the two known mitochondrion division machineries and their distribution in eukaryotic groups. Colourful proteins have recognised homologue in prokaryotes and were likely inherited from the eubacterial endosymbiont. Proteins in the shades of grey are eukaryotic inventions. Abbreviations: ERMES, ER mitochondrion encounter structure; ETC, electron transport chain; LSU, large subunit; MICOS, mitochondrial inner membrane complex; MPP, mitochondrial processing peptidase; mtSRP, mitochondrial signal recognition particle-based system; SAM, sorting and assembly machinery; SSU, small subunit; TIM, translocase of the inner mitochondrial membrane; TOM, translocase of the outer mitochondrial membrane; T2SS, mitochondrial type 2 secretion system

two receptor recognition system of Tom40 thus represents an example of convergent evolution at the molecular level [59]. The Tim17 family of proteins plays a key role in the protein transport across, and protein maturation within, the inner membrane. Tim17 and Tim23 form a TIM23 translocase complex delivering the presequence-containing proteins into the mitochondrial matrix. In this complex, the cavity of Tim17 forms the translocation path and Tim23 has a structural role [63]. Tim22 forms another complex responsible for the lateral insertion of MCF proteins and other members of this family are responsible for the maturation of complex I of the electron transport chain. There is no convincing evidence for prokaryotic ancestry of this family and the current view proposes that the last eukaryotic common ancestor (LECA) contained all three key members of this family (Tim17/Tim22/Tim23). Interestingly, in some lineages, this system underwent simplifications, and some of these proteins likely modified their functions. Only Tim22 is present in the minimalistic mitosomes of microsporidia but also in the aerobic mitochondrion of *T. brucei*, where it likely performs the roles of both Tim22 and Tim17 [64]. In trypanosomes, this change has been accompanied by the replacement of the ancestral PAM motor subunit Pam18 with an unrelated functional analogue, TbPam27 [65]. In *Giardia intestinalis* and *Cryptosporidium* spp., it is only the Tim17 member of the family that is present and likely took over all relevant functions in these reduced mitochondria, while for *T. vaginalis* the same holds for Tim23 encoded in multiple variants [66, 67]. In this context, it may look rather surprising that all three members are retained in the reduced mitochondrion of *Paratrimastix pyriformis* [43]. The reduced and modified mitochondria of anaerobes and microaerophiles always contain minimalistic sets of proteins involved in protein import [43, 66, 68, 69], the extreme example represented by Archamoebae (e.g. *E. histolytica*), where the outer membrane inventory reduced to Tom40, Sam50 and the lineage-specific receptor Tom60 while not a single component is known for the inner membrane [70–73]. This view should, however, be taken with a grain of salt as it may be an artefact of the approach by which gene inventories are determined. Typically, omics data sets are screened using sequence similarity searches with query sequences taken from a few model organisms (e.g. yeast, human, *A. thaliana* and trypanosomes) in which the proteins were determined experimentally. In addition to the difficulties with the identification of distant homologues clearly exemplified by Tom40 in trypanosomes [74, 75] and Tim17 in *Giardia* [76], this methodology precludes the identification of lineage-specific components which are very likely to be present. Identification of such proteins must therefore await functional or proteomic studies in these non-model organisms.

Another interesting aspect of the evolution of the protein import system concerns the acquisition of mitochondrion-targeting signals on the proteins being imported. For the sake of simplicity, we narrow our focus to the largest class of proteins imported into the mitochondrial matrix—those using N-terminal pre-sequences as mitochondrial-targeting sequences (MTS). These pre-sequences have been well characterised for model species with aerobic mitochondria [77] and their key characteristic is the formation of amphipathic alpha helix containing positively charged residues on one side and hydrophobic on the other. This helix binds to the smaller Tom40-associated receptor (Tom20 in yeast) and unfolded proteins are directed to the Tom40 pore. Upon specific interaction with the TIM complex in the intermembrane space, the N-terminal alpha-helix reaches the matrix in a proton motive force driven process. Then, polypeptide translocation is mediated by active pulling by the Hsp70 chaperone of the PAM complex at the cost of ATP hydrolysis. The situation in non-model organisms, particularly those with mitochondria adapted to low-oxygen environments, is poorly understood. For some of these species, such as the metamonads *G. intestinalis* and *T. vaginalis,* the N-terminal pre-sequences are shorter and they have lost positively charged amino acids. The latter may have occurred concomitantly with the loss of the electron transport chain and therefore the electrochemical gradient on the inner mitochondrial membrane [78, 79]. In some *T. vaginalis* proteins, it has been shown that the import is independent of the N-terminal pre-sequences and likely relies on hypothetical internal targeting signals [80–83].

From an evolutionary point of view, it is reasonable to assume that the appearance of signals on targeted proteins and their fine-tuning happened simultaneously with the formation and precision of the import system itself. Once both targeting signals and importers were already present, though ineffective and imprecise, it is not difficult to imagine natural selection acting to enhance their efficiency and selectivity. Like in many other cases, it can be very difficult to conceive of the very beginning of the evolutionary process when the components of a system first arose. What did the protein translocases do initially, when there were no common targeting features on proteins to recognize and, vice versa, why should proteins contain targeting labels if the system cannot recognise them yet? The proposed solutions for this chicken or egg conundrum mostly assume that before the emergence of mitochondrion, some proteins in the cell exhibited features that could in low efficiency drive their import inside providing a platform for natural selection to take off. Indeed, the presence of simple features facilitating the import, e.g. a positive charge at the N-terminus [78], in a subpopulation of proteins is highly probable. This minor mistargeting mechanism [84] could indeed help to populate the emerging organelle with partially mistargeted cytosolic proteins or proteins destined for other compartments (nucleus, ER or secretory system). Consistent with this hypothesis is the observation that 14 proteins from the cytosolic extract of amitochondriate eukaryote *M. exilis* entered the hydrogenosomes of *T. vaginalis* after co-incubation [85]. In the same vein, it was proposed that some classes of prokaryotic proteins, tail-anchored proteins [86] and antimicrobial peptides [87] may be preferentially driven to the mitochondrial membranes or matrix. It has been predicted that approximately 5% of *Escherichia coli* proteins are predisposed to targeting yeast mitochondria because they possess N-terminal extensions with properties similar to mitochondrion-targeting signals; however, this prediction was experimentally confirmed only for a single protein, YhaR [88]. Finally, the evidence for protein import into kleptoplasts (temporary and incompletely integrated plastids from a prey) of Ross Sea Dinoflagellate and mixotrophic euglenid *Rapaza viridis* indicates that the evolution of a protein import system may be relatively fast and could predate the full integration of the symbiont into the cell [89, 90]. It should be noted, however, that the import to higher order (i.e. secondary or tertiary) plastids or kleptoplasts, unlike mitochondria and primary plastids, involves the ready-to-use secretory and endomembrane systems, which may simplify the evolutionary process leading to the establishment of the protein import.

The protein import systems described above are present in some form in all mitochondria. Research on non-model eukaryotes from diverse lineages has also revealed mitochondria endowed with vestiges of systems for transporting proteins in the other direction, i.e. from the matrix across or into the inner membrane, or across both membranes to the cytosol [91]. Although their taxonomic distribution and our understanding of their functional importance are limited, they are implicated in the origin of mitochondria because they relate to bacterial Sec, YidC, Tat and T2SS secretory systems that were likely present in the ancestral mitochondrial symbiont (Fig. 1) [91].

Sec is a fundamental prokaryotic system enabling both the co-translational and post-translational transport of proteins across the plasma membrane. It is centred around the trimeric SecYEG translocase, in which SecY serves as the protein-conducting channel. Eukaryotes use evolutionary derivates of this system to transport proteins from cytosol into the endoplasmic reticulum via Sec61 and into thylakoids and the intermembrane space of plastids via SEC1 and SEC2. Most mitochondria have lost the Sec system, yet some preserved a few of its parts. Mitochondrial genomes of jakobids encode a gene for the SecY channel component which, based on structural modelling, exhibits a conserved structure [33, 91, 92]. At the same time, jakobids lack the genes for SecE and SecG proteins. They also lack the signal recognition particle (SRP) binding to the nascent peptide synthesised at the ribosome as well as the docking protein FtsY, which would bring the ribosome-SRP complex to the membrane channel. The function of SecY in these organisms is thus unclear. Interestingly, homologues of SRP and FtsY designated as mtFfh and mtFtsY, respectively, have been discovered in mitochondria of several other protist groups, namely Heterolobosea (including *Naegleria gruberi*), Alveida (recently referred to as Nebulidia), *Goniomonas* and Hemimastigophora [93]. Consistent with the presence of these proteins, a subset of mitochondrion-encoded proteins in *N. gruberi* contain N-terminal signal peptides necessary for mtFfh binding. Since no homologues of the SecYEG translocase can be found in these species, it is speculated that the system uses the inner mitochondrial membrane protein Oxa1 instead. Oxa1 is a homologue of the second prokaryotic co-translational secretory system YidC and, unlike Sec, it is present in all genome-containing mitochondria where it is responsible for the insertion of proteins from the matrix into the inner mitochondrial membrane. It represents one of only a few widely conserved components of the inner mitochondrial membrane with a clear prokaryotic ancestry [91].

The Tat secretory system present in some mitochondria also has prokaryotic roots and seems to be ancestral to the mitochondrial symbiont. Unlike the Sec and YidC systems, it is able to transport folded proteins across the plasma membrane. The Tat system consists of TatA, TatB and TatC subunits, of which the TatA forms a membrane pore of variable diameter by oligomerisation around the substrate bound to TatC [94]. Homologues of these proteins are patchily distributed across eukaryotes and, except for TatB, are encoded in mitochondrial genomes. Most eukaryotes contain only TatC, the function of which is unknown, however, when accompanied by TatA as in jakobids, Alveida, cryptophytes and ochrophytes, it probably forms a functional Tat system. This was demonstrated experimentally for *Andaluci godoyi* proteins, which were able to substitute for the *E. coli* TatABC complex in vivo [95]. Similarly, the TatBC complex in plant mitochondria was shown to be responsible for the translocation of the Rieske FeS subunit of complex III into the intermembrane space [96].

Finally, a group of unrelated protist lineages, most of them mentioned in various contexts in previous paragraphs (malawimonads, jakobids and heteroloboseans), were shown to contain a reduced version of the prokaryotic type 2 secretion system (T2SS) [97]. This well-characterised system allows bacteria to secrete proteins simultaneously across both membranes by the formation of filaments (pseudopili) stretching from the plasma to the periplasmic membrane. Evidence for mitochondrial localisation and/or pore formation and/or mutual interactions exists for some of these proteins. Substrates for this system are not known and it is hypothesised that it may export mature forms of nucleus-encoded proteins, which were imported into mitochondrion to undergo a final maturation step such as the addition of a prosthetic group. Despite the patchy distribution of TS2II, Tat and Sec system homologues across eukaryotes and our limited understanding of their functions, the emerging picture shows the ancestral mitochondria of LECA exhibited more bacterial features and were equipped with a richer set of communication tools with the host cell cytoplasm than most of today's mitochondria (Fig. 1).

5 Division of Mitochondria

Another necessary condition for the successful integration of the endosymbiont is the synchronisation of its division with the host cell cycle. As such synchronisation is observed in other bacterial endosymbionts of eukaryotes, the origin of the synchronisation system likely occurred in the early phase of the mitochondrial endosymbiosis. The bacterial division system is dependent on the inventory of filamentous temperature-sensitive (fts) genes of which FtsZ is a GTPase forming a ring at the division site responsible for the constriction of the cell [98]. The location of the FtsZ ring is controlled by the antagonistic action of Min genes (MinC, D and E). Although the mitochondrial endosymbiont must have contained the bacterial division system, the mitochondria in humans, yeast but also Apicomplexa (e.g., *Plasmodium*), plants, *T. vaginalis* and many other species were shown to divide using the host-derived constriction system acting on the external surface of the organelle. The scission force here is provided by a ring composed of dynamins (Drp1/Dnm1 in human/yeast) or dynamin-related proteins and GTP hydrolysis [99]. This ring is assembled on the outer surface of the organelle upon the dynamin monomers that are recruited from the cytosol by receptors; the most conserved of these receptors is Fis1 [100]. Dynamins and dynamin-related proteins are a diverse family involved in membrane remodelling in various contexts—'strangling' of vesicles during endocytosis, division of peroxisomes and other vesicles—and so coopting this machinery for the division of endosymbiotic organelles was apparently a straightforward and plausible step.

A broader inspection of gene inventories led to the discovery of a functional core of the ancestral prokaryotic cell division system (FtsZ and Min genes) in species of jakobids and heteroloboseans, a malawimonad, many stramenopiles, amoebozoans, a breviate, an apusomonad and a red alga [101–103]. For some of these cases, the association of proteins with mitochondria has been demonstrated experimentally. This strongly supports the hypothesis that the prokaryotic division system was primordial in mitochondria and was substituted by a host-derived dynamin system later in some lineages. With the potential exception of the red alga *Cyanidioschyzon merolae* [102], the presence of both division rings in one organism has not been reported, so it is not clear whether the dynamin-like proteins were recruited for the mitochondrial division prior to LECA or multiple times later (Fig. 1).

6 Diversification of Mitochondrial Morphology and Function—Cristae Morphology, Adaptation to Anaerobiosis

The presence of the electron transport chain (ETC) and F_0F_1 ATPase in all eukaryotic major clades and the relationship of its core subunits to the eubacterial ETC and ATPases clearly suggest that the mitochondrion of LECA was capable of oxidative phosphorylation coupled with oxygen respiration that it had inherited from its proteobacterial ancestor [104]. In eukaryotes, the presence of ETC and F_0F_1 ATPase almost strictly correlates with conspicuous morphological structures—cristae [105]. Cristae are extensions of the inner mitochondrial membrane into the interior of the mitochondrion forming pockets of various shapes—thin flat lamellae, blob-like vesicles, discs or tubes [105]. Except for aberrant mitochondria, the cristae are connected with the inner membrane by one or more crista junctions. These form a partial barriers separating the cristae lumen from the space between the inner and outer membrane. This increases the efficacy of oxidative phosphorylation because the ETC complexes are primarily present in the crista membranes, pumped protons accumulate mainly in the crista lumen and the proton motive force is harvested by F_0F_1 ATPase present also there. Interestingly, the eukaryotic F_0F_1 ATPase directly contributes to cristae formation by forming dimers in which membrane F_0 moieties make a species-specific angle enforcing the positive membrane curvature at the tips of cristae. The negative membrane curvature at the crista junction is enforced by the MICOS complex [106–108]. Mic10, one of the core components, is present in all eukaryotes with cristae-forming mitochondria and it very likely possesses membrane-deforming activity as shown both in vitro and in vivo [109, 110]. The larger core subunit Mic60 (mitofilin) contains an N-terminal transmembrane domain, while its C-terminal domain extends to intermembrane space as is likely to form contacts with several potential partners there (VDAC/Porin, TOM complex, the SAM complex or Ugo1 involved in mitochondrial fusion) [111]. These contacts likely anchor the cristae to the mitochondrial envelope and prevent them from budding off (the latter is known to occur in Mic60 knockouts) [109]. It has been also demonstrated in yeast that the MICOS complexes assemble in the proximity of ER–mitochondrion contact sites facilitated by outer membrane complex ERMES and distantly from mitochondrion–vacuole contact sites facilitated by the vCLAMP complex [112, 113]. In summary, cristae positioning and shaping are affected by multiple factors including proteins in the outer mitochondrial membrane and contacts of other organelles with mitochondria [114].

An interesting evolutionary finding accompanied the identification of prokaryotic homologues of Mic60 in most investigated α-proteobacteria [115, 116]. These proteins contain all of the important features of Mic60—N-terminal transmembrane domain, middle coiled-coil domain and C-terminal mitofilin domain—and seem to be likely derived from a gene in the haem synthesis operon (HemX) [116]. Strikingly, many alphaproteobacterial species are able to build intracytoplasmic compartments morphologically resembling cristae and Mic60 has been localised into these compartments in several species including *Rhodobacter sphaeroides* [117, 118], *Rhodospirillum rubrum* [119] and *Rhodopseudomonas palustris* [120]. Moreover, the Mic60 homologue in *Rhodobacter sphaeroides* has been shown to physically interact with Bam50, the homologue of Sam50 [121], which is a very strong indication that core molecular and ultrastructural features of cristae have a presymbiotic origin [122].

Adaptation to environments with low oxygen concentration typically leads to changes in mitochondrial metabolism associated with transient or permanent exposure to hypoxia. Four major categories (anaerobic mitochondrion, hydrogen-producing mitochondrion, hydrogenosome and mitosome) of such adapted mitochondria have been defined based on the combination of the presence–absence of ETC, nature of electron acceptors and the production of hydrogen and ATP [123]. The actual evolutionary plasticity of these organelles is, however, wider precluding the inclusion of all known mitochondria into these defined categories. This is why the somewhat complicated-sounding term 'mitochondrion-related organelle' (MRO) was coined to encompass all of them. We suggest that the term mitochondrion be extended to encompass all of these organelles that are derived from the same ancestral symbiosis and limit the use of MRO to discussions focused only on the permanently anaerobically functioning organelles. In the remainder of this text, we will use the term mitochondrion in this broader sense.

A large body of literature including books and reviews describing mitochondria in anaerobic and microaerophilic organisms had accumulated in the previous decades [123, 124] which we do not intend to summarise here but will point only to some general aspects of this adaptive process. The adaptation to this environment starts with using alternative terminal electron acceptors [125], shortening of the ETC by loss (or downregulation) of complexes III and IV [126], loss of mitochondrial DNA, acquisition of the ability to convert ubiquinone into rhodoquinone [127] and the loss of F_oF_1 ATP synthase [128–131]. The loss of the latter complex is almost always accompanied by the loss of cristae, although some exceptions in ciliates have been reported [128, 129]. Further adaptations in some lineages involved the acquisitions of the specific repertoire of enzymes to produce acetyl-CoA (e.g. pyruvate:ferredoxin oxidoreductase, PFO; pyruvate:NADH oxidoreductase, PNO; pyruvate formate lyase, PFL), electron carriers (ferredoxins) and electron sink ([FeFe]hydrogenase). Even though these proteins are present in a wide diversity of unrelated eukaryotes living in low-oxygen environments, their phylogenies are inconsistent with the host eukaryote relationships and, in some cases (e.g. [FeFe]hydrogenase), there are multiple distinct eukaryote clades embedded within different bacterial groups. This suggests that rather than being ancestral to mitochondria, they represent more recent adaptive acquisitions via horizontal gene transfer from other prokaryotic and eukaryotic organisms living in the same niche. However, in some cases, the enzyme phylogenies are poorly resolved (e.g. PFO), so it is not possible to definitively rule out the possibility that they are ancestral to extant eukaryotes [128, 132, 133].

Studies on the most highly reduced mitochondria, some of them designated as mitosomes, are particularly interesting because they demonstrate the extent to which mitochondrial functions can be stripped down and, by extension, which functions are the hardest to lose. The first function to mention in this context is the synthesis of FeS clusters, which are essential cofactors of dozens of enzymes found in different compartments of eukaryotic cells [134]. For unclear reasons, the canonical pathway serving this purpose—the 'iron-sulphur cluster' (ISC) system—which has been inherited from the mitochondrial endosymbiont is always localised within mitochondria suggesting that it cannot function in the cytosol. In several extant mitochondria, the need to retain this pathway represents probably the single *raison d'être* for their existence [135, 136]. However, it is important to note that the synthesis of FeS clusters in the cytosol is possible in principle because we know of two groups of anaerobic protists—the Preaxostyla and Archamoebae—which do it using other FeS cluster biogenesis pathways acquired by horizontal gene transfer—the SUF pathway and the NIF-like pathway, respectively [137–139]. The answer to the obvious question about what the reduced mitochondria in these two groups provide to the cell differs between the two cases. The mitochondria of Archamoebae contain three enzymes synthesising the activated sulphate derivative (PAPS) crucial for sulpholipid synthesis and encystation [72, 73, 140, 141]. This situation is rather unusual because no other eukaryote is known to carry out these reactions in the mitochondrion; in animals and plants, they typically occur in the cytosol [142, 143]. In Preaxostyla, specifically *Paratrimastix pyriformis*, the reason for maintaining mitochondria is its involvement in one-carbon metabolism producing formate and folate derivatives essential for the methionine cycle in the cytosol [43]. Although the involvement of mitochondria in one-carbon metabolism is well known in model organisms [144], not much attention has been paid to this pathway in MROs. Nevertheless, the presence of a glycine cleavage system in many MROs of free-living anaerobes suggests that these organelles may also provide this functions [69]. Consistent with this, mitochondria were completely lost in oxymonads [145–147] after the acquisition of a cytosolic SUF system for the FeS cluster assembly and the loss of one-carbon metabolism. Along similar lines, an independent instance of complete (or near-complete) mitochondrial loss was reported in a free-living anaerobic flagellate *Skoliomonas litria* that is a member of a newly described deeply branching metamonad group. In this case, the loss was likely facilitated by the previous acquisition of an archaeal-type minimal SUF-like FeS cluster biogenesis system and the loss of all of the components of the glycine cleavage system [148].

7 Timeframe of the Mitochondrial Endosymbiosis

Timing mitochondrial origins involves two different aspects: (i) the 'absolute time' in Earth's history when the mitochondrial symbiosis first occurred and (ii) the chronology of the origin of mitochondria versus other events that occurred during eukaryogenesis and delineating the role mitochondria played in this evolutionary transition.

7.1 *The Absolute Timing of the Mitochondrial Symbiosis*

The age of the mitochondrial symbiosis is difficult to establish with any precision because it doesn't correspond to a specific node in any molecular phylogenetic tree (Fig. 2). With the exception of a recent paper [149] in which aerobic mitochondria were suggested to originate later and independently from metamonad MROs, ignoring abundant evidence for the common origin of these organelles (see [26, 43, 79] and references therein), there is a broad consensus that the last eukaryote common

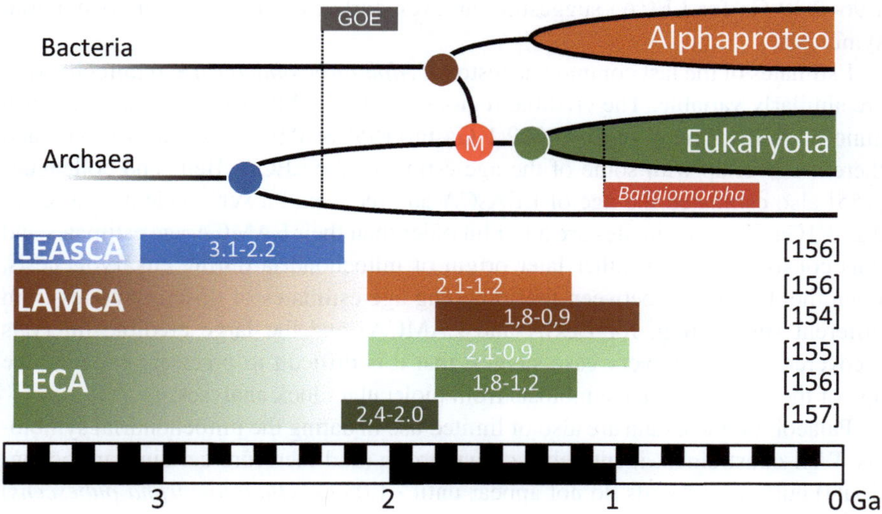

Fig. 2 Schematic representation of nodes in the tree of life, which are relevant to the mitochondrial endosymbiosis and their dating based on the molecular clock. The mitochondrial endosymbiosis is shown by the circle marked with the letter M. The lower bars in the respective colours represent confidence intervals for the ages of these nodes from the literature. References are given on the right-hand side. Two other important events, the Great Oxidation Event (GOE) and the appearance of the first identifiable fossil of the crown group of eukaryotes (the rhodophyte Bangiomorpha pubescens), are also indicated. The time scale in billions of years (Ga) is shown at the bottom. Other abbreviations: LEAsCA, last common ancestor of eukaryotes/Asgard (LEAsCA); LAMCA, last common ancestor of alphaproteobacteria/mitochondria; LECA, last common ancestor of eukaryotes

ancestor (LECA) possessed a mitochondrion [150]. Assuming this is correct, the mitochondrial endosymbiosis must have happened prior to LECA but it must be younger than the last common ancestor of the eukaryote nucleocytoplasmic lineage and its closest related asgard archaeal lineage; here we call this the 'last eukaryote/asgard common ancestor': LEAsCA [151, 152] and note that, in this definition, LEAsCA is younger than the last asgard common ancestor. Finally, the mitochondrial endosymbiosis must also be younger than the last common ancestor of the mitochondrial lineage and extant Alphaproteobacteria (the 'last alphaproteobacterial/mitochondrial common ancestor': LAMCA). All these nodes on the tree of life are very difficult to precisely date. Recent relaxed molecular clock analyses and considerations of the paleobiological record suggest that LECA likely existed between ~1.3 and 2.3 Ga [153–156], but the credible regions on most of the molecular clock estimates of this date are very wide and the age estimates vary greatly depending on what fossil data are used as calibrations, how bounds on calibrations are treated in the analyses (and the root prior) and the specific relaxed molecular clock and sequence evolution models are used. For example, a quite recent molecular clock estimate of the ages of plastid symbioses [156], which uses similar methods but a larger data set and more calibration points than many earlier analyses [155], recovers substantially older ages for crown-group eukaryotes clades and LECA (i.e. 1.81–2.14 Ga for the primary photosynthetic eukaryote group and 1.96–2.39 Ga for LECA) suggesting an even earlier origin for the mitochondrial symbiosis.

Estimates of the last common ancestor of *Alphaproteobacteria* and mitochondria are similarly variable. The credible regions for the LAMCA node sometimes span almost a billion years (e.g. ~1.2–2.1 Ga in [155] or 0.91–1.84 Ga in [153]) and therefore overlap with some of the age estimates for LECA. Betts and colleagues [155] also estimated the age of LEAsCA and recovered a very wide age range of 2.2–3.1Ga. These estimates are a fair bit older than their LAMCA age estimates and thus consistent with a rather later origin of mitochondria during eukaryogenesis. Even the differences between the foregoing age estimates of different nodes from different studies (e.g. for LECA and LAMCA) and the large credible intervals recovered in almost every case suggest that it is difficult to precisely estimate the age of the mitochondrial symbiosis from molecular clock analyses.

Palaeobiological data are also of limited use in dating the mitochondrial symbiosis. Well-constrained, identifiably crown group (and therefore certainly mitochondriate) eukaryote fossils do not appear until ~1.05 Ga (*Bangiomorpha pubescens*) [157]. Although earlier Mesoproterozoic microfossils may correspond to stem group lineages (i.e. extinct lineages that diverged prior to LECA), it is not possible to determine whether they possessed mitochondria [158]. Furthermore, few of these putative stem eukaryote fossils from the Mesoproterozoic have accompanying data bearing on redox conditions, so it is difficult to know what ranges of conditions such organisms lived in (see [158] and references therein). Nevertheless, we do know that wherever the symbiosis occurred, oxygen must have been in sufficient concentrations to support aerobic respiration (i.e. ~1 μM O_2 or higher, see [159]) as this pathway was certainly in LECA and clearly traces its origin to the alphaproteobacterial endosymbiont [104]. Although it's difficult to pinpoint the exact origin of oxygenic photosynthesis, by the early Proterozoic (~2.3 Ga) oxygen was present only in shallow marine environments [160], likely associated with photosynthetic microbial mats. Such environments are therefore almost certainly where the mitochondrial endosymbiosis first occurred and their emergence in the Paleoproterozoic serves as a rough upper bound on its age. On the other hand, attempts to relate the mitochondrial symbiosis to the scarcity of oxygen in the Proterozoic in general and the strict anoxic conditions in deep oceanic environments at the time (e.g [161].) make little sense; the host-mitochondrial symbiont system could not have persisted over many generations in an anoxic environment and still maintain a fully functional oxygen-dependent electron transport chain. Even if the host and/or symbiont were metabolically adapted to fluctuating oxygen concentrations which seems possible (i.e. they were facultative aerobes/anaerobes as is often suggested, e.g. [159, 162]), shallow marine photosynthesizing microbial mats are a much more likely habitat, as they feature steep redox gradients with oxic and anoxic layers in close proximity and diurnal fluctuations in oxygen tension (e.g. [163, 164]). Finally, there has been much speculation about this supposed facultatively anaerobic nature of the protoeukaryotic holobiont and the possible role of anaerobic H_2-based syntrophy in eukaryogenesis [159, 165–167]. However, there is currently no direct evidence supporting the view that the mitochondrial symbiont possessed H_2-producing or -consuming metabolic pathways or that such interactions occurred between the host and endosymbiont.

7.2 The Relative Timing of the Mitochondrial Symbiosis and Other Events in Eukaryogenesis

The timing of the mitochondrial symbiosis relative to other innovations that occurred during eukaryogenesis has been an extremely contentious topic in the past few decades. There are many possible orderings of the events in eukaryogenesis, but the most frequently articulated alternatives are: (i) the mitochondrial symbiosis occurred very early and the energetic boost it provided was critically important to setting eukaryogenesis in motion (these are sometimes called 'mitochondria-early' hypotheses) and (ii) the mitochondrial symbiosis occurred later after the cytoskeletal and endomembrane-related machinery for phagocytosis and intracellular digestion had already evolved (i.e. 'phagocytosis-early/mitochondria-later' hypotheses that usually assume a lesser role of the mitochondrial symbiosis in eukaryogenesis). There are many different versions of these specific hypotheses including some that directly concern the origin of other eukaryote features (e.g. the origin of the nucleus and endomembrane system as per the 'inside-out' hypothesis of Baum and Baum [168]). A common theme in many of the mitochondrial early ideas is that anaerobic H_2-based syntrophy cemented the initial interactions between the symbiont and host. In the 'Hydrogen Hypothesis', the host was an H_2 consumer and the mitochondrial symbiont an H_2 producer with modern-day hydrogen-producing mitochondria in diverse eukaryote lineages directly inheriting these original anaerobic metabolic pathways (see [165]). A recent twist on this idea—the 'E3 hypothesis'—features an asgard archaeal host that is the hydrogen producer and adds in a sulphate-reducing bacterium as a syntrophic partner consuming hydrogen facilitating the eventual adaptation to low levels of oxygen [167]. In contrast, the 'reverse-flow' hypothesis of Spang et al. [166] holds that the original syntrophic interaction was also based on H_2-exchange, but the roles of H_2-producer and H_2-consumer were reversed relative to the Hydrogen Hypothesis. They suggested a syntrophic interaction between an asgard 'host' cell that disposed of electrons by reducing protons to H_2 using membrane-associated [NiFe]-hydrogenases and excreted acetate produced by the Wood-Llungdahl pathway with the proto-mitochondrial symbiont actively consuming these two metabolites. A common theme in some of the foregoing hypotheses is that the genomically predicted metabolic features of some anaerobic Asgard Archaea and the cellular features of the two lokiarchaeal taxa so far cultivated [167, 169] were conserved features of the original host cell that first interacted with the proto-mitochondrial endosymbiont >1.5–2 Ga ago. Most other eukaryote traits (e.g. an endomembrane system, a nucleus, phagocytosis, etc.) are suggested to have evolved after this syntrophy was established. Several authors have even suggested that phagocytosis not only had nothing to do with mitochondrial origins but was not even a trait of LECA, having evolved later polyphyletically [170, 171]. Yet, candidate root positions of the eukaryotic tree [172, 173], combined with detailed ancestral reconstructions of the flagellar apparatus across eukaryotic diversity [174, 175], indicate that LECA was almost certainly a unicellular heterotrophic flagellate. Since the vast majority of known unicellular heterotrophic flagellates are phagotrophs, it is very likely that LECA was too.

Some mitochondria-early narratives have been justified on energetic grounds. That is, the ATP surplus eventually afforded by the mitochondrial endosymbiont was a pre-requisite for a massive expansion of gene families and cell size, and the subsequent 'evolutionary experimentation' that occurred to evolve so many novel functions, processes, as well as molecular and cellular structures during eukaryogenesis. These claims have been critically reviewed in detail elsewhere (see [176–178]) and will not be discussed in depth here. However, we are aware of no strong evidence supporting the idea that surplus ATP provided by aerobically respiring mitochondria was necessary for the origin of the novel genes, gene families or canonical eukaryotic cellular structures of LECA.

Mitochondria-late or -later hypotheses, in contrast, generally hold that most events that occurred during eukaryogenesis were not a consequence of mitochondrial symbiosis. In these scenarios, other features of eukaryote cells—especially the ability to take up and digest cells for food by phagocytosis or a simple precursor process—evolved prior to the mitochondrial symbiosis. The idea that the origin of phagocytosis-based feeding and nutrient uptake was the first step in the origin of eukaryotic cells was originally suggested by de Duve and Wattiaux [179] prior to Margulis' first serial endosymbiosis eukaryogenesis theory paper [9] and therefore prior to subsequent widespread recognition that a symbiotic origin of mitochondria was possible. Interestingly, the same idea and selective rationale was separately proposed by Roger Stanier in 1970 [180]. Although neither of these proposals explicitly suggested that phagocytosis was a necessary precondition for a symbiotic origin of mitochondria, Stanier argued that bacterial endosymbioses would be vastly more likely after phagocytosis had evolved. Most subsequent phagocytosis-early/mitochondria-later hypotheses [181] have assumed that the protomitochondrial symbiont was taken up by a phagocytic host cell in much the same way as many other post-LECA endosymbioses were established, such as those that gave rise to primary (and secondary) plastids and myriad bacterial endosymbionts known in protists across the tree of life (see [182]).

If mitochondria post-dated the origins of phagocytosis, then many orderings of the other innovations in eukaryogenesis are possible. Thankfully, studies of asgard metagenome data and experimental analyses have recently helped to constrain possible scenarios somewhat by demonstrating that LEAsCA likely had a dynamic actin-based cytoskeleton (with profilin, gelsolin and ARP complex subunits, [169]), possibly tubulin-like filaments, some subunits of eukaryotic-type protein targeting systems, a ubiquitin system, and a number of the molecular features likely involved in membrane bending, scission and intracellular vesicle trafficking such as homologues of ESCRT I, II and III complexes as well as TRAPP, COPII, retromer, HOPS and CORVET complexes [151, 152, 183]. Based on this, it seems clear that a simple cytoskeleton and endomembrane system was the first canonically eukaryotic feature to have evolved, although experimental evidence for internalized membrane structures and their trafficking is currently lacking. The finding of a planctomycete cell [184] that appears to possess an actin-based filament system that consumes and digests other cells (both prokaryote and eukaryote) in a phagocytosis-like process indicates that the basic requirements of cellular engulfing and digestion may not require much beyond an ability to deform the cell membrane to surround prey, to secrete hydrolytic enzymes into the prey-enclosing space and import the resulting nutrients. The existence of a bacterium with this feeding mode that lacks any mitochondrion-like organelle invalidates arguments [185] that phagocytosis could not evolve prior to the origin of mitochondria. It will be of great interest to observe, cultivate and experimentally dissect the closest known Asgard archaeal relatives of eukaryotes such as the Hodarchaeota [152] to see if they also can phagocytose prey and determine what structures and functions had already evolved prior to the splitting of the eukaryote nucleocytoplasmic lineage from its Asgard archaeal relatives prior to the mitochondrial symbiosis.

Two studies [186, 187] have attempted to order events in eukaryogenesis using a completely different kind of evidence. Both constructed phylogenetic trees of proteins that were apparently in LECA that had prokaryotic homologues and extracted the length of the 'stem branch'—the branch separating prokaryote and eukaryote clades. Vosseberg et al. [187] also analysed the lengths of the branches between the eukaryotic clades of proteins that duplicated during eukaryogenesis ('duplication lengths'). They then used the distributions of normalised stem lengths and duplication lengths to attempt to order the times of origin of the various proteins. Both studies found distributions of archaeal-origin eukaryotic protein stem lengths to be much larger on average than genes of bacterial origin which were longer than those of mitochondrial origin. This was interpreted as evidence that mitochondrial symbiosis took place substantially after the eukaryote lineage split from Asgard Archaea. Vosseberg and colleagues [187] analyses extended this to gene family expansions and concluded those related to cytoskeleton, membrane trafficking a few other systems occurred prior to the mitochondrial symbiosis, whereas many of the other eukaryotic gene family duplications occurred simultaneously or after this symbiosis. These kinds of analyses have not been universally accepted, with the 'stem length distribution' methodology receiving heavy criticism [186]. Although some of the latter criticisms are unfounded, the methods do rely on some debatable assumptions, such as the requirement that no systematic differences in sequence evolutionary processes affect proteins of different functional classes or different origins or that genes in particular functional classes should all have duplicated at roughly the same time (see [188, 189] and for more discussion).

8 Future Directions

Even after decades of study, the topics of mitochondrial origins and evolution continue to increase in visibility because of the recent flood of comparative genomic, proteomic and cell biological data recently available from diverse eukaryote lineages. While many questions are still not satisfactorily answered (e.g. the precise physiological properties of the mitochondrial endosymbiont), others are much better understood (e.g. the origin of mitochondrial cristae), and still new questions continue to arise. Past research has shown how important it is to study the diversity of organisms on all sides of the mitochondrial endosymbiosis—bacteria, archaea and eukaryotes—and opportunities for new insights are expanding as new lineages in each of these groups are discovered. For example, it is quite possible that further exploration of bacterial diversity using metagenomic approaches will identify the mitochondrial endosymbiont's immediate relatives (i.e. sister group), thereby enabling huge improvements in our estimates of the properties of the original endosymbiont. On the other hand, research on the deepest-level diversity of protistan eukaryotes and their mitochondria will most likely reveal further unexpected features of these organelles absent in model system organisms, as exemplified by the discovery of unexpected secretory and transport systems (TAT, T2SS, SRP). Importantly, the molecular and cell biology of these non-model organisms should continue to develop including the design of protocols and tools for their culturing, imaging and genetic manipulation to clarify how the discovered systems and features actually function in the cells. Only then will we be able to get a glimpse of the properties of the earliest 'primitive' mitochondria and the physiological nature of their early interactions with the host cell. Likewise, continued investigation of known eukaryotes with very simple mitochondria (e.g. metamonads and Archamoebae) or without mitochondria (oxymonads and *Skoliomonas litria*) identifies mitochondrial functions dispensable under certain conditions and reveals others—such as FeS cluster synthesis, one-carbon metabolism and sulphate activation—whose importance was previously underappreciated. These organisms with minimalist mitochondria also represent potentially useful systems in which hypotheses on the integration of mitochondria into the eukaryotic cell can be tested including the evolution of protein import. Development of tools to genetically manipulate these and other novel model organisms for investigation of mitochondrial functions is therefore a promising area of future research. We should not forget that every extant eukaryote is the outcome of a >1.5 billion year evolutionary experiment, so the more diverse lineages we study, the more we learn about the principles, constraints and forces shaping the evolution of their mitochondria.

Acknowledgements We thank Michelle Leger, Alastair Simpson and Pavel Doležal for constructive criticism and discussion. The salary of Vladimír Hampl was supported by the European Research Council (ERC) under the European Union's Horizon 2020 research and innovation programme (grant agreement No. 771592 to VH), from the Ministry of Education, Youth and Sports (MEYS) of the Czech Republic (CR) in the Centre for Research of pathogenicity and virulence of parasites (project No. CZ.02.1.01/0.0/0.0/16_019/0000759). Andrew J. Roger's research in this area is supported by Moore Simons Project on the Origin of the Eukaryotic Cell, Simons Foundation grant 735923LPI (DOI: 10.46714/735923LPI) and by a Discovery Grant from the Natural Sciences and Engineering Research Council of Canada nr. RGPIN-2022-05430.

References

1. Scheffler IE (2008) Mitochondria. Wiley-Liss, Hoboken, New Jersey
2. Altmann R (1894) Die Elementarorganismen und ihre Beziehungen zu den Zellen. 2nd edn. de Gruyter, Boston
3. Archibald JM (2014) One Plus One Equals One: symbiosis and the evolution of complex life. Oxford University Press, New York
4. Sapp J (2016) The symbiotic self. Evol Biol 43:596–603. https://doi.org/10.1007/s11692-016-9378-3
5. Sapp J (1994) Evolution by association: a history of symbiosis. Oxford University Press, New York
6. McLean JR, Cohn GL, Brandt IK, Simpson MV (1958) Incorporation of labeled amino acids into the protein of muscle and liver mitochondria. J Biol Chem 233:657–663. https://doi.org/10.1016/s0021-9258(18)64722-2
7. Kužela S, Wielburski A, Nelson DB (1981) Translation of mitochondrial proteins in digitonin-treated rat hepatocytes. FEBS Lett 135:89–92. https://doi.org/10.1016/0014-5793(81)80950-7
8. Nass MM, Nass S (1963) Intramitochondrial fibers with DNA characteristics: I. Fixation and electron staining reactions. J Cell Biol 19:593. https://doi.org/10.1083/JCB.19.3.593
9. Sagan L (1967) On the origin of mitosing cells. J Theor Biol 14:255–274
10. Margulis L (1970) Origin of eukaryotic cells. Yale University Press, New Haven, Connecticut
11. Gray MW, Doolittle WF (1982) Has the endosymbiont hypothesis been proven? Microbiol Rev 46:1–42. https://doi.org/10.1128/MR.46.1.1-42.1982
12. Anderson S, Bankier AT, Barrell BG, De Bruijn MHL, Coulson AR, Drouin J, Eperon IC, Nierlich DP, Roe BA, Sanger F, Schreier PH, Smith AJH, Staden R, Young IG (1981) Sequence and organization of the human mitochondrial genome. Nature 290:457–465. https://doi.org/10.1038/290457A0
13. Woese CR (1977) Endosymbionts and mitochondrial origins. J Mol Evol 10:93–96. https://doi.org/10.1007/BF01751802
14. John P, Whatley FR (1975) *Paracoccus denitrificans* and the evolutionary origin of the mitochondrion. Nature 254:495–498. https://doi.org/10.1038/254495A0

15. Bonen L, Cunningham RS, Gray MW, Doolittle WF (1977) Wheat embryo mitochondrial 18S ribosomal RNA: evidence for its prokaryotic nature. Nucleic Acids Res 4:663. https://doi.org/10.1093/NAR/4.3.663
16. Schwartz RM, Dayhoff MO (1978) Origins of prokaryotes, eukaryotes, mitochondria, and chloroplasts. Science 199:395–403. https://doi.org/10.1126/SCIENCE.202030
17. Yang D, Oyaizu Y, Oyaizu H, Olsen GJ, Woese CR (1985) Mitochondrial origins. Proc Natl Acad Sci U S A 82:4443. https://doi.org/10.1073/PNAS.82.13.4443
18. Olsen GJ, Woese CR, Overbeek R (1994) The winds of (evolutionary) change: breathing new life into microbiology. J Bacteriol 176:1–6. https://doi.org/10.1128/JB.176.1.1-6.1994
19. Viale AM, Arakaki AK (1994) The chaperone connection to the origins of the eukaryotic organelles. FEBS Lett 341:146–151. https://doi.org/10.1016/0014-5793(94)80446-X
20. Andersson SGE, Zomorodipour A, Andersson JO, Sicheritz-Ponten T, Alsmark UCM, Podowski RM, Naslund AK, Eriksson AS, Winkler HH, Kurland CG (1998) The genome sequence of *Rickettsia prowazekii* and the origin of mitochondria. Nature 396:133–140. https://doi.org/10.1038/24094
21. Thrash JC, Boyd A, Huggett MJ, Grote J, Carini P, Yoder RJ, Robbertse B, Spatafora JW, Rappé MS, Giovannoni SJ (2011) Phylogenomic evidence for a common ancestor of mitochondria and the SAR11 clade. Sci Rep 1:13. https://doi.org/10.1038/srep00013
22. Georgiades K, Madoui MA, Le P, Robert C, Raoult D (2011) Phylogenomic analysis of *Odyssella thessalonicensis* fortifies the common origin of Rickettsiales, *Pelagibacter ubique* and *Reclimonas americana* mitochondrion. PloS One 6:e24857. https://doi.org/10.1371/JOURNAL.PONE.0024857
23. Rodríguez-Ezpeleta N, Embley TM (2012) The SAR11 group of alpha-proteobacteria is not related to the origin of mitochondria. PloS One 7:e30520. https://doi.org/10.1371/JOURNAL.PONE.0030520
24. Brindefalk B, Ettema TJG, Viklund J, Thollesson M, Andersson SGE (2011) A phylometagenomic exploration of oceanic alphaproteobacteria reveals mitochondrial relatives unrelated to the SAR11 clade. PloS One 6:e24457. https://doi.org/10.1371/journal.pone.0024457
25. Viklund J, Ettema TJG, Andersson SGE (2012) Independent genome reduction and phylogenetic reclassification of the oceanic SAR11 clade. Mol Biol Evol 29:599–615. https://doi.org/10.1093/MOLBEV/MSR203
26. Roger AJ, Muñoz-Gómez SA, Kamikawa R (2017) The origin and diversification of mitochondria. Curr Biol 27:R1177–R1192. https://doi.org/10.1016/j.cub.2017.09.015
27. Wang Z, Wu M (2015) An integrated phylogenomic approach toward pinpointing the origin of mitochondria. Sci Rep 51(5):1–12. https://doi.org/10.1038/srep07949
28. Martijn J, Vosseberg J, Guy L, Offre P, Ettema TJG (2018) Deep mitochondrial origin outside the sampled alphaproteobacteria. Nature 557:101–105. https://doi.org/10.1038/s41586-018-0059-5
29. Fan L, Wu D, Goremykin V, Xiao J, Xu Y, Garg S, Zhang C, Martin WF, Zhu R (2020) Phylogenetic analyses with systematic taxon sampling show that mitochondria branch within Alphaproteobacteria. Nat Ecol Evol 4:1213–1219. https://doi.org/10.1038/S41559-020-1239-X
30. Martijn J, Vosseberg J, Guy L, Offre P, Ettema TJG (2022) Phylogenetic affiliation of mitochondria with Alpha-II and Rickettsiales is an artefact. 6:1829–1831. https://doi.org/10.1038/s41559-022-01871-3
31. Muñoz-gómez SA, Susko E, Williamson K, Eme L, Slamovits CH, Moreira D, López-garcía P, Roger AJ (2022) Site-and-branch-heterogeneous analyses of an expanded dataset favour mitochondria as sister to known Alphaproteobacteria. Nat Ecol Evol 6:253–262. https://doi.org/10.1038/s41559-021-01638-2
32. Boussau B, Karlberg EO, Frank AC, Legault BA, Andersson SGE (2004) Computational inference of scenarios for α-proteobacterial genome evolution. Proc Natl Acad Sci U S A 101:9722–9727. https://doi.org/10.1073/pnas.0400975101
33. Burger G, Gray MW, Forget L, Lang BF (2013) Strikingly bacteria-like and gene-rich mitochondrial genomes throughout jakobid protists. Genome Biol Evol 5:418–438. https://doi.org/10.1093/gbe/evt008

34. Janouškovec J, Tikhonenkov DV, Burki F, Howe AT, Rohwer FL, Mylnikov AP, Keeling PJ (2017) A new lineage of eukaryotes illuminates early mitochondrial genome reduction. Curr Biol 27:3717–3724.e5. https://doi.org/10.1016/J.CUB.2017.10.051
35. García Pascual B, Nordbotten JM, Johnston IG (2023) Cellular and environmental dynamics influence species-specific extents of organelle gene retention. Proc Biol Sci 290:20222140. https://doi.org/10.1098/RSPB.2022.2140
36. Giannakis K, Arrowsmith SJ, Richards L, Gasparini S, Chustecki JM, Røyrvik EC, Johnston IG (2022) Evolutionary inference across eukaryotes identifies universal features shaping organelle gene retention. Cell Syst 13:874–884.e5. https://doi.org/10.1016/J.CELS.2022.08.007
37. Thorsness PE, Fox TD (1990) Escape of DNA from mitochondria to the nucleus in *Saccharomyces cerevisiae*. Nature 346:376–379. https://doi.org/10.1038/346376a0
38. Wei W, Schon KR, Elgar G, Orioli A, Tanguy M, Giess A, Tischkowitz M, Caulfield MJ, Chinnery PF (2022) Nuclear-embedded mitochondrial DNA sequences in 66,083 human genomes. Nature 611:105–114. https://doi.org/10.1038/s41586-022-05288-7
39. Hazkani-Covo E, Zeller RM, Martin W (2010) Molecular poltergeists: Mitochondrial DNA copies (numts) in sequenced nuclear genomes. PLoS Genet 6:e1000834. https://doi.org/10.1371/journal.pgen.1000834
40. Björkholm P, Ernst AM, Hagström E, Andersson SGE (2017) Why mitochondria need a genome revisited. FEBS Lett 591:65–75. https://doi.org/10.1002/1873-3468.12510
41. Allen JF (2015) Why chloroplasts and mitochondria retain their own genomes and genetic systems: colocation for redox regulation of gene expression. Proc Natl Acad Sci U S A 112:10231–10238. https://doi.org/10.1073/PNAS.1500012112
42. Johnston IG, Williams BP (2016) Evolutionary inference across eukaryotes identifies specific pressures favoring mitochondrial gene retention. Cell Syst 2:101–111. https://doi.org/10.1016/J.CELS.2016.01.013
43. Zítek J, Füssy Z, Treitli SC, Peña-Diaz P, Vaitová Z, Zavadska D, Harant K, Hampl V (2022) Reduced mitochondria provide an essential function for the cytosolic methionine cycle. Curr Biol 32:5057–5068.e5. https://doi.org/10.1016/j.cub.2022.10.028
44. Flegontov P, Michálek J, Janouškovec J, Lai D-H, Jirků M, Hajdušková E, Tomčala A, Otto TD, Keeling PJ, Pain A, Oborník M, Lukeš J (2015) Divergent mitochondrial respiratory chains in phototrophic relatives of apicomplexan parasites. Mol Biol Evol 32:1115–1131. https://doi.org/10.1093/molbev/msv021
45. John U, Lu Y, Wohlrab S, Groth M, Janouškovec J, Kohli GS, Mark FC, Bickmeyer U, Farhat S, Felder M, Frickenhaus S, Guillou L, Keeling PJ, Moustafa A, Porcel BM, Valentin K, Glöckner G (2019) An aerobic eukaryotic parasite with functional mitochondria that likely lacks a mitochondrial genome. Sci Adv 5:1–11. https://doi.org/10.1126/sciadv.aav1110
46. Kayal E, Smith DR (2021) Is the dinoflagellate *Amoebophrya* really missing an mtDNA? Mol Biol Evol 38:2493–2496. https://doi.org/10.1093/MOLBEV/MSAB041
47. Formaggioni A, Luchetti A, Plazzi F (2021) Mitochondrial genomic landscape: a portrait of the mitochondrial genome 40 years after the first complete sequence. Life (Basel, Switzerland) 11:663. https://doi.org/10.3390/LIFE11070663
48. Zardoya R (2020) Recent advances in understanding mitochondrial genome diversity. F1000 Research 9:F1000 Faculty Rev-270. https://doi.org/10.12688/F1000RESEARCH.21490.1
49. Feagin JE (1994) The extrachromosomal DNAs of apicomplexan parasites. Annu Rev Microbiol 48:81–104. https://doi.org/10.1146/ANNUREV.MI.48.100194.000501
50. Burger G, Forget L, Zhu Y, Gray MW, Lang BF (2003) Unique mitochondrial genome architecture in unicellular relatives of animals. Proc Natl Acad Sci U S A 100:892–897. https://doi.org/10.1073/pnas.0336115100
51. Swart EC, Nowacki M, Shum J, Stiles H, Higgins BP, Doak TG, Schotanus K, Magrini VJ, Minx P, Mardis ER, Landweber LF (2012) The *Oxytricha trifallax* mitochondrial genome. Genome Biol Evol 4:136–154. https://doi.org/10.1093/GBE/EVR136
52. Sloan DB, Alverson AJ, Chuckalovcak JP, Wu M, McCauley DE, Palmer JD, Taylor DR (2012) Rapid evolution of enormous, multichromosomal genomes in flowering plant mitochondria with exceptionally high mutation rates. PLoS Biol 10:e1001241. https://doi.org/10.1371/JOURNAL.PBIO.1001241

53. Smith DR, Keeling PJ (2015) Mitochondrial and plastid genome architecture: reoccurring themes, but significant differences at the extremes. Proc Natl Acad Sci U S A 112:10177–10184. https://doi.org/10.1073/pnas.1422049112
54. Chevigny N, Schatz-Daas D, Lotfi F, Gualberto JM (2020) DNA repair and the stability of the plant mitochondrial genome. Int J Mol Sci 21:328. https://doi.org/10.3390/IJMS21010328
55. Lukeš J, Guilbride DL, Votýpka J, Zíková A, Benne R, Englund P (2002) Kinetoplast DNA network : evolution of an improbable structure. Eukaryot Cell 1:495–502. https://doi.org/10.1128/EC.1.4.495-502.2002
56. Vlcek C, Marande W, Teijeiro S, Lukeš J, Burger G (2011) Systematically fragmented genes in a multipartite mitochondrial genome. Nucleic Acids Res 39:979–988. https://doi.org/10.1093/nar/gkq883
57. Lukeš J, Wheeler R, Jirsová D, David V, Archibald JM (2018) Massive mitochondrial DNA content in diplonemid and kinetoplastid protists. IUBMB Life 70:1267–1274. https://doi.org/10.1002/iub.1894
58. Hammond M, Dorrell RG, Speijer D, Lukeš J (2022) Eukaryotic cellular intricacies shape mitochondrial proteomic complexity. Bioessays 44:1–13. https://doi.org/10.1002/bies.202100258
59. Schneider A (2022) Evolution and diversification of mitochondrial protein import systems. Curr Opin Cell Biol 75:102077. https://doi.org/10.1016/j.ceb.2022.102077
60. Fukasawa Y, Oda T, Tomii K, Imai K (2017) Origin and evolutionary alteration of the mitochondrial import system in eukaryotic lineages. Mol Biol Evol 34:1–13. https://doi.org/10.1093/molbev/msx096
61. Schneider A (2020) Evolution of mitochondrial protein import – lessons from trypanosomes. Biol Chem 401:663–676. https://doi.org/10.1515/HSZ-2019-0444
62. Wiedemann N, Pfanner N (2017) Mitochondrial machineries for protein import and assembly. Annu Rev Biochem 86:685–714. https://doi.org/10.1146/ANNUREV-BIOCHEM-060815-014352
63. Sim SI, Chen Y, Lynch DL, Gumbart JC, Park E (2023) Structural basis of mitochondrial protein import by the TIM23 complex. Nature 621:620–626. https://doi.org/10.1038/s41586-023-06239-6
64. Harsman A, Oeljeklaus S, Wenger C, Huot JL, Warscheid B, Schneider A (2016) The non-canonical mitochondrial inner membrane presequence translocase of trypanosomatids contains two essential rhomboid-like proteins. Nat Commun 7:13707. https://doi.org/10.1038/NCOMMS13707
65. von Känel C, Muñoz-Gómez SA, Oeljeklaus S, Wenger C, Warscheid B, Wideman JG, Harsman A, Schneider A (2020) Homologue replacement in the import motor of the mitochondrial inner membrane of trypanosomes. Elife 9:e52560. https://doi.org/10.7554/ELIFE.52560
66. Rada P, Doležal P, Jedelský PL, Bursac D, Perry AJ, Šedinová M, Smíšková K, Novotný M, Beltrán NC, Hrdý I, Lithgow T, Tachezy J (2011) The core components of organelle biogenesis and membrane transport in the hydrogenosomes of *Trichomonas vaginalis*. PloS One 6:e24428. https://doi.org/10.1371/JOURNAL.PONE.0024428
67. Žárský V, Doležal P, Dole P (2016) Evolution of the Tim17 protein family. Biol Direct 11:1–13
68. Martincová E, Voleman L, Pyrih J, Žárský V, Vondráčková P, Kolísko M, Tachezy J, Doležal P (2015) Probing the biology of *Giardia intestinalis* mitosomes using in vivo enzymatic tagging. Mol Cell Biol 35:2864–2874. https://doi.org/10.1128/MCB.00448-15
69. Leger MM, Kolisko M, Kamikawa R, Stairs CW, Kume K, Čepička I, Silberman JD, Andersson JO, Xu F, Yabuki A, Eme L, Zhang Q, Takishita K, Inagaki Y, Simpson AGBB, Hashimoto T, Roger AJ (2017) Organelles that illuminate the origins of *Trichomonas* hydrogenosomes and *Giardia* mitosomes. Nat Ecol Evol 1:0092. https://doi.org/10.1038/s41559-017-0092
70. Doležal P, Dagley MJ, Kono M, Wolynec P, Likić VA, Foo JH, Šedinová M, Tachezy J, Bachmann A, Bruchhaus I, Lithgow T (2010) The essentials of protein import in the degenerate mitochondrion of *Entamoeba histolytica*. PLoS Pathog 6:e1000812. https://doi.org/10.1371/journal.ppat.1000812

71. Makiuchi T, Mi-Ichi F, Nakada-Tsukui K, Nozaki T (2013) Novel TPR-containing subunit of TOM complex functions as cytosolic receptor for *Entamoeba* mitosomal transport. Sci Rep 3:1–7. https://doi.org/10.1038/srep01129
72. Nývltová E, Stairs CW, Hrdý I, Rídl J, Mach J, Pačes J, Roger AJ, Tachezy J (2015) Lateral gene transfer and gene duplication played a key role in the evolution of *Mastigamoeba balamuthi* hydrogenosomes. Mol Biol Evol 32:1039–1055. https://doi.org/10.1093/MOLBEV/MSU408
73. Záhonová K, Treitli SC, Le T, Škodová-Sveráková I, Hanousková P, Čepička I, Tachezy J, Hampl V (2022) Anaerobic derivates of mitochondria and peroxisomes in the free-living amoeba *Pelomyxa schiedti* revealed by single-cell genomics. BMC Biol 20:56. https://doi.org/10.1186/S12915-022-01247-W
74. Pusnik M, Schmidt O, Perry AJ, Oeljeklaus S, Niemann M, Warscheid B, Lithgow T, Meisinger C, Schneider A (2011) Mitochondrial preprotein translocase of trypanosomatids has a bacterial origin. Curr Biol 21:1738–1743. https://doi.org/10.1016/j.cub.2011.08.060
75. Zarsky V, Tachezy J, Dolezal P (2012) Tom40 is likely common to all mitochondria. Curr Biol 22:R479–R481. https://doi.org/10.1016/J.CUB.2012.03.057
76. Pyrihová E, Motyčková A, Voleman L, Wandyszewska N, Fišer R, Seydlová G, Roger A, Kolísko M, Doležal P (2018) A single tim translocase in the mitosomes of *Giardia intestinalis* illustrates convergence of protein import machines in anaerobic eukaryotes. Genome Biol Evol 10:2813–2822. https://doi.org/10.1093/gbe/evy215
77. Kunze M, Berger J (2015) The similarity between N-terminal targeting signals for protein import into different organelles and its evolutionary relevance. Front Physiol 6:1–27. https://doi.org/10.3389/fphys.2015.00259
78. Garg SG, Gould SB (2016) The role of charge in protein targeting evolution. Trends Cell Biol 26:894–905. https://doi.org/10.1016/j.tcb.2016.07.001
79. Stairs CW, Táborský P, Salomaki ED, Kolisko M, Pánek T, Eme L, Hradilová M, Vlček Č, Jerlström-Hultqvist J, Roger AJ, Čepička I (2021) Anaeramoebae are a divergent lineage of eukaryotes that shed light on the transition from anaerobic mitochondria to hydrogenosomes. Curr Biol 31:5605–5612.e5. https://doi.org/10.1016/j.cub.2021.10.010
80. Mentel M, Zimorski V, Haferkamp P, Martin W, Henze K (2008) Protein import into hydrogenosomes of *Trichomonas vaginalis* involves both N-terminal and internal targeting signals: a case study of thioredoxin reductases. Eukaryot Cell 7:1750–1757. https://doi.org/10.1128/EC.00206-08
81. Zimorski V, Major P, Hoffmann K, Brás XP, Martin WF, Gould SB (2013) The N-terminal sequences of four major hydrogenosomal proteins are not essential for import into hydrogenosomes of *Trichomonas vaginalis*. J Eukaryot Microbiol 60:89–97. https://doi.org/10.1111/jeu.12012
82. Burstein D, Gould SB, Zimorski V, Kloesges T, Kiosse F, Major P, Martin WF, Pupko T, Dagan T (2012) A machine learning approach to identify hydrogenosomal proteins in *Trichomonas vaginalis*. Eukaryot Cell 11:217–228. https://doi.org/10.1128/EC.05225-11
83. Rada P, Makki AR, Zimorski V, Garg S, Hampl V, Hrdý I, Gould SB, Tachezy J (2015) N-terminal presequence-independent import of phosphofructokinase into hydrogenosomes of *Trichomonas vaginalis*. Eukaryot Cell 14:1264–1275. https://doi.org/10.1128/EC.00104-15/SUPPL_FILE/ZEK999094508SO1.PDF
84. Martin W (2010) Evolutionary origins of metabolic compartmentalization in eukaryotes. Philos Trans R Soc B Biol Sci 365:847–855. https://doi.org/10.1098/rstb.2009.0252
85. Fang YK, Vaitová Z, Hampl V (2023) A mitochondrion-free eukaryote contains proteins capable of import into an exogenous mitochondrion-related organelle. Open Biol 13:220238. https://doi.org/10.1098/RSOB.220238
86. Lutfullahoğlu-Bal G, Seferoğlu AB, Keskin A, Akdoğan E, Dunn CD (2018) A bacteria-derived tail anchor localizes to peroxisomes in yeast and mammalian cells. Sci Rep 8:1–11. https://doi.org/10.1038/s41598-018-34646-7
87. Garrido C, Caspari OD, Choquet Y, Wollman FA, Lafontaine I (2020) Evidence supporting an antimicrobial origin of targeting peptides to endosymbiotic organelles. Cells 9:1795. https://doi.org/10.3390/CELLS9081795

88. Lucattini R, Likić VA, Lithgow T (2004) Bacterial proteins predisposed for targeting to mitochondria. Mol Biol Evol 21:652–658. https://doi.org/10.1093/molbev/msh058
89. Karnkowska A, Yubuki N, Maruyama M, Yamaguchi A, Kashiyama Y, Suzaki T, Keeling PJ, Hampl V, Leander BS (2023) Euglenozoan kleptoplasty illuminates the early evolution of photoendosymbiosis. Proc Natl Acad Sci U S A 120:e2220100120. https://doi.org/10.1073/PNAS.2220100120
90. Hehenberger E, Gast RJ, Keeling PJ (2019) A kleptoplastidic dinoflagellate and the tipping point between transient and fully integrated plastid endosymbiosis. Proc Natl Acad Sci 116:17934–17942. https://doi.org/10.1073/pnas.1910121116
91. Petrů M, Dohnálek V, Füssy Z, Doležal P (2021) Fates of Sec, Tat, and YidC translocases in mitochondria and other eukaryotic compartments. Mol Biol Evol 38:5241–5254. https://doi.org/10.1093/molbev/msab253
92. Lang BF, Burger G, O'Kelly CJ, Cedergren R, Golding GB, Lemieux C, Sankoff D, Turmel M, Gray MW (1997) An ancestral mitochondrial DNA resembling a eubacterial genome in miniature. Nature 387:493–497. https://doi.org/10.1038/387493A0
93. Pyrih J, Pánek T, Durante IM, Rašková V, Cimrhanzlová K, Kriegová E, Tsaousis AD, Eliáš M, Lukeš J (2021) Vestiges of the bacterial signal recognition particle-based protein targeting in mitochondria. Mol Biol Evol 38:3170–3187. https://doi.org/10.1093/molbev/msab090
94. Alcock F, Stansfeld PJ, Basit H, Habersetzer J, Baker MA, Palmer T, Wallace MI, Berks BC (2016) Assembling the Tat protein translocase. Elife 5:20718. https://doi.org/10.7554/ELIFE.20718
95. Petrů M, Wideman J, Moore K, Alcock F, Palmer T, Doležal P (2018) Evolution of mitochondrial TAT translocases illustrates the loss of bacterial protein transport machines in mitochondria. BMC Biol 16:1–14. https://doi.org/10.1186/s12915-018-0607-3
96. Schäfer K, Künzler P, Klingl A, Eubel H, Carrie C (2020) The plant mitochondrial TAT pathway is essential for complex III biogenesis. Curr Biol 30:2206. https://doi.org/10.1016/J.CUB.2020.05.057
97. Horváthová L, Žárský V, Pánek T, Derelle R, Pyrih J, Motyčková A, Klápšťová V, Vinopalová M, Marková L, Voleman L, Klimeš V, Petrů M, Vaitová Z, Čepička I, Hryzáková K, Harant K, Gray MW, Chami M, Guilvout I, Francetic O, Franz Lang B, Vlček Č, Tsaousis AD, Eliáš M, Doležal P (2021) Analysis of diverse eukaryotes suggests the existence of an ancestral mitochondrial apparatus derived from the bacterial type II secretion system. Nat Commun 12:1–18. https://doi.org/10.1038/s41467-021-23046-7
98. Errington J, Daniel RA, Scheffers D-J (2003) Cytokinesis in bacteria. Microbiol Mol Biol Rev 67:52–65. https://doi.org/10.1128/MMBR.67.1.52-65.2003
99. Antonny B, Burd C, De Camilli P, Chen E, Daumke O, Faelber K, Ford M, Frolov VA, Frost A, Hinshaw JE, Kirchhausen T, Kozlov MM, Lenz M, Low HH, McMahon H, Merrifield C, Pollard TD, Robinson PJ, Roux A, Schmid S (2016) Membrane fission by dynamin: what we know and what we need to know. EMBO J 35:2270–2284. https://doi.org/10.15252/embj.201694613
100. Kalia R, Wang RYR, Yusuf A, Thomas PV, Agard DA, Shaw JM, Frost A (2018) Structural basis of mitochondrial receptor binding and constriction by DRP1. Nature 558:401–405. https://doi.org/10.1038/s41586-018-0211-2
101. Beech PL, Nheu T, Schultz T, Herbert S, Lithgow T, Gilson PR, McFadden GI (2000) Mitochondrial FtsZ in a chromophyte alga. Science 287:1276–1279. https://doi.org/10.1126/science.287.5456.1276
102. Kuroiwa T, Nishida K, Yoshida Y, Fujiwara T, Mori T, Kuroiwa H, Misumi O (2006) Structure, function and evolution of the mitochondrial division apparatus. Biochim Biophys Acta – Mol Cell Res 1763:510–521. https://doi.org/10.1016/j.bbamcr.2006.03.007
103. Leger MM, Petrů M, Žárský V, Eme L, Vlček Č, Harding T, Lang BF, Eliáš M, Doležal P, Roger AJ (2015) An ancestral bacterial division system is widespread in eukaryotic mitochondria. Proc Natl Acad Sci U S A 112:10239–10246. https://doi.org/10.1073/pnas.1421392112
104. Wang Z, Wu M (2014) Phylogenomic reconstruction indicates mitochondrial ancestor was an energy parasite. PloS One 9:e110685

105. Pánek T, Eliáš M, Vancová M, Lukeš J, Hashimi H (2020) Returning to the fold for lessons in mitochondrial crista diversity and evolution. Curr Biol 30:R575–R588. https://doi.org/10.1016/j.cub.2020.02.053
106. von der Malsburg K, Müller JM, Bohnert M, Oeljeklaus S, Kwiatkowska P, Becker T, Loniewska-Lwowska A, Wiese S, Rao S, Milenkovic D, Hutu DP, Zerbes RM, Schulze-Specking A, Meyer HE, Martinou JC, Rospert S, Rehling P, Meisinger C, Veenhuis M, Warscheid B, van der Klei IJ, Pfanner N, Chacinska A, van der Laan M (2011) Dual role of mitofilin in mitochondrial membrane organization and protein biogenesis. Dev Cell 21:694–707. https://doi.org/10.1016/J.DEVCEL.2011.08.026
107. Harner M, Körner C, Walther D, Mokranjac D, Kaesmacher J, Welsch U, Griffith J, Mann M, Reggiori F, Neupert W (2011) The mitochondrial contact site complex, a determinant of mitochondrial architecture. EMBO J 30:4356–4370. https://doi.org/10.1038/EMBOJ.2011.379
108. Hoppins S, Collins SR, Cassidy-Stone A, Hummel E, DeVay RM, Lackner LL, Westermann B, Schuldiner M, Weissman JS, Nunnari J (2011) A mitochondrial-focused genetic interaction map reveals a scaffold-like complex required for inner membrane organization in mitochondria. J Cell Biol 195:323–340. https://doi.org/10.1083/JCB.201107053
109. Stephan T, Brüser C, Deckers M, Steyer AM, Balzarotti F, Barbot M, Behr TS, Heim G, Hübner W, Ilgen P, Lange F, Pacheu-Grau D, Pape JK, Stoldt S, Huser T, Hell SW, Möbius W, Rehling P, Riedel D, Jakobs S (2020) MICOS assembly controls mitochondrial inner membrane remodeling and crista junction redistribution to mediate cristae formation. EMBO J 39:1–24. https://doi.org/10.15252/embj.2019104105
110. Barbot M, Jans DC, Schulz C, Denkert N, Kroppen B, Hoppert M, Jakobs S, Meinecke M (2015) Mic10 oligomerizes to bend mitochondrial inner membranes at cristae junctions. Cell Metab 21:756–763. https://doi.org/10.1016/J.CMET.2015.04.006
111. van der Laan M, Horvath SE, Pfanner N (2016) Mitochondrial contact site and cristae organizing system. Curr Opin Cell Biol 41:33–42. https://doi.org/10.1016/j.ceb.2016.03.013
112. Tirrell PS, Nguyen KN, Luby-Phelps K, Friedman JR (2020) MICOS subcomplexes assemble independently on the mitochondrial inner membrane in proximity to ER contact sites. J Cell Biol 219:e202003024. https://doi.org/10.1083/JCB.202003024
113. Hönscher C, Mari M, Auffarth K, Bohnert M, Griffith J, Geerts W, van der Laan M, Cabrera M, Reggiori F, Ungermann C (2014) Cellular metabolism regulates contact sites between vacuoles and mitochondria. Dev Cell 30:86–94. https://doi.org/10.1016/j.devcel.2014.06.006
114. Friedman JR (2022) Mitochondria from the outside in: the relationship between inter-organelle crosstalk and mitochondrial internal organization. Contact 5:251525642211332. https://doi.org/10.1177/25152564221133267
115. Muñoz-Gómez SA, Slamovits CH, Dacks JB, Baier KA, Spencer KD, Wideman JG (2015) Ancient homology of the mitochondrial contact site and cristae organizing system points to an endosymbiotic origin of mitochondrial cristae. Curr Biol 25:1489–1495. https://doi.org/10.1016/j.cub.2015.04.006
116. Huynen MA, Mühlmeister M, Gotthardt K, Guerrero-Castillo S, Brandt U (2016) Evolution and structural organization of the mitochondrial contact site (MICOS) complex and the mitochondrial intermembrane space bridging (MIB) complex. Biochim Biophys Acta - Mol Cell Res 1863:91–101. https://doi.org/10.1016/j.bbamcr.2015.10.009
117. D'Amici GM, Rinalducci S, Murgiano L, Italiano F, Zolla L (2010) Oligomeric characterization of the photosynthetic apparatus of *Rhodobacter sphaeroides* R26.1 by nondenaturing electrophoresis methods. J Proteome Res 9:192–203. https://doi.org/10.1021/PR9005052
118. Jackson PJ, Lewis HJ, Tucker JD, Hunter CN, Dickman MJ (2012) Quantitative proteomic analysis of intracytoplasmic membrane development in *Rhodobacter sphaeroides*. Mol Microbiol 84:1062–1078. https://doi.org/10.1111/J.1365-2958.2012.08074.X
119. Selao TT, Branca R, Chae PS, Lehtiö J, Gellman SH, Rasmussen SGF, Nordlund S, Norén A (2011) Identification of chromatophore membrane protein complexes formed under different nitrogen availability conditions in *Rhodospirillum rubrum*. J Proteome Res 10:2703–2714. https://doi.org/10.1021/PR100838X

120. Fejes AP, Yi EC, Goodlett DR, Beatty JT (2003) Shotgun proteomic analysis of a chromatophore-enriched preparation from the purple phototrophic bacterium *Rhodopseudomonas palustris*. Photosynth Res 78:195–203. https://doi.org/10.1023/B:PRES.0000006752.81486.74
121. Muñoz-Gómez SA, Cadena LR, Gardiner AT, Leger MM, Sheikh S, Connell LB, Bilý T, Kopejtka K, Beatty JT, Koblížek M, Roger AJ, Slamovits CH, Lukeš J, Hashimi H (2023) Intracytoplasmic-membrane development in alphaproteobacteria involves the homolog of the mitochondrial crista-developing protein Mic60. Curr Biol 33:1099–1111.e6. https://doi.org/10.1016/j.cub.2023.02.059
122. Muñoz-Gómez SA, Wideman JG, Roger AJ, Slamovits CH, Muñoz-Gómez SA, Wideman JG, Roger AJ, Slamovits CH, Agashe D (2017) The origin of mitochondrial cristae from alphaproteobacteria. Mol Biol Evol 34:943–956. https://doi.org/10.1093/molbev/msw298
123. Müller M, Mentel M, van Hellemond JJ, Henze K, Woehle C, Gould SB, Yu R-Y, van der Giezen M, Tielens AGM, Martin WF (2012) Biochemistry and evolution of anaerobic energy metabolism in eukaryotes. Microbiol Mol Biol Rev 76:444–495. https://doi.org/10.1128/MMBR.05024-11
124. Tachezy J (2019) Hydrogenosomes and mitosomes: Mitochondria of anaerobic eukaryotes. 2nd edn. Springer, Cham
125. Kamp A, Høgslund S, Risgaard-Petersen N, Stief P (2015) Nitrate storage and dissimilatory nitrate reduction by eukaryotic microbes. Front Microbiol 6:1–15. https://doi.org/10.3389/fmicb.2015.01492
126. Gawryluk RMR, Kamikawa R, Stairs CW, Silberman JD, Brown MW, Roger AJ (2016) The earliest stages of mitochondrial adaptation to low oxygen revealed in a novel rhizarian. Curr Biol 26:2729–2738. https://doi.org/10.1016/j.cub.2016.08.025
127. Stairs CW, Eme L, Muñoz-Gómez SA, Cohen A, Dellaire G, Shepherd JN, Fawcett JP, Roger AJ (2018) Microbial eukaryotes have adapted to hypoxia by horizontal acquisitions of a gene involved in rhodoquinone biosynthesis. Elife 7:1–23. https://doi.org/10.7554/eLife.34292
128. Lewis WH, Lind AE, Sendra KM, Onsbring H, Williams TA, Esteban GF, Hirt RP, Ettema TJG, Embley TM (2020) Convergent evolution of hydrogenosomes from mitochondria by gene transfer and loss. Mol Biol Evol 37:524–539. https://doi.org/10.1093/molbev/msz239
129. Rotterová J, Salomaki E, Pánek T, Bourland W, Žihala D, Táborský P, Edgcomb VP, Beinart RA, Kolísko M, Čepička I (2020) Genomics of new ciliate lineages provides insight into the evolution of obligate anaerobiosis. Curr Biol 30:2037–2050.e6. https://doi.org/10.1016/j.cub.2020.03.064
130. Salomaki ED, Terpis KX, Rueckert S, Kotyk M, Varadínová ZK, Čepička I, Lane CE, Kolisko M (2021) Gregarine single-cell transcriptomics reveals differential mitochondrial remodeling and adaptation in apicomplexans. BMC Biol 19:1–19. https://doi.org/10.1186/s12915-021-01007-2
131. Mathur V, Wakeman KC, Keeling PJ (2021) Parallel functional reduction in the mitochondria of apicomplexan parasites. Curr Biol 31:2920–2928.e4. https://doi.org/10.1016/j.cub.2021.04.028
132. Hug LA, Stechmann A, Roger AJ (2010) Phylogenetic distributions and histories of proteins involved in anaerobic pyruvate metabolism in eukaryotes. Mol Biol Evol 27:311–324. https://doi.org/10.1093/molbev/msp237
133. Stairs CW, Roger AJ, Hampl V (2011) Eukaryotic pyruvate formate lyase and its activating enzyme were acquired laterally from a firmicute. Mol Biol Evol 28:2087–2099. https://doi.org/10.1093/molbev/msr032
134. Braymer JJ, Freibert SA, Rakwalska-Bange M, Lill R (2021) Mechanistic concepts of iron-sulfur protein biogenesis in biology. Biochim Biophys Acta - Mol Cell Res 1868:118863. https://doi.org/10.1016/j.bbamcr.2020.118863
135. Burki F, Corradi N, Sierra R, Pawlowski J, Meyer GR, Abbott CL, Keeling PJ (2013) Phylogenomics of the intracellular parasite *Mikrocytos mackini* reveals evidence for a mitosome in rhizaria. Curr Biol 23:1541–1547. https://doi.org/10.1016/j.cub.2013.06.033
136. Tovar J, Leon-Avila G, Sanchez LB, Sutak R, Tachezy J, van der Giezen M, Hernandez M, Muller M, Lucocq JM (2003) Mitochondrial remnant organelles of *Giardia* function in iron-sulphur protein maturation. Nature 426:172–176. https://doi.org/10.1038/nature01945

137. Vacek VV, Novák LVF, Treitli SC, Táborský P, Čepička I, Kolísko M, Keeling PJ, Hampl VV (2018) Fe-S cluster assembly in oxymonads and related protists. Mol Biol Evol 35:2712–2718. https://doi.org/10.1093/molbev/msy168
138. Ali V, Shigeta Y, Tokumoto U, Takahashi Y, Nozaki T (2004) An intestinal parasitic protist, *Entamoeba histolytica*, possesses a non-redundant nitrogen fixation-like system for iron-sulfur cluster assembly under anaerobic conditions. J Biol Chem 279:16863–16874
139. Peña-Diaz P, Braymer JJ, Vacek V, Zelená M, Lometto S, Hrdý I, Treitli SC, Hochberg G, Py B, Lill R, Hampl V (2023) Characterisation of the SUF FeS cluster machinery in the amitochondriate eukaryote *Monocercomonoides exilis*. bioRxiv 20230330534840. https://doi.org/10.1101/2023.03.30.534840
140. Mi-ichi F, Yousuf MA, Nakada-Tsukui K, Nozaki T, Abu Yousuf M, Nakada-Tsukui K, Nozaki T (2009) Mitosomes in *Entamoeba histolytica* contain a sulfate activation pathway. Proc Natl Acad Sci 106:21731–21736. https://doi.org/10.1073/pnas.0907106106
141. Mi-ichi F, Makiuchi T, Furukawa A, Sato D, Nozaki T (2011) Sulfate activation in mitosomes plays an important role in the proliferation of *Entamoeba histolytica*. PLoS Negl Trop Dis 5:1–7. https://doi.org/10.1371/journal.pntd.0001263
142. Venkatachalam KV, Akita H, Strott CA (1998) Molecular cloning, expression, and characterization of human bifunctional 3′-phosphoadenosine 5′-phosphosulfate synthase and its functional domains. J Biol Chem 273:19311–19320. https://doi.org/10.1074/JBC.273.30.19311
143. Koprivova A, Kopriva S (2016) Sulfation pathways in plants. Chem Biol Interact 259:23–30. https://doi.org/10.1016/J.CBI.2016.05.021
144. Pan S, Fan M, Liu Z, Li X, Wang H (2021) Serine, glycine and one-carbon metabolism in cancer (Review). Int J Oncol 58:158–170. https://doi.org/10.3892/ijo.2020.5158
145. Karnkowska A, Vacek V, Zubáčová Z, Treitli SC, Petrželková R, Eme L, Novák L, Žárský V, Barlow LD, Herman EK, Soukal P, Hroudová M, Doležal P, Stairs CW, Roger AJ, Eliáš M, Dacks JB, Vlček Č, Hampl V (2016) A eukaryote without a mitochondrial organelle. Curr Biol 26:1274–1284. https://doi.org/10.1016/j.cub.2016.03.053
146. Karnkowska A, Treitli SC, Brzoň O, Novák L, Vacek V, Soukal P, Barlow LD, Herman EK, Pipaliya SV, Pánek T, Žihala D, Petrželková R, Butenko A, Eme L, Stairs CW, Roger AJ, Eliáš M, Dacks JB, Hampl V (2019) The oxymonad genome displays canonical eukaryotic complexity in the absence of a mitochondrion. Mol Biol Evol 36:2292–2312. https://doi.org/10.1093/molbev/msz147
147. LVF N, Treitli SC, Pyrih J, Hałakuc P, Pipaliya SV, Vacek V, Brzoň O, Soukal P, Eme L, Dacks JB, Karnkowska A, Eliáš M, Hampl V (2023) Genomics of preaxostyla flagellates illuminates the path towards the loss of mitochondria. PLoS Genet 19:e1011050. https://doi.org/10.1371/journal.pgen.1011050
148. Williams SK, Hultqvist JJ, Eglit Y, Salas-Leiva DE, Curtis B, Orr R, Stairs CW, Simpson AGB, Roger AJ (2023) Extreme mitochondrial reduction in a novel group of free-living metamonads. bioRxiv 20230503539051. https://doi.org/10.1101/2023.05.03.539051
149. Jewari CA, Baldauf SL (2023) An excavate root for the eukaryote tree of life. Sci Adv 9. https://doi.org/10.1126/SCIADV.ADE4973
150. Koumandou VL, Wickstead B, Ginger ML, van der Giezen M, Dacks JB, Field MC (2013) Molecular paleontology and complexity in the last eukaryotic common ancestor. Crit Rev Biochem Mol Biol 48:373–396. https://doi.org/10.3109/10409238.2013.821444
151. Zaremba-Niedzwiedzka K, Caceres EF, Saw JH, Bäckström D, Juzokaite L, Vancaester E, Seitz KW, Anantharaman K, Starnawski P, Kjeldsen KU, Stott MB, Nunoura T, Banfield JF, Schramm A, Baker BJ, Spang A, Ettema TJG, Vancaester E, Kiley W, Nunoura T, Banfield JF, Schramm A, Baker BJ, Spang A, Ettema TJG, Seitz KW, Anantharaman K, Starnawski P, Kjeldsen KU, Stott MB, Nunoura T, Banfield JF, Schramm A, Baker BJ, Spang A, Ettema TJG (2017) Asgard archaea illuminate the origin of eukaryotic cellular complexity. Nature 541:353–358. https://doi.org/10.1038/nature21031
152. Eme L, Tamarit D, Caceres EF, Stairs CW, De Anda V, Schön ME, Seitz KW, Dombrowski N, Lewis WH, Homa F, Saw JH, Lombard J, Nunoura T, Li W-J, Hua Z-S, Chen L-X, Banfield JF, John ES, Reysenbach A-L, Stott MB, Schramm A, Kjeldsen KU, Teske AP, Baker BJ,

Ettema TJG (2023) Inference and reconstruction of the heimdallarchaeial ancestry of eukaryotes. Nature 618:992–999. https://doi.org/10.1038/s41586-023-06186-2
153. Shih PM, Matzke NJ (2013) Primary endosymbiosis events date to the later Proterozoic with cross-calibrated phylogenetic dating of duplicated ATPase proteins. Proc Natl Acad Sci U S A 110:12355–12360. https://doi.org/10.1073/pnas.1305813110
154. Eme L, Sharpe SC, Brown MW, Roger AJ (2014) On the age of eukaryotes: evaluating evidence from fossils and molecular clocks. Cold Spring Harb Perspect Biol 6. https://doi.org/10.1101/cshperspect.a016139
155. Betts HC, Puttick MN, Clark JW, Williams TA, Donoghue PCJ, Pisani D (2018) Integrated genomic and fossil evidence illuminates life's early evolution and eukaryote origin. Nat Ecol Evol 2:1556–1562. https://doi.org/10.1038/S41559-018-0644-X
156. Strassert JFH, Irisarri I, Williams TA, Burki F (2021) A molecular timescale for eukaryote evolution with implications for the origin of red algal-derived plastids. Nat Commun 12:1879. https://doi.org/10.1038/S41467-021-22044-Z
157. Gibson TM, Shih PM, Cumming VM, Fischer WW, Crockford PW, Hodgskiss MSW, Wörndle S, Creaser RA, Rainbird RH, Skulski TM, Halverson GP (2018) Precise age of *Bangiomorpha pubescens* dates the origin of eukaryotic photosynthesis. Geology 46:135–138. https://doi.org/10.1130/G39829.1
158. Porter SM (2020) Insights into eukaryogenesis from the fossil record. Interface Focus 10. https://doi.org/10.1098/RSFS.2019.0105
159. Mills DB, Boyle RA, Daines SJ, Sperling EA, Pisani D, Donoghue PCJ, Lenton TM (2022) Eukaryogenesis and oxygen in Earth history. Nat Ecol Evol 6:520–532
160. Fischer WW, Hemp J, Johnson JE (2016) Evolution of oxygenic photosynthesis. Annu Rev Earth Planet Sci 44:647–683. https://doi.org/10.1146/ANNUREV-EARTH-060313-054810
161. Mentel M, Martin W (2008) Energy metabolism among eukaryotic anaerobes in light of Proterozoic ocean chemistry. Philos Trans R Soc Lond B Biol Sci 363:2717–2729. https://doi.org/10.1098/rstb.2008.0031
162. Zimorski V, Mentel M, Tielens AGM, Martin WF (2019) Energy metabolism in anaerobic eukaryotes and Earth's late oxygenation. Free Radic Biol Med 140:279–294. https://doi.org/10.1016/j.freeradbiomed.2019.03.030
163. Wong HL, Visscher PT, White RA, Smith DL, Patterson MM, Burns BP (2017) Dynamics of archaea at fine spatial scales in Shark Bay mat microbiomes. Sci Rep 71(7):1–12. https://doi.org/10.1038/srep46160
164. López-García P, Moreira D (2020) The Syntrophy hypothesis for the origin of eukaryotes revisited. Nat Microbiol 5:655–667. https://doi.org/10.1038/S41564-020-0710-4
165. Martin WF, Neukirchen S, Zimorski V, Gould SB, Sousa FL (2016) Energy for two: new archaeal lineages and the origin of mitochondria. Bioessays 38:850–856. https://doi.org/10.1002/BIES.201600089
166. Spang A, Stairs CW, Dombrowski N, Eme L, Lombard J, Caceres EF, Greening C, Baker BJ, Ettema TJG (2019) Proposal of the reverse flow model for the origin of the eukaryotic cell based on comparative analyses of Asgard archaeal metabolism. Nat Microbiol 47(4):1138–1148. https://doi.org/10.1038/s41564-019-0406-9
167. Imachi H, Nobu MK, Nakahara N, Morono Y, Ogawara M, Takaki Y, Takano Y, Uematsu K, Ikuta T, Ito M, Matsui Y, Miyazaki M, Murata K, Saito Y, Sakai S, Song C, Tasumi E, Yamanaka Y, Yamaguchi T, Kamagata Y, Tamaki H, Takai K (2020) Isolation of an archaeon at the prokaryote-eukaryote interface. Nature 577:519–525. https://doi.org/10.1038/S41586-019-1916-6
168. Baum DA, Baum B (2014) An inside-out origin for the eukaryotic cell. BMC Biol 12:1–22. https://doi.org/10.1186/S12915-014-0076-2
169. Rodrigues-Oliveira T, Wollweber F, Ponce-Toledo RI, Xu J, Rittmann SKMR, Klingl A, Pilhofer M, Schleper C (2023) Actin cytoskeleton and complex cell architecture in an Asgard archaeon. Nature 613:332–339. https://doi.org/10.1038/S41586-022-05550-Y
170. Bremer N, Tria FDK, Skejo J, Garg SG, Martin WF (2022) Ancestral state reconstructions trace mitochondria but not phagocytosis to the last eukaryotic common ancestor. Genome Biol Evol 14:evac079. https://doi.org/10.1093/GBE/EVAC079

171. Mills DB (2020) The origin of phagocytosis in Earth history. Interface Focus 10:20200019. https://doi.org/10.1098/RSFS.2020.0019
172. Derelle R, Torruella G, Klimeš V, Brinkmann H, Kim E, Vlček Č, Lang BF, Eliáš M (2015) Bacterial proteins pinpoint a single eukaryotic root. Proc Natl Acad Sci 112:201420657. https://doi.org/10.1073/pnas.1420657112
173. Al Jewari C, Baldauf SL (2022) Conflict over the eukaryote root resides in strong outliers, mosaics and missing data sensitivity of site-specific (CAT) mixture models. Syst Biol syac 029. https://doi.org/10.1093/SYSBIO/SYAC029
174. Yubuki N, Leander BS (2013) Evolution of microtubule organizing centers across the tree of eukaryotes. Plant J 75:230–244
175. Heiss AA, Walker G, Simpson AGB (2013) The microtubular cytoskeleton of the apusomonad *Thecamonas*, a sister lineage to the opisthokonts. Protist 164:598–621. https://doi.org/10.1016/J.PROTIS.2013.05.005
176. Muñoz-Gómez SA (2023) Energetics and evolution of anaerobic microbial eukaryotes. Nat Microbiol 8:197–203. https://doi.org/10.1038/S41564-022-01299-2
177. Schavemaker PE, Muñoz-Gómez SA (2022) The role of mitochondrial energetics in the origin and diversification of eukaryotes. Nat Ecol Evol 6:1307–1317. https://doi.org/10.1038/s41559-022-01833-9
178. Hampl V, Čepička I, Eliáš M (2018) Was the mitochondrion necessary to start eukaryogenesis? Trends Microbiol 27:1–9. https://doi.org/10.1016/j.tim.2018.10.005
179. De Duve C, Wattiaux R (1966) Functions of lysosomes. Annu Rev Physiol 28:435–492. https://doi.org/10.1146/ANNUREV.PH.28.030166.002251
180. Stanier R (1970) Some aspects of the biology of cells and their possible evolutionary significance. Symp Soc Gen Microbiol 20:1–38
181. Cavalier-Smith T (1987) The origin of eukaryotic and archaebacterial cells. Ann N Y Acad Sci 503:17–54. https://doi.org/10.1111/J.1749-6632.1987.TB40596.X
182. Husnik F, Tashyreva D, Boscaro V, George EE, Lukeš J, Keeling PJ (2021) Bacterial and archaeal symbioses with protists. Curr Biol 31:R862–R877. https://doi.org/10.1016/j.cub.2021.05.049
183. Liu Y, Makarova KS, Huang WC, Wolf YI, Nikolskaya AN, Zhang X, Cai M, Zhang CJ, Xu W, Luo Z, Cheng L, Koonin EV, Li M (2021) Expanded diversity of Asgard archaea and their relationships with eukaryotes. Nature 593:553–557. https://doi.org/10.1038/S41586-021-03494-3
184. Shiratori T, Suzuki S, Kakizawa Y, Kichiro I (2019) Phagocytosis-like cell engulfment by a planctomycete bacterium. Nat Commun 10:5529. https://doi.org/10.1038/S41467-019-13499-2
185. Martin WF, Tielens AGM, Mentel M, Garg SG, Gould SB (2017) The physiology of phagocytosis in the context of mitochondrial origin. Microbiol Mol Biol Rev 81:e00008–e00017. https://doi.org/10.1128/MMBR.00008-17
186. Pittis AA, Gabaldón T (2016) Late acquisition of mitochondria by a host with chimaeric prokaryotic ancestry. Nature 531:101–104. https://doi.org/10.1038/NATURE16941
187. Vosseberg J, van Hooff JJE, Marcet-Houben M, van Vlimmeren A, van Wijk LM, Gabaldón T, Snel B (2021) Timing the origin of eukaryotic cellular complexity with ancient duplications. Nat Ecol Evol 5:92–100. https://doi.org/10.1038/S41559-020-01320-Z
188. Susko E, Steel M, Roger AJ (2021) Conditions under which distributions of edge length ratios on phylogenetic trees can be used to order evolutionary events. J Theor Biol 526:110788. https://doi.org/10.1016/J.JTBI.2021.110788
189. Roger AJ, Susko E, Leger MM (2021) Evolution: reconstructing the timeline of eukaryogenesis. Curr Biol 31:R193–R196. https://doi.org/10.1016/J.CUB.2020.12.035

The Extent, Role, and Timing of Endosymbiotic Gene Transfer in Plastids

Elisabeth Hehenberger and Fabien Burki

Abstract Endosymbiotic gene transfer (EGT), the transfer of endosymbiont genes to the host nucleus, is considered a fundamental process of plastid endosymbiosis. Together with retargeting of the protein products of EGTs to the plastid where they function, EGTs are viewed as a hallmark of plastids as genetically integrated organelles. Despite its central role in one of the biggest evolutionary transitions on our planet, and a long history of inquiry into plastid evolution, our knowledge about the extent of EGTs, their roles in the host cell and timing of acquisition, is still patchy. This chapter summarizes our current knowledge about EGT, framing the discussion in the more general context of horizontal gene transfer (HGT), and highlighting the issues that research in this field is facing. While the need to investigate gene transfer in the context of plastid endosymbiosis is universally acknowledged, there is no consensus on the methodology used to research EGT and HGT, making comparisons between studies difficult. However, some patterns are beginning to emerge and the central role of EGT in plastid establishment is now being shifted toward a shared role between EGT, HGT, and contributions by the host.

Keywords EGT · HGT · Plastid · Evolution · Endosymbiosis · Algae · Primary plastids · Complex plastids · Kleptoplastidy · Phylogeny · Bioinformatics · Plastid loss

E. Hehenberger (✉)
Institute of Parasitology, Biology Centre, CAS, České Budějovice, Czech Republic
e-mail: elisabeth.hehenberger@paru.cas.cz

F. Burki
Department of Organismal Biology, Program in Systematic Biology, Uppsala University, Uppsala, Sweden
e-mail: fabien.burki@ebc.uu.se

S. D. Schwartzbach et al. (eds.), *Endosymbiotic Organelle Acquisition*,
https://doi.org/10.1007/978-3-031-57446-7_4

1 Introduction

It is generally acknowledged that the majority of plastids in eukaryotes originated from a single endosymbiotic event: the uptake and integration of a β-cyanobacterium by a eukaryotic cell. This event, termed primary endosymbiosis, took place in an ancestor of the supergroup Archaeplastida, comprising green algae and plants, red algae, and glaucophytes as well as the recently placed lineages Rhodelphidia and Picozoa [1–3]. Many important transformations happened during the integration of a free-living cyanobacterium into a plastid, leading to a general reduction of the organelle compared to their free-living relatives. One striking difference is the genome size, with the largest plastid genomes encoding at most about 10% of the proteins found in cyanobacteria [4]. In this chapter, we discuss one of the key events in plastid establishment that largely contributed to their genome size reduction: the transfer of genes from the plastid genome to the nuclear genome of the host cell—a process termed endosymbiotic gene transfer or EGT (Fig. 1). The existence of EGTs

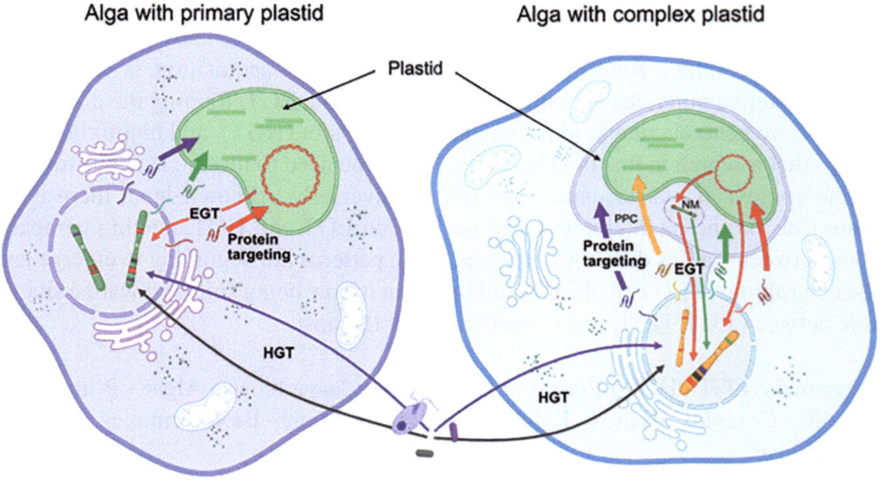

Fig. 1 Plastid endosymbiosis in algae with primary and complex plastids. In algae with primary plastids, genes from the cyanobacterial endosymbiont (=EGTs) have been transferred to the host genome and the products of some of these genes are targeted back to the plastid (slim and bold red arrows, respectively). In addition, genes from sources other than the endosymbiont (prokaryotic and/or eukaryotic) have been integrated into the host genome (=HGTs), some encoding for proteins functioning in the plastid (slim and bold purple arrows, respectively). Finally, host-derived proteins also contribute to the plastid proteome (bold green and yellow arrows in primary and complex plastid-containing algae, respectively). Algae with complex plastids resulted from the uptake of primary plastid-bearing alga by a nonphotosynthetic eukaryote. As a result, cyanobacterial as well as eukaryotic genes have been transferred from the endosymbiont nucleus to the complex algal nucleus, representing EGTs (red and green arrows, respectively). Like in algae with primary plastids, also HGTs and host-derived proteins can function in the plastid (purple and yellow arrows, respectively). In chlorarachniophytes and cryptophytes, that retained the remnant nucleus ("nucleomorph"(NM)) and cytoplasm ("periplastidial compartment" (PPC)) of the endosymbiont, proteins encoded by all types of genes transfers as well as host-encoded proteins can also be targeted to the PPC [8].

is not a new notion, and it has been unambiguously confirmed in many algae by phylogenetic analyses as well as on experimental level [5–11].

Intracellular life, relaxing the selective pressure on numerous plastid genes, has led to plastid genome erosion and the concomitant transfer of many endosymbiotic genes to the host genome. EGTs have also been proposed to be energetically beneficial to the host cell, thus providing a selective advantage toward transferring genes out of the plastids [12]. In today's plastids, most biochemical pathways are encoded in the host genome, requiring tightly coordinated plastid import of hundreds of proteins to sustain proper plastid functions. But the products of EGTs are probably not all functioning in plastids; a proportion of EGTs—the extent of which remains unknown—acquired new functions in the host. Plastids are also supported by many noncyanobacterial proteins, most of which are derived from the repurposing of host proteins but also from genes acquired horizontally from eukaryotes and prokaryotes other than the cyanobacterial endosymbiont; we will refer to these genes as horizontal gene transfers (HGTs) (Fig. 1). Bioinformatically recognizing EGT comes with many challenges, a matter not made easier when trying to distinguish EGTs from HGTs. As both EGT and HGT have likely played central roles during the establishment of plastids, our discussion in this chapter will encompass both types of gene transfer.

Interestingly, varied origins of plastid proteins have also been shown in another lineage with primary plastids—the only other known lineage that acquired a plastid directly from cyanobacteria. Indeed, photosynthetic members of the rhizarian amoeba *Paulinella* possess primary plastids—termed chromatophores—that were acquired not only independently from a different group of cyanobacteria (α-cyanobacteria) but also about a billion years later than the plastids in Archaeplastida. As such, *Paulinella* has become a fascinating study system in plastid evolution that provides invaluable insights into the role of gene transfer during plastid establishment. Interestingly, the results of these studies do not necessarily align with the findings from studies using archaeplastid models, a discrepancy we will discuss in this chapter.

Following primary endosymbiosis in Archaeplastida, several additional endosymbioses have taken place where both interacting partners were eukaryotes—one of which was an alga. These higher-order endosymbioses resulted in plastids with additional membranes, often integrated within the host endomembrane system and sometimes even retaining a highly reduced nuclear genome (called the nucleomorph). Due to their complexity, these plastids are often referred to as "complex plastids" to differentiate them from primary plastids (Fig. 1). Both green and red algae have been involved in these events, giving rise to two groups with complex green plastids (euglenids and chlorarachniophytes) and four groups with complex red plastids. For simplicity, we will collectively refer here to the complex red algal plastid-bearing lineages as MOCHa, which stands for myzozoans (comprising dinoflagellates as well as the parasitic apicomplexans, chromerids, and colpodellids), ochrophytes (photosynthetic stramenopiles such as diatoms), cryptophytes (e.g., the model alga *Guillardia theta*), and haptophytes (e.g., the bloom-forming *Emiliania huxleyi*) [13]. By avoiding descriptors commonly used such as secondary or tertiary endosymbiosis, we acknowledge the ongoing debate about the number of endosymbiotic events that led to the current distribution of plastids in MOCHa [13, 14]. In general, systems with complex plastids are less well studied than Archaeplastida, but EGT has been shown to occur in these lineages too. Phylogenetics has been the tool of choice to assess the extent and timing of EGT and will be discussed in this chapter in the context of lineages hosting primary and complex plastids, as well as in taxa only transiently maintaining plastids stolen from algal preys (known as kleptoplasts).

2 Evidence for EGTs

The existence of EGT has been proposed more than 50 years ago. Although not termed EGT then, Goksøyr [15] hypothesized that mitochondrial and plastidial endosymbionts lost DNA during their integration and transferred at least some of it to the host nuclear genome to sustain proper organellar functions. About 15 years later, Weeden [16] examined the implications of endosymbiosis on plastids, focusing on the genome reduction of the organelle. In his treatise, Weeden discussed two possibilities for the small genome size of plastids: outright gene loss (and replacement of their function by host genes) or the transfer of genes from the endosymbiont to the nucleus and the continued function of their protein product in the plastid after import. He showed that the majority of studies on plastidial enzymatic isoforms and metabolic pathways support the latter postulate. He also predicted that plastidial isoforms of enzymes (e.g., glycolytic enzymes) are more similar to cyanobacterial than cytosolic homologs (and thus precluding their host origin); however, data available at the time did not allow to prove this hypothesis unequivocally. Almost two decades later, a phylogenetic study confirmed the cyanobacterial origin of over 40 nucleus-encoded genes in Archaeplastida [5]. This was soon followed by a landmark genome-wide analysis of gene origin in the plant model *Arabidopsis thaliana*. This study predicted ~1700 *Arabidopsis* genes with cyanobacterial origin among the genes amenable for phylogenetic reconstruction (about 37% of all *Arabidopsis* genes); when extrapolated to the whole genome, the astounding number of ~4500 genes was predicted to have been acquired from the cyanobacterial ancestor of plastids via EGT [17]. The process of EGT was experimentally demonstrated in another plant model, tobacco, selecting for gene transfer from the plastid to the nucleus and functional activation of the gene with antibiotics [6, 18]. However, similar experiments in the green algal model *Chlamydomonas reinhardtii* did not lead to any detectable DNA transfer from plastid to nucleus and the authors speculated that the presence of only a single plastid in this species might preclude this process [19].

The discovery of primary plastids in *Paulinella* widely opened up the field of plastid endosymbiosis research by showing that a second photosynthetic organelle has been independently and much more recently established directly from cyanobacteria [20]. This comparatively recent plastid uptake provides a rare opportunity to investigate in detail which transformations occur early in plastid establishment, including EGT. Not surprisingly, phylogenetic analyses indicated α-cyanobacterial origin for a number of nuclear genes in *Paulinella chromatophora* that likely resulted from EGT [7, 21, 22]. However, as we will discuss in more detail below, a distinctly lower number of EGT events is predicted for *Paulinella* as compared to the numbers predicted for a range of Archaeplastida representatives.

Identifying EGT in complex plastids that resulted from eukaryote-to-eukaryote endosymbioses requires additional considerations. Although the number of these endosymbioses and the identity of the partners involved remain vividly debated, it is unequivocally accepted that both red and green algal endosymbionts contributed to the evolutionary success of algae across the tree of eukaryotes. In all studied cases, numerous putative EGTs were reported from red or green algal sources, respectively [8, 10, 23, 24]. However, detecting EGTs in these lineages is even more complicated than in primary plastid-containing lineages because genes have been transferred not only from the plastid genome but also from the endosymbiont nuclear genome to the host nuclear genome, resulting in a more blurry EGT signal due to the shorter evolutionary distance between two eukaryotic genomes (Fig. 1). EGTs in complex plastids are thus less straightforward to distinguish from vertically inherited genes in phylogenetic reconstructions. EGT analyses are based on single-protein trees (amino acid, not DNA sequences, are usually used for EGT tree reconstructions), and depending on the length of the protein and degree of sequence conservation, the resolution of these single-protein trees is often low, making unambiguous calls about the origin of the EGT candidate impossible. This issue can be partially compensated by comprehensive and carefully balanced taxon sampling, which has become more and more attainable with the wealth of data for a large diversity of eukaryotes (and prokaryotes) that are now available.

Initial EGT surveys in algae with complex red algal plastids have observed, besides the expected contribution from red algae, high numbers of genes of green algal origin [25, 26]. This was interpreted as evidence for a so-called cryptic green algal endosymbiosis possibly preceding the establishment of the complex red plastids [25]. However, reanalyses of these data using a more comprehensive taxon sampling in combination with manual tree inspection called these results into question, indicating that the expected red algal contribution to putative EGTs is the dominant signal in algae with complex red plastids [23, 24]. The interpretation of Moustafa et al. [25] that green algal genes support a cryptic endosymbiosis is based on the general assumption that the integration of plastids leaves a genomic "footprint" in the genome even after it has been lost or replaced by another plastid. This notion was introduced by the *Arabidopsis* study that reported that the majority of the thousands of putative cyanobacterial EGTs function outside of the plastid [17]. Note, however, that these results have never been reassessed in light of the mass of new data that has accumulated since that publication, and newer comprehensive, functional EGT studies in other lineages are few (see also Sect. 6).

One recent study using curated phylogenies based exclusively on cyanobacterial genes with essential plastid functions reinforced the observation that the EGT signal is dominated by red algal contributions in lineages with complex red plastids; this was shown for all MOCHa lineages [10]. Conversely, in the same study, an unexpectedly large fraction of putative transfers from red algae in euglenids and chlorarachniophytes, that is, lineages with complex green plastids, was recovered. Whether these foreign genes are "true" EGTs suggesting a cryptic red algal endosymbiosis, or rather represent HGTs from sources other than the endosymbiont, could not be distinguished—an issue that will follow us throughout this chapter.

Some dinoflagellates represent a special case of eukaryote-to-eukaryote endosymbiosis because they have replaced, on multiple occasions, their ancestral complex red plastid (called the "peridinin" plastid in dinoflagellates) with a different plastid that can be derived from lineages with primary or complex plastids. Due to the younger age of these endosymbioses, these dinoflagellates allow glimpses into events of plastid establishment that may have been obscured by time in older endosymbioses, much like in *Paulinella* for primary endosymbiosis. In several of these dinoflagellates, the presence of putative EGTs from their current "replacement" plastid has been studied. For example, the species *Lepidodinium chlorophorum*, which has replaced its peridinin plastid with a plastid from pedinophyte green algae, contains at least six nucleus-encoded plastid-targeted genes (out of ~4700 randomly picked clones of a cDNA library) of green algal origin, interpreted to have arisen via EGT [27]. Similarly, in a more recent transcriptome-wide study of the unrelated dinoflagellate strains MGD and TGD, which also carry pedinophyte plastids but still retain a relic endosymbiont nucleus, green algal transcripts (164 and 201, respectively) were found to be encoded in the host nucleus and interpreted as gene transfers from the current endosymbiont [11]. Since several homologs were found to be encoded in both the endosymbiont nucleus as well as in the host nuclear genome, it was suggested that the process of EGT is still ongoing in these lineages. In *Karenia brevis* and *Karlodinium veneficum*, both belonging to the family Kareniaceae that has replaced the ancestral peridinin plastid with plastids of haptophyte origin, 90 genes were identified with affinity to haptophytes in phylogenetic reconstructions [9]. The same study also investigated gene transfers in the so-called dinotoms, which contain minimally reduced but permanently integrated diatom endosymbionts. It identified only 14 putative cases of EGT based on the affinity of these genes to diatoms. However, a follow-up study benefiting from a more extensive taxon sampling for dinoflagellates and diatoms inferred these genes as affiliated with dinoflagellates instead, rendering the currently known EGTs in dinotoms zero [28]. The absence of genetic integration in dinotoms is likely correlated to the almost "complete" state of the endosymbiont, precluding gene transfer to the host nucleus.

3 Problems with Identifying EGTs and HGTs

Although the presence of EGT is considered an integral part of plastid evolution, the extent of EGT in plastid-bearing lineages remains contentious and often uninvestigated. There are numerous reasons for the absence of clear-cut EGT estimates, from challenges inherent to the data to the variation between methodological approaches, as we will explain below. As almost all genome-wide EGT assessment studies have been performed in Archaeplastida (and within this group predominantly in green lineages), and the rhizarian *Paulinella*, we focus our discussion in this and the next sections on results from studies on primary plastid-containing lineages if not otherwise specified. All the aspects of EGT analysis discussed below can also be applied to the analysis of HGTs, but for the sake of simplicity, we use the term EGT throughout.

One of the main aspects preventing straightforward EGT assessment is the lack of consensus about the methodological approach. EGT estimates have been performed either using only BLAST searches, defining top hits from cyanobacteria as EGTs [29] or more often employing superior phylogenetic approaches. Only phylogenetics allows tracing of the evolutionary history of an EGT candidate and thus determining whether the suspected gene transfer, for example, in the case of primary plastids, actually originated from *within* cyanobacteria as expected. The usage of the term "within" here is crucial as only the location of an EGT candidate within a clade of cyanobacteria—and not as sister to cyanobacteria—can be considered unambiguous evidence for gene transfer. To analyze the overall extent of EGT for a given lineage using phylogenetics, trees for *all* genes of that lineage should be inferred and examined. In reality, there is usually a substantial fraction of genes for which it is difficult to infer clear orthology, precluding their analysis. To overcome this issue, EGT numbers that can be determined via phylogenies have been extrapolated to take all predicted genes of a given species into account, i.e., the number of EGTs found in the genes amenable to phylogenetic analysis was extrapolated to the total number of genes in an organism. This approach is based, however, on the untested assumption that conservation among EGT and the other nuclear genes is distributed in a similar manner.

In addition to the issue of conservation, several other aspects influence the outcome of EGT analysis, including the quality of the input data, the database used to search for orthologs of each gene, and the method used to sort through the generated phylogenies. In general, the genome quality of plants has been better compared to that of many algae. Moreover, for many algae there are few or no genomes available, so analyses have relied heavily on transcriptomic data instead. As the majority of phylogenetic EGT studies have been performed over 10 years ago, the data used for ortholog inference did not cover a large taxonomic diversity and were strongly biased toward well-studied model organisms. But as already pointed out above (Sect. 2), lately this issue has been largely overcome by the increased data availability across eukaryotic and prokaryotic diversity. However, there is no established method for sorting through thousands of trees, resulting in EGT estimates that are not directly comparable across studies. Finally, not every EGT event results in an intact nuclear-encoded cyanobacterial gene (in the case of primary plastids): retention of only partial genes, domain fusion, repeated transfers, rearrangements, and mutations impede unambiguous identification as EGT.

4 Impact of EGTs and HGTs on the Host Genomes

Even when taking into consideration all the hurdles in correctly identifying EGTs as well as the methodological bias described above, based on the limited number of studies available, an overall trend toward higher EGT numbers in land plants compared to other Archaeplastida lineages has emerged. Depending on the study, the EGT estimates based on phylogenetic reconstructions (before extrapolation to the whole genome) range between 801 and 1700 in *Arabidopsis* [17, 30], 637 in the rice *Oryza sativa* [30], and 903 in the moss *Physcomitrella patens* [30]. In unicellular algae, the inferred numbers of EGTs are noticeably lower, for example, 478 and 403 in the green algae *Chlamydomonas reinhardtii and Ostreococcus tauri,* respectively [30], 307 in the red alga *Cyanidioschyzon merolae* [30], and 274 in the glaucophyte alga *Cyanophora paradoxa* [31]. The higher EGT number estimates in plants, if not biased by methodology and/or data quality, may simply be due to plant-specific gene family expansions and duplication of EGTs resulting in larger plant genome sizes compared to algae. It is also possible that the uptake of additional cyanobacterial genes via HGT took place during plant evolution. These processes have important implications for the estimation of EGT numbers in general, as neither can be clearly distinguished from bona fide EGTs. It is likely that the number of cyanobacterial genes identified in a given archaeplastid genome is an overestimation of the actual number of EGT events that took place during evolution.

Compared to the several hundred cases of putative EGT estimated in Archaeplastida, only 72 genes were found in the host nucleus of *Paulinella chromatophora* to be of α-cyanobacterial origin and therefore inferred as EGT [7, 21, 22]. Several of these genes likely originated from gene duplications after the EGT events, possibly reducing the number of actual transfers to as few as ~30 [7, 21]. Distinguishing EGT from cyanobacterial HGT in *Paulinella* is as challenging as in Archaeplastida, but the distinctly younger age of the chromatophore makes it more straightforward to infer cyanobacterial origin. Interestingly, EGT analysis in *Paulinella* also revealed that two photosynthesis-related genes are encoded and expressed by both the nuclear and chromatophore genomes [21]. In combination with the low number of putative EGT, this observed dual location of plastid-associated genes suggests that EGT in *P. chromatophora* is an ongoing process. Since the chromatophore is estimated to be between 90 and 140 million years old [32], this provides a minimum time frame for genetic integration by EGT during primary endosymbiosis.

In contrast to the relatively low EGT numbers in *P. chromatophora*, at least 171 nuclear genes have been acquired from bacterial lineages other than α-cyanobacteria [22], suggesting a predominance of HGT over EGT in chromatophore evolution. Not all cases of HGT need to be connected with the establishment of the chromatophore, but none of these HGTs was found in a—very partial—genome of a heterotrophic sister lineage (*P. ovalis*) and many HGT gene products seem to function in the chromatophore where they complement organelle-encoded metabolic pathways [22]. The impact of HGT has also been investigated in Archaeplastida, in particular, prokaryotic HGT but also eukaryotic and viral HGT [33, 34]; however, few studies put these gene transfers in context with plastid establishment, preventing valid comparisons with *Paulinella*. The draft genome of the glaucophyte *C. paradoxa* contains 15 putative cases of bacterial HGT shared among other Archaeplastida [31]. In an improved genome assembly of the same organism, specifically targeting the chlamydial contribution to bacterial HGT, 35 candidates for chlamydial HGTs were found, with about half of them shared by other Archaeplastida [35]. Chlamydial HGT is thought to be directly associated with plastid establishment under the "menage a trois" hypothesis (MATH) [36]. Accordingly, several chlamydial HGTs function in metabolic pathways that are postulated to represent ancient interactions between the cyanobacterial symbiont and a chlamydial parasite co-existing during this tripartite symbiosis scenario [35]. Interestingly, *P. chromatophora* also encodes several chlamydial HGTs in its nuclear genome that are annotated to function in genetic information processing and the bacterial cell envelope [22]. It remains to be investigated whether Chlamydiales in *Paulinella* could have also acted as a facilitator of cyanobacterial endosymbiosis as proposed for the Archaeplastida plastid.

As described in Sect. 2, EGT has been shown in various lineages with both green and red algal complex plastids, but genome (or transcriptome)-wide studies were performed only in two diatoms [25], one chlorarachniophyte and one cryptophyte [8], and two dinoflagellates [11]. In the case of the diatoms, the results were later revised by Deschamps and Moreira [24] as described above, but as this follow-up study only re-investigated a subset of the diatom genes, the resulting number of 126 EGTs from red algae (the source of the plastid in diatoms) represents only an approximation of the overall EGT signal in diatoms. The same study also reestimated the number of green algal genes encoded in diatom genomes and found only 16 genes of green algal origin which were interpreted as HGTs. In the chlorarachniophyte *Bigelowiella natans* (green algal plastid) and the cryptophyte *Guillardia theta* (red algal plastid), 207 green and 100 red algal genes, respectively, were identified and considered possible cases of EGT. Conversely, 45 red algal genes were found in *B. natans* and 252 green algal genes in *G. theta*, representing possible cases of HGT [8]. The authors of this study, however, caution that methodological artifacts, especially the fact that the taxon sampling available at the time was heavily biased toward green algae, may have resulted in artificial inflation of the green signal. In the dinoflagellates MGD and TGD, 164 and 201 putative green algal EGTs, respectively, have been identified by a combination of BLAST searches and codon usage analysis [11]. As this study did not target any putative gene transfer sources other than from the green lineage, there are currently no estimates of HGT in these dinoflagellates. Taking into account the many uncertainties introduced by different approaches in different studies, we see that the EGT counts in lineages with complex plastids are roughly similar to the numbers reported in Archaeplastida, putatively rendering the land plants outliers with overall higher EGT numbers.

Unlike for algae with permanent plastids, there is still very little genomic data for organisms that harbor temporary endosymbionts (photosymbiosis) or stolen plastids (kleptoplasts). Kleptoplastidy is particularly fascinating because this type of interaction has been proposed as one scenario explaining the initial stages of plastid endosymbiosis [37]. The retention time of kleptoplasts, which are overall rare but found across a range of eukaryotes, varies from hours to several months. The Ross Sea Dinoflagellate (RSD) represents one of the better-studied kleptoplastidial examples regarding EGT. In this system, the plastids are stolen from haptophyte prey and are retained for at least 30 months, representing the longest kleptoplastidic association known to date [38]. In addition, RSD is highly specific in its choice of kleptoplasts, suggesting that RSD might be on the path of establishing a new, stable complex plastid of haptophyte origin [39]. A transcriptome-wide search for genes of haptophyte origin in RSD—potentially representing gene transfers from the kleptoplast—revealed just 13 host-encoded haptophyte genes. Strikingly, however, only one of these genes could be traced back directly to the current kleptoplast source, the rest being from other haptophyte sources [40]. In addition, this study identified several cases of kleptoplast-targeted genes that originated from diverse, non-haptophyte sources not only in RSD but also in close algal relatives with permanent haptophyte plastids such as *K. brevis* and *K. veneficum*. This suggests that in a not-so-distant past, an ancestor of these lineages interacted with several different endosymbionts and/or food sources, which has left traces in the host nuclear genomes. These observations are in line with hypotheses on plastid origin such as the "shopping bag" [41] or the "targeting-ratchet" models [42], that envision plastid endosymbiosis as a process driven by transient associations with various endosymbionts (or preys) before fixation and transformation of the final endosymbiont into a plastid. But these models also challenge our understanding of EGT, since they blur the threshold during the endosymbiotic integration, where we can define a gene transfer as endosymbiotic versus horizontal. In the case of RSD, can the only kleptoplast-derived gene identified be considered an EGT, assuming the current source of kleptoplast is eventually going to be the source of the final, stably integrated plastid? Or broadening the question to all algae: which of the inferred EGTs are the result of gene transfers after the plastids were established and which are the results from transient associations with endosymbionts closely related to the current plastid? These are distinctions that are impossible to make in most cases, complicating EGT assessments even further.

5 Impact of EGTs and HGTs on Plastid Proteomes

To investigate the impact of EGTs and HGTs on plastid functions with more confidence, it is appealing to catalog plastid proteomes, that is, the full set of proteins located in the plastid. Mass spectrometry-derived plastid proteomes are available only for a few organisms mainly with primary plastids [43–45]. Experimentally derived plastid proteomes are also available for a couple of lineages with complex plastids [46, 47]; however, in these organisms most data currently come from computational localization prediction approaches only. Although plastid proteomes have yet to be comprehensively analyzed in the context of EGT/HGT, it is worth noting that in *A. thaliana*, 24% of an experimentally verified nucleus-encoded primary plastid proteome is derived from cyanobacteria and thus considered prime examples of EGTs [48]. This cyanobacterial fraction of the plastid proteome corresponds to 391 genes, while the lowest bioinformatic estimate for overall EGT in *Arabidopsis* is more than double this number ($n = 801$ [30]). Interestingly, a subset of these EGT candidates does not seem to function in the plastid itself, an observation we will explore in the next section. The same study also revealed that the experimentally derived nucleus-encoded cyanobacterial contribution to the plastid proteomes of two primary algae, the green alga *C. reinhardtii* and the glaucophyte *C. paradoxa,* follows a similar trend as in *Arabidopsis* [48]. The fraction of the host-encoded plastid proteome of cyanobacterial origin in these two algae ($n = 197$ for Cr and $n = 73$ for Cp) is also much lower than the bioinformatically predicted EGT estimates ($n = 478$ for Cr [30] and $n = 274$ for Cp [31]), amounting to only 19% and 12% of the plastid proteome in *C. paradoxa* and *C. reinhardtii*, respectively. Qiu et al. [48] also investigated the impact of bacterial but non-cyanobacterial HGT in the plastid proteome of these three taxa and estimated that between 7% ($n = 40$; *C. paradoxa*) and 15% ($n = 241$; *A. thaliana*) of the organellar content is derived from bacterial sources other than the endosymbionts. Together these numbers for the EGT and HGT contributions to the plastid proteome suggest that in Archaeplastida at least, both EGT and HGT have comparable impact on the protein composition of the organelle and that the majority of protein contribution to the plastid does not necessarily stem from these gene transfers but originated from the host.

A similar trend can be observed in the plastid proteome of *Paulinella*, for which data are available from both mass spectrometry and in silico prediction [49]. This combined approach identified 433 nucleus-encoded proteins imported into the chromatophore. Out of these, only 17 (4%) were determined to be of α-cyanobacterial origin and inferred as EGT. Another 26 plastid-targeted proteins (6%) were of other bacterial origin, likely representing HGT candidates localizing to the chromatophore. However, the vast majority of import candidates were shown to be derived from the host cell, mirroring the results described above for archaeplastid primary plastids. These interesting data support the notion that EGT only plays a small role during the initial integration of the chromatophore, which may still be ongoing. This conclusion stands in stark contrast to the almost 400 EGT candidates (24% of the plastid proteome) targeted to the plastids in *Arabidopsis* or the somewhat lower but still substantial EGT contributions to the plastid proteome estimated for archaeplastid algae. Assuming a convergent evolutionary path between *Paulinella* and Archaeplastida, these differences in the EGT content of the plastid proteomes likely reflect the >1 billion-year difference between the establishment of both organelles, which would have allowed for more EGTs to accumulate in the Archaeplastida genomes. By extension, this suggests that EGTs accumulate not in one burst at the onset of the endosymbiotic process but over an extended time span and that, in fact, most EGTs occur in later stages of plastid integration. We will come back to this idea in the last part of this chapter, as these considerations have large implications not only on how we view plastid establishment but also plastid loss.

In the complex plastid-bearing chlorarachniophyte *B. natans,* the plastid proteome has also been analyzed on experimental and phylogenetic levels [46]. Of the

302 nucleus-encoded proteins localizing to the plastid, only 22 (~7%) could unambiguously be traced to green algae (the plastid donor in chlorarachniophytes), indicating their origin via EGT. *B. natans* is a special case among lineages with complex plastids, as it also retains the remnant nucleus (the nucleomorph) and the cytoplasm (the periplastidial compartment; PPC) of the green algal endosymbiont, which were co-purified with the plastid compartment in this study. Some of the green algal EGT identified may therefore be targeted to either the plastids, the PPC, or the nucleomorph, but biological function and the presence of specific targeting peptides strongly suggested localization of the gene products of these EGTs to the plastid. Almost the same number of proteins ($n = 24$) showed affinities to red algae or other algae with complex red-algal plastids, which were interpreted as plastid-targeted HGTs. Thus, the contribution of EGT and HGT to the complex plastid proteome in *B. natans* is of similar size and only corresponds to a small fraction of the plastid proteome, reflecting the observations made for primary plastids. The euglenozoan *Euglena gracilis*, like *B. natans*, also harbors a complex green plastid, but its plastid proteome is significantly larger than the one of chlorarachniophyte. Experimental validation determined that almost 1300 nucleus-encoded proteins localize to the *E. gracilis* plastid [50]. Again, only 95 (~7%) of these were found to associate with green algae in phylogenies, representing putative EGTs functioning in the plastid. Another 251 genes likely originated via transfer from other phototrophic eukaryotes, 20 from nonphototrophic eukaryotes and 13 from prokaryotes, adding up to 284 (~22%) HGT candidates targeted to the plastid. Unlike the roughly equal contribution of EGT and HGT in all plastid proteomes discussed above, the relative HGT contribution to the plastid proteome of *E. gracilis* exceeds the relative EGT content of its plastid proteome by a factor of three. Nevertheless, the dominant contribution of the host to the plastid proteome seems to be a hallmark of all investigated plastids.

Finally, the composition of a complex "plastid" proteome was bioinformatically analyzed in a kleptoplastidic lineage, the euglenozoan *Rapaza viridis*, which harbors kleptoplasts from the green algal genus *Tetraselmis* [51]. A comparable analysis was not performed in the kleptoplastidic lineage RSD, as the RSD sequence data were generated from a mixture of host and prey cells, precluding analysis of the kleptoplast proteome. Based on in silico predictions employing the presence/absence of plastid-targeting peptides, 274 putative kleptoplast-targeted proteins were identified. Out of these, 18 were found to group with or within the *Tetraselmis* lineage, but interestingly no candidate was identical to the prey strain *Tetraselmis* sp., indicating that no transfer from the current kleptoplast was detected. While the contribution of core chlorophytes (*Tetraselmis* and others) to the plastid proteome in *Rapaza* dominates, almost 30% of the proteins determined to be kleptoplast-targeted clustered with various primary and complex red-plastid lineages. Overall, *Rapaza* shows striking similarities to the kleptoplastidic dinoflagellate RSD in several ways: (1) no clear case of EGT was identified, (2) only a few plastid-targeted transfers were detected from lineages closely related to the prey (18 in *Rapaza*, 13 in RSD), and (3) plastid-targeted HGTs involve gene transfers from a variety of donors.

These two kleptoplastidic lineages do not possess a stably integrated plastid; however, both lineages are obligatory kleptoplastidic and obtain their plastids only from one specific type of prey. As these two lineages are not evolutionarily related, the observed similarities are best viewed as convergent and point to the possibility that plastids originated in similar ways. It is thus tempting to compare or even try to reconcile the observations made in these kleptoplastidic lineages with the current knowledge about EGT and HGT in vertically inherited plastids. In particular, the various sources of plastid-targeted HGT in RSD and *Rapaza* and the dominance of HGT over "EGT" (= sequences closely related to prey) in the *Rapaza* kleptoplast proteome (a similar analysis was not done in RSD) invite comparisons with *Paulinella*, in which HGT from a variety of bacterial donors contributed more than EGT to plastid function [22, 49]. Interestingly, a similar observation was made in yet another system with putative kleptoplasts, this time in the centrohelid *Meringosphaera* with dictyochophyte endosymbionts [52]. Although the search for putative EGT in this species only included a small set of nuclear-encoded proteins known to function in plastids, it was found that the putative origin of these proteins was varied.

6 Function of EGT and HGT in Primary and Complex Plastid Endosymbiosis

Assuming we can get to a reasonable estimate of EGT and HGT, what are the functions of the products of these genes and in which cell compartment do they fulfill them? For EGTs, the simplest explanation would be that the majority of proteins are targeted to the plastid where they perform their original role in the endosymbionts. However, as is often the case in science, the simplest explanation is not necessarily correct. Indeed, the limited number of species where the function and location of putative EGT products were analyzed shows a complex pattern. The first study exploring the function of EGTs was undertaken in the plant model *Arabidopsis* and determined that more than half of the ~1700 proteins of cyanobacterial origin do not, in fact, function in the plastid. Instead, these (putative) endosymbiotic proteins were predicted to localize to various compartments such as mitochondria, the cytosol, or to enter the secretory pathway [17]. Moreover, the predicted functions of many of these proteins were atypical for cyanobacteria or plastids, falling into categories such as intracellular transport or disease resistance, which suggests the acquisition of novel functions after transfer from the cyanobacterial endosymbiont to the host genome. In contrast, an EST-based analysis of the glaucophyte *C. paradoxa* showed that EGTs in this alga are mostly limited to plastidial functions [53]. In another analysis and as already discussed above, the relatively low numbers of cyanobacterial-derived proteins functioning in the plastids of *Arabidopsis*, *C. paradoxa*, and *C. reinhardtii* suggested that the majority of EGTs function outside of the organelle they have originated from [48].

In *Paulinella*, the ~32 identified EGT candidates displayed an evident bias toward proteins functioning in photosynthesis and photoprotection [7], but here also the chromatophore seems to be serviced by only a subset of these EGTs (17 proteins of α-cyanobacterial origin were found to be imported into the plastid [49]). *Paulinella* also represents one of the few taxa in which the functions of HGTs were analyzed. Of the 171 HGTs reported in *P. chromatophora* [22], gene products for 26 of these were found to be imported into the chromatophore [49]. Interestingly, many of these proteins fill specific gaps in metabolic pathways encoded by the organelle, suggesting that they have replaced functions that were lost from the chromatophore during endosymbiont reduction [22].

The observation that EGTs are not exclusively targeted back to the plastid also applies to complex plastid-bearing lineages. For example, in the chlorarachniophyte *Bigelowiella natans,* fewer green algal genes (56)—interpreted as EGTs—were predicted to be targeted to the plastid than to other cellular compartments (207) [8]. In addition, in both the cryptophyte *G. theta* and in *B. natans*, a subset of genes of algal origin (71 and 32, respectively) have undergone duplication events and in several cases, the resulting protein products are targeted to multiple compartments, suggesting that EGTs and HGTs can take up new functions and/or function in novel locations. Thus, both primary and complex plastid-bearing lineages seem to routinely retarget and re-functionalize their EGTs, with at least half of the EGT candidates taking on roles outside of the plastid.

Fewer studies have looked at kleptoplastidic associations, but in general, as mentioned above the number of EGTs that can be precisely traced back to the kleptoplast source are low. In the dinoflagellate RSD, only a small number of nuclear genes ($n = 13$) have been found to be related to haptophytes, the source of stolen plastids [40]. The majority of these gene transfers have obvious plastidial functions and/or are targeted to the kleptoplast (or the remnant ancestral dinoflagellate plastid still present in RSD), with only one candidate not having a clear plastid association. It is tempting to speculate that for these temporary associations, gene transfers with plastid function are being preferentially retained and utilized by the host to control the kleptoplast early on, while targeting of other proteins, for example host-derived, evolves only later in the host–symbiont relationship. In the euglenid *Rapaza*, kleptoplast-targeted proteins are, like in *Paulinella*, dominated by HGTs from various sources [51] and it will be interesting to investigate whether some of these HGTs have similar functions to the "metabolic rescue genes" in *Paulinella* [22].

Currently, there are more questions than answers regarding the functions of transferred genes. Future expansion of functional data for EGTs and HGTs will provide the exciting opportunity to infer patterns in functions gained by gene transfers, for example, whether specific functions are predominantly transferred via either type of gene transfer. In addition, linking the results of EGT and HGT analyses to the results of proteomic and targeting analyses will allow us to build a more comprehensive understanding of where in the cell these transferred genes function.

7 Fate of EGTs when Photosynthesis and/or Plastids Are Lost

Because host nuclear genomes have been the recipient of endosymbiotic genes during the course of plastid integration, which may take on new roles unrelated to plastid functions, it is often assumed that functional EGTs are retained even after the plastid is lost. A corollary of this assumption is that the presence of algal genes—but not necessarily plastid pathways—constitute a genetic "footprint" of endosymbioses, the evidence that ancestors of living plastid-lacking lineages were once photosynthetic [25, 54–56]. This reasoning was for example used to interpret 16 genes in plastid-lacking ciliates shared with algae and plants, some with a cyanobacterial origin and/or plastid targeting signal, as of putative endosymbiotic origin [54]. The assumption of a pervasive EGT footprint was also used to suggest an ancient, cryptic endosymbiosis in an ancestor of diatoms and related lineages with a different endosymbiont—from green algae—than the red algal endosymbiont related to the current red complex plastids in MOCHa groups [25].

Understanding the fate of EGTs after an endosymbiont is lost is crucial to avoid circularity if this footprint is to be used to infer plastid losses and more generally the punctuated distribution of photosynthesis and plastids in the tree of eukaryotes. Plastid loss has been viewed as a central force shaping the evolution of algal lineages, most famously in the chromalveolate hypothesis [57]. This hypothesis postulated that all plastids of red algal origin—those found in MOCHa groups—originated by a single ancestral secondary endosymbiosis with a red alga, implying numerous plastid losses (at least ten based on our current knowledge of the tree of eukaryotes) in the plastid-lacking lineages related to MOCHas (e.g., ciliates, nonochrophyte stramenopiles, katablepharids, and centrohelids) [13]. The chromalveolate hypothesis, however, has lost much of its traction mainly because empirical data indicate that complete plastid losses are extremely uncommon, unlike plastid endosymbioses. Indeed, the possibility that similar plastids can derive independently from similar endosymbionts, even in closely related hosts, is substantiated by several examples of dinoflagellates that have replaced their ancestral photosynthetic plastid with a new plastid of primary or complex algal origin, such as in the Kareniaceae, dinotoms, or the yet unnamed species MDG and TGD [11, 40, 58].

At the core of the difficulties in interpreting the penetrating impact of EGTs on the host genomes, besides the methodological difficulties discussed in the previous sections, is the lack of a reliable expectation to distinguish a lineage where a plastid once existed but was lost from a lineage that never possessed a plastid. We do not have a clear baseline to test the notion that the establishment of plastid necessarily makes a detectable impact in the host genome even after a cell re-engineers its metabolism to cope with plastid loss [2, 59]. To assess the lasting mark of plastids on host genomes, study systems are needed where plastid loss can be inferred independently of EGTs. These study systems are rare occurrences, because they ideally require enough morphological evidence to be convincing that not even a cryptic plastid is present, and, unlike the MOCHa groups, that they are phylogenetically nested within a plastid-containing clan for which it can be clearly established that the plastids have a common origin.

Demonstrable cases of plastid loss are not only few, they are also *not* far between. Only 3–4 relatively clear cases of plastid loss are known in groups for which photosynthetic ancestry by eukaryote-to-eukaryote endosymbiosis can be demonstrated. The losses of these complex plastids all occurred in the same group of eukaryotes—in Myzozoa—and, importantly, are only known in parasites: in *Cryptosporidium* [60], two losses in gregarines [61], and in the syndinian dinoflagellates *Hematodinium* [62] and *Amoebophrya* [63]. The underlying reason for the scarcity of plastid loss probably lies in the essential biochemical functions of plastids besides photosynthesis, including the biosynthesis of proteins, fatty acids, and isoprenoids. When the plastid is lost, these functions need re-engineering—for example, by metabolite scavenging from the host—or circumvented entirely, which appears extremely difficult to achieve for a cell. Therefore, while the loss of photosynthesis is common in algae, outright plastid loss seems not. The lasting EGT footprint in these taxa seems at best minimal with for example only one putative EGT in *Hematodinium* (the hemD gene of the heme biosynthesis pathway), which was re-evaluated as an HGT when including more taxon sampling in the phylogeny [59]. Even in other myzozoans where cryptic nonphotosynthetic plastids were suggested by genomic analyses, for example in *Oxyrrhis*, perkinsinds, or eleftherids, the impact of EGTs is likely restricted to core nonphotosynthetic metabolic plastid pathways [59, 64].

Outside of myzozoans, the eukaryvorous heterotrophic Actinophryds (stramenopiles) have been proposed as candidates for plastid loss due to their possible placement in ochrophytes. These are free-living cells, so the demonstration of plastid loss in this group would be of great significance. But a phylogenomic analysis showed that the actinophryd *Actinophrys* form a robust *sister* group to ochrophytes, thus weakening the criterion of phylogenetic derivation described above needed to facilitate the inference of plastid loss [65]. The analyses of several biosynthetic pathways based on transcriptomic data also revealed no evidence of plastid-type proteins, supporting the absence of plastid in *Actinophrys*. Yet, a dozen genes with algal origin were reported, all encoding organellar aminoacyl-tRNA synthases (aaRSs) now targeted to the mitochondria in *Actinophrys* but shared with ochrophytes and other algae where they can be dually targeted to mitochondria and plastids [65]. Whereas these aaRS-coding genes represent EGTs from an ancient endosymbiont common to actinophryds and ochrophytes but lost in actinophryds, or instead are HGTs independent of endosymbiosis, remains unclear. Regardless, *Actinophrys* represents an excellent example of the low level of the algal signal that can be expected in heterotrophic species with somewhat close evolutionary connection to algae. Similar low algal signals were reported in oomycetes (stramenopiles) [55], or as already mentioned in ciliates (alveolates) [54].

While the examples discussed so far in this section all concern the EGT footprint related to the loss of complex plastids, especially that of red algal-derived plastids, much less data are available for archaeplastid plastids. This is because the origin of these primary plastids has been generally considered much less controversial and, until recently, there was no known archaeplastid lineage without a plastid. The discovery that the tiny plastid-lacking marine flagellates Picozoa are closely related to red algae opens up the possibility of evaluating the EGT signal after primary plastid loss, under the dominant hypothesis of a single plastid origin from cyanobacteria in an ancestor of Archaeplastida [2]. A common assumption is that this archaeplastid host of primary endosymbiosis was already the recipient of extensive EGT from the cyanobacterial endosymbiont. This assumption derives from the substantial—although variable—number of EGTs inferred in all studied archaeplastids, ranging from ~1713 in *C. paradoxa* to as high as 4500 in *Arabidopsis* [17, 31]. Thus, lineages like Picozoa would have also inherited at least some of these genes despite that they have since lost their plastids.

What is the EGT footprint in Picozoa? Leveraging newly produced single-cell genomes (SAGs), the cyanobacterial signal in Picozoa was evaluated by phylogenetic analysis [2]. Importantly, the signal in Picozoa was directly compared to the signal inferred using the same approach in various species with photosynthetic and nonphotosynthetic plastids, and species with plastid loss or no plastid ancestry. This analysis showed, in a standardized way, that the cyanobacterial signal, while easily recoverable in photosynthetic species, drops by a good margin when photosynthesis is lost. This signal reduction continues after the plastid is lost. Only three and ten cyanobacterial genes were detected in the plastid-lacking myzozoan *Cryptosporidium* and *Hematodinium*, respectively, which is less than in all included species with plastids. In Picozoa, 16 cyanobacterial genes were detected, placing this signal in a comparable zone as for *Rhodelphis* and *Toxoplasma* (both with nonphotosynthetic plastids; 14 and 17 genes, respectively) or *Telonema* (plastid-lacking; 15 genes). The observation that plastid-lacking species have comparatively very few putative EGT from cyanobacteria becomes even more apparent when also taking into account HGTs from other prokaryotic sources [2]. In this case, while species with plastids (including nonphotosynthetic plastids) continue to show a stronger putative EGT signal, nearly all included plastid-lacking species show more putative HGTs than EGTs (with the exception of *Cryptosporidium*, but only three cyanobacterial genes were found in this species). It should be noted that this analysis represents a coarse attempt to contrast the putative EGT signal in plastid-lacking species with the different plastid "types" included in the analysis of Schön et al. [2], but further work is needed in particular to more systematically exclude dubious gene transfers that could simply be contamination. Still, it shows that with no plastid, even when the plastid was lost, detectable EGTs are few and cannot be differentiated from the background of possible HGTs. Plastids do not seem to necessarily have a lasting impact on their host genome and, in turn, the EGT footprint is likely not a reliable measure to reconstruct the history of plastid gains and losses.

If so little EGT signal is left in Picozoa, and pronounced EGT signal in Archaeplastida and other algae only comes from lineages that have evolved for a very long evolutionary time—in some cases more than 1.5 billion years—in the presence of plastids, then the next important question is when these EGTs occurred during plastid integration. Obvious reasons for the lack of signal are that we failed to detect it, or that Picozoa never had a plastid, or, as discussed above, that the transferred genes have been lost with the plastid. But another interesting possibility, which will require investigation, is that EGTs accumulate slowly during the endosymbiotic integration of plastids, or at least that the bulk of transfers happens "late" during the process. This is contrary to the common assumption that the bulk of EGTs and retargeting of their products to the plastids is a cornerstone of early plastid establishment [48]. Picozoa is interesting in this regard because it diverged relatively early in the evolution of Archaeplastida. The hypothesis that few EGTs occur at the onset of plastid endosymbiosis is consistent with the low level of EGT seen in photosynthetic *Paulinella*, which possesses relatively more recent primary plastids [4]. It is also consistent with the few putative EGTs found in *Actinophrys*, under the assumption that it has lost its red algal plastid early in ochrophyte evolution [65]. Importantly, this hypothesis is fully compatible with the notion that (some) EGT precedes organelle stabilization but is complemented by gene transfers from sources other than the endosymbiont, as demonstrated for the kleptoplastidic dinoflagellate RSD or the euglenid *Rapaza* [40, 51]. The initial exposure to foreign DNA, be it from a series of temporary endosymbionts or food sources, forms the basis of the "shopping bag" and "targeting-ratchet" models of plastid origin [41, 42]. By placing the emphasis on HGTs for the origin of plastids (in this case also including transfers from transient endosymbionts that may have been of different origins than the plastid), these models provide a fuller framework to interpret the fate of EGTs during plastid gain and loss.

8 Concluding Remarks

Numerous studies have sought to measure the impact of EGTs on various eukaryotic genomes and the roles that these foreign genes play in the cell. Although the inference of EGTs is a challenging problem, and accordingly EGT estimates vary widely, our current knowledge suggests a possible, yet uninvestigated role of EGT not only in plastid but also in other cellular functions. Increasingly, the importance of HGT not only on nuclear genomes but also on plastid proteomes has been highlighted, in several cases even surpassing EGT. Nevertheless, the extent to which gene transfers associated with plastid endosymbiosis have impacted nuclear genomes and when they took place are still open questions. To answer these important questions, the examination of a wide breadth of algae, including secondarily nonphotosynthetic lineages, together with the reexamination of previous findings in light of new data need to be tackled. This is a challenging task and the constant growth of available data will likely lead to additional complexity in our understanding of the role and timing of EGT. But ultimately these undertakings will not only allow us to more confidently estimate the impact of plastid-associated gene transfers but will also help us pinpoint their emergence. Finally, for both, EGT and HGT analysis, the emergence of spatial proteomic approaches promises the exciting possibility to study both mechanisms at subcellular levels. Combining (spatial) proteomic studies—which will provide answers to the question of where in the cell EGTs and HGTs function—with phylogenetic investigations—that will identify the source of the transferred gene—will contribute to constructing a more complete model of the role of gene transfers in plastid endosymbiosis.

Acknowledgments EH would like to acknowledge support from the Czech Academy of Sciences (Lumina Quaeruntur grant LQ200962204). FB would like to acknowledge support from Science for Life Laboratory, the Swedish Research Council VR (2021-04055), and the European Research Council (ERC consolidator grant 101044505).

References

1. Gawryluk RMR, Tikhonenkov DV, Hehenberger E, Husnik F, Mylnikov AP, Keeling PJ (2019) Non-photosynthetic predators are sister to red algae. Nature 572:240–243. https://doi.org/10.1038/s41586-019-1398-6
2. Schön ME, Zlatogursky VV, Singh RP, Poirier C, Wilken S, Mathur V (2021) Single cell genomics reveals plastid-lacking Picozoa are close relatives of red algae. Nat Commun 12. https://doi.org/10.1038/s41467-021-26918-0
3. Irisarri I, Strassert JFH, Burki F (2021) Phylogenomic insights into the origin of primary plastids. Syst Biol:2–45. https://doi.org/10.1093/sysbio/syab036
4. Nowack ECM, Weber APM (2018) Genomics-informed insights into endosymbiotic organelle evolution in photosynthetic eukaryotes. Annu Rev Plant Biol 69:51–84. https://doi.org/10.1146/annurev-arplant-042817-040209
5. Martin W, Goremykin V, Hansmann S (1998) Gene transfer to the nucleus and the evolution of chloroplasts. Curr Opin Plant Biol 1:276. https://doi.org/10.1016/1369-5266(88)80011-6
6. Stegemann S, Hartmann S, Ruf S, Bock R (2003) High-frequency gene transfer from the chloroplast genome to the nucleus. Proc Natl Acad Sci U S A 100:8828–8833. https://doi.org/10.1073/pnas.1430924100
7. Nowack ECM, Vogel H, Groth M, Grossman AR, Melkonian M, Glöckner G (2011) Endosymbiotic gene transfer and transcriptional regulation of transferred genes in *Paulinella chromatophora*. Mol Biol Evol 28:407–422. https://doi.org/10.1093/molbev/msq209
8. Curtis BA, Tanifuji G, Maruyama S, Gile GH, Hopkins JF, Eveleigh RJM, Nakayama T, Malik SB, Onodera NT, Slamovits CH et al (2012) Algal genomes reveal evolutionary mosaicism and the fate of nucleomorphs. Nature 492:59–65. https://doi.org/10.1038/nature11681
9. Burki F, Imanian B, Hehenberger E, Hirakawa Y, Maruyama S, Keeling PJ (2014) Endosymbiotic gene transfer in tertiary plastid-containing dinoflagellates. Eukaryot Cell 13:246–255. https://doi.org/10.1128/EC.00299-13
10. Ponce-Toledo RI, Moreira D, López-García P, Deschamps P (2018) Secondary plastids of euglenids and chlorarachniophytes function with a mix of genes of red and green algal ancestry. Mol Biol Evol 35:2198–2204. https://doi.org/10.1093/molbev/msy121
11. Sarai C, Tanifuji G, Nakayama T, Kamikawa R, Takahashi K, Yazaki E, Matsuo E, Miyashita H, Ishida KI, Iwataki M et al (2020) Dinoflagellates with relic endosymbiont nuclei as models for elucidating organellogenesis. Proc Natl Acad Sci U S A 117:5364–5375. https://doi.org/10.1073/pnas.1911884117

12. Kelly S (2021) The economics of organellar gene loss and endosymbiotic gene transfer. Genome Biol 22:345. https://doi.org/10.1186/s13059-021-02567-w
13. Burki F (2017) The convoluted evolution of eukaryotes with complex plastids, 1st edn. Elsevier Ltd. https://doi.org/10.1016/bs.abr.2017.06.001
14. Sibbald SJ, Archibald JM (2020) Genomic insights into plastid evolution. Genome Biol Evol 12:978–990. https://doi.org/10.1093/GBE/EVAA096
15. Goksøyr J (1967) Evolution of eucaryotic cells. Nature 214:1967
16. Weeden NF (1981) Genetic and biochemical implications of the endosymbiotic origin of the chloroplast. J Mol Evol 17:133–139. https://doi.org/10.1007/BF01733906
17. Martin W, Rujan T, Richly E, Hansen A, Cornelsen S, Lins T, Leister D, Stoebe B, Hasegawa M, Penny D (2002) Evolutionary analysis of *Arabidopsis, cyanobacterial, and chloroplast genomes reveals plastid phylogeny and thousands of cyanobacterial genes in the nucleus. Proc Natl Acad Sci U S A 99:12246–12251.* https://doi.org/10.1073/pnas.182432999
18. Stegemann S, Bock R (2006) Experimental reconstruction of functional gene transfer from the tobacco plastid genome to the nucleus. Plant Cell 18:2869–2878. https://doi.org/10.1105/tpc.106.046466
19. Lister DL, Bateman JM, Purton S, Howe CJ (2003) DNA transfer from chloroplast to nucleus is much rarer in *Chlamydomonas* than in tobacco. Gene 316:33–38. https://doi.org/10.1016/S0378-1119(03)00754-6
20. Marin B, Nowack ECM, Melkonian M (2005) A plastid in the making: evidence for a second primary endosymbiosis. Protist 156:425–432. https://doi.org/10.1016/j.protis.2005.09.001
21. Zhang R, Nowack ECM, Price DC, Bhattacharya D, Grossman AR (2017) Impact of light intensity and quality on chromatophore and nuclear gene expression in *Paulinella chromatophora*, an amoeba with nascent photosynthetic organelles. Plant J 90:221–234. https://doi.org/10.1111/tpj.13488
22. Nowack ECM, Price DC, Bhattacharya D, Singer A, Melkonian M, Grossman AR (2016) Gene transfers from diverse bacteria compensate for reductive genome evolution in the chromatophore of *Paulinella chromatophora*. Proc Natl Acad Sci U S A 113:12214–12219. https://doi.org/10.1073/pnas.1608016113
23. Burki F, Flegontov P, Oborník M, Cihlář J, Pain A, Lukeš J, Keeling PJ (2012) Re-evaluating the green versus red signal in eukaryotes with secondary plastid of red algal origin. Genome Biol Evol 4:626–635. https://doi.org/10.1093/gbe/evs049
24. Deschamps P, Moreira D (2012) Reevaluating the green contribution to diatom genomes. Genome Biol Evol 4:683–688. https://doi.org/10.1093/gbe/evs053
25. Moustafa A, Beszteri B, Maier UG, Bowler C, Valentin K, Bhattacharya D (2009) Genomic footprints of a cryptic plastid endosymbiosis in diatoms. Science (80-) 324:1724–1726. https://doi.org/10.1126/science.1172983
26. Woehle C, Dagan T, Martin WF, Gould SB (2011) Red and problematic green phylogenetic signals among thousands of nuclear genes from the photosynthetic and apicomplexa-related Chromera velia. Genome Biol Evol 3:1220–1230. https://doi.org/10.1093/gbe/evr100
27. Minge MA, Shalchian-Tabrizi K, Tørresen OK, Takishita K, Probert I, Inagaki Y, Klaveness D, Jakobsen KS (2010) A phylogenetic mosaic plastid proteome and unusual plastid-targeting signals in the green-colored dinoflagellate *Lepidodinium chlorophorum*. BMC Evol Biol 10. https://doi.org/10.1186/1471-2148-10-191
28. Hehenberger E, Burki F, Kolisko M, Keeling PJ (2016) Functional relationship between a dinoflagellate host and its diatom endosymbiont. Mol Biol Evol 33:2376–2390. https://doi.org/10.1093/molbev/msw109
29. Shih PM, Matzke NJ (2013) Primary endosymbiosis events date to the later Proterozoic with cross-calibrated phylogenetic dating of duplicated ATPase proteins. Proc Natl Acad Sci U S A 110:12355–12360. https://doi.org/10.1073/pnas.1305813110

30. Dagan T, Roettger M, Stucken K, Landan G, Koch R, Major P, Gould SB, Goremykin VV, Rippka R, De Marsac NT et al (2013) Genomes of stigonematalean cyanobacteria (subsection V) and the evolution of oxygenic photosynthesis from prokaryotes to plastids. Genome Biol Evol 5:31–44. https://doi.org/10.1093/gbe/evs117
31. Price DC, Chan CX, Yoon HS, Yang EC, Qiu H, Weber APM, Schwacke R, Gross J, Blouin NA, Lane C et al (2012) *Cyanophora paradoxa* genome elucidates origin of photosynthesis in algae and plants. Science (80-) 335:843–847. https://doi.org/10.1126/science.1213561
32. Delaye L, Valadez-Cano C, Pérez-Zamorano B (2016) How really ancient is *Paulinella chromatophora*? PLoS Curr 8. https://doi.org/10.1371/currents.tol.e68a099364bb1a1e129a17b4e06b0c6b
33. Huang J, Yue J (2013) Horizontal gene transfer in the evolution of photosynthetic eukaryotes. J Syst Evol 51:13–29. https://doi.org/10.1111/j.1759-6831.2012.00237.x
34. Ma J, Wang S, Zhu X, Sun G, Chang G, Li L, Hu X, Zhang S, Zhou Y, Song CP et al (2022) Major episodes of horizontal gene transfer drove the evolution of land plants. Mol Plant 15:857–871. https://doi.org/10.1016/j.molp.2022.02.001
35. Price DC, Goodenough UW, Roth R, Lee JH, Kariyawasam T, Mutwil M, Ferrari C, Facchinelli F, Ball SG, Cenci U et al (2019) Analysis of an improved *Cyanophora paradoxa* genome assembly. DNA Res 26:287–299. https://doi.org/10.1093/dnares/dsz009
36. Cenci U, Bhattacharya D, Weber APM, Colleoni C, Subtil A, Ball SG (2017) Biotic host–pathogen interactions as major drivers of plastid endosymbiosis. Trends Plant Sci 22:316–328. https://doi.org/10.1016/j.tplants.2016.12.007
37. Bodył A (2018) Did some red alga-derived plastids evolve via kleptoplastidy? A hypothesis. Biol Rev 93:201–222. https://doi.org/10.1111/brv.12340
38. Gast RJ, Moran DM, Dennett MR, Caron DA (2007) Kleptoplasty in an Antarctic dinoflagellate: caught in evolutionary transition? Environ Microbiol 9:39–45. https://doi.org/10.1111/j.1462-2920.2006.01109.x
39. Sellers CG, Gast RJ, Sanders RW (2014) Selective feeding and foreign plastid retention in an Antarctic dinoflagellate. J Phycol 50:1081–1088. https://doi.org/10.1111/jpy.12240
40. Hehenberger E, Gast RJ, Keeling PJ (2019) A kleptoplastidic dinoflagellate and the tipping point between transient and fully integrated plastid endosymbiosis. Proc Natl Acad Sci U S A 116:17934–17942. https://doi.org/10.1073/pnas.1910121116
41. Larkum AWD, Lockhart PJ, Howe CJ (2007) Shopping for plastids. Trends Plant Sci 12:189–195. https://doi.org/10.1016/j.tplants.2007.03.011
42. Keeling PJ (2013) The number, speed, and impact of plastid endosymbioses in eukaryotic evolution. Annu Rev Plant Biol 64:583–607. https://doi.org/10.1146/annurev-arplant-050312-120144
43. Facchinelli F, Pribil M, Oster U, Ebert NJ, Bhattacharya D, Leister D, Weber APM (2013) Proteomic analysis of the *Cyanophora paradoxa* muroplast provides clues on early events in plastid endosymbiosis. Planta 237:637–651. https://doi.org/10.1007/s00425-012-1819-3
44. Terashima M, Specht M, Naumann B, Hippler M (2010) Characterizing the anaerobic response of *Chlamydomonas reinhardtii* by quantitative proteomics. Mol Cell Proteomics 9:1514–1532. https://doi.org/10.1074/mcp.M900421-MCP200
45. van Wijk KJ, Baginsky S (2011) Plastid proteomics in higher plants: current state and future goals. Plant Physiol 155:1578–1588. https://doi.org/10.1104/pp.111.172932
46. Hopkins JF, Spencer DF, Laboissiere S, Neilson JAD, Eveleigh RJM, Durnford DG, Gray MW, Archibald JM (2012) Proteomics reveals plastid- and periplastid-targeted proteins in the chlorarachniophyte alga *Bigelowiella natans*. Genome Biol Evol 4:1391–1406. https://doi.org/10.1093/gbe/evs115
47. Schober AF, Río Bártulos C, Bischoff A, Lepetit B, Gruber A, Kroth PG (2019) Organelle studies and proteome analyses of mitochondria and plastids fractions from the diatom *Thalassiosira pseudonana*. Plant Cell Physiol 60:1811–1828. https://doi.org/10.1093/pcp/pcz097

48. Qiu H, Price DC, Weber APM, Facchinelli F, Yoon HS, Bhattacharya D (2013) Assessing the bacterial contribution to the plastid proteome. Trends Plant Sci 18:680–687. https://doi.org/10.1016/j.tplants.2013.09.007
49. Singer A, Poschmann G, Mühlich C, Valadez-Cano C, Hänsch S, Hüren V, Rensing SA, Stühler K, Nowack ECM (2017) Massive protein import into the early-evolutionary-stage photosynthetic organelle of the amoeba *Paulinella chromatophora*. Curr Biol 27:2763–2773.e5. https://doi.org/10.1016/j.cub.2017.08.010
50. Novák Vanclová AMG, Zoltner M, Kelly S, Soukal P, Záhonová K, Füssy Z, Ebenezer TGE, Lacová Dobáková E, Eliáš M, Lukeš J et al (2020) Metabolic quirks and the colourful history of the *Euglena gracilis* secondary plastid. New Phytol 225:1578–1592. https://doi.org/10.1111/nph.16237
51. Karnkowska A, Yubuki N, Maruyama M, Yamaguchi A, Kashiyama Y, Suzaki T, Keeling PJ, Hampl V, Leander BS (2023) Euglenozoan kleptoplasty illuminates the early evolution of photoendosymbiosis. Proc Natl Acad Sci 120:e2220100120. https://doi.org/10.1073/pnas.2220100120
52. Sørensen MES, Zlatogursky VV, Onuţ-Brännström I, Walraven A, Foster RA, Burki F (2023) A novel kleptoplastidic symbiosis revealed in the marine centrohelid *Meringosphaera* with evidence of genetic integration. Curr Biol 33:3571–3584.e6. https://doi.org/10.1016/j.cub.2023.07.017
53. Reyes-Prieto A, Hackett JD, Soares MB, Bonaldo MF, Bhattacharya D (2006) Cyanobacterial contribution to algal nuclear genomes is primarily limited to plastid functions. Curr Biol 16:2320–2325. https://doi.org/10.1016/j.cub.2006.09.063
54. Reyes-Prieto A, Moustafa A, Bhattacharya D (2008) Multiple genes of apparent algal origin suggest ciliates may once have been photosynthetic. Curr Biol 18:956–962. https://doi.org/10.1016/j.cub.2008.05.042
55. Stiller JW, Huang J, Ding Q, Tian J, Goodwillie C (2009) Are algal genes in nonphotosynthetic protists evidence of historical plastid endosymbioses? BMC Genomics 10:484. https://doi.org/10.1186/1471-2164-10-484
56. Tyler BM, Tripathy S, Zhang X, Dehal P, Jiang RHY, Aerts A, Arredondo FD, Baxter L, Bensasson D, Beynon JL et al (2006) Phytophthora genome sequences uncover evolutionary origins and mechanisms of pathogenesis. Science (80-) 313:1261–1266. https://doi.org/10.1126/science.1128796
57. Cavalier-Smith T (1999) Principles of protein and lipid targeting in secondary symbiogenesis: Euglenoid, dinoflagellate, and sporozoan plastid origins and the eukaryote family tree. J Eukaryot Microbiol 46:347–366. https://doi.org/10.1111/j.1550-7408.1999.tb04614.x
58. Yamada N, Sym SD, Horiguchi T (2017) Identification of highly divergent diatom-derived chloroplasts in dinoflagellates, including a description of *Durinskia kwazulunatalensis* sp. nov. (Peridiniales, Dinophyceae). Mol Biol Evol 34:1335–1351. https://doi.org/10.1093/molbev/msx054
59. Holt CC, Hehenberger E, Tikhonenkov DV, Jacko-Reynolds VKL, Okamoto N, Cooney EC, Irwin NAT, Keeling PJ (2023) Multiple parallel origins of parasitic Marine Alveolates. Nat Commun 14. https://doi.org/10.1038/s41467-023-42807-0
60. Xu P, Widmer G, Wang Y, Ozaki LS, Alves JM, Serrano MG, Pulu D, Manque P, Akiyoshi D, Mackey AJ et al (2004) The genome of *Cryptosporidium hominis*. Nature 431:1107–1112. https://doi.org/10.1038/nature02977
61. Mathur V, Salomaki ED, Wakeman KC, Na I, Kwong WK, Kolisko M, Keeling PJ (2023) Reconstruction of plastid proteomes of apicomplexans and close relatives reveals the major evolutionary outcomes of cryptic plastids. Mol Biol Evol 40:1–12. https://doi.org/10.1093/molbev/msad002
62. Gornik SG, Febrimarsa CAM, MacRae JI, Ramaprasad A, Rchiad Z, McConville MJ, Bacic A, McFadden GI, Pain A et al (2015) Endosymbiosis undone by stepwise elimination of the plastid in a parasitic dinoflagellate. Proc Natl Acad Sci U S A 112:5767–5772. https://doi.org/10.1073/pnas.1423400112

63. John U, Lu Y, Wohlrab S, Groth M, Janouškovec J, Kohli GS, Mark FC, Bickmeyer U, Farhat S, Felder M et al (2019) An aerobic eukaryotic parasite with functional mitochondria that likely lacks a mitochondrial genome. Sci Adv 5:1–12. https://doi.org/10.1126/sciadv.aav1110
64. Janouskovec J, Gavelis GS, Burki F, Dinh D, Bachvaroff TR, Gornik SG, Bright KJ, Imanian B, Strom SL, Delwiche CF et al (2017) Major transitions in dinoflagellate evolution unveiled by phylotranscriptomics. Proc Natl Acad Sci U S A 114:E171–E180. https://doi.org/10.1073/pnas.1614842114
65. Azuma T, Pánek T, Tice AK, Kayama M, Kobayashi M, Miyashita H, Suzaki T, Yabuki A, Brown MW, Kamikawa R (2022) An enigmatic stramenopile sheds light on early evolution in Ochrophyta plastid organellogenesis. Mol Biol Evol 39:1–12. https://doi.org/10.1093/molbev/msac065

Easier Lost than Found? What We Know about Plastid Genome Reduction

Anežka Konupková, Lucia Tomečková, Kristína Záhonová, Miroslav Oborník, and Zoltán Füssy

Abstract Plastids are eukaryotic organelles that evolved from a photobiotic symbiont, imparting photosynthetic abilities to heterotrophic hosts. Plastids lose much of their complexity during the endosymbiont-to-organelle transition, reflecting the need of the symbiotic partners to synchronize reproduction and streamline metabolism. This is obvious from genome size reduction, and while the genomes of plastid predecessors, cyanobacteria, typically range 1.6–7.8 Mbp, most plastid genomes range 110–190 kbp. In some lineages, plastid genomes depart from convention, which manifests two-way. Whereas in rhodophytes, chlorophytes, plants, and euglenids this leads to the expansion of noncoding DNA, in dinoflagellates, the plastid

---

A. Konupková · L. Tomečková
Faculty of Science, BIOCEV, Charles University, Prague, Czech Republic
e-mail: anezka.konupkova@natur.cuni.cz; lucia.hadariova@natur.cuni.cz

K. Záhonová
Faculty of Science, BIOCEV, Charles University, Prague, Czech Republic

Institute of Parasitology, Biology Centre, Czech Academy of Sciences, České Budějovice, Czech Republic

Life Science Research Centre, Department of Biology and Ecology, Faculty of Science, University of Ostrava, Ostrava, Czech Republic

Division of Infectious Diseases, Department of Medicine, Faculty of Medicine and Dentistry, University of Alberta, Edmonton, AB, Canada

M. Oborník
Institute of Parasitology, Biology Centre, Czech Academy of Sciences, České Budějovice, Czech Republic

Faculty of Science, University of South Bohemia, České Budějovice, Czech Republic
e-mail: obornik@paru.cas.cz

Z. Füssy (✉)
Faculty of Science, BIOCEV, Charles University, Prague, Czech Republic

Faculty of Science, University of South Bohemia, České Budějovice, Czech Republic

Scripps Institution of Oceanography, University of California San Diego, La Jolla, CA, United States of America

S. D. Schwartzbach et al. (eds.), *Endosymbiotic Organelle Acquisition*,
https://doi.org/10.1007/978-3-031-57446-7_5

genome is fragmented into single-gene minicircles, and in one chlorophyte lineage into linear single-stranded hairpin chromosomes. Yet, plastids may later enter the dark phase of their "life history." Driven by competition, even established phototrophic organisms sometimes revert to heterotrophy or parasitism, leading to further impairment or complete loss of photosynthesis. Here, we recapitulate the history of plastids from early acquisition to their disappearance in nonphotosynthetic algae and plants. We compare how molecular functions encoded by plastids vary in diverse eukaryotic lineages that acquired them, and how they vary in lineages about to lose them. We highlight how genome reduction accompanies plastid life cycles and how evolutionary history shapes their ultimate future.

Keywords Comparative genomics · Reductive evolution · Plastid genome topology · Intron expansion · Repeat expansion · Minicircles · Endosymbiotic gene transfer · Horizontal gene transfer · Nonphotosynthetic plastids · Serial endosymbiosis · Complete plastid loss

1 Introduction

Chloroplasts have been of immense importance to the evolution and perpetuation of complex life. Plants and algae that carry these photosynthetic organelles are responsible for ~90% of global primary production [1, 2]. Through harvesting the energy of light to reduce inorganic carbon (CO_2), they provide a major input of biomass that drives food chains in virtually all of Earth's surface environments. The molecular machinery for oxygenic photosynthesis was invented before the Great Oxidation Event ca. 2.4–2.3 billion years ago (reviewed in [3]) by the ancestors of modern cyanobacteria. Still, it was not until ca. 1.5–0.9 billion years ago that eukaryotes first took hold of this invention via endosymbiotic uptake of a cyanobacterium [4–6]. Somewhat later yet, during the Neoproterozoic Oxidation Event (800–600 million years ago), atmospheric oxygen levels reached todays' values, possibly coinciding with a radiation of algae in the ocean [7], which allowed oxygen-dependent lineages such as animals to flourish into their contemporary diversity.

Albeit much reduced, chloroplasts are very reminiscent of their cyanobacterial ancestors. Chloroplasts are enclosed by at least two membranes hypothesized to originate from the cyanobacterial cell envelope [8]. In the stroma, their equivalent of cytosol, chloroplasts possess additional membranous structures termed the thylakoids where the photosynthetic apparatus resides; more or less conspicuous pyrenoids that harbor the enzymes of the carbon-concentrating mechanism (CCM), functionally analogous to cyanobacterial carboxysomes; a reduced circular-mapping genome with a typical bacterial-like organization localized into nucleoids [9]; and a translation subcompartment dedicated to the biosynthesis of photosystems [10]. Indeed, among the emerging diversity of eukaryotic endosymbionts, it is not uncommon to see a substantial reduction in coding capacity and function (reviewed in, e.g., [11]), and chloroplasts are excellent examples of maintaining vital functions over the course of their evolution while losing functions that were disposable [12].

Despite many apparent similarities, a closer look at chloroplast ultrastructure and molecular equipment suggests a convoluted distribution among eukaryotic lineages (Fig. 1). A single primary endosymbiosis event is thought to have enabled photosynthesis in Archaeplastida, the supergroup consisting of rhodophytes (red algae), glaucophytes, and Viridiplantae (green algae and plants). Higher-order (also called eukaryote-to-eukaryote or complex) endosymbiotic events then grafted chloroplasts into other nonrelated eukaryotic lineages, including cryptophytes, alveolates, stramenopiles, and haptophytes (or CASH lineages [13]) as well as chlorarachniophytes and euglenids. Whereas the latter two provide more clear-cut examples of endosymbiosis with green algae, CASH lineages formed endosymbioses with rhodophyte-derived chloroplasts in an order that is still being debated (e.g., [14–16]; see Chap. 2). More recently, another primary endosymbiosis event occurred in Paulinellidae (Rhizaria), where a *Synechococcus*-related endosymbiont slowly evolves into an organelle termed the cyanelle or chromatophore [17]. Yet other examples of cyanobacterial symbionts known as the UCYN-A (i.e., unicellular N_2-fixing cyanobacteria group A) have been found in freshwater diatoms and a marine haptophyte, albeit these are apparently on a different evolutionary route toward becoming dinitrogen (N_2)-fixing organelles, analogous to those known from plant mycorrhizal symbioses [18].

In this chapter, we track the evolution of chloroplast genomes across the diversity of eukaryotes that have acquired them, from their origin in endosymbiosis to their demise in algae and plants that no longer photosynthesize. We look at which functions they retain most conservatively and what changes they undergo when they lose control. To present a more rigorous picture, we used ortholog grouping, phylogenetic analyses, and hidden Markov model (HMM) profile searches with gene annotations extracted from >3500 complete reference genomes available in GenBank to double-check their annotations and gene family homologies. Although this approach is not comprehensive, it outlines the challenges in plastid genome biology for the next era.

2 Dynamics of Plastid Genome Reduction

Plastids, referring to both photosynthetic and nonphotosynthetic organelles, contain genomes with a characteristic architecture. These genomes are circular, typically 110–190 kbp long, carrying around 70 protein-coding genes in the Viridiplantae lineage and 100–200 genes in the rhodophyte lineage, 25–32 tRNA, and 3–4 rRNA genes (Fig. 2). The number of genes is lower in nonphotosynthetic plastids compared with their photosynthetic siblings, ranging from zero, in plants that lost plastid genomes entirely, to 90 in nonphotosynthetic cryptomonads. Some plant leucoplasts are an exception, as they retain a full genome but are physiologically nonphotosynthetic via gene repression. There are two large inverted repeats (IRs) containing the rRNA operons, and unique regions of different sizes, referred to as

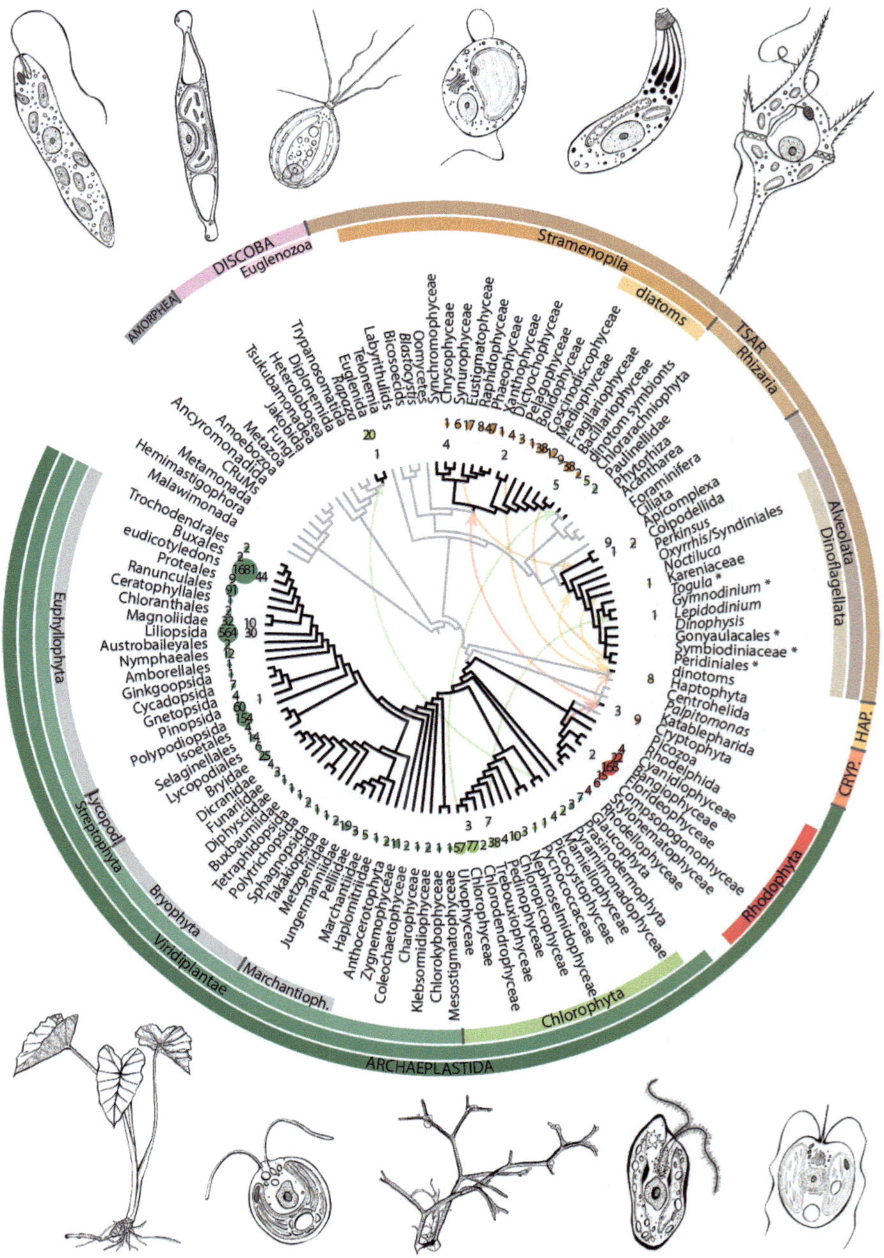

Fig. 1 The evolution of eukaryotic photosynthesis and the taxonomic distribution of plastid genomes have been revisited in this chapter. The cladogram depicts the relationships of major nonphotosynthetic and photosynthetic lineages (gray and black branches, respectively). Higher taxonomic classification is shown over the arcs joining related lineages. The arrows illustrate our current understanding of chloroplast evolution (according to [16]), showing endosymbioticFig.

the Small Single Copy (SSC) and Large Single Copy (LSC) regions. This organization has been implicated in helping maintain genome integrity through DNA recombination [19]. Most of the protein-coding genes code for subunits of photosystems I and II (*psa* and *psb*, respectively) and ribosomal proteins (*rps* and *rpl* for proteins of the small and large subunit, respectively). This is in striking contrast with the genomes of cyanobacteria, which typically are 1.6–7.8 Mbp long and encode between 1700 and 12,300 proteins [20, 21], with the closest known relative to plastids, *Gloeomargarita lithophora*, encoding ~3000 proteins [4].

Indeed, chloroplast functions require hundreds of proteins (996 in *Chlamydomonas*, [22]; 1856 in *Arabidopsis,* [23]; 1345 in *Euglena*, [24]; and 433 imported plus 876 chromatophore-encoded in *Paulinella,* [25]), and most of them are encoded by the nuclear genome and need to be imported into the organelle. The establishment of a protein translocation machinery in the chloroplast envelope is thought to facilitate large-scale functional replacement of chloroplast-encoded proteins with homologs newly encoded by the nuclear genome and retargeted back to the organelle [26]. Every time a gene is transferred from the symbiont genome into the host nucleus and acquires a functional plastid targeting presequence [e.g., [27]], the copy in the plastid genome becomes exposed to genetic drift or adaptive streamlining and is lost. This process persists during complex endosymbioses, as complex plastids encode, on average, fewer genes than their primary ancestors (Fig. 2) [28]. Few genes transferred from the endosymbiont appear to be essential for photosynthesis, as they are usually important under stress conditions. For instance, they assist in avoiding oxidative stress that would otherwise prevent long-term plastid retention [29].

Homologs of host origin, or horizontally acquired from entirely different sources, can also be (re)targeted to the plastid and functionally replace a plastid-encoded homolog, which translates into a chimeric evolutionary origin of plastid proteomes [24, 25, 30–35]. This phenomenon seems to affect primary and complex chloroplast endosymbioses alike, although the sources of plastid-targeting genes and the magnitude of gene influx might differ from lineage to lineage. For instance, only 17 of the 433 proteins imported into the chromatophore seem to be of cyanobacterial origin, 26 of other bacterial origin, and the vast majority are of host or uncertain origin [25]. In comparison, while many genes encoding core plastid proteins of Archaeplastida seem to originate in bacteria or eukaryotes, most of them can be traced to cyanobacterial ancestors [34, 36]. In complex algae, considerably more genes are derived from other algae than bacteria [24, 32, 33], reflecting that recruitment of pre-existing algal solutions could facilitate endosymbiont acquisition in

1 (continued) transfers from donor lineages (arrowtail) to their hosts (arrowhead). Peridinin-pigmented dinoflagellates, as opposed to those with chlorophyte, haptophyte or diatom endosymbionts, are marked by asterisks. For each lineage, the number of available photosynthetic (colored) and nonphotosynthetic plastid sequences (gray) is marked at branch tips with a number and a circle in scale. Note that some of plastid sequence redundancy was removed using average nucleotide identity filtering (ANI distance <0.02). Drawings of clade representatives are shown around the cladogram. Abbreviations: CRYP Cryptophyceae, HAP Haptophyceae, Lycopod Lycopodiopsida

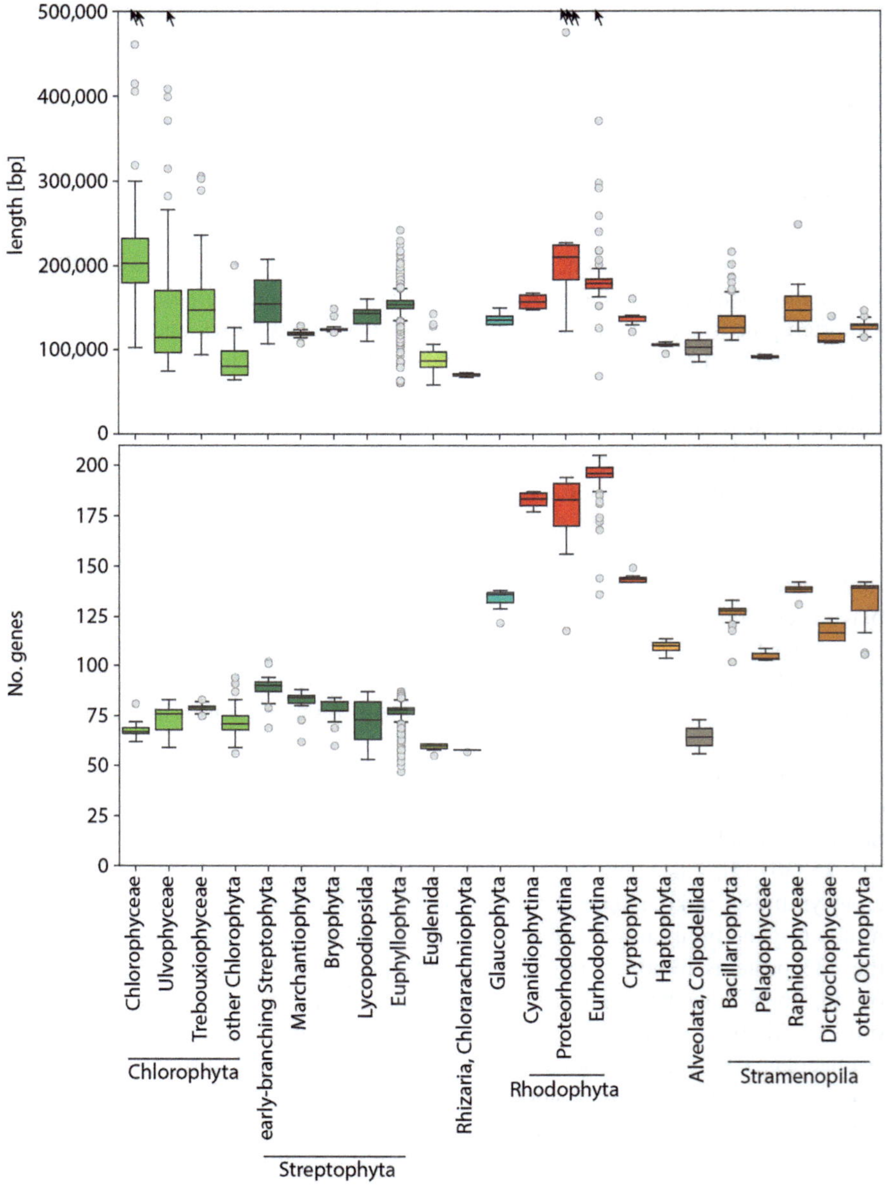

Fig. 2 Variations in chloroplast genome size and gene content across photosynthetic lineages. Box-and-whisker plots characterize each data set, with outliers shown as gray circles. Arrows mark genomes larger than 500 kbp: *Floydiella terrestris* (521 kbp), *Haematococcus lacustris* (1.35 Mbp) (Chlorophyceae), *Acetabularia acetabulum* (estimated ca. 1 Mbp) (Ulvophyceae), *Dixoniella grisea* (516 kbp) (Bangiophyceae), *Bulboplastis apyrenoidosa* (610 kbp), *Corynoplastis japonica* (1.12 Mbp), and *Rhodella violacea* (903 kbp) (Rhodellophyceae)

another eukaryotic lineage [29, 37]. If the host attempting to establish a symbiotic relationship is a predator or forms other strong biotic relationships, its biotic interactions are also likely to affect horizontal gene flow into its nuclear genome [35, 38]. Needless to add, despite endosymbiotic gene transfer (EGT), that is, horizontal gene transfer (HGT) specifically from the endosymbiont prevails over HGT from other donors across algal and plant lineages [39], not all transferred genes encode proteins targeted to plastids, and host molecular functions become chimeric as well [40, 41].

Closely linked with the genetic innovations of the host nucleus, the evolution of the endosymbiont/plastid genome has a clear dynamic [36, 42]. Following a sudden increase in the cellular complexity of the endosymbiotic couple, there is a lengthy reductive phase when the resulting union gradually consolidates. This happens for several major reasons: (1) regulatory or life cycle-related functions of the nascent symbiont, such as sensing of nutrients or biological threats, might diminish, as host-enclosed symbionts find themselves in a relatively more stable environment compared with their free-living siblings; (2) the symbiont loses independence as the host gains control over its division and metabolism; (3) metabolic redundancy, that is, a situation when analogous biochemical pathways run concurrently in the host and the symbiont, is adaptively streamlined through metabolite exchange and losses of either host or endosymbiont pathways or their parts; and (4) near-neutral selective pressure allows gene loss by genetic drift. Arguably, reductive evolution is much faster at the initiation of endosymbiosis, which is when selection on the plastid genome transiently relaxes as well [28, 29]. Consistently, chlorophyte-derived plastids of euglenids, chlorarachniophytes, and the dinoflagellate *Lepidodinium* lost noticeably more plastid genes at their stem branch compared with their chlorophyte sister lineages or descendant branches [28]. Major losses of plastid genes are also apparent in rhodophyte-derived plastids with respect to their rhodophyte counterparts, mainly those encoding phycocyanin pigment proteins and enzymes. Later, when endosymbiosis is functionally well established, selection takes over to enforce the coordination of plastid metabolic and housekeeping processes, and the rate of gene loss from the plastid genome decreases. This dynamic is not limited to plastids and many endosymbionts face similar reductive trends [11].

The situation might change for organisms that are not obligately phototrophic, that is, exercise other trophic strategies such as predation, osmotrophy, or facultative parasitism. The maintenance of the photosynthetic machinery is metabolically expensive [43, 44], and depending on how much mixotrophic algae rely on nutrients acquired heterotrophically, the selective pressure on photosynthetic genes may become relaxed [28]. In fact, many algal lineages lose photosynthesis secondarily and return to heterotrophy or become parasitic. In these cases, photosynthesis-related genes are lost quickly from both nuclear and plastid genomes [45–48], corresponding to the early phases of endosymbiosis when excess complexity is eliminated. In contrast, the sets of genes that remain in nonphotosynthetic algal genomes are usually well conserved, which applies to both evolutionarily old lineages, such as apicomplexans, and younger lineages that independently evolved from a common ancestor, such as nonphotosynthetic chlorophytes, suggesting convergent evolution (see below).

One of the important factors influencing the rate of organelle-to-nucleus DNA transfer, and hence potentially EGT and plastid genome reduction, is the number of plastids per cell. In organisms having a single plastid (i.e., monoplastidic), the total length of nuclear plastid DNA-like sequences (NUPTs) is reduced compared with polyplastidic organisms [49]. This is possibly owing to a pressure on monoplastidic species to keep their plastid intact, which is known as the "limited transfer window" [50]. The tendency of large plant genomes to accumulate noncoding DNA helps us understand how often NUPTs are transferred to the host nucleus if they are not eliminated by selection [49]. NUPTs have been found in hundreds-to-thousands of genomic loci in plants. Although the majority of NUPTs are short noncoding sequences, functional plastid genes are probably transferred to the nucleus by similar mechanisms. The early stages of endosymbioses could have been analogous to polyplastidy, in a sense that the hosts more likely experienced endosymbiont lysis, which would expose the endosymbiont DNA for a potential transfer into the host nucleus [50]. Therefore, the ability of hosts to recruit genes with plastid-targeted functions into their nuclear genome strongly shapes their plastids' genome evolution as well.

In several algal lineages, the gene transfer window closed much sooner than in others, resulting in remarkable dead stops. The plastids of cryptophytes and chlorarachniophytes not only contain genomes of cyanobacterial origin but also vestigial nuclei of the algal ancestors of their plastids, that is, a rhodophyte and a chlorophyte, respectively [51, 52]. These organelles are known as nucleomorphs and harbor genomes that are much smaller than those of free-living rhodophytes and chlorophytes (estimated ~550–845 kbp in cryptophytes and ~330–610 kbp in chlorarachniophytes) but encode dozens of housekeeping genes essential for plastid maintenance, which failed to be transferred to the host nucleus (reviewed in [53]). With data currently at hand, it is difficult to conclude whether nucleomorphs are still slowly evolving and will be eventually lost, or whether there are mechanistic or functional barriers preventing their elimination.

This illustrates that despite all plastids trace back to cyanobacteria, there are subtle structural and functional differences among plastid-bearing lineages. These changes supposedly stem from different modes of acquisition of their immediate ancestors, and perhaps random chance alike. With that in mind, we next elaborate on some of these differences.

3 Chloroplasts in Algae and Plants

Archaeplastida The chloroplasts in Archaeplastida originated from a primary endosymbiosis ca. 1.5–0.9 billion years ago. They are surrounded by two membranes and exhibit many ancestral features that we do not see in more derived complex chloroplasts. The glaucophyte chloroplasts, for instance, contain a peptidoglycan wall between the inner and outer membrane, and their thylakoids are concentric around a carboxysome. Both glaucophyte and rhodophyte thylakoids are unstacked and decorated by large light-harvesting structures called phycobilisomes. Only chlorophyll (Chl) *a* is present. In Viridiplantae, which unites prasinodermophytes, chlorophytes, and streptophytes, the lineage ultimately evolving into plants, phycobilisomes have been lost and the light-harvesting complexes are membrane-embedded, allowing the formation of tighter and more efficient thylakoid arrangements, called stacks. Their photosynthetic pigments are Chl *a* and *b* and zeaxanthin. In a sense, their chloroplasts appear more physiologically advanced. Notably, glaucophytes and rhodophytes keep their polysaccharide storage in the cytosol whereas viridiplants in their chloroplast. These marked differences, conserved within each lineage, led to hypotheses about multiple origins of chloroplasts in Archaeplastida, which has not been unambiguously shown [39, 54]; see also Chap. 1.

Many of these features are performed by proteins encoded by chloroplast genomes (Fig. 3). Therefore, rhodophyte and glaucophyte chloroplast genomes are more gene-rich, coding for many proteins and complexes lost from the plastid genomes of Viridiplantae. These include apoproteins of photosynthetic pigments

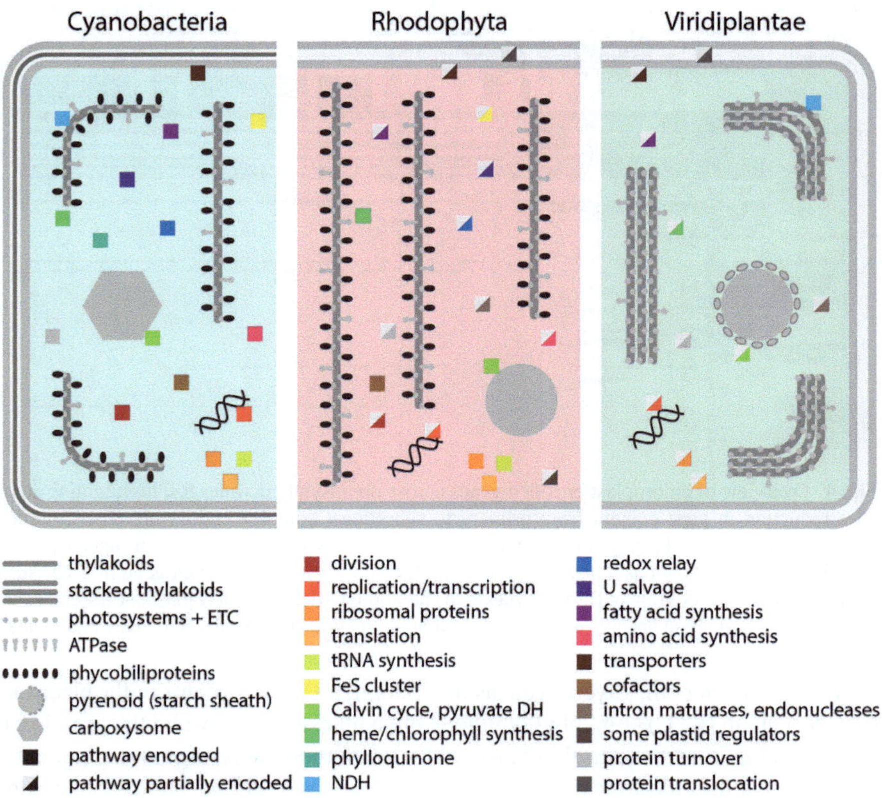

Fig. 3 Comparison of cyanobacteria and chloroplasts of Rhodophyta and Viridiplantae. Major chloroplast pathways with genome-encoded components are shown in color, partially or entirely relocated pathways as (half) gray

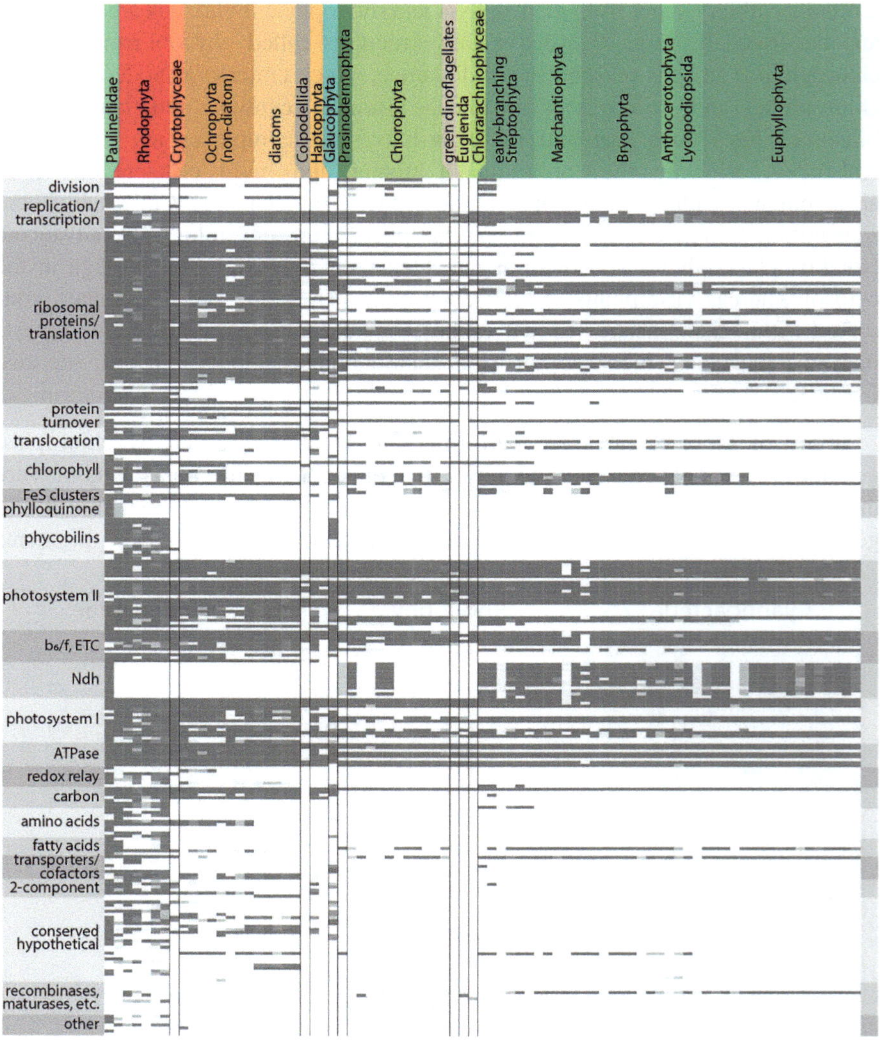

Fig. 4 Overview of chloroplast-encoded genes across photosynthetic lineages, highlighting stepwise reduction in gene content from primary to subsequent complex endosymbioses. Columns represent taxonomic subgroups of shown lineages, whereas rows represent genes grouped by function. The figure is shown as a heatmap, reflecting the conservation of a given gene in each (sub) lineage. The second column in Haptophyta are the haptophyte-derived chloroplasts of Kareniaceae. More information about specific events are found in the section "Chloroplasts in algae and plants"

(phycocyanins in both, phycoerythrins in rhodophytes), components of the electron-transport chain (ETC) (e.g., phylloquinone biosynthesis pathway and the ferredoxin relay system), enzymes partaking in synthesizing amino acids (arginine, glutamate, isoleucine/valine, tryptophan), and some of the ribosomal and photosystem subunits, not to mention several unknown-function conserved proteins (named Ycf's, Fig. 4). "The greens," on the other hand, encode in their chloroplast genomes subunits of the proton-pumping NADH:plastoquinone oxidoreductase Ndh having a dual role in cyclic (light) and respiratory (dark) electron transport [55], although its importance and implications for its conservation across Viridiplantae have not been clarified [56].

The structure of the chloroplast genomes seems ancestrally canonical in Archaeplastida, with two direct or inverted repeats and SSC/LSC regions present in most, and perhaps no introns in genes. The number and phylogenetic distribution of syntenic blocks and rDNA operons, however, suggest that genomic rearrangements and repeat inversions and losses are in fact quite common [57].

At a higher taxonomic resolution, rhodophyte classes differ very little as to their plastid genome content. Only Cyanidiophyceae encode two additional iron-sulfur (Fe-S) cluster assembly proteins CysA/CysT and the enzymes of the phylloquinone pathway. Among rhodophyte classes, there are mainly differences in the retention of genes encoding subunits of the fatty acid synthesis enzyme acetyl-CoA carboxylase (*accA, -B, -D*), subunits of the chlorophyll biosynthesis enzyme known as light-independent protochlorophyllide oxidoreductase (LIPOR; *chlL, -N, -B*), and several unknown-function proteins (*ycf*). Most classes of the subphylum Proteorhodophytina encode plastid intron maturases, consistent with their intron richness and size expansion [58], although Florideophyceae have been shown to encode one or two maturases in the *trnM-e* gene as well [59]. Plastid introns are considered selfish DNA and have been probably acquired by horizontal transfer.

Glaucophytes possess most of the plastid genes found in rhodophytes with a few exceptions. They lack the genes for DNA helicase (*dnaB*), RNA polymerase subunit Z (*rpoZ*), seven ribosomal proteins (*rpl4, -13, -24, -29, -31, -32*, and *rps1*), translation factors (*infB, -C, tsf*), protein translocase subunits (*secA, tatC*), phycoerythrin apoproteins and their tetrapyrrole-attachment protein (*cpeA, -B, -S*), acetyl-CoA carboxylase, and five photosystem and ETC subunits (*ycf20, petJ, ccs1, psaD, -L*). Some of the unknown-function *ycf* genes are shared with rhodophytes. Interestingly, glaucophyte plastid genomes also lack the otherwise well-conserved gene for the photosystem quality control and division protease *ftsH*, though they contain other division-related genes *ftsQ* and *ftsW*, the latter shared with some early-branching Viridiplantae. Glaucophytes and a handful of Viridiplantae encode a rubrerythrin (*rbr*A) that functions as a ferredoxin:NADPH reductase-dependent peroxidase.

Viridiplantae chloroplasts ancestrally contain the lowest number of genes among Archaeplastida. The most apparent difference is the number of ribosomal proteins (25 in at least three classes, as opposed to 47 in rhodophytes), photosystems, ETC, and ATPase subunits (25, 6, and 6, respectively, as opposed to 36, 10, and 8 in rhodophytes). The small subunit and regulatory proteins (*rbcS, rbcR, cbbX*) of the key carbon fixation enzyme, ribulose-1,5-bisphosphate carboxylase/oxygenase (RuBisCO), were also transferred to the host nucleus. Multiple housekeeping genes were transferred or lost as well, including ones encoding translation factors (*infB, -C, tsf*), protein turnover chaperones (*dnaK, clpC, groEL*), translocases (*secA, -Y, tatC*), and ferredoxin-thioredoxin relay (*ftrC, dsbD, trxA, bas1*). None of the amino acid biosynthesis enzymes remained encoded in the genome, and only the gene coding for acetyl-CoA carboxylase subunit D is widely found. LIPOR subunits are conserved to various levels and lost in flowering plants and two chlorophyte lineages (Picocystophyceae and Pedinophyceae). Notable synapomorphies of the chlorophyte+streptophyte lineage include the Ycf1 and Ycf2 inner-membrane translocase components, which are not easily found in cyanobacteria, rhodophytes, or glaucophytes. In fact, these genes represent members of the Tic20/Ycf60 and FtsH protein families, respectively, so divergent that they were considered horizontal gene transfers [60, 61]. Intron maturases have been acquired by early Streptophyta, Pyramimonadophyceae, and a few other chlorophyte classes, leading to various levels of intron accumulation and consequent genome size expansion [62].

Within Viridiplantae, gene losses follow five major patterns. There are 60 highly conserved protein-coding genes (Fig. 4). Several genes are only found in unicellular algae and absent in multicellular streptophytes (*minD*, *ftsH*, *rps9*), or lost in vascular or higher plants (*tilS, rpl5, -12, -19, -36, tufA, chlI, -L, -N, -B, psb30, ycf20*). There are also genes only retained in streptophytes (*rpl21, -33, rps15, -16, -36, psbG, odpB, ycf66*), and the distribution of some genes is patchy (*ftsI, -W, rpl22, secG, cysA, -T, petN, ndhA-L, rbrA*). We did not find genes that would be commonly present in Chlorophyta and absent in Streptophyta. The loss of the Ndh complex genes in various lineages could be attributed to the presence of alternative pathways, such as type II NAD(P)H dehydrogenase or PGR5/PGRL1 [63, 64]. Similarly, the loss of LIPOR has been linked to functional replacement by its light-dependent analog POR [65]. The distribution of other genes suggests gene transfers and random losses.

Paulinellidae Outside Archaeplastida, there is another case of a more recent primary plastid endosymbiosis. Genus *Paulinella*, photosynthetic amoebae from Rhizaria, possess a primary plastid, also called the chromatophore, from an endosymbiotic event that happened 90–140 million years ago [66]. The cyanobacterial origin of the endosymbiont still manifests as the bacterial peptidoglycan cell wall, located between its two surrounding membranes, and as carboxysomes and phycobilisomes. On the other hand, *Paulinella*'s chromatophore divides with the host cell and is unable to survive outside the host (reviewed in [67]). The majority of the proteins imported into the chromatophore are of ancestral host origin [25], which ensures the dependence on the host cell.

The main reduction of the chromatophore genome occurred before the diversification of photosynthetic *Paulinella* species from their common ancestor ca. 60 million years ago. Only ~35% of gene families remained in the genome [68], corresponding to a reduction from ~2–5 Mbp genomic DNA in free-living alfa-cyanobacteria to 0.97–1.02 Mbp with 867–915 protein-coding genes of *Paulinella* endosymbionts' single circular chromosomes [17, 68]. Gene reduction mainly concerns losses of biosynthetic pathways of some amino acids and cofactors, the tricarboxylic acid cycle, environmental information processing, and ABC transporters [25, 68].

Cryptophytes, Haptophytes, Stramenopiles, and Chromerids Four of the rhodophyte-derived "CASH" algal lineages (cryptophytes, stramenopiles, haptophytes, and chromerids) contain plastids with four-membrane envelopes, so the first and second outermost membranes represent novel barriers for proteins targeted to the core plastid. Topologically, the endosymbiont/plastid (enclosed by three membranes – two innermost membranes of the former cyanobacterium and the plasma membrane of the engulfed alga) resides inside the host's endomembrane system, as the outermost membrane is contiguous with the ER and studded with ribosomes. The translocation across the first plastid membrane is therefore typically facilitated by the addition of a signal peptide to a protein's N-terminus, which is a common motif for proteins entering the endomembrane system to many final cellular destinations. Notably, the translocation across the second membrane was resolved in cryptophytes by the recruitment of the endosymbiont's ERAD complex (ER-associated protein degradation) for membrane translocation. The neofunctionalized machinery was named SELMA (symbiont-specific ERAD-like machinery) and the phylogeny of its central component, the Cdc48-ATPase, suggests it was repeatedly recruited also during subsequent red-derived endosymbioses [69].

Cryptophytes were therefore likely the first to obtain a complex plastid of rhodophyte origin. The plastids of these unicellular algae harbor a nucleomorph located between the outermost and innermost pairs of plastid membranes, with a genome organized into three small linear chromosomes [51, 70, 71]. Cryptophytes encode a phycobiliprotein (phycoerythrin), which constitutes a low-molecular-weight secondary light-harvesting complex, though phycobilisomes are not formed as in rhodophytes or glaucophytes, and thylakoids are stacked in doublets. Photosynthetic pigments include Chl *a* and c_2 and alloxanthin. The (cyanobacterial) plastid genome features of photosynthetic cryptophytes vary quite little. The genome size ranges from 121 kbp in *Guillardia theta* to 161 kbp in *Cryptomonas borealis*, with 32–38.1% GC. There are 142–149 protein-coding genes in photosynthetic species, the main differences being the presence of LIPOR only in *Storeatula* sp. and *Cryptomonas borealis* (Fig. 4) [72]. Losses compared with Rhodophytes, the assumed direct plastid donors, include most of the genes for the synthesis of chlorophyll, amino acids, fatty acids or cofactors, and a handful of genes involved in translation (tRNA synthetases, *rpl9*, *-28*, *rps1*, *infC*). Notably, several genes were retained from rhodophytes that encode two-component His-to-Asp sensory kinases and effector proteins of the Tsg1/Trg1 type (*dfr*, *ompR*, *ycf19*, *-29*, *-55*), however, their regulatory cues have not been determined [73–75].

Haptophyte plastids are possibly most akin to those of cryptophytes [76] and similarly rather uniform. Haptophytes usually possess two chloroplasts that are broadly pigmented with Chl *a* and $c_{1\text{-}2}$, fucoxanthin, and its haptophyte-specific derivatives 19′-hexanoyl- and 19′-butanoyloxyfucoxanthin (reviewed in [77]). The chloroplast genome lengths are in the range 95–107 kbp in *Diacronema lutheri* and *Phaeocystis globosa*, respectively, with 35.4–37.2% GC and 104–114 protein-coding genes (Fig. 2) [74]. Consistent with their descent from cryptophyte chloroplasts, haptophyte plastid genomes encode fewer genes, with losses involving genes for DNA replication, translation (ten *rpl*'s, *rps20, infB, tsf*), photosystem subunits (*psaE, -K, psbW, petF, ycf33, -37*), phycobiliproteins, and primary metabolite synthesis (*ilvB, -H, acpP*). The two-component system, as found in cryptophytes, is retained.

Ochrophytes, that is, the plastid-bearing subgroup of Stramenopila, employ in their plastids a significant fraction of proteins that have a chlorophyte phylogenetic signal, although their genomes are clearly rhodophyte derived [32, 78]. This led to speculations about a serial green-then-red endosymbiosis deep in the evolution of the lineage. Ochrophytes encompass a wide diversity of marine and freshwater algae, including diatoms, kelp, golden algae, and pelagophytes, many with commercial or ecological significance. Accordingly, their chloroplasts differ in their genome architecture and biological complexity. Photosynthetic plastids of ochrophytes utilize Chl *a* and c_{1-2} and fucoxanthin, similarly to haptophytes. Their genome size is in a broader range than in the other CASH lineages—it varies from 90 kbp in *Aureococcus anophagefferens* (Pelagophyceae) to 249 kbp in *Fibrocapsa japonica* (Raphidophyceae), with GC content 28–36% and 101–142 protein-coding genes (Fig. 2). This is largely attributed to differences in the extent of EGT and the length of introns and intergenic regions, although a comprehensive ochrophyte-wide analysis has not been conducted. Eustigmatophyceae and Synurophyceae are the most divergent in phylogenomic analyses, but genome size, architecture, and content are comparable within these lineages [79, 80]. Similarly, Phaeophyceae are quite uniform in genome architecture, although some rearrangements have occurred [81]. Dictyochophyceae genomes are more diverse in genome size and content and show numerous rearrangements involving IR loss and flip to tandem repeats [82]. Chrysophyceae are sister to Synurophyceae but their genomes are highly rearranged, probably owing to their various trophic strategies [83]. Plastid genomes of diatoms were found particularly labile in terms of size, structure, and gene content [84]. Compared with rhodophytes, multiple genes were lost in all ochrophytes, encoding proteins partaking in translation (*rne, tilS, syh, rpl28, infB, -C*), metabolism (*ycf53, pbsA, pgmA, odpA, -B, cemA*, amino acid synthesis except *ilvB* and *-H*, fatty acid synthesis except *acpP*), transporters (*mntA, -B, bioY*), photosynthesis (phycobilin cluster, *ycf20, psaK*), and multiple *ycf*'s (Fig. 4). There are 105 highly conserved genes in ochrophytes, largely overlapping the gene complement in the most gene-poor class Pelagophyceae. Several genes were differentially lost across various classes including housekeeping (*ftsH, dnaB, rpoZ, syfB, rpl9, -18, -32, rps1, -6, ycf65, tsf, clpN, secG, ycf60*), metabolic (*acsF, ycf54, chlL, -N, -B, rbcR, ilvB, -H, acpP*), cofactor (*thiG, -S*), and photosynthesis genes (*psbW, -Z, ycf39, petF, -J, psaI, ycf33, -37, ftrB, bas1*). Three *ycf* genes are broadly specific to ochrophytes (*ycf66, -88, -90*, although the first one is also found in Streptophyta). Of the two-component system, only *ycf19* was retained in multiple classes, and some Raphidophyceae retained the *dfr-ompR* pair. The tRNA synthetase *syfB* gene is particularly persistent in multiple classes, possibly because a nuclear replacement is unsuitable or unavailable [84]. The acquisition of sequence-specific recombinases (*serC, tyrC*, and *xerC*) and intron maturases in Raphidophyceae, Dictyochophyceae, and diatoms coincides with genome expansions. Notable is a lateral transfer of a streptophyte-related gene *cemA* into Synurophyceae [80].

"Chromerids" refers to a polyphyletic grouping of two photosynthetic algae within Apicomplexa sensu lato [85]. Ultrastructurally, the plastids of *Vitrella brassicaformis* and *Chromera velia* are similar, with four-membrane envelopes and triplet-stacked thylakoids [86]. The chloroplast genomes of chromerids are, however, strikingly different in both architecture and gene content [87]. The genome of *V. brassicaformis* is compactly circular and canonical in structure, 85 kbp long, unusually GC-rich (47.7%), and encoding 74 standard-length protein-coding genes, 35 tRNAs, and two ribosomal operons as IRs. The genome of *C. velia* has an unusual linear topology with terminal inverted repeats, standard GC content (36.6%), and in 120 kbp encodes only 57 proteins, 28 tRNAs, and a single rDNA operon. Genes missing in *C. velia* encode ribosomal proteins, LIPOR, and photosystems/ETC subunits. Uniquely among algae, except dinoflagellates with which they share ancestry, chromerids employ transcript 3′-end polyuridylation, which is apparently not biased toward a specific gene function.

Dinoflagellates Arguably, dinoflagellates showcase the possibilities in endosymbiont acquisition. The sister group of extant dinoflagellates, Perkinsozoa, retain a non-photosynthetic plastid, and so do their earliest branching lineages *Oxyrrhis marina* and Syndiniales, except *Hematodinium* that lost its plastid altogether. Although dinoflagellates are considered one of the CASH lineages [13], their plastids are remarkably different. The majority of "core" dinoflagellates contain a peridinin-type plastid, named after a signature dinoflagellate apocarotenoid that enhances the absorption capacity of their light-harvesting complexes in the blue-green spectrum [88]. They contain triplet-stacked thylakoids and a pyrenoid. Storage polysaccharides are synthesized in the host cytosol. Peridinin plastids are surrounded by three membranes, as opposed to four in other rhodophyte-derived CASH plastids, and the outermost membrane is not continuous with the ER. The mechanism for protein translocation across the outermost membrane is largely unknown, but after signal peptide cleavage on the ER membrane, plastid-targeted proteins remain docked in the membrane by a transmembrane domain, with a leading transit peptide-like sequence exposed to the lumen of the transport vesicle [89].

On the molecular level, peridinin plastids are extremely divergent from other conventional plastids. All but ~17 genes have been lost from the plastid genome, highlighting the importance of host control in the plastid's biology. Of the 17 genes, 12 encode photosynthesis proteins (*psaA, -B, psbA, -B, -C, -D, -E, -I, petB, -D*, and *atpA, -B*), three are tRNAs (*trnM, trnP* and *trnW*) and two are rRNAs (LSU, SSU) [90]. These genes are dispersed on small circular chromosomes, or minicircles, which are 2–3 kbp in size, each usually encoding a single gene, although some contain more genes, while others are "empty" or otherwise aberrant (reviewed in [91]). Besides genes, minicircles have a conserved core region considered to be the origin of replication and transcription [90]. More complicated, digenic ~10 kbp minicircles with multiple hairpin structures have also been found, raising questions about recombination playing a role in their replication [92]. Two notable post-transcriptional modifications occur in peridinin plastids, 3′-end polyuridylation and RNA editing. Poly(U) tails are added to all protein-coding transcripts, possibly marking full-length transcripts. The machinery is probably homologous to that of chromerids. Transcript editing seems limited to dinoflagellates (except one basal lineage, *Amphidinium*; [93]) but its extent (hundreds of edited sites) and range (up to nine different substitutions, including transversions) are unmatched in plastid-bearing lineages (reviewed in [94]). There are apparently no strict mechanisms to coordinate the replication and partition of minicircles into daughter chloroplasts with the division of the host, and their numbers can lower drastically in fast-growing cells [95]. Although peridinin plastids have been around for >250 My [5, 96] because of their extraordinary fragmentation they are thought to be prone to losing photosynthesis, particularly if the host inclines to mixotrophy.

Many dinoflagellates resolved the problems of maintaining the peridinin plastid by kleptoplasty, that is, a strategy when the plastid is stolen from a photosynthetic alga and transiently maintained (reviewed in [97]). Best-known examples of kleptoplastic dinoflagellates include the obligately mixotrophic *Dinophysis* spp., *Noctiluca scintillans*, and the as-yet undescribed "Ross Sea Dinoflagellate" (RSD), which cannot survive without regularly replacing their kleptoplasts at least once in several generations. *Dinophysis* uses cryptophyte plastids stolen from a ciliate *Mesodinium rubrum* [98]; in *Noctiluca*, the kleptoplast is an intact green alga *Pedinomonas noctilucae* [99]; and in RSD, the kleptoplast is from a haptophyte *Phaeocystis antarctica* [100].

In other dinoflagellate lineages, stable endosymbionts were recruited as a stand-in for their deteriorating peridinin plastid (Fig. 1). Serial replacement events have been documented for three distantly related lineages. *Durinskia*, *Kryptoperidinium*, and *Peridiniopsis* (Peridiniales:Kryptoperidiniaceae), also called dinotoms, possess a diatom endosymbiont (hence the name) that divides synchronously with the host. The endosymbiont is bound by a single membrane and remarkably intact, besides a typical four membrane-bound diatom plastid also retaining a reduced nucleus and mitochondria [101]. The structural integrity of the dinotom endosymbionts and their frequent replacement by other diatom lineages [102] suggest that this is an early-stage endosymbiosis. *Lepidodinium* spp. (Gymnodiniales: Gymnodiniaceae) and two other taxonomically unassigned dinoflagellates possess green algal endosymbionts that form a monophyletic group sister to Pedinophyceae [103, 104]. The endosymbionts have four surrounding membranes, with the innermost two corresponding to green algal plastid membranes. Reduced nuclei and free ribosomes have been observed in the endosymbionts' cytoplasm [103, 105]. *Karenia* and *Karlodinium*, which are related to RSD (Gymnodiniales:Kareniaceae), possess an endosymbiont of haptophyte origin. This endosymbiont is reduced to a "naked" plastid, that is, it is devoid of other organelles and has three membranes, analogously to peridinin plastids (reviewed in [106]). The pigment composition of serial replacement plastids in dinotoms, "green" dinoflagellates, and Kareniaceae is more similar to each respective endosymbiont donor than to peridinin plastids [107–109]. A vestigial peridinin plastid is nevertheless often retained in the hosts as the eyespot [110].

Genome reduction in dinoflagellates with replacement chloroplasts mirrors these structural observations. The mitochondrial and chloroplast genomes in dinotoms are hardly reduced and without any apparent gene losses, consistent with early-stage endosymbiosis [111, 112]. The chloroplast genomes of green endosymbionts lost 12–18 genes compared with Pedinophyceae, eight of which are missing in all three (*minD, tilS, infA, cysA, -T, ccsA, accD, cemA*) (Fig. 4). Transcript editing has been observed in the green plastids, most likely grafted into the plastids by the host-encoded machinery previously operating in the peridinin plastid [104]. From Kareniaceae, only the plastid genome of *Karlodinium veneficum* is available as a long contig. At least 39 genes have been lost from this genome since the haptophyte's acquisition, and an additional 22 have in-frame stop codons or frameshifts, some of which are known to undergo transcript editing to compensate for these mutations [113, 114]. At least one gene, *dna*K, is encoded by a minicircle [115]. All and all, dinoflagellate hosts have an unusually strong impact on the evolution of their plastids. This impact is imposed by host-encoded proteins that both allow and compensate for deleterious effects on the plastid genomic DNA, which is why serial plastid replacements are facing similar fragmentation pressure as the peridinin plastids [94].

Euglenids The plastid of euglenids formed ca. 539 million years ago from a green alga related to Pyramimonadophyceae [116, 117]. Thylakoids are triple-stacked and a pyrenoid is often present. Photosynthetic pigments are Chl *a* and *b*. The plastid envelope consists of three membranes, the outermost not continuous with the ER. Protein import into euglenid plastids includes a vesicular route similar to that of peridinin dinoflagellates, possibly via the Golgi but not involving SNARE proteins [118]. The process of translocation is not well described; however, in *Euglena gracilis, E. longa,* and *Eutreptiella* spp., the TIC/TOC pathway was found greatly reduced, with no TOC subunits observed in the middle membrane, and only a limited number of TIC subunits in the innermost membrane [24, 48].

Some euglenids have smaller plastid genomes, such as *Phacus inflexus* with 58 kbp, while other genomes are larger, such as *Euglena gracilis* with 143 kbp, although they consistently encode 55–61 proteins (Fig. 4). The length range is wider in comparison with known free-living relatives of their plastid, Pyramimonadophyceae (85 and 102 kbp, 75 and 83 protein-coding genes, in *Cymbomonas tetramitiformis* and *Pyramimonas parkeae*, respectively). However, euglenid plastid genomes contain many introns, whereas that of *P. parkeae* contains only one, and *C. tetramitiformis* lacks introns altogether [116]. The loss of the NDH complex accounts for most of the differences between euglenid and pyramimonad chloroplast genomes. Further losses include genes encoding housekeeping functions (*ftsH, infA, clpP, ycf1B*), photosystem subunits (*ycf20, ycf3*), and electron transfer (*petA, -D, -N*). From chlorophyll synthesis, only *chlI* remains in euglenid plastid genomes.

Chlorarachniophytes The morphologically diverse chlorarachniophytes contain a green-derived chloroplast surrounded by four-membrane envelopes with a nucleomorph between the second and the third membranes. The specific origin of this green plastid, acquired ca. 318 million years ago, remains unclear; however, the endosymbiont ancestor was a relative of the siphonous algae Bryopsidales (Ulvophyceae) [117, 119]. Chloroplasts usually contain many appressed thylakoids, a pyrenoid, and Chl *a* and *b*. The outermost membrane is not continuous with the ER and the plastid protein import is not well characterized. Eleven subunits of the TOC/TIC complex have been identified to localize in the plastid of *Bigelowiella natans*; however, how proteins translocate through the two outermost membranes remains uncertain [120].

The size of chlorarachniophyte plastid genomes varies quite little, from 67 to 73 kbp in *Gymnochlora stellata* and *Partenskyella glossopodia*, respectively, with 57 or 58 protein-coding genes. The chlorarachniophyte plastid genomes are quite compacted due to shorter intergenic regions and fewer genes and introns, especially in comparison with the average in their ancestor group, Ulvophyceae, which is 150 kbp with 74 protein-coding genes [119]. During the endosymbiosis, gene losses occurred in the synthesis of chlorophyll (*chlL, chlN, chlB, ccsA*), fatty acids (*accD*), and FeS clusters (*cysA, cysT*). Some reduction is observed in photosystems and electron transfer (*psbZ, psb30, ycf20, psaI, psaM, petL*), ribosomal proteins (*rpl12, rpl32, rps9*), translation (*infA*), and division (*ftsH*).

Many Shades of Plastids Above, we elaborated on how different eukaryotic lineages acquired their chloroplasts. The convoluted evolutionary history of these chloroplasts left genetic footprints in their genomes, some obvious, some more subtle, allowing us to reconstruct these events. Yet, not all of these partnerships endured in the long term. In the next part, we explore the diversity of organisms, for which photosynthesis was no longer advantageous.

4 Plastid Genome Reduction in Nonphotosynthetic Lineages

With the notable exception of glaucophytes, haptophytes, and chlorarachniophytes, most plastid-bearing lineages include secondarily nonphotosynthetic representatives. As discussed above, loss of photosynthesis is tightly linked to changes in feeding habits and therefore constrained by the ability of the host to obtain nutrients via predation, parasitism, or osmotrophy. Ultrastructurally, nonphotosynthetic plastids are markedly reduced, appearing as inconspicuous vesicles lacking thylakoids, and they are primarily recognized by the number of envelope membranes, which seems to correspond to their photosynthetic ancestors [86, 110, 121]. On the molecular level, this is a continuation of the reductive process, with large numbers of photosynthesis-related genes quickly lost or pseudogenized, thus similar to the initial phase of endosymbiosis [45–48]. The remaining plastid gene content seems strongly lineage-dependent, and so is the tendency of nonphotosynthetic descendants to remain heterotrophic or parasitic (Fig. 5a). The extent of plastid genome reduction therefore varies, and a division into four nonphotosynthetic classes ("CHER") was suggested by Maciszewski and Karnkowska [122], named after species representing each class: *Cornospumella*, possessing an almost full set of plastid genes, including photosynthesis-related genes; *Helicosporidium*, whose plastid genes were partially transferred to the nucleus; *Eimeria*, with most of the genes lost;

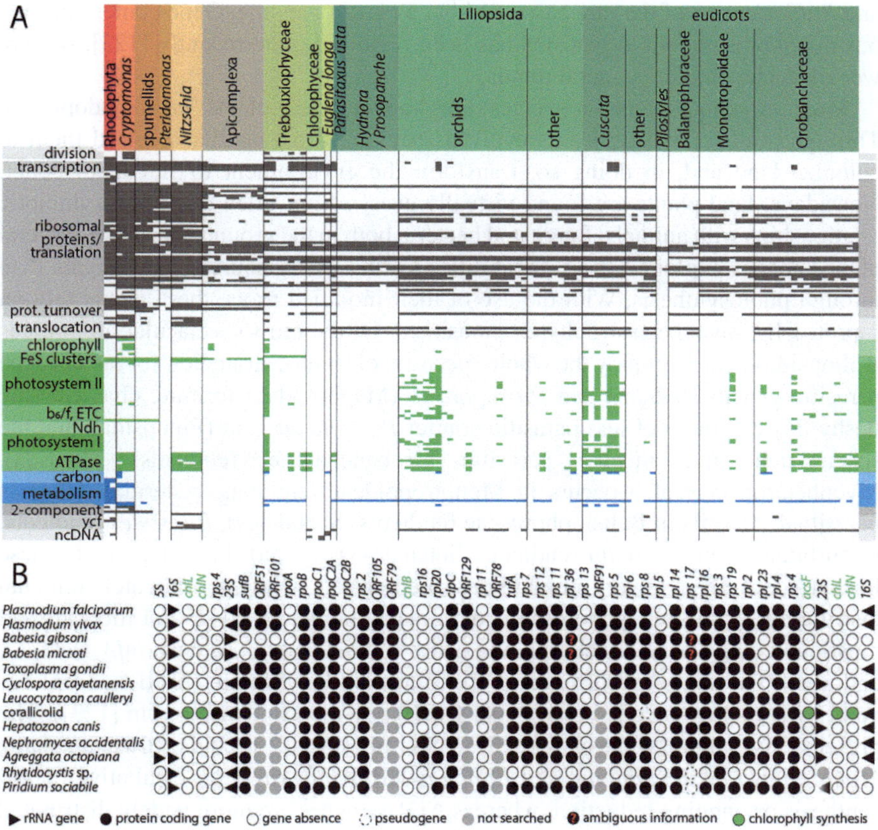

Fig. 5 Nonphotosynthetic plastid gene complements. Panel A shows the distribution of all plastid genes found at least once in nonphotosynthetic algae and plants. Columns represent individual species. Only two genomes are available for Rhodophytes (*Choreocolax polysiphoniae* and *Pterocladiophila hemisphaerica*) as many parasitic rhodophytes are kleptoplastidic. Nonphotosynthetic Trebouxiophyceae include the genera *Auxenochlorella* (a facultative heterotroph), *Helicosporidium,* and *Prototheca*, nonphotosynthetic Chlorophyceae include *Leontynka*, *Polytoma*, and Volvocales sp. NrCl902. Magnoliids are represented by the conspicuous parasites *Hydnora* and *Prosopanche*. Liliopsida include many species of orchids and mycoheterotrophic plants *Burmannia*, Corsiaceae, *Geosiris*, *Petrosavia*, and *Sciaphila*. Eudicots include (semi)parasitic and mycoheterotrophic groups *Cuscuta*, Balanophoraceae, Monotropoideae, Orobanchaceae, and less-known genera *Cynomorium*, *Cytinus*, *Epirixanthes*, *Lennoa*, *Pholisma,* and *Pilostyles*. Photosynthesis-related and metabolic genes are shown in green and blue, respectively. This highlights the occurrence of hemiparasitic and parasitic plants (orchids, *Cuscuta*, *Aphyllon californicum*, *Monotropastrum*, *Lathraea squamaria*), suggesting ongoing reduction or perhaps cryptic photosynthesis. The widespread occurrence of ATPase genes could be conditioned by the need to maintain thylakoids and/or membrane potential, similarly to *Cryptomonas* and *Nitzschia*. Metabolic genes *rbcL* and *accD* are also quite widespread in nonphotosynthetic algae and plants. Among noncoding nucleic acid-interacting proteins (ncDNA), MatK is most frequently encoded. Panel B shows a presence–absence matrix for genes in Apicomplexa. Genes are ordered according to *Plasmodium*

and *Rafflesia*, one of the rare cases without a plastid genome. Gene content in various nonphotosynthetic organisms has been reviewed quite recently [123], so here we will focus on more general trends.

Most nonphotosynthetic species are known from plants and rhodophytes. Through multicellularity, plants evolved specialized organs that allowed them to colonize land and, in doing so, transform the environment to suit them better. Nowadays, land plants are found virtually everywhere, often exploring symbiotic relationships with animals, fungi, and bacteria both in the ground and above. Several plant lineages exploited these capabilities to find easier sources of nutrients than through photosynthesis. With the use of their modified roots, these plants rely on saprotrophy, mycoheterotrophy, or parasitism. Of the most spectacular are orchids (Liliopsida), which employ the whole spectrum of trophic strategies, for instance the parasitic genera *Hydnora* and *Prosopanche* (Magnoliidae) forming characteristic fleshy flowers, the root-less parasitic conifer *Parasitaxus usta* (Pinopsida), and the species-rich and mostly parasitic Orobanchaceae (eudicots: Asterida). Mycoheterotrophy also occurs in Monotropoideae (eudicots: Asterida), whereas parasitism prevails in Balanophoraceae (eudicots: Santalales), *Pilostyles* (eudicots: Cucurbitales), and *Cuscuta* (eudicots: Solanales) (Fig. 5a). Plastid genes in these lineages converged on very similar housekeeping sets (with approximately bimodal distribution, 16–18 in Hydnoroideae and Balanophoraceae, 20–25 in Liliopsida and most eudicots). These largely include genes for ribosomal proteins, *infA*, *ycf1*, *-2*, and *clpC*. The sparse occurrence of *rpo* genes is likely due to functional replacement by a nuclear-encoded phage-like (NEP) polymerase (reviewed in [123]). The presence of photosynthesis-related genes in several orchids, *Cuscuta* spp., *Monotropastrum sciaphilum*, and *Epifagus virginiana* suggests facultative photosynthesis or ongoing reduction, whereas ATPase genes are more widely distributed and perhaps required to maintain membrane potential for the Tat and Sec protein import systems [124]. From metabolic pathways, *accD* is highly conserved owing to the localization of de novo fatty acid synthesis in plant plastids, whereas *rbcL* largely matches the distribution of photosynthesis-related genes (Fig. 5a), except for *Aphyllon* spp., *Petrosavia stellaris*, and *Pholisma arenarium*, where its function remains unknown. Intron maturase MatK is commonly encoded by less reduced genomes. The absence of a plastid genome in *Rafflesia* (eudicot) is controversial, as one of the essential pathways localized in plastids, tetrapyrrole synthesis, uses plastid-encoded tRNA-Glu as a precursor [50].

Rhodophyte parasites have evolved dozens of times independently. They are quite diverse but exploit the molecular machinery of closely or more distantly related relatives within Florideophyceae [46]. As a result, they do not typically possess plastids because they steal them from their host. Only two plastid sequences are available from rhodophyte parasites that appear to have their own plastids (*Choreocolax polysiphoniae* and *Pterocladiophila hemisphaerica*). These are rich in housekeeping genes (54 in total, encoding RNA-polymerase subunits, ribosomal proteins, components of protein turnover and translocation, and two-component-type regulators) and retain several metabolic genes of amino acid and fatty acid pathways. Three genes have photosynthesis-related functions, although they are probably involved in metabolic redox reactions; these are coding for ferredoxin (*petF*) and proteins that provide it with an Fe-S cluster (*sufB, -C*) (Fig. 5a).

The plastid gene content of nonphotosynthetic *Cryptomonas* spp. (Cryptophyceae), spumellids (Ochrophyta: Chrysophyceae), and *Nitzschia* spp. (Ochrophyta: diatoms) largely matches that of the two rhodophyte parasites. Most genes encode housekeeping factors. ATPase subunits are conserved in *Cryptomonas* spp. and *Nitzschia* spp., consistently with the presence of genes encoding Tat and Sec subunits (*tatC, secA, -G, -Y*), and *sufB/C* are conserved in all three groups, though only cryptomonads and spumellids encode *petF* (Fig. 5a). Cryptomonads consistently encode four metabolic proteins (*ilvB*, *-H*, *acpP*, *cemA*), *Cryptomonas paramecium* uniquely encodes both RuBisCO subunits and one of its regulators (*cbbX*), whereas two other *Cryptomonas* spp. encode LIPOR subunits, which *C. paramecium* does not. Importantly, individual strains of *Cryptomonas* spp., spumellids, and likely also *Nitzschia* spp. are not directly related, and lost their photosynthetic abilities independently, arriving at equivalent gene complements by convergence [121, 125, 126]. *Pteridomonas* spp. (Ochrophyta: Dictyochophyceae) are more distantly related to *Nitzschia* spp. and spumellids, and were only found to encode housekeeping genes, suggesting that their plastid genome is retained solely to provide tRNA-Glu for tetrapyrrole synthesis [127].

Similarly to their rhodophyte-derived counterparts, nonphotosynthetic green algae and euglenids converge on a similar housekeeping gene content as higher plants (23–32 genes). The absence of *rpo* genes in some Trebouxiophyceae (*Prototheca bovis* and *P. ciferrii*) calls for a targeted search for a nuclear-encoded replacement in these species. The only other conserved genes are *accD* and *cysT* in Trebouxiophyceae. Photosynthesis-related genes are only present in the facultatively heterotrophic *Auxenochlorella protothecoides*, whereas ATPase genes are found in about half of the available species. *Euglena longa* lost all photosynthesis-related genes except *rbcL*, for which a unique redox-balancing function has been proposed [128].

Apicomplexa sensu Cavalier-Smith [2018] consist of two main subgroups. Sporozoa, or Apicomplexa sensu stricto, are obligate symbionts, typically parasites of animals, with >6000 described species including medically and economically important representatives (e.g., *Cryptosporidium*, *Eimeria*, *Plasmodium*, *Theileria*, and *Toxoplasma*) in four major clades (Hematozoa, Coccidia, Marosporida, and gregarines). Apicomonada, or chrompodellids, include photosynthetic chromerids, predatory colpodellids, and symbiotic species, such as parasitic *Piridium sociabile* [129] and the presumably mutualistic symbiont X [130]. Most apicomplexans contain complex plastids, which are photosynthetic in chromerids, whereas others harbor vestigial plastids with (Sporozoa, *P. sociabile*) or without (colpodellids) genomes [47, 131], pointing to a lasting dependency on plastid-localized pathways for tetrapyrrole, isoprenoid or fatty acid synthesis [132–134] (see also Chap. 7). Meanwhile, *Cryptosporidium* spp. and some gregarines lost their plastids entirely [135, 136].

According to their phylogeny, apicomplexans lost photosynthesis multiple times, and multiple times also became parasitic. The similarity of plastid genomes and inferred proteomes of Sporozoa and *P. sociabile* suggests strong convergent evolution (Fig. 5b). The plastids of Apicomplexa, or apicoplasts, have circular genomes of ca. 35 kbp with minor lineage-specific gene losses and rearrangements (Fig. 5b). Apicoplast-to-nucleus gene transfer might be reduced by the "limited transfer window," as only one apicoplast is present per cell. The main differences are in genes coding for short ribosomal proteins, *sufB* (missing in *Babesia*), *rpoA* (only present in *Piridium*), rpoC2B (a copy of rpoC2 specific to *Cyclospora* and *Leucocytozoon*), and several ORFs with unclear homologies. Most apicoplasts have two ribosomal RNA operons encoding 16S and 23S rRNA, though *Babesia* spp. and *Rhytidocystis* sp. have one. 5S RNA is rarely encoded, only in corallicolids and two Marosporida (*Aggregata octopiana* and *Merocystis kathae*). The most gene-rich are the genomes of corallicolids and *P. sociabile* (32 and 27 protein-coding genes, respectively), whereas the most gene-poor is the genome of *Rhytidocystis* sp. (Marosporida) with merely 8 protein-coding genes [137]. Notably, corallicolids also encode four genes involved in chlorophyll synthesis (Fig. 5b), although biochemical evidence for chlorophyll synthesis is missing, and their function remains unknown [138].

Secondary loss of photosynthesis is quite common compared to chloroplast endosymbiosis, although both are strongly lineage-specific and likely affected by trophic strategies. For instance, dinoflagellates have mastered predation for algae and increased their prey's longevity, whereas apicomplexans sensu lato rarely keep photosynthetic plastids. Achlorophyllous plants acquire nutrients through mycorrhizal interactions or parasitism, parasitic rhodophytes exploit the vulnerability of their photosynthetic relatives, and unicellular nonphotosynthetic algae ultimately choose between predation and parasitism. From what we know, it has become clear that general rules apply to both photosynthetic and nonphotosynthetic plastids, as well as other endosymbionts. Genome reduction occurs more suddenly during transitions to photosynthetic or nonphotosynthetic lifestyle, followed by later sporadic gene losses and transfers. Accordingly, despite their common origin, plastids from various eukaryotic lineages underwent numerous host-specific modifications reflected by losses and, sometimes, gains of genes. Although there is no clear evidence for the independent origin of plastids in rhodophytes and Viridiplantae, and therefore the role of convergence in shaping primary plastid genomes, all nonphotosynthetic plastid genomes arrived at similar gene composition, driven by plastid-mediated dependency, with fine differences that highlight the stochastic dimension of the process. In the next part, we delve into the phenomena particularly connected to randomness, as they represent unique departures from general rules of plastid architecture and biology.

5 Special Cases

Noncanonical Topologies Even though plastid genomes usually assemble as circular quadripartite chromosomes (with two single copy regions and two repeats), in vivo they appear in various supramolecular conformations. In developing plant leaves, for instance, chloroplast DNA concentrates in entangled multigenomic bodies, which become progressively fragmented and degraded by the end of leaf maturation [139]. In *Acetabularia* (Chlorophyceae), small minicircular molecules of DNA coexist with the large chromosomal plastid genome [140]. In diatom and rhodophyte chloroplasts, plasmids occasionally occur as independently replicating molecules, and their fragments are found integrated into chloroplast genomes [111, 141, 142]. Nevertheless, there are several examples of plastid genomes that substantially differ in their architecture and never occur in a conventional form. These include minicircles in dinoflagellates, the linear chromosome in *C. velia*, as discussed above, and fragmented genomes that have been found in green algae. The first example is *Koshicola spirodelophila*, which has a genome split into three smaller circular chromosomes, totaling 385 kbp [143]. *Koshicola* belongs to the clade Chaetopeltidales and its closest relative with available sequence data, *Floydiella terrestris*, has a massively expanded plastid genome of 521 kbp, although their gene complement is almost identical [143]. Even dinoflagellates' minicircles however appear quite regular in comparison with another example, the hairpin chromosomes found in the chloroplasts of Cladophorales [144]. The sequencing of *Boodlea composita* revealed the presence of at least 34 short contigs (1.2–6.9 kbp) with very high GC content (54–60%), representing single-stranded DNA hairpins that encode 21 proteins and 16S rRNA, or their fragments. Fully photosynthetic, therefore, *Boodlea* converged on a gene set highly overlapping that of peridinin dinoflagellates. Similar fragmentation was found in nine other Cladophorales and seems to have occurred early in their evolution. Retrotransposon-like repeats in *Boodlea* suggest that this arrangement could have resulted from an introduction of nuclear retrotransposons into the plastid genome, leading to expansion and fragmentation by repeat recombination and cleavage [144]. Importantly, the chloroplast genomes of Trentepohliales, the sister group of Cladophorales, were found highly inflated and rearranged already, albeit still maintaining a single circular chromosome architecture [145].

Expanded Genomes Chloroplast genome deterioration often manifests as size inflation, generally due to the expansion of introns or repeats. Two types of introns are found in chloroplast genomes, group I, with self-splicing properties, and group II, which requires intron maturases for splicing. Both groups of introns may encode proteins that determine their mobility. Group I introns often encode a sequence-specific homing endonuclease, whereas group II introns employ reverse transcriptases, intron maturases, and endonucleases, but less frequently encode them (e.g., [62]). The largest known chloroplast genomes disclose the consequences of "junk" DNA inflation. In rhodophytes *Corynoplastis japonica* and *Bulboplastis apyrenoidosa* (1.127 Mbp and 610 kbp, respectively), the expansion is fueled by group II introns and transposable elements of possibly bacterial origin [58]. In chlorophytes, the expansion is mainly driven by intergenic repeats. In *Floydiella terrestris* (521 kbp) these repeats are >30 bp, dispersed, and heterogeneous [146], in *Volvox carteri* (525 kbp) they are palindromic and on average 50 bp long [147], and in *Acetabularia acetabulum* (estimated ca. 1 Mbp) they are tandemly arranged copies of a 10-kbp element [148]. Repeat expansions also occur in nonphotosynthetic Chlamydomonadales, and *Leontynka pallida* and *Polytoma uvella* have comparable plastid genomes (362 and 230 kbp, respectively) despite their reduced coding capacity [149, 150]. Notably, some repeats exhibit conserved positions at the 3′-ends of specific coding sequences and have been linked to modulation of transcript expression [74, 151]. Repeat expansion should not be confused with IR expansion, which refers to extending the IRs to include neighboring genes, possibly induced by recombination or double-strand breakage repair and gene conversion [62, 152]. As there is a wide range in IR gene content and size, they substantially contribute to genome size variation [62]. Yet, IRs have been only shown to expand in some lineages, for example, Viridiplantae and diatoms, whereas in others the IR boundaries remain conserved [74, 84].

Horizontal Gene Transfers HGT into plastid genomes are very rare, although they are a major cause of the acquisition of selfish introns. In 5 billion years of combined vertical descent, however, there are several documented cases of adaptive HGT, which are sometimes linked to thought-provoking evolutionary synapomorphies. Rhodophytes encode a proteobacterial RuBisCO distantly related to the gene found in glaucophytes and Viridiplantae [153]. This event implies an HGT if Archaeplastida are monophyletic, although it is also consistent with an independent origin of plastids in Rhodophyta [39, 54] (see also Chap. 1). Similarly, a horizontally transferred *rpl36* links the history of plastid genomes of cryptophytes and haptophytes [76]. More lineage-specific HGTs include a spirochete *clpC* and three originally mitochondrial ribosomal protein-coding genes transferred into the apicoplast genome; a firmicute *dnaX* in the genomes of cryptophytes; a batch of genes from Bacteroidetes located in the minicircles of dinoflagellates *Pyrocystis lunula* and *Ceratium horridum* (reviewed in [154]); the Bacteroidetes-derived *ebo* operon in Eustigmatophyceae [155]; a proteobacterial *leuC/leuD* gene pair in *Gracilaria tenuistipitata* [59]; *cemA* in Synurophyceae acquired from within Viridiplantae [80]; and large stretches of intergenic DNA in diatoms that have not been functionally characterized but appear to be of foreign origin based on elevated GC content [84]. Although evolutionary odds are clearly against stable HGT into plastid genomes [54], the growing list of adaptive transfers documents that HGT does occur, and indeed there are multiple events of horizontal transfer of selfish sequences [62, 156].

Easier Lost Than Found? Ultimately, reductive evolution leads to a complete loss of plastid genomes and plastids. Plastid genome losses are nevertheless quite uncommon, and only three undeniable plastid losses have been identified so far, cryptosporidia, gregarines (both Apicomplexa), and *Hematodinium* (Dinoflagellata) [135, 136, 157]. This resistance to a complete loss is connected to the host's dependency on the plastid itself rather than its genome. During endosymbiosis, many pathways have been grafted into this membrane-bound compartment (reviewed in Chap. 7), and their relocation to another would require concerted retargeting of all the protein components of that pathway. This could be insurmountable on evolutionary timescales. However, peridinin plastid minicircles only encode photosynthesis-related proteins and RNA required for their production, and therefore nonphotosynthetic dinoflagellates lose plastid genomes relatively more easily than the organelle. In chrompodellids, plastid genome dependency was lost with the relocation of *sufB* into the nucleus, and predatory colpodellids lack plastid genomes as well [47]. Plastid genome losses were also suggested in *Polytomella* (Chlorophyceae) and *Rafflesia* (eudicot), although their potential dependency on plastid tRNA-Glu for priming tetrapyrrole biosynthesis has not been resolved [158]. Clearly, algal hosts form with their plastids a deeply raveled entity, which makes later severance of this bond extremely difficult.

6 Future Prospects

Extensive comparative research of plastid genomes challenged our views on many aspects of this remarkable organelle and translated into a better understanding of other endosymbionts as well. More questions remain to be answered regarding genome architecture, gene acquisition and loss, and the history of the rhodophyte lineage itself. To fill these gaps, we need to look beyond the current scope, biased toward land plants and multicellular rhodophytes, and target plastid-bearing lineages that have been overlooked (Fig. 1). Many of these lineages are in fact just as ecologically and economically important, including diatoms, chrysophytes, dictyochophytes, marine chlorophytes, euglenids, and haptophytes [159].

In most cases, the physical structure of plastid genomes complicates the assembly of their full-length sequences. With short reads, it is often impossible to correctly assemble and place both IRs, particularly if they contain polymorphisms, and additional efforts using long PCR and Sanger sequencing are needed to circularize plastid sequences. With the wide use of long-read sequencing, plastid assemblies should become more feasible and accurate, which will in turn allow a better understanding of how IRs are stabilized on evolutionary timescales. For instance, haptophyte IRs are not identical and accumulate polymorphisms, questioning the stringency of IR recombination in this lineage [74].

Consistent gene prediction also remains a major problem, despite the use of specialized annotation tools such as MFannot. Some of the confusion comes from the dual naming of genes, for example, *psbN* is also known as *pbf1*, *secG* as *ycf47*, *ycf3* as *pafI*, *ycf4* as *pafII*, and *cp30* as *ycf60*. Alternative genetic codes may sometimes prevent the prediction of coding sequences, although these have been known only from a handful of plastids, notably apicomplexans and *Chromera velia*, Cladophorales, a chlorophyte *Jenufa minuta*, and a holoparasitic plant *Balanophora* [62, 86, 144, 160]. In addition, there are several conserved unknown-function proteins (Ycf's), along with less widespread predicted ORFs, encoded by plastid genomes that await characterization. Identifying distant homologies between these sequences and their counterparts in free-living organisms could help us understand which conserved or novel functions are important for plastid genomes to retain. For instance, the function and origins of essential *ycf1* and *ycf2* of Viridiplantae in *tic20/ycf60* and *ftsH* families, respectively, have only been unveiled recently [61]. Finally, a comprehensive determination of plastid proteomes in various algae is necessary to fully appreciate their complexity and bring more insights into the specific roles plastids have within their hosts and how they are integrated into regulatory networks.

7 Conclusions

The evolutionary history of plastids is truly colorful. Regardless if they arose from cooperation or manipulation, willingness or theft, with each endosymbiotic event, mechanisms had to be implemented to develop and maintain these organelles. This contributes substantially to the diversity of plastids in eukaryotes on structural and functional levels. We see that genome reduction closely accompanies plastid "life cycles" from their birth to disappearance. This process is partly adaptive and partly stochastic, dictated by selection on both functionality and streamlining, as well as randomness in gene replacement. The evolutionary histories of plastids and their recipients have a major role in shaping this reduction. All hosts independently developed a tool set to stabilize their endosymbionts and provide hundreds of proteins required for their function but not encoded in their genomes. Many of these symbionts suffered further accidental mutations in the process or carried a perilous burden represented by mobile DNA. As much as we have learned about plastids, however, there is more to be accomplished to evenly map the landscape of plastid biology in eukaryotes.

Acknowledgments The authors acknowledge funding from the Charles University institutional funding (Cooperatio Biology, SVV 260432/2018), the Czech Science Foundation (grant nos. 21-03224S and 23-07277S), and the Ministry of Education, Youth and Sports of the Czech Republic (MEYS-CR) within the Centre for Research of Pathogenicity and Virulence of Parasites (project No. CZ.02.1.01/0.0/0.0/16_019/0000759).

References

1. Flombaum P et al (2013) Present and future global distributions of the marine cyanobacteria *Prochlorococcus* and *Synechococcus*. Proc Natl Acad Sci U S A 110:9824–9829. https://doi.org/10.1073/pnas.1307701110
2. Rousseaux CS, Gregg WW (2014) Interannual variation in phytoplankton primary production at a global scale. Remote Sens (Basel) 6:1–19. https://doi.org/10.3390/rs6010001
3. Sánchez-Baracaldo P, Cardona T (2019) On the origin of oxygenic photosynthesis and cyanobacteria. New Phytol 225:1440–1446. https://doi.org/10.1111/nph.16249
4. Ponce-Toledo RI, Deschamps P, López-García P, Zivanovic Y, Benzerara K, Moreira D (2017) An early-branching freshwater cyanobacterium at the origin of plastids. Curr Biol 27:386–391. https://doi.org/10.1016/j.cub.2016.11.056

5. Parfrey LW, Lahr DJG, Knoll AH, Katz LA (2011) Estimating the timing of early eukaryotic diversification with multigene molecular clocks. Proc Natl Acad Sci U S A 108:13624–13629. https://doi.org/10.1073/pnas.1110633108
6. Eme L, Sharpe SC, Brown MW, Roger AJ (2014) On the age of eukaryotes: evaluating evidence from fossils and molecular clocks. Cold Spring Harb Perspect Biol 6:a016139. https://doi.org/10.1101/cshperspect.a016139
7. Strassert JFH, Jamy M, Mylnikov AP, Tikhonenkov DV, Burki F (2019) New phylogenomic analysis of the enigmatic phylum Telonemia further resolves the eukaryote tree of life. Mol Biol Evol 36:757–765. https://doi.org/10.1093/molbev/msz012
8. Jarvis P, Soll J (2001) Toc, Tic, and chloroplast protein import. Biochim Biophys Acta - Mol Cell Res 1541:64–79. https://doi.org/10.1016/S0167-4889(01)00147-1
9. Kamimura Y, Tanaka H, Kobayashi Y, Shikanai T, Nishimura Y (2018) Chloroplast nucleoids as a transformable network revealed by live imaging with a microfluidic device. Commun Biol 1:47. https://doi.org/10.1038/s42003-018-0055-1
10. Sun Y, Valente-Paterno M, Bakhtiari S, Law C, Zhan Y, Zerges W (2019) Photosystem biogenesis is localized to the translation zone in the chloroplast of *Chlamydomonas*. Plant Cell 31:3057–3072. https://doi.org/10.1105/tpc.19.00263
11. Husnik F, Keeling PJ (2019) The fate of obligate endosymbionts: reduction, integration, or extinction. Curr Opin Genet Dev 58–59:1–8. https://doi.org/10.1016/j.gde.2019.07.014
12. Oborník M (2022) Organellar evolution: a path from benefit to dependence. Microorganisms 10:122. https://doi.org/10.3390/microorganisms10010122
13. Petersen J, Ludewig AK, Michael V, Bunk B, Jarek M, Baurain D, Brinkmann H (2014) *Chromera velia*, endosymbioses and the rhodoplex hypothesis – Plastid evolution in cryptophytes, alveolates, stramenopiles, and haptophytes (CASH lineages). Genome Biol Evol 6:666–684. https://doi.org/10.1093/gbe/evu043
14. Stiller JW, Schreiber J, Yue J, Guo H, Ding Q, Huang J (2014) The evolution of photosynthesis in chromist algae through serial endosymbioses. Nat Commun 5:5764. https://doi.org/10.1038/ncomms6764
15. Bodył A (2017) Did some red alga-derived plastids evolve via kleptoplastidy? A hypothesis. Biol Rev 93:201–222. https://doi.org/10.1111/brv.12340
16. Penot M, Dacks JB, Read B, Dorrell RG (2022) Genomic and meta-genomic insights into the functions, diversity and global distribution of haptophyte algae. Appl Phycol 3:340–359. https://doi.org/10.1080/26388081.2022.2103732
17. Nowack ECM, Melkonian M, Glöckner G (2008) Chromatophore genome sequence of *Paulinella* sheds light on acquisition of photosynthesis by eukaryotes. Curr Biol 18:410–418. https://doi.org/10.1016/j.cub.2008.02.051
18. Nakayama T, Kamikawa R, Tanifuji G, Kashiyama Y, Ohkouchi N, Archibald JM, Inagaki Y (2014) Complete genome of a nonphotosynthetic cyanobacterium in a diatom reveals recent adaptations to an intracellular lifestyle. Proc Natl Acad Sci U S A 111:11407–11412. https://doi.org/10.1073/pnas.1405222111
19. Maréchal A, Brisson N (2010) Recombination and the maintenance of plant organelle genome stability. New Phytol 186:299–317. https://doi.org/10.1111/j.1469-8137.2010.03195.x
20. Dagan T et al (2013) Genomes of stigonematalean cyanobacteria (subsection V) and the evolution of oxygenic photosynthesis from prokaryotes to plastids. Genome Biol Evol 5:31–44. https://doi.org/10.1093/gbe/evs117
21. Simm S, Keller M, Selymesi M, Schleiff E (2015) The composition of the global and feature specific cyanobacterial core-genomes. Front Microbiol 6:219. https://doi.org/10.3389/fmicb.2015.00219
22. Terashima M, Specht M, Hippler M (2011) The chloroplast proteome: a survey from the *Chlamydomonas reinhardtii* perspective with a focus on distinctive features. Curr Genet 57:151–168. https://doi.org/10.1007/s00294-011-0339-1
23. Ferro M et al (2010) AT-CHLORO, a comprehensive chloroplast proteome database with subplastidial localization and curated information on envelope proteins. Mol Cell Proteomics 9:1063–1084. https://doi.org/10.1074/mcp.M900325-MCP200

24. Novák Vanclová AMG et al (2019) Metabolic quirks and the colourful history of the *Euglena gracilis* secondary plastid. New Phytol 225:1578–1592. https://doi.org/10.1111/nph.16237
25. Singer A, Poschmann G, Mühlich C, Valadez-Cano C, Hänsch S, Hüren V, Rensing SA, Stühler K, Nowack ECM (2017) Massive protein import into the early-evolutionary-stage photosynthetic organelle of the amoeba *Paulinella chromatophora*. Curr Biol 27:2763–2773. https://doi.org/10.1016/j.cub.2017.08.010
26. Gross J, Bhattacharya D (2009) Mitochondrial and plastid evolution in eukaryotes: an outsiders' perspective. Nat Rev Genet 10:495–505. https://doi.org/10.1038/nrg2610
27. Jiroutová K, Kořený L, Bowler C, Oborník M (2010) A gene in the process of endosymbiotic transfer. PloS One 5:e13234. https://doi.org/10.1371/journal.pone.0013234
28. Uthanumallian K, Iha C, Repetti SI, Chan CX, Bhattacharya D, Duchene S, Verbruggen H (2021) Tightly constrained genome reduction and relaxation of purifying selection during secondary plastid endosymbiosis. Mol Biol Evol 39:msab295. https://doi.org/10.1093/molbev/msab295
29. Dorrell RG, Howe CJ (2012) What makes a chloroplast? Reconstructing the establishment of photosynthetic symbioses. J Cell Sci 125:1865–1875. https://doi.org/10.1242/jcs.102285
30. Archibald JM, Rogers MB, Toop M, Ken-Ichiro I, Keeling PJ (2003) Lateral gene transfer and the evolution of plastid-targeted proteins in the secondary plastid-containing alga *Bigelowiella natans*. Proc Natl Acad Sci U S A 100:7678–7683. https://doi.org/10.1073/pnas.1230951100
31. Suzuki K, Miyagishima SY (2009) Eukaryotic and eubacterial contributions to the establishment of plastid proteome estimated by large-scale phylogenetic analyses. Mol Biol Evol 27:581–590. https://doi.org/10.1093/molbev/msp273
32. Dorrell RG et al (2017) Chimeric origins of ochrophytes and haptophytes revealed through an ancient plastid proteome. Elife 6:e23717. https://doi.org/10.7554/eLife.23717
33. Ponce-Toledo RI, Moreira D, López-García P, Deschamps P (2018) Secondary plastids of euglenids and chlorarachniophytes function with a mix of genes of red and green algal ancestry. Mol Biol Evol 35:2198–2204. https://doi.org/10.1093/molbev/msy121
34. Sato N (2019) Complex origins of chloroplast membranes with photosynthetic machineries: multiple transfers of genes from divergent organisms at different times or a single endosymbiotic event? J Plant Res 133:15–33. https://doi.org/10.1007/s10265-019-01157-z
35. Dorrell RG et al (2021) Phylogenomic fingerprinting of tempo and functions of horizontal gene transfer within ochrophytes. Proc Natl Acad Sci U S A 118:e2009974118. https://doi.org/10.1073/PNAS.2009974118
36. Ku C, Nelson-Sathi S, Roettger M, Garg S, Hazkani-Covo E, Martin WF (2015) Endosymbiotic gene transfer from prokaryotic pangenomes: inherited chimerism in eukaryotes. Proc Natl Acad Sci U S A 112:10139–10146. https://doi.org/10.1073/pnas.1421385112
37. Larkum AWD, Lockhart PJ, Howe CJ (2007) Shopping for plastids. Trends Plant Sci 12:189–195. https://doi.org/10.1016/j.tplants.2007.03.011
38. Ford DW (1998) You are what you eat: a gene transfer ratchet could account for bacterial genes in eukaryotic nuclear genomes. Trends Genet 14:307–311. https://doi.org/10.1016/S0168-9525(98)01494-2
39. Schön ME et al (2021) Single cell genomics reveals plastid-lacking Picozoa are close relatives of red algae. Nat Commun 12:6651. https://doi.org/10.1038/s41467-021-26918-0
40. Martin W et al (2002) Evolutionary analysis of *Arabidopsis*, cyanobacterial, and chloroplast genomes reveals plastid phylogeny and thousands of cyanobacterial genes in the nucleus. Proc Natl Acad Sci U S A 99:12246–12251. https://doi.org/10.1073/pnas.182432999
41. Kořený L, Oborník M (2011) Sequence evidence for the presence of two tetrapyrrole pathways in *Euglena gracilis*. Genome Biol Evol 3:359–364. https://doi.org/10.1093/gbe/evr029
42. Wolf YI, Koonin EV (2013) Genome reduction as the dominant mode of evolution. Bioessays 35:829–837. https://doi.org/10.1002/bies.201300037
43. Wilken S, Huisman J, Naus-Wiezer S, Van Donk E (2013) Mixotrophic organisms become more heterotrophic with rising temperature. Ecol Lett 16:225–233. https://doi.org/10.1111/ele.12033

44. Jones RI (2000) Mixotrophy in planktonic protists: an overview. Freshw Biol 45:219–226. https://doi.org/10.1046/j.13652427.2000.00672.x
45. Borza T, Popescu CE, Lee RW (2005) Multiple metabolic roles for the nonphotosynthetic plastid of the green alga *Prototheca wickerhamii*. Eukaryot Cell 4:253–261. https://doi.org/10.1128/EC.4.2.253-261.2005
46. Salomaki ED, Lane CE (2014) Are all red algal parasites cut from the same cloth? Acta Soc Bot Pol 83:369–375. https://doi.org/10.5586/asbp.2014.047
47. Janouškovec J, Tikhonenkov DV, Burki F, Howe AT, Kolísko M, Mylnikov AP, Keeling PJ (2015) Factors mediating plastid dependency and the origins of parasitism in apicomplexans and their close relatives. Proc Natl Acad Sci 112:10200–10207. https://doi.org/10.1073/pnas.1423790112
48. Záhonová K, Füssy Z, Birčák E, Novák Vanclová AMG, Klimeš V, Vesteg M, Krajčovič J, Oborník M, Eliáš M (2018) Peculiar features of the plastids of the colourless alga *Euglena longa* and photosynthetic euglenophytes unveiled by transcriptome analyses. Sci Rep 8:17012. https://doi.org/10.1038/s41598-018-35389-1
49. Smith DR, Crosby K, Lee RW (2011) Correlation between nuclear plastid DNA abundance and plastid number supports the limited transfer window hypothesis. Genome Biol Evol 3:365–371. https://doi.org/10.1093/gbe/evr001
50. Barbrook AC, Howe CJ, Purton S (2006) Why are plastid genomes retained in non-photosynthetic organisms? Trends Plant Sci 11:101–108. https://doi.org/10.1016/j.tplants.2005.12.004
51. Douglas S et al (2001) The highly reduced genome of an enslaved algal nucleus. Nature 410:1091–1096. https://doi.org/10.1038/35074092
52. Gilson PR, Su V, Slamovits CH, Reith ME, Keeling PJ, McFadden GI (2006) Complete nucleotide sequence of the chlorarachniophyte nucleomorph: nature's smallest nucleus. Proc Natl Acad Sci U S A 103:9566–9571. https://doi.org/10.1073/pnas.0600707103
53. Archibald JM, Lane CE (2009) Going, going, not quite gone: nucleomorphs as a case study in nuclear genome reduction. J Hered 100:582–590. https://doi.org/10.1093/jhered/esp055
54. Stiller JW (2014) Toward an empirical framework for interpreting plastid evolution. J Phycol 50:462–471. https://doi.org/10.1111/jpy.12178
55. Sazanov LA, Burrows PA, Nixon PJ (1998) The plastid ndh genes code for an NADH-specific dehydrogenase: isolation of a complex I analogue from pea thylakoid membranes. Proc Natl Acad Sci U S A 95:1319–1324. https://doi.org/10.1073/pnas.95.3.1319
56. Shikanai T (2016) Chloroplast NDH: a different enzyme with a structure similar to that of respiratory NADH dehydrogenase. Biochim Biophys Acta - Bioenerg 1857:1015–1022. https://doi.org/10.1016/j.bbabio.2015.10.013
57. Lee JM, Cho CH, Park SI, Choi JW, Song HS, West JA, Bhattacharya D, Yoon HS (2016) Parallel evolution of highly conserved plastid genome architecture in red seaweeds and seed plants. BMC Biol 14:75. https://doi.org/10.1186/s12915-016-0299-5
58. Muñoz-Gómez SA, Mejía-Franco FG, Durnin K, Colp M, Grisdale CJ, Archibald JM, Slamovits CH (2017) The new red algal subphylum Proteorhodophytina comprises the largest and most divergent plastid genomes known. Curr Biol 27:1677–1684. https://doi.org/10.1016/j.cub.2017.04.054
59. Janouškovec J, Liu SL, Martone PT, Carré W, Leblanc C, Collén J, Keeling PJ (2013) Evolution of red algal plastid genomes: ancient architectures, introns, horizontal gene transfer, and taxonomic utility of plastid markers. PloS One 8:e59001. https://doi.org/10.1371/journal.pone.0059001
60. de Vries J, Sousa FL, Bölter B, Soll J, Goulda SB (2015) YCF1: a green TIC? Plant Cell 27:1827–1833. https://doi.org/10.1105/tpc.114.135541
61. Kikuchi S et al (2018) A Ycf2-FtsHi heteromeric AAA-ATPase complex is required for chloroplast protein import. Plant Cell 30:2677–2703. https://doi.org/10.1105/tpc.18.00357
62. Turmel M, Lemieux C (2018) Evolution of the plastid genome in green algae. Adv Bot Res 85:157–193. https://doi.org/10.1016/bs.abr.2017.11.010

63. Jans F et al (2008) A type II NAD(P)H dehydrogenase mediates light-independent plastoquinone reduction in the chloroplast of *Chlamydomonas*. Proc Natl Acad Sci U S A 105:20546–20551. https://doi.org/10.1073/pnas.0806896105
64. Ruhlman TA et al (2015) NDH expression marks major transitions in plant evolution and reveals coordinate intracellular gene loss. BMC Plant Biol 15:100. https://doi.org/10.1186/s12870-015-0484-7
65. Hunsperger HM, Randhawa T, Cattolico RA (2015) Extensive horizontal gene transfer, duplication, and loss of chlorophyll synthesis genes in the algae. BMC Evol Biol 15:16. https://doi.org/10.1186/s12862-015-0286-4
66. Delaye L, Valadez-cano C, Pérez-Zamorano B (2016) How really ancient is *Paulinella chromatophora*? PLOS Curr 8:ecurrents.tol.e68a099364bb1a1e129a17b4e06b0c6b. https://doi.org/10.1371/currents.tol.e68a099364bb1a1e129a17b4e06b0c6b
67. Gabr A, Grossman AR, Bhattacharya D (2020) *Paulinella*, a model for understanding plastid primary endosymbiosis. J Phycol 56:837–843. https://doi.org/10.1111/jpy.13003
68. Lhee D, Ha J, Kim S, Park MG, Bhattacharya D, Yoon HS (2019) Evolutionary dynamics of the chromatophore genome in three photosynthetic *Paulinella* species. Sci Rep 9:2560. https://doi.org/10.1038/s41598-019-38621-8
69. Felsner G, Sommer MS, Gruenheit N, Hempel F, Moog D, Zauner S, Martin W, Maier UG (2010) ERAD components in organisms with complex red plastids suggest recruitment of a preexisting protein transport pathway for the periplastid membrane. Genome Biol Evol 3:140–150. https://doi.org/10.1093/gbe/evq074
70. Lane CE, Van Den Heuvel K, Kozera C, Curtis BA, Parsons BJ, Bowman S, Archibald JM (2007) Nucleomorph genome of *Hemiselmis andersenii* reveals complete intron loss and compaction as a driver of protein structure and function. Proc Natl Acad Sci U S A 104:19908–19913. https://doi.org/10.1073/pnas.0707419104
71. Kim JI, Tanifuji G, Jeong M, Shin W, Archibald JM (2022) Gene loss, pseudogenization, and independent genome reduction in non-photosynthetic species of *Cryptomonas* (Cryptophyceae) revealed by comparative nucleomorph genomics. BMC Biol 20:227. https://doi.org/10.1186/s12915-022-01429-6
72. Kim JI, Moore CE, Archibald JM, Bhattacharya D, Yi G, Yoon HS, Shin W (2017) Evolutionary dynamics of cryptophyte plastid genomes. Genome Biol Evol 9:1859–1872. https://doi.org/10.1093/gbe/evx123
73. Duplessis MR, Karol KG, Adman ET, Choi LYS, Jacobs MA, Cattolico RA (2007) Chloroplast His-to-Asp signal transduction: a potential mechanism for plastid gene regulation in *Heterosigma akashiwo* (Raphidophyceae). BMC Evol Biol 7:70. https://doi.org/10.1186/1471-2148-7-70
74. Hovde BT, Starkenburg SR, Hunsperger HM, Mercer LD, Deodato CR, Jha RK, Chertkov O, Monnat RJ, Cattolico RA (2014) The mitochondrial and chloroplast genomes of the haptophyte *Chrysochromulina tobin* contain unique repeat structures and gene profiles. BMC Genomics 15:604. https://doi.org/10.1186/1471-2164-15-604
75. Zverkov OA, Seliverstov AV, Lyubetsky VA (2016) Regulation of expression and evolution of genes in plastids of rhodophytic branch. Life 6:7. https://doi.org/10.3390/life6010007
76. Rice DW, Palmer JD (2006) An exceptional horizontal gene transfer in plastids: gene replacement by a distant bacterial paralog and evidence that haptophyte and cryptophyte plastids are sisters. BMC Biol 4:31. https://doi.org/10.1186/1741-7007-4-31
77. Andersen RA (2004) Biology and systematics of heterokont and haptophyte algae. Am J Bot 91:1508–1522. https://doi.org/10.3732/ajb.91.10.1508
78. Moustafa A, Beszteri B, Maier UG, Bowler C, Valentin K, Bhattacharya D (2009) Genomic footprints of a cryptic plastid endosymbiosis in diatoms. Science 324:1724–1726. https://doi.org/10.1126/science.1172983
79. Ševčíková T, Yurchenko T, Fawley KP, Amaral R, Strnad H, Santos LMA, Fawley MW, Eliáš M (2019) Plastid genomes and proteins illuminate the evolution of eustigmatophyte algae and their bacterial endosymbionts. Genome Biol Evol 11:362–379. https://doi.org/10.1093/gbe/evz004

80. Kim JI, Shin H, Škaloud P, Jung J, Yoon HS, Archibald JM, Shin W (2019) Comparative plastid genomics of Synurophyceae: inverted repeat dynamics and gene content variation. BMC Evol Biol 19:20. https://doi.org/10.1186/s12862-018-1316-9
81. Starko S, Bringloe TT, Soto Gomez M, Darby H, Graham SW, Martone PT (2021) Genomic rearrangements and sequence evolution across brown algal organelles. Genome Biol Evol 13:evab124. https://doi.org/10.1093/gbe/evab124
82. Han KY, Maciszewski K, Graf L, Yang JH, Andersen RA, Karnkowska A, Yoon HS (2019) Dictyochophyceae plastid genomes reveal unusual variability in their organization. J Phycol 55:1166–1180. https://doi.org/10.1111/jpy.12904
83. Graupner N, Jensen M, Bock C, Marks S, Rahmann S, Beisser D, Boenigk J (2018) Evolution of heterotrophy in chrysophytes as reflected by comparative transcriptomics. FEMS Microbiol Ecol 94:fiy039. https://doi.org/10.1093/FEMSEC/FIY039
84. Ruck EC, Nakov T, Jansen RK, Theriot EC, Alverson AJ (2014) Serial gene losses and foreign DNA underlie size and sequence variation in the plastid genomes of diatoms. Genome Biol Evol 6:644–654. https://doi.org/10.1093/gbe/evu039
85. Cavalier-Smith T (2018) Kingdom Chromista and its eight phyla: a new synthesis emphasising periplastid protein targeting, cytoskeletal and periplastid evolution, and ancient divergences. Protoplasma 255:297–357. https://doi.org/10.1007/s00709-017-1147-3
86. Moore RB et al (2008) A photosynthetic alveolate closely related to apicomplexan parasites. Nature 451:959–963. https://doi.org/10.1038/nature06635
87. Janouškovec J, Horák A, Oborník M, Lukeš J, Keeling PJ (2010) A common red algal origin of the apicomplexan, dinoflagellate, and heterokont plastids. Proc Natl Acad Sci U S A 107:10949–10954. https://doi.org/10.1073/pnas.1003335107
88. Hofmann E, Wrench PM, Sharples FP, Hiller RG (1996) Structural basis of light harvesting by carotenoids. Science 272:1788–1791
89. Patron NJ, Waller RF (2007) Transit peptide diversity and divergence: a global analysis of plastid targeting signals. Bioessays 29:1048–1058. https://doi.org/10.1002/bies.20638
90. Barbrook AC, Santucci N, Plenderleith LJ, Hiller RG, Howe CJ (2006) Comparative analysis of dinoflagellate chloroplast genomes reveals rRNA and tRNA genes. BMC Genomics 7:1–15. https://doi.org/10.1186/1471-2164-7-297
91. Howe CJ, Barbrook AC, Nisbet RER, Lockhart PJ, Larkum AWD (2008) The origin of plastids. Philos Trans R Soc B Biol Sci 363:2675–2685. https://doi.org/10.1098/rstb.2008.0050
92. Nelson MJ, Green BR (2005) Double hairpin elements and tandem repeats in the non-coding region of *Adenoides eludens* chloroplast gene minicircles. Gene 358:102–110. https://doi.org/10.1016/j.gene.2005.05.024
93. Barbrook AC, Dorrell RG, Burrows J, Plenderleith LJ, Nisbet RER, Howe CJ (2012) Polyuridylylation and processing of transcripts from multiple gene minicircles in chloroplasts of the dinoflagellate *Amphidinium carterae*. Plant Mol Biol 79:347–357. https://doi.org/10.1007/s11103-012-9916-z
94. Dorrell RG, Howe CJ (2015) Integration of plastids with their hosts: lessons learned from dinoflagellates. Proc Natl Acad Sci U S A 112:10247–10254. https://doi.org/10.1073/pnas.1421380112
95. Koumandou VL, Howe CJ (2007) The copy number of chloroplast gene minicircles changes dramatically with growth phase in the dinoflagellate *Amphidinium operculatum*. Protist 158:89–103. https://doi.org/10.1016/j.protis.2006.08.003
96. Janouskovec J et al (2017) Major transitions in dinoflagellate evolution unveiled by phylotranscriptomics. Proc Natl Acad Sci U S A 114:E171–E180. https://doi.org/10.1073/pnas.1614842114
97. Park MG, Kim M, Kim S (2014) The acquisition of plastids/phototrophy in heterotrophic dinoflagellates. Acta Protozool 53:39–50. https://doi.org/10.4467/16890027AP.14.005.1442
98. Park GM, Kim S, Kim SH, Myung G, Kang GY, Yih W (2006) First successful culture of the marine dinoflagellate *Dinophysis acuminata*. Aquat Microb Ecol 45:101–106. https://doi.org/10.3354/ame045101

99. Sweeney BM (1976) *Pedinomonas noctilucae* (Prasinophyceae), the flagellate symbiotic in *Noctiluca* (Dinophyceae) in Southeast Asia. J Phycol 12:460–464. https://doi.org/10.1111/j.1529-8817.1976.tb02874.x
100. Gast RJ, Moran DM, Dennett MR, Caron DA (2007) Kleptoplasty in an Antarctic dinoflagellate: caught in evolutionary transition? Environ Microbiol 9:39–45. https://doi.org/10.1111/j.1462-2920.2006.01109.x
101. Tomas RN, Cox ER, Steidinger KA (1973) *Peridinium balticum* (Levander) Lemmermann, an unusual dinoflagellate with a mesocaryotic and an eucaryotic nucleus. J Phycol 9:91–98. https://doi.org/10.1111/j.0022-3646.1973.00091.x
102. Yamada N, Sym SD, Horiguchi T (2017) Identification of highly divergent diatom-derived chloroplasts in dinoflagellates, including a description of *Durinskia kwazulunatalensis* sp. nov. (Peridiniales, Dinophyceae). Mol Biol Evol 34:1335–1351. https://doi.org/10.1093/molbev/msx054
103. Sarai C et al (2020) Dinoflagellates with relic endosymbiont nuclei as models for elucidating organellogenesis. Proc Natl Acad Sci U S A 117:5364–5375. https://doi.org/10.1073/pnas.1911884117
104. Matsuo E, Morita K, Nakayama T, Yazaki E, Sarai C, Takahashi K, Iwataki M, Inagaki Y (2022) Comparative plastid genomics of green-colored dinoflagellates unveils parallel genome compaction and RNA editing. Front Plant Sci 13:918543. https://doi.org/10.3389/fpls.2022.918543
105. Elbrächter M, Schnepf E (1996) *Gymnodinium chlorophorum*, a new, green, bloom-forming dinoflagellate (Gymnodiniales, Dinophyceae) with a vestigial prasinophyte endosymbiont. Phycologia 35:381–393. https://doi.org/10.2216/i0031-8884-35-5-381.1
106. Schnepf E, Elbrächter M (1999) Dinophyte chloroplasts and phylogeny - A review. Grana 38:81–97. https://doi.org/10.1080/713786928
107. Tengs T, Dahlberg OJ, Shalchian-Tabrizi K, Klaveness D, Rudi K, Delwiche CF, Jakobsen KS (2000) Phylogenetic analyses indicate that the 19'hexanoyloxy-fucoxanthin-containing dinoflagellates have tertiary plastids of haptophyte origin. Mol Biol Evol 17:718–729. https://doi.org/10.1093/oxfordjournals.molbev.a026350
108. Hansen G, Botes L, De Salas M (2007) Ultrastructure and large subunit rDNA sequences of *Lepidodinium viride* reveal a close relationship to *Lepidodinium chlorophorum* comb. nov. (= *Gymnodinium chlorophorum*). Phycol Res 55:25–41. https://doi.org/10.1111/j.1440-1835.2006.00442.x
109. Yamada N, Tanaka A, Horiguchi T (2015) Pigment compositions are linked to the habitat types in dinoflagellates. J Plant Res 128:923–932. https://doi.org/10.1007/s10265-015-0745-4
110. Dodge JD (1984) The functional and phylogenetic significance of dinoflagellate eyespots. Biosystems 16:259–267. https://doi.org/10.1016/0303-2647(83)90009-6
111. Imanian B, Pombert JF, Keeling PJ (2010) The complete plastid genomes of the two 'Dinotoms' *Durinskia baltica* and *Kryptoperidinium foliaceum*. PloS One 5:e10711. https://doi.org/10.1371/journal.pone.0010711
112. Imanian B, Pombert JF, Dorrell RG, Burki F, Keeling PJ (2012) Tertiary endosymbiosis in two dinotoms has generated little change in the mitochondrial genomes of their dinoflagellate hosts and diatom endosymbionts. PloS One 7:e43763. https://doi.org/10.1371/journal.pone.0043763
113. Gabrielsen TM et al (2011) Genome evolution of a tertiary dinoflagellate plastid. PloS One 6:e19132. https://doi.org/10.1371/journal.pone.0019132
114. Jackson CJ, Gornik SG, Waller RF (2013) A tertiary plastid gains RNA editing in its new host. Mol Biol Evol 30:788–792. https://doi.org/10.1093/molbev/mss270
115. Richardson E, Dorrell RG, Howe CJ (2014) Genome-wide transcript profiling reveals the coevolution of plastid gene sequences and transcript processing pathways in the fucoxanthin dinoflagellate *Karlodinium veneficum*. Mol Biol Evol 31:2376–2386. https://doi.org/10.1093/molbev/msu189

116. Turmel M, Gagnon MC, O'Kelly CJ, Otis C, Lemieux C (2009) The chloroplast genomes of the green algae *Pyramimonas*, *Monomastix*, and *Pycnococcus* shed new light on the evolutionary history of prasinophytes and the origin of the secondary chloroplasts of euglenids. Mol Biol Evol 26:631–648. https://doi.org/10.1093/molbev/msn285
117. Jackson C, Knoll AH, Chan CX, Verbruggen H (2018) Plastid phylogenomics with broad taxon sampling further elucidates the distinct evolutionary origins and timing of secondary green plastids. Sci Rep 8:1523. https://doi.org/10.1038/s41598-017-18805-w
118. Slávikovά S, Vacula R, Fang Z, Ehara T, Osafune T, Schwartzbach SD (2005) Homologous and heterologous reconstitution of Golgi to chloroplast transport and protein import into the complex chloroplasts of *Euglena*. J Cell Sci 118:1651–1661. https://doi.org/10.1242/jcs.02277
119. Suzuki S, Hirakawa Y, Kofuji R, Sugita M, Ishida K, ichiro. (2016) Plastid genome sequences of *Gymnochlora stellata*, *Lotharella vacuolata*, and *Partenskyella glossopodia* reveal remarkable structural conservation among chlorarachniophyte species. J Plant Res 129:581–590. https://doi.org/10.1007/s10265-016-0804-5
120. Hirakawa Y, Burki F, Keeling PJ (2012) Genome-based reconstruction of the protein import machinery in the secondary plastid of a chlorarachniophyte alga. Eukaryot Cell 11:324–333. https://doi.org/10.1128/EC.05264-11
121. Kamikawa R et al (2015) Multiple losses of photosynthesis in *Nitzschia* (Bacillariophyceae). Phycol Res 63:19–28. https://doi.org/10.1111/pre.12072
122. Maciszewski K, Karnkowska A (2019) Should I stay or should I go? Retention and loss of components in vestigial endosymbiotic organelles. Curr Opin Genet Dev 58–59:33–39. https://doi.org/10.1016/j.gde.2019.07.013
123. Hadariová L, Vesteg M, Hampl V, Krajčovič J (2017) Reductive evolution of chloroplasts in non-photosynthetic plants, algae and protists. Curr Genet 64:365–387. https://doi.org/10.1007/s00294-017-0761-0
124. Kohzuma K, Froehlich JE, Davis GA, Temple JA, Minhas D, Dhingra A, Cruz JA, Kramer DM (2017) The role of light–dark regulation of the chloroplast ATP synthase. Front Plant Sci 8:1248. https://doi.org/10.3389/fpls.2017.01248
125. Tanifuji G, Kamikawa R, Moore CE, Mills T, Onodera NT, Archibald JM, Inagaki Y, Hashimoto T (2020) Comparative plastid genomics of *Cryptomonas* species reveals fine-scale genomic responses to loss of photosynthesis. Genome Biol Evol 12:3926–3937. https://doi.org/10.1093/gbe/evaa001
126. Grossmann L, Bock C, Schweikert M, Boenigk J (2016) Small but manifold - Hidden diversity in '*Spumella*-like flagellates'. J Eukaryot Microbiol 63:419–439. https://doi.org/10.1111/jeu.12287
127. Kayama M et al (2020) A non-photosynthetic green alga illuminates the reductive evolution of plastid electron transport systems. BMC Biol 18:126. https://doi.org/10.1186/s12915-020-00853-w
128. Füssy Z, Záhonová K, Tomčala A, Krajčovič J, Yurchenko V, Oborník M, Eliáš M (2020) The cryptic plastid of *Euglena longa* defines a new type of nonphotosynthetic plastid organelle. mSphere 5:e00675-20. https://doi.org/10.1128/msphere.00675-20
129. Mathur V, Kolísko M, Hehenberger E, Irwin NAT, Leander BS, Kristmundsson Á, Freeman MA, Keeling PJ (2019) Multiple independent origins of apicomplexan-like parasites. Curr Biol 29:2936–2941. https://doi.org/10.1016/j.cub.2019.07.019
130. Janouskovec J, Paskerova GG, Miroliubova TS, Mikhaiiov KV, Birley T, Aieoshin VV, Simdyanov TG (2019) Apicomplexan-like parasites are polyphyletic and widely but selectively dependent on cryptic plastid organelles. Elife 8:e49662. https://doi.org/10.7554/eLife.49662
131. Waller RF et al (1998) Nuclear-encoded proteins target to the plastid in *Toxoplasma gondii* and *Plasmodium falciparum*. Proc Natl Acad Sci U S A 95:12352–12357. https://doi.org/10.1073/pnas.95.21.12352

132. Yeh E, DeRisi JL (2011) Chemical rescue of malaria parasites lacking an apicoplast defines organelle function in blood-stage *Plasmodium falciparum*. PLoS Biol 9:e1001138. https://doi.org/10.1371/journal.pbio.1001138
133. Vaughan AM, O'neill MT, Tarun AS, Camargo N, Phuong TM, Aly ASI, Cowman AF, SHI K (2009) Type II fatty acid synthesis is essential only for malaria parasite late liver stage development. Cell Microbiol 11:506–520. https://doi.org/10.1111/j.1462-5822.2008.01270.x
134. Ke H et al (2014) The heme biosynthesis pathway is essential for *Plasmodium falciparum* development in mosquito stage but not in blood stages. J Biol Chem 289:34827–34837. https://doi.org/10.1074/jbc.M114.615831
135. Abrahamsen MS et al (2004) Complete genome sequence of the apicomplexan, *Cryptosporidium parvum*. Science 304:441–445. https://doi.org/10.1126/science.1094786
136. Toso MA, Omoto CK (2007) *Gregarina niphandrodes* may lack both a plastid genome and organelle. J Eukaryot Microbiol 54:66–72. https://doi.org/10.1111/j.1550-7408.2006.00229.x
137. Mathur V, Kwong WK, Husnik F, Irwin NAT, Kristmundsson Á, Gestal C, Freeman M, Keeling PJ (2021) Phylogenomics identifies a new major subgroup of apicomplexans, *Marosporida* class nov., with extreme apicoplast genome reduction. Genome Biol Evol 13:evaa244. https://doi.org/10.1093/gbe/evaa244
138. Kwong WK, del Campo J, Mathur V, Vermeij MJA, Keeling PJ (2019) A widespread coral-infecting apicomplexan with chlorophyll biosynthesis genes. Nature 568:103–107. https://doi.org/10.1038/s41586-019-1072-z
139. Shaver JM, Oldenburg DJ, Bendich AJ (2006) Changes in chloroplast DNA during development in tobacco, *Medicago truncatula*, pea, and maize. Planta 224:72–82. https://doi.org/10.1007/s00425-005-0195-7
140. Green B (1976) Covalently closed minicircular DNA associated with *Acetabularia* chloroplasts. Biochim Biophys Acta 447:156–166
141. Brembu T, Winge P, Tooming-Klunderud A, Nederbragt AJ, Jakobsen KS, Bones AM (2014) The chloroplast genome of the diatom *Seminavis robusta*: new features introduced through multiple mechanisms of horizontal gene transfer. Mar Genomics 16:17–27. https://doi.org/10.1016/j.margen.2013.12.002
142. Wang L et al (2013) Complete sequence and analysis of plastid genomes of two economically important red algae: *Pyropia haitanensis* and *Pyropia yezoensis*. PloS One 8:e65902. https://doi.org/10.1371/journal.pone.0065902
143. Watanabe S, Fučíková K, Lewis LA, Lewis PO (2016) Hiding in plain sight: *Koshicola spirodelophila* gen. et sp. nov. (Chaetopeltidales, Chlorophyceae), a novel green alga associated with the aquatic angiosperm *Spirodela polyrhiza*. Am J Bot 103:865–875. https://doi.org/10.3732/ajb.1500481
144. Del Cortona A et al (2017) The plastid genome in Cladophorales green algae is encoded by hairpin chromosomes. Curr Biol 27:3771–3782. https://doi.org/10.1016/j.cub.2017.11.004
145. Fang J, Liu B, Liu G, Verbruggen H, Zhu H (2021) Six newly sequenced chloroplast genomes from Trentepohliales: the inflated genomes, alternative genetic code and dynamic evolution. Front Plant Sci 12:780054. https://doi.org/10.3389/fpls.2021.780054
146. Brouard JS, Otis C, Lemieux C, Turmel M (2010) The exceptionally large chloroplast genome of the green alga *Floydiella terrestris* illuminates the evolutionary history of the Chlorophyceae. Genome Biol Evol 2:240–256. https://doi.org/10.1093/gbe/evq014
147. Smith DR, Lee RW (2009) The mitochondrial and plastid genomes of *Volvox carteri*: bloated molecules rich in repetitive DNA. BMC Genomics 10:132. https://doi.org/10.1186/1471-2164-10-132
148. Tymms MJ, Schweiger HG (1985) Tandemly repeated nonribosomal DNA sequences in the chloroplast genome of an *Acetabularia mediterranea* strain. Proc Natl Acad Sci U S A 82:1706–1710. https://doi.org/10.1073/pnas.82.6.1706
149. Figueroa-Martinez F, Nedelcu AM, Smith DR, Reyes-Prieto A (2017) The plastid genome of *Polytoma uvella* is the largest known among colorless algae and plants and reflects contrasting evolutionary paths to nonphotosynthetic lifestyles. Plant Physiol 173:932–943. https://doi.org/10.1104/pp.16.01628

150. Pánek T et al (2022) A new lineage of non-photosynthetic green algae with extreme organellar genomes. BMC Biol 20:66. https://doi.org/10.1186/s12915-022-01263-w
151. Adams CC, Stern DB (1990) Control of mRNA stability in chloroplasts by 3′ inverted repeats: effects of stem and loop mutations on degradation of PsbA mRNA in vitro. Nucleic Acids Res 18:6003–6010. https://doi.org/10.1093/nar/18.20.6003
152. Zhu A, Guo W, Gupta S, Fan W, Mower JP (2016) Evolutionary dynamics of the plastid inverted repeat: the effects of expansion, contraction, and loss on substitution rates. New Phytol 209:1747–1756. https://doi.org/10.1111/nph.13743
153. Delwiche CF, Palmer JD (1996) Rampant horizontal transfer and duplication of Rubisco genes in eubacteria and plastids. Mol Biol Evol 13:873–882. https://doi.org/10.1093/oxfordjournals.molbev.a025647
154. Mackiewicz P, Bodył A, Moszczyński K (2013) The case of horizontal gene transfer from bacteria to the peculiar dinoflagellate plastid genome. Mob Genet Elem 3:e25845. https://doi.org/10.4161/mge.25845
155. Yurchenko T, Ševčíková T, Strnad H, Butenko A, Eliáš M (2016) The plastid genome of some eustigmatophyte algae harbours a bacteria-derived six-gene cluster for biosynthesis of a novel secondary metabolite. Open Biol 6:160249. https://doi.org/10.1098/rsob.160249
156. Cattolico R et al (2008) Chloroplast genome sequencing analysis of *Heterosigma akashiwo* CCMP452 (West Atlantic) and NIES293 (West Pacific) strains. BMC Genomics 9:211. https://doi.org/10.1186/1471-2164-9-211
157. Gornik SG et al (2015) Endosymbiosis undone by stepwise elimination of the plastid in a parasitic dinoflagellate. PNAS 112:5767–5772. https://doi.org/10.1073/pnas.1423400112
158. Smith DR, Asmail SR (2014) Next-generation sequencing data suggest that certain nonphotosynthetic green plants have lost their plastid genomes. New Phytol 204:7–11. https://doi.org/10.1111/nph.12919
159. De Vargas C et al (2015) Eukaryotic plankton diversity in the sunlit ocean. Science 348:1261605. https://doi.org/10.1126/science.1261605
160. Su HJ et al (2019) Novel genetic code and record-setting AT-richness in the highly reduced plastid genome of the holoparasitic plant *Balanophora*. Proc Natl Acad Sci U S A 116:934–943. https://doi.org/10.1073/pnas.1816822116

Dinotoms Illuminate Early Pathways to the Stable Acquisition of Photosynthetic Endosymbionts

Norico Yamada, William H. Lewis, Takeo Horiguchi, and Ross F. Waller

Abstract It is a widely accepted hypothesis that plastids and mitochondria arose by endosymbiosis; however, many of the evolutionary processes of organellar acquisition and development are still unknown. While many organism models represent points in a possible transition from facultative photosynthetic endosymbionts to highly reduced plastid organelles, only 'dinotoms' can be applied to answer the question: How do free-living microalgae transform into stable plastid-bearing endosymbionts? Many of the proposed models for plastidogenesis are either still in a kleptoplastidic stage requiring ongoing endosymbiont gain or have already transitioned to fixed endosymbionts with dramatic reduction of the symbiont nuclei and other organelles. Dinotom dinoflagellates, on the other hand, provide a missing link. Their plastids are derived from diatoms at various stages of association and permanence within dinoflagellate hosts. Some are at the kleptoplastid stage, where ingested diatoms are maintained only for months; some are at a stage in which multiple diatoms are stably maintained; and some are in a permanent partnership with only a single diatom endosymbiont. Dinotoms maintain all core diatom organelles regardless of their evolutionary stage, except for the plasma membrane and the frustule. In this chapter, we summarise the current knowledge of the dinotom system with respect to cell biology and molecular biology and discuss the applicability of dinotoms as model organisms for endosymbiosis studies.

Keywords Diatom · Dinoflagellate · Dinotom · Kleptoplastid · Mitochondria · Evolutionary intermediate plastid · Kleptokaryon · Kryptoperidiniaceae · Nucleomorph · Organelle redundancy · Plastid diversity · Tertiary plastid

N. Yamada (✉)
Department of Biology, University of Konstanz, Konstanz, Germany
e-mail: norico.yamada@uni-konstanz.de

W. H. Lewis · R. F. Waller
Department of Biochemistry, University of Cambridge, Cambridge, UK
e-mail: whl30@cam.ac.uk; rfw26@cam.ac.uk

T. Horiguchi
Department of Biological Sciences, Hokkaido University, Sapporo, Japan

S. D. Schwartzbach et al. (eds.), *Endosymbiotic Organelle Acquisition*,
https://doi.org/10.1007/978-3-031-57446-7_6

1 Introduction

The acquisition of phototrophy through endosymbiotic gain of a plastid organelle is not an event unique to Archaeplastida. Plastids have been acquired independently many times in different eukaryotic groups. Unlike the primary gain of a plastid from a photosynthetic cyanobacterium, many plastids are derived from secondary, tertiary, or higher-order endosymbioses, where heterotrophic ancestors engulfed eukaryotic microalgae, which themselves possessed primary, secondary, or higher-order plastids [1–3]. Dinoflagellates are a group of such eukaryotes with red-algal-derived plastids acquired by secondary or higher-order endosymbioses [4, 5]. Approximately 2500 extant dinoflagellate species have been recognised so far [6], although metabarcoding ocean surveys suggest that this is a substantial underestimate of the full dinoflagellate diversity [7]. A distinctive light-harvesting pigment, peridinin, evolved early in dinoflagellate plastid evolution and is retained throughout most of the photosynthetic members of dinoflagellates. However, approximately half of them have lost their photosynthetic ability returning to a purely heterotrophic mode of nutrition. This process of photoautotrophic loss has occurred multiple times independently across dinoflagellate taxa revealing a puzzling frequency for loss of such a seemingly useful trait [8, 9]. Furthermore, some dinoflagellates have regained or substituted their photosynthetic capacity by acquiring new plastids from different eukaryotic microalgal sources. These new plastids in dinoflagellates represent several different stages of endosymbiont integration, from kleptoplastids through to stable organelles consisting only of the photosynthetic plastid [10–17]. Dinoflagellates, therefore, offer rich opportunities to investigate plastidogenesis processes, as well as the loss of these organelles.

Since the late 1960s, it has been known that in place of the peridinin pigment, some dinoflagellates have plastids that contain fucoxanthin as the principal light-harvesting pigment, which is a pigment more typically found in diatoms [18, 19], and led to the coining of the name 'dinotoms' for these anomalous dinoflagellates [20]. The first direct evidence of the presence of diatom endosymbionts in these cells was provided by transmission electron microscopy (TEM) by Dodge [21], who reported the presence of two different types of eukaryotic nuclei within a single dinoflagellate cell in the species *Glenodinium foliaceum* (:*Kryptoperidinium triquetrum*). Ground-breaking dinotom studies by Tomas and Cox [22, 23] provided ultrastructural detail of another dinotom, *Peridinium balticum* (:*Durinskia baltica*), revealing that this organism possessed a structurally independent endosymbiont. Subsequently, it was verified by molecular data that the endosymbionts of *K. triquetrum* and *D. baltica* are actually derived from diatoms [24, 25]. Since these discoveries, five genera containing more than 20 species of dinotoms have been recognised: *Blixaea*, *Dinothrix*, *Durinskia*, *Kryptoperidinium*, and *Unruhdinium* (Table 1). These genera all belong to a monophyletic clade classified as the family Kryptoperidiniaceae [26].

Table 1 A list of dinotom taxa, their endosymbiont's nearest free-living taxa, and the availability of genomic and transcriptomic data

Host dinoflagellates	Endosymbiotic diatoms	Whole genomic and transcriptomic data
Blixaea		
Blixaea quinquecornis	*Chaetoceros tenuissimus/simplex*	–
Dinothrix		
Dinothrix paradoxa	*Nitzschia* sp. clade 6B	–
D. phymatodea	*Nitzschia* sp. clade 6B	–
D. pseudoparadoxa	*Nitzschia* sp. clade 6B	–
D. quadrilobata	*Nitzschia* sp. clade 6B	–
D. rugata	*Nitzschia* sp. clade 6B	In progress (genome, both diatom and dinoflagellate) In progress (transcriptome)
Durinskia		
Durinskia agilis	Unknown	–
D. cf. *baltica* strain CS-38	*Nitzschia palea* clade 6B	PRJNA173778 (transcriptome)
Other *D.* cf. *baltica*	Several species of *Nitzschia* clade 6B	–
D. capensis	*Nitzschia captiva* clade 6B and a close relative of *N. captiva*	In progress (genome, diatom) PRJDB15210 (transcriptome, diatom)
D. dybowskii	Unknown	–
D. kwazulunatalensis	*Simonsenia* sp. clade W	In progress (transcriptome)
D. oculata	*Nitzschia* sp. clade 6B	PRJEB62010 (assembled genome, diatom) PRJEB61219 (draft genome, both) In progress (transcriptome)
Kryptoperidinium		
Kryptoperidinium sp. strain CCMP/NCMA1326	*Nitzschia* sp. clade 6B	PRJNA178060 (transcriptome)
Other *Kryptoperidinium* spp.	Several species of *Nitzschia* clade 6B or clade 3	–
K. triquetrum strain GeoB 459/VGO	*Nitzschia* sp. clade 6B	–
K. triquetrum strains W1/W4/G etc.	*Nitzschia* sp. clade 6B	PRJEB63457 (assembled genome, diatom) PRJEB61217 (draft genome, both) In progress (transcriptome)
Unruhdinium		
Unruhdinium jiulongense	*Discostella* sp. LO4-2	–
U. cf. *kevei*		–

(continued)

Table 1 (continued)

Host dinoflagellates	Endosymbiotic diatoms	Whole genomic and transcriptomic data
U. minima	*Cyclotella* sp.	–
U. niei	*Cyclostephanos* sp.	–
U. penardii	*Discostella* sp. LO4-2	–
U. penardii var. *robustum*	*Discostella lacus-karluki*	–

Each diatom clade is classified based on [41]. Endosymbiotic *Nitzschia* spp. clade 6B in different host dinoflagellates are phylogenetically distinct from each other [31, 33, 41, 42, 51]

2 'Dinotoms' are Dinoflagellates that have Acquired Diatom Endosymbionts for Photosynthesis Multiple Times

Initially, the endosymbiotic diatoms of dinotoms were believed to be from a single species of diatoms belonging to the genus *Nitzschia* [25, 27, 28]. It was subsequently discovered, however, that *Blixaea* and *Unruhdinium* possess *Chaetoceros* sp. and *Discostella*/*Cyclotella*/*Cyclostephanos* spp. endosymbionts, respectively [29–31]. These taxonomically distinct endosymbionts of dinotoms challenged the notion that the establishment of stable endosymbionts is a rare event that happens infrequently, even at evolutionary timescales. Furthermore, it was subsequently discovered that even morphologically and systematically identical host dinoflagellates possess distinct diatom endosymbionts [31–33]. Currently, endosymbiotic diatoms are inferred to be derived from more than 20 species, 6 genera, and 3 families (Bacillariaceae, Chaetocerotaceae, and Stephanodiscaceae) of free-living diatoms. Thus, there is strong evidence that the plastids of dinotoms did not evolve from a single endosymbiont species. A similar trend of multiple independent gains of related but distinct plastids is apparent in other dinoflagellates that have exchanged their peridinin plastids for those derived from other algae, namely *Lepidodinium* spp., MGD, and TGD, which have plastids originated from different species of green algae [11, 16], as well as members of the family Kareniaceae, which have plastids derived from haptophytes [17, 34, 35].

3 Multiple Acquisitions of Endosymbionts have Occurred in Dinotoms Resulting in their Plastids Representing Different Evolutionary Stages

The study of at least 20 distinct diatom endosymbionts in dinotoms has revealed that the stage of endosymbiotic integration is not the same in all cases (Fig. 1). Moreover, the recent revelation that some dinotoms are still at a stage of hosting temporary endosymbionts shows that dinotoms represent a wide range of endosymbiotic stages. *Durinskia capensis* is recently confirmed as a kleptoplastidic dinotom that acquires diatoms but can only maintain them as endosymbionts for a

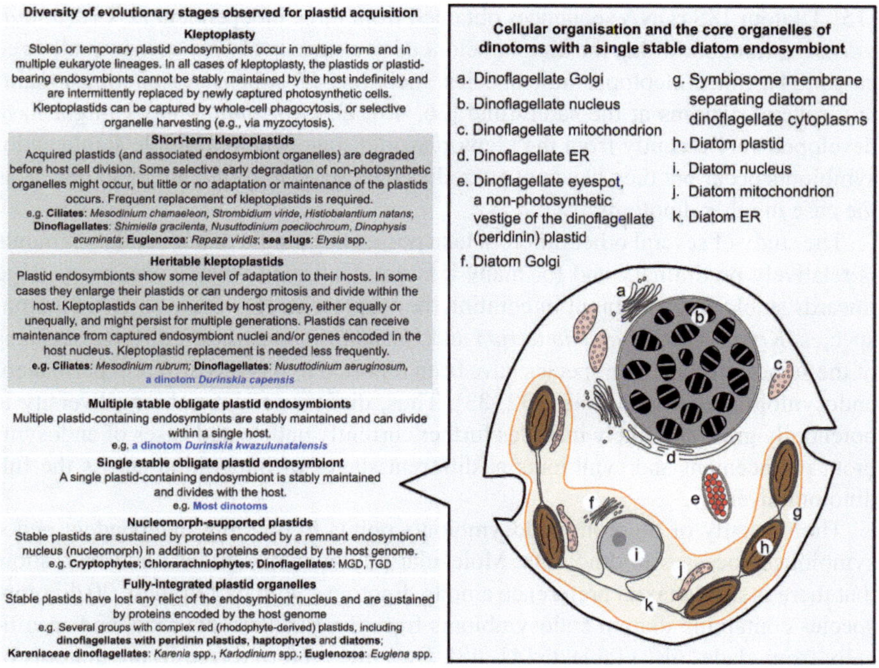

Fig. 1 The kleptoplastid to plastid organelle continuum and the stages of dinotom plastid evolution. A schematic of plastid evolution highlighting stages of plastid integration, transformation and maintenance observed from diverse cases of photosynthetic eukaryotes. Not all these points are necessarily required to achieve full plastid integration, nor might all endosymbionts be capable of doing so. A diagram of dinotoms that have a fixed endosymbiont, representing the evolutionary stage of a host cell having a single stable obligate plastid endosymbiont, is shown. These dinotoms retain the diatom endosymbiont organelles in addition to dinoflagellate versions, meaning that single dinotom cells contain functionally overlapping organelle copies that have distinct evolutionary histories. So far, separate dinotom species have been identified that represent three different evolutionary stages of plastid evolution (highlighted in blue text)

few months [15]. This dinoflagellate is cultivable only when specific essential diatoms are present: *Nitzschia captiva* or an undescribed close relative of *N. captiva*, although it can use many species of diatoms as partially functional kleptoplastids also [36]. The possible existence of further examples of kleptoplastidic dinotoms has been indicated by several reports of dinotoms that were observed to possess only the dinoflagellate nucleus [37–39]. Loss of the diatom nucleus, or nuclei, has also often been observed in starved *D. capensis* that have been cultured without free-living diatoms [15].

Durinskia kwazulunatalensis is the closest known relative of *D. capensis* and it reveals a further tantalising piece of the plastid evolutionary puzzle. *D. kwazulunatalensis* has progressed to the 'fixed' endosymbiotic stage as seen for other well-studied stable dinotoms. Unusually, however, it contains multiple (three to eight) endosymbiotic diatom nuclei in a single host dinoflagellate cell, rather than one

[15]. Diatom 18S rDNA sequences obtained from three independent *D. kwazulunatalensis* strains show that the diatom belonged to a species of *Simonsenia* in all three cases [31]. Since kleptoplastidic species, including *D. capensis*, sometimes maintain several diatoms at the same time [36, 40], *D. kwazulunatalensis* might have developed very recently from the kleptoplastidic stage where multiple stable endosymbionts occur but they have not yet reduced these *Simonsenia* nuclei to one, as is the case in other dinotoms.

The study of several other dinotom taxa possessing single diatom endosymbionts is relatively preliminary and for many it is not confirmed what evolutionary stage towards stable endosymbiont integration they represent. However, for two dinotom species, *Kryptoperidinium triquetrum* and *Unruhdinium penardii*, different isolates of the same dinoflagellate species have been reported to possess different permanent endosymbiotic diatom species [32, 33]. Thus, dinotom endosymbiont diversity is potentially great and likely includes further currently undetected cases of endosymbiont replacement and symbionts at different stages of integration across the full dinotom diversity.

The diversity of dinotom endosymbionts points to ongoing independent endosymbiotic processes in dinotoms. Molecular phylogenetic studies, however, show that there is strong taxon preference among dinotoms, with half of these 20 dinotom species containing diatom endosymbionts from the Bacillariophyceae, and specifically from clade '6B' (Table 1) [41, 42]. Even the preferred free-living diatoms of kleptoplastidic *D. capensis* belong to the diatom clade 6B [42]. Although *D. capensis* is able to use many diatom species as partially functional kleptoplastids, in which only the photosystems II and I are still active, carbon fixation in the endosymbiont is only seen with the favourable diatoms [36]. This constrained taxon choice of suitable partners for many dinotoms indicates that there have been preadaptations, likely through earlier kleptoplastidic interactions with diatoms of this clade, where some level of kleptoplastid support has developed that is specifically tailored to this clade.

4 Dinotom Endosymbionts are Surrounded by a Single Membrane and Possess Apparent Redundancy of Common Organelles

Eukaryotic endosymbiosis theory often implies the acquisition of extra membranes surrounding newly acquired plastids, with two extra membranes being the norm. This includes retention of the endosymbiont's plasma membrane, as well as a phagosome membrane from the host forming the new outermost membrane of the plastids. Dinotom endosymbionts, however, are surrounded by only one additional membrane leaving the cytoplasms of the host dinoflagellate and the diatom endosymbionts separated by this single membrane barrier (Figs. 1 and 2) [43]. This membrane must, therefore, be the point of exchange of any materials that are shared

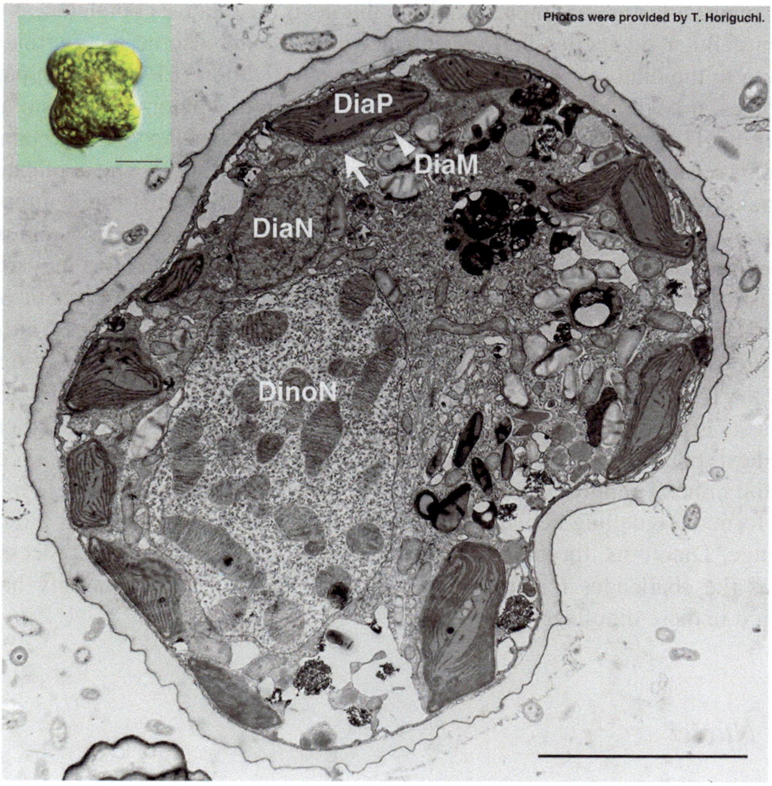

Fig. 2 A dinotom *Dinothrix quadrilobata* by light and transmission electron microscopy. (inset light microscope image) *D. quadrilobata* is a clover-shaped cyst-like immotile dinotom. The swimming cells appear only for a few hours after cell division. (Transmission electron microscope image) *D. quadrilobata* in an immotile cyst-like phase. The diatom nucleus (DiaN), diatom mitochondria (arrowhead; DiaM), and diatom plastids (DiaP) are observed. These diatom organelles are separated from the dinoflagellate cytoplasm by a single symbiosome membrane (arrow). DinoN = dinoflagellate nucleus. bar = 10 μm.

between the host and endosymbiont. The origin of this 'symbiosome' membrane is still controversial. Eschbach et al. [44] posited from freeze-fracture TEM data that this membrane is derived from the diatom plasma membrane. These authors speculated that the endosymbiotic diatoms were engulfed via phagocytosis but that the phagosomal membrane and the diatom cellular wall (frustule) were subsequently lost. However, later it was observed that dinotoms feed on diatoms using myzocytosis [15], a method in which the prey's organelles along with the cytoplasm are directly transferred to the dinoflagellate through a feeding tube called a peduncle [45]. This form of phagocytosis eliminates the prey's plasma membrane (and frustule) in dinotoms and the single symbiosome membrane is therefore likely derived from the phagosomal membrane of the host dinoflagellate.

An unusual feature of dinotom endosymbionts is the relative completeness of the diatom cellular organelles that are retained in the symbiosome compartment. In addition to the diatom plastids, the endoplasmic reticulum (ER), Golgi apparatus, ribosomes, mitochondria, and diatom nucleus within the diatom cytoplasm are all evident by TEM (Fig. 2) [21–23, 43, 46, 47]. These organelles are all retained in dinotoms regardless of which stage of endosymbiosis they are at. The duplication of core eukaryotic organelles in a single dinotom cell implies that the symbiosome membrane remains a very restrictive barrier between the host and endosymbiont. To date, no reports of protein targeting across it have been made, and this may explain the lack of diatom nuclear reduction [48–50]. The duplication of biosynthetic organelles and pathways implies that the exchange of metabolites is also limited, although there are some reports of metabolite transporters and vesicular export systems in dinotoms [50, 51]. This presents a conundrum because the raison d'etre for the symbiosis is presumably for the endosymbiont to provide a supply of fixed carbon to the host. In turn, the endosymbiont must at least be dependent on the host cell for essential mineral supply. Moreover, for coordination of cell division, there must be some form of signalling between the two which is presumably based on molecular exchange. Dinotoms, therefore, present opportunities to study these processes as well as the challenges for further molecular integration that seemingly has not occurred in these dinoflagellates.

4.1 Nuclei

Dinotom cells possess two nuclei with different ancestries, one belonging to the dinoflagellate and the other to the diatom endosymbiont. As with other elements of the host and endosymbiont, these two nuclei are spatially compartmentalised and separated by the symbiosome membrane. Vestigial endosymbiont nuclei, commonly called nucleomorphs, remain in three unrelated eukaryotic groups: cryptophytes, chlorarachniophytes, and some dinoflagellates possessing green-algal-derived plastids [16, 52, 53]. These nucleomorph genomes are substantially reduced relative to the nuclear genomes of free-living eukaryotes, with most endosymbiont genes transferred to the host nuclear genome and the protein products of these genes targeted back to the endosymbiont compartment [54, 55]. Unlike nucleomorphs, the nucleus of the diatom endosymbionts found in dinotoms maintains and expresses approximately the full coding capacity of their free-living diatom relatives [50]. The assembled genome sizes of endosymbiotic diatoms in *Durinskia oculata* and *K. triquetrum* are, respectively, 52.13 Mb or 39.99 Mb (Table 1), which are comparable to some free-living diatoms such as *Thalassiosira pseudonana* (31.30 Mb) [56] and *Fistulifera solaris* (49.73 Mb) [57]. Only nine genes so far have been identified from transcriptomic data that have likely been transferred from the diatom nucleus to the dinoflagellate nucleus, but it is unclear if these are functionally expressed [48, 50]. Thus, dinotoms apparently demonstrate that stable endosymbiosis can precede the establishment of protein-trafficking systems from the host to the endosymbiont.

In contrast to the coding capacity of dinotom endosymbiont nuclei, which has shown little change compared to free-living diatoms, the morphology and behaviour of these diatom nuclei have developed some unique properties within their dinoflagellate hosts. While the nucleus of free-living diatoms is typically spherical in shape during interphase, the nuclei of diatom endosymbionts in dinotoms are frequently extended, lobed, and can have a branched structure [10, 15, 21, 58]. During cell division and mitosis, the diatom nucleus then reverts to the round shape of free-living diatoms and divides synchronously with the host cell. However, rather than generating a clear mitotic spindle that is typically very conspicuous in diatom nuclei, dinotom endosymbiont nuclei lack ultrastructural evidence of a spindle and it is even suggested that they are 'amitotic' [10].

Although this nuclear dynamic has been known since early dinotom studies, it has been unclear why the diatom nucleus changes shape during interphase. *D. kwazulunatalensis* provides a clue for this unique nuclear dynamic [15]: in this dinotom species, the diatom nuclei become string- or dot-shaped and spatially move closer towards the cell surface of the host dinoflagellate, where the diatom plastids are localised by the host dinoflagellate, whilst still remaining within the symbiosome membrane. These diatom nuclei then revert to a round shape and return to the centre of the dinoflagellate cell prior to karyokinesis, and are then inherited by the two host daughter cells, respectively. A possible explanation for these morphological and spatial dynamics of endosymbiont nuclei is that host dinoflagellate may facilitate the positioning of the diatom nuclei to be closer to the diatom plastids to support plastid activity. Interestingly, this morphological change has not been observed in the kleptoplastidic dinotom *D. capensis*, where the nucleus of ingested diatoms retains the round shape of free-living diatoms [15, 36], suggesting that nuclear remodelling could be an integral part of the development of dinotom endosymbiont stability.

4.2 Mitochondria

In contrast to endosymbiotic nuclei, which can persist as nucleomorphs, the retention of endosymbiotic mitochondria is an uncommon feature of microalgae. This apparent rarity could indicate that mitochondria are typically lost at very early stages of plastid evolution, or are directly digested in the host cells during initial ingestion of endosymbionts [59–61]. The latter scenario often seems to be the case for many kleptoplastidic species, which typically only sequester prey plastids during prey ingestion. However, the retention of prey mitochondria is known from a handful of kleptoplastidic species, including the ciliates *Mesodinium* spp. (sequestered from cryptophyte prey) [62], *Perispira ovum* (sequestered from euglena prey) [63], the dinoflagellate *Nusuttodinium* spp. (sequestered from cryptophyte prey) [61], and dinotoms (sequestered from diatom prey). In dinotoms with stable diatom endosymbionts, it has been shown that the two endosymbiont mitochondrial genomes of *Durinskia baltica* and *Kryptoperidinium foliaceum* (currently,

Kryptoperidinium sp. according to [33]) retain a high degree of synteny with one another, as well as with the mitochondrial genome of the free-living diatom *Fragilariopsis cylindrus* [64]. The transcription of several mitochondria genes of the endosymbiont, including NADH-dehydrogenase subunits 2 (*nad2*) and 5 (*nad5*) of Complex I; cytochrome b (*cob*) of Complex III; and cytochrome c oxidase 1 (*cox1*), cytochrome c oxidase 2 (*cox2*) and cytochrome c oxidase 3 (*cox3*) of Complex IV; as well as a gene block encoding the mitochondrial proteins rps19-rps3-rpl16; and a gene for the mitochondrial large subunit ribosomal RNA (LSU rRNA) were also detected in RT-qPCR and RNA-seq data, suggesting that the diatom mitochondria in dinotoms are functional [65, 66]. Despite the maintenance of these functional diatom mitochondrial genomes, the *cox1* and *cob* genes are also expressed from host dinoflagellate mitochondria [65]. Thus, the dinotom cell possesses functionally overlapping, seemingly redundant, mitochondrial organelles.

Electron microscopy data have also shown that the diatom mitochondria are structurally well conserved and maintained typical features of diatom mitochondria: enclosed by a double-membrane, possess tubular cristae, and are located in close proximity to plastids [28, 46]. The kleptoplastidic dinotom *D. capensis* has also been shown to retain structurally intact diatom mitochondria from ingested prey, and these were maintained for at least 21 days after collection from natural habitats [15, 47]. These observations demonstrate that the mitochondria of diatom endosymbionts remain functionally intact within dinoflagellate host cells, and substantial reduction of these organelles and their genomes has not occurred.

4.3 Plastids

Another surprising case of organelle redundancy in dinotoms is the presence of two evolutionarily distinct plastids. The first of these, the diatom plastid, is the more conspicuous and drives the autotrophic capacity of dinotoms. These plastids are surrounded by four bounding membranes as is also the case in free-living diatoms [43, 47]. The diatom-derived plastids are functionally identical to free-living diatoms in their photosynthetic ability [15, 36]. Their thylakoid and pyrenoids structures are preserved and they express all the expected photosynthesis-related genes [21, 50]. The structure, size, and gene content of the plastid genome in the diatom endosymbiont of *D. baltica* are remarkably similar to those of a free-living model diatom *Phaeodactylum tricornutum*: only three genes (tsf, acpP, and ycf42) are absent from this dinotom plastid [20]. Interestingly, the plastid genome of *Kryptoperidinium* sp. is larger than those of *D. baltica* and *P. tricornutum* with more non-coding DNA than has been found even in free-living diatoms. Furthermore, some dinotom plastid genomes code several additional genes (tyrC, serC1, and serC2) which have not been found in *P. tricornutum* [20].

The second, less conspicuous plastid is a relic of the peridinin plastid that has lost its photosynthetic capacity, instead serving as a phototactic eyespot and which has been noted in dinotoms since their first observation. Eyespots are widely distributed in motile cells of many groups of algae and in general, they function in phototaxis [67]. All dinotom species possess the same structure of eyespot, defined as Type D, which is unique to dinotoms (Figs. 2 and 3) [68]. The dinotom eyespot is always located near the longitudinal flagellum and consists of rows of red-pigmented plastoglobules surrounded by a triple membrane layer. The more common dinoflagellate eyespot is classified as Type A [68], which is formed by a row of red-pigmented lipid globules located within the plastid lumen amongst the thylakoids. As early as 1984, Dodge speculated that the Type D eyespot is dervied from the dinoflagellate peridinin plastid due to the conspicuous retention of three bounding membranes, which is a distinctive characteristic of peridinin plastids [69]. Dodge therefore inferred that the thylakoid membranes had been lost from this organelle, and only the plastoglobules appeared to have been retained. Later, this hypothesis was supported by Hehenberger et al. [70] who showed that the peridinin plastid protein translocon, Tic110, is retained in dinotoms. Moreover, they observed that genes for metabolic functions—the tetrapyrrole pathway and methylerythritol phosphate/1-deoxy-D-xylulose-5-phosphate (MEP/DOXP) pathway for isoprene precursors [70]—with peridinin-type features are also present and likely provide biosynthetic functions in this reduced plastid. These pathways are also present in the diatom plastid, and therefore, as for the mitochondria, there is redundancy and evidence of a surprising lack of sharing of some common metabolites between host and symbiont in dinotoms. It is intriguing to note that in continuous cultures some dinotoms have been observed to lose their eyespots (or their pigmentation at least), potentially due to the lack of need for phototaxis in these artificial light-saturating conditions [71]. This may indicate that the tetrapyrrole and isoprenoid precursor synthesis is primarily required for eyespot function and not exported to the rest of the cell.

5 Dinotoms as Models for Early Plastid Organelle Development and/or an Alternative Path to Stable Photosynthetic Endosymbiosis

To understand the processes of organellogenesis, we need model systems that represent the separate key stages in this process, particularly examples at early points of symbiont establishment. The photosynthetic and heterotrophic amoeba, *Paulinella* spp., have provided exciting models for studying an independent primary endosymbiosis of a cyanobacterium [72, 73]. However, the processes of eukaryotic endosymbioses are likely different with the coordination and merger of two eukaryotic-type systems required for these long-lived partnerships. Many model systems have been proposed for this purpose, including sea slugs *Elysia* spp. [74, 75], ciliates *Mesodinium* spp. [62, 76, 77], the cryptophyte *Guillardia theta* [54], the chlororarachniophyte *Bigelowiella natans* [55], and undescribed dinoflagellates MGD and TGD with green algal endosymbionts [16]. But these all represent cases

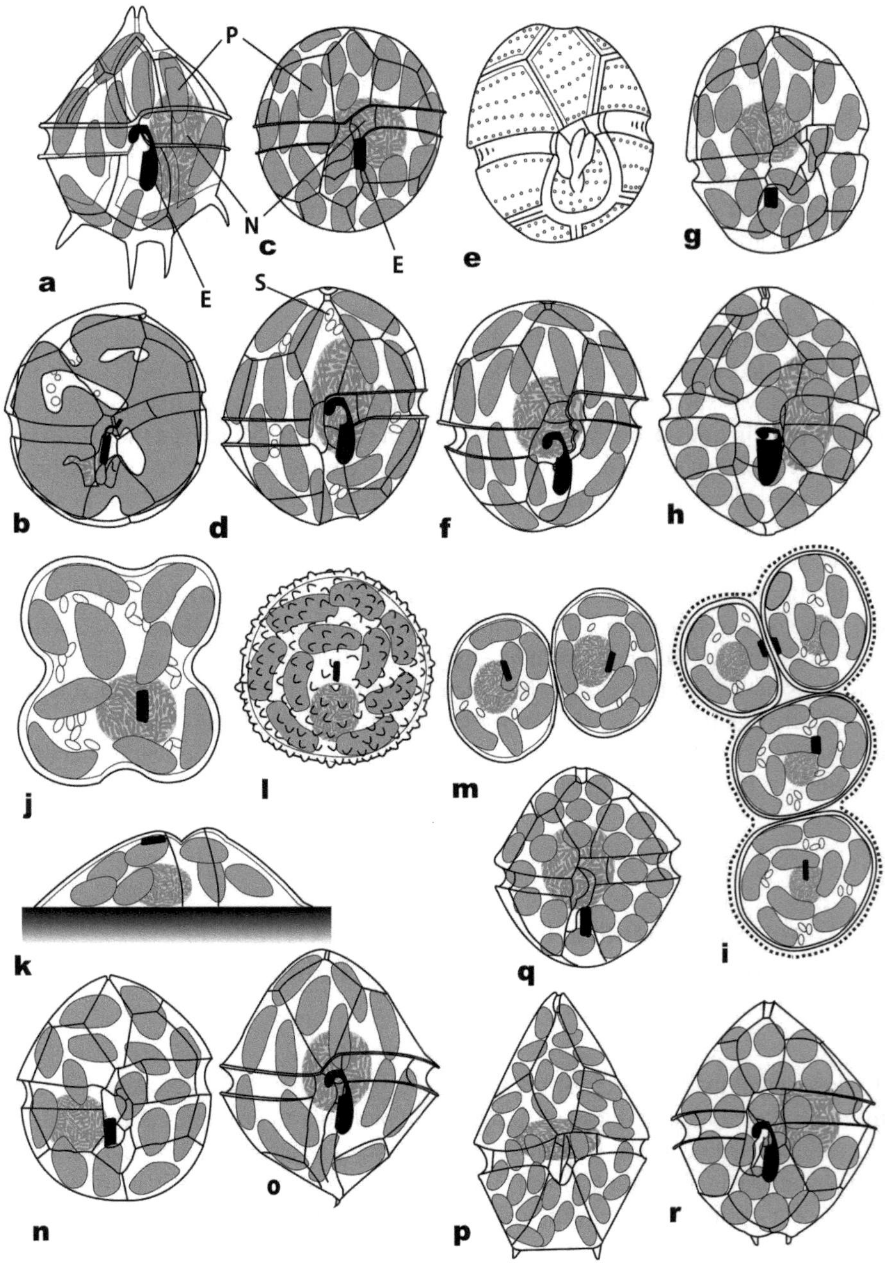

Fig. 3 The eyespot is a well-conserved morphological trait of dinotoms. Dinotoms are morphologically diverse and in many cases it is difficult to identify common morphological features between species. One exception to this is D-type eyespot, present in most dinotoms, which seems to be either hook-shaped or rectangular depending on the species. The eyespot is located near the sulcus of the dinoflagellate. It is typically observable during the motile swimming stage but

of either unestablished kleptoplastids or permanent endosymbionts already with substantial nuclear gene transfer such that only nucleomorphs remain in the endosymbiont. Dinotoms offer unique models for endosymbiosis because they represent a point in endosymbiont fixation apparently before gene relocation to the host nucleus and targeting of proteins back to the symbiont has occurred. They offer multiple lineages with different endosymbionts allowing the reproducibility of processes and adaptations to be examined, and they also contain examples prior to this where kleptoplasty is still maintained.

Dinotoms, as models for organelle origins, also raise an important question: do all endosymbionts follow the same path during organellogenesis? Protein trafficking to endosymbionts is considered a key step to becoming an organelle. It requires two essential processes; (1) the expression of nucleus-encoded proteins with appropriate regulation and (2) sorting mechanisms to direct proteins both to and across the membranes that surround the compartments where the proteins function (e.g. plastid stroma). In many cases of organellogenesis, some of the existing membrane translocons of the endosymbiont are employed: the Translocons of the Outer and Inner Chloroplast membranes (TOCs and TICs) that transfer proteins across the inner two membranes, and a repurposed ER-Associated Degradation (ERAD) translocon that transfers proteins across a third membrane in many algae with red-algal-derived complex plastids [78, 79]. Thus, with endosymbiont gene transfer (EGT) from the endosymbiont to the host, much of the machinery required for the development of protein trafficking might often already be in place. The 'shopping bag' model for EGT suggests that gene transfer and protein targeting are established before endosymbionts become fixed as stable symbionts and some models for early endosymbiont establishment do show evidence of protein targeting at the kleptoplasty stage [80]. Both the euglenoid *Rapaza viridis* and the RSD dinoflagellates have seemingly acquired genes in their nuclei to support their kleptoplastids [81, 82]. While it is yet to be demonstrated, these genes are presumed to target as proteins back to the kleptoplastid to sustain key symbiont functions and kleptoplastid longevity [82, 83]. In a slightly different arrangement, the ciliate *Mesodinium* spp. separately retains its kleptoplastids' nuclei (so-called kleptokaryons) which express genes for kleptoplastid maintenance [62]. Indeed, so successful is *Mesodinium* in maintaining its kleptoplastids that they can even divide these plastids, and the host

Fig. 3 (continued) becomes difficult to identify by light microscopy when the dinotoms are in the cyst stage [51, 58]. With the exception of *Durinskia dybowskii* (e) and *Unruhdinium niei* (p), all dinotoms have been reported to possess the D-type eyespot. The dinotoms depicted are (**a**) *Blixaea quinquecornis*, (**b**) *Durinskia agilis* [modified from [86]]; (**c**) *Durinskia baltica*, (**d**) *Durinskia capensis*, (**e**) *Durinskia dybowskii* [after [87]], (**f**) *Durinskia kwazulunatalensis*, (**g**) *Durinskia oculata* [modified from [88]], (**h**) *Kryptoperidinium triquetrum* [modified from [33]], (**i**) *Dinothrix paradoxa*, (**j**) *Dinothrix quadrilobata*, (**k**) *Dinothrix rugata*, l: *Dinothrix phymatodea*, (**m**) *Dinothrix pseudoparadoxa*, n: *Unruhdinium jiulongense* [modified from [89]], (**o**) *Unruhdinium kevei*, (**p**) *Unruhdinium niei* [modified from [90]], (**q**) *Unruhdinium minima* [based on figures in [91]], and (**r**) *Unruhdinium penardii*. The drawings without references are all made using private dinotom photos taken by T. Horiguchi. Grey object = Plastid (P), Black object = Eyespot (E), White circle = Starch granule (S), and Spherical grey object with linear structures = Dinoflagellate nucleus (N)

is required to acquire kleptokaryons for their maintenance more often than replacing the plastids themselves [40]. These cases all suggest that effective protein targeting from the host cytoplasm to the symbiont has been achieved before the fixation of a permanent endosymbiont. But this is seemingly the opposite for dinotoms that show the repeated establishment of stable endosymbionts before protein targeting for maintenance is developed. The reasons for the lack of protein targeting are unclear, but it might be that the symbiosome membrane, in addition to the four membranes around the diatom plastid, has created too great a series of obstacles for protein trafficking processes to overcome. Whether this challenge for protein targeting is simply yet to be met, and dinotoms will ultimately follow a similar route of other alga to genetically integrated plastid organelles, or whether dinotoms are set on their own path where their symbionts will be locked in this more complete eukaryotic stage, is unclear (Fig. 1).

The absence of protein targeting to dinotom stable endosymbionts nevertheless provides a further opportunity for endosymbiosis studies—to understand the forms of communication and coordination that exist before (or without) direct genetic control from the host. Many photosynthetic endosymbionts show radical morphological changes upon uptake in their hosts, for example *Nephroselmis* as a temporary symbiont of *Hatena arenicola* expands massively to fill the host's cytoplasm following uptake [84]. This implies that the host can rapidly direct developmental programmes in the endosymbiont. In dinotoms, the coordination of endosymbiont volume, metabolism, and division with host cytokinesis all must occur through unknown, but finely controlled, signalling processes. The unique point of dinotom endosymbiont evolution thus provides opportunities to unveil these mechanisms.

Establishing dinotoms as experimental model systems is showing great promise with increasing amounts of transcriptomic and genomic data becoming available, and more in the pipeline (Table 1). Moreover, while experimental genetic tools in dinoflagellates are still in their early days, GFP expression and some success with the CRISPR/Cas9 genome editing tool achieved for the diatom symbiont *N. captiva* provide opportunities to manipulate and study kleptoplasty in *D. capensis* [85]. These tools offer exciting opportunities for exploiting these new endosymbiotic models.

6 Conclusions

Dinotoms have demonstrated that organelle origins involve the early development of symbiont taxon preferences likely through species-specific preadaptations; that within these emerging host–symbiont relationships, the fixation of stable endosymbionts can happen multiple times and potentially repeatedly; and that while protein targeting may precede endosymbiont stability, this is not essential and that multiple routes might lead to the gain of new plastids. These lessons shape the way we must think about the origins of all of plastid diversity and are likely not the last discoveries that will be made by the fascinating cases of dinotoms.

References

1. Delwiche CF (1999) Tracing the thread of plastid diversity through the tapestry of life. Am Nat 154:164–177. https://doi.org/10.1086/303291
2. Archibald JM (2015) Endosymbiosis and eukaryotic cell evolution. Curr Biol 25:911–921. https://doi.org/10.1016/j.cub.2015.07.055
3. Keeling PJ (2010) The endosymbiotic origin, diversification and fate of plastids. Philos Trans R Soc B 365:729–748. https://doi.org/10.1098/rstb.2009.0103
4. Zhang Z, Green BR, Cavalier-Smith T (1999) Single gene circles in dinoflagellate chloroplast genomes. Nature 400:155–159. https://doi.org/10.1038/220999
5. Waller RF, Koreny L (2017) Plastid complexity in dinoflagellates: a picture of gains, losses, replacements and revisions. Secondary Endosymbioses 84:105–143. https://doi.org/10.1016/bs.abr.2017.06.004
6. Hoppenrath M (2017) Dinoflagellate taxonomy - a review and proposal of a revised classification. Mar Biodiv 47:381–403. https://doi.org/10.1007/s12526-016-0471-8
7. De Vargas CD, Audic S, Henry N, Decelle J, Mahé F, Logares R et al (2015) Eukaryotic plankton diversity in the sunlit ocean. Science 348:1261605. https://doi.org/10.1126/science.1261605
8. Saldarriaga JF, Taylor FJR, Keeling PJ, Cavalier-Smith T (2001) Dinoflagellate nuclear SSU rRNA phylogeny suggests multiple plastid losses and replacements. J Mol Evol 53:204–213. https://doi.org/10.1007/s002390010210
9. Janouškovec J, Gavelis GS, Burki F, Dinh D, Bachvaroff TR, Gornik SG et al (2017) Major transitions in dinoflagellate evolution unveiled by phylotranscriptomics. Proc Natl Acad Sci U S A 114:171–180. https://doi.org/10.1073/pnas.1614842114
10. Tippit DH, Pickett-Heaps JD (1976) Apparent amitosis in the binucleate dinoflagellate *Peridinium balticum*. J Cell Sci 21:273–289. https://doi.org/10.1242/jcs.21.2.273
11. Watanabe MM, Suda S, Inouye I, Sawaguchi T, Chihara M (1990) *Lepidodinium viride* gen. et sp. nov. (Gymnodiniales, Dinophyta), a green dinoflagellate with a chlorophyll *a*- and *b*-containing endosymbiont. J Phycol 26:741–751. https://doi.org/10.1111/j.0022-3646.1990.00741.x
12. Gast RJ, Moran DM, Beaudoin DJ, Blythe JN, Dennett MR, Caron DA (2006) Abundance of a novel dinoflagellate phylotype in the Ross Sea, Antarctica. J Phycol 42:233–242. https://doi.org/10.1111/j.1529-8817.2006.00183.x
13. Park MG, Kim S, Kim HS, Myung G, Kang YG, Yih W (2006) First successful culture of the marine dinoflagellate *Dinophysis acuminata*. Aquat Microb Ecol 45:101–106. https://doi.org/10.3354/ame045101
14. Park MG, Kim M, Kang M (2013) A dinoflagellate *Amylax triacantha* with plastids of the cryptophyte origin: phylogeny, feeding mechanism, and growth and grazing responses. J Eukaryot Microbiol 60:363–376. https://doi.org/10.1111/jeu.12041
15. Yamada N, Bolton JJ, Trobajo R, Mann DG, Dąbek P, Witkowski A et al (2019) Discovery of a kleptoplastic 'dinotom' dinoflagellate and the unique nuclear dynamics of converting kleptoplastids to permanent plastids. Sci Rep 9:10474. https://doi.org/10.1038/s41598-019-46852-y
16. Sarai C, Tanifuji G, Nakayama T, Kamikawa R, Takahashi K, Yazaki E et al (2020) Dinoflagellates with relic endosymbiont nuclei as models for elucidating organellogenesis. Proc Natl Acad Sci U S A 117:5364–5375. https://doi.org/10.1073/pnas.1911884117
17. Ok JH, Jeong HJ, Lee SY, Park SA, Noh JH (2021) *Shimiella* gen. nov. and *Shimiella gracilenta* sp. nov. (Dinophyceae, Kareniaceae), a kleptoplastidic dinoflagellate from Korean waters and its survival under starvation. J Phycol 57:70–91. https://doi.org/10.1111/jpy.13067
18. Riley JP, Wilson TRS (1967) The pigments of some marine phytoplankton species. J Mar Biol Assoc U K 47:351–361. https://doi.org/10.1017/S0025315400056459
19. Mandelli E (1968) Carotenoid pigments of the dinoflagellate *Glenodinium foliaceum* Stein. J Phycol 4:347–348. https://doi.org/10.1111/j.1529-8817.1968.tb04706.x
20. Imanian B, Pombert J, Keeling PK (2010) The complete plastid genomes of the two 'dinotoms' *Durinskia baltica* and *Kryptoperidinium foliaceum*. PloS One 5:e10711. https://doi.org/10.1371/journal.pone.0010711

21. Dodge JD (1971) A dinoflagellate with both a mesocaryotic and a eucaryotic nucleus I. Fine structure of the nuclei. Protoplasma 73:145–157. https://doi.org/10.1007/BF01275591
22. Tomas RN, Cox ER, Steidinger KA (1973) *Peridinium balticum* (Levander) Lemmermann, an usual dinoflagellate with a mesocaryotic and an eucaryotic nucleus. J Phycol 9:91–98. https://doi.org/10.1111/j.0022-3646.1973.00091.x
23. Tomas RN, Cox ER (1973) Observations on the symbiosis of *Peridinium balticum* and its intracellular alga. I. ultrastracture. J Phycol 9:304–323. https://doi.org/10.1111/j.1529-8817.1973.tb04098.x
24. Chesnick JM, Morden CW, Schmieg AM (1996) Identity of the endosymbiont of *Peridinium foliaceum* (Pyrrophyta): analysis of the *rbcLS* operon. J Phycol 32:850–857. https://doi.org/10.1111/j.0022-3646.1996.00850.x
25. Chesnick JM, Kooistra WHCF, Wellbrock U, Medlin LK (1997) Ribosomal RNA analysis indicates a benthic pennate diatom ancestry for the endosymbionts of the dinoflagellates *Peridinium foliaceum* and *Peridinium balticum* (Pyrrhophyta). J Eukaryot Microbiol 44:314–320. https://doi.org/10.1111/j.1550-7408.1997.tb05672.x
26. Gottschling M, Čalasan AŽ, Kretschmann J (2017) Two new generic names for dinophytes harbouring a diatom as an endosymbiont, *Blixaea* and *Unruhdinium* (Kryptoperidiniaceae, Peridiniales). Phytotaxa 306:296–300. https://doi.org/10.11646/PHYTOTAXA.306.4.6
27. Inagaki Y, Dacks JB, Doolittle WF, Watanabe KI, Ohama T (2000) Evolutionary relationship between dinoflagellates bearing obligate diatom endosymbionts: insight into tertiary endosymbiosis. Int J Syst Evol Microbiol 50:2075–2081. https://doi.org/10.1099/00207713-50-6-2075
28. Tamura M, Shimada S, Horiguchi T (2005) *Galeidinium rugatum* gen. et sp. nov. (Dinophyceae), a new coccoid dinoflagellate with a diatom endosymbiont. J Phycol 41:658–671. https://doi.org/10.1111/j.1529-8817.2005.00085.x
29. Horiguchi T, Takano Y (2006) Serial replacement of a diatom endosymbiont in the marine dinoflagellate *Peridinium quinquecorne* (Peridiniales, Dinophyceae). Phycol Res 54:193–200. https://doi.org/10.1111/j.1440-1835.2006.00426.x
30. Takano Y, Hansen G, Fujita D, Horiguchi T (2008) Serial replacement of diatom endosymbionts in two freshwater dinoflagellates, *Peridiniopsis* spp. (Peridiniales, Dinophyceae). Phycologia 47:41–53. https://doi.org/10.2216/07-36.1
31. Yamada N, Sym SD, Horiguchi T (2017) Identification of highly divergent diatom-derived chloroplasts in dinoflagellates, including a description of *Durinskia kwazulunatalensis* sp. nov. (Peridiniales, Dinophyceae). Mol Biol Evol 34:1335–1351. https://doi.org/10.1093/molbev/msx054
32. Mertens KN, Takano Y, Meyvisch P, Carbonell-Moore MC, Chomérat N, Bogus K, Leitão M (2021) Morpho-molecular and spectroscopic characterization of the freshwater dinoflagellate *Unruhdinium penardii* var. *robustum* (Kryptoperidiniaceae, Peridiniales), blooming in the Loir River, France. Nova Hedwigia 112:283–306. https://doi.org/10.1127/nova_hedwigia/2021/0633
33. Tillmannn U, Wietkamp S, Kretschmann J, Chacón J, Gottschling M (2023) Spatial fragmentation in the distribution of diatom endosymbionts from the taxonomically clarified dinophyte *Kryptoperidinium triquetrum* (= *Kryptoperidinium foliaceum*, Peridiniales). Sci Rep 13:8593. https://doi.org/10.1038/s41598-023-32949-y
34. Tengs T, Dahlberg OJ, Shalchian-Tabrizi K, Klaveness D, Rudi K, Delwiche CF et al (2000) Phylogenetic analyses indicate that the 19' hexanoyloxy-fucoxanthin-containing dinoflagellates have tertiary plastids of haptophyte origin. Mol Biol Evol 17:718–729. https://doi.org/10.1093/oxfordjournals.molbev.a026350
35. Takahashi K, Benico G, Lum WM, Iwataki M (2019) *Gertia stigmatica* gen. et sp. nov. (Kareniaceae, Dinophyceae), a new marine unarmored dinoflagellate possessing the peridinin-type chloroplast with an eyespot. Protist 170:125680. https://doi.org/10.1016/j.protis.2019.125680
36. Yamada N, Lepetit B, Mann DG, Sprecher BN, Buck JM, Bergmann P et al (2023) Prey preference in a kleptoplastic dinoflagellate is linked to photosynthetic performance. ISME J 17:1578–1588. https://doi.org/10.1038/s41396-023-01464-3

37. Kempton JW, Wolny J, Tengs T, Rizzo P, Morris R, Tunnell J et al (2002) *Kryptoperidinium foliaceum* blooms in South Carolina: a multi-analytical approach to identification. Harmful Algae 1:383–392. https://doi.org/10.1016/S1568-9883(02)00051-3
38. Zhang Q, Liu G, Hu Z (2011) Morphological differences and molecular phylogeny of freshwater blooming species, *Peridiniopsis* spp. (Dinophyceae) from China. Eur J Protistol 47:149–160. https://doi.org/10.1016/j.ejop.2011.03.001
39. Okolodkov YB, Merino-Virgilio FC, Huerta-Quintanilla DA, Gárate-Lizárraga I, Steidinger KA, Aguilar-Trujillo AC et al (2020) A Kryptoperidiniaceae species (Dinophyceae: Peridiniales) blooming in coastal Yucatan waters, Gulf of Mexico. Protistology 14:58–69
40. Kim M, Drumm K, Daugbjerg N, Hansen PJ (2017) Dynamics of sequestered cryptophyte nuclei in *Mesodinium ruburum* during starvation and refeeding. Front Microbiol 8. https://doi.org/10.3389/fmicb.2017.00423
41. Mann DG, Trobajo R, Sato S, Li C, Witkowski A, Rimet F et al (2021) Ripe for reassessment: a synthesis of available molecular data for the speciose diatom family Bacillariaceae. Mol Phylogenet Evol 158:106985. https://doi.org/10.1016/j.ympev.2020.106985
42. Mann DG, Yamada N, Bolton JJ, Witkowski A, Trobajo R (2023) *Nitzschia captiva* sp. nov., the essential prey diatom of the kleptoplastic dinoflagellate *Durinskia capensis*, compared with *N. agnita*, *N. kuetzingioides* and other species. Phycologia 62:136–151. https://doi.org/10.1080/00318884_2023.2169024
43. Jeffery SW, Vesk M (1976) Further evidence for a membrane-bound endosymbiont within the dinoflagellate *Peridinium foliaceum*. J Phycol 12:450–455. https://doi.org/10.1111/j.1529-8817.1976.tb02872.x
44. Eschbach S, Speth V, Hansmann P, Sitte P (1990) Freeze-fracture study of the single membrane between host cell and endosymbiont in the dinoflagellates *Glenodinium foliaceum* and *Peridinium balticum*. J Phycol 26:324–328
45. Hansen PJ, Calado AJ (1999) Phagotrophic mechanisms and prey selection in free-living dinoflagellates. J Eukaryot Microbiol 46:382–389. https://doi.org/10.1111/j.1550-7408.1999.tb04617.x
46. Horiguchi T, Pienaar RN (1991) Ultrastructure of a marine dinoflagellate, *Peridinium quinquecorne* Abé (Peridiniales) from South Africa with particular reference to its chrysophyte endosymbiont. Botanica Marina 34:123–131. https://doi.org/10.1515/botm.1991.34.2.123
47. Pienaar RN, Sakai H, Horiguchi T (2007) Description of a new dinoflagellate with a diatom endosymbiont, *Durinskia capensis* sp. nov. (Peridiniales, Dinophyceae) from South Africa. J Plant Res 120:247–258. https://doi.org/10.1007/s10265-006-0047-y
48. Burki F, Imanian B, Hehenberger E, Hirakawa Y, Maruyama S, Keeling PJ (2014) Endosymbiotic gene transfer in tertiary plastid-containing dinoflagellates. Eukaryot Cell 13:246–255. https://doi.org/10.1128/ec.00299-13
49. Imanian B, Keeling PJ (2014) Horizontal gene transfer and redundancy of tryptophan biosynthetic enzymes in dinotoms. Genome Biol Evol 6:333–343. https://doi.org/10.1093/gbe/evu014
50. Hehenberger E, Burki F, Kolisko M, Keeling PJ (2016) Functional relationship between a dinoflagellate host and its diatom endosymbiont. Mol Biol Evol 33:2376–2390. https://doi.org/10.1093/molbev/msw109
51. Yamada N, Sakai H, Onuma R, Horiguchi T (2020) Five non-motile dinotom dinoflagellates of the genus *Dinothrix*. Front Plant Sci 11. https://doi.org/10.3389/fpls.2020.591050
52. Hibberd D, Norris R (1984) Cytology and ultrastructure of *Chlorarachnion reptans* (Chlorarachniophyta divisio nova, Chlorarachniophyceae classis nova). J Phycol 20:310–330. https://doi.org/10.1111/j.0022-3646.1984.00310.x
53. Douglas SE, Murphy C, Spencer D, Gray M (1991) Cryptomonad algae are evolutionary chimaeras of two phylogenetically distinct unicellular eukaryotes. Nature 350:149–151. https://doi.org/10.1038/350148a0
54. Douglas S, Zauner S, Fraunholz M, Beaton M, Penny S, Deng LT et al (2001) The highly reduced genome of an enslaved algal nucleus. Nature 410:1091–1096

55. Gilson PR, Su V, Slamovits CH, Reith ME, Keeling PJ, McFadden GI (2006) Complete nucleotide sequence of the chlorarachniophyte nucleomorph: nature's smallest nucleus. Proc Natl Acad Sci U S A 103:9566–9571. https://doi.org/10.1073/pnas.0600707103
56. Armbrust V, Berges JA, Bowler C, Green BR, Martinez D, Putnam NH et al (2004) The genome of the diatom *Thalassiosira pseudonana*: ecology, evolution, and metabolism. Science 306:79–86. https://doi.org/10.1126/science.1101156
57. Tanaka T, Maeda Y, Veluchamy A, Tanaka M, Abida H, Maréchal E et al (2015) Oil accumulation by the oleaginous diatom *Fistulifera solaris* as revealed by the genome and transcriptome. Plant Cell 27:162–176. https://doi.org/10.1105/tpc.114.135194
58. Figueroa RI, Bravo I, Fraga S, Garcés E, Llaveria G (2009) The life history and cell cycle of *Kryptoperidinium foliaceum*, a dinoflagellate with two eukaryotic nuclei. Protist 160:285–300. https://doi.org/10.1016/j.protis.2008.12.003
59. Kim M, Nam SW, Shin W, Coats DW, Park MG (2012) *Dinophysis caudata* (Dinophyceae) sequesters and retains plastids from the mixotrophic ciliate prey *Mesodinium rubrum*. J Phycol 48:569–579. https://doi.org/10.1111/j.1529-8817.2012.01150.x
60. Yamaguchi A, Yubuki N, Leander BS (2012) Morphostasis in a novel eukaryote illuminates the evolutionary transition from phagotrophy to phototrophy: description of *Rapaza viridis* n. gen. et sp. (Euglenozoa, Euglenida). BMC Evol Biol 12:1–16. https://doi.org/10.1186/1471-2148-12-29
61. Onuma R, Horiguchi T (2013) Morphological transition in kleptochloroplasts after ingestion in the dinoflagellates *Amphidinium poecilochroum* and *Gymnodinium aeruginosum* (Dinophyceae). Protist 164:622–642. https://doi.org/10.1016/j.protis.2013.06.003
62. Johnson MD, Oldach D, Delwiche CF, Stoecker DK (2007) Retention of transcriptionally active cryptophyte nuclei by the ciliate *Myrionecta rubra*. Nature 445:426–428. https://doi.org/10.1038/nature05496
63. Johnson PW, Donaghay PL, Small EB, Sieburth JM (1995) Ultrastructure and ecology of *Perispira ovum* (Ciliophora: Litostomatea): an aerobic, planktonic ciliate that sequesters the chloroplasts, mitochondria, and paramylon of *Euglena proxima* in a micro-oxic habitat. J Eukaryot Microbiol 42:323–335. https://doi.org/10.1111/j.1550-7408.1995.tb01588.x
64. Imanian B, Pombert JF, Dorrell RG, Burki F, Keeling PJ (2012) Tertiary endosymbiosis in two dinotoms has generated little change in the mitochondrial genomes of their dinoflagellate hosts and diatom endosymbionts. PloS One 7:e43763. https://doi.org/10.1371/journal.pone.0043763
65. Imanian B, Carpenter KJ, Keeling PJ (2007) Mitochondrial genome of a tertiary endosymbiont retains genes for electron transport proteins. J Eukaryot Microbiol 54:146–153. https://doi.org/10.1111/j.1550-7408.2007.00245.x
66. Imanian B, Keeling PJ (2007) The dinoflagellates *Durinskia baltica* and *Kryptoperidinium foliaceum* retain functionally overlapping mitochondria from two evolutionarily distinct lineages. BMC Evol Biol 7:1–11. https://doi.org/10.1186/1471-2148-7-172
67. Kawai H, Kreimer G (2000) Sensory mechanisms. In: The flagellates: unity, diversity and evolution. Taylor & Francis, New York, p 124. https://doi.org/10.2307/3285296
68. Moestrup Ø, Calado AJ (2018) Süßwasserflora von Mitteleuropa, Bd. 6-freshwater flora of Central Europe, Vol. 6: Dinophyceae, vol 6. Springer. https://doi.org/10.1007/978-3-662-56269-7
69. Dodge JD (1984) The functional and phylogenetic significance of dinoflagellate eyespots. Biosystems 16:259–267. https://doi.org/10.1016/0303-2647(83)90009-6
70. Hehenberger E, Imanian B, Burki F, Keeling PJ (2014) Evidence for the retention of two evolutionary distinct plastids in dinoflagellates with diatom endosymbionts. Genome Biol Evol 6:2321–2334. https://doi.org/10.1093/gbe/evu182
71. Moldrup M, Moestrup Ø, Hansen PJ (2013) Loss of phototaxis and degeneration of an eyespot in long-term algal cultures: evidence from ultrastructure and behaviour in the dinoflagellate *Kryptoperidinium foliaceum*. J Eukaryot Microbiol 60:327–334. https://doi.org/10.1111/jeu.12036
72. Gabr A, Grossman AR, Bhattacharya D (2020) *Paulinella*, a model for understanding plastid primary endosymbiosis. J Phycol 56:837–843. https://doi.org/10.1111/jpy.13003

73. Bhattacharya D, Etten JV, Benites LF, Stephens TG (2023) Endosymbiotic ratchet accelerates divergence after organelle origin: the *Paulinella* model for plastid evolution. Bioessays 45:2200165. https://doi.org/10.1002/bies.202200165
74. Vries JD, Christa G, Gould SB (2014) Plastid survival in the cytosol of animal cells. Trends Plant Sci 19:347–350. https://doi.org/10.1016/j.tplants.2014.03.010
75. Cruz S, Paulo C (2022) Kleptoplasty: getting away with stolen chloroplasts. PLoS Biol 20:e3001857. https://doi.org/10.1371/journal.pbio.3001857
76. Gustafson DE, Stoecker DK, Johnson MD, Heukelem WFV, Sneider K (2000) Cryptophyte algae are robbed of their organelles by the marine ciliate *Mesodinium rubrum*. Nature 405:1049–1052. https://doi.org/10.1038/35016570
77. Moestrup Ø, Garcia-Cuetos L, Hansen PJ, Fenchel T (2012) Studies on the genus *Mesodinium* I: ultrastructure and description of *Mesodinium chamaeleon* n. sp., a benthic marine species with green or red chloroplasts. J Eukaryot Microbiol 59:20–39. https://doi.org/10.1111/j.1550-7408.2011.00593.x
78. Bolte K, Bullmann L, Hempel F, Bozarth A, Zauner S, Maier UG (2009) Protein targeting into secondary plastids. J Eukaryot Microbiol 56:9–15. https://doi.org/10.1111/j.1550-7408.2008.00370.x
79. Bodył A (2018) Did some red alga-derived plastids evolve *via* kleptoplastidy? A hypothesis. Biol Rev 93:201–222. https://doi.org/10.1111/brv.12340
80. Larkum AWD, Lockhart PJ, Howe CJ (2007) Shopping for plastids. Trends Plant Sci 12:189–195. https://doi.org/10.1016/j.tplants.2007.03.011
81. Hehenberger E, Gast RJ, Keeling PJ (2019) A kleptoplastidic dinoflagellate and the tipping point between transient and fully integrated plastid endosymbiosis. Proc Natl Acad Sci U S A 116:17934–17942. https://doi.org/10.1073/pnas.1910121116
82. Karnkowska A, Yubuki N, Maruyama M, Yamaguchi A, Kashiyama Y, Suzaki T et al (2023) Euglenozoan kleptoplasty illuminates the early evolution of photoendosymbiosis. Proc Natl Acad Sci U S A 120:e2220100120. https://doi.org/10.1073/pnas.2220100120
83. Maruyama M, Kagamoto T, Matsumoto Y, Onuma R, Miyagishima S, Tanifuji G et al (2023) Horizontally acquired nitrate reductase realized kleptoplastic photoautotrophy of *Rapaza viridis*. Plant Cell Physiol 64:pcad044. https://doi.org/10.1093/pcp/pcad044
84. Okamoto N, Inouye I (2006) *Hatena arenicola* gen. et sp. nov., a katablepharid undergoing probable plastid acquisition. Protist 157:401–419. https://doi.org/10.1016/j.protis.2006.05.011
85. Sprecher BN, Buck JM, Ropella LL, Ramsperger A, Kroth PG, Yamada N (2023) Genetic transformation methods for diatom *Nitzschia captiva*: new tools to better understand dinotom endosymbiosis. Algal Res 72:103136. https://doi.org/10.1016/j.algal.2023.103136
86. Saburova M, Chomérat N, Hoppenrath M (2012) Morphology and SSU rDNA phylogeny of *Durinskia agilis* (Kofoid & Swezy) comb. nov. (Peridiniales, Dinophyceae), a thecate, marine, sand-dwelling dinoflagellates formerly classified within *Gymnodinium*. Phycologia 51:287–302. https://doi.org/10.2216/10-22.1
87. Woloszynska J (1916) Polnische Süsswasser-Peridineen. Bull Int Acad Sci Crocovie B 1915:260–285
88. Kretschmann J, Čalasan AŽ, Gottschling M (2018) Molecular phylogenetics of dinophytes harboring diatoms as endosymbionts (Kryptoperidiniaceae, Peridiniales), with evolutionary interpretations and a focus on the identity of *Durinskia oculata* from Prague. Mol Phylogenet Evol 118:392–402. https://doi.org/10.1016/j.ympev.2017.10.011
89. You X, Luo Z, Su Y, Gu L, Gu H (2015) *Peridiniopsis jiulongensis*, a new freshwater dinoflagellate with a diatom endosymbiont from China. Nova Hedwigia 101:313–326
90. Liu GX, Pei GF, Hu ZY (2008) *Peridiniopsis niei* sp. nov. (Dinophyceae), a new species of freshwater red tide dinoflagellates from China. Nova Hedwigia 87:487. https://doi.org/10.1127/0029-5035/2008/0087-0487
91. Hameed HA, Saburova M, Chomérat N (2016) Taxonomic characterization of a brackish water, bloom-forming *Peridiniopsis* species (Peridiniales, Dinophyceae) from Iraq. Mar Biodiv Rec 9:40. https://doi.org/10.1186/s41200-016-0042-5

Plastid Evolution in Non-photosynthetic Lineages

Martin Kolisko, Kacper Maciszewski, and Anna Karnkowska

Abstract Plastids are endosymbiotic organelles widespread among eukaryotic lineages. They were pivotal in the evolution of eukaryotes, as they allowed host eukaryotic cells to perform photosynthesis. From an evolutionary point of view, there are two main types of plastids, primary and complex (higher-order) plastids. Primary plastids originated through a single event, the engulfment of a cyanobacteria by a eukaryotic host cell, giving rise to the group Archaeplastida (which includes plants). On the other hand, complex plastids, which occur when a primary plastid containing eukaryote is enslaved by another host cell, have originated multiple times within eukaryotes, and have given rise to most other plastid-bearing lineages.

The most common function of plastids is photosynthesis, which facilitates carbon fixation using energy from sunlight. However, plastids participate in many essential metabolic processes such as fatty acid metabolism, as well as iron–sulfur cluster and isoprenoid synthesis. Similar to mitochondria, where there are many lineages with reduced mitochondrial organelles incapable of oxidative phosphorylation, there are many eukaryotic lineages with plastids that have lost photosynthesis. In this chapter, we will discuss these lineages as well as the genomic changes associated with photosynthesis loss and its evolutionary consequences.

Keywords Non-photosynthetic plastid · Plastid loss · Plastid genome evolution · Plastid metabolism evolution · Plastid evolution

M. Kolisko (✉) · K. Maciszewski
Institute of Parasitology, Biology Centre, Czech Academy of Sciences, České Budějovice, Czech Republic
e-mail: kolisko@paru.cas.cz; kacper.maciszewski@paru.cas.cz

A. Karnkowska (✉)
Institute of Evolutionary Biology, University of Warsaw, Warsaw, Poland
e-mail: a.karnkowska@uw.edu.pl

S. D. Schwartzbach et al. (eds.), *Endosymbiotic Organelle Acquisition*,
https://doi.org/10.1007/978-3-031-57446-7_7

1 Introduction

The function of typical plastids is fixation of carbon using energy from sunlight through the process of photosynthesis. However, there are many other biochemical pathways and reactions localized to the plastid, including the biosynthesis of iron–sulfur clusters, isoprenoids, fatty acids, and nucleotides [1]. While photosynthesis could be considered the plastid's primary function, it is the other plastid-localized biochemical pathways that appear to create a dependency on this organelle [2]. It is not uncommon for plastid-bearing algae to become non-photosynthetic, but it is very uncommon for plastids to be lost entirely, suggesting that targeting of non-photosynthesis pathways to the plastid is responsible for its retention [3] (Fig. 1). So far, complete plastid loss has been observed in only a few cases, including gregarines, Cryptosporidia, and *Hematodinium* [4, 5]. This is similar to the situation with mitochondria, where again it is not uncommon to find organisms with highly reduced mitochondria, even down to one or two biochemical cellular functions, but it is very hard to identify a eukaryote that has lost mitochondria completely (until the recently discovered secondarily amitochondriate *Monocercomonoides* [6]). In this chapter, we will catalog and describe lineages across most of the known diversity of eukaryotes that have lost photosynthesis and most of the photosynthetic genes. Much of the current research efforts have been enabled by the recent massive growth of available sequence data made possible by next-generation sequencing and single-cell-based techniques [7]. While colorless unicellular algae may be considerably more difficult to find and cultivate than land plants, studying the functions of their remnant plastids using transcriptomics is much easier, as it is not hampered by tissue-specific gene expression.

This chapter will emphasize the similarities that occur during independent instances of reductive plastid genome evolution and commonalities in the retention of plastid functions, as well as highlight the differences that have brought about a stunning mosaic of retained plastid functions and plastid genome composition across these diverse lineages [8].

2 Archaeplastida

2.1 *Chlorophyta:* Prototheca *(Trebouxiophyceae)*

The class Trebouxiophyceae is a vast assemblage of cosmopolitan, unicellular green algae, including both free-living and symbiotic taxa [9]. However, despite their diversity, loss of photosynthetic capabilities has thus far been documented only in the so-called AHP clade, including genera *Auxenochlorella, Helicosporidium,* and *Prototheca*. While *Auxenochlorella* is photosynthetic, both *Helicosporidium* and the paraphyletic *Prototheca* comprise exclusively non-photosynthetic species, exhibiting a range of lifestyles, from free-living osmotrophs, through opportunistic pathogens of various animals, to highly specialized gut parasites of insects [10].

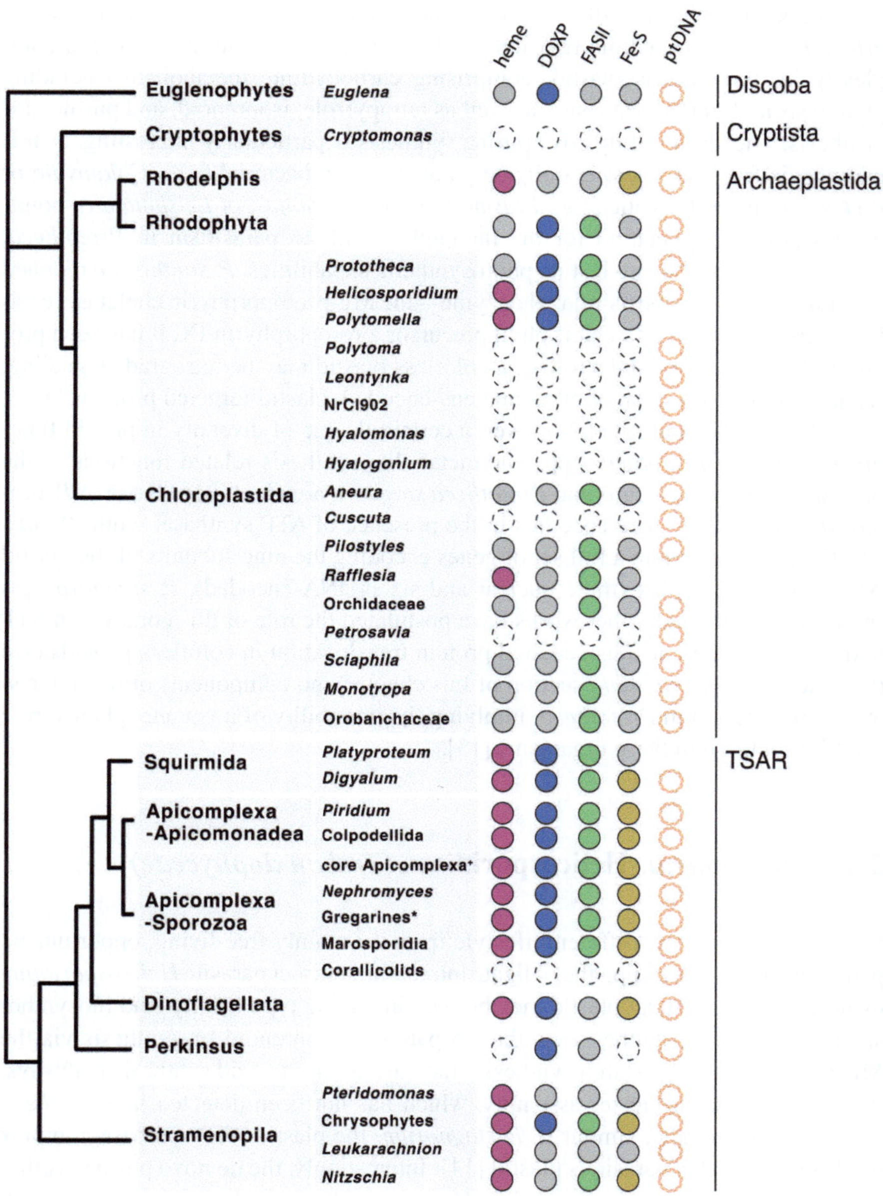

Fig. 1 Cartoon of lineages that have lost photosynthesis and their evolutionary relationship to one another. The circles show putative presence (filled color circle) or absence (grey circle; empty circle with dashed outline means unknown) of metabolic processes in the plastids of non-photosynthetic representatives: heme—heme biosynthesis, DOXP—isoprenoid biosynthesis, FASII—fatty acid biosynthesis, and Fe–S—iron–sulfur cluster biosynthesis. A symbol of circular DNA denotes the presence of plastid genomes in non-photosynthetic representatives. * In gregarines there are several types of plastids, ranging from complete plastid loss, to plastids with reduced metabolism and no plastid genome, to plastids with genomes and all four pathways present. + 'core' Apicomplexa include the taxa Hemosporidia, Piroplasmida, and Coccidia, whose apicoplasts are very similar to one another (see Sect. 9 for details)

A transcriptomic study of *Prototheca xanthoriae* (classified at the time as *P. wickerhamii*; see the taxonomic revision in [11]) revealed substantial metabolic complexity in its colorless plastid, comprising carbohydrate metabolism (including glycolysis and starch synthesis), as well as tetrapyrrole, isoprenoid, and purine biosynthesis. The ability of de novo purine synthesis is particularly interesting, as this pathway is frequently lost in obligate parasites, both bacterial (e.g., *Chlamydia* or *Rickettsia*) and eukaryotic (e.g., *Plasmodium, Trichomonas,* or *Leishmania*), pointing toward an explanation for the incomplete shift to parasitism in *Prototheca.* Curiously, despite having lost its photosynthetic capabilities, *P. xanthoriae* retained a hallmark photosynthesis-related enzyme—the Mg-protoporphyrin chelatase, catalyzing the formation of a chlorophyll precursor, protoporphyrin IX. It has been proposed that this compound's role in a colorless plastid may be retrograde signaling, regulating the expression of other nucleus-encoded, plastid-targeted proteins [12].

Further studies indicated, however, a certain degree of diversity in plastid functionality among *Prototheca* spp. The metabolite synthesis-related functions of the plastids of *Prototheca cutis* and *Prototheca stagnora* are identical to those of *P. xanthoriae*, yet the key difference lies in the presence of ATP synthase: while *P. cutis* and *P. xanthoriae* retain a full set of genes encoding the nine subunits of the plastid ATP synthase complex (three nuclear and six ptDNA-encoded), *P. stagnora* possesses none. Although other works have postulated the role of this complex in ATP hydrolysis to facilitate Tat-mediated protein translocation in colorless plastids (see the Stramenopila: *Nitzschia* section of this chapter), no components of the Tat system were detected in *Prototheca,* implying the possibility of a yet unexplained role for ATP synthase in these organisms [13].

2.2 *Chlorophyta:* Helicosporidium *(Trebouxiophyceae)*

Despite its completely different lifestyle from the mainly free-living, opportunistic pathogens *Prototheca* spp., the obligate intracellular insect parasite *Helicosporidium* exhibits nearly identical plastid metabolism, including type II fatty acid biosynthesis, tetrapyrrole biosynthesis via the C5 pathway, isoprenoid biosynthesis via the MEP/DOXP pathway, starch synthesis, and various amino acid synthesis pathways, but also iron–sulfur cluster assembly, which has not been detected in *Prototheca* [14, 15]. Additionally, similar to *P. stagnorum,* the plastid ATP synthase complex has been lost in this parasite's plastid [14]. Interestingly, the de novo purine synthesis pathway has not been detected in *Helicosporidium*, further supporting its link with the transition to obligate parasitism [12].

Moreover, the gene repertoire of *Helicosporidium*'s ptDNA is almost identical to those of *Prototheca* spp., including a transcription and translation-related gene set, as well as the acyl carrier protein gene *accD,* involved in fatty acid biosynthesis; the sulfur transporter gene *cysT,* involved in cysteine biosynthesis; and the protein translocation complex (TIC) component *ycf1* [16]. The key difference between *Helicosporidium* and *Prototheca* spp. plastid metabolism lies in carbon fixation, which has been almost entirely retained in the former, despite the lack of photosystems and the ribulose-1,5-bisphosphate carboxylase/oxygenase (RuBisCO). It has therefore been proposed that the carbon fixation pathway in *Helicosporidium* is linked to the synthesis of storage materials, such as starch [15].

2.3 *Chlorophyta:* Polytomella *(Chlorophyceae)*

Chlorophyceae are another of the four main classes of green algae, similarly widespread and diverse as Trebouxiophyceae, encompassing extensively studied model organisms, such as *Chlamydomonas reinhardtii* [9]. Among Chlorophyceae, photosynthesis losses have been documented in multiple distantly related lineages, and their extant non-photosynthetic representatives exhibit substantial divergence. The first among colorless chlorophyceans to undergo genomic studies was *Polytomella parva,* and the results were rather astonishing. As no nuclear genes involved in plastome expression, replication, or repair were identified, and as the plastid rDNA/rRNA could neither be detected via Northern blot nor PCR, it was concluded that *P. parva* has lost its entire plastid genome—the first case of such an occurrence documented at the time [17].

However, the absence of key metabolically relevant ptDNA-encoded genes in *P. parva,* previously identified in *Prototheca* and *Helicosporidium*, has since found an explanation. The *accD* gene, involved in plastid fatty acid biosynthesis, has been relocated to the nucleus of *P. parva*, while *ycf1,* although not identified in the nuclear genome, is known to be a fast-evolving gene that eludes detection via homology-based search. Most interestingly, however, the plastid of *P. parva* retained a considerably large portion of its ancestral metabolism, including isoprenoid, fatty acid, tetrapyrrole, heme, and nucleotide biosynthesis [17]. This intriguing case study may serve as a perfect example of the disparities in complexity between ptDNA and plastid metabolism, underscoring the importance of studying the nuclear component of the multifaceted phenomenon of organellar evolution.

2.4 *Chlorophyta:* Polytoma *(Chlorophyceae)*

In terms of ptDNA evolution following the loss of photosynthesis, *Polytoma uvella* could be considered the polar opposite of *Polytomella parva.* This organism, whose photosynthetic capabilities were lost independently of *P. parva* (as they are, contrary to their names, quite distantly related), was found to carry a giant plastid genome of approximately 230 kbp, mostly consisting of non-coding short repeated regions. Its genetic repertoire, however, includes only 25 protein-coding genes, which is not only quantitatively comparable to *Helicosporidium* and *Prototheca stagnora*, but functionally as well, as only six genes differ between the groups—three of its genes (*rps9*, *rps18*, *clpP*) are absent in the colorless AHP, while three others, present in AHP (*rpl12, rpl32, cysT*), are missing in *P. uvella* [18]. Although the full extent of metabolic roles of the plastid in *P. uvella* remains unknown due to the lack of transcriptomic and nuclear genomic data, it would be logical to assume that they are rather similar to the non-photosynthetic AHP, with the exception of cysteine biosynthesis being likely missing in *P. uvella* due to the lack of *cysT*.

2.5 Chlorophyta: Leontynka *(Chlorophyceae)*

A recent study revealed yet another unique non-photosynthetic lineage to reside within Chlorophyceae: a newly described genus *Leontynka*, encompassing only two species—*L. pallida* and *L. elongata*. Detailed ultrastructural and genomic analyses have shown that *Leontynka* has retained almost all the hallmark features of photosynthetic algae, save for photosynthesis itself—it has an eyespot (sometimes even two) and its plastids possess thylakoids, as evidenced both by microscopic observations and identification of the thylakoid biogenesis machinery in its transcriptome. The plastid genome of *L. pallida* is extremely inflated, even compared to *Polytoma uvella* (over 360 kbp), but hosts only 34 protein-coding genes, including ribosomal protein-coding genes (families *rps* and *rpl*), RNA polymerase subunits (family *rpo*), and a handful of miscellaneous genes (*tufA, clpP, ftsH,* and *ycf1*). Furthermore, *L. pallida* has retained its plastid ATP synthase, with six subunits (*atpABEFHI*) remaining encoded within the ptDNA, while the other three (*atpCDG*) have been relocated to the nuclear genome. All subunits of the Tat translocase have also been identified, supporting the possible connection between ATP synthase retention and the Tat protein translocation system in plastids. However, as an alternative explanation for ATP synthase presence, the authors of the study suggest that it may be involved in the maintenance of basic stromal pH, which is optimal for a variety of enzymes in the plastid stroma [19].

Although not all plastid-localized secondary metabolite biosynthetic pathways have been investigated, it has been demonstrated that *L. pallida* performs carotenoid and plastoquinone biosynthesis in its plastids, and utilizes the plastid terminal oxidase (PTOX) in plastoquinone recycling, implying the presence of a form of electron transport chain, possibly involved in other redox reactions. It is also worth noting that *Leontynka* is one of the rare cases of a non-photosynthetic lineage without any close photosynthetic relatives, as its phylogenetic position is between two large clades of Chlamydomonadales. While this may be the result of incomplete taxon sampling and the genus might have as yet undiscovered photosynthetic relatives, the possibility that *Leontynka* represents an ancient loss of photosynthesis cannot be ruled out [19].

2.6 *Chlorophyta: Strain NrCl902 (Chlorophyceae)*

A further example of photosynthesis losses leading to similar outcomes in Chlorophyceae is the chlamydomonadalean strain NrCl902. This organism, currently lacking a full systematic name, possesses a slightly smaller (176 kbp) plastome than *Polytoma uvella* but contains exactly identical gene content. Notably, it branches among the photosynthetic relatives of *Polytoma*, implying that its photosynthesis loss is an independent event from that of *P. uvella* [20].

A transcriptome-based analysis of the plastid metabolism of this strain has demonstrated the presence of the pentose phosphate pathway, likely supplying erythrose-4-phosphate, a substrate for aromatic amino acid biosynthesis, as well as the biosynthesis and turnover of plastoquinone via PTOX (as in *Leontynka*). The role of plastoquinone synthesis and recycling in this organism's plastid is likely associated with the need for an electron sink in carotenoid biosynthesis. However, this finding is rather puzzling, as plastoquinone activity is associated with the thylakoid membrane, and the plastids of strain NrCl902, unlike *Leontynka*, do not have discernible thylakoids [20]. Nonetheless, the possible redox homeostasis constraint for the retention of plastoquinone in non-photosynthetic plastids is a very important finding for our understanding of the evolution of these organelles, which should not be neglected in future studies.

2.7 *Chlorophyta:* **Hyalomonas oviformis** *(Chlorophyceae)*

Superficially similar genome contents in non-photosynthetic plastids are not necessarily required to follow the same evolutionary paths. A case of possible non-obvious divergence has been observed in the ptDNA of *Hyalomonas oviformis*—the type species of its newly-described genus, previously classified as a representative of *Polytoma*. Its plastid genome, small (almost 132 kbp) compared to other non-photosynthetic Chlorophyceae, carries the same set of ATP synthase subunit genes as *Leontynka pallida*. These genes are not only spread across the entire ptDNA molecule (as in the case of *L. pallida*) as opposed to clustered together like in most other lineages, but they are also subjected to relaxed selection, which is evident from their accelerated substitution rate. What is more, the twin-arginine translocation system, proposed to be functionally linked to the hydrolytic activity of ATP synthase, has only been partially identified in *H. oviformis*, pointing towards an even greater dissimilarity of plastid functions between this species and *L. pallida* [21].

2.8 *Chlorophyta:* Hyalogonium fusiforme *(Chlorophyceae)*

In the study mentioned above, a colorless representative of a different chlamydomonadalean genus underscores further important insights into the evolution of non-photosynthetic plastids. The plastid genome of *Hyalogonium fusiforme* not only sets another record for ptDNA size in a non-photosynthetic alga at an astonishing 442 kbp, but in sharp contrast with its non-photosynthetic cousins among Chlorophyceae, it displays almost no signs of coding content reduction concomitant with the loss of photosynthesis. In fact, it carries a nearly complete set of photosynthesis-related genes, along with cytochrome *b6f* components, chlorophyll biosynthesis machinery, and the large RuBisCO subunit. Disruptions in a majority of the aforementioned genes suggests their loss of function, responsible for the observable non-photosynthetic phenotype of *H. fusiforme*; however, these sequences are not conventional pseudogenes with frameshifts or internal stop codons (as is often observed in the ptDNA of non-photosynthetic land plants; see in the following sections), but instead, they are mostly fragmented into smaller, possibly expressible ORFs [21].

2.9 *Streptophyta:* Aneura mirabilis *(Marchantiophyta)*

Although photosynthesis is usually considered the hallmark feature of land plants, its loss is actually quite frequent among this group, leading to considerable variability among secondarily heterotrophic plants, whose lifestyles always involve some form of parasitism [22]. Interestingly, this phenomenon is not exclusive to the so-called higher plants (i.e., gymnosperms and conifers), as the loss of photosynthesis has also been observed in a parasitic liverwort, *Aneura mirabilis*. As a mycoheterotroph, *A. mirabilis* obtains organic carbon from a fungus (genus *Tulasnella*) living in mycorrhizal association with photosynthetic plants, such as pines or birches [23].

An analysis of the plastid genome of *A. mirabilis* has shown it has a substantially different evolutionary path than in the case of non-photosynthetic chlorophytes mentioned earlier. Its ptDNA exhibits no discernible reduction (nor inflation) in size, and only a handful of its photosynthesis-related genes have been lost. Many genes, however, have been pseudogenized by frameshift mutations and stop codon insertions, but this applies only to photosystems, cytochrome *b6f*, chlororespiration, and cysteine biosynthesis components; the ATP synthase components, the large RuBisCO subunit gene, chlorophyll biosynthesis-related genes, and all housekeeping genes are intact. The detectability of pseudogenes may be indicative of a very recent loss of photosynthesis in *A. mirabilis*. Regardless, the colorless plastids in *A. mirabilis* likely carry out chlorophyll and fatty acid biosynthesis, as indicated by the presence of these pathways' components in its ptDNA [23].

2.10 *Streptophyta:* Cuscuta *(Convolvulaceae)*

One of the first cases of a transition to parasitism in vascular plants to be studied using genomic techniques is the genus *Cuscuta,* commonly known as dodder. This plant holoparasite has retained its photosynthetic capabilities, however, analyses of the plastid genome contents of four of its species—*C. reflexa, C. gronovii, C. exaltata,* and *C. obtusiflora*—have revealed crucial insights into ptDNA reduction in parasites. While all photosystem and cytochrome *b6f* components, as well as all constituents of the chlorophyll biosynthesis pathway remain, plastomes of *Cuscuta* spp. have lost a broad array of other genes, including all NADH dehydrogenase complex genes (*ndh*), required for carrying out chlororespiration, the translation initiation factor gene, *infA*, the lysine tRNA gene, the intron maturase gene, *matK*, and in three out of four species studied (excluding *C. exaltata*), the plastid RNA polymerase operon [24, 25].

This variety of missing genes, together with the increased substitution rate observed in the plastid genomes of *Cuscuta,* points toward a very important conclusion: that the gradual ptDNA decay associated with lifestyle transition in ancestrally photosynthetic organisms does not begin with the loss of photosynthesis itself. Instead, ptDNA reduction may already affect a major part of the plastid genome's genetic content before photosynthesis is eventually lost. Furthermore, this suggests it is most likely the NADH dehydrogenase, not the photosystems, that constitute the first step (or the first point of no return) towards the parasitism-associated reductive evolution of plastids [25, 26].

2.11 *Streptophyta:* Pilostyles *(Apodanthaceae)*

In parasitic plants, the variability of forms and functions of plastids and their genomes can be quite considerable, and a perfect counterpart to the aforementioned *Cuscuta* spp. is the genus *Pilostyles.* These endoparasitic members of Apodanthaceae, inhabiting the tissues of photosynthetic legumes (Fabaceae), possess the smallest plastid genomes documented thus far—approximately 11 kbp for *P. aethiopica* and 15 kbp for *P. hamiltoni*—with a total repertoire of only four functional protein-coding genes (*accD, rps3, rps4, and rps12*), two rRNA genes, and either one or two pseudogenes (*rpl2* and *rpoC2* in *P. hamiltoni* and only *rpl2* in *P. aethiopica*). While the retention of *accD*—the gene encoding an acyl carrier protein, involved in plastid fatty acid biosynthesis—is not entirely surprising, as it is one of the most often retained ptDNA-encoded genes in non-photosynthetic plastids, it is rather puzzling that intact genes can be found in a plastome that is missing almost all of the organellar gene expression machinery, including ribosomal protein-coding genes and the entire tRNA gene complement [22].

2.12 *Streptophyta:* Rafflesia *(Rafflesiaceae)*

The reductive genome evolution of non-photosynthetic plastids in parasitic plants can progress even further, as evidenced by the genomic and transcriptomic studies of the genus *Rafflesia*. In a curious case of convergent evolution with the green alga *Polytomella*, these spectacular holoparasites of photosynthetic plants, bearing the largest flowers ever documented in nature, have most likely lost their entire plastomes. Among the high-throughput sequencing datasets obtained from two representatives of this genus—*R. cantleyi* and *R. lagascae*—several plastid-derived pseudogene sequences were identified; however, their phylogeny indicates they are either contamination or traces of lateral gene transfer from the genus *Tetrastigma,* the photosynthetic host plant of *Rafflesia* spp. Still, cellular imaging of yet another species of *Rafflesia, R. philippensis,* via TEM has revealed the presence of plastid-like compartments, prompting further studies [27].

A detailed transcriptomic analysis of *R. cantleyi* has confirmed the retention of a plastid which, despite lacking its genome, possesses clearly discernible metabolism. While photosynthesis-related proteins, including those involved in the biosynthesis of chlorophyll (and tetrapyrroles in general), have not been found and isoprenoid synthesis takes place in the cytosol, it has been demonstrated that *R. cantleyi* performs amino acid and fatty acid biosynthesis, as well as partial heme biosynthesis, in its vestigial plastids. This finding shows that despite superficial similarity in the form of ptDNA loss, *Rafflesia* and *Polytomella* have retained vastly different plastid functions [28].

2.13 *Streptophyta: Orchidaceae*

Orchidaceae are one of the most distinctive and speciose land plant families, characterized by a wide variety of fascinating adaptations for the exploitation of other organisms, such as fungi and insects [29–31]. What's more, it has been estimated that photosynthesis loss has taken place approximately 30 times independently among orchids, which makes this occurrence more frequent in this family than in all other land plant taxa combined [32]. Therefore, one might expect that the plastome contents and plastid functions, in general, would be highly diverse among the Orchidaceae; however, the representatives studied so far display only superficial differences, underlined by deeper functional similarities.

Among orchids, the genus with the least reduced plastomes is *Corallorhiza*. In one representative, *C. striata*, the ptDNA contents resemble those of the colorless liverwort *Aneura mirabilis*—its size of 137.5 kbp is comparable with photosynthetic plants. Numerous genes encoding photosystem subunits and assembly proteins, cytochrome *b6f* components, NADH dehydrogenase complex components, and the large subunit of RuBisCO have been pseudogenized, while only a few (e.g. *psbC*, *psbD*, *psbI*) have undergone complete deletion. On the other hand, the entire housekeeping machinery (except the pseudogenized RNA polymerase subunit genes) is fully intact, as are the ATP synthase subunit genes, the fatty acid biosynthesis-associated gene *accD*, the protein translocation system component genes *ycf1* and *ycf2*, and the protease subunit gene *clpP* [33]. This pattern of gene degradation and conservation has been corroborated by a follow-up study of nine other species of *Corallorhiza*. Surprisingly, this study additionally demonstrated that many of the investigated organisms retain chlorophyll biosynthesis [34]. Without the ability to photosynthesize, it has been proposed that *Corallorhiza* spp. utilize chlorophyll as a plastid-to-nucleus signaling component or as a protective factor against light-induced damage to DNA.

An intermediary level of ptDNA degradation in Orchidaceae has been observed in *Neottia nidus-avis*. Its approximately 92 kbp plastome is missing the *matK* intron maturase gene and the entire ATP synthase-associated gene family, but otherwise, it is functionally similar to *Corallorhiza* spp. Instead of being pseudogenized, the photosystem subunits, cytochrome *b6f* components, and NADH dehydrogenase complex genes have mostly been deleted, which accounts for the difference in plastid genome size. Like in *Corallorhiza* spp., the housekeeping genes, as well as *accD, ycf1, ycf2,* and *clpP* genes are retained in the plastome of *N. nidus-avis* [35].

Finally, the most reduced orchid plastomes belong to *Rhizanthella gardneri* and two representatives of the genus *Epipogium—E. aphyllum and E. roseum.* Although all of them retain the aforementioned set of four essential non-housekeeping genes, found, for example, in the ptDNA of *N. nidus-avis*, the drastic reduction of their ptDNA contents encompasses even the housekeeping machinery: *R. gardneri* carries only nine out of over 30 tRNA genes usually present in plant plastomes, while *E. roseum* carries seven, and *E. aphyllum* only six. The ribosomal protein repertoire is also partially lost, with only 12 of the approximately 25–30 of these genes that are usually present in photosynthetic species being retained in *E. aphyllum.* As the gene content is, to some extent, proportional to the total ptDNA size (more so in *Epipogium* than in *Rhizanthella*, whose plastome is largely non-coding due to the remnant pseudogenes, which are absent in *Epipogium*), the plastid genome of *E. roseum* is only 19 kbp long, which is substantially closer to the previously mentioned smallest ptDNA-bearing genus *Pilostyles* than any other orchid [36, 37]. Most importantly, taken together, the findings outlined above show that all Orchidaceae share the same set of indispensable plastome components, and despite their independent shifts to obligate heterotrophy, the functions of their plastid organelles are likely very similar.

2.14 Streptophyta: Petrosavia stellaris *(Petrosaviaceae)*

Petrosavia stellaris is among the few examples of photosynthesis loss in monocots, apart from Orchidaceae. The non-photosynthetic plastid genome of this species was found to be even less reduced than those of *Corallorhiza* spp., having retained not only the ATP synthase subunit genes and the set of four non-housekeeping genes retained in all orchids but also an intact large RuBisCO subunit gene, as well as the genes for some cytochrome *b6f* and photosystem components, such as *psbZ* and *petG*. Interestingly, while the missing photosystem subunit genes have been fully deleted, the NADH dehydrogenase genes are pseudogenized, suggesting that their degradation might have followed, rather than preceded the loss of photosynthesis, in stark contrast to the previously mentioned case of *Cuscuta* spp. Furthermore, *P. stellaris* is among the very few colorless species to retain partial inverted repeats in their ptDNA, which, despite being conserved in the plastomes of photosynthetic plants and green algae, are almost always lost along with the photosynthetic apparatus [38].

2.15 Streptophyta: Sciaphila densiflora *(Triuridaceae)*

Regrettably, large groups of non-photosynthetic plants are often not as well-studied as the captivating Orchidaceae, even though their evolutionary history might not be any less unusual. One such understudied lineage is the entirely mycoheterotrophic and colorless family Triuridaceae, which encompasses around 50 species of small herbs with reduced leaves, yet only one—*Sciaphila densiflora*—has been the subject of genomic studies. Its petite 21 kbp plastid genome carries 18 protein-coding genes, an rDNA operon, and a set of six tRNA genes, which at first sight, might seem similar to the plastomes of the *Epipogium* orchids. However, the non-housekeeping gene repertoire retained by *S. densiflora* is quite different: the *matK* intron maturase gene is retained (although as a stand-alone gene, as opposed to photosynthetic plants where it is usually encoded within an intron), while the protein trafficking-related genes (*ycf1, ycf2*) are lost [39]. As nearly all other known taxa with non-photosynthetic primary plastids retain the aforementioned transport machinery components, their loss in *S. densiflora* is exceptionally peculiar. Further studies of the Triuridaceae family will certainly help our understanding of how these achlorophyllous plants managed to overcome the common constraint against the loss of transport-related plastid-encoded genes.

2.16 Streptophyta: Monotropa hypopitys *(Ericaceae)*

Interestingly, the atypical gene composition of ptDNA in *S. densiflora* outlined above brings us to yet another rare case of convergence across the entire phylogeny of angiosperms. The plastid genome of a mycoheterotrophic representative of Ericaceae, *Monotropa hypopitys*, has been found to carry a slightly more complete set of housekeeping genes and a more than twice as large tRNA complement as compared to *S. densiflora*; however, apart from the transcription and translation apparatus, its gene content includes only *accD, clpP,* and a non-intron-encoded *matK*. It has not retained *ycf1, ycf2*, RNA polymerase subunits, ATP synthase components, nor any of the other non-photosynthesis-related genes that are usually present in non-photosynthetic plastids [40]. Although the convergence demonstrated here involves only a handful of genes, further investigation of the nucleus-encoded portions of plastid metabolism in *S. densiflora* and *M. hypopitys* might prove to be more illuminating for evaluating the similarities between these species.

2.17 Streptophyta: Orobanchaceae

Single, ancient losses of photosynthesis in large plant taxa are rather uncommon; what is more often observed is multiple independent losses among the already radiated lineages, leading to a patchy distribution of photosynthetic capabilities on the phylogenetic tree, as in orchids and broomrapes (Orobanchaceae). This family comprises interwoven lineages of autotrophs, as well as hemi- and holoparasites, often inhabiting extreme environments, such as deserts [26, 41]. Analyses of their plastid genomes indicate substantially greater variability of retained functions than in the case of Orchidaceae. One example, a hemiparasitic and photosynthetic species *Schwalbea americana*, was found to carry a large (almost 161 kbp) plastome with the almost complete gene repertoire of its photosynthetic relatives (e.g., *Lindenbergia philippensis*), save for the pseudogenized *ndh* (NADH dehydrogenase components) family. This would be an example of near-perfect convergence with *Cuscuta* spp., if not for the pseudogenization of the *accD* gene, which is found intact in almost all other plant plastomes, including the non-photosynthetic ones [26].

In contrast, two among the eleven holoparasitic (and non-photosynthetic) species investigated thus far have retained intact photosystem II subunit genes—*psbM* in the case of *Cistanche deserticola*, and *psbF, psbI, psbJ, psbK,* and *psbL* in *Myzorrhiza californica,* which also carries an intact *rbcL* (a large subunit of RuBisCO) gene, postulated to be involved in serine and glycine biosynthesis [26, 41]. Finally, the most unusual ptDNA content among Orobanchaceae has been identified in *Phelipanche ramosa*. The plastid genome of this holoparasitic species possesses pseudogenized *accD, clpP,* and even *ycf1,* but it also carries a full set of intact ATP synthase subunit genes [26]. In brief, the plastid genome repertoires of the non-photosynthetic Orobanchaceae have very little in common, save for the invariable pseudogenization or loss of the NADH dehydrogenase and RNA polymerase components [26, 41, 42]. Unfortunately, transcriptomic data from Orobanchaceae remain scarce; our knowledge of the nucleus-encoded portion of their plastid metabolism is currently limited to the presence of the chlorophyll biosynthesis pathway in *Phelipanche aegyptiaca* and its lack, but with the retention of heme biosynthesis, in *Aeginetia indica* [42].

It is also worth mentioning that like *Petrosavia stellaris* (Petrosaviaceae), almost all of the holoparasitic Orobanchaceae, with the exception of only two species, retain the quadripartite structure typical for photosynthetic plant plastomes. Even though the inverted repeats evolved into atypical forms in some broomrape species—for example, in *A. indica*, where their inflation by non-coding regions led to immense size, covering almost 75% of the entire plastome—it is rather unusual for them to predominate in the genomes of non-photosynthetic plastids, especially when derived from independent photosynthesis losses [26, 42].

3 Rhodophyta

Red algae represent one of the lineages of Archaeplastida that possess primary plastids. Although the loss of photosynthesis in red algae is generally linked to parasitism and it is estimated to have happened over a hundred times independently, there are actually very few molecular evolutionary studies of the red algal non-photosynthetic plastids. Most parasitic red algae are represented by the so-called neoplastic parasites that have completely lost their plastids, compensating for this by utilizing their hosts' plastids. Alternatively, archaeplastic parasites have been shown to retain their own reduced plastid [43].

The plastids of two alloparasitic red algae — *Choreocolax polysiphoniae* and *Harveyella mirabilis* — have been studied, revealing that they have both lost a majority of the genes linked to photosynthesis, except the *petF* gene, which can function in other pathways. Their plastid genomes are similar in size: that of *C. polysiphoniae* contains ±90 genes and that of *H. mirabilis* contains ±100 genes. Based on the presence of genes, both plastids are likely involved in fatty acid, amino acid, and isoprenoid biosynthesis [43].

3.1 Rhodelphis *spp.*

Recently, a pair of novel eukaryovorous organisms that form a sister clade to red algae have been discovered: *Rhodelphis limneticus* and *Rhodelphis marinus*. Both of these species are predatory flagellates; they do not perform photosynthesis. Nevertheless, as a sister lineage to red algae, they must have once possessed a primary plastid, and indeed, they do appear to contain a cryptic plastid. These plastids have not been observed with TEM, and no plastid genome sequences have been found in a genomic survey; however, transcriptomic analyses have identified proteins with putative functions in primary plastid import (TIC20, TIC22, TIC32, and TOC75). No homologs of the SELMA complex have been identified, which is in agreement with the presence of a primary plastid. Furthermore, the identification of proteins with a primary plastid targeting presequence has detected around 50 other proteins that are likely imported into these cryptic plastids, suggesting that at least iron–sulfur cluster and part of the heme biosynthesis pathway is localized in the plastid organelle [44]. The first precursor of heme, ALA, is putatively synthetized in their mitochondria via the Shemin pathway (typical for heterotrophs), while the rest of the pathway is likely localized to the plastid. This arrangement is identical to *Chromera velia* and very similar to Apicomplexa, where the pathway terminates in the mitochondria [45]. *Rhodelphis* has most likely completely lost the plastid genome, as so far there has been no evidence of its presence.

4 Euglenophyta: *Euglena longa*

Euglenophytes, a group of algae with secondary plastids of green-algal origin, are predominantly photosynthetic [46]. However, several colorless species have been described within different genera, suggesting they lost photosynthetic ability independently. At present, only one colorless species—*Euglena longa*—has been analyzed in detail and was shown to possess a reduced plastid genome [47]. *E. longa* is of particular interest due to its lack of photosynthesis and its close relationship with *Euglena gracilis*, a model euglenophycean. Early studies revealed that the *E. longa* plastid genome is around half the size of the chloroplast genome of *E. gracilis* and lacks genes for all photosynthesis-related proteins, except for *rbcL* (a large subunit of RuBisCO). The surprising retention of this gene prompted further studies on the possible role of the RbcL protein; it was demonstrated that it is highly divergent compared to its homologs from photosynthetic relatives but still transcribed. Similar to photosynthetic species, *E. longa* encodes the small subunit of RuBisCO (RbcS) in the nuclear genome. Both RbcL and RbcS proteins are synthesized in *E. longa*; yet, their abundance is very low, and processing of the precursor polyprotein has been shown to be deficient [48].

Studies of transcriptomic data revealed that nearly all Calvin–Benson cycle enzymes (assembling a putative linear CB pathway) have been retained in *E. longa* [49]. While a subset of CB cycle enzymes has been reported from unrelated colorless algae and plants, the repertoire identified in *E. longa* is unique. Furthermore, the ferredoxin/thioredoxin (Fd/Trx) system was identified despite the absence of photosynthesis, which suggests a possible role for the CB pathway in *E. longa*. Füssy and colleagues hypothesized that the Fd/Trx system activates the linear CB pathway in response to the redox status of the *E. longa* cell, and therefore, its primary role is maintaining cellular redox balance [49]. This explanation is in line with previously shown adaptations of *E. gracilis* to anaerobiosis [50].

The only other unexpected feature of the *E. longa* plastid genome is the loss of the gene for plastid ribosomal protein S18, which is otherwise broadly conserved in photosynthetic eukaryotes [51]. However, *rps18* is missing from some other colorless plastid genomes, suggesting that the protein S18 is not essential for translation in the plastid.

The plastid of *E. longa* seems to be indispensable: experiments with bleaching agents (streptomycin and ofloxacin) leading to the loss of functional plastids, ultimately inhibited the growth of *E. longa* [52]. Dependency on the plastid most likely lies in crucial non-redundant metabolic pathways; but which ones specifically are yet to be identified. Transcriptome analyses have revealed that the *E. longa* plastid, in contrast to many other non-photosynthetic plastids, plays no role in the biosynthesis of isoprenoid precursors and fatty acids, and only partially in heme biosynthesis, which is restricted to the synthesis of siroheme. On the other hand, the plastid is surprisingly involved in isoprenoid metabolism, including the synthesis of tocopherols and a phylloquinone derivative [49].

5 Cryptophyta: *Cryptomonas parvum*

Cryptophyte plastids are derived from a red-algal secondary endosymbiosis, but in contrast to many other red-originated lineages, in addition to plastid genomes, they also contain nucleomorphs. The genus *Cryptomonas* encompasses photosynthetic and non-photosynthetic species that have lost the ability to photosynthesize at least three separate times [53]. The recurrent loss of photosynthesis in this genus might be related to a mixotrophic lifestyle [54]. A detailed evolutionary scenario explaining recurrent losses of photosynthesis in *Cryptomonas* was recently proposed by Suzuki and colleagues [55]. In this scenario, the common ancestor of cryptophytes possessed mobile group II introns, which expanded and are maintained in mixotrophic species but degenerated and decreased in obligate photosynthetic species. Under heterotrophic conditions, a large number of group II introns could have been retained and might have triggered frequent structural variants in mixotrophic species. Consequently, this could result in the disruption of the inverted repeat structure (IRs) due to intron insertions or homologous recombination. Over time, this could lead to the loss of IRs and the destabilization of genomic structures among the examined non-photosynthetic cryptophytes. Finally, photosynthesis-related genes would be lost in non-photosynthetic species.

The plastid genome of *C. paramecium* CCAP977/2a, the first one sequenced among non-photosynthetic cryptomonads, is approximately half the size of the plastomes of photosynthetic species. The reduction in genome size is mainly due to the loss of photosynthesis-related genes, such as members of the *psa* and *psb* gene families [56]. As expected from the aforementioned model of non-photosynthetic plastid evolution, it lacks IRs and bears only one rRNA operon. Despite lacking most of the genes involved in photosynthesis, the *C. paramecium* plastid genome possesses a nearly complete complement of the ATP synthase subunit genes found in photosynthetic cryptomonads. These six genes are also present in other non-photosynthetic plastids—such as several species of *Prototheca* (Chlorophyta, see above), *Nitzschia* sp. (Stramenopila, see below), and some parasitic plants—but the exact roles of the ATP synthase complex in non-photosynthetic plastids are a matter of much debate [13, 57]. Although most photosynthesis-related genes have been lost in all non-photosynthetic cryptomonad strains investigated thus far, some differences have been observed in the variety of carbon fixation– and chlorophyll synthesis–related genes that have been retained.

Analysis of the photosynthetic cryptomonad genomes suggests that nuclear protochlorophyllide reductase (POR) is the main enzyme participating in the chlorophyll biosynthetic pathway. In contrast, the plastid-encoded light-independent protochlorophyllide reductase (LIPOR) is dispensable. Surprisingly, LIPOR comprising *chlB*, *L*, and *N* is present in two non-photosynthetic strains, SAG977-2f and CCAC1634B [58]. This may be because these strains are still undergoing the process of gene loss; however, genes encoding those proteins usually disappear at an early stage of photosynthesis loss because they can encode proteins that participate in the generation of reactive oxygen species (ROS). Retention of genes involved in chlorophyll biosynthesis was also noted in the corallicolid plastid, in contrast to the closely related parasitic Apicomplexa, which lack those genes in their apicoplasts [59] and photosynthetic *Chromera* [60]. Taken together, the presence of genes involved in chlorophyll biosynthesis in non-photosynthetic cryptomonads and corallicolids may represent evolutionary intermediates, or the retention of these genes could have a biological purpose that is as yet unknown.

One more oddity of the non-photosynthetic cryptomonad plastid genomes is the retention of RuBisCO-related genes (*rbcL*, *rbcS*, and *cbbX*) in strain CCAP977/2a, while two other investigated strains have lost the RuBisCO protein-coding genes [58]. One proposed explanation for the presence of RuBisCO in non-photosynthetic strains is that it has functions other than the Calvin–Benson cycle: for example, it could improve carbon efficiency by converting hexose to pyruvate in lipid biosynthesis. However, if RuBisCO plays an essential role in *C. paramecium* CCAP977/2a, other RuBisCO-lacking strains would have to replace its role with other enzymes. Another plausible explanation is that RuBisCO is in the process of being lost [58].

In addition to their reduced plastid genomes, non-photosynthetic cryptomonads have a nucleomorph, which has also lost genes associated with plastid functions [61]. Two genes (*cbb*X and *ycf*20) show differential presence/absence patterns among non-photosynthetic strains of cryptomonads. The cbbX protein is a red-algal type ATPase enzyme involved in activating RuBisCO and may serve as a molecular chaperone for RuBisCO subunit assembly. Typically, the *cbb*X, *rbc*L, and *rbc*S genes are arranged as an operon in the plastid genomes of red algae and cryptophytes. In cryptophytes, two distinct *cbb*X genes are present: one in the nucleomorph genome and one in the RuBisCO operon of the plastid genome. Interestingly, the canonical RuBisCO operon is present in the plastid genome of the colorless species *C. paramecium* CCAP977/2a; yet, *cbbX* is missing from both the plastid and nucleomorph genomes of the very closely related strain *C. paramecium* KR, while it still retains the *rbc*L and *rbc*S genes in its plastid genome. In contrast, the plastid RuBisCO operon and the nucleomorph *cbb*X gene are both missing in the colorless species *Cryptomonas* sp. CCAC1634B [62]. The patchy distribution of *ycf*20 suggests that its function might not be essential in at least some cryptomonads.

6 Stramenopila

6.1 *Stramenopila:* Nitzschia *(Bacillariophyta)*

Bacillariophytes, commonly known as diatoms, are an ochrophytean lineage constituting key planktonic primary producers of the global ocean. Despite their tremendous diversity, estimated to encompass up to 200,000 distinct species [63], losses of photosynthesis are very rare among diatoms. Thus far, they have been described in only seven species of two closely-related genera—*Nitzschia* and *Hantzschia* [64]. The non-photosynthetic diatoms exhibit a saprotrophic lifestyle, and have been found thriving on many continents in mangroves, intertidal sands, and as epiphytes of macroalgae, such as *Sargassum* or *Fucus* spp. [64–66]. Still, as their sampling remains limited and phylogenetic reconstructions have been based on only a few marker sequences, it has been vigorously disputed whether the extant apochlorotic diatoms are descendants of multiple organisms that forfeited photosynthesis independently or a single ancestor that lost its photosynthetic apparatus, with the latter hypothesis currently having stronger phylogenetic support [64, 67]. Nonetheless, genomic and/or transcriptomic data have been obtained for six colorless diatom strains—*Nitzschia putrida, Nitzschia* sp. NIES-3581, NIES-3576, PL1-4, PL3-2, and Nitz4—providing interesting insights into their plastid metabolism [57, 65, 66, 68].

Unsurprisingly, the ptDNA of all studied non-photosynthetic *Nitzschia* spp. have lost the entire photosynthesis-related portion of their genetic repertoire, including photosystem subunits, cytochrome *b6f* complexes, as well as Calvin–Benson cycle and chlorophyll biosynthesis components. Curiously though, all of them have retained seven out of eight plastid-encoded ATP synthase subunits, but not the same ones. While strain NIES-3581 is missing the *atpE* gene, encoding a regulator/inhibitor of the hydrolytic activity of ATP synthase, the strains NIES-3576, PL1-4, and PL3-2 are missing *atpF*—a subunit forming the stator stalk between the F0 and F1 components of the ATP synthase complex. As other studies have demonstrated that both of these subunits are crucial for efficient ATP synthesis activity, but not for ATP hydrolysis, it has been suggested that the plastid ATP synthase of the apochlorotic *Nitzschia* spp. acts, in fact, as an ATP hydrolase. Its role would therefore be in maintaining the proton gradient across the thylakoid membrane, which is required for protein translocation into the lumen by the twin-arginine translocator (Tat) system [57, 65].

As shown by transcriptomic analyses, the apochlorotic *Nitzschia* spp. possess a rather rich, although predominantly nucleus-encoded, plastid metabolism. In *Nitzschia* sp. NIES-3581, the only conventional non-photosynthesis-related plastid pathway identified as lost is the MEP pathway of isoprenoid synthesis, the function of which is fully compensated by its cytosolic counterpart, the MVA pathway [69]. On the other hand, this strain has retained its plastid pathways for heme, iron–sulfur cluster, fatty acid, and amino acid biosynthesis, as well as glycolysis/gluconeogenesis and the reductive/non-oxidative pentose phosphate pathway (rPPP).

Surprisingly, *Nitzschia putrida* has additionally been shown to retain the plastid ornithine cycle—a nitrogen-recycling pathway that had not been identified in a non-photosynthetic plastid before [68, 69]. The non-obvious role of rPPP in the absence of RuBisCO and PRK (and therefore CO_2 fixation) is supplying the substrates, such as erythrose-4-phosphate, pyruvate, and phosphoenolpyruvate, for amino acid synthesis [69].

Interestingly, a different apochlorotic strain, *Nitzschia* sp. Nitz4, has been demonstrated to lack cytosolic enolase, indicating that, in contrast with most model eukaryotes, these organisms do not perform glycolysis in their cytosol, but exclusively in organelles. Additionally, the authors of this study proposed a NADPH constraint as the explanation for the scarcity of photosynthesis losses in diatoms. A variety of metabolic pathways in plastids, outlined earlier, utilize NADPH as a reducing factor. The turnover of NADP+/NADPH is thought to be carried out either by light-dependent photosynthetic reactions or by the oxidative pentose phosphate pathway, which happens to be ancestrally absent in all diatoms. Naturally, the very existence of the apochlorotic *Nitzschia* spp. and their complex plastid metabolism indicates that another mechanism—possibly using an isoform of the glycolytic enzyme GAPDH, or an NAD(P) transhydrogenase using the proton motive force for the interconversion of NADH and NADPH—must be in place. Nevertheless, the necessity to conserve NADPH might explain not only the rarity of non-photosynthetic plastids in diatoms, but also the elimination of isoprenoid biosynthesis from these organelles [66].

6.2 Stramenopila: **Pteridomonas** *(Dictyochophyceae)*

Dictyochophyceae are a lineage of unicellular planktonic algae closely related to diatoms [70, 71]. Despite being ancestrally capable of photosynthesis thanks to their secondary red alga-derived plastids, shifting away from the autotrophic lifestyle has been documented in many of their representatives, such as the mixotrophic, bacterivorous genus *Pseudopedinella* [72] and the fully heterotrophic genera *Pteridomonas* and *Ciliophrys* [73]. The evolutionary path of the plastid in the latter genus remains enigmatic, as an ultrastructural study of *Ciliophrys infusionum* did not identify any presence of this organelle. In contrast, the plastid of *Pteridomonas danica* has not only been identified morphologically, but its unique physiology has been revealed by genomic and transcriptomic studies [73, 74].

The vestigial plastids of *Pteridomonas* spp. have been demonstrated to carry out glycolysis/gluconeogenesis and the non-oxidative pentose phosphate pathway, as well as heme biosynthesis. However, no components of the plastid iron–sulfur cluster assembly system have been detected, indicating loss of this pathway in *Pteridomonas*. Furthermore, two complete plastid genome sequences of *Pteridomonas* spp. revealed no metabolite synthesis-related protein-coding genes; instead, it has been proposed that the key function of their plastomes is encoding the glutamyl–tRNA molecule, which is a substrate of the plastid heme biosynthesis pathway [74]. Surprisingly, the *rbcL* gene has not been found in any of the two investigated *Pteridomonas* plastomes or transcriptomes, even though this gene has been identified via PCR in a different strain of *P. danica* [73, 74]. Nonetheless, the unprecedented (although previously theorized [75]) finding of a tRNA gene, without additional protein-coding genes, being the main constraint against ptDNA loss in a non-photosynthetic plastid holds great importance for our understanding of the convoluted paths of reductive evolution of organelles.

6.3 *Stramenopila:* Leukarachnion salinum *(Synchromophyceae)*

Interestingly, not long after the study of *Pteridomonas* plastid genomes was published, a convergently similar case of ptDNA containing no apparent protein-coding genes involved directly in secondary metabolite synthesis was identified in a distantly related ochrophytean lineage—a secondarily non-photosynthetic amoeboflagellate, *Leukarachnion salinum*. While the remnant plastid is difficult to observe using TEM, its existence in *L. salinum* has been confirmed by the detection of a plastid genome, as well as transcripts for numerous plastid-specific proteins, such as components of the TIC and SELMA protein trafficking systems and the plastid division machinery (e.g., FtsZ). What's more, the organelle retains not only the function of heme biosynthesis, like in the case of *Pteridomonas,* but also the synthesis of folate, tocotrienol, and amino acid precursors [76], which further underscores the often-underestimated gap between the extent of ptDNA reduction and the complexity of plastid functionality.

6.4 *Stramenopila: Chrysophytes*

Chrysophytes are a diverse group of protists belonging to Stramenopila, with varying trophic modes: heterotrophic, mixotrophic, and phototrophic. Similar to other stramenopiles, they possess complex plastids of red algal origin. The non-photosynthetic species are collectively called *Spumella*-like flagellates and belong to several genera — *Poteriospumella, Cornospumella, Spumella, Pedospumella,* and *Paraphysomonas* — which do not form a monophyletic group. This suggests that non-photosynthetic chrysophytes evolved several times independently, which makes them an ideal model to study the impact of the loss of photosynthesis on the metabolic pathways and structure of plastids [77]. Each of the non-photosynthetic clades is sister to photosynthetic or mixotrophic relatives within chrysophytes, allowing detailed comparative studies [54, 77]. All of the known strains of non-photosynthetic chrysophytes have been shown to retain plastid structures and most of them, except *Paraphysomonas*, have also kept their plastid genomes [54].

Detailed studies of the plastid genomes of non-photosynthetic *Spumella*-like flagellates, as expected, show loss of the photosynthesis-related genes found in their photosynthetic relatives [78]. Besides the most commonly lost genes, such as those encoding photosystem and ATP synthase subunits, the investigated genomes have also lost genes for carbon fixation (RuBisCO), chlorophyll biosynthesis (*chlI*), and those encoding cytochrome c biogenesis proteins (*ccs1* and *ccsA*). Although the *pet* (photosynthetic electron transport) genes are almost entirely absent from the plastid genomes of all non-photosynthetic *Spumella*-like flagellates, surprisingly, the *petF* (ferredoxin) gene has been identified and is suggested to encode a protein involved in various redox reactions. Unsurprisingly, besides genes encoding proteins for translation, transcription, and proteolysis, there are also *sufB* and *sufC* genes present in the investigated plastid genomes, which are some of the most commonly retained genes among non-photosynthetic red-originated plastid genomes. Several other genes—such as *dnaK, secA, secY*, and *tufA*—have more complex distribution patterns among colorless chrysophytes. The *Spumella*-like flagellates have lost the vast majority of genes from their plastid genomes. Additionally, they have lost thylakoid membranes in their plastids, except for *Cornospumella fuschlensis*, which retains visible thylakoid structures [77, 78].

Analysis of the plastid genomes from chrysophyte lineages that independently evolved non-photosynthetic plastids shows almost identical gene content and structure for most of the investigated species, suggesting that non-photosynthetic chrysophytes have experienced parallel gene losses as they independently transitioned from phototrophy to heterotrophy. Parallel evolution is further corroborated by the shared conserved set of plastid-targeted proteins, such as proteins associated with oxidative carbon metabolism, plastid glycolysis, heme synthesis, fatty acid and steroid metabolism, cysteine and Fe–S cluster synthesis, as well as plastid protein import, division, and genome expression proteins [54].

However, two exceptions should be noted: the least reduced non-photosynthetic plastids of *Cornospumella fuschlensis* and the dramatically reduced plastids of *Paraphysomonas. C. fuschlensis*, in contrast to other *Spumella*-like flagellates, retains multiple plastid-targeted proteins associated with photosystems and light-harvesting, as well as the majority of other typical plastid functions, except reductive carbon metabolism and amino acid biosynthesis. At the opposite end of the spectrum, the vestigial plastids of *Paraphysomonas* are likely only involved in heme, lipid and steroid synthesis, glycolysis, and plastid protein import and division [54].

7 Dinoflagellata

Dinoflagellate plastids have several unusual features, as this lineage ancestrally contains peridinin plastids. Peridinin is a secondary, carotenoid pigment specific to dinoflagellates. Uniquely, dinoflagellate plastids, along with those of chromerids (see below), use the II RuBisCO gene that has been obtained via lateral gene transfer from bacteria and has replaced the I RuBisCO of cyanobacterial origin [79, 80]. Moreover, the peridinin plastids are enveloped in three membranes, which is likely reflected in protein import mechanisms that do not use SELMA, since the outer membrane is not connected to the endoplasmic reticulum, and it appears to utilize the Golgi apparatus for protein import [81–84]. Both dinoflagellates and chromerids also use a polyuridylate tail for their plastid mRNA [79]. Dinoflagellates represent an interesting lineage for studying the evolution of plastids, as many of them have replaced the peridinin plastid through higher-order endosymbiosis. Many dinoflagellate lineages also contain a so-called kleptoplastid, a plastid that originated through higher-order symbiosis that is temporary and is serially replaced by another higher-order plastid [84, 85].

Many dinoflagellate lineages are non-photosynthetic (up to 50% [86]) and phylogenetic analyses have clearly shown that these non-photosynthetic plastids do not form a monophylum, suggesting multiple independent losses of photosynthesis [84]. However, for most lineages there is relatively little known about the exact metabolic potential of these non-photosynthetic plastids or the status of their genomes. Data from the dinoflagellates, *Abedinium*, *Noctiluca scintillans* (and other Noctilucales), *Crypthecodinium cohnii,* and *Oxyrrhis marina,* suggest that they have preserved isoprenoid, heme, and iron-sulfur cluster biosynthesis pathways in their plastids [87–89]. Little is known about their plastid genomes, yet we can surmise they were likely lost, as the *sufB* gene has been transferred to the nuclear genome, which commonly corresponds with genome loss in these lineages [3]. *Hematodinium* and *Amoebophrya* [90] represent a rare case within dinoflagellates, where complete plastid loss has been shown [4]. They are sister lineages to MALV-II and MALV-IV, and it is assumed that the plastid was completely lost in their common ancestor.

8 Perkinsozoa

Perkinsozoa is a group of parasitic species closely related to dinoflagellates [91]. *Perkinsus* sp. is an oyster parasite and any evidence pointing to a plastid-like organelle is spurious at best [92]. Sequence data have failed to identify the presence of a plastid genome, but did provide evidence for isoprenoid biosynthesis in the cryptic organelle [93]. Somewhat less conclusive evidence exists for the presence of FASII and iron–sulfur cluster biosynthesis pathways, as only a few of their components have been identified in sequence data; however, sensitivity to inhibitors of FASII has been observed in *Perkinsus* [92, 94]. Our understanding of this putative relic plastid organelle in *Perkinsus* is very likely to change as more data and experimental evidence become available.

9 Apicomplexa and Their Relatives

Apicomplexa and their relatives could almost be considered the model group for plastid reduction and loss, as this group is comprised of organisms that span the entire spectrum of plastid/plastome diversity: there are lineages with fully functioning photosynthetic plastids, those that have lost photosynthesis but still carry out chlorophyll biosynthesis, those with reduced plastids and highly reduced plastid genomes, those with highly reduced plastids without genomes, and finally several lineages that have lost their plastids entirely [95]. The traditional 'core' Apicomplexa (Hemosporidia, Piroplasmida, Coccidia) are obligatory intracellular parasites that contain highly reduced plastid organelles called apicoplasts [95, 96]. The taxonomy of Apicomplexa and their relatives has been relatively dynamic, with many new lineages being discovered in the past decade. Here, we will be referring to Apicomplexa sensu Cavalier-Smith [97], where the Apicomplexa taxon contains the group Sporozoa, which includes apicomplexan parasites and represents what would traditionally be considered Apicomplexa. It also contains the group Apicomonadea, which also harbors Sporozoa's photosynthetic relatives *Chromera* and *Vitrella* as well as predatory colpodellids and parasitic *Piridium*. These lineages form monophyletic group in recent phylogenomic studies and are often classified as chrompodellids [3, 5, 98].

The apicoplast is a very small organelle and it took several years of research to convincingly demonstrate that it is indeed a reduced plastid [99–101]. The apicoplasts of different 'core' apicomplexan lineages are more or less identical, are surrounded by four membranes, and have lost most of the original plastid functions, including photosynthesis. The four main metabolic functions preserved in these apicoplasts are isoprenoid, heme, iron–sulfur cluster, and fatty acid biosynthesis [102–104], the exception being the apicoplast of the Piroplasmida group, which seemingly only functions in isoprenoid synthesis. Generally, apicoplast genomes are less than 40 kbp long, with most of the gene content involved in replication and translation; this includes genes for the 16S and 23S ribosomal subunits, ribosomal proteins, tRNAs, RNA polymerases, and elongation factor Tu (*tufA*), as well as the *sufB* and *clpC* genes, which are considered the key for the existence of the genome [3]. SufB functions in iron–sulfur cluster synthesis, while the function of ClpC in apicomplexan metabolism remains unknown. Here again, the exception is Piroplasmida, where *sufB* has been lost from the apicoplast genome; therefore, retention of the genome is likely due to the *clpC* gene in this lineage [3].

Recently, several novel lineages related to Apicomplexa have been studied at the transcriptomic/genomic level. While most of this work has been done using "only" comparative genomics and transcriptomics methods, it has greatly expanded our understanding of the diverse mosaic of functional and genomic reduction of the plastids within these organisms. However, it is important to note that further investigation, whether wet lab or genomics based, will be necessary to fully understand and test our current explanations of the data.

9.1 *Sporozoa*: Nephromyces

Nephromyces are symbionts of the renal sacs of tunicates, where they likely degrade urate as a source of carbon and nitrogen [105]. Recent metagenomic and metatranscriptomic analyses have shown that *Nephromyces* are in fact an internal lineage of Apicomplexa, where they likely branch sister to the Hemosporidia and Piroplasmida clade. They also seem to possess a plastid organelle very similar to the typical apicoplast: its genome appears to be 35 kbp long, with a structure and gene content much like other Apicomplexa [106]. Analysis of *Nephromyces*-enriched metatranscriptomic data has suggested the presence of typical isoprenoid, heme, iron–sulfur cluster, and fatty acid biosynthetic pathways in the plastid [106].

9.2 *Sporozoa: Gregarines*

Gregarines are an invertebrate-infecting, incredibly diverse, group of Apicomplexa. Until recently, very little has been known about their plastid-bearing status; however, with the development of single-cell transcriptomic sequencing, the number of species that have been sequenced and analyzed has greatly increased [5, 96, 107]. Analyses of their transcriptomic and occasionally plastid genome data have shown that plastid evolution within gregarines is quite convoluted, as there are lineages with apicoplast-like organelles, those with cryptic, highly reduced organelles, and lineages that have likely lost their plastid completely. In a recent publication of phylogenomic and plastid analyses, it was shown that Archigregarines contain a plastid that functions in isoprenoid, heme, fatty acid, and iron–sulfur cluster biosynthesis [98]. Their plastid also appears to contain a plastid genome resembling that of the apicoplast. Within Eugregarines, the genera *Lankesteria, Lecudina,* and *Pterospora*, likely contain a cryptic plastid only functioning in fatty acid biosynthesis. The remaining gregarine species appear to have lost their plastids entirely [98]. Interestingly, the current phylogenomic tree suggests two independent instances of complete plastid loss within gregarines; however, it is important to note that these are mostly based on incomplete transcriptomic datasets, and with larger taxon sampling or more complete sequence data, this picture might still change.

9.3 *Sporozoa: Marosporidia*

Marosporidia is the only recently described subgroup of Apicomplexa, which currently consists of four somewhat understudied genera that infect marine invertebrates—*Aggregata*, *Merocystis*, *Margolisiella*, and *Rhytidocystis* [107, 108]. These four genera have been studied at the genomic/transcriptomic level, with phylogenomic analyses revealing that they form a novel lineage within Apicomplexa. Like other apicomplexans, they possess non-photosynthetic plastids that appear to function in the four typical apicoplast pathways—isoprenoid, heme, fatty acid, and iron–sulfur cluster biosynthesis. Based on the available data, *Merocystis* possesses only a few genes to support the existence of these pathways in its plastid; however, this is most likely the result of a poor transcriptomic dataset and will be clarified in the future when more comprehensive data are available. While the plastid genomes of *Aggregata* and *Merocystis* are very similar to traditional apicoplast genomes, those of *Margolisiella* and *Rhytidocystis* are tiny in comparison. They are ±18 kbp long, containing only the 16S and 23S ribosomal subunit genes, a few ribosomal proteins, and tRNA genes, as well as the *clpC* and *sufB* genes (also *tufA* in the case of *Margolisiella*). Notably, the plastid genomes of the latter two lineages are completely missing the RNA polymerase genes, which is striking as they are considered essential for the transcription of plastid-encoded proteins. It has been proposed that the T7 phage-derived polymerase that usually functions in mitochondria could either be duplicated or dually targeted to function in *Margolisiella* and *Rhytidocystis* plastids [108].

9.4 *Sporozoa: Corallicolids*

Surveys of coral microbiomes have revealed a large diversity of plastid-derived SSU sequences that branch closely with their apicomplexan counterparts [109, 110]. In 2019, Kwong and colleagues identified the actual organisms in coral tissue and called them informally corallicolids [59]. Corallicolids are the second most common microbial eukaryotes in corals and appear to be present across a large diversity of coral species. Very little is known about the biochemical functions of their plastids, but they are very interesting in regards to the evolution of non-photosynthetic plastid genomes, since they appear to exist in a middle ground. Their plastid genomes seem to be similar to that of a traditional apicoplast, but they also contain genes for chlorophyll biosynthesis. These genes appear to be transcribed and under strong selection, suggesting their importance for the organism. However, the precise function of these genes remains unknown [59].

9.5 *Apicomonadea:* Piridium

Piridium sociabile is an intracellular parasite of whelk, which was first described as an apicomplexan in the early twentieth century [111]. However, a recent transcriptomic and plastid genomic study has shown that it is in fact related to a photosynthetic algae *Vitrella brassicaformis*, which is closely related to the Apicomplexa [5]. Its plastid seems to be functionally involved in heme, isoprenoid, iron–sulfur cluster, and fatty acid biosynthesis. *Piridium*'s plastid genome is very similar to the traditional apicoplast genome of core Apicomplexa, including the presence of *sufB* and *clpC* genes. However, it has lost inverted ribosomal operon repeats; in fact, it possesses one complete operon and remnants of the second operon (specifically part of the 28S rRNA gene), which are in tandem orientation. Overall, this is significant because it means that *Piridium's* plastid has independently evolved to the same form as the plastid of most core apicomplexans, possibly suggesting a "predefined" path in plastid reduction, at least within these lineages [5].

9.6 Apicomonadea: Colpodellids

Colpodella, *Voromonas,* and *Alphamonas* are predatory flagellates that hunt other microbial eukaryotes. Phylogenetically, they do not form a monophyletic group but are related to Apicomplexa, their closest relatives are *Chromera*, *Vitrella,* and *Piridium* and they are generally referred to as colpodellids [3]. While they are not photosynthetic, based on their phylogenetic position, they are expected to possess a plastid. The molecular data remains scarce, but based on what is available, they appear to possess a cryptic plastid, without a genome, that likely functions in isoprenoid, heme, iron–sulfur cluster, and fatty acid biosynthesis [3]. Interestingly, the *sufB* gene, which is normally encoded on the plastid genome, has been relocated to the nucleus. This suggests that *sufB* may create a dependency on the plastid genome in Apicomplexa and their relatives and that once *sufB* is in the nucleus, the genome can be lost [3].

9.7 Platyproteum *and* Digyalum

Two other lineages that have only recently been studied using genome/transcriptome scale molecular data are *Platyproteum* and *Digyalum*, which in the latest phylogenomic analysis form a monophyletic group [98] together with *Filipodium*. This grouping is classified as Squirmida [112], which forms an independent basal lineage closely related to other Apicomplexa [98]. The best-studied genera are *Platyproteum* and *Digyalum*, both obligatory parasites of animals. These lineages possess non-photosynthetic plastids akin to the apicoplast, but their plastid genomes differ. *Digyalum's* plastid is very similar to the typical apicoplast in both function (isoprenoid, heme, iron–sulfur cluster, and fatty acid biosynthesis) and gene content/genome size. In contrast, the *Platyproteum* plastid is most likely completely lacking a genome, but it is functionally comparable, as it does appear to be involved in isoprenoid, heme, and iron–sulfur cluster biosynthesis [5, 107].

9.8 *Apicomplexa and Their Relatives: Conclusions*

In their recent study, Mathur et al. used the "complete" apicoplast proteome of the apicomplexan, *Toxoplasma gondii*, and searched those proteins across the available diversity of Apicomplexa and their relatives [98]. Their analyses have shown what can be inferred from the text above, that the retention of plastid genomes and plastid pathways within this group is very mosaic. It appears that different lineages have formed alternative dependencies on their plastids, most likely due to their varied ability to scavenge metabolites (e.g., from the host) based on their lifestyle.

10 Conclusions

There are many primary and complex plastid-bearing lineages across the eukaryotic tree. With exception of glaucophytes, haptophytes, and chlorarachniophytes, all of the plastid-bearing lineages contain organisms with non-photosynthetic plastids (Fig. 1). Although giving up a near-infinite source of energy might seem like a surprising strategy, it is most commonly associated with considerable changes in the lifestyle of the plastid-bearing organism, such as the shift toward parasitism. Most of the secondarily non-photosynthetic lineages still contain plastids, as there are only a handful of known cases of complete plastid loss. The known non-photosynthetic plastid-bearing lineages usually retain plastid genomes; nonetheless, there are far more examples of plastid genome loss (with concomitant retention of the organelle itself) than complete plastid loss. Both plastids and their genomes, often with diverse contents, are retained due to the divergent patterns of dependency on the organelle, or its genome, even in non-photosynthetic settings. It is important to note that while complete loss of the plastid appears to be rare, as we have only a small number of examples, it is also somewhat tricky to detect this because the "historical plastid existence" can sometimes leave no footprint in the nuclear genome, which would make identification of plastid loss using current techniques almost impossible. Very often we actually need strong phylogenetic context to show plastid loss, like in the case of *Cryptosporidium*, which branches deep within plastid-bearing lineages.

Plastid Genome Dependency During the evolution of non-photosynthetic plastids, most, if not all, of the photosynthetic genes are lost from the plastid genome. The occasional exceptions are usually represented by typically photosynthesis-associated genes that additionally function in other metabolic pathways or processes, as is evident from the discoveries of fascinating non-photosynthetic lineages that have retained the chlorophyll biosynthesis pathway within their plastid genomes, such as corallicolids, *Cryptomonas parvum*, and several non-photosynthetic representatives of green algae. The biological function that has facilitated this retention remains unknown.

Most genes that remain encoded within reduced plastid genomes are involved in gene transcription and protein translation, which, at first sight, might seem rather redundant. However, in nearly all non-photosynthetic plastid genomes known to date, there are also genes encoding "utility" proteins (i.e., not involved in DNA replication, transcription, or translation), whose roles are described as essential, thus forcing the retention of the plastid gene expression machinery. Consequently, several genes, such as *rbcL*, *clpC*, and *sufB*, appear to be consistently retained in the non-photosynthetic plastid genomes of diverse lineages; for example, it has been postulated that the plastid-encoded *sufB* gene is one of the reasons for plastid genome retention in apicomplexans. Support for this hypothesis comes from studies of the colpodellids—a lineage of free-living predatory flagellates related to Apicomplexa—which are hypothesized to contain cryptic plastids devoid of genomes and have been observed to carry the *sufB* gene in the nucleus.

Plastid Dependency While photosynthesis is considered the primary function of "standard" plastids, other metabolic pathways located in the plastid can still be essential for the organism's survival. This dependency on the plastid likely has its source in the very process of establishing the endosymbiotic organelle, namely the resolution of pathway duplication, which results from a fully self-sufficient host obtaining an organelle with complex metabolic capabilities. The pattern of preservation of host-derived versus symbiont-derived functions is mosaic, with different lineages maintaining different original plastid roles, but the retained organellar pathways are almost exclusively related to the biosynthesis of secondary metabolites, such as heme, isoprenoids, fatty acids, iron–sulfur clusters, purines, or amino acids.

The last two decades have witnessed an explosion in the availability of next-generation sequencing data, which has immensely enriched our understanding of the molecular evolution of organelles and their genomes. Consequently, sequence data has become the foremost source of evidence for the presence or absence of plastid-encoded and plastid-targeted proteins, thus constituting the basis for our knowledge about non-photosynthetic plastids. Genomics and transcriptomics have, unfortunately, one important caveat: the commonplace incompleteness of obtained datasets, which might lead to inaccurate conclusions. A solution to this issue comes from the rapidly-developing novel sequencing technologies, which enable considerable improvement in the quality of produced data at minimal increased cost.

However, as evidenced by a number of studies outlined in this chapter, important insights often come from entirely different approaches than classical genomics. These include, but are not limited to: mass-spectrometry-based proteomics, experimental verification of protein functions via genetic modification (e.g., gene knockouts or heterologous expression), bioenergetic analyses, or measurement of organellar influx or efflux of specific elements and compounds via nanoscale mass spectrometry. With the recent discoveries of fascinating and still enigmatic novel non-photosynthetic lineages, such as the corallicolids or *Rhodelphis*, it is reasonable to expect that studies on the diversity of microbial eukaryotes and their dynamically evolving organelles will lead to exciting findings, or even become part of the foundation for paradigm shifts in evolutionary biology in the upcoming decades.

Acknowledgments We would like to thank the editors for the opportunity to contribute to this book as well as for their valuable comments on the text. We would also like to thank our colleagues Elisabeth Hehenberger, Eric Salomaki, and Rachel Kolisko for their help and suggestions for improving the text.

References

1. Hadariová L, Vesteg M, Hampl V, Krajčovič J (2018) Reductive evolution of chloroplasts in non-photosynthetic plants, algae and protists. Curr Genet 64:365–387. https://doi.org/10.1007/s00294-017-0761-0
2. Oborník M (2022) Organellar evolution: a path from benefit to dependence. Microorganisms 10:122. https://doi.org/10.3390/microorganisms10010122
3. Janouškovec J, Tikhonenkov DV, Burki F, Howe AT, Kolísko M, Mylnikov AP, Keeling PJ (2015) Factors mediating plastid dependency and the origins of parasitism in apicomplexans and their close relatives. Proc Natl Acad Sci U S A 112:10200–10207. https://doi.org/10.1073/pnas.1423790112
4. Gornik SG, Febrimarsa CAM, MacRae JI, Ramaprasad A, Rchiad Z, McConville MJ, Bacic A, McFadden GI, Pain A, Waller RF (2015) Endosymbiosis undone by stepwise elimination of the plastid in a parasitic dinoflagellate. Proc Natl Acad Sci U S A 112:5767–5772. https://doi.org/10.1073/pnas.1423400112
5. Mathur V, Kolísko M, Hehenberger E, Irwin NAT, Leander BS, Kristmundsson Á, Freeman MA, Keeling PJ (2019) Multiple independent origins of apicomplexan-like parasites. Curr Biol 29:2936–2941.e5. https://doi.org/10.1016/j.cub.2019.07.019
6. Karnkowska A, Vacek V, Zubáčová Z, Treitli SC, Petrželková R, Eme L, Novák L, Žárský V, Barlow LD, Herman EK, Soukal P, Hroudová M, Doležal P, Stairs CW, Roger AJ, Eliáš M, Dacks JB, Vlček Č, Hampl V (2016) A eukaryote without a mitochondrial organelle. Curr Biol 26:1274–1284. https://doi.org/10.1016/j.cub.2016.03.053
7. Picelli S, Faridani OR, Björklund ÅK, Winberg G, Sagasser S, Sandberg R (2014) Full-length RNA-seq from single cells using Smart-seq2. Nat Protoc 9:171–181. https://doi.org/10.1038/nprot.2014.006
8. Maciszewski K, Karnkowska A (2019) Should I stay or should I go? Retention and loss of components in vestigial endosymbiotic organelles. Curr Opin Genet Dev 58:33–39. https://doi.org/10.1016/j.gde.2019.07.013
9. Lemieux C, Otis C, Turmel M (2014) Chloroplast phylogenomic analysis resolves deep-level relationships within the green algal class Trebouxiophyceae. BMC Evol Biol 14:211. https://doi.org/10.1186/s12862-014-0211-2
10. Shave CD, Millyard L, May RC (2021) Now for something completely different: *Prototheca*, pathogenic algae. PLoS Pathog 17:e1009362. https://doi.org/10.1371/journal.ppat.1009362
11. Jagielski T, Bakuła Z, Gawor J, Maciszewski K, Kusber WH, Dyląg M, Nowakowska J, Gromadka R, Karnkowska A (2019) The genus *Prototheca* (Trebouxiophyceae, Chlorophyta) revisited: implications from molecular taxonomic studies. Algal Res 43:101639. https://doi.org/10.1016/j.algal.2019.101639
12. Borza T, Popescu CE, Lee RW (2005) Multiple metabolic roles for the nonphotosynthetic plastid of the green alga *Prototheca wickerhamii*. Eukaryot Cell 4:253–261. https://doi.org/10.1128/ec.4.2.253-261.2005
13. Suzuki S, Endoh R, Manabe RI, Ohkuma M, Hirakawa Y (2018) Multiple losses of photosynthesis and convergent reductive genome evolution in the colourless green algae *Prototheca*. Sci Rep 8:940. https://doi.org/10.1038/s41598-017-18378-8
14. de Koning AP, Keeling PJ (2004) Nucleus-encoded genes for plastid-targeted proteins in *Helicosporidium*: functional diversity of a cryptic plastid in a parasitic alga. Eukaryot Cell 3:1198–1205. https://doi.org/10.1128/ec.3.5.1198-1205.2004

15. Pombert JF, Blouin NA, Lane C, Boucias D, Keeling PJ (2014) A lack of parasitic reduction in the obligate parasitic green alga *Helicosporidium*. PLoS Genet 10:1004355. https://doi.org/10.1371/journal.pgen.1004355
16. de Koning AP, Keeling PJ (2006) The complete plastid genome sequence of the parasitic green alga *Helicosporidium* sp. is highly reduced and structured. BMC Biol 4:12. https://doi.org/10.1186/1741-7007-4-12
17. Smith DR, Lee RW (2014) A plastid without a genome: evidence from the nonphotosynthetic green algal genus *Polytomella*. Plant Physiol 164:1812–1819. https://doi.org/10.1104/pp.113.233718
18. Figueroa-Martinez F, Nedelcu AM, Smith DR, Reyes-Prieto A (2016) The plastid genome of *Polytoma uvella* is the largest known among colorless algae and plants and reflects contrasting evolutionary paths to nonphotosynthetic lifestyles. Plant Physiol 173:932–943. https://doi.org/10.1104/pp.16.01628
19. Pánek T, Barcytė D, Treitli SC, Záhonová K, Sokol M, Ševčíková T, Zadrobílková E, Jaške K, Yubuki N, Čepička I, Eliáš M (2022) A new lineage of non-photosynthetic green algae with extreme organellar genomes. BMC Biol 20:66. https://doi.org/10.1186/s12915-022-01263-w
20. Kayama M, Chen JF, Nakada T, Nishimura Y, Shikanai T, Azuma T, Miyashita H, Takaichi S, Kashiyama Y, Kamikawa R (2020) A non-photosynthetic green alga illuminates the reductive evolution of plastid electron transport systems. BMC Biol 18:126. https://doi.org/10.1186/s12915-020-00853-w
21. DeShaw AE, Figueroa-Martinez F, Pröschold T, Lorenz M, Nedelcu AM, Smith DR, Reyes-Prieto A (2022) The plastomes of *Hyalomonas oviformis* and *Hyalogonium fusiforme* evolved dissimilar architecture after the loss of photosynthesis. Plant Direct 6:e454. https://doi.org/10.1002/pld3.454
22. Bellot S, Renner SS (2016) The plastomes of two species in the endoparasite genus *Pilostyles* (Apodanthaceae) each retain just five or six possibly functional genes. Genome Biol Evol 8:189–201. https://doi.org/10.1093/gbe/evv251
23. Wickett NJ, Zhang Y, Hansen SK, Roper JM, Kuehl JV, Plock SA, Wolf PG, DePamphilis CW, Boore JL, Goffinet B (2008) Functional gene losses occur with minimal size reduction in the plastid genome of the parasitic liverwort *Aneura mirabilis*. Mol Biol Evol 25:393–401. https://doi.org/10.1093/molbev/msm267
24. Funk HT, Berg S, Krupinska K, Maier UG, Krause K (2007) Complete DNA sequences of the plastid genomes of two parasitic flowering plant species, *Cuscuta reflexa* and *Cuscuta gronovii*. BMC Plant Biol 7:45. https://doi.org/10.1186/1471-2229-7-45
25. McNeal JR, Kuehl JV, Boore JL, DePamphilis CW (2007) Complete plastid genome sequences suggest strong selection for retention of photosynthetic genes in the parasitic plant genus *Cuscuta*. BMC Plant Biol 7:57. https://doi.org/10.1186/1471-2229-7-57
26. Wicke S, Müller KF, DePamphilis CW, Quandt D, Wickett NJ, Zhang Y, Renner SS, Schneeweiss GM (2013) Mechanisms of functional and physical genome reduction in photosynthetic and nonphotosynthetic parasitic plants of the broomrape family. Plant Cell 25:3711–3725. https://doi.org/10.1105/tpc.113.113373
27. Molina J, Hazzouri KM, Nickrent D, Geisler M, Meyer RS, Pentony MM, Flowers JM, Pelser P, Barcelona J, Inovejas SA, Uy I, Yuan W, Wilkins O, Michel CI, Locklear S, Concepcion GP, Purugganan MD (2014) Possible loss of the chloroplast genome in the parasitic flowering plant *Rafflesia lagascae* (Rafflesiaceae). Mol Biol Evol 31:793–803. https://doi.org/10.1093/molbev/msu051
28. Ng SM, Lee XW, Mat-Isa MN, Aizat-Juhari MA, Adam JH, Mohamed R, Wan KL, Firdaus-Raih M (2018) Comparative analysis of nucleus-encoded plastid-targeting proteins in *Rafflesia cantleyi* against photosynthetic and non-photosynthetic representatives reveals orthologous systems with potentially divergent functions. Sci Rep 8:17258. https://doi.org/10.1038/s41598-018-35173-1
29. Houlihan PR, Stone M, Clem SE, Owen M, Emmel TC (2019) Pollination ecology of the ghost orchid (*Dendrophylax lindenii*): a first description with new hypotheses for Darwin's orchids. Sci Rep 9:12850. https://doi.org/10.1038/s41598-019-49387-4

30. Jąkalski M, Minasiewicz J, Caius J, May M, Selosse MA, Delannoy E (2021) The genomic impact of mycoheterotrophy in orchids. Front Plant Sci 12:632033. https://doi.org/10.3389/fpls.2021.632033
31. Kim YK, Jo S, Cheon SH, Joo MJ, Hong JR, Kwak M, Kim KJ (2020) Plastome evolution and phylogeny of Orchidaceae, with 24 new sequences. Front Plant Sci 11:22. https://doi.org/10.3389/fpls.2020.00022
32. Barrett CF, Sinn BT, Kennedy AH, Pupko T (2019) Unprecedented parallel photosynthetic losses in a heterotrophic orchid genus. Mol Biol Evol 36:1884–1901. https://doi.org/10.1093/molbev/msz111
33. Barrett CF, Davis JI (2012) The plastid genome of the mycoheterotrophic *Corallorhiza striata* (Orchidaceae) is in the relatively early stages of degradation. Am J Bot 99:1513–1523. https://doi.org/10.3732/ajb.1200256
34. Barrett CF, Freudenstein JV, Li J, Mayfield-Jones DR, Perez L, Pires JC, Santos C (2014) Investigating the path of plastid genome degradation in an early-transitional clade of heterotrophic orchids, and implications for heterotrophic angiosperms. Mol Biol Evol 31:3095–3112. https://doi.org/10.1093/molbev/msu252
35. Logacheva MD, Schelkunov MI, Penin AA (2011) Sequencing and analysis of plastid genome in mycoheterotrophic orchid *Neottia nidus-avis*. Genome Biol Evol 3:1296–1303. https://doi.org/10.1093/gbe/evr102
36. Delannoy E, Fujii S, Colas Des Francs-Small C, Brundrett M, Small I (2011) Rampant gene loss in the underground orchid *Rhizanthella gardneri* highlights evolutionary constraints on plastid genomes. Mol Biol Evol 28:2077–2086. https://doi.org/10.1093/molbev/msr028
37. Schelkunov MI, Shtratnikova VY, Nuraliev MS, Selosse MA, Penin AA, Logacheva MD (2015) Exploring the limits for reduction of plastid genomes: a case study of the mycoheterotrophic orchids *Epipogium aphyllum* and *Epipogium roseum*. Genome Biol Evol 7:1179–1191. https://doi.org/10.1093/gbe/evv019
38. Logacheva MD, Schelkunov MI, Nuraliev MS, Samigullin TH, Penin AA (2014) The plastid genome of mycoheterotrophic monocot *Petrosavia stellaris* exhibits both gene losses and multiple rearrangements. Genome Biol Evol 6:238–246. https://doi.org/10.1093/gbe/evu001
39. Lam VKY, Gomez MS, Graham SW (2015) The highly reduced plastome of mycoheterotrophic *Sciaphila* (Triuridaceae) is colinear with its green relatives and is under strong purifying selection. Genome Biol Evol 7:2220–2236. https://doi.org/10.1093/gbe/evv134
40. Ravin NV, Gruzdev EV, Beletsky AV, Mazur AM, Prokhortchouk EB, Filyushin MA, Kochieva EZ, Kadnikov VV, Mardanov AV, Skryabin KG (2016) The loss of photosynthetic pathways in the plastid and nuclear genomes of the non-photosynthetic mycoheterotrophic eudicot *Monotropa hypopitys*. BMC Plant Biol 16:238. https://doi.org/10.1186/s12870-016-0929-7
41. Li X, Zhang TC, Qiao Q, Ren Z, Zhao J, Yonezawa T, Hasegawa M, Crabbe MJC, Li J, Zhong Y (2013) Complete chloroplast genome sequence of holoparasite *Cistanche deserticola* (Orobanchaceae) reveals gene loss and horizontal gene transfer from its host *Haloxylon ammodendron* (Chenopodiaceae). PloS One 8:e58747. https://doi.org/10.1371/journal.pone.0058747
42. Chen J, Yu R, Dai J, Liu Y, Zhou R (2020) The loss of photosynthesis pathway and genomic locations of the lost plastid genes in a holoparasitic plant *Aeginetia indica*. BMC Plant Biol 20:199. https://doi.org/10.1186/s12870-020-02415-2
43. Salomaki ED, Lane CE (2019) Molecular phylogenetics supports a clade of red algal parasites retaining native plastids: taxonomy and terminology revised. J Phycol 55:279–288. https://doi.org/10.1111/jpy.12823
44. Gawryluk RMR, Tikhonenkov DV, Hehenberger E, Husnik F, Mylnikov AP, Keeling PJ (2019) Non-photosynthetic predators are sister to red algae. Nature 572:240–243. https://doi.org/10.1038/s41586-019-1398-6
45. Kořený L, Sobotka R, Janouškovec J, Keeling PJ, Oborník M (2011) Tetrapyrrole synthesis of photosynthetic chromerids is likely homologous to the unusual pathway of apicomplexan parasites. Plant Cell 23:3454–3462. https://doi.org/10.1105/tpc.111.089102

46. Kostygov AY, Karnkowska A, Votýpka J, Tashyreva D, Maciszewski K, Yurchenko V, Lukeš J (2021) Euglenozoa: taxonomy, diversity and ecology, symbioses and viruses. Open Biol 11:200407. https://doi.org/10.1098/rsob.200407
47. Gockel G, Hachtel W (2000) Complete gene map of the plastid genome of the nonphotosynthetic euglenoid flagellate *Astasia longa*. Protist 151:347–351. https://doi.org/10.1078/S1434-4610(04)70033-4
48. Záhonová K, Füssy Z, Oborník M, Eliáš M, Yurchenko V (2016) RuBisCO in nonphotosynthetic alga *Euglena longa*: divergent features, transcriptomic analysis and regulation of complex formation. PloS One 11:e0158790. https://doi.org/10.1371/journal.pone.0158790
49. Füssy Z, Záhonová K, Tomčala A, Krajčovič J, Yurchenko V, Oborník M, Eliáš M (2020) The cryptic plastid of *Euglena longa* defines a new type of nonphotosynthetic plastid organelle. mSphere 5:e00675–e00620. https://doi.org/10.1128/msphere.00675-20
50. Zimorski V, Rauch C, Van JJ, Tielens AGM, Martin WF (2017) The Mitochondrion of *Euglena gracilis*. In: Schwartzbach SD, Shigeoka S (eds) Euglena: biochemistry, cell and molecular biology. Springer International Publishing, pp 19–37. https://doi.org/10.1007/978-3-319-54910-1_2
51. Záhonová K, Füssy Z, Birčák E, Novák Vanclová AMG, Klimeš V, Vesteg M, Krajčovič J, Oborník M, Eliáš M (2018) Peculiar features of the plastids of the colourless alga *Euglena longa* and photosynthetic euglenophytes unveiled by transcriptome analyses. Sci Rep 8:17012. https://doi.org/10.1038/s41598-018-35389-1
52. Hadariová L, Vesteg M, Birčák E, Schwartzbach SD, Krajčovič J (2017) An intact plastid genome is essential for the survival of colorless *Euglena longa* but not *Euglena gracilis*. Curr Genet 63:331–334. https://doi.org/10.1007/s00294-016-0641-z
53. Hoef-Emden K, Tran HD, Melkonian M (2005) Lineage-specific variations of congruent evolution among DNA sequences from three genomes, and relaxed selective constraints on rbcL in *Cryptomonas* (Cryptophyceae). BMC Evol Biol 5:56. https://doi.org/10.1186/1471-2148-5-56
54. Dorrell RG, Azuma T, Nomura M, de Kerdrel GA, Paoli L, Yang S, Bowler C, Ishii KI, Miyashita H, Gile GH, Kamikawa R (2019) Principles of plastid reductive evolution illuminated by nonphotosynthetic chrysophytes. Proc Natl Acad Sci U S A 116:6914–6923. https://doi.org/10.1186/1471-2148-5-56
55. Suzuki S, Matsuzaki R, Yamaguchi H, Kawachi M (2022) What happened before losses of photosynthesis in cryptophyte algae? Mol Biol Evol 39:msac001. https://doi.org/10.1093/molbev/msac001
56. Donaher N, Tanifuji G, Onodera NT, Malfatti SA, Chain PSG, Hara Y, Archibald JM (2009) The complete plastid genome sequence of the secondarily nonphotosynthetic alga *Cryptomonas paramecium*: reduction, compaction, and accelerated evolutionary rate. Genome Biol Evol 1:439–448. https://doi.org/10.1093/gbe/evp047
57. Kamikawa R, Tanifuji G, Ishikawa SA, Ishii KI, Matsuno Y, Onodera NT, Ishida KI, Hashimoto T, Miyashita H, Mayama S, Inagaki Y (2015) Proposal of a twin arginine translocator system-mediated constraint against loss of ATP synthase genes from nonphotosynthetic plastid genomes. Mol Biol Evol 32:2598–2604. https://doi.org/10.1093/molbev/msv134
58. Tanifuji G, Kamikawa R, Moore CE, Mills T, Onodera NT, Kashiyama Y, Archibald JM, Inagaki Y, Hashimoto T (2020) Comparative plastid genomics of *Cryptomonas* species reveals fine-scale genomic responses to loss of photosynthesis. Genome Biol Evol 12:3926–3937. https://doi.org/10.1093/gbe/evaa001
59. Kwong WK, del Campo J, Mathur V, Vermeij MJA, Keeling PJ (2019) A widespread coral-infecting apicomplexan with chlorophyll biosynthesis genes. Nature 568:103–107. https://doi.org/10.1038/s41586-019-1072-z
60. Oborník M, Lukeš J (2015) The organellar genomes of *Chromera* and *Vitrella*, the phototrophic relatives of apicomplexan parasites. Annu Rev Microbiol 69:129–144. https://doi.org/10.1146/annurev-micro-091014-104449

61. Tanifuji G, Onodera NT, Wheeler TJ, Dlutek M, Donaher N, Archibald JM (2011) Complete nucleomorph genome sequence of the nonphotosynthetic alga *Cryptomonas paramecium* reveals a core nucleomorph gene set. Genome Biol Evol 3:44–54. https://doi.org/10.1093/gbe/evq082
62. Kim JI, Tanifuji G, Jeong M, Shin W, Archibald JM (2022) Gene loss, pseudogenization, and independent genome reduction in non-photosynthetic species of *Cryptomonas* (Cryptophyceae) revealed by comparative nucleomorph genomics. BMC Biol 20:227. https://doi.org/10.1186/s12915-022-01429-6
63. Malviya S, Scalco E, Audic S, Vincent F, Veluchamy A, Poulain J, Wincker P, Iudicone D, De Vargas C, Bittner L, Zingone A, Bowler C (2016) Insights into global diatom distribution and diversity in the world's ocean. Proc Natl Acad Sci U S A 113:E1516–E1525. https://doi.org/10.1073/pnas.1509523113
64. Kamikawa R, Yubuki N, Yoshida M, Taira M, Nakamura N, Ishida KI, Leander BS, Miyashita H, Hashimoto T, Mayama S, Inagaki Y (2015) Multiple losses of photosynthesis in *Nitzschia* (Bacillariophyceae). Phycol Res 63:19–28. https://doi.org/10.1111/pre.12072
65. Kamikawa R, Azuma T, Ishii KI, Matsuno Y, Miyashita H (2018) Diversity of organellar genomes in non-photosynthetic diatoms. Protist 169:351–361. https://doi.org/10.1016/j.protis.2018.04.009
66. Onyshchenko A, Roberts WR, Ruck EC, Lewis JA, Alverson AJ (2021) The genome of a nonphotosynthetic diatom provides insights into the metabolic shift to heterotrophy and constraints on the loss of photosynthesis. New Phytol 232:1750–1764. https://doi.org/10.1111/nph.17673
67. Onyshchenko A, Ruck EC, Nakov T, Alverson AJ (2019) A single loss of photosynthesis in the diatom order Bacillariales (Bacillariophyta). Am J Bot 106:560–572. https://doi.org/10.1002/ajb2.1267
68. Kamikawa R, Mochizuki T, Sakamoto M, Tanizawa Y, Nakayama T, Onuma R, Cenci U, Moog D, Speak S, Sarkozi K, Toseland A, van Oosterhout C, Oyama K, Kato M, Kume K, Kayama M, Azuma T, Ishii KI, Miyashita H, Henrissat B, Lombard V, Win J, Kamoun S, Kashiyama Y, Mayama S, Miyagishima SY, Tanifuji G, Mock T, Nakamura Y (2022) Genome evolution of a nonparasitic secondary heterotroph, the diatom *Nitzschia putrida*. Sci Adv 8:5075. https://doi.org/10.1126/sciadv.abi5075
69. Kamikawa R, Moog D, Zauner S, Tanifuji G, Ishida KI, Miyashita H, Mayama S, Hashimoto T, Maier UG, Archibald JM, Inagaki Y (2017) A non-photosynthetic diatom reveals early steps of reductive evolution in plastids. Mol Biol Evol 34:2355–2366. https://doi.org/10.1093/molbev/msx172
70. Derelle R, López-García P, Timpano H, Moreira D (2016) A phylogenomic framework to study the diversity and evolution of stramenopiles (=heterokonts). Mol Biol Evol 33:2890–2898. https://doi.org/10.1093/molbev/msw168
71. Han KY, Maciszewski K, Graf L, Yang JH, Andersen RA, Karnkowska A, Yoon HS (2019) Dictyochophyceae plastid genomes reveal unusual variability in their organization. J Phycol 55:1166–1180. https://doi.org/10.1111/jpy.12904
72. Gerea M, Saad JF, Izaguirre I, Queimaliños C, Gasol JM, Unrein F (2016) Presence, abundance and bacterivory of the mixotrophic algae *Pseudopedinella* (Dictyochophyceae) in freshwater environments. Aquat Microb Ecol 76:219–232. https://doi.org/10.3354/ame01780
73. Sekiguchi H, Moriya M, Nakayama T, Inouye I (2022) Vestigial chloroplasts in heterotrophic stramenopiles *Pteridomonas danica* and *Ciliophrys infusionum* (Dictyochophyceae). Protist 153:157–167. https://doi.org/10.1078/1434-4610-00094
74. Kayama M, Maciszewski K, Yabuki A, Miyashita H, Karnkowska A, Kamikawa R (2020) Highly reduced plastid genomes of the non-photosynthetic dictyochophyceans *Pteridomonas* spp. (Ochrophyta, SAR) are retained for tRNA-Glu-based organellar heme biosynthesis. Front. Plant Sci 11:602455. https://doi.org/10.3389/fpls.2020.602455
75. Barbrook AC, Howe CJ, Purton S (2006) Why are plastid genomes retained in non-photosynthetic organisms? Trends Plant Sci 11:101–108. https://doi.org/10.1016/j.tplants.2005.12.004

76. Jaške K, Barcytė D, Pánek T, Ševčíková T, Eliášová A, Eliáš M (2022) The net-like heterotrophic amoeba *Leukarachnion salinum* sp. nov. (Ochrophyta, Stramenopiles) has a cryptic plastid. bioRxiv. https://doi.org/10.1101/2022.04.05.487141
77. Grossmann L, Bock C, Schweikert M, Boenigk J (2016) Small but manifold—hidden diversity in "*Spumella*-like flagellates". J Eukaryot Microbiol 63:419–439. https://doi.org/10.1111/jeu.12287
78. Kim JI, Jeong M, Archibald JM, Shin W (2020) Comparative plastid genomics of non-photosynthetic chrysophytes: genome reduction and compaction. Front Plant Sci 11:572703. https://doi.org/10.3389/fpls.2020.572703
79. Janouškovec J, Horák A, Oborník M, Lukeš J, Keeling PJ (2010) A common red algal origin of the apicomplexan, dinoflagellate, and heterokont plastids. Proc Natl Acad Sci U S A 107:10949–10954. https://doi.org/10.1073/pnas.1003335107
80. Whitney SM, Shaw D, Yellowlees D (1995) Evidence that some dinoflagellates contain a ribulose-1,5-bisphosphate carboxylase/oxygenase related to that of the α-proteobacteria. Proc R Soc B: Biol Sci 259:271–275. https://doi.org/10.1098/rspb.1995.0040
81. Nassoury N, Cappadocia M, Morse D (2013) Plastid ultrastructure defines the protein import pathway in dinoflagellates. J Cell Sci 116:2867–2874. https://doi.org/10.1242/jcs.00517
82. Patron NJ, Waller RF, Archibald JM, Keeling PJ (2005) Complex protein targeting to dinoflagellate plastids. J Mol Biol 348:1015–1024. https://doi.org/10.1016/j.jmb.2005.03.030
83. Schnepf E, Elbrächter E (1999) Dinophyte chloroplasts and phylogeny—a review. Grana 38:81–97. https://doi.org/10.1080/00173139908559217
84. Waller RF, Kořený L (2017) Plastid complexity in dinoflagellates: a picture of gains, losses, replacements and revisions. Adv Bot Res 84:105–143. https://doi.org/10.17863/CAM.16938
85. Sibbald SJ, Archibald JM (2020) Genomic insights into plastid evolution. Genome Biol Evol 12:978–990. https://doi.org/10.1093/gbe/evaa096
86. Taylor FJR, Hoppenrath M, Saldarriaga JF (2008) Dinoflagellate diversity and distribution. Biodivers Conserv 17:407–418. https://doi.org/10.1007/s10531-007-9258-3
87. Cooney EC, Leander BS, Keeling PJ (2022) Phylogenomics shows unique traits in Noctilucales are derived rather than ancestral. PNAS Nexus 1:pgac202. https://doi.org/10.1093/pnasnexus/pgac202
88. Cooney EC, Okamoto N, Cho A, Hehenberger E, Richards TA, Santoro AE, Worden AZ, Leander BS, Keeling PJ (2020) Single-cell transcriptomics of *Abedinium* reveals a new early-branching dinoflagellate lineage. Genome Biol Evol 12:2417–2428. https://doi.org/10.1093/gbe/evaa196
89. Janouškovec J, Gavelis GS, Burki F, Dinh D, Bachvaroff TR, Gornik SG, Bright KJ, Imanian B, Strom SL, Delwiche CF, Waller RF, Fensome RA, Leander BS, Rohwer FL, Saldarriaga JF (2017) Major transitions in dinoflagellate evolution unveiled by phylotranscriptomics. Proc Natl Acad Sci U S A 114:E171–E180. https://doi.org/10.1073/pnas.1614842114
90. John U, Lu Y, Wohlrab S, Groth M, Janouškovec J, Kohli GS, Mark FC, Bickmeyer U, Farhat S, Felder M, Frickenhaus S, Guillou L, Keeling PJ, Moustafa A, Porcel BM, Valentin K, Glöckner G (2019) An aerobic eukaryotic parasite with functional mitochondria that likely lacks a mitochondrial genome. Sci Adv 5:eaav1110. https://doi.org/10.1126/sciadv.aav111
91. Jeon BS, Park MG (2021) A novel parasitoid of marine dinoflagellates, *Pararosarium dinoexitiosum* gen. et sp. nov. (Perkinsozoa, Alveolata), showing characteristic beaded sporocytes. Front Microbiol 12:748092. https://doi.org/10.3389/fmicb.2021.748092
92. Fernández Robledo JA, Caler E, Matsuzaki M, Keeling PJ, Shanmugam D, Roos DS, Vasta GR (2011) The search for the missing link: a relic plastid in *Perkinsus*? Int J Parasitol 41:1217–1229. https://doi.org/10.1016/j.ijpara.2011.07.008
93. Matsuzaki M, Kuroiwa H, Kuroiwa T, Kita K, Nozaki H (2008) A cryptic algal group unveiled: a plastid biosynthesis pathway in the oyster parasite *Perkinsus marinus*. Mol Biol Evol 25:1167–1179. https://doi.org/10.1093/molbev/msn064
94. Stelter K, El-Sayed NM, Seeber F (2007) The expression of a plant-type ferredoxin redox system provides molecular evidence for a plastid in the early dinoflagellate *Perkinsus marinus*. Protist 158:119–130. https://doi.org/10.1016/j.protis.2006.09.003

95. Salomaki ED, Kolísko M (2019) There is treasure everywhere: reductive plastid evolution in Apicomplexa in light of their close relatives. Biomolecules 9:378. https://doi.org/10.3390/biom9080378
96. Waller RF, McFadden G (2005) The apicoplast: a review of the derived plastid of apicomplexan parasites. Curr Issues Mol Biol 7:57–80. https://doi.org/10.21775/cimb.007.057
97. Cavalier-Smith T (2018) Kingdom Chromista and its eight phyla: a new synthesis emphasizing periplastid protein targeting, cytoskeletal and periplastid evolution, and ancient divergences. Protoplasma 255:297–357. https://doi.org/10.1007/s00709-017-1147-3
98. Mathur V, Salomaki ED, Wakeman KC, Na I, Kwong WK, Kolisko M, Keeling PJ (2023) Reconstruction of plastid proteomes of apicomplexans and close relatives reveals the major evolutionary outcomes of cryptic plastids. Mol Biol Evol 40:msad002. https://doi.org/10.1093/molbev/msad002
99. Elaagip A, Absalon S, Florentin A (2022) Apicoplast dynamics during *Plasmodium* cell cycle. Front Cell Infect Microbiol 12:864819. https://doi.org/10.3389/fcimb.2022.864819
100. McFadden GI, Reith ME, Munholland J, Lang-Unnasch N (1996) Plastid in human parasites. Nature 381:482. https://doi.org/10.1038/381482a0
101. Palmer J (1992) Green ancestry of malarial parasites? Curr Biol 2:318–320. https://doi.org/10.1016/0960-9822(92)90887-G
102. Diniz JAP, Silva EO, Lainson R, De Souza W (2000) The fine structure of *Garnia gonadati* and its association with the host cell. Parasitol Res 86:971–977. https://doi.org/10.1007/PL00008528
103. Köhler S, Delwiche CF, Denny PW, Tilney LG, Webster P, Wilson RJM, Palmer JD, Roos DS (1997) A plastid of probable green algal origin in Apicomplexan parasites. Science 275:1485–1489. https://doi.org/10.1126/science.275.5305.1485
104. Lemgruber L, Kudryashev M, Dekiwadia C, Riglar DT, Baum J, Stahlberg H, Ralph SA, Frischknecht F (2013) Cryo-electron tomography reveals four-membrane architecture of the *Plasmodium* apicoplast. Malar J 12:25. https://doi.org/10.1186/1475-2875-12-25
105. Paight C, Slamovits CH, Saffo MB, Lane CE (2019) *Nephromyces* encodes a urate metabolism pathway and predicted peroxisomes, demonstrating that these are not ancient losses of apicomplexans. Genome Biol Evol 11:41–53. https://doi.org/10.1093/gbe/evy251
106. Muñoz-Gómez SA, Durnin K, Eme L, Paight C, Lane CE, Saffo MB, Slamovits CH (2019) *Nephromyces* represents a diverse and novel lineage of the apicomplexa that has retained apicoplasts. Genome Biol Evol 11:2727–2740. https://doi.org/10.1093/gbe/evz155
107. Janouškovec J, Paskerova GG, Miroliubova TS, Mikhaiiov KV, Birley T, Aieoshin VV, Simdyanov TG (2019) Apicomplexan-like parasites are polyphyletic and widely but selectively dependent on cryptic plastid organelles. Elife 8:e49662. https://doi.org/10.7554/eLife.49662
108. Mathur V, Kwong WK, Husnik F, Irwin NAT, Kristmundsson Á, Gestal C, Freeman M, Keeling PJ (2021) Phylogenomics identifies a new major subgroup of apicomplexans, Marosporida class nov., with extreme apicoplast genome reduction. Genome Biol Evol 13:evaa244. https://doi.org/10.1093/gbe/evaa244
109. Janouškovec J, Horák A, Barott KL, Rohwer FL, Keeling PJ (2013) Environmental distribution of coral-associated relatives of apicomplexan parasites. ISME J 7:444–447. https://doi.org/10.1038/ismej.2012.129
110. Janouškovec J, Horák A, Barott KL, Rohwer FL, Keeling PJ (2012) Global analysis of plastid diversity reveals apicomplexan-related lineages in coral reefs. Curr Biol 22:R518–R519. https://doi.org/10.1016/j.cub.2012.04.047
111. Patten R (1936) Notes on a new Protozoon, *Piridium sociabile* n. gen, n. sp, from the foot of *Buccinum undatum*. Parasitology 28:502–516. https://doi.org/10.1017/S003118200002268X
112. Cavalier-Smith T (2014) Gregarine site-heterogeneous 18S rDNA trees, revision of gregarine higher classification, and the evolutionary diversification of Sporozoa. Eur J Protistol 50:472–495. https://doi.org/10.1016/j.ejop.2014.07.002

Targeting Plastids in an Animal Cell

Jenny Melo Clavijo, Sven B. Gould, and Gregor Christa

Abstract Sacoglossa display one of nature's most curious photosymbiotic systems. Instead of establishing a symbiosis with an algal partner, like other photosymbiotic animals, these slugs sequester only the plastids of their macroalgal prey. The stolen plastids—kleptoplasts—are phagocytized by epithelial cells of the digestive gland system and are initially encapsulated by a phagosomal membrane. In some sacoglossan species, the kleptoplasts remain photosynthetically active over weeks or months, autonomously and separated from the genetic support the algal nucleus offers. The photosynthetic activity of the kleptoplasts resulted in a decade-long debate on whether the slugs support the kleptoplasts with proteins the animals encode due to genes horizontally transferred from the alga. Transcriptome and genome analyses have revealed that slugs do not encode algal-derived genes, and other unknown mechanisms that explain kleptoplast longevity in Sacoglossa are at work. In algae and plants, dual-targeting nuclear-encoded proteins to mitochondria and plastids is common. The correct targeting of the proteins is based on a sequence and its charge, and a median charge could allow the translocation of mitochondrial slug proteins to the chloroplasts. Whether dual targeting is involved in plastid longevity in sacoglossan sea slugs is unknown. A major challenge would be first to translocate the proteins across the phagosomal membrane, if not removed, before they can enter the kleptoplasts. The system is even more complex as some sacoglossan species sequester heterokontophyte plastids that are surrounded by four membranes in their native host. However, only the inner two of those four membranes appear to be retained in the animal cells. In this chapter, we summarize the current knowledge on the slug-kleptoplast system from a cell biology point of view and provide a theoretical road map on how animal proteins could be translocated to the kleptoplasts—or what would hinder such transport.

J. Melo Clavijo · G. Christa (✉)
Institute for Zoology and Didactics of Biology, University of Wuppertal, Wuppertal, Germany
e-mail: meloclavijo@uni-wuppertal.de; christa@uni-wuppertal.de

S. B. Gould
Institute for Molecular Evolution, HHU Düsseldorf, Düsseldorf, Germany
e-mail: gould@hhu.de

S. D. Schwartzbach et al. (eds.), *Endosymbiotic Organelle Acquisition*,
https://doi.org/10.1007/978-3-031-57446-7_8

Keywords Crawling leaves · *Elysia* · Kleptoplasty · Photosymbiosis · *Plakobranchus* · Sacoglossa · Sea slugs · Starch metabolism

1 Introduction

Oxygen-developing photosynthesis of cyanobacteria, algae, and plants is one of the most fundamental biological processes that drives life on Earth [1]. Typically, heterotrophic organisms consume phototrophs to obtain their photosynthetic assimilates, which are stored mainly as glucose bonds, but also in the form of other organic molecules [2]. In some protists and a few animals, however, symbiotic relationships with phototrophs have evolved that allow the heterotrophs to harness photosynthetic assimilates directly [3–6]. The association of corals and photosynthetic dinoflagellates from the Symbiodiniaceae family is probably the most well-known example of such photosymbiosis. Corals harbor their partner algae intracellularly surrounded by an inhibited phagosomal membrane (the symbiosome) [7] and supply them with ammonium and carbon dioxide, while the algae provide glucose, lipids, and oxygen in return [8–11]. Algae can supply more energy than the corals' demand for the calcification processes to build reefs [12, 13]. The nutrient exchange is critical in maintaining the photosymbiotic relationship. Its disruption results in the photosymbiosis breakdown, which leads to what we observe as coral bleaching [14–17].

Heterotrophic hosts most commonly establish a photosymbiosis with unicellular algae or cyanobacteria [6]. Several protist lineages and two animal taxa, microturbellarian flatworms and sacoglossan sea slugs, display unusual photosymbiotic systems [18–21]. These organisms sequester the chloroplasts from their algal prey and retain them within their cells. The ciliate *Mesodinium rubrum* is the only plastid-stealing organism that retains along with the chloroplasts the prey's nucleus (kleptokaryon) and even the mitochondria (Fig. 1). In the ciliate, the kleptoplasts and the mitochondria are surrounded by a phagosomal membrane, while the kleptokaryon resides without an additional membrane inside the cytosol. The kleptokaryon

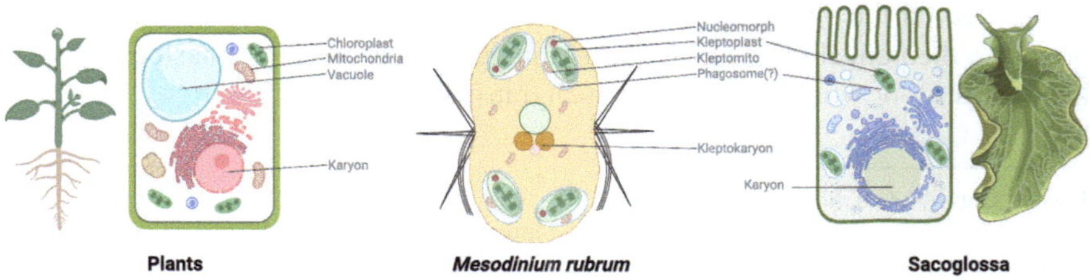

Fig. 1 Chloroplasts in plants, ciliates, and animals. Schematic model of a plant cell, the ciliate *M. rubrum,* and the epithelial cell of the digestive gland system in sacoglossan sea slugs. The presence of a phagosome membrane surrounding the stolen organelles in the slugs remains unclear. Only the ciliate *M. rubrum* can sequester the nucleus of its prey, *Geminigera cryophila* (kleptokaryon). (The figure was created with BioRender.com)

remains transcriptionally active for up to 10 days [18] and is hypothesized to support the kleptoplasts by expressing photosynthetic relevant genes for kleptoplast function and fission [22]. This unique mode of kleptoplasty enables *M. rubrum* to thrive as a mixotroph for up to 2 weeks, combining autotrophic and heterotrophic modes of nutrition [23].

Microturbellarian worms and Sacoglossa sea slugs sequester only the chloroplasts of their algal prey. While the kleptoplasts rapidly lose their functionality in the worms [19], they continue to photosynthesize and build up organic compounds for several weeks to months in a few of the 500 known Sacoglossa species [24–29]. What is fascinating is the longevity of the organelle inside the slugs in the absence of the genetic information (the algal nuclear DNA) encoding many hundreds of proteins that are otherwise considered essential. A canonical plastid imports these proteins for its maintenance through translocons that evolved for this very purpose [30]. This fascinating ability of the slugs raises many questions about how they can incorporate chloroplasts in their cytosol for so long.

2 Functional Kleptoplasty in Sacoglossa

The ability of Sacoglossa to incorporate algal chloroplasts has been known for decades and was first described in 1876 as green granules [31, 32]. Since then, dozens of studies have tried to shed light on the diversity and functionality of the kleptoplasts in Sacoglossa [20, 32–34]. More recently, mechanisms enabling kleptoplast maintenance [9, 35, 36] and properties of the chloroplasts of the food algae are moving into focus [37–40]. Most Sacoglossa feed on green macroalgae of the class Ulvophyceae, but some prey on heterokontophyte algae [27, 41–47]. During the feeding process, the chloroplasts are selectively phagocytized, probably recognized by specific pattern recognition receptors, and reside intracellularly [4, 35, 48, 49]. Once inside the animals' cells, the slugs are categorized based on the functionality of the kleptoplasts during these starvation periods.

Most Sacoglossa cannot retain functional kleptoplasts but digest them quickly and are thus called nonretention species (NR). Functional short-term retention (StR) kleptoplasty refers to maintaining kleptoplasts functional for about 2 weeks [44, 50]. In Sacoglossa, StR evolved twice independently [51, 52] (Fig. 2). During starvation in StR species, the abundance of kleptoplasts decreases along with the photosynthetic activity of the kleptoplasts (measured as the maximum quantum yield (F_v/F_m)), the oxygen generation, and the CO_2 fixation [24, 28, 29, 37, 53, 54]. These StR species can generally starve for much longer than their kleptoplasts remain functional, which is why StR is not considered to represent a truly beneficial photosymbiosis—the slugs are not mixotrophs. However, there are five long-term retention (LtR) species known in which the kleptoplasts remain photosynthetically active for several months (Fig. 2): *Costasiella ocellifera*, *Elysia timida*, *Elysia crispata*, *Elysia chlorotica*, and *Plakobranchus ocellatus*. Functional LtR kleptoplasty evolved independently in each of these species. Yet, the duration of photosynthetic

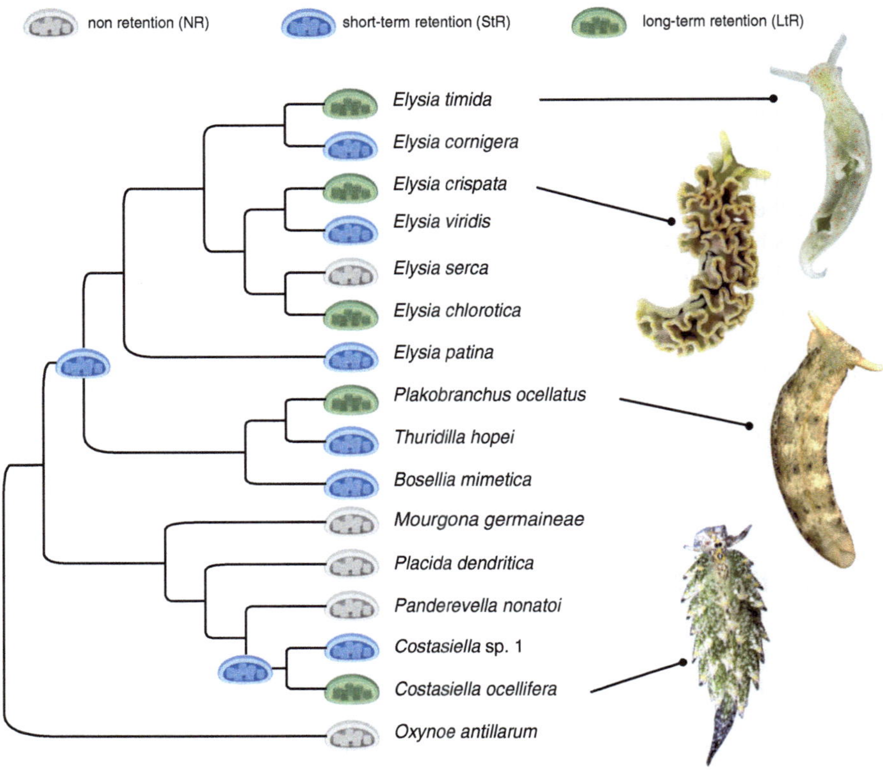

Fig. 2 Evolution of functional kleptoplasty in Sacoglossa after [145]. Four of the five long-term retention (LtR) species are shown. (The figure was created with BioRender.com) (Photos: G. Christa)

kleptoplasts maintenance differs between the species, spanning from 2 to 6 months. Noteworthy, there is no apparent connection between the food algae and the ability for short- or long-term retention, and many NR species feed on the same algae as StR or LtR species [42, 44, 50].

How kleptoplasts are kept functional for months in Sacoglossa is not well understood and remains an active field of research. It was long thought that horizontal gene transfer (HGT) from the algal nucleus to the slug's genome was the source of plastid longevity [55–57]. Among the genes discussed were genes for chlorophyll synthesis and the photosystem II. However, these reports turned out to be the result of PCR artifacts and, likely, some algal traces of mRNA from the food source remaining in the prepped slug tissue due to slow digestion. More recent RNA-Seq and genomic data find no evidence for HGT from alga to the animal [28, 36, 48, 49, 58–63]. If there are no genes of algal origin in the slugs to support kleptoplasty, other mechanisms must be considered to explain the extended function of the alien organelles.

One idea is based on the (dual) targeting of slug proteins to kleptoplasts that are otherwise destined for native cell compartments, such as the animal's mitochondria [20, 32, 61]. Because several physiological processes in mitochondria and chloroplasts pose similar problems—for example, reactive oxygen species (ROS) accumulation associated with respiration and photosynthesis—one can imagine that certain proteins, such as thioredoxins, are dually targeted and aid in kleptoplast longevity [64]. This book chapter provides insight into if and how dual targeting could be involved in kleptoplast longevity in the slugs, and which other explanations should be considered.

3 Protein Translocation in Algae and Plants

To investigate protein targeting to animal kleptoplasts, it is necessary to understand how it works in algae and plants first. In green algae and higher plants, nuclear-encoded and chloroplast-targeted proteins are synthesized in the cytosol and are subsequently imported to the chloroplasts. The newly synthesized proteins destined for the chloroplast stroma usually contain an N-terminal transit peptide (NTS) with specific characteristics [65, 66]. These NTSs are typically 20–100 amino acids long, rich in hydroxylated and hydrophobic amino acids, and are net positively charged [67–69]. Two key protein complexes facilitate the subsequent translocation of proteins into the chloroplast: the translocon at the outer and inner chloroplast membrane (TOC/TIC, respectively). The main TOC components are the TOC75 channel and the two TOC34 and TOC159 GTPase receptor subunits [70]. TOC34 and TOC159 recognize the NTS on the incoming precursor protein that might be assisted by a guidance complex [66, 71]. It is hypothesized that net positive charges on the NTS interact with negatively charged residues on the TOC receptors [72]. After passing through the TOC complex, the precursor protein is transferred to the TIC complex. The core components of the TIC complex consist of TIC20, TIC56, TIC100, and TIC214 [73], except for grasses, where an evolutionary more ancient cross-species conserved TIC complex is assembled around TIC110 and TIC214 is absent [74, 75]. The translocation process through these complexes is energy-dependent, requiring GTP hydrolysis at the TOC and ATP hydrolysis at the TIC complex [76]. After the translocation into the chloroplast, a stromal processing peptidase (SPP) typically cleaves the transit peptide, which leads to protein folding, often assisted by chaperones such as HSP70 and HSP60, into its functional conformation [77].

A TOC-independent protein translocation system that appears to be used by a few substrates has also been identified [78, 79]. Some nuclear-encoded chloroplast proteins contain a glycosylated N-terminal signal peptide, which triggers vesicle transport to the chloroplasts through the endoplasmic reticulum (ER) and the Golgi apparatus. The vesicles fuse with the outer envelope membrane of the chloroplasts and release their cargo into the intermembrane space. This transportation mode is considered the ancestral mechanism for protein trafficking to chloroplasts [80] but is not yet fully understood, and some evidence suggests otherwise, positioning the vesicular trafficking as a derived state [81, 82].

The protein translocation in algae with plastids of secondary endosymbiotic origin (i.e., the plastid originates from an engulfed alga and not a cyanobacterium) differs [83, 84]. Nuclear-encoded proteins destined for these complex plastids must cross a minimum of three, in most cases four membranes: an outer membrane that sometimes continues to the ER, a second membrane of ER origin or the plasma membrane of the engulfed alga, and the canonical outer and inner chloroplast envelope [84]. Hence, in addition to a chloroplast transit peptide, proteins targeted to complex chloroplasts are characterized by bipartite or even tripartite leader sequences [85]. For complex plastids of red algal origin (such as the food alga of the sea slug *Elysia chlorotica*), an N-terminal signal peptide precedes a plastid NTS and serves the co-translational transport across the outer membrane via Sec61 [85, 86]. Within the ER, the proteins are translocated across the second membrane via SELMA, a complex derived from ERAD [87]. Protein translocation from the periplasmatic space across the remaining two membranes to reach the stroma is facilitated by the TOC and TIC complexes like that of primary plastids.

4 Protein Translocation into Sacoglossan Kleptoplasts

Whether and how kleptoplasts in sacoglossans actively import proteins remains unknown. In Sacoglossa, the plastids are phagocytized by the cells lining the digestive gland system (DGS), and they are hence, at least initially, surrounded by a phagosomal membrane [88, 89]. In other photosynthetic endosymbiosis, this membrane is called symbiosome and is crucial for establishing a stable photosymbiosis. The symbiosome contains proton pumps to acidify the interior, bicarbonate and ammonium transporters to provide the symbiont with sufficient inorganic and organic compounds to enhance photosynthesis, and glucose transporters to translocate sugars into the animal's cytosol [10, 90, 91].

The fate of the phagosomal membrane in the slugs, and if it resembles a symbiosome, remains a controversial topic. Based on transmission electron microscopy (TEM), some studies confirm the presence of this membrane surrounding (at least some of) the kleptoplasts in different sacoglossan species [43, 88, 92, 93]. Other studies report its removal and the kleptoplasts appear to be released into the cytosol surrounded by only their native double membrane envelope [58, 94]. The fate of membranes is even more complicated in the case of *Elysia chlorotica*, which feeds on the heterokontophyte *Vaucheria litorea. V. litorea* carries plastids of secondary endosymbiotic origin that are surrounded by four membranes. In *E. chlorotica*, however, the kleptoplasts' outer two membranes are lost during phagocytosis [89]. How this selective membrane removal is managed remains unknown, but it has implications on how we need to investigate the potential targeting of slug proteins to kleptoplasts. So far, no study has investigated whether any of the transporters known to be important to establish a symbiosome in cnidarians also surround the kleptoplasts in slugs, which would be clear evidence for the presence of the phagosomal membrane (or a symbiosome).

A phagosomal membrane surrounding kleptoplasts presents itself as a significant obstacle. The TOC complex would not be in direct contact with the cytosol, and an inadvertent import of, for example, mitochondrial proteins is impossible. In this case, delivery to the outer phagosomal membrane might happen through vesicles that release the proteins for recognition by the TOC complex [78, 79]. Usually, proteins are transported to phagosomes through an initial integration into lysosomal vesicles [95]. Lysosomal enzymes are partly synthesized in the ER, and most of these proteins are tagged on their passage through the Golgi with a specific marker, mannose-6-phosphate [96]. The tagged lysosomal proteins are packaged into transport vesicles that bud off the trans-Golgi network [97]. These transport vesicles then fuse with early endosomes, which are sorting compartments within the cell. In the early endosome, lysosomal enzymes are sorted out from other proteins and lipids [98]. The early endosomes then mature into late endosomes, which continue the sorting process and begin to acidify, a necessary condition for the lysosomal enzymes to function [99]. Finally, the late endosomes transform into fully formed lysosomes, capable of degrading a wide range of substrates [100]. The transformation includes further acidification and concentration of lysosomal enzymes. The lysosome then fuses with phagosomes that require the recruitment and function of various proteins, such as Rab GTPases (i.e., Rab7), SNAREs (soluble N-ethylmaleimide-sensitive factor attachment protein receptors), and tethering factors (like the HOPS complex) [101–104].

Rab7, a small GTPase, plays a crucial role in phagosome maturation by converting early phagosomes to late phagosomes and promoting fusion with lysosomes. In Sacoglossa, there is evidence that Rab7 is down-regulated during starvation and feeding [28, 62], which would at least slow down the fusion of lysosomes with phagosomes and, therefore, subsequent phagosome maturation. This inhibition could be relevant for kleptoplast maintenance, as it is important in other photosymbiotic associations, such as cnidarians and their symbiotic dinoflagellate algae [9, 105]. The inhibition of lysosome fusion would simultaneously affect a mechanism that could theoretically deliver proteins to the phagosome to support kleptoplast functionality.

The ciliate *Mesodinium rubrum* might illuminate the issue, given that the kleptokaryon expresses proteins for kleptoplasts, which are also surrounded by a phagosomal membrane (Fig. 1). The kleptokaryon changes gene expression when captured by *M. rubrum*, particularly down-regulating genes associated with ER protein processing and transport [22, 106]. Yet, the details of this process, including how proteins are translocated into the kleptoplasts, remain unknown. However, the ciliate's endomembrane system is hypothesized to be involved in targeting cryptophyte proteins to the enslaved kleptoplast complexes. In the sacoglossan *Elysia chlorotica*, the gene expression changes during the establishment of kleptoplasts in developing juveniles but does not involve the endomembrane system as in the ciliate [48]. Instead, mainly genes associated with development are up-regulated, which is to be expected for developing juveniles. In starving slugs, especially metabolism-related genes are down-regulated and autophagy is induced [28, 36, 62]. However, the cellular mechanisms that would allow the translocation of proteins into algal and plant plastids, either through the TIC/TOC complex, vesicles, dual targeting, or the endomembrane system, have not yet been investigated in sea slugs. It remains to be uncovered whether or not any of these mechanisms could account for plastid longevity in these marine animals.

5 Functional Kleptoplasty Resembles Leaf Senescence in Plants

With all these uncertainties in mind, let us assume a phagosomal membrane surrounding the kleptoplasts is present to establish a symbiosome-like environment. If so, animal proteins would need specific phagosomal targeting information that precedes, for example, native mitochondrial targeting sequences that could theoretically lead to dual targeting of a protein to kleptoplasts. There is no indication this is the case, so how else could the sequestered organelles maintain photosynthetic activity for a long time? One theory has it that certain plastids are naturally robust, a concept first put forward many decades ago [107], and for which some supporting clues are emerging [32, 38, 39, 108–110]. Among the factors that account for the plastid robustness are photoprotection mechanisms that would prevent or reduce photoinhibition in the kleptoplasts. As mentioned above, kleptoplast functionality is typically assessed by chlorophyll fluorescence measurements and determining the maximum quantum yield of the photosystem II (F_v/F_m). Indeed, in the five LtR species, the functionality of the kleptoplasts declines during starvation [44, 46, 50, 111–115], which is likely the consequence of insufficient genetic support. The decline of F_v/F_m depends on the species, population, and temperature. But in some LtR species, F_v/F_m exceeds values higher than 0.5 for at least 2 months, which is considered a photosynthetically active state. In addition to F_v/F_m, the CO_2 fixation rate of the kleptoplast has been measured and used for efficiency calculations [24, 28, 29, 116–118]. Unlike the F_v/F_m, the decline of the CO_2 fixation rate is observed to be a lot faster; only about a third of the initial CO_2 fixation capacity is retained after a single week of starvation. Oxygen evolution is less commonly studied, but its decline is slower than CO_2 fixation [27, 36]. The observed decrease of the F_v/F_m, the reduction of oxygen evolution, and the decline in CO_2 fixation of kleptoplasts in starving slugs share remarkable similarities with leaf senescence in plants (Fig. 3).

Leaf senescence is a final, intricate stage of leaf development, embodying a suite of systemically coordinated and strictly regulated biochemical and physiological changes, culminating in organized cell death [119]. This process is carefully modulated by an assortment of endogenous signals (developmental age, phytohormones) and environmental cues such as light intensity, temperature, and nutrient availability [120]. Early signs of leaf senescence include reduced chlorophyll content. Several enzymes, including chlorophyllase, degrade chlorophyll, resulting in a decreased

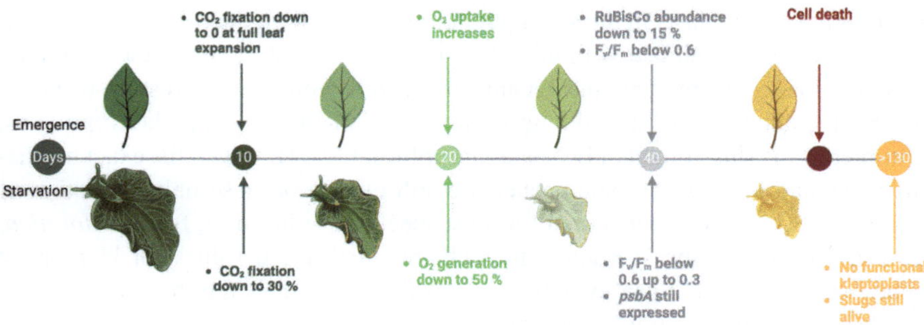

Fig. 3 Functional kleptoplasty resembles leaf senescence. Comparison between the leaf senescence (above) and kleptoplasty in slugs (below). CO_2 fixation, O_2 generation, photosynthetic activity measured by F_v/F_m, and gene expression of the main genes involved in photosynthesis are shown. (The figure was created with BioRender.com)

ability of plants to absorb and utilize light for photosynthesis [121]. Consequently, the reduction in chlorophyll levels leads to a decrease in chlorophyll fluorescence [122], similar to the decline in the F_v/F_m of kleptoplasts in the slugs [50, 111, 113, 114, 123, 124]. Along with chlorophyll reduction, secondary pigments, such as carotenoids and anthocyanins, become more prominent, causing the characteristic shift in leaf color [125]. A similar color change can be observed in starving slugs, which become yellow or slightly brownish and eventually pale with ongoing starvation [29, 113, 124] (Fig. 3).

In plants, the reduction in chlorophyll content accompanies a reduction in RuBisCO (Ribulose-1,5-bisphosphat-carboxylase/−oxygenase) abundance [126], oxygen evolution, and F_v/F_m [122]. In *Arabidopsis thaliana,* CO_2 fixation and oxygen evolution decrease shortly after the beginning of senescence to nearly one-third of the initial values [127–129], which is similar to what is observed in Sacoglossa after 10 days of starvation [28, 29, 118]. The decline in *Arabidopsis* is also induced by the reduction of the gene expression of key PSII components such as psbA, psbC, and psbD, resulting in the disassembly of the photosystems. Along with chlorophyll degradation and reduction in photosynthetic capacity, proteolytic activity targeting the abundant photosynthetic proteins within chloroplasts is initiated in plants. This proteolytic activity is facilitated by a complex of enzymes in the chloroplast stroma and the central vacuole, leading to the systematic degradation of proteins into their constituent amino acids. These amino acids are then mobilized to other tissues via the phloem, supporting new growth and development [130].

Except for active catabolic activities, like chlorophyll and protein degradations, which likely do not occur in the slugs, functional kleptoplasty resembles leaf senescence in plants. Hence, the maintenance of the kleptoplasts is likely based mainly on the autonomy of the kleptoplasts, especially concerning protein longevity. For protein longevity, particularly photoprotection mechanisms will be helpful [37–39, 110, 124, 131], at least at the beginning of the starvation periods and nonstressful light conditions. Eventually, these protection mechanisms also cease when the slugs are starving, which might induce plastid digestion via autophagy. Long-term retention of kleptoplasts in the slugs might thus be based on the robustness of kleptoplast proteins, combined with inhibiting phagosome maturation at the beginning of starvation to ensure kleptoplast maintenance, host immune suppression response against the alien organelle, and long-lasting functionality of kleptoplasts' proteins. Ultimately, the slugs might only bear aging chloroplasts. However, the exact mechanisms behind kleptoplast maintenance are still elusive because only a fraction of experiments have been performed for most species. For instance, for *E. chlorotica,* no F_v/F_m values during starvation are available, and initial values for kleptoplast gene expression or oxygen evolution in freshly fed animals are missing [26, 27]. These knowledge gaps hinder gaining a complete insight into the system.

Still, the slugs evidently use the kleptoplasts as an energy source. Whether this is by an active export or by accessing the photosynthates through digestion is a matter of open debate. Yet, the lack of genetic support of the kleptoplasts and a phagosomal membrane potentially surrounding the alien organelles not only inhibits the targeting of proteins to the kleptoplasts. Likewise, communication between the host nucleus and the kleptoplasts would be unlikely, inhibiting the export of photosynthates to the slugs. This retrograde signaling is essential in plants to regulate nuclear gene expression, ensuring a coordinated cellular response to changing environmental and developmental conditions as well as starch synthesis and degradation [132]. The missing retrograde signaling in kleptoplasts could help to explain observations on starch accumulation in Sacoglossa and favors the hypothesis that the kleptoplasts function as a larder accessible by digestion.

6 Starch Synthesis Eventually Declines, but Its Accumulation Increases at the Outset of Starvation in Slugs

The regulatory mechanisms of starch synthesis and degradation are thought to be tied to the circadian rhythm (the internal 24-hour clock of the plant) and light exposure [133]. These mechanisms likely regulate enzymatic activity at the transcriptional and posttranslational levels. A key signal triggering starch degradation is the fall in the light level at the end of the day, signaling the transition from the light period (photosynthesis and starch synthesis) to the dark period (respiration and starch degradation) [134, 135]. Starch degradation is also thought to be regulated by the availability of end products of photosynthesis, such as ATP and NADPH. When these products are abundant, the plant effectively converts light energy into chemical energy, and less starch degradation is needed. Furthermore, starch degradation might also be regulated by redox signals (changes in the balance between oxidized and reduced molecules in the cell), and by the metabolic status of the cell, that is, whether the cell has sufficient energy and resources at its disposal [136, 137].

In sacoglossan slugs, starch accumulates at the beginning of starvation [138], pointing towards a lack of starch breakdown in the kleptoplasts. How the slugs may access and degrade the accumulated starch remains to be determined. But because the amount of starch and the abundance of kleptoplasts decline with prolonged starvation, digestion via autophagy seems the most likely scenario. Amylases are needed to break down starch into maltose which is then exported into the cytosol in plants. In the starving LtR species *Elysia timida*, amylases do not change their expression pattern upon starvation, while in the StR slug *E. cornigera,* maltase is upregulated [28]. In the StR slug *E. viridis*, however, no maltase regulation was determined [62]. But even if starch-degrading enzymes were synthesized inside the slug cell's cytosol, as previously discussed, chances are low that they could reach the kleptoplast's stroma. There is an alternative route of starch breakdown in plants, though. Research on *Nicotiana benthamiana* leaves revealed an autophagy's role in starch granule degradation. This process involves autophagy-mediated transport of starch granules by structures termed small starch granule-like structures (SSGLs) [139]. Eventually, the SSGLs fuse with the vacuole for starch degradation [140]. In the slugs *E. timida* and *E. viridis,* starch granules have been observed at the proximity of the kleptoplasts and appeared to be in a phase of budding off [141, 142]. Hence, this offers a possibility of how the slugs might profit from accumulated starch in the kleptoplasts gradually over time (Fig. 4).

Other mechanisms regulating starch degradation are posttranslational modifications such as reversible glucan phosphorylation and redox regulation [133]. The addition of phosphate disrupts the surface of the starch grain, facilitating access by starch-degradation enzymes. Two enzymes are crucial for phosphate addition to the starch granules: glucan water dikinase and phosphoglucan water dikinase. Evidence suggests that the starch degradation pathway may be regulated by the enzymatic activity of these water dikinases [133, 143, 144], but the presence of both genes in the slugs' kleptoplasts has not yet been explored. These genes are not plastid encoded, so they are unlikely to be present in the slugs. So far, no evidence exists for activating the regulatory mechanisms for starch degradation at the posttranslational level in Sacoglossa. The absence of a regulated starch degradation mechanism might thus be based on the lack of nuclear-encoded enzymes responsible for starch breakdown in the kleptoplasts after phagocytosis, as well as the missing feedback system occurring in the algal prey, which, if not present, would directly lead to the observed starch accumulation in kleptoplasts. Hence, the sea slugs could only access the starch by autophagy and kleptoplast digestion/degradation.

7 Conclusion

Functional kleptoplasty in sacoglossan sea slugs stands as an intriguing model for studying photosymbiosis and photosynthesis maintenance in nonplant organisms. Especially since several fundamental aspects surrounding sea slug kleptoplasty remain shrouded in controversy. Several questions remain unanswered, and some prevailing notions remain speculative. Based on cellular processes such as

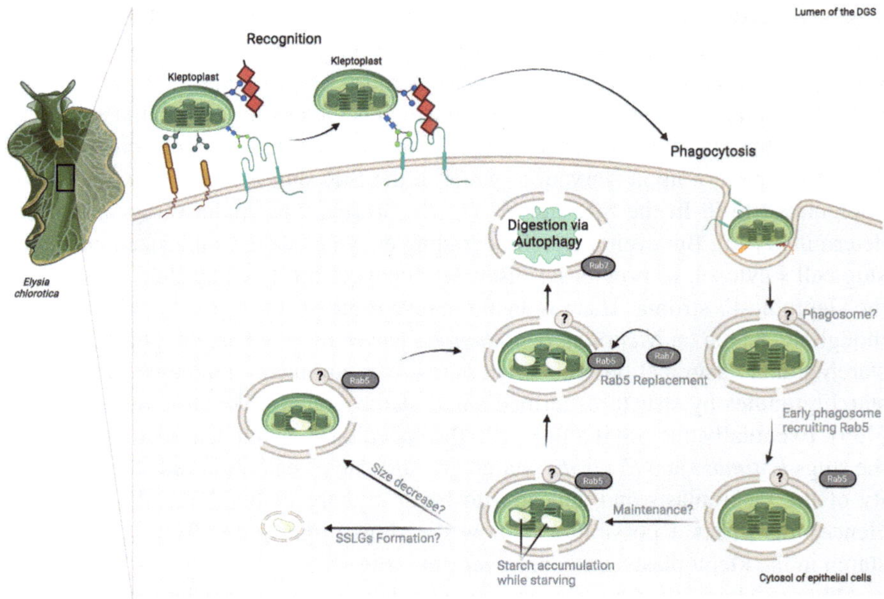

Fig. 4 Potential fate of kleptoplasts inside the slugs' cells. Epithelial cells of the digestive gland system (DGS) likely recognize the chloroplasts with different pattern recognition receptors and initiate phagocytosis. A theory in cnidarians states that the phagosome maturation is inhibited by recruiting Rab5 and inhibiting the replacement by Rab7, inducing symbiont maintenance. It is still unknown if this occurs or if the phagosomal membrane gets lost during this process in the slugs. Once the kleptoplasts are non-functional, for instance, damaged through excessive light, the kleptoplasts might be digested by autophagy. A not yet investigated possibility is that small starch-like granules (SSLGs) could be formed before autophagy, and the starch content of these granules gets digested by the animal. As a result, the size of kleptoplasts would decrease, which can be occasionally observed. (The figure was created with BioRender.com)

phagocytosis for sequestering kleptoplasts, the disruption of ER membranes in complex red plastids of *E. chlorotica*, and the accumulation of photosynthates, active ongoing support by the sea slugs to maintain the photosynthetic activity of kleptoplasts appears unlikely or, at least, far less important than often perceived. Nevertheless, determining the presence or absence of the phagosomal membrane is critical for a clearer understanding of how kleptoplasts might be supported and maintained. After all, the proteins introduced by the kleptoplasts appear to have exceptional stability. Functional kleptoplasty bears striking similarities to mechanisms involved in plant leaf senescence, except for the probable absence of induced active catabolic processes for chlorophyll degradation or proteolysis. Therefore, the slugs seem to capture functional chloroplasts that eventually age within their cytosol and then digest them, or only starch-containing parts, when additional energy is required.

References

1. Hohmann-Marriott MF, Blankenship RE (2011) Evolution of photosynthesis. Annu Rev Plant Biol 62:515–548. https://doi.org/10.1146/annurev-arplant-042110-103811
2. Blankenship RE (2010) Early evolution of photosynthesis. Plant Physiol 154:434–438. https://doi.org/10.1104/pp.110.161687
3. Venn AA, Loram JE, Douglas AE (2008) Photosynthetic symbioses in animals. J Exp Bot 59:1069–1080. https://doi.org/10.1093/jxb/erm328
4. Rumpho ME, Pelletreau KN, Moustafa A, Bhattacharya D (2011) The making of a photosynthetic animal. J Exp Biol 214:303–311. https://doi.org/10.1242/jeb.046540
5. Decelle J, Colin S, Foster RA (2015) Photosymbiosis in Marine Planktonic protists. In: Ohtsuka S, Suzaki T, Horiguchi T et al (eds) Marine protists. Springer Japan, Tokyo, pp 465–500
6. Melo Clavijo J, Donath A, Serôdio J, Christa G (2018) Polymorphic adaptations in metazoans to establish and maintain photosymbioses. Biol Rev 93:2006–2020. https://doi.org/10.1111/brv.12430
7. Hill M, Hill A (2012) The magnesium inhibition and arrested phagosome hypotheses: new perspectives on the evolution and ecology of *Symbiodinium* symbioses. Biol Rev 87:804–821. https://doi.org/10.1111/j.1469-185X.2012.00223.x
8. Barott KL, Venn AA, Perez SO et al (2015) Coral host cells acidify symbiotic algal microenvironment to promote photosynthesis. Proc Natl Acad Sci U S A 112:607–612. https://doi.org/10.1073/pnas.1413483112
9. Mohamed AR, Cumbo V, Harii S et al (2016) The transcriptomic response of the coral *Acropora digitifera* to a competent *Symbiodinium* strain: the symbiosome as an arrested early phagosome. Mol Ecol 25:3127–3141. https://doi.org/10.1111/mec.13659
10. Matthews JL, Crowder CM, Oakley CA et al (2017) Optimal nutrient exchange and immune responses operate in partner specificity in the cnidarian-dinoflagellate symbiosis. Proc Natl Acad Sci U S A 114:13194–13199. https://doi.org/10.1073/pnas.1710733114
11. Thies AB, Quijada-Rodriguez AR, Zhouyao H et al (2022) A Rhesus channel in the coral symbiosome membrane suggests a novel mechanism to regulate NH_3 and CO_2 delivery to algal symbionts. Sci Adv 8:eabm0303. https://doi.org/10.1126/sciadv.abm0303
12. Stanley GD Jr (2006) Photosymbiosis and the evolution of modern coral reefs. Science 312:857–858. https://doi.org/10.1126/science.1123701
13. Stanley GD, Helmle KP (2010) Middle triassic coral growth bands and their implication for photosymbiosis. PALAIOS 25:754–763. https://doi.org/10.2110/palo.2010.p10-039r
14. Brown BE (1997) Coral bleaching: causes and consequences. Coral Reefs 16:129–138
15. Hughes TP, Anderson KD, Connolly SR et al (2018) Spatial and temporal patterns of mass bleaching of corals in the Anthropocene. Science 359:80–83. https://doi.org/10.1126/science.aan8048
16. Blackstone NW, Golladay JM (2018) Why do Corals Bleach? conflict and conflict mediation in a host/symbiont community. Bioessays 40:1800021. https://doi.org/10.1002/bies.201800021
17. Rädecker N, Pogoreutz C, Gegner HM et al (2021) Heat stress destabilizes symbiotic nutrient cycling in corals. Proc Natl Acad Sci U S A 118:e2022653118. https://doi.org/10.1073/pnas.2022653118
18. Hansen PJ, Nielsen LT, Johnson M et al (2013) Acquired phototrophy in *Mesodinium* and *Dinophysis* – A review of cellular organization, prey selectivity, nutrient uptake and bioenergetics. Harmful Algae 28:126–139. https://doi.org/10.1016/j.hal.2013.06.004
19. Van Steenkiste NWL, Stephenson I, Herranz M et al (2019) A new case of kleptoplasty in animals: marine flatworms steal functional plastids from diatoms. Sci Adv 5:eaaw4337. https://doi.org/10.1126/sciadv.aaw4337
20. Cruz S, Cartaxana P (2022) Kleptoplasty: getting away with stolen chloroplasts. PLoS Biol 20:e3001857. https://doi.org/10.1371/journal.pbio.3001857

21. Christa G (2023) Kleptoplasty. Curr Biol 33:R465–R467. https://doi.org/10.1016/j.cub.2023.03.036
22. Johnson MD, Moeller HV, Paight C et al (2023) Functional control and metabolic integration of stolen organelles in a photosynthetic ciliate. Curr Biol S0960982223000611. https://doi.org/10.1016/j.cub.2023.01.027
23. Gustafson DE, Stoecker DK, Johnson MD et al (2000) Cryptophyte algae are robbed of their organelles by the marine ciliate *Mesodinium rubrum*. Nature 405:1049–1052. https://doi.org/10.1038/35016570
24. Hinde R, Smith DC (1975) The role of photosynthesis in the nutrition of the mollusc *Elysia viridis*. Biol J Linn Soc 7:161–171. https://doi.org/10.1111/j.1095-8312.1975.tb00738.x
25. Hinde R (1978) The metabolism of photosynthetically fixed carbon by isolated chloroplasts from *Codium fragile* (Chlorophyta: Siphonales) and by *Elysia viridis* (Mollusca: Sacoglossa). Biol J Linn Soc 10:329–342. https://doi.org/10.1111/j.1095-8312.1978.tb00019.x
26. Mujer CV, Andrews DL, Manhart JR et al (1996) Chloroplast genes are expressed during intracellular symbiotic association of *Vaucheria litorea* plastids with the sea slug Elysia chlorotica. Proc Natl Acad Sci U S A 93:12333–12338. https://doi.org/10.1073/pnas.93.22.12333
27. Green BJ, Li W-Y, Manhart JR et al (2000) Mollusc-algal chloroplast endosymbiosis. Photosynthesis, thylakoid protein maintenance, and chloroplast gene expression continue for many months in the absence of the algal nucleus. Plant Physiol 124:331–342. https://doi.org/10.1104/pp.124.1.331
28. de Vries J, Woehle C, Christa G et al (2015) Comparison of sister species identifies factors underpinning plastid compatibility in green sea slugs. Proc R Soc B 282:20142519. https://doi.org/10.1098/rspb.2014.2519
29. Rauch C, Tielens AGM, Serôdio J et al (2018) The ability to incorporate functional plastids by the sea slug *Elysia viridis* is governed by its food source. Mar Biol 165:82. https://doi.org/10.1007/s00227-018-3329-8
30. Cavalier-Smith T (2000) Membrane heredity and early chloroplast evolution. Trends Plant Sci 5:174–182. https://doi.org/10.1016/S1360-1385(00)01598-3
31. de Negri A, de Negri G (1876) Farbstoffe aus *Elysia viridis*. Ber Deut Chem Gesell 9:84
32. de Vries J, Rauch C, Christa G, Gould SB (2014a) A sea slug's guide to plastid symbiosis. Acta Soc Bot Pol 83:415–421
33. Wägele H, Raupach MJ, Burghardt I et al (2010) Solar powered seaslugs (Opisthobranchia, Gastropoda, Mollusca): incorporation of photosynthetic units: a key character enhancing radiation? In: Glaubrecht M (ed) Evolution in action. Springer, Berlin Heidelberg, pp 263–282
34. Wägele H, Martin WF (2014) Endosymbioses in Sacoglossan Seaslugs: plastid-bearing animals that keep photosynthetic organelles without borrowing genes. In: Löffelhardt W (ed) Endosymbiosis. Springer Vienna, Vienna, pp 291–324
35. Melo Clavijo J, Frankenbach S, Fidalgo C et al (2020) Identification of scavenger receptors and thrombospondin-type-1 repeat proteins potentially relevant for plastid recognition in Sacoglossa. Ecol Evol:ece3.6865. https://doi.org/10.1002/ece3.6865
36. Maeda T, Takahashi S, Yoshida T et al (2021) Chloroplast acquisition without the gene transfer in kleptoplastic sea slugs, *Plakobranchus ocellatus*. Elife 10:e60176. https://doi.org/10.7554/eLife.60176
37. Vieira S, Calado R, Coelho H, Serôdio J (2009) Effects of light exposure on the retention of kleptoplastic photosynthetic activity in the sacoglossan mollusc *Elysia viridis*. Mar Biol 156:1007–1020. https://doi.org/10.1007/s00227-009-1144-y
38. Cartaxana P, Morelli L, Jesus B et al (2019) The photon menace: kleptoplast protection in the photosynthetic sea slug *Elysia timida*. J Exp Biol 222:jeb202580. https://doi.org/10.1242/jeb.202580
39. Havurinne V, Handrich M, Antinluoma M et al (2021a) Genetic autonomy and low singlet oxygen yield support kleptoplast functionality in photosynthetic sea slugs. J Exp Bot erab216. https://doi.org/10.1093/jxb/erab216
40. Christa G, Cruz S, Jahns P et al (2017) Photoprotection in a monophyletic branch of chlorophyte algae is independent of energy-dependent quenching (qE). New Phytol 214:1132–1144. https://doi.org/10.1111/nph.14435

41. Jensen KR (1997) Evolution of the Sacoglossa (Mollusca, Opisthobranchia) and the ecological associations with their food plants. Evol Ecol 11:301–335. https://doi.org/10.1023/A:1018468420368
42. Christa G, Händeler K, Schäberle TF et al (2014a) Identification of sequestered chloroplasts in photosynthetic and non-photosynthetic sacoglossan sea slugs (Mollusca, Gastropoda). Front Zool 11:15. https://doi.org/10.1186/1742-9994-11-15
43. Curtis NE, Massey SE, Pierce SK (2006) The symbiotic chloroplasts in the sacoglossan *Elysia clarki* are from several algal species. Inverteb Biol 125:336–345. https://doi.org/10.1111/j.1744-7410.2006.00065.x
44. Curtis NE, Middlebrooks ML, Schwartz JA, Pierce SK (2015) Kleptoplastic sacoglossan species have very different capacities for plastid maintenance despite utilizing the same algal donors. Symbiosis 65:23–31. https://doi.org/10.1007/s13199-015-0317-3
45. Händeler K, Wägele H, Wahrmund U et al (2010) Slugs' last meals: molecular identification of sequestered chloroplasts from different algal origins in Sacoglossa (Opisthobranchia, Gastropoda). Mol Ecol Resour 10:968–978. https://doi.org/10.1111/j.1755-0998.2010.02853.x
46. Maeda T, Hirose E, Chikaraishi Y et al (2012) Algivore or Phototroph? *Plakobranchus ocellatus* (Gastropoda) continuously acquires kleptoplasts and nutrition from multiple algal species in nature. PloS One 7:e42024. https://doi.org/10.1371/journal.pone.0042024
47. Wade RM, Sherwood AR (2018) Updating *Plakobranchus cf. ianthobapsus* (Gastropoda, Sacoglossa) host use: diverse algal-animal interactions revealed by NGS with implications for invasive species management. Mol Phylogenet Evol 128:172–181. https://doi.org/10.1016/j.ympev.2018.07.010
48. Chan CX, Vaysberg P, Price DC et al (2018) Active host response to algal symbionts in the sea slug *Elysia chlorotica*. Mol Biol Evol 35:1706–1711. https://doi.org/10.1093/molbev/msy061
49. Mendoza M, Rocha S, Troncoso J et al (2022) Transcriptome landscape of kleptoplastic sea slug *Elysia viridis*. J Molluscan studies 89: eyad001. https://doi.org/10.1093/mollus/eyad001
50. Händeler K, Grzymbowski YP, Krug PJ, Wägele H (2009) Functional chloroplasts in metazoan cells – a unique evolutionary strategy in animal life. Front Zool 6:28. https://doi.org/10.1186/1742-9994-6-28
51. Christa G, Händeler K, Kück P et al (2015) Phylogenetic evidence for multiple independent origins of functional kleptoplasty in Sacoglossa (Heterobranchia, Gastropoda). Org Divers Evol 15:23–36. https://doi.org/10.1007/s13127-014-0189-z
52. Hirokane Y, Miyamoto A, Kitaura J et al (2022) Phylogeny and evolution of functional chloroplast retention in sacoglossan sea slugs (Gastropoda: Heterobranchia). Org Divers Evol 22:419–429. https://doi.org/10.1007/s13127-021-00532-w
53. Hinde R, Smith DC (1972) Persistence of functional chloroplasts in *Elysia viridis* (Opisthobranchia, Sacoglossa). Nat New Biol 239:30–31. https://doi.org/10.1038/newbio239030a0
54. Serôdio J, Cruz S, Cartaxana P, Calado R (2014) Photophysiology of kleptoplasts: photosynthetic use of light by chloroplasts living in animal cells. Philos Trans R Soc B 369:20130242. https://doi.org/10.1098/rstb.2013.0242
55. Rumpho ME, Worful JM, Lee J et al (2008) Horizontal gene transfer of the algal nuclear gene *psbO* to the photosynthetic sea slug *Elysia chlorotica*. Proc Natl Acad Sci U S A 105:17867–17871. https://doi.org/10.1073/pnas.0804968105
56. Pierce SK, Curtis NE, Schwartz JA (2009) Chlorophyll a synthesis by an animal using transferred algal nuclear genes. Symbiosis 49:121–131. https://doi.org/10.1007/s13199-009-0044-8
57. Pierce SK, Fang X, Schwartz JA et al (2012) Transcriptomic evidence for the expression of horizontally transferred algal nuclear genes in the photosynthetic sea slug, *Elysia chlorotica*. Mol Biol Evol 29:1545–1556. https://doi.org/10.1093/molbev/msr316
58. Wägele H, Deusch O, Handeler K et al (2011) Transcriptomic evidence that longevity of acquired plastids in the photosynthetic slugs *Elysia timida* and *Plakobranchus ocellatus* does not entail lateral transfer of algal nuclear genes. Mol Biol Evol 28:699–706. https://doi.org/10.1093/molbev/msq239

59. Eastman KE, Pendleton AL, Shaikh MA et al (2023) A reference genome for the long-term kleptoplast-retaining sea slug *Elysia crispata* morphotype clarki. G3: Genes, Genomes. Genetics 13:jkad234. https://doi.org/10.1093/g3journal/jkad234
60. Cai H, Li Q, Fang X et al (2019) A draft genome assembly of the solar-powered sea slug *Elysia chlorotica*. Sci Data 6:190022. https://doi.org/10.1038/sdata.2019.22
61. Pelletreau KN, Bhattacharya D, Price DC et al (2011) Sea Slug Kleptoplasty and plastid maintenance in a Metazoan. Plant Physiol 155:1561–1565. https://doi.org/10.1104/pp.111.174078
62. Frankenbach S, Melo Clavijo J, Brück M et al (2023) Shedding light on starvation in darkness in the plastid-bearing sea slug *Elysia viridis* (Montagu, 1804). Mar Biol 170:89. https://doi.org/10.1007/s00227-023-04225-0
63. Rauch C, de Vries J, Rommel S et al (2015) Why it is time to look beyond algal genes in photosynthetic slugs. Genome Biol Evol 7:2602–2607. https://doi.org/10.1093/gbe/evv173
64. Rumpho ME, Summer EJ, Green BJ et al (2001) Mollusc/algal chloroplast symbiosis: how can isolated chloroplasts continue to function for months in the cytosol of a sea slug in the absence of an algal nucleus? Zoology 104:303–312. https://doi.org/10.1078/0944-2006-00036
65. Knopp M, Garg SG, Handrich M, Gould SB (2020) Major changes in plastid protein import and the origin of the chloroplastida. iScience 23:100896. https://doi.org/10.1016/j.isci.2020.100896
66. Sun Y, Jarvis RP (2023) Chloroplast proteostasis: import, sorting, ubiquitination, and proteolysis. Annu Rev Plant Biol 74:259–283. https://doi.org/10.1146/annurev-arplant-070122-032532
67. Franzén L-G, Rochaix J-D, Von Heijne G (1990) Chloroplast transit peptides from the green alga *Chlamydomonas reinhardtii* share features with both mitochondrial and higher plant chloroplast presequences. FEBS Lett 260:165–168. https://doi.org/10.1016/0014-5793(90)80094-Y
68. Bruce BD (2000) Chloroplast transit peptides: structure, function and evolution. Trends Cell Biol 10:440–447. https://doi.org/10.1016/S0962-8924(00)01833-X
69. Garg SG, Gould SB (2016) The role of charge in protein targeting evolution. Trends Cell Biol 26:894–905. https://doi.org/10.1016/j.tcb.2016.07.001
70. Kessler F, Schnell DJ (2006) The function and diversity of plastid protein import pathways: a multilane GTPase highway into plastids: chloroplast protein import. Traffic 7:248–257. https://doi.org/10.1111/j.1600-0854.2005.00382.x
71. May T, Soll J (2000) 14-3-3 proteins form a guidance complex with chloroplast precursor proteins in plants. Plant Cell 12:53–63. https://doi.org/10.1105/tpc.12.1.53
72. Pool MR (2005) Signal recognition particles in chloroplasts, bacteria, yeast and mammals. Mol Membr Biol 22:3–15. https://doi.org/10.1080/09687860400026348
73. Kikuchi S, Bédard J, Hirano M et al (2013) Uncovering the protein translocon at the chloroplast inner envelope membrane. Science 339:571–574. https://doi.org/10.1126/science.1229262
74. De Vries J, Sousa FL, Bölter B et al (2015) YCF1: a green TIC? Plant Cell 27:1827–1833. https://doi.org/10.1105/tpc.114.135541
75. Rochaix J (2022) Chloroplast protein import machinery and quality control. FEBS J 289:6908–6918. https://doi.org/10.1111/febs.16464
76. Inaba T, Schnell DJ (2008) Protein trafficking to plastids: one theme, many variations. Biochem J 413:15–28. https://doi.org/10.1042/BJ20080490
77. Zhao Q, Liu C (2018) Chloroplast Chaperonin: an intricate protein folding machine for photosynthesis. Front Mol Biosci 4:98. https://doi.org/10.3389/fmolb.2017.00098
78. Villarejo A, Burén S, Larsson S et al (2005) Evidence for a protein transported through the secretory pathway en route to the higher plant chloroplast. Nat Cell Biol 7:1224–1231. https://doi.org/10.1038/ncb1330
79. Radhamony RN, Theg SM (2006) Evidence for an ER to Golgi to chloroplast protein transport pathway. Trends Cell Biol 16:385–387. https://doi.org/10.1016/j.tcb.2006.06.003
80. Bhattacharya D, Archibald JM, Weber APM, Reyes-Prieto A (2007) How do endosymbionts become organelles? Understanding early events in plastid evolution. Bioessays 29:1239–1246. https://doi.org/10.1002/bies.20671

81. Bodył A, Mackiewicz P, Stiller JW (2009) Early steps in plastid evolution: current ideas and controversies. Bioessays 31:1219–1232. https://doi.org/10.1002/bies.200900073
82. Gagat P, Bodył A, Mackiewicz P (2013) How protein targeting to primary plastids via the endomembrane system could have evolved? a new hypothesis based on phylogenetic studies. Biol Direct 8:18. https://doi.org/10.1186/1745-6150-8-18
83. Keeling PJ (2013) The number, speed, and impact of plastid endosymbioses in eukaryotic evolution. Annu Rev Plant Biol 64:583–607. https://doi.org/10.1146/annurev-arplant-050312-120144
84. Gould SB, Maier U-G, Martin WF (2015) Protein import and the origin of red complex plastids. Curr Biol 25:R515–R521. https://doi.org/10.1016/j.cub.2015.04.033
85. Maier UG, Zauner S, Hempel F (2015) Protein import into complex plastids: cellular organization of higher complexity. Eur J Cell Biol 94:340–348. https://doi.org/10.1016/j.ejcb.2015.05.008
86. Agrawal S, Striepen B (2010) More membranes, more proteins: complex protein import mechanisms into secondary plastids. Protist 161:672–687. https://doi.org/10.1016/j.protis.2010.09.002
87. Stork S, Moog D, Przyborski JM et al (2012) Distribution of the SELMA Translocon in secondary plastids of red algal origin and predicted uncoupling of Ubiquitin-dependent translocation from degradation. Eukaryot Cell 11:1472–1481. https://doi.org/10.1128/EC.00183-12
88. Evertsen J, Johnsen G (2009) In vivo and in vitro differences in chloroplast functionality in the two north Atlantic sacoglossans (Gastropoda, Opisthobranchia) *Placida dendritica* and *Elysia viridis*. Mar Biol 156:847–859. https://doi.org/10.1007/s00227-009-1128-y
89. Pelletreau KN, Weber APM, Weber KL, Rumpho ME (2014) Lipid accumulation during the establishment of kleptoplasty in *Elysia chlorotica*. PloS One 9:e97477. https://doi.org/10.1371/journal.pone.0097477
90. Armstrong EJ, Roa JN, Stillman JH, Tresguerres M (2018) Symbiont photosynthesis in giant clams is promoted by V-type H^+-ATPase from host cells. J Exp Biol 221:jeb177220. https://doi.org/10.1242/jeb.177220
91. Roberty S, Béraud E, Grover R, Ferrier-Pagès C (2020) Coral productivity is co-limited by bicarbonate and ammonium availability. Microorganisms 8:640. https://doi.org/10.3390/microorganisms8050640
92. Muscatine L, Pool RR, Trench RK (1975) Symbiosis of algae and invertebrates: aspects of the symbiont surface and the host-symbiont interface. Trans Am Microsc Soc 94:450. https://doi.org/10.2307/3225518
93. Mondy WL, Pierce SK (2005) Apoptotic-like morphology is associated with annual synchronized death in kleptoplastic sea slugs (*Elysia chlorotica*). Inverteb Biol 122:126–137. https://doi.org/10.1111/j.1744-7410.2003.tb00078.x
94. Martin R, Walther P, Tomaschko K-H (2013) Phagocytosis of algal chloroplasts by digestive gland cells in the photosynthesis-capable slug *Elysia timida* (Mollusca, Opisthobranchia, Sacoglossa). Zoomorphology 132:253–259. https://doi.org/10.1007/s00435-012-0184-x
95. Corona Velazquez AF, Jackson WT (2018) So many roads: the multifaceted regulation of autophagy induction. Mol Cell Biol 38. https://doi.org/10.1128/MCB.00303-18
96. Braulke T, Bonifacino JS (2009) Sorting of lysosomal proteins. Biochimica et Biophysica Acta (BBA) - Molecular Cell Research 1793:605–614. https://doi.org/10.1016/j.bbamcr.2008.10.016
97. Bonifacino JS, Glick BS (2004) The mechanisms of vesicle budding and fusion. Cell 116:153–166. https://doi.org/10.1016/S0092-8674(03)01079-1
98. Maxfield FR, McGraw TE (2004) Endocytic recycling. Nat Rev Mol Cell Biol 5:121–132. https://doi.org/10.1038/nrm1315
99. Huotari J, Helenius A (2011) Endosome maturation: endosome maturation. EMBO J 30:3481–3500. https://doi.org/10.1038/emboj.2011.286
100. Saftig P, Klumperman J (2009) Lysosome biogenesis and lysosomal membrane proteins: trafficking meets function. Nat Rev Mol Cell Biol 10:623–635. https://doi.org/10.1038/nrm2745

101. Bucci C, Thomsen P, Nicoziani P et al (2000) Rab7: a key to lysosome biogenesis. MBoC 11:467–480. https://doi.org/10.1091/mbc.11.2.467
102. Jones S, Farquhar M, Martin A, Howl J (2005) Intracellular translocation of the decapeptide carboxyl terminal of Gi3α induces the dual phosphorylation of p42/p44 MAP kinases. Biochimica et Biophysica Acta (BBA) – Molecular Cell Research 1745:207–214. https://doi.org/10.1016/j.bbamcr.2005.05.006
103. Balderhaar HJK, Ungermann C (2013) CORVET and HOPS tethering complexes – coordinators of endosome and lysosome fusion. J Cell Sci 126:1307–1316. https://doi.org/10.1242/jcs.107805
104. Luzio JP, Hackmann Y, Dieckmann NMG, Griffiths GM (2014) The biogenesis of lysosomes and lysosome-related organelles. Cold Spring Harb Perspect Biol 6:a016840–a016840. https://doi.org/10.1101/cshperspect.a016840
105. Fransolet D, Roberty S, Plumier J-C (2012) Establishment of endosymbiosis: the case of cnidarians and *Symbiodinium*. J Exp Mar Biol Ecol 420–421:1–7. https://doi.org/10.1016/j.jembe.2012.03.015
106. Lasek-Nesselquist E, Wisecaver JH, Hackett JD, Johnson MD (2015) Insights into transcriptional changes that accompany organelle sequestration from the stolen nucleus of *Mesodinium rubrum*. BMC Genomics 16:805. https://doi.org/10.1186/s12864-015-2052-9
107. Giles KL, Sarafis V (1972) Chloroplast survival and division in vitro. Nat New Biol 236:56–58. https://doi.org/10.1038/newbio236056a0
108. de Vries J, Christa G, Gould SB (2014b) Plastid survival in the cytosol of animal cells. Trends Plant Sci 19:347–350. https://doi.org/10.1016/j.tplants.2014.03.010
109. de Vries J, Habicht J, Woehle C et al (2013) Is ftsH the key to plastid longevity in Sacoglossan Slugs? Genome Biol Evol 5:2540–2548. https://doi.org/10.1093/gbe/evt205
110. Havurinne V, Tyystjärvi E (2020) Photosynthetic sea slugs induce protective changes to the light reactions of the chloroplasts they steal from algae. Elife 9:e57389. https://doi.org/10.7554/eLife.57389
111. Casalduero FG, Muniain C (2008) The role of kleptoplasts in the survival rates of *Elysia timida* (Risso, 1818): (Sacoglossa: Opisthobranchia) during periods of food shortage. J Exp Mar Biol Ecol 357:181–187. https://doi.org/10.1016/j.jembe.2008.01.020
112. Christa G, Gould SB, Franken J et al (2014b) Functional kleptoplasty in a limapontioidean genus: phylogeny, food preferences and photosynthesis in *Costasiella*, with a focus on *C. ocellifera* (Gastropoda: Sacoglossa). J Moll Stud 80:499–507. https://doi.org/10.1093/mollus/eyu026
113. Laetz EMJ, Rühr PT, Bartolomaeus T et al (2017a) Examining the retention of functional kleptoplasts and digestive activity in sacoglossan sea slugs. Org Divers Evol 17:87–99. https://doi.org/10.1007/s13127-016-0308-0
114. Donohoo SA, Wade RM, Sherwood AR (2020) Finding the sweet spot: sub-ambient light increases fitness and Kleptoplast survival in the sea slug *Plakobranchus* cf. *ianthobaptus* Gould, 1852. Biol Bull 238:154–166. https://doi.org/10.1086/709371
115. Christa G, Wescott L, Schäberle TF et al (2013) What remains after 2 months of starvation? Analysis of sequestered algae in a photosynthetic slug, *Plakobranchus ocellatus* (Sacoglossa, Opisthobranchia), by barcoding. Planta 237:559–572. https://doi.org/10.1007/s00425-012-1788-6
116. Marín A, Ros JD (1992) Dynamics of a peculiar plant-herbivore relationship: the photosynthetic ascoglossan *Elysia timida* and the chlorophycean *Acetabularia acetabulum*. Mar Biol 112:677–682. https://doi.org/10.1007/BF00346186
117. Raven JA, Walker DI, Jensen KR et al (2001) What fraction of the organic carbon in sacoglossans is obtained from photosynthesis by kleptoplastids? An investigation using the natural abundance of stable carbon isotopes. Mar Biol 138:537–545. https://doi.org/10.1007/s002270000488
118. Christa G, Zimorski V, Woehle C et al (2014c) Plastid-bearing sea slugs fix CO_2 in the light but do not require photosynthesis to survive. Proc R Soc B 281:20132493. https://doi.org/10.1098/rspb.2013.2493

119. Lim PO, Kim HJ, Gil Nam H (2007) Leaf senescence. Annu Rev Plant Biol 58:115–136. https://doi.org/10.1146/annurev.arplant.57.032905.105316
120. Woo HR, Kim HJ, Lim PO, Nam HG (2019) Leaf senescence: systems and dynamics aspects. Annu Rev Plant Biol 70:347–376. https://doi.org/10.1146/annurev-arplant-050718-095859
121. Hörtensteiner S (2013) Update on the biochemistry of chlorophyll breakdown. Plant Mol Biol 82:505–517. https://doi.org/10.1007/s11103-012-9940-z
122. Wingler A, Marès M, Pourtau N (2004) Spatial patterns and metabolic regulation of photosynthetic parameters during leaf senescence. New Phytol 161:781–789. https://doi.org/10.1111/j.1469-8137.2004.00996.x
123. Cartaxana P, Trampe E, Kühl M, Cruz S (2017) Kleptoplast photosynthesis is nutritionally relevant in the sea slug *Elysia viridis*. Sci Rep 7:7714. https://doi.org/10.1038/s41598-017-08002-0
124. Christa G, Pütz L, Sickinger C et al (2018) Photoprotective Non-photochemical quenching does not prevent kleptoplasts from net photoinactivation. Front Ecol Evol 6:121. https://doi.org/10.3389/fevo.2018.00121
125. Esteban R, Moran JF, Becerril JM, García-Plazaola JI (2015) Versatility of carotenoids: an integrated view on diversity, evolution, functional roles and environmental interactions. Environ Exp Bot 119:63–75. https://doi.org/10.1016/j.envexpbot.2015.04.009
126. Breeze E, Harrison E, McHattie S et al (2011) High-resolution temporal profiling of transcripts during *Arabidopsis* leaf senescence reveals a distinct chronology of processes and regulation. Plant Cell 23:873–894. https://doi.org/10.1105/tpc.111.083345
127. Betsche T, Eising R (1989) CO_2-fixation, glycolate formation, and enzyme activities of photorespiration and photosynthesis during greening and senescence of sunflower cotyledons. J Exp Bot 40:1037–1043. https://doi.org/10.1093/jxb/40.9.1037
128. Hensel LL, Grbić V, Baumgarten DA, Bleecker AB (1993) Developmental and age-related processes that influence the longevity and senescence of photosynthetic tissues in arabidopsis. Plant Cell 5:553–564. https://doi.org/10.1105/tpc.5.5.553
129. Patro L, Mohapatra PK, Biswal UC, Biswal B (2014) Dehydration induced loss of photosynthesis in Arabidopsis leaves during senescence is accompanied by the reversible enhancement in the activity of cell wall β-glucosidase. J Photochem Photobiol B Biol 137:49–54. https://doi.org/10.1016/j.jphotobiol.2014.03.018
130. Gregersen PL, Holm PB, Krupinska K (2008) Leaf senescence and nutrient remobilisation in barley and wheat. Plant Biol 10:37–49. https://doi.org/10.1111/j.1438-8677.2008.00114.x
131. Havurinne V, Aitokari R, Mattila H et al (2021b) Ultraviolet screening by slug tissue and tight packing of plastids protect photosynthetic sea slugs from photoinhibition. Photosynth Res. https://doi.org/10.1101/2021.07.15.452583
132. Zeeman SC, Kossmann J, Smith AM (2010) Starch: its metabolism, evolution, and biotechnological modification in plants. Annu Rev Plant Biol 61:209–234. https://doi.org/10.1146/annurev-arplant-042809-112301
133. Stitt M, Zeeman SC (2012) Starch turnover: pathways, regulation and role in growth. Curr Opin Plant Biol 15:282–292. https://doi.org/10.1016/j.pbi.2012.03.016
134. Smith SM, Fulton DC, Chia T et al (2004) Diurnal changes in the transcriptome encoding enzymes of starch metabolism provide evidence for both transcriptional and posttranscriptional regulation of starch metabolism in arabidopsis leaves. Plant Physiol 136:2687–2699. https://doi.org/10.1104/pp.104.044347
135. Farré EM, Weise SE (2012) The interactions between the circadian clock and primary metabolism. Curr Opin Plant Biol 15:293–300. https://doi.org/10.1016/j.pbi.2012.01.013
136. Santelia D, Trost P, Sparla F (2015) New insights into redox control of starch degradation. Curr Opin Plant Biol 25:1–9. https://doi.org/10.1016/j.pbi.2015.04.003
137. Skryhan K, Gurrieri L, Sparla F et al (2018) Redox regulation of starch metabolism. Front Plant Sci 9:1344. https://doi.org/10.3389/fpls.2018.01344
138. Laetz EMJ, Moris VC, Moritz L et al (2017b) Photosynthate accumulation in solar-powered sea slugs – starving slugs survive due to accumulated starch reserves. Front Zool 14:4. https://doi.org/10.1186/s12983-016-0186-5

139. Wang Y, Yu B, Zhao J et al (2013) Autophagy contributes to leaf starch degradation. Plant Cell 25:1383–1399. https://doi.org/10.1105/tpc.112.108993
140. Izumi M, Nakamura S, Li N (2019) Autophagic turnover of chloroplasts: its roles and regulatory mechanisms in response to sugar starvation. Front Plant Sci 10:280. https://doi.org/10.3389/fpls.2019.00280
141. Cruz S, LeKieffre C, Cartaxana P et al (2020) Functional kleptoplasts intermediate incorporation of carbon and nitrogen in cells of the Sacoglossa sea slug *Elysia viridis*. Sci Rep 10:10548. https://doi.org/10.1038/s41598-020-66909-7
142. Laetz EMJ, Wägele H (2019) Comparing amylose production in two solar-powered sea slugs: the sister taxa *Elysia timida* and *E. cornigera* (Heterobranchia: Sacoglossa). J Moll Stud 85:166–171. https://doi.org/10.1093/mollus/eyy047
143. Ritte G, Heydenreich M, Mahlow S et al (2006) Phosphorylation of C6- and C3-positions of glucosyl residues in starch is catalysed by distinct dikinases. FEBS Lett 580:4872–4876. https://doi.org/10.1016/j.febslet.2006.07.085
144. Zeeman SC, Delatte T, Messerli G et al (2007) Starch breakdown: recent discoveries suggest distinct pathways and novel mechanisms. Funct Plant Biol 34:465. https://doi.org/10.1071/FP06313
145. Kodžoman K, Clavijo JM, Sickinger C, Christa G (2023) Funktionale Kleptoplastie in Meeresnacktschnecken. Biologie in unserer Zeit. https://doi.org/10.11576/BIUZ-6349

Part II
Host and Endosymbiont Processes Utilized for Acquisition and Evolution of an Endosymbiont into an Organelle

Membrane Delivery to the Vacuole and the Multifunctional Roles of Vacuoles

Ann-Kathrin Rößling, Jürgen Kleine-Vehn, and Kai Dünser

Abstract The vacuole is the largest organelle in plants and provides various crucial functions to plant cells. This book chapter recapitulates the multifunctional roles of plant vacuoles and mainly focuses on the highly dynamic vacuolar morphology. Vacuolar appearance and function differ depending on the cell type and its developmental stage. Lytic vacuoles (LVs) are the most common vacuoles in vegetative tissues. They display an acidic pH and contain proteinases as well as nucleases to fulfil their function as a part of the degradation machinery. The so-called protein storage vacuoles (PSVs) mainly occur in seed tissues and can be converted into LVs during development.

Besides discussing the multiple roles of the vacuole, we review how distinct vesicle trafficking routes define the identity and, thereby, function of the vacuolar membrane (also called tonoplast). Sorting signals and receptors define the specific delivery of cargo vesicles, which require a complex interplay of vesicle coats, teth-

A.-K. Rößling (✉)
Faculty of Biology, Chair of Molecular Plant Physiology (MoPP), University of Freiburg, Freiburg, Germany
e-mail: ann-kathrin.roessling@biologie.uni-freiburg.de

J. Kleine-Vehn
Faculty of Biology, Chair of Molecular Plant Physiology (MoPP), University of Freiburg, Freiburg, Germany

Center for Integrative Biological Signalling Studies (CIBSS), University of Freiburg, Freiburg, Germany
e-mail: juergen.kleine-vehn@biologie.uni-freiburg.de

K. Dünser
Faculty of Biology, Chair of Molecular Plant Physiology (MoPP), University of Freiburg, Freiburg, Germany

Center for Integrative Biological Signalling Studies (CIBSS), University of Freiburg, Freiburg, Germany

Institute of Plant Biotechnology and Cell Biology, University of Natural Resources and Life Sciences, Wien, Austria
e-mail: kai.duenser@boku.ac.at

S. D. Schwartzbach et al. (eds.), *Endosymbiotic Organelle Acquisition*,
https://doi.org/10.1007/978-3-031-57446-7_9

ering factors, as well as membrane fusion machinery. Vacuolar size and the overall size of the cell are coordinated, which plays a pivotal role in cell size determination. The vacuolar volume increases during cellular elongation and thereby allows the cytosolic volume of the cell to remain relatively constant. Here we highlight that plant cells use a complex network of vesicle trafficking routes to precisely coordinate cargo delivery to the tonoplast, as well as, the overall size of the vacuole.

Keywords Trafficking · Vacuole · Growth control · Vesicles · Cell wall · Elongation · Vacuolar biogenesis · Protein storage vacuole · Lytic vacuole · pH · Homeostasis

1 Introduction to the Multifunctional Role of the Vacuole

The word *vacuole* originates from the Latin word *vacuus*, meaning vacuum. Therefore, the vacuole was first thought of as an empty and possibly even functionless space within the cell. Over the years, quite the opposite was proven to be true—the vacuole has a plethora of different and even contradicting functions, such as ensuring lytic degradation as well as storage of cellular material. The vacuole plays a crucial role among others in cellular growth, immune responses against pathogens, control of turgor pressure and pH homeostasis [1–3]. In order to accommodate storage and lytic functions, plants evolved different types of vacuoles (see Sect. 2), such as the protein storage vacuoles (PSVs) and the lytic vacuoles (LVs) [4]. PSVs mainly reside in seed tissues whereas LVs are the predominant vacuoles in vegetative tissues [5].

Vacuole biogenesis seems to be a cell-type-specific process in plants, and multiple models and mechanisms for vacuole biogenesis have been proposed (see Sect. 3). Mechanistically, vesicle trafficking routes are central to defining and dynamically altering the multitude of vacuolar functions (see Sect. 4). Within this cellular frame, vacuoles receive many membranes and proteins from various subcellular sources [3, 6]. Vacuolar cargo delivery depends on sorting motifs present in the cargo proteins and the specific machinery that recognises such motifs. Some of these cargos are hydrolytic enzymes, such as proteinases and nucleases, allowing LVs to fulfil their function as a part of the degradation machinery. These enzymes also require the maintenance of an acidic pH, mediated by vacuolar proton pumps [7, 8] proton gradient provides the necessary energy for secondary active transport across the tonoplast [9].

The vacuole is also the final destination of autophagic cargo and can also trigger self-destructing cellular programs [6, 10, 11], but these aspects will not be explicitly covered in this review. Instead, a focus will be given here on the pivotal role of vacuoles in cellular homeostasis (see Sect. 5) and cell size determination (see Sect. 6). The vacuolar volume increases during cellular elongation, which allows the cytosolic content of the cell to remain relatively constant, enabling fast and energy-efficient cellular elongation in plants [12–14]. Besides occupying space, the vacuole

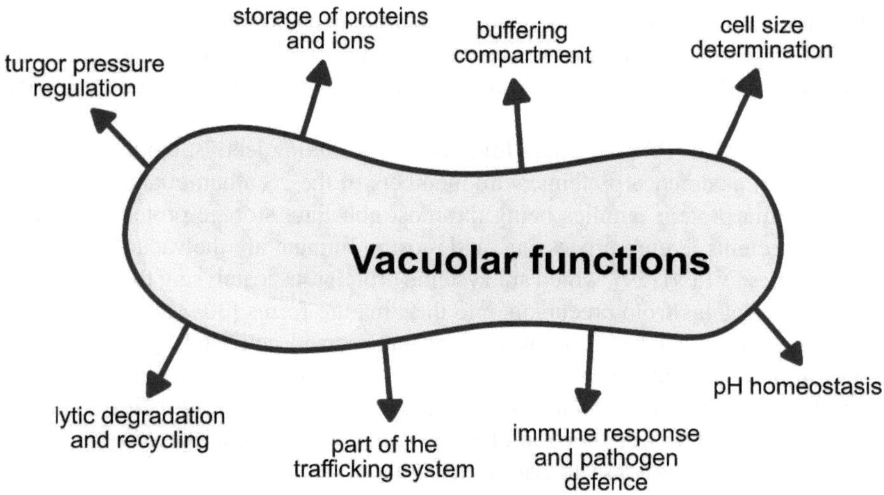

Fig. 1 The vacuole is a multifunctional cellular compartment within plant cells, contributing to various aspects of the lifecycle of plants

is also involved in regulating the osmotic potential of a cell. This contributes to the so-called cell turgor pressure, which is a prerequisite of cell expansion (see Fig. 1) [15, 16].

2 Distinct Types of Vacuoles

As briefly introduced above, plants contain PSVs and LVs mainly in developing seeds and vegetative cells, respectively. Both types are relatively easy to distinguish from other organelles due to their large size but differ from each other in a separate set of tonoplast intrinsic proteins (TIPs), cargo proteins, as well as luminal pH. In terms of membrane markers, α-TIP and δ-TIP are indicating PSVs, whereas γ-TIPs are typically present in LVs [4, 17, 18]. PSVs and LVs were further shown to co-exist during seed development in legumes [19], requiring precise tuning of the sorting machinery.

As the name suggests, PSVs accumulate large amounts of proteins but also sugars, and thereby function as a storage compartment in seeds, providing sufficient nutrient supply for the developing embryo (reviewed in [20]). They are further capable of a transient accumulation of storage compounds, which can be observed for instance in bark, tubers, leaves or pods [21, 22]. In general, PSVs were shown to contain three morphological regions: the matrix, crystalloid and globoid [20, 23, 24]. While the matrix and crystalloid contain storage proteins, the globoid store phosphate in the form of phytic acid crystals [24]. During the maturation of the seed, storage proteins are synthesised in huge amounts in the ER and delivered into the vacuoles, where they are kept until consumption during germination [22, 25].

PSVs are mainly present in cotyledons, aleurone or endosperm of seeds, but may also be present as small PSVs in vegetative tissues, such as root tip or leaf mesophyll cells. PSVs typically possess higher pH and lower hydrolytic activities compared to LVs [26].

The contents of PSVs are manifold, usually including lectins, albumins, globulins, as well as defence proteins, with members of the 2S albumin and the 7S and 12S globulin protein families being the most abundant storage proteins [27, 28]. Storage proteins require processing, and most prominent are the vacuolar processing enzymes (VPEs) [29], which are cysteine proteinases, catalysing the conversion of storage proteins from precursors into their mature forms [30–32]. The enzymes themselves are synthesised as inactive precursors and activated by self-catalytic removal of an autoinhibitory domain at their C-terminus [33]. There are four VPE homologs present in the *Arabidopsis* genome ($\alpha - \delta VPE$). Besides their role in protein storage, another key function of VPEs is the promotion of the vacuolar membrane collapse in response to pathogens. This proteolytic cascade aids pathogen defence, ultimately leading to programmed cell death (PCD) [31, 34, 35].

During seed development, precursors of storage proteins are synthesised at the endoplasmic reticulum (ER) and reach the PSVs via at least three different vesicle trafficking pathways: (1) direct transport from the Golgi to the PSVs, (2) the autophagic pathway and (3) Golgi-independent transport utilising dense vesicles (DVs) (see Fig. 2) [36]. DVs were first discovered in developing pea cotyledons [17]. They obtained their name due to their dense appearance in electron microscopy, which is caused by their highly aggregated storage globulins. DVs appear small and uniform, 150–200 nm in diameter, and form at the periphery of the very first *cis*-cisterna of the Golgi apparatus, carrying on to the *trans* face via maturation [37]. They typically do not contain any sort of protein coat, but clathrin-coated vesicles (CCVs) are shown to bud off the surface of DVs later on [36]. At the end of their journey, DVs fuse with multivesicular bodies (MVBs), also known as prevacuolar compartments (PVCs) in plants [38], delivering the contents further to the PSVs.

The processing of the storage proteins starts at the MVBs, demonstrated by the fact that the 2S albumin was shown to be present as a precursor as well as in the processed state in this compartment [28]. A shift in pH could initiate the processing of the storage proteins at the MVBs, because its progressive acidification during organelle maturation [39] facilitates the activity of processing enzymes with a low pH optimum, such as VPE. Further, the low pH level could initiate the detachment of the ligands from their receptors [40, 41].

Storage proteins are initially stored followed by a rapid breakdown during germination to provide nutrients. Notably, storage proteins and their corresponding processing proteases are delivered via separated vesicle trafficking routes to the MVBs [27, 28], which may prevent premature digestion. Seed imbibition will initiate PSVs to be converted into LVs [42, 43]. During the conversion process, vacuolar identity is altered by the degradation of regulatory proteins as well as by biosynthetic replacement.

Lytic vacuoles typically contain high levels of proteinases and nucleases requiring an acidic pH for their enzymatic functions [44]. There are indications that lytic

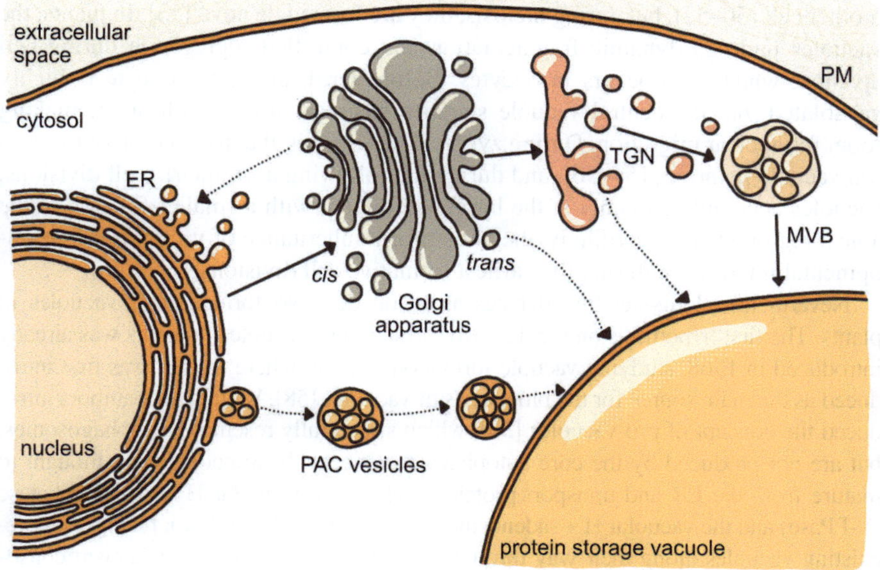

Fig. 2 Overview displaying the trafficking pathways transporting cargo towards protein storage vacuoles in plant cells. Canonically, storage proteins start their journey at the ER and reach the Golgi apparatus via anterograde vesicle trafficking. A Golgi-bypassing route involves precursor accumulating (PAC) vesicles as intermediate compartments for direct transport of cargo to the vacuole. Through the Golgi apparatus as a major sorting station, cargo proteins get transported to the *trans*-Golgi network (TGN). From the *cis* face of the Golgi apparatus, aggregation-based sorting using dense vesicles (DVs) is taking place and these vesicles are able to bypass the TGN trafficking route. Exiting the TGN, cargo is transported to the vacuole via multivesicular bodies (MVBs) as the final destination of the cargo. This step can be bypassed by AP3-dependent trafficking from the TGN directly to the vacuole.

vacuoles can further differentiate in content and function. In 2005, Otegui and colleagues demonstrated the existence of senescence-associated vacuoles (SAVs) in mesophyll and guard cells in Arabidopsis as well as soybean. They are distinguishable from central vacuoles in terms of a lower pH level and a different composition of the tonoplast, especially a lack of γ-TIP aquaporins [45]. SAVs are further linked to autophagic processes due to their involvement in the degradation of chloroplasts [45].

The accumulation of a variety of secondary metabolites, such as organic acids, alkaloids and flavonoids (including anthocyanins) can furthermore define the vacuolar function (reviewed in [46]. Flavonoids are among the most abundant compounds, usually accumulating highly in seeds, fruits, and flowers, where they contribute to an enormous range of colours, but also provide UV protection. Flavonoids also function as developmental cues and their vacuolar accumulation is used as a readout against plant-feeding insects and herbivores [47–49].

3 Vacuolar Biogenesis

In the yeast species *S. cerevisiae* vacuoles are inherited during mitosis from the mother cell [50–52], but during meiosis, they are formed de novo [53]. In plants, the vacuoles undergo dynamic fragmentation and controlled segregation during cell division, which also occurs in oocytes. Before fertilisation, the oocyte is highly vacuolated, but the central vacuole starts to fragment upon fertilisation, making room for nuclear migration. During zygote development, the distribution of the central vacuole polarizes [54–56], and during the following asymmetric cell divisions, vacuoles accumulate mainly at the basal cell region, with a smaller fraction in the apical region [57]. Accordingly, the nonuniform inheritance of vacuoles has developmental importance during the earliest formative cell divisions.

Nevertheless, long-standing debates arose on de novo formation of vacuoles in plants. The first hypothesis on the de novo formation of vacuoles in plants was already introduced in 1968, studying vacuole formation in corn, where the ER was first introduced as the main source for the biogenesis of vacuoles [58]. Viotti and coauthors introduced the concept of pro-vacuoles [59], which structurally resemble autophagosomes, but are not produced by the core autophagy machinery. Provacuoles were thought to mature from the ER and transport proteins such as the vacuolar H^+-pyrophosphatase (V-PPase) and the vacuolar H + -adenosine triphosphatase (V-ATPase), fusing with preexisting vacuoles along their way through the cell periphery. This mechanism constitutes a direct ER-to-vacuole pathway, bypassing the Golgi apparatus [59].

On the contrary, an alternative model based on endosomal fusion was supported by 2D and 3D TEM data, showing that small vacuoles mature and derive from MVB fusions in the root cortex of Arabidopsis [3, 38, 60]. By using whole-cell electron tomography, it was shown that small vacuoles (SVs), 400–1000 nm in diameter, can be formed by maturation and fusion of MVBs [38, 61]. MVBs are originating from the fusion of DVs and small vesicles (SVs). Further, it was suggested that MVBs are able to fuse with each other [28]. This scenario is somewhat reminiscent of mammalian cells, which contain a large number of lysosomes of a rather small size. These lytic compartments are passed on from the mother cells during mitosis but were also shown to be formed from the maturation and fusion of late endosomes [62].

In essence, plant vacuoles can be inherited as well as newly synthesised (reviewed in [63]) and the underlying mechanisms (ER- vs. MVB-derived) could exist in parallel or might be tied to tissue- and/or context-specific cues.

4 Trafficking Routes Towards the Vacuole

4.1 *Membrane Sorting Towards the Vacuole*

The sorting of cargo towards the vacuole is initiated at the ER. This network of tubules and cisternae around the nuclear envelope represents the first compartment of the secretory pathway and the entry point into the endomembrane system. Further, it is an important checkpoint to ensure correct protein folding, so that only proteins of proper quality can exit the ER [64, 65]. Proteins with an ER-localisation are retained in the ER or recycled back from the Golgi apparatus [64]. So far, there is no evidence of an intermediate compartment between the ER and the Golgi [66].

Canonical trafficking towards the vacuole can be implemented via conventional anterograde transport from the ER to the Golgi and then further via post-Golgi transport [67, 68]. In plants, the *trans*-Golgi network (TGN) functions as an early endosome [69], receiving cargo from endocytosis events from the PM as well as from the Golgi apparatus. While the secretory and the endocytic pathway converge at the TGN, their respective cargoes are forwarded to the MVB by different sorting mechanisms and it further serves as a site for recycling [69, 70]. Protein sorting at subdomains as well as organelle maturation allows the cargo delivery to and formation of multivesicular endosomes. These MVBs display the ability to fuse with vacuoles (reviewed in [71]).

Vacuolar trafficking may have evolved specific features to deal with storage proteins since they are mainly insoluble, a potentially problematic issue in a rather aqueous environment [72]. Further, they are synthesised in such large amounts, that the classical transport route does not appear to suffice [73]. Plant cells evolved a transport route for a bulk of material towards the PSVs, which can bypass the Golgi [74]. Experimental data demonstrated that membrane cargos could still be delivered, even when Golgi and post-Golgi trafficking were chemically blocked [75]. A Golgi-bypassing route was also discovered for precursor-accumulating (PAC) vesicles, delivering storage proteins to PSVs [76]. PAC vesicles morphologically resemble characteristics of PVCs and are distinguishable from DVs due to their electron-translucent outer layer, containing glycoproteins. These types of vesicles are formed by protein aggregation in the ER, known at least from studies in pea, developing pumpkin, and castor bean seeds [77]. An equivalent structure termed dark intrinsic proteins (DIP) organelle, based on the presence of DIP, a receptor-like protein of the RMR family, was reported in tomato, snapdragon, tobacco and Arabidopsis [23].

On the other hand, not only storage proteins were shown to bypass the Golgi compartment. A way to transport cargo proteins towards LVs, which seems to be unique to the plant kingdom, is a direct transport from the ER to the LVs via ER-derived organelles termed ER bodies [78, 79], highlighting the enormous plasticity of the ER in plant cells. The ER bodies are <1 μm in diameter with a dense core surrounded by a membrane originating from the ER itself [73]. This system was thought to mediate the transport of otherwise transport-incompetent insoluble material within the cell. Proteins that are transported via this route are, for instance, the vacuolar H^+-ATPase VHA-a3, the H^+-pyrophosphatase AVP1 and the α-TIP aquaporin [59, 80, 81]. However, there could be other Golgi-independent routes of transport to the vacuole, since the sucrose transporter SUC4 and the tonoplast *myo*-inositol transporter INT1 were shown to be trafficked in an AP-3-dependent manner [82, 83]. It is conceivable that novel trafficking routes and sorting mechanisms will come to light in future research.

Besides canonical vesicle trafficking, also autophagy provides a route to deliver storage proteins from the ER towards the vacuole [84–86]. In general, autophagy encompasses the catabolic degradation of proteins as well as whole organelles, also acting in environmental stress responses and pathogen defence, in the end releasing the autophagic body into the vacuolar lumen for subsequent degradation [87]. Inducing ER stress in Arabidopsis was shown to lead to the delivery of ER components to the vacuole via autophagy, underlining its involvement in trafficking [88]. Further, it was proposed, that autophagy is utilised as a transport pathway to deliver storage proteins to PSVs in developing seeds [89, 90]. Normally, vesicles containing cargo proteins are engulfed by the phagophore membrane and autophagosomes are later fused with the tonoplast for the lytic degradation in the vacuole.

4.2 MVB Formation and Cellular Waste Management

Transmembrane proteins that are marked for lytic degradation initially accumulate in MVBs or late endosomes. Ubiquitination displays one of the most common post-translational modifications, designating proteins for the degradation pathway. Ubiquitination affects protein stability by acting as a sorting signal [91]. The process of ubiquitination is strongly regulated [92] by a consecutive set of E1, E2, and E3 ubiquitin ligases that ultimately attach ubiquitin to target proteins [93, 94]. The abundance of many proteins is regulated through this degradation mechanism, which is for instance the case for the auxin efflux facilitator PIN2 [95].

The sorting of cargo marked by ubiquitination into intraluminal vesicles (ILVs) for the transport towards the vacuole/lysosome [96, 97] is taken over by the ESCRT complex, a multi-subunit machinery for membrane deformation and scission. It is involved in the biogenesis of MVBs [98] and present in various isoforms, including ESCRT-I, -II, -III and the VACUOLAR PROTEIN SORTING4/SUPPRESSOR OF K^+ TRANSPORT GROWTH DEFECT 1 (VPS4/SKD1) subcomplex [100]. So far, no homolog to the canonical ESCRT-0 subunits has been detected in plants (reviewed in [100]), but this function seems to be taken over by TOL proteins in plants [101, 102]. The ESCRT machinery appears evolutionary conserved in fungi and animal cells; however, plants seem to display unique ESCRT components, such as FYVE DOMAIN PROTEIN REQUIRED FOR ENDOSOMAL SORTING 1 (FREE1) as well as POSITIVE REGULATOR OF SKD1 (PROS) [103, 104], suggesting plant-specific features.

4.3 Membrane Tethering and Fusion at the Tonoplast

Vesicle trafficking ensures the delivery of vacuolar cargo proteins. The underlying complex sorting mechanisms ensure the correct delivery and require a plethora of tethering factors, vesicle coats, GTPases, and SNAREs (reviewed in [68]). The canonical delivery of cargo to the vacuole requires molecular maturation of MVBs. Rab GTPases regulate many steps of membrane trafficking and a replacement of RAB5 with RAB7 is crucial for MVB maturation. This is in part mediated by the dimeric complex formed by MONESIN SENSITIVITY1 (MON1) and CALCIUM CAFFEINE ZINC SENSITIVITY1 (CCZ1), which function as guanine nucleotide exchange factors (GEFs) for RAB7 [38, 98, 105].

Ultimately, MVBs and other vesicles require the soluble *N*-ethylmaleimide-sensitive fusion attachment protein receptors (SNAREs) to fuse with vacuoles (reviewed in [106]). SNARE proteins are the initiators of membrane fusion by forming a *trans*-complex utilising their R- or Q-SNARE motifs [107]. The underlying vacuolar SNARE complex consists of the R-SNARE VAMP711, the Qa-SNARE SYP22, the Qb-SNARE VTI11 and the two Qc-SNAREs SYP51 and SYP52 [108], however, multiple isoforms of this complex have evolved over time. Despite functional redundancy, loss of *VTI11* leads to a strong vacuolar phenotype with rounded and partially fragmented vacuoles [108–111].

The first mutant described in *Arabidopsis*, which completely fails to build a vacuole is *vacuoleless1* (*vcl1*). The mutant is embryo lethal, arresting at the late torpedo stage, highlighting the essential role of vacuoles for plant development [112, 113]. In the *vcl1* mutant, cells fail to form a central vacuole, instead, accumulate autophagosome-like structures and display strong defects in cell division orientation and elongation [112]. The *vcl1* mutant carries a T-DNA insertion in a gene encoding a core subunit of the homotypic fusion and vacuole protein sorting (HOPS) complex [114]. The HOPS complex mediates with VAMP713-containing SNARE and RAB7 primarily homotypic fusion between vacuoles. In contrast, the CORVET complex facilitates heterotypic membrane fusion between MVBs and the vacuole with RAB5 and VAMP727-containing SNARE complex [115]. Both, the HOPS and the CORVET complex are heterohexameric tethering complexes. They share four core subunits (VPS11, VPS16, VPS18 and VPS33), while VPS3 and VPS8 are CORVET specific and VPS39 and VPS41 only belong to the HOPS complex [114, 116].

4.4 Sorting Signals and Receptors for Vacuolar Sorting

Vacuolar sorting signals are crucial for cargo to get recruited to the delivery system [117, 118]. Two types of receptors are currently known to be involved in sorting soluble cargos to the vacuole: the vacuolar sorting receptor (VSR) family and the receptor membrane ring-H2 (RMR) family (reviewed in [25]). Members of the VSR family were first described in CCVs preparations of pea cotyledon [119], where they were originally termed BP80 (binding protein 80 kDa). VSRs were hence postulated to function in sorting proteins to lytic vacuoles by recruiting clathrin via their cytosolic motif. VSRs provide nucleation sites required for protein aggregation at the edge of Golgi cisterna [120, 121], but cargo association appears very early in the secretory pathway (reviewed in [122]). Structurally they are type I membrane proteins with a cytosolic motif, a large luminal domain, recognising and binding the NPIRL-(like) consensus motifs in cargo proteins, as well as a short C-terminal cytosolic tail, that interacts with trafficking components for vesicle formation and interacts with factors required for their recycling [123, 124]. VSRs themselves are able to be bound by the μ1 subunit of the AP-1 complex by the YMPL motif in their C-terminus, leading to transport via CCVs [67, 125]. They are thought to mediate sorting to LVs, but experimental data also showed VSRs to be involved in sorting storage proteins in other plant species, suggesting that there might be different isoforms for different sorting events [126]. Here VSRs may have overlapping functions with the RMR family, which is primarily involved in cargo sorting to PSVs [123, 127].

In contrast to mammals, soluble vacuolar proteins in plants do not require modified glycans as sorting signals [128]. Instead, they utilise short peptide sequences for vacuolar sorting, which is reminiscent of vacuolar sorting of proteases in yeast [129, 130]. Specific vacuolar sorting determinants (VSDs) mark proteins for delivery to the LVs and PSVs [118, 120]. Proteins without any of those motifs would simply follow the default pathway [131]. Initially, VSDs are recognised by VSRs, then the motif is typically removed during the process of protein maturation by vacuolar proteases after sorting [25].

Ligands dissociate from the VSR receptor at the lower pH of the TGN and/or MVB. The receptor is then recycled, which is mediated by the retromer coat protein [132–134]. The retromer coat complex is a pentameric complex with two subcomplexes: the core retromer and the sorting nexin (SNX). The core retromer complex consists of the subunits VPS26, VPS29 and VPS35, whereas VPS35 functions as the receptor-binding one [135, 136]. The core complex is known to function in plant development as well as biotic stress responses since it enables retrieval of transmembrane or membrane-associated proteins from the MVBs and/or TGN (see Fig. 3) [134, 135].

5 Vacuoles and Cell Homeostasis

The vacuole is an acidic organelle. A precisely tuned luminal pH is necessary because many of the enzymes contained in the vacuolar lumen, typically display a pH optimum at lower pH levels, which is critical for the lysis of delivered compounds [29, 76, 137]. pH control depends on the proton pumps at the tonoplast, which ensures the lytic function but also energises secondary active transport processes across the tonoplast. The V-type H^+-ATPases and an H^+-pyrophosphatase (V-PPase, also called AVP1) jointly control the vacuolar pH [138]. Furthermore, AHA10, a P-type H^+-ATPase contributes in a cell-type-specific manner to vacuolar pH [139]. Additionally, VHA-a1 activity at the TGN adds to the vesicle-based delivery of protons to the vacuole [137]. A finding from yeast, which might apply to plants as well, is that there is a collaborative mechanism of regulation for PM-localised and vacuolar proton pumps involved in cytosolic pH homeostasis

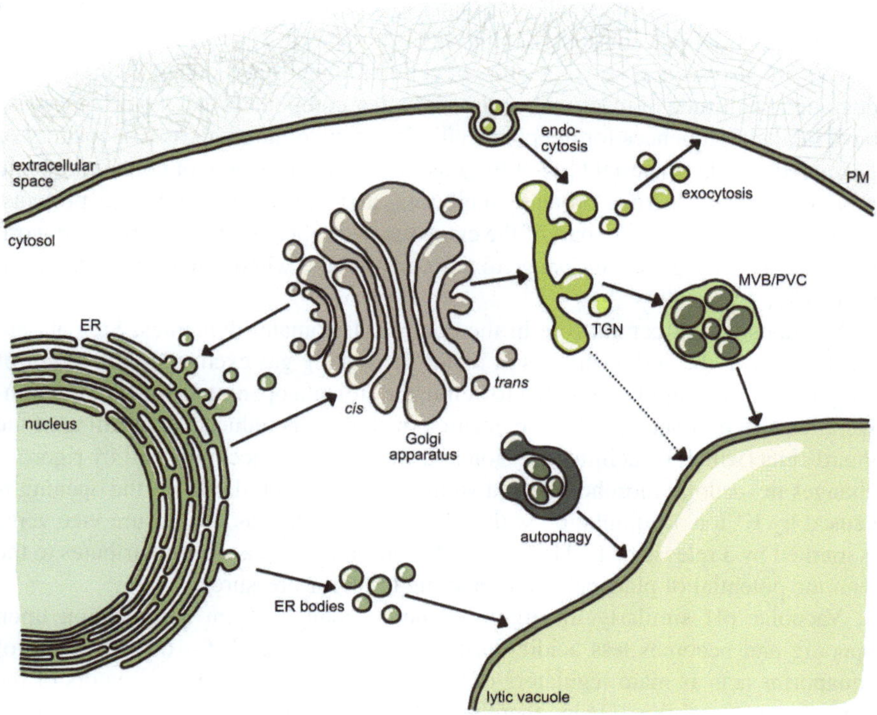

Fig. 3 Overview of the different trafficking pathways ending up in the lytic vacuole in plant cells. Proteins destined for the LV enter the secretory pathway at the ER, where anterograde transport delivers them towards the *cis* face of the Golgi apparatus. The Golgi route can be bypassed by vesicle transport via ER bodies, that directly go to the vacuole. Passing through the Golgi, the cargo reaches the *trans*-Golgi network (TGN), where also cargo from endocytic events is entering. Trafficking from and to the cell surface takes place through exo- and endocytosis, which are also responsible for recycling. From the TGN vesicles further mature towards the MVB compartment, which leads to the vacuole. Vesicles transported in an AP3-dependent manner from the TGN as well as autophagic processes can bypass the MVBs

[140, 141]. The proton pump activity is precisely regulated through 14-3-3 proteins, phosphorylation, ion concentration as well as redox conditions to maintain the functionality of the cytosol [140].

The membrane potential difference between cytosol and the vacuolar lumen is relatively low (~30 mV), suggesting that anion fluxes are involved in the discharge created by the proton pumps [142].

The vacuole plays a rather complex role in metal ion homeostasis. Besides accumulating metal ions in the vacuolar lumen, it also defines metal ion uptake by contributing to the turnover of its uptake components [6, 143]. A prominent example is the borate transporter BOR1, which is located in the inner polar domain of root cells. It mediates the loading of boron into the xylem. Under surplus boron concentrations, the BOR1 transporter is getting ubiquitinated, internalised and targeted for degradation to the vacuoles, whereas under low-boron amounts it recycles between the PM and TGN in a constant loop [144–146].

LVs are also seen as key regulators of cellular homeostasis due to their ability to degrade a multitude of macromolecules and other compounds that would otherwise have fatal consequences for the plant cell. The vacuole functions as a depository of unwanted cellular material [147–149], but they are also capable of containing toxic proteins, such as lectins, protease inhibitors, and ribosome-inactivating proteins. Thereby, they are not only part of the cytosolic detoxification process but also work in defence strategies, releasing toxic compounds when physically damaged [150, 151].

Vacuoles play a central role in the control of stomatal dynamics. Stomata are small pores on the aerial surfaces of plants facilitating gas exchange. The turgor of the guard cells is tightly regulated to define the stomata opening and hence the transpiration rate of a plant [152]. The opening of stomata is induced by swelling of the guard cells (which is facilitated by ion and water fluxes), accompanied by rigorous changes in vacuolar morphology and volume. On a subcellular level, the opening is caused by K^+ ion accumulation within the vacuole, whereas the closure vice versa is marked by a release of K^+ [153–155]. Accordingly, the vacuole contributes to the osmotic potential of plant cells, contributing to turgor pressure.

Vacuolar pH similarly undergoes dynamic changes from acidification upon opening and becomes less acidic during stomata closure [156–160]. A family of aquaporins acts as main regulators of water intake and hence as facilitators of the tonoplast permeability [161], thereby buffering the water content of the cytoplasm [162].

6 The Role of Vacuoles During Cell Expansion

Plant cell elongation is a highly dynamic process, where the vacuole undergoes drastic morphological changes to facilitate enormously fast cellular elongation. The primary root displays highly coordinated cellular expansion dynamics, which take place in three phases (see Fig. 4) [12, 163].

(1) Cell expansion rates are relatively slow in the meristematic zone where the cells enlarge rather gradually after cytokinesis.
(2) When the cells enter the elongation zone, cell wall loosening and turgor pressure drive cell expansion. These cells are highly vacuolated, which in part can be explained by the high uptake of water, supported by aquaporins [80, 161, 162]. Notably, *TIP1;1* is a prominent member of the tonoplast-localised aquaporins and its expression correlates with the onset of cellular elongation and the corresponding water movement [161, 164].
(3) The fast cell expansion phase is ceasing with the rigidification of the cell wall, the deposition of new cell wall material by the action of cell wall modifying enzymes, and the promotion of cross-linking within the wall [12, 165]. In meri-

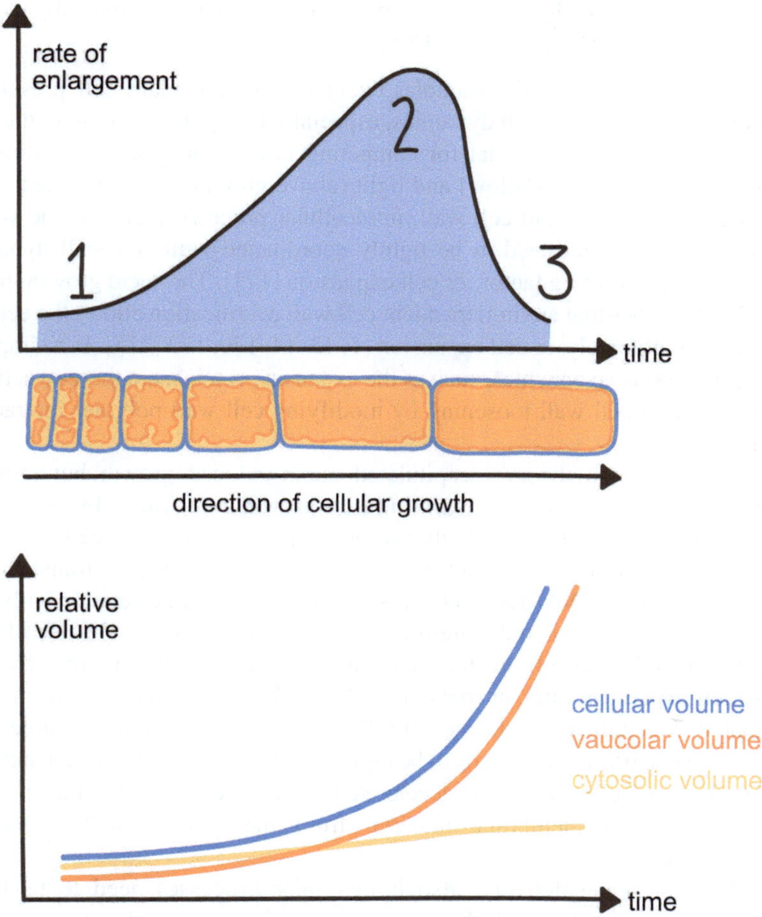

Fig. 4 During growth, plant cells gradually go through different phases. The cells start with rather slow (1) meristematic growth, then transition to rapid expansion during (2) vacuolated growth. The cell elongation process ends with (3) growth cessation (in part adapted after Cosgrove, 2000). While the overall cellular volume increases during growth, the vacuole follows by increasing its size, whereas the cytosolic volume stays relatively constant. (Adapted after Dünser et al., 2019 [13])

stematic cells, the vacuolar volumes are relatively small and cells can contain multiple vacuoles, which take up only 20–30% of the cellular volume [13, 166]. Small and possibly fragmented vacuoles seem to be a hallmark of meristematic cells. It is conceivable that bigger vacuoles would physically limit cell plate formation. In the so-called transition zone, cells stop dividing and contain typically a single, central vacuole [167]. The size of the vacuole drastically changes at the onset of elongation: when cells are increasing in size, the vacuole gradually occupies more cellular space, leading to a vacuolar occupancy of up to 90% [13]. This dynamic control of vacuolar volume explains why vacuolar size is tightly correlated with cell volume [168]. The relative increase in vacuolar size allows plant cells to drastically expand without the need to dramatically increase their cytosolic content [13, 167, 169].

The steady increase in the vacuolar occupancy of the plant cell presumably allows energy-efficient growth dynamics, especially to rapidly increase in the overall cell size. The latter is crucial for competing in covering the surroundings for mining essential nutrients (below) and light (above ground). As every plant cell is surrounded by a rather rigid cell wall, intracellular processes, such as the volume increase of the vacuole, need to be tightly coordinated with cell wall dynamics, since it displays a limiting factor for cell expansion [171]. The "acid growth theory" [171, 172] describes that auxin-dependent cell wall acidification allows the cell wall to loosen, and ultimately, to cell expansion [1, 13, 14, 170, 173–175]. In a simplified model, pH-responsive enzymes, such as the expansin proteins, are thereby activated and contribute to cell wall loosening by modifying cell wall polymer interactions [163, 176].

The acidic growth theory recapitulated auxin-induced growth but remained debated in tissues that display auxin-induced growth repression. In roots, auxin transiently induces apoplastic alkalinisation, required to repress cell elongation [173]. In these cells, auxin also impacts vacuolar morphology by post-translationally modulating vacuolar SNAREs [169]. It is conceivable that interference with vacuolar SNARE complexes is reducing the vesicle fusion and hence the size of vacuoles. Consequently, auxin limits the vacuolar occupancy of cells and thereby limits the fast cell expansion rates in roots [167, 169]. Vacuolar shape and size are also coordinated by actin/myosin-dependent folding and unfolding of the vacuole [167]. However, vesicle trafficking seems to be highly relevant to coordinate cell and vacuolar size. Endocytic trafficking is accelerated during the onset of cellular elongation, implying that membrane material from the PM is fuelling vacuolar enlargement [109].

Besides membrane delivery, also intracellular processes need to be highly coordinated with the extracellular space. A prominent example of a molecular integrator is FERONIA (FER), a receptor-like kinase from the *Catharanthus roseus* receptor-like kinases 1-like (*Cr*RLK1L) family. FER presumably functions as a mechanosensor, which feeds back on the extensibility of cell walls [176, 177]. The malectin-like domains of FER can bind the peptide ligand RAPID ALKALINISATION FACTOR1 (RALF1), which consequently leads to FER-dependent apoplast alkalinisation and repression of cellular expansion (see Fig. 5) [178, 179].

The *fer-4* mutant shows a strong vacuolar phenotype with enlarged and more roundish vacuoles [13]. The interaction of FER with the extracellular cell wall binding protein leucine-rich repeat extensins (LRXs) allows to sense changes in the cell wall and to relay this information into the cell. This sensing mechanism coordinates the loosening of the cell wall and the size increase of the vacuole, enabling the onset of cell expansion [13]. The extensin domain of the LRX proteins interacts with cell wall components, but the precise mechanism is unknown [180, 181].

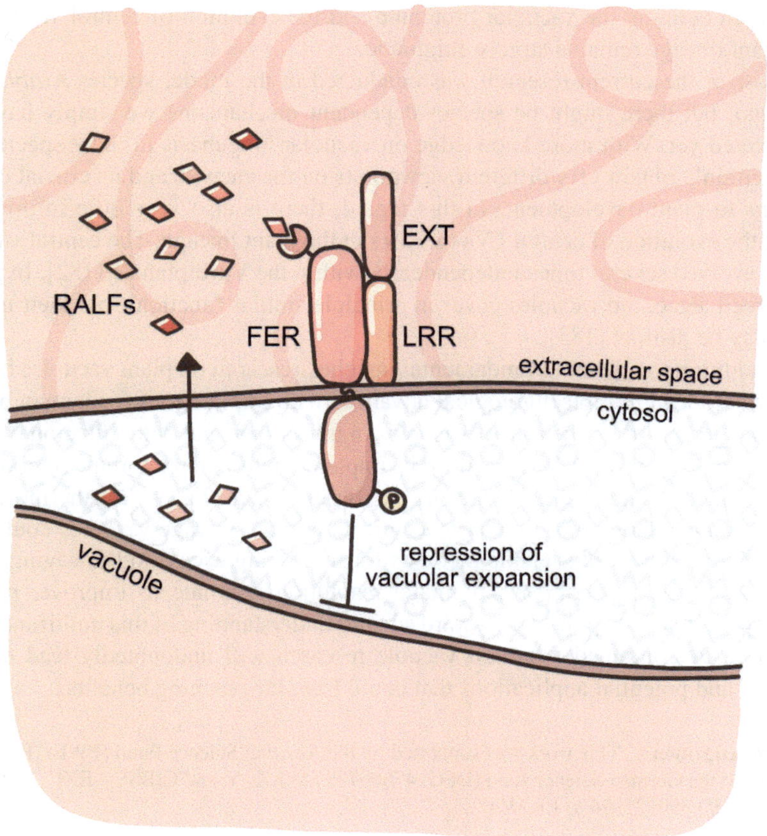

Fig. 5 Schematic overview of the impact of cell wall sensing on vacuolar occupancy. Signals from the cell wall are integrated via the LRX-FER-sensing module. The extensin (EXT) domain of the LRX proteins is able to perceive changes within the extracellular matrix and transfers them onto the receptor-like kinase FERONIA (FER) through their leucine-rich repeat (LRR) domain for protein–protein interactionsThe signal is then passed to the vacuole leading to the repression of vacuolated growth. The precise mechanism is currently unknown. FER is further known as a receptor for RALF peptides, which could further be involved in cell wall sensing and signal transductionAll figure images were created using the Concepts App for iOS (Version 6.9.2, TopHatch, Inc. (Turku, Finnland)) and edited using AffinityDesigner (version 1, Serif (West Bridgford, UK))

7 Outlook and Future Directions

This review has highlighted the critical role of the vacuole in plant development and the challenges that remain in understanding its complex biology. While we have made significant progress in characterising the mechanisms of vesicle trafficking and cargo sorting to the vacuole, we still lack a complete understanding of how the cell integrates and monitors the different routes of membrane flows to ensure that vacuole identity and size meet cellular needs. Additionally, the role of external signals in defining the vacuolar proteome and the evolution of central vacuoles in the plant lineage remains largely enigmatic.

Most of the current research was conducted in the model species *Arabidopsis thaliana*, but there might be species-dependent mechanisms we simply have not discovered yet. With more knowledge on vacuolar biogenesis in other species, we will certainly obtain very different viewpoints on the vacuole and its crucial contributions to plant development. In this regard, there is also very little information about the evolution of central LVs or PSVs in the plant lineage. The central vacuole likely evolved several times independently within the Viridiplantae [182]. In plants and green algae, the vacuoles cover in principle similar functions, but their utilisation may be distinct [183].

In addition to obtaining fundamental insights, research on plant vacuoles has the potential to lead to biotechnological advances with implications for human health. Knowledge gained on the role of vacuoles in the storage and detoxification of compounds could be further developed for its implementation in phytoremediation strategies, which could reduce exposure to harmful substances. Furthermore, the production and storage of biopharmaceuticals in plant storage vacuoles could provide cost-effective and scalable methods for improving access to life-saving medicines. As tools, procedures and image resolution continue to improve, method development is paving the way toward a better understanding of this multifunctional organelle. A continued focus on vacuole research will undoubtedly lead to new insights and potential applications that could have far-reaching benefits.

Acknowledgements This work was supported by the Austrian Science Fund (FWF) (P33044 to J.KV.,) and the German Science fund (DFG; 470007283 to J.K.-V. and CIBSS—EXC-2189. Project ID 390939984 to J.K.-V.).

References

1. Kaiser S, Scheuring D (2020) To lead or to follow: contribution of the plant vacuole to cell growth. Front Plant Sci 11. Frontiers Media S.A. https://doi.org/10.3389/fpls.2020.00553
2. Krüger F, Schumacher K (2018) Pumping up the volume—vacuole biogenesis in *Arabidopsis thaliana*. Semin Cell Dev Biol 80:106–112). Elsevier Ltd. https://doi.org/10.1016/j.semcdb.2017.07.008
3. Marty F (1999) Plant vacuoles. Plant Cell 11. www.plantcell.org
4. Paris N, Stanley CM, Jones RL (1996) Plant cells contain two functionally distinct vacuolar compartments. Cell 85:563-572. https://doi.org/10.1016/s0092-8674(00)81256-8
5. Herman EM, Larkins BA (1999) Protein storage bodies and vacuoles. Plant Cell 11:601–613. https://doi.org/10.1105/tpc.11.4.601
6. Shimada T, Takagi J, Ichino T, Shirakawa M, Hara-Nishimura I (2018) Plant vacuoles. Annu Rev Plant Biol 14(6):123–168. https://doi.org/10.1146/annurev-arplant-042817
7. Briskin DP, Hanson JB (1992) How does the plant plasma membrane H^+-ATPase pump protons? J Exp Bot 43(248). https://academic.oup.com/jxb/article/43/3/269/526911
8. Schumacher K (2014) pH in the plant endomembrane system—an import and export business. Curr Opin Plant Biol 22:71–76. https://doi.org/10.1016/j.pbi.2014.09.005

9. Seidel T, Siek M, Marg B, Dietz KJ (2013) Energization of vacuolar transport in plant cells and its significance under Stress. Int Rev Cell Mol Biol 304:57–131). Elsevier Inc. https://doi.org/10.1016/B978-0-12-407696-9.00002-6
10. Hara-Nishimura I, Hatsugai N (2011) The role of vacuole in plant cell death. Cell Death Differ 18:1298–1304
11. Hatsugai N, Hara-Nishimura I (2010) Hatsugai and Hara-Nishimura 2010 (Plant Signaling & Behavior). Plant Signal Behav 5(12):1568–1570
12. Cosgrove DJ (1993) Do plant cell walls extend? Plant Physiol 102:1–6. www.plantphysiol.org
13. Dünser K, Gupta S, Herger A, Feraru MI, Ringli C, Kleine-Vehn J (2019) Extracellular matrix sensing by FERONIA and Leucine-Rich Repeat Extensins controls vacuolar expansion during cellular elongation in Arabidopsis thaliana. EMBO J 38(7). https://doi.org/10.15252/embj.2018100353
14. Dünser K, Kleine-Vehn J (2015) Differential growth regulation in plants-the acid growth balloon theory. Curr Opin Plant Biol 28:55–59). Elsevier Ltd. https://doi.org/10.1016/j.pbi.2015.08.009
15. Beauzamy L, Nakayama N, Boudaound A (2014) Flowers under pressure: ins and outs of turgor regulation in development. Ann Bot 114:1517–1533
16. Haines TH (1994) Water transport across biological membranes. FEBS Lett 346(1):115–122. https://doi.org/10.1016/0014-5793(94)00470-6
17. Hoh B, Hinz G, Jeong B-K, Robinson DG (1995) Protein storage vacuoles form de novo during pea cotyledon development. J Cell Sci 108:299–310
18. Jauh G-Y, Phillips TE, Rogers JC (1999) Tonoplast intrinsic protein isoforms as markers for vacuolar functions. Plant Cell 11. www.plantcell.org
19. Frigerio L, Hinz G, Robinson DG (2008) Multiple vacuoles in plant cells: rule or exception? Traffic 9(10):1564–1570. https://doi.org/10.1111/j.1600-0854.2008.00776.x
20. Jiang L, Phillips TE, Hamm CA, Drozdowicz YM, Rea PA, Maeshima M, Rogers SW, Rogers JC (2001) The protein storage vacuole: a unique compound organelle. J Cell Biol 155(6):991–1002. https://doi.org/10.1083/jcb.200107012
21. Jiang YT, Yang LH, Ferjani A, Lin WH (2021) Multiple functions of the vacuole in plant growth and fruit quality. Mol Hortic 1(1):1–13. https://doi.org/10.1186/s43897-021-00008-7
22. Vitale A, Hinz G (2005) Sorting of proteins to storage vacuoles: how many mechanisms? Trends Plant Sci 10(7):316–323. https://doi.org/10.1016/j.tplants.2005.05.001
23. Jiang L, Phillips TE, Rogers SW, Rogers JC (2000) Biogenesis of the protein storage vacuole crystalloid. J Cell Biol 150:755–769
24. Weber E, Neumann D (1980) Protein bodies, storage organelles in plant seeds. Biochem Physiol Pflanzen 175:247–306
25. Xiang L, Etxeberria E, van den Ende W (2013) Vacuolar protein sorting mechanisms in plants. FEBS J 280(4):979–993. https://doi.org/10.1111/febs.12092
26. Swanson SJ, Jones RL (1996) Gibberellic acid Induces vacuolar acidification in Barley Aleurone. Plant Cell 8:2211–2221
27. Robinson DG, Oliviusson P, Hinz G (2005) Protein sorting to the storage vacuoles of plants: a critical appraisal. Traffic 6(8):615–625. https://doi.org/10.1111/j.1600-0854.2005.00303.x
28. Otegui MS, Herder R, Schulze J, Jung R, Staehelin LA (2006) The proteolytic processing of seed storage proteins in Arabidopsis embryo cells starts in the multivesicular bodies. Plant Cell 18(10):2567–2581. https://doi.org/10.1105/tpc.106.040931
29. Kinoshita T, Yamada K, Hiraiwa N, Kondo M, Nishimura M, Hara-Nishimura I (1999) Vacuolar processing enzyme is up-regulated in the lytic vacuoles of vegetative tissues during senescence and under various stressed conditions. Plant J 19(1):43–53. https://doi.org/10.1046/j.1365-313X.1999.00497.x
30. Hara-Nishimura I, Takeuchi Y, Nishimura M (1993) Molecular characterization of a vacuolar processing enzyme related to a Putative Cysteine Proteinase of *Schistosoma mansoni*. Plant Cell 5. https://academic.oup.com/plcell/article/5/11/1651/5984613

31. Nakaune S, Yamada K, Kondo M, Kato T, Tabata S, Nishimura M, Hara-Nishimura I (2005) A vacuolar processing enzyme, delta-VPE, is involved in seed coat formation at the early stage of seed development. Plant Cell 17:876–887
32. Shimada T, Yamada K, Kataoka M, Nakaune S, Koumoto Y, Kuroyanagi M, Tabata S, Kato T, Shinozaki K, Seki M, Kobayashi M, Kondo M, Nishimura M, Hara-Nishimura I (2003) Vacuolar processing enzymes are essential for proper processing of seed storage proteins in *Arabidopsis thaliana*. J Biol Chem 278(34):32292–32299. https://doi.org/10.1074/jbc.M305740200
33. Kuroyanagi M, Nishimura M, Hara-Nishimura I (2002) Activation of arabidopsis vacuolar processing enzyme by self-catalytic removal of an auto-inhibitory domain of the C-terminal propeptide. Plant Cell Physiol 43(2):143
34. Hatsugai N, Kuroyanagi M, Yamada K, Meshi T, Tsuda S, Kondo M, Nishimura M, Hara-Nishimura I (2004) A plant vacuolar protease, VPE, mediates virus-induced hypersensitive cell death. Science 305:855–858. www.sciencemag.org
35. Kuroyanagi M, Yamada K, Hatsugai N, Kondo M, Nishimura M, Hara-Nishimura I (2005) Vacuolar processing enzyme is essential for mycotoxin-induced cell death in *Arabidopsis thaliana*. J Biol Chem 280(38):32914–32920. https://doi.org/10.1074/jbc.M504476200
36. Hohl I, Robinson DG, Chrispeels MJ, Hinz G (1996) Transport of storage proteins to the vacuole is mediated by vesicles without a clathrin coat. J Cell Sci 109:2539–2550
37. Hillmer S, Movafeghi A, Robinson DG, Hinz G (2001) Vacuolar storage proteins are sorted in the cis-Cisternae of the Pea Cotyledon Golgi Apparatus. J Cell Biol 152(1). http://www.jcb.org/cgi/content/full/152/1/41
38. Cui Y, Cao W, He Y, Zhao Q, Wakazaki M, Zhuang X, Gao J, Zeng Y, Gao C, Ding Y, Wong HY, Wong WS, Lam HK, Wang P, Ueda T, Rojas-Pierce M, Toyooka K, Kang BH, Jiang L (2019) A whole-cell electron tomography model of vacuole biogenesis in Arabidopsis root cells. Nature Plants 5(1):95–105. https://doi.org/10.1038/s41477-018-0328-1
39. Scheuring D, Viotti C, Krüger F, Künzl F, Sturm S, Bubeck J, Hillmer S, Frigerio L, Robinson DG, Pimpl P, Schumacher K (2011) Multivesicular bodies mature from the Trans-Golgi network/early endosome in arabidopsis. Plant Cell 23:3463–3481
40. Dautry-Varsat A, Ciechanover A, Lodish HF (1983) Cell Biology pH and the recycling of transferrin during receptor-mediated endocytosis (endocytic vesicle/apotransferrin/transferrin receptor). Proc Natl Acad Sci U S A 80. https://www.pnas.org
41. Januliene D, Andersen JL, Nielsen JA, Quistgaard EM, Hansen M, Strandbygaard D, Moeller A, Petersen CM, Madsen P, Thirup SS (2017) Acidic environment induces dimerization and ligand binding site collapse in the Vps10p domain of Sortilin. Structure 25(12):1809–1819.e3. https://doi.org/10.1016/j.str.2017.09.015
42. Olbrich A, Hillmer S, Hinz G, Oliviusson P, Robinson DG (2007) Newly formed vacuoles in root meristems of barley and pea seedlings have characteristics of both protein storage and lytic vacuoles. Plant Physiol 145(4):1383–1394. https://doi.org/10.1104/pp.107.108985
43. Zheng H, Staehelin LA (2011) Protein storage vacuoles are transformed into lytic vacuoles in root meristematic cells of germinating seedlings by multiple, cell type-specific mechanisms. Plant Physiol 155(4):2023–2035. https://doi.org/10.1104/pp.110.170159
44. Zouhar J, Rojo E (2009) Plant vacuoles: where did they come from and where are they heading? Curr Opin Plant Biol 12(6):677–684. https://doi.org/10.1016/j.pbi.2009.08.004
45. Otegui MS, Noh YS, Martínez DE, Vila Petroff MG, Staehelin LA, Amasino RM, Guiamet JJ (2005) Senescence-associated vacuoles with intense proteolytic activity develop in leaves of *Arabidopsis* and soybean. Plant J 41(6):831–844. https://doi.org/10.1111/j.1365-313X.2005.02346.x
46. Grotewold E (2006) The genetics and biochemistry of floral pigments. Annu Rev Plant Biol 57:761–780. https://doi.org/10.1146/annurev.arplant.57.032905.105248
47. Brunetti C, di Ferdinando M, Fini A, Pollastri S, Tattini M (2013) Flavonoids as antioxidants and developmental regulators: relative significance in plants and humans. Int J Mol Sci 14:3540–3555

48. Harborne JB, Williams CA (2000) Advances in flavonoid research since 1992. Phytochemistry 55:481–504
49. Taylor LP, Grotewold E (2005) Flavonoids as developmental regulators. Curr Opin Plant Biol 8(3 SPEC. ISS):317–323). Elsevier Ltd. https://doi.org/10.1016/j.pbi.2005.03.005
50. Armstrong J (2010) Yeast vacuoles: more than a model lysosome. Trends Cell Biol 20(10):580–585. https://doi.org/10.1016/j.tcb.2010.06.010
51. Roeder AD, Shaw JM (1996) Vacuole partitioning during meiotic division in yeast. Genetics 144:445–458
52. Weisman LS (2003) Yeast vacuole inheritance and dynamics. Annu Rev Genet 37:435–460. https://doi.org/10.1146/annurev.genet.37.050203.103207
53. Bryant NJ, Stevens TH (1998) Vacuole biogenesis in *Saccharomyces cerevisiae*: protein transport pathways to the yeast vacuole. Microbiol Mol Biol Rev 62(1):230–247
54. Souter M, Lindsey K (2000) Polarity and signalling in plant embryogenesis. J Exp Bot 51(347):971. https://about.jstor.org/terms
55. Ueda M, Zhang Z, Laux T (2011) Transcriptional activation of Arabidopsis axis patterning genes WOX8/9 links zygote polarity to embryo development. Dev Cell 20(2):264–270. https://doi.org/10.1016/j.devcel.2011.01.009
56. Zhao P, Begcy K, Dresselhaus T, Sun MX (2017) Does early embryogenesis in eudicots and monocots involve the same mechanism and molecular players? Plant Physiol 173(1):130–142. https://doi.org/10.1104/pp.16.01406
57. Kimata Y, Kato T, Higaki T, Kurihara D, Yamada T, Segami S, Morita MT, Maeshina M, Hasezawa S, Higashiyama T, Tasaka M, Ueda M (2019) Polar vacuolar distribution is essential for accurate asymmetric division of Arabidopsis zygotes. PNAS 116(6):2338–2343
58. Matile P, Moor H (1968) Vacuolation: origin and development of the lysosomal apparatus in root-tip cells. Planta 80:159–175
59. Viotti C, Krüger F, Krebs M, Neubert C, Fink F, Lupanga U, Scheuring D, Boutté Y, Frescatada-Rosa M, Wolfenstetter S, Sauer N, Hillmer S, Greb M, Schumacher K (2013) The endoplasmic reticulum is the main membrane source for biogenesis of the lytic vacuole in *Arabidopsis*. Plant Cell 25(9):3434–3449. https://doi.org/10.1105/tpc.113.114827
60. Marty F (1978) Cytochemical studies on GERL, provacuoles, and vacuoles in root meristematic cells of Euphorbia (electron microscopy/enzyme localizations/lysosome/cellular autophagy/membrane flow). Cell Biol 75(2):852
61. Tse YC, Mo B, Hillmer S, Zhao M, Lo SW, Robinson DG, Jiang L (2004) Identification of multivesicular bodies as prevacuolar compartments in *Nicotiana tabacum* BY-2 Cells. Plant Cell 16:672–693
62. Luzio JP, Hackmann Y, Dieckmann NMG, Griffiths GM (2014) The biogenesis of lysosomes and lysosome-related organelles. Cold Spring Harb Perspect Biol 6(9). https://doi.org/10.1101/cshperspect.a016840
63. Cui Y, Zhao Q, Hu S, Jiang L (2020) Vacuole biogenesis in plants: how many vacuoles, how many models? Trends Plant Sci 25(6):538–548). Elsevier Ltd. https://doi.org/10.1016/j.tplants.2020.01.008
64. Vitale A, Denecke J (1999) The endoplasmic reticulum-gateway of the secretory pathway. Plant Cell 11. www.plantcell.org
65. Brandizzi F, Hanton S, Pinto DaSilva LL, Boevink P, Evans D, Oparka K, Denecke J, Hawes C (2003) ER quality control can lead to retrograde transport from the ER lumen to the cytosol and the nucleoplasm in plants. Plant J 34(3):269–281. https://doi.org/10.1046/j.1365-313X.2003.01728.x
66. Neumann U, Brandizzi F, Hawes C (2003) Protein transport in plant cells: in and out of the Golgi. Ann Bot 92:167–180
67. Gershlick DC, Marcos Lousa C, Foresti O, Lee AJ, Pereira EA, da Silva LLP, Bottanelli F, Denecke J (2014) Golgi-dependent transport of vacuolar sorting receptors is regulated by COPII, AP1, and AP$ protein complexes in tobacco. Plant Cell 26:1308–1329

68. Jürgens G (2004) Membrane trafficking in plants. Annu Rev Cell Dev Biol 20:481–504. https://doi.org/10.1146/annurev.cellbio.20.082503.103057
69. Dettmer J, Hong-Hermesdorf A, Stierhof YD, Schumacher K (2006) Vacuolar H+-ATPase activity is required for endocytic and secretory trafficking in *Arabidopsis*. Plant Cell 18(3):715–730. https://doi.org/10.1105/tpc.105.037978
70. Heard W, Sklenar J, Tomé DFA, Robatzek S, Jones AME (2015) Identification of regulatory and cargo proteins of endosomal and secretory pathways in *Arabidopsis thaliana* by proteomic dissection. Mol Cell Proteomics 14(7):1796–1813
71. Rosquete MR, Davis DJ, Drakakaki G (2018) The plant Trans-Golgi network: not just a matter of distinction. Plant Physiol 176:187–198
72. Herman E, Schmidt M (2004) Endoplasmic reticulum to vacuole trafficking of endoplasmic reticulum bodies provides an alternate pathway for protein transfer to the vacuole. Plant Physiol 136:3440–3446
73. Herman EM (2008) Endoplasmic reticulum bodies: solving the insoluble. Curr Opin Plant Biol 11(6):672–679. https://doi.org/10.1016/j.pbi.2008.08.004
74. Galili G (2004) ER-derived compartments are formed by highly regulated processes and have special functions in plants. Plant Physiol 136(3):3411–3413). American Society of Plant Biologists. https://doi.org/10.1104/pp.104.900125
75. Ebine K, Inoue T, Ito J, Ito E, Uemura T, Goh T, Abe H, Sato K, Nakano A, Ueda T (2014) Plant vacuolar trafficking occurs through distinctly regulated pathways. Curr Biol 24(12):1375–1382. https://doi.org/10.1016/j.cub.2014.05.004
76. Hara-Nishimura I, Inoue K, Nishimura M (1991) A unique vacuolar processing enzyme responsible for conversion of several proprotein precursors into the mature forms. FEBS Lett 294(1–2):89–93. https://doi.org/10.1016/0014-5793(91)81349-D
77. Hara-Nishimura I, Nishimura M (1987) Proglobulin processing enzyme in vacuoles isolated from developing Pumkin Cotyledons. Plant Physiol 85:440–445
78. Matsushima R, Hayashi Y, Kondo M, Shimada T, Nishimura M, Hara-Nishimura I (2002) An endoplasmic reticulum-derived structure that is induced under stress conditions in *Arabidopsis*. Plant Physiol 130:1807–1814
79. Matsushima R, Hayashi Y, Yamada K, Shimada T, Nishimura M, Hara-Nishimura I (2003) The ER body, a novel endoplasmic reticulum-derived structure in *Arabidopsis*. Plant Cell Physiol 44(7):661–666. https://academic.oup.com/pcp/article/44/7/661/1855686
80. Maurel C, Santoni V, Luu DT, Wudick MM, Verdoucq L (2009) The cellular dynamics of plant aquaporin expression and functions. Curr Opin Plant Biol 12(6):690–698. https://doi.org/10.1016/j.pbi.2009.09.002
81. Viotti C (2014) ER and vacuoles: never been closer. Front Plant Sci 5(20):1–7
82. Wolfenstetter S, Wirsching P, Dotzauer D, Schneider S, Sauer N (2012) Routes to the tonoplast: the sorting of tonoplast transporters in Arabidopsis mesophyll protoplasts. Plant Cell 24(1):215–232. https://doi.org/10.1105/tpc.111.090415
83. Zwiewka M, Feraru E, Möller B, Hwang I, Feraru M, Kleine-Vehn J, Weijers D, Friml J (2011) The AP-3 adaptor complex is required for vacuolar function in *Arabidopsis*. Cell Res 21:1711–1722
84. Bassham DC, Laporte M, Marty F, Moriyasu Y, Ohsumi Y, Olsen LJ, Yoshimoto K (2006) Autophagy in development and stress responses of plants. Autophagy 2(1):2–11). Taylor and Francis Inc. https://doi.org/10.4161/auto.2092
85. Merkulova EA, Guiboileau A, Naya L, Masclaux-Daubresse C, Yoshimoto K (2014) Assessment and optimization of autophagy monitoring methods in Arabidopsis roots indicate direct fusion of autophagosomes with vacuoles. Plant Cell Physiol 55(4):715–726. https://doi.org/10.1093/pcp/pcu041
86. Phillips AR, Suttangkakul A, Vierstra RD (2008) The ATG12-conjugating enzyme ATG10 Is essential for autophagic vesicle formation in *Arabidopsis thaliana*. Genetics 178:1339–1353
87. Yoshimoto K, Ohsumi Y (2018) Unveiling the molecular mechanisms of plant autophagy-from autophagosomes to vacuoles in plants. Plant Cell Physiol 59(7):1337–1344). Oxford University Press. https://doi.org/10.1093/pcp/pcy112

88. Liu Y, Burgos JS, Deng Y, Srivastava R, Howell SH (2012) Degradation of the endoplasmic reticulum by autophagy during endoplasmic reticulum stress in Arabidopsis. Plant Cell 24(11):4635–4651
89. Li F, Vierstra RD (2012) Autophagy: a multifaceted intracellular system for bulk and selective recycling. Trends Plant Sci 17(9):526–537. https://doi.org/10.1016/j.tplants.2012.05.006
90. Liu Y, Bassham DC (2012) Autophagy: pathways for self-eating in plant cells. Annu Rev Plant Biol 63:215–237. https://doi.org/10.1146/annurev-arplant-042811-105441
91. Tanno H, Komada M (2013) The ubiquitin code and its decoding machinery in the endocytic pathway. J Biochem 153(6):497–504. https://doi.org/10.1093/jb/mvt028
92. Vierstra RD (2009) The ubiquitin-26S proteasome system at the nexus of plant biology. Nat Rev Mol Cell Biol 10(6):385–397. https://doi.org/10.1038/nrm2688
93. Mazzucotelli E, Belloni S, Marone D, de Leonardis AM, Guerra D, di Fonzo N, Cattivelli L, Mastrangelo AM (2006) The E3 Ubiquitin Ligase gene family in plants: regulation by degradation. Curr Genomics 7:509–522
94. Miricescu A, Goslin K, Graciet E (2018) Ubiquitylation in plants: signaling hub for the integration of environmental signals. J Exp Bot 69(19):4511–4527). Oxford University Press. https://doi.org/10.1093/jxb/ery165
95. Leitner J, Petrásek J, Tomanov K, Retzer K, Parezova M, Korbei B, Bachmair A, Zazimalova E, Luschnig C (2012) Lysine63-linked ubiquitylation of PIN2 auxin carrier protein governs hormonally controlled adaptation of Arabidopsis root growth. PNAS 109(21):8322–8327
96. Henne WM, Buchkovich NJ, Emr SD (2011) The ESCRT Pathway. Dev Cell 21(1):–77–91. https://doi.org/10.1016/j.devcel.2011.05.015
97. Raiborg C, Stenmark H (2009) The ESCRT machinery in endosomal sorting of ubiquitylated membrane proteins. Nature 458(7237):445–452. https://doi.org/10.1038/nature07961
98. Cui Y, Shen J, Gao C, Zhuang X, Wang J, Jiang L (2016) Biogenesis of plant prevacuolar multivesicular bodies. Mol Plant 9:774–786
99. Gao C, Zhuang X, Shen J, Jiang L (2017) Plant ESCRT complexes: moving beyond endosomal sorting. Trends Plant Sci 22(11):986–998). Elsevier Ltd. https://doi.org/10.1016/j.tplants.2017.08.003
100. Winter V, Hauser MT (2006) Exploring the ESCRTing machinery in eukaryotes. Trends Plant Sci 11(3):115–123. https://doi.org/10.1016/j.tplants.2006.01.008
101. Yoshinari A, Korbei B, Takano J (2018) TOL proteins mediate vacuolar sorting of the borate transporter BOR1 in *Arabidopsis thaliana*. Soil Sci Plant Nutr 64(5):598–605. https://doi.org/10.1080/00380768.2018.1504322
102. Moulinier-Anzola J, Schwihla M, De-Araújo L, Artner C, Jörg L, Konstantinova N, Luschnig C, Korbei B (2020) TOLs function as Ubiquitin receptors in the early steps of the ESCRT pathway in higher plants. Mol Plant 13(5):717–731. https://doi.org/10.1016/j.molp.2020.02.012
103. Gao C, Luo M, Zhao Q, Yang R, Cui Y, Zeng Y, Xia J, Jiang L (2014) A Unique plant ESCRT component, FREE1, regulates multivesicular body protein sorting and plant growth. Curr Biol 24(21):2556–2563. https://doi.org/10.1016/j.cub.2014.09.014
104. Reyes FC, Buono RA, Roschzttardtz H, di Rubbo S, Yeun LH, Russinova E, Otegui MS (2014) A novel endosomal sorting complex required for transport (ESCRT) component in *Arabidopsis thaliana* controls cell expansion and development. J Biol Chem 289(8):4980–4988. https://doi.org/10.1074/jbc.M113.529685
105. Nordmann M, Cabrera M, Perz A, Bröcker C, Ostrowicz C, Engelbrecht-Vandré S, Ungermann C (2010) The Mon1-Ccz1 complex is the GEF of the late Endosomal Rab7 Homolog Ypt7. Curr Biol 20:1654–1659
106. Kim SJ, Brandizzi F (2012) News and views into the SNARE complexity in *Arabidopsis*. Front Plant Sci 3(FEB). https://doi.org/10.3389/fpls.2012.00028
107. Jahn R, Lang T, Südhof TC (2003) Review Membrane Fusion dynamic supramolecular assemblies involving con. Cell 112:519
108. Fujiwara M, Uemura T, Ebine K, Nishimori Y, Ueda T, Nakano A, Sato MH, Fukao Y (2014) Interactomics of Qa-SNARE in *Arabidopsis thaliana*. Plant Cell Physiol 55(4):781–789. https://doi.org/10.1093/pcp/pcu038

109. Dünser K, Schöller M, Rößling A-K, Löfke C, Xiao N, Pařízková B, Melnik S, Rodriguez-Franco M, Stöger E, Novák O, Kleine-Vehn J (2022) Endocytic trafficking promotes vacuolar enlargements for fast cell expansion rates in plants. Elife 11. https://doi.org/10.7554/eLife.75945
110. Yano D, Sato M, Saito C, Sato MH, Terao Morita M, Tasaka M (2003) A SNARE complex containing SGR3AtVAM3 and ZIGVTI11 in gravity-sensing cells is important for Arabidopsis shoot gravitropism. PNAS 100(14):8589–8594. www.pnas.orgcgi. https://doi.org/10.1073/pnas.1430749100
111. Zheng J, Han SW, Rodriguez-Welsh MF, Rojas-Pierce M (2014) Homotypic vacuole fusion requires VTI11 and is regulated by phosphoinositides. Mol Plant 7(6):1026–1040. https://doi.org/10.1093/mp/ssu019
112. Rojo E, Gillmor CS, Kovaleva V, Somerville CR, Raikhel NV (2001) VACUOLELESS1 is an essential gene required for vacuole formation and morphogenesis in *Arabidopsis*. Dev Cell 1:303–310
113. Rojo E, Zouhar J, Kovaleva V, Hong S, Raikhel N, v. (2003) The AtC-VPS Protein Complex Is Localized to the Tonoplast and the Prevacuolar Compartment in *Arabidopsis*. Mol Biol Cell 14:361–369. https://doi.org/10.1091/mbc.E02-08
114. Balderhaar HJ, kleine, & Ungermann, C. (2013) CORVET and HOPS tethering complexes—coordinators of endosome and lysosome fusion. In. J Cell Sci 126(6):1307–1316. https://doi.org/10.1242/jcs.107805
115. Takemoto K, Ebine K, Askani JC, Krüger F, Gonzalez ZA, Ito E, Goh T, Schumacher K, Nakano A, Ueda T (2018) Distinct sets of tethering complexes, SNARE complexes, and Rab GTPases mediate membrane fusion at the vacuole in *Arabidopsis*. PNAS 115(10):E2457–E2466. https://doi.org/10.1073/pnas.1717839115
116. Vukašinovic N, Žárský V (2016) Tethering complexes in the Arabidopsis endomembrane system. Front Cell Dev Biol 4(MAY)). Frontiers Media S.A. https://doi.org/10.3389/fcell.2016.00046
117. Hadlington JL, Denecke J (2000) Sorting of soluble proteins in the secretory pathway of plants. Curr Opin Plant Biol 3:461–468
118. Kang H, Hwang I (2014) Vacuolar sorting receptor-mediated trafficking of soluble vacuolar proteins in plant cells. Plants 3(3):392–408). MDPI AG. https://doi.org/10.3390/plants3030392
119. Kirsch T, Parist N, Butler JM, Beevers L, Rogerst JC (1994) Purification and initial characterization of a potential plant vacuolar targeting receptor. Proc Nati Acad Sci USA 91:3403
120. Marcos Lousa C, Gershlick DC, Denecke J (2012) Mechanisms and concepts paving the way towards a complete transport cycle of plant vacuolar sorting receptors. Plant Cell 24:1714–1732
121. Kim H, Kang H, Jang M, Chang JH, Miao Y, Jiang L, Hwang I (2010) Homomeric interaction of AtVSR1 is essential for its function as a vacuolar sorting receptor. Plant Physiol 154(1):134–148. https://doi.org/10.1104/pp.110.159814
122. Robinson DG, Neuhaus JM (2016) Receptor-mediated sorting of soluble vacuolar proteins: myths, facts, and a new model. J Exp Bot 67(15):4435–4449). Oxford University Press. https://doi.org/10.1093/jxb/erw222
123. Cao X, Rogers SW, Butler J, Beevers L, Rogers JC (2000) Structural requirements for ligand binding by a probable plant vacuolar sorting receptor. Plant Cell 12:493. https://academic.oup.com/plcell/article/12/4/493/6008799
124. Luo F, Fong YH, Zeng Y, Shen J, Jiang L, Wong K-B (2014) How vacuolar sorting receptor proteins interact with their cargo proteins: crystal structures of Apo and Cargo-Bound Forms of the protease-associated domain from an Arabidopsis vacuolar sorting receptor. Plant Cell 26:3693–3708
125. Fuji K, Shirakawa M, Shimono Y, Kunieda T, Fukao Y, Koumoto Y, Takahashi H, Hara-Nishimura I, Shimada T (2016) The adaptor complex AP-4 regulates vacuolar protein sorting

at the trans-Golgi network by interacting with VACUOLAR SORTING RECEPTOR1. Plant Physiol 170:211–219
126. Zouhar J, Muñoz A, Rojo E (2010) Functional specialization within the vacuolar sorting receptor family: VSR1, VSR3 and VSR4 sort vacuolar storage cargo in seeds and vegetative tissues. Plant J 64(4):577–588. https://doi.org/10.1111/j.1365-313X.2010.04349.x
127. Park M, Lee D, Lee GJ, Hwang I (2005) AtRMR1 functions as a cargo receptor for protein trafficking to the protein storage vacuole. J Cell Biol 170(5):757–767. https://doi.org/10.1083/jcb.200504112
128. Voelker TA, Herman EM, Chrispeels MJ (1989) In vitro mutated Phytohemagglutinin genes expressed in tobacco seeds: role of Glycans in protein targeting and stability. Plant Cell 1:95-104. https://10.1105/tpc.1.1.95
129. Bednarek SY, Wilkins TA, Dombrowski JE, Raikhel NV (1990) A carboxyl-terminal propeptide necessary for proper sorting of Barley Lectin to vacuoles of tobacco. Plant Cell 2:1145. American Society of Plant Physiologists
130. Holwerdaya BC, Padgettyb HS, Rogersaibi JC (1992) Proaleurain vacuolar targeting mediated by short contiguous peptide lnteractions. Plant Cell 4:307-18. https://doi.org/10.1105/tpc.4.3.307
131. Hwang I (2008) Sorting and anterograde trafficking at the Golgi Apparatus. Plant Physiol 148:673–683
132. Pourcher M, Santambrogio M, Thazar N, Thierry AM, Fobis-Loisy I, Miège C, Jaillais Y, Gaude T (2010) Analyses of SORTING NEXINs reveal distinct Retromer-Subcomplex functions in development and protein sorting in *Arabidopsis thaliana*. Plant Cell 22(12):3980–3991. https://doi.org/10.1105/tpc.110.078451
133. Shimada T, Koumoto Y, Li L, Yamazaki M, Kondo M, Nishimura M, Hara-Nishimura I (2006) AtVPS29, a putative component of a retromer complex, is required for the efficient sorting of seed storage proteins. Plant Cell Physiol 47(9):1187–1194. https://doi.org/10.1093/pcp/pcj103
134. Yamazaki M, Shimada T, Takahashi H, Tamura K, Kondo M, Nishimura M, Hara-Nishimura I (2008) Arabidopsis VPS35, a retromer component, is required for vacuolar protein sorting and involved in plant growth and leaf senescence. Plant Cell Physiol 49(2):142–156. https://doi.org/10.1093/pcp/pcn006
135. Heucken N, Ivanov R (2018) The retromer, sorting nexins and the plant endomembrane protein trafficking. J Cell Sci 131(2) Company of Biologists Ltd. https://doi.org/10.1242/jcs.203695
136. Robinson DG, Pimpl P, Scheuring D, Stierhof YD, Sturm S, Viotti C (2012) Trying to make sense of retromer. Trends Plant Sci 17(7):431–439. https://doi.org/10.1016/j.tplants.2012.03.005
137. Kriegel A, Andrés Z, Medzihradszky A, Krüger F, Scholl S, Delang S, Görkem Patir-Nebioglu M, Gute G, Yang H, Murphy AS, Peer WA, Pfeiffer A, Krebs M, Lohmann JU, Schumacher K (2015) Job sharing in the endomembrane system: vacuolar acidification requires the combined activity of V-ATPase and V-Ppase. Plant Cell 27(12):3383–3396. https://doi.org/10.1105/tpc.15.00733
138. Faraco M, Spelt C, Bliek M, Verweij W, Hoshino A, Espen L, Prinsi B, Jaarsma R, Tarhan E, deBoer AH, DiSansebastiano G, Pietro Koes R, Quattrocchio FM (2014) Hyperacidification of vacuoles by the combined action of two different P-ATPases in the tonoplast determines flower color. Cell Rep 6(1):32–43. https://doi.org/10.1016/j.celrep.2013.12.009
139. Appelhagen I, Nordholt N, Seidel T, Spelt K, Koes R, Quattrochio F, Sagasser M, Weisshaar B (2015) TRANSPARENT TESTA 13 is a tonoplast P3A-ATPase required for vacuolar deposition of proanthocyanidins in *Arabidopsis thaliana* seeds. Plant J 82(5):840–849. https://doi.org/10.1111/tpj.12854
140. Cosse M, Seidel T (2021) Plant proton pumps and cytosolic pH-homeostasis. Front Plant Sci 12. Frontiers Media S.A. https://doi.org/10.3389/fpls.2021.672873

141. Martínez-Munoz GA, Kane P (2008) Vacuolar and plasma membrane proton pumps collaborate to achieve cytosolic pH homeostasis in yeast. J Biol Chem 283(29):20309–20319
142. Martinoia E, Maeshima M, Neuhaus HE (2007) Vacuolar transporters and their essential role in plant metabolism. J Exp Bot 58(1):83–102. https://doi.org/10.1093/jxb/erl183
143. Clemens S (2001) Molecular mechanisms of plant metal tolerance and homeostasis. Planta 212:475–486
144. Kasai K, Takano J, Miwa K, Toyoda A, Fujiwara T (2011) High boron-induced ubiquitination regulates vacuolar sorting of the BOR1 borate transporter in *Arabidopsis thaliana*. J Biol Chem 286(8):6175–6183. https://doi.org/10.1074/jbc.M110.184929
145. Takano J, Miwa K, Yuan L, von Wiré N, Fujiwara T (2005) Endocytosis and degradation of BOR1, a boron transporter of *Arabidopsis thaliana*, regulated by boron availability. PNAS 102(34):12276–12281. https://www.pnas.org
146. Takano J, Tanaka M, Toyoda A, Miwa K, Kasai K, Fuji K, Onouchi H, Naito S, Fujiwara T (2010) Polar localization and degradation of Arabidopsis boron transporters through distinct trafficking pathways. Proc Natl Acad Sci U S A 107(11):5220–5225. https://doi.org/10.1073/pnas.0910744107
147. Callis J, Carpenter T, Sun C-W, Vierstrat RD (1995) Structure and evolution of genes encoding Polyubiquitin and Ubiquitin-Like proteins in *Arabidopsis thaliana* Ecotype Columbia. Genetics 139:921–939
148. Chen Q, Yu F, Xie Q (2020) Insights into endoplasmic reticulum-associated degradation in plants. New Phytol 226:345–350
149. Vierstra RD (1993) Protein degradation in plants. Annu Rev Plant Physiol Plant Mol Biol 44:385–410. www.annualreviews.org
150. de Brito Francisco R, Martinoia E (2018) The vacuolar transportome of plant specialized metabolites. Plant Cell Physiol 59(7):1326–1336). Oxford University Press. https://doi.org/10.1093/pcp/pcy039
151. Martinoia E, Meyer S, de Angeli A, Nagy R (2012) Vacuolar transporters in their physiological context. Annu Rev Plant Biol 63:183–213. https://doi.org/10.1146/annurev-arplant-042811-105608
152. Pillitteri LJ, Dong J (2013) Stomatal development in *Arabidopsis*. In: The Arabidopsis book, pp 1–26
153. Barragán V, Leidi EO, Andrés Z, Rubio L, de Luca A, Fernández JA, Cubero B, Pardo JM (2012) Ion exchangers NHX1 and NHX2 mediate active potassium uptake into vacuoles to regulate cell turgor and stomatal function in arabidopsis. Plant Cell 24:1127–1142
154. de Angeli A, Zhang J, Meyer S, Martinoia E (2013) AtALMT9 is a malate-activated vacuolar chloride channel required for stomatal opening in *Arabidopsis*. Nat Commun 4:1–10
155. Gobert A, Isayenkov S, Voelker C, Czempinski K, Maathuis FJM (2007) The two-pore channel TPK1 gene encodes the vacuolar K+ conductance and plays a role in K+ homeostasis. PNAS 104(25):10726–10731
156. Bak G, Lee EJ, Lee Y, Kato M, Segami S, Sze H, Maeshima M, Hwang JU, Lee Y (2013) Rapid structural changes and acidification of guard cell vacuoles during stomatal closure require phosphatidylinositol 3,5-bisphosphate. Plant Cell 25(6):2202–2216. https://doi.org/10.1105/tpc.113.110411
157. Eisenach C, de Angeli A (2017) Ion transport at the vacuole during stomatal movements. Plant Physiol 174(2):520–530. https://doi.org/10.1104/pp.17.00130
158. Franks PJ, Buckley TN, Shope JC, Mott KA (2001) Guard cell volume and pressure measured concurrently by confocal microscopy and the cell pressure probe. Plant Physiol 125:1577–1584. http://bioweb.usu.edu/kmott
159. Shope JC, DeWald DB, Mott KA (2003) Changes in surface area of intact guard cells are correlated with membrane internalization. Plant Physiol 133:1314–1321
160. Tanaka Y, Kutsuna N, Kanazawa Y, Kondo N, Hasezawa S, Sano T (2007) Intra-vacuolar reserves of membranes during stomatal closure: the possible role of guard cell vacuoles esti-

mated by 3-D reconstruction. Plant Cell Physiol 48(8):1159–1169. https://doi.org/10.1093/pcp/pcm085

161. Ludevid D, Höfte H, Himelblau E, Chrispeels MJ (1992) The expression pattern of the Tonoplast Intrinsic Protein y-TIP in *Arabidopsis thaliana* is correlated with cell enlargement. Plant Physiol 100:1633–1639
162. Kaldenhoff R, Ribas-Carbo M, Sans JF, Lovisolo C, Heckwolf M, Uehlein N (2008) Aquaporins and plant water balance. Plant Cell Environ 31(5):658–666. https://doi.org/10.1111/j.1365-3040.2008.01792.x
163. Cosgrove DJ (2000) Expansive growth of plant cell walls. Plant Physiol Biochem 38:109–124
164. Beebo A, Thomas D, Der C, Sanchez L, Leborgne-Castel N, Marty F, Schoefs B, Bouhidel K (2009) Life with and without AtTIP1;1, an Arabidopsis aquaporin preferentially localized in the apposing tonoplasts of adjacent vacuoles. Plant Mol Biol 70(1–2):193–209. https://doi.org/10.1007/s11103-009-9465-2
165. Cosgrove DJ (2018) Diffuse growth of plant cell walls. Plant Physiol 176(1):16–27). American Society of Plant Biologists. https://doi.org/10.1104/pp.17.01541
166. Kang BH, Anderson CT, Arimura SI, Bayer E, Bezanilla M, Botella MA, Brandizzi F, Burch-Smith TM, Chapman KD, Dünser K, Gu Y, Jaillais Y, Kirchhoff H, Otegui MS, Rosado A, Tang Y, Kleine-Vehn J, Wang P, Zolman BK (2022) A glossary of plant cell structures: current insights and future questions. Plant Cell 34(1):10–52. https://doi.org/10.1093/plcell/koab247
167. Scheuring D, Löfke C, Krüger F, Kittelmann M, Eisa A, Hughes L, Smith RS, Hawes C, Schumacher K, Kleine-Vehn J (2016) Actin-dependent vacuolar occupancy of the cell determines auxin-induced growth repression. Proc Natl Acad Sci U S A 113(2):452–457. https://doi.org/10.1073/pnas.1517445113
168. Owens T, Poole RJ (1979) Regulation of cytoplasmic and vacuolar volumes by plant cells in suspension culture. Plant Physiol 64:900–904
169. Löfke CL, Dünser K, Scheuring D, Kleine-Vehn J (2015) Auxin regulates SNARE-dependent vacuolar morphology restricting cell size. Elife 1–16. https://doi.org/10.7554/eLife.05868.001
170. Wolf S, Hématy K, Höfte H (2012) Growth control and cell wall signaling in plants. Annu Rev Plant Biol 63:381–407. https://doi.org/10.1146/annurev-arplant-042811-105449
171. Rayle DL, Cleland R (1970) Enhancement of wall loosening and elongation by acid solutions. Plant Physiol 46. https://academic.oup.com/plphys/article/46/2/250/6091254
172. Rayle DL, Cleland RE (1992) The acid growth theory of Auxin-induced cell elongation is alive and well. Plant Physiol 99:1271–1274
173. Barbez E, Dünser K, Gaidora A, Lendl T, Busch W (2017) Auxin steers root cell expansion via apoplastic pH regulation in *Arabidopsis thaliana*. PNAS:E4884–E4893. https://doi.org/10.5061/dryad.sq7s3
174. Fendrych M, Leung J, Friml J (2016) TIR1/AFB-Aux/IAA auxin perception mediates rapid cell wall acidification and growth of Arabidopsis hypocotyls. Elife 1–19. https://doi.org/10.7554/eLife.19048.001
175. Spartz AK, Ren H, Park MY, Grandt KN, Lee SH, Murphy AS, Sussman MR, Overvoorde PJ, Gray WM (2014) SAUR inhibition of PP2C-D phosphatases activates plasma membrane H+-ATPases to promote cell expansion in Arabidopsis. Plant Cell 26(5):2129–2142. https://doi.org/10.1105/tpc.114.126037
176. Hématy K, Sado PE, van Tuinen A, Rochange S, Desnos T, Balzergue S, Pelletier S, Renou JP, Höfte H (2007) A receptor-like kinase mediates the response of Arabidopsis cells to the inhibition of cellulose synthesis. Curr Biol 17:922–931
177. Shih HW, Miller ND, Dai C, Spalding EP, Monshausen GB (2014) The receptor-like kinase FERONIA is required for mechanical signal transduction in Arabidopsis seedlings. Curr Biol 24(16):1887–1892. https://doi.org/10.1016/j.cub.2014.06.064
178. Blackburn MR, Haruta M, Moura DS (2020) Twenty years of progress in physiological and biochemical investigation of RALF peptides. Plant Physiol 182(4):1657–1666). American Society of Plant Biologists. https://doi.org/10.1104/PP.19.01310

179. Haruta M, Sabat G, Stecker K, Minkoff BB, Sussman MR (2014) A peptide hormone and its receptor protein kinase regulate plant cell expansion. Science 343(6169):408–411. https://doi.org/10.1126/science.1244454
180. Herger A, Dünser K, Kleine-Vehn J, Ringli C (2019) Leucine-Rich repeat extensin proteins and their role in cell wall sensing. Curr Biol 29(17):R851–R858). Cell Press. https://doi.org/10.1016/j.cub.2019.07.039
181. Herger A, Gupta S, Kadler G, Franck CM, Boisson-Dernier A, Ringli C (2020) Overlapping functions and protein-protein interactions of LRR-extensins in *Arabidopsis*. PLoS Genet 16(6). https://doi.org/10.1371/journal.pgen.1008847
182. Becker B (2007) Function and evolution of the vacuolar compartment in green algae and land plants (Viridiplantae). Int Rev Cytol 264:1–24. https://doi.org/10.1016/S0074-7696(07)64001-7
183. Mitsuhashi N, Ohnishi M, Sekiguchi Y, Kwon YU, Chang YT, Chung SK, Inoue Y, Reid RJ, Yagisawa H, Mimura T (2005) Phytic acid synthesis and vacuolar accumulation in suspension-cultured cells of Catharanthus roseus induced by high concentration of inorganic phosphate and cations. Plant Physiol 138(3):1607–1614. https://doi.org/10.1104/pp.105.060269

Bacterial Protein Transport Pathways and Analogous Conserved Pathways in Eukaryotes

Philip Kauffman, Sharbani Kaushik, Andreas Kuhn, and Ross E. Dalbey

Abstract Approximately 30% of the proteins in bacteria are exported to the cell surface. The export process is crucial for bacterial growth and survival, protein biogenesis, membrane function, signaling, lipid synthesis, and metabolism. This review will present recent advances in the export of proteins to the periplasm and outer membrane, primarily in Gram-negative bacteria and discuss how proteins are targeted and inserted into the inner membrane. Many systems are used for this process, which include the secretion (Sec) system, the insertase YidC, the twin-arginine translocation (Tat) system, the β-barrel assembly machinery (BAM) insertase, and the localization of lipoprotein (Lol) system, in Gram-negative bacteria. The accumulating evidence reveals that many of the fascinating insertion/translocation components used in bacteria have found their way into eukaryotic cells through evolution, now playing crucial roles in ER, mitochondria, and chloroplasts.

Keywords Sec · Tat · YidC · Bam · Lol · Protein export · Membrane insertion · β-barrel proteins · Oxa1 · Mitochondria · Chloroplast · Endoplasmic reticulum

P. Kauffman · R. E. Dalbey (✉)
Department of Chemistry and Biochemistry, The Ohio State University, Columbus, OH, USA
e-mail: Kauffman.210@osu.edu; dalbey.1@osu.edu

S. Kaushik
Department of Chemistry and Biochemistry, The Ohio State University, Columbus, OH, USA

Department of Chemical Engineering at Dr. Vishwanath Karad MIT, World Peace University, Pune, India
e-mail: sharbani.kaushik@mitwpu.edu.in

A. Kuhn
Institute of Biology, University of Hohenheim, Stuttgart, Germany
e-mail: andreas.Kuhn@uni-hohenheim.de

S. D. Schwartzbach et al. (eds.), *Endosymbiotic Organelle Acquisition*,
https://doi.org/10.1007/978-3-031-57446-7_10

1 Introduction

The cell surface of Gram-negative bacteria contains three compartments, the inner membrane, the outer membrane, and the periplasmic space localized between the two membranes (Fig. 1). The inner membrane contains α-helical integral membrane proteins, whereas the outer membrane contains mainly β-barrel proteins. Both contain lipoproteins, which are peripheral proteins that are anchored to the membrane by their lipid moieties. The inner membrane is made up entirely of phospholipids whereas the outer membrane contains a phospholipid inner leaflet and lipopolysaccharides (LPS) located in the outer leaflet.

The bacterial cell wall consists of the inner membrane, periplasm with peptidoglycan and the outer membrane, and performs many important functions [1]. The outer membrane LPS protects the bacterial cell from e.g., antimicrobial compounds and creates a permeability barrier that controls the entrance of substrates. Porins in the outer membrane function to mediate the transport of ions and polar molecules. The peptidoglycan is localized in the periplasmic space and is important for maintaining the cell shape and stabilizes the cell. Other important structures on the

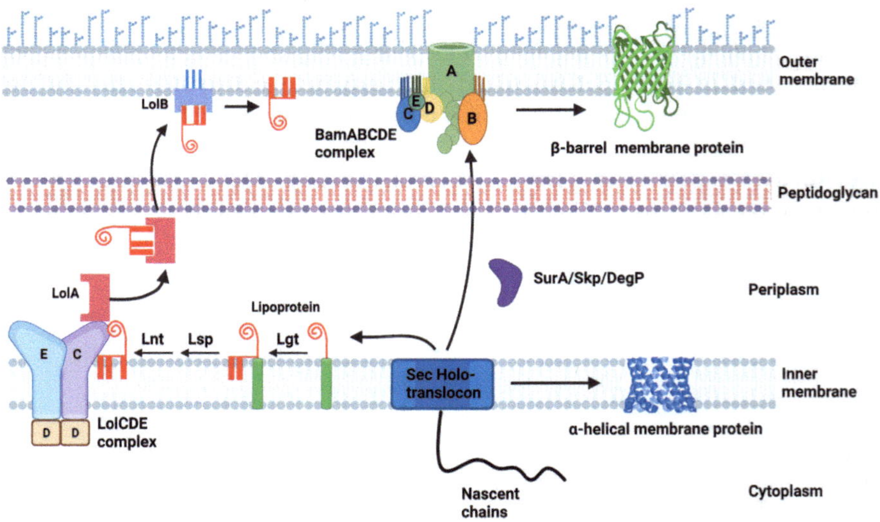

Fig. 1 The biomachineries involved in the inner membrane insertion, protein export and trafficking, and assembly of proteins into the outer membrane in *E. coli*. SecYEG holocomplex (SecYEGDF–YajC–YidC) is primarily responsible for protein export to the periplasm or insertion into the inner membrane. Following translocation, precursors possessing a lipoprotein signal sequence undergo maturation and processing where the cysteine at the +1 position is modified by a diacylglycerol catalyzed by Lgt, followed by cleavage of the signal sequence by Lsp (SPase 2 or lipoprotein signal peptidase), and modification of the amino-terminus with a fatty acid carried out by the Lnt enzyme. The Lol pathway directs lipoproteins destined to the outer membrane and involves the inner membrane LolCDE transporter, the periplasmic chaperone LolA and outer membrane receptor LolB. The Bam machinery (BamABCDE) mediates the insertion and assembly of β-barrel proteins into the outer membrane. The unfolded Omps, following membrane translocation, are escorted by the SurA/Skp/DegP chaperones to the BamABCDE complex for insertion and assembly into the outer membrane. (Created with BioRender.com)

cell surface are the fimbriae and the flagella that allow the bacteria to bind to surfaces and move, respectively.

In recent years, there have been remarkable advances in understanding the biomachineries for insertion of proteins into the inner [2–7] and outer membranes [8–12] of bacteria and for protein export to the periplasmic space [4, 7, 13–17]. In this review, we will discuss the roles of the Sec translocase and YidC insertase in the integration of proteins into the cytoplasmic membrane, in addition to the function of Sec and the Tat translocases in membrane transport. We will describe the recent work on the role of the Lol machinery in the trafficking of lipoproteins to the outer membrane, and the Bam system in inserting β-barrel proteins into the outer membrane. Finally, we will provide an update on the evolutionarily conserved Bam, Sec, Tat, and YidC systems found in the endoplasmic reticulum (ER), mitochondria, and chloroplasts of eukaryotes.

2 Insertion of Proteins into the Inner Membrane

The principal route for targeting proteins to the inner membrane of *Escherichia coli* is the Signal Recognition particle (SRP) pathway [18, 19]. SRP, comprised of the subunit Ffh and the 4.5 S RNA [20], typically binds cotranslationally to the first hydrophobic segment of a membrane protein as it leaves the ribosome [21] and targets it to FtsY, the SRP receptor [22]. Following targeting, the nascent protein chain is transferred to the membrane-bound Sec machinery or YidC insertase [19, 23] (Fig. 2).

While the majority of membrane proteins are targeted by the SRP/FtsY system, some proteins are directed to the membrane without their aid [24]. For example, the M13 coat protein does not require SRP for targeting. SecA can also cotranslationally target some membrane proteins such as RodZ directly to the membrane-bound SecYEG insertase [25, 26]. In addition, a subset of Tat-dependent integral membrane proteins possessing a folded domain and a C-terminal transmembrane (TM) segment can be directly targeted to the TatABC translocase (Fig. 3a).

2.1 *Sec System*

Following the delivery to the membrane, most bacterial SRP substrates are transferred to the SecYEG machinery [18]. SecYEG functions as a protein conducting channel [27] and associates with the trimeric membrane protein SecDF–YajC [28] and the insertase YidC (Fig. 2a) [29]. SecDF facilitates translocation of the polypeptide through the Sec channel using the energy of the proton motive force (pmf) [30] and YidC plays an important role in membrane integration and folding steps (see below) [2]. Some membrane proteins also require SecA, the motor ATPase [31–33], particularly for membrane translocation of large loops. Interestingly, the components may assemble to form a SecYEGDF–YajC–YidC holotranslocon (HTL)

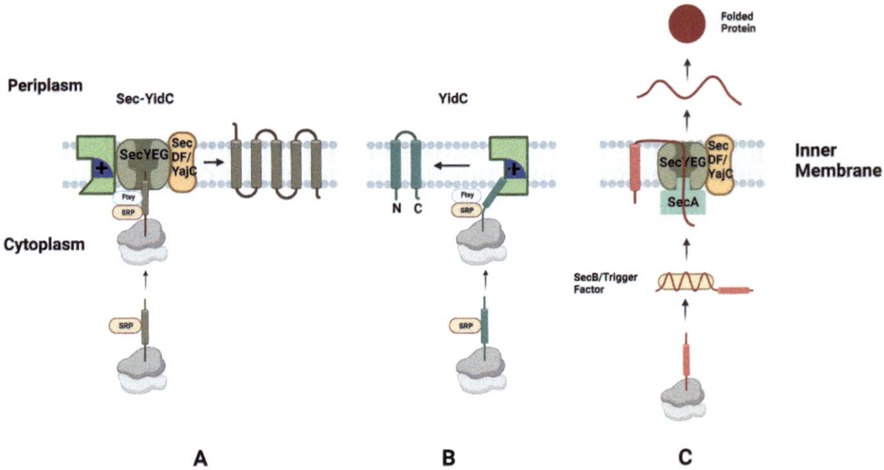

Fig. 2 Protein export and inner membrane insertion mediated by SecYEG and YidC. (**a**) The SecYEGDF–YajC–YidC complex is the main translocase to insert nascent proteins into the inner membrane. The YidC insertase can function with the Sec complex or (**b**) independently in membrane protein insertion. The signal recognition particle (SRP) and SRP receptor (FtsY) are involved in targeting Sec and YidC proteins. (**c**) The SecYEGDF–YajC complex is also responsible for exporting the majority of proteins to the periplasmic space. The motor ATPase SecA energizes export through the Sec channel. These exported proteins often are kept in an unfolded state after release from the ribosome by interaction with chaperones such as SecB and trigger factor. (Created with BioRender.com

complex (Fig. 2a). The HTL has been reported to be more efficient than SecYEG alone in cotranslational membrane protein insertion [34] and can also promote SecA-dependent translocation.

Clues to the function of the SecYEG complex in inserting membrane proteins were revealed by the structure of the SecY protein [35]. Along the center of the channel at its narrowest point is the pore ring comprised of six nonpolar aliphatic residues. This pore ring is sealed at the periplasmic side by the so-called plug helix. Interestingly, when the SecYEG channel is activated, the plug helix is displaced from the pore ring [36, 37], allowing the nascent chain to move through it to the periplasmic side of the membrane [35]. Another important functional domain of the SecYEG complex is the lateral gate, which, when open, provides a gateway for the newly inserted hydrophobic segments to exit the channel and to enter the lipid bilayer. Thereby, the TM segments of multispanning membrane proteins can be inserted into the lipid bilayer (via the lateral gate) one by one. In some cases, the flanking surface helices, or re-entrant loops that go roughly halfway through the membrane and turn back toward the side it commences, can advance or delay insertion of the TM segments of multispanning membrane proteins [38]. Other variations can also occur with pairs of transmembrane helices exiting the channel during insertion of multispanning membrane proteins [39, 40].

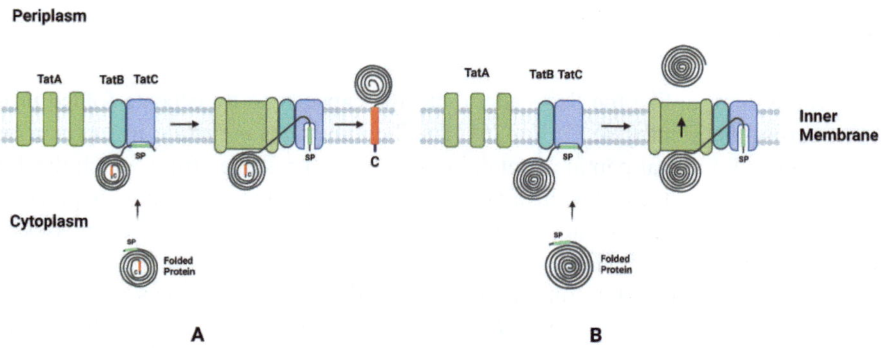

Fig. 3 Membrane insertion and protein export by the Tat pathway. (**a**) The Tat signal peptide targets the folded protein possessing a C-terminal TM segment to the TatBC receptor complex. Following deep insertion of the signal peptide into the TatBC complex, many TatA subunits are recruited to form the translocation channel of the Tat complex in the presence of the pmf. After translocation of the folded protein through the translocation site, the signal peptide is removed, and the C-terminal hydrophobic segment anchors the protein in the membrane. Following translocation, the active Tat complex is disassembled. (**b**) The Tat exported protein uses the same mechanism for translocation across the inner membrane into the periplasmic space. Due to lacking a C-terminal hydrophobic membrane anchor, the protein is translocated completely across the membrane into the periplasm. (Created with BioRender.com)

2.2 *YidC Insertase*

YidC can cooperate with the SecYEG machinery to perform important functions in membrane protein biogenesis (Fig. 2a) [41–44]. For LacY, which spans the membrane 12 times, it promotes the folding of the protein while the SecYEG machinery is exclusively involved in translocation of the six periplasmic loops of the protein [45, 46]. YidC functions as a foldase/assembly site for the Sec-dependent multi-spanning proteins, MalF and mannitol permease [47, 48]. In other cases, YidC, along with SecYEG, drives the membrane insertion of subunit a of ATP synthase [49, 50], CyoA [51], and TatC [52, 53].

YidC can also insert proteins independent of the Sec system (Fig. 2b) [23, 49, 54–60]. YidC-only substrates typically have 1 or 2 TM segments with short periplasmic regions, in line with results showing that it has a minimal translocase function [61–64]. In contrast to SecYEG, YidC does not possess a membrane-spanning aqueous channel. Rather it contains a hydrophilic groove localized to the inner leaflet region, which is open to the cytoplasm and the lipid bilayer [65–69]. The hydrophilic groove has a strictly conserved positively charged residue that transiently accepts hydrophilic regions of the substrate membrane protein that are to be translocated across the membrane [70]. In addition, there is a hydrophobic slide region where TM segments of YidC substrates move through YidC before moving laterally to integrate into the lipid bilayer [71]. Finally, YidC family members possess a cytoplasmic helical hairpin-like region that has been implicated in substrate binding and interaction with the SecY and the targeting components SRP and FtsY [72, 73].

2.3 TatABC

The Tat system, unlike the Sec system, has the capacity to translocate folded proteins across the membrane [7, 74–76]. The Tat substrates are made in a precursor form with a Tat signal peptide containing two arginyl residues from which the Tat (*t*win *a*rginine *t*ranslocase) pathway gets its name [77]. Of the thirty known proteins that are membrane translocated by the Tat pathway in *E. coli*, five of them possess a C-terminal TM segment [78].

The Gram-negative Tat complex is assembled by multiple copies of the TatA, TatB, and TatC subunits. Briefly, in *E. coli*, the Tat proteins are encoded in an operon comprised of the *tatA*, *tatB*, *tatC*, and also *tatD* genes [79]. Another gene, *tatE*, encodes a protein homologous to TatA, but is expressed at low levels in the cell and is not needed for Tat function [79]. Both TatA and TatB have similar features, consisting of a small periplasmic N-terminal tail, a short TM segment, an amphipathic helix and a cytoplasmic domain [80]. TatC consist of 6 TM segments [81, 82] and has a cupped hand shape based on the crystal structure [83, 84]. TatD is a cytoplasmic protein and plays a role in quality control for FeS-containing exported proteins [85].

Export of a Tat substrate is initiated by the binding of the signal peptide to the TatBC module of the Tat resting complex (Fig. 3a) [86, 87]. Afterwards, there is a deep insertion of the Tat signal peptide within the Tat complex. This triggers the oligomerization of many TatA protomers to form an active translocase, a process which requires the proton motive force (pmf) [88–90] (Fig. 3a) [7]. The TatA protomers are recruited from a separate pool of TatA molecules in the membrane. After translocation of the folded protein domain, the signal peptide is cleaved off by a signal peptidase [91] and the active Tat complex is disassembled. In the case of a Tat membrane protein, the C-terminal hydrophobic segment acts as an anchor sequence, which halts translocation so that the TM segment can laterally integrate into the membrane bilayer.

TatB and TatC play a role in the recognition of the Tat signal peptide and the transfer of the substrate to the TatA oligomers [87]. TatC plays the primary role in the initial recognition of the signal peptide, while TatB is proposed to be involved in the delivery of the substrate from TatC to TatA protomers at the translocation site [92]. Many details of this process and how TatBC subunits are configured to produce the active translocation state have been explored using crosslinking methods [86, 90, 92–96], but the structure of the active translocation complex is still needed [17, 97, 98].

3 Export of Proteins to the Periplasm

Proteins exported across the inner membrane into the periplasmic space are made in a precursor form with an amino-terminal signal sequence that directs the protein to the translocase (Figs. 2 and 3). The signal sequence that directs a protein to the Sec translocase has three domains: an amino-terminal positively charged region, a central hydrophobic domain, and a hydrophilic region that possesses the sequence motifs for signal peptidase processing [99]. For Tat-dependent protein export, the twin-arginine signal peptides play a crucial role [16, 100].

Targeting of exported proteins to the Sec complex can occur by three different routes involving chaperones such as SecB, the ribosome-associated trigger factor or the SecA motor ATPase [3]. In the first route, after release from the ribosome, the preprotein binds to the chaperone SecB, which functions as a holdase and maintains the preprotein's unfolded state [101]. The SecB–preprotein complex is then delivered to SecA that is localized at the SecYEG translocation site [102]. Alternatively, SecA binds directly to the nascent chain as it is being synthesized by the ribosome and then directs the exported protein to the SecYEG [103]. In a third targeting pathway, the preprotein can be released from the ribosome with the bound trigger factor and, after it is recruited to SecB, it is targeted to the SecAYEG (Fig. 2c) [104].

Tat preproteins are folded in the cytoplasm. The signal peptide of the folded preprotein binds directly to the TatBC components at the inner membrane. Many of the Tat substrates possess redox cofactors that need to be introduced into the cytoplasm [105]. Notably, there are chaperones called REMPs (for redox enzyme maturation proteins) that are involved in maturation of complex iron sulfur molybdoenzymes. REMPs facilitate the insertion of the cofactors and perform proof reading functions in the maturation and assembly of the cofactor [106, 107]. Some of the REMPs appear to bind to the Tat signal peptide until the Tat protein is fully folded [108–110]. This way, the chaperones protect the signal peptide and prevent premature targeting of the Tat substrate to the translocation site.

3.1 Sec System

The preprotein–SecA complex targeted to the membrane activates the SecYEG complex for translocation (Fig. 2c) [111]. Activation is achieved by SecA binding to SecYEG, which leads to partial opening of the lateral gate. There is an expansion of the pore ring and dissociation of the plug helix after the signal peptide of the preprotein binds to the partially open lateral gate [33, 112]. The flanking C-terminal region of the mature protein chain is then able to pass through the center of the channel within the pore ring [113]. After multiple rounds of ATP hydrolysis by the peripheral subunit SecA, the polypeptide is moved completely across the SecYEG channel [14, 32, 114, 115]. The signal peptide is removed from the preprotein by signal peptidase, allowing the release of the chain into the periplasmic space [116]. Folding of the exported protein then occurs in the periplasmic space.

The SecA motor ATPase has 5 functional domains [117]: the nucleotide binding domains 1 and 2 (NBD1 and NBD2); preprotein crosslinking domain (PPXD); an α- helical scaffold domain (HSD); the helical wing domain (HWD); and carboxyl-terminal linker domain (CTL). ATP binds at the interface of NBD1 and NBD2 while the precursor substrate and the SecYEG channel are recruited to the PPXD domain. The SecA clamp region, located at the interface of the PPXD and NBD2, can capture the substrate polypeptide chain [118]. The SecA groove formed at the interface of PPXD and the 2-helix finger (2HF) has been proposed to constitute the signal peptide recognition site [119]. The 2HF in the HSD domain is important for SecA function and for Sec-dependent translocation [120]. In the SecA–SecYEG complex, it is located at the entrance of the Sec channel [118] and is positioned close to the preprotein [120]. It may move backwards and forwards to drive the translocation of the substrate through the SecYEG channel [121].

Several models have been proposed for how SecA uses ATP to energize the stepwise movement of the polypeptide through SecYEG [14, 31]. In the push and slide model, SecA pushes the preprotein into the SecYEG channel. Specifically, the 2HF interacting with the preprotein may push the preprotein into the channel when SecA binds ATP [121]. After ATP hydrolysis, the 2HF dissociates from the preprotein chain after insertion and retracts back to the cytoplasmic side of the channel. During the retraction of the 2HF, the clamp region of SecA closes around the polypeptide to hold it and prevent it from back sliding. Evidence supporting this mechanism was provided by a single molecule FRET study [122]. When ADP is bound to SecA, the polypeptide clamp will reopen and then the protein chain can be translocated through the SecYEG channel by a passive sliding [121]. In this mechanism, the chain can slide in the forward or backward direction but net translocation is achieved when it occurs in the forward direction for certain protein chains moving across the SecYEG pore. Forward translocation is facilitated, in one scenario, by the folding of the translocated chain on the periplasmic side of the membrane. The pushing step repeats after ATP binds again to SecA. A somewhat surprising result is that the protein can still be translocated across the SecYEG channel even when 2HF is crosslinked to a cytoplasmic loop of SecY [123].

Another mechanism is the ratcheting model proposed by Allen et al. [124]. In this model, the polypeptide with small amino acid side chains can slide through a partially open state of SecYEG when ADP is bound to SecA. When bulky amino acid sidechains in the polypeptide chain are encountered in the SecYEG channel, a conformational change occurs in SecA, leading to the binding of ATP. This causes further opening of the SecYEG channel resulting in the polypeptide sliding through the channel. Backsliding is prevented because of the bulky residues trapped on the periplasmic side of the membrane. After ATP hydrolysis, the channel closes again and the cycle repeats. In this model, 2HF is proposed to trigger SecA nucleotide exchange on sensing the amino acids with large sidechains, allowing ATP to bind, opening the channel. Other models of SecA-mediated translocation are the catch and release mechanism for translocation of the preprotein that is coupled to the intrinsic dynamics of SecA [125] as well as the hand-switching model for movement by the clamp that drives translocation [126].

The SecYEG accessory component SecDF also promotes translocation of the polypeptide chain most likely late in the export process using the energy of the proton motive force (pmf) [30, 127]. Indeed, a late stage of translocation was shown to be SecDF and pmf-dependent and did not require ATP [127]. The crystal structure of SecDF showed it forms a 12-helix transmembrane domain with two large periplasmic domains, P1 and P4 [127]. P1 was proposed to bind the partially translocated chain, and then SecDF would pull the chain toward the periplasmic side of the membrane, resulting in the forward movement of the protein chain as P1 undergoes PMF-dependent conformational changes [128].

3.2 *Tat System*

Tat-dependent export of folded proteins to the periplasmic space occurs by the same mechanism put forward for the integral membrane proteins containing C-terminal tail-anchor sequences (Fig. 3b). How the assembled TatA oligomers form the active substrate translocation complex [16] is not clear. One model proposed that after cargo docks onto TatBC, TatA oligomers are recruited to form different sized proteinaceous pores that accommodate the different sized folded proteins to be translocated [129, 130]. A second model proposes that the TatA oligomers cause membrane thinning due to the short TatA TM segment and the presence of the conserved glutamic acid/glutamine in the oligomer. Thinning would occur such that the membrane transiently breaks down at the TatA translocation site [131, 132]. The membrane destabilization may be controlled by the TatA amphipathic helix which can switch its orientation relative to the membrane surface upon the recruitment of substrate [132]. The bilayer may disrupt altogether and transiently form a lipid lined toroidal pore that would allow the translocation of the folded domain [133]. Notably, hydrophobic mismatch of TatA and TatB is crucial for Tat transport across the membrane [134, 135].

4 Trafficking to the Outer Membrane

Outer membrane destined β-barrel proteins use typical signal peptides and are exported across the inner membrane by the Sec translocase into the periplasmic space. In the periplasm, chaperones are involved in escorting the still unfolded proteins through the loose meshwork of the peptidoglycan to the Bam machinery for outer membrane insertion (Fig. 1).

Lipoprotein precursors, on the other hand, are synthesized with signal peptides similar to Sec-peptides except the carboxyl-terminal region is different. There, the lipoprotein signal sequences contain a four-residue motif (Leu^{-3}-Ala/Ser^{-2}-Gly/Ala^{-1}-Cys^{+1} at the −3 to +1 position) called the lipobox [136]. The invariant +1 Cys at the amino-terminus of the mature protein gets modified by diacylglycerol by the enzyme preprolipoprotein diacylglycerol transferase (Lgt) (Fig. 1) [136]. The lipid modified preprotein is then cleaved between the −1 and +1 amino acids by the signal peptidase 2 (Lsp) (also called lipoprotein signal peptidase) [137]. Following cleavage, the mature lipoprotein undergoes further maturation by the addition of a lipid at its amino-terminus. This later modification is performed by an N-acyl transferase (Lnt) [138].

4.1 Lol Transport Movement for Lipoproteins to the Outer Membrane

Lipoproteins typically have several fates after lipid modification. They can remain within the inner membrane or they are directed to the outer membrane after being extracted out of the inner membrane by the LolCDE transporter (Fig. 1). Although in *E. coli*, these outer membrane lipoproteins typically face the periplasmic space, in some bacteria, they cross the outer membrane to the exterior of the cell surface.

The Lol (*l*ocalization *o*f *l*ipoproteins) machinery directing lipoproteins to the outer membrane was discovered by Tokuda and coworkers. The Lol system is comprised of five proteins: LolA, LolB, LolC, LolD, and LolE (Fig. 1) [139–141]. LolA is localized in the periplasmic space and functions to escort the lipoprotein from the inner membrane to the outer membrane (Fig. 1) [139]. After transiting the periplasmic space, LolA delivers the lipoprotein to LolB protruding from the outer membrane [140]. This occurs via a mouth-to-mouth mechanism, as determined by photocrosslinking studies [142]. In the last step, the lipoprotein is inserted by partitioning of the triacylated lipid moiety into the periplasmic leaflet of the outer membrane.

LolCDE functions as an ABC transporter to extract the inner membrane lipoproteins destined for the outer membrane. In this complex, LolC and LolE are integral membrane subunits, each having a large periplasmic domain and 4 transmembrane spans (Fig. 1) [143]. LolD, located on the cytoplasmic side of the membrane as a dimer (Fig. 1), is the ATPase subunit of the transporter, and possesses the conserved Walker A and Walker B motifs. The structure of LolCDE [144–146] revealed that the protein has a V-shaped substrate-binding cavity open to the periplasm, which is created by the bent TM1 and TM2 of the LolC and LolE subunits at their interface [146]. The lipoprotein enters the V-shaped cavity with its three acyl chains located within two hydrophobic pockets [144–146]. The first three amino acids of the lipoprotein are within the substrate binding cavity and the remainder of the protein protrudes out of this region toward the LolE periplasmic domain. It is striking that the three acyl chains of the lipoprotein are already localized to the level of the periplasmic surface of the membrane. The structure of the LolCDE intermediate with bound ATP shows the two LolD subunits come together, leading to the collapse of the V-shaped cavity within LolCE [144–146]. This collapse of the cavity pushes the lipoprotein out of the cavity toward the LolE periplasmic domain [147]. Hydrolysis of ATP then leads to a conformational change that promotes the further movement of the lipoprotein from LolE to a LolC-bound LolA. Finally, incorporation of the lipoprotein into LolA causes dissociation of the protein from the inner membrane.

Insight into how LolA is recruited to the LolC component of the LolCDE transporter was provided by the structure of LolA bound to the LolC periplasmic domain [148], and verified by the cryoEM structure of a LolACDE complex by cryoEM [144]. LolA is recruited to LolC by a β-hairpin loop that functions as a hook and a pad region comprised of three surface residues. Recruitment of LolA to the LolC is facilitated by the binding of lipoprotein to the LolCDE complex [142] and does not depend on ATP binding or hydrolysis by the LolCDE complex [148]. The recruitment of LolA to LolC also enlarges the mouth of the LolA, priming it to bind to the lipid moiety of the lipoprotein substrate [148]. When it binds lipoprotein, the LolA mouth size increases even further and the three acyl chains are incorporated [149]. Because there is substantial overlap of the lipoprotein and the LolC binding site within the LolA cavity, it is expected that the binding of the lipoprotein to LolA, which is bound to LolC, would initiate the dissociation of LolA–lipoprotein complex from the LolCDE transporter complex. Lipoproteins that remain in the inner membrane contain a Lol avoidance sequence that is characterized with an aspartic acid at the +2 position and certain residues at +3 of the mature region of the protein [150, 151]. Lol substrates with the Lol avoidance signal do not enter the LolCDE V-shaped cavity [146].

4.2 Bam Pathway

Outer membrane β-barrel proteins are exported initially across the inner membrane by the Sec pathway (Fig. 1). In the periplasm, the Omps (outer membrane proteins) bind to chaperones that direct their transport to the outer membrane (Fig. 1). Chaperone SurA associates with unfolded proteins, escorting them to their destined outer membrane [152]. Another chaperone pathway that helps in the trafficking of Omps to the outer membrane involves the chaperones, Skp and DegP [153–155].

The *b*arrel *a*ssembly *m*achinery (BAM) catalyzes the insertion and folding of β-barrel proteins into the outer membrane (Fig. 1). In *E. coli*, Bam is comprised of 5 components [156, 157]. BamA, a β-barrel protein itself, and four lipoproteins, BamB, BamC, BamD, and BamE (Fig. 1). BamA and BamD are essential for bacterial cell viability and for β-barrel assembly in the outer membrane [156, 158–160]. While the BamB, BamC, and BamE subunits can be individually deleted without compromising cell viability, these proteins make the assembly of β-barrel proteins more efficient [156, 157, 161].

The essential BamA is a member of the Omp85 superfamily, whose members catalyze both the membrane insertion of β-barrel proteins into the outer membrane and the translocation across the outer membrane [162, 163]. BamA is characterized by a β-barrel at its C-terminus that is joined to five polypeptide-transporter associated (POTRA) domains located in the periplasm [163, 164]. The BamA barrel plays a crucial role in the membrane insertion of Omps and its POTRA domains function to form a scaffold for binding the Bam lipoprotein subunits [165, 166]. In addition, the POTRA domains are important for interacting with the escorted unfolded Bam substrate and the chaperone, SurA. Specifically, the BamA POTRA domain 1 acts as a receptor for SurA [167]. The BamD subunit has been shown to bind to substrate client proteins [168, 169].

A big advance in the field was made with the determination of the structure of BamA [170]. Strikingly, in one of the structures, the C-terminal β-strand 16 of the β-barrel was twisted and interacted with the β-strand 1 with only two hydrogen bonds [170]. This would allow this part of BamA barrel region to easily open, allowing the incoming polypeptide to insert into the interior of the BamA cavity to the outer membrane. The opening of the lateral gate/seam between the β-1 and β-16 strands is crucial for Bam function, since locking the two strands together by a disulfide bond inactivates BamA [171]. Another interesting finding is that BamA distorts the membrane around the β1-β16 lateral gate to cause significant thinning of the membrane, which may promote the folding of the Omps into the lipid phase [170].

Since the first structure of BamA appeared in 2013, there have been many structures of BamA and Bam complexes that have been determined by X-ray crystallography and cryo-EM in detergent or in nanodiscs [172–175]. More recently, structures of Bam-substrate folding intermediates have been determined, shedding light on the mechanism of β-barrel biogenesis [176–178] (see below). There are currently several models proposed for how BamA accelerates the folding of β-barrel proteins.

The BamA budding model [179, 180] hypothesizes that the substrates β-strands assemble at the β1/16 seam of BamA. For example, the BamA β-strand 1 binds to the incoming C-terminal β-signal of the nascent outer membrane protein (Omp). This allows the exposed β-strand 1 region of the BamA β-barrel to function as a template for catalyzing the formation of a hybrid BamA-Omp barrel via β-augmentation. Each added pair of strands then nucleate the folding of the next β-strand until the folding of the new Omp is terminated. The release of the new Omp is accomplished by the interaction of the substrate amino-terminal β-strand with its C-terminal β-signal after it is released from the β-strand 1 of BamA. This is favored because binding of its β-signal to its own first strand is of higher affinity because more hydrogen bonds can form [12, 176].

The swing model [181] proposes that the initial stages go by a budding model with the folding of the nascent protein occurring at the membrane interface, and then membrane integration of the partially folded β-barrel occurs. In this model, BamA facilitates the folding of an open β-sheet to a barrel-like structure in a mechanism that that is facilitated by outer-membrane tension [178].

The BamA-assisted model [171, 182, 183] proposes that BamA functions to destabilize and distort the adjacent membrane, allowing the incoming OmpA to insert into the membrane. Moreover, Bam lipoproteins themselves also cause lipid disordering and membrane destabilization [184].

5 Evolutionary Conserved Pathways in the ER, Mitochondria, and Chloroplasts

Eukaryotic cells are thought to have evolved from the symbiosis of an archaeal host cell with an alphaproteobacterial (mitochondrial) endosymbiont. This is why many of the proteins found in the ER (which is part of the endomembrane system) show similarities to proteins located in the plasma membrane of archaeal host cells, and mitochondrial proteins show similarity to those proteins found in Gram-negative bacteria [185]. Later in evolution, plant cells originated from a eukaryotic cell with mitochondria that then engulfed an ancient cyanobacterium. This explains why some of the prokaryotic-like biomachineries are found in the chloroplast envelope and thylakoid membrane [185].

5.1 ER

The principal translocation machinery in the ER is Sec61 that is a descendant of the archaeal Sec channel located in the plasma membrane (Fig. 4). The ER Sec61αβγ complex, like the archaeal SecYEβ, forms the channel where protein substrates move across. Sec61α is homologous to the SecY in archaea and bacteria; Secγ is related to the SecE in archaea and bacteria; Sec61β is similar to the Secβ in archaea but is unrelated to SecG in bacteria [186]. In addition to the Sec61 channel, there are the accessory components Sec62/Sec63/Sec71/Sec72, translocation-associated membrane protein (TRAM), and the lumenal ATPase BiP.

The eukaryotic SRP system that evolved from the archaeal SRP is an essential targeting pathway that couples the synthesis of nascent proteins with delivery to the ER membrane. The eukaryotic cytosolic SRP is composed of six protein subunits (Srp54, Srp19, Srp9/14, Srp68/72) and a 7S RNA [187] whereas the archaeal SRP is composed of Srp54, Srp19, and 7S RNA. The Srp54 performs the main function and binds the nascent chain and the SRP receptor. Srp54 is also homologous to the Ffh subunit in the *E. coli* SRP (Ffh and 4.5S RNA). The ER SRP receptor (SR) contains two subunits (SRα and SRβ) [188], of which SRα is homologous to the FtsY subunit in archaea and bacteria.

During cotranslational export into the ER lumen, the signal peptide of the nascent chain binds to SRP as it emerges from the ribosome. SRP targets the ribosome nascent chains (RNC), via its membrane associated SR, to the Sec translocon [189] where the hydrophilic region of the nascent exported protein or the hydrophilic domain of membrane proteins move through the channel cotranslationally as the protein is being synthesized in the ribosome. The TM segments of membrane proteins enter and exit the channel sideways through the lateral gate to integrate into the membrane bilayer. In some cases, TRAM facilitates membrane insertion [190, 191].

Alternatively, exported proteins can be translocated across the ER membrane post-translationally [192, 193]. These precursors contain a less hydrophobic signal peptide and are not recognized by SRP. They are targeted to the posttranslational protein translocation machinery, comprised of Sec61αβγ associated with the Sec62/63/71/72

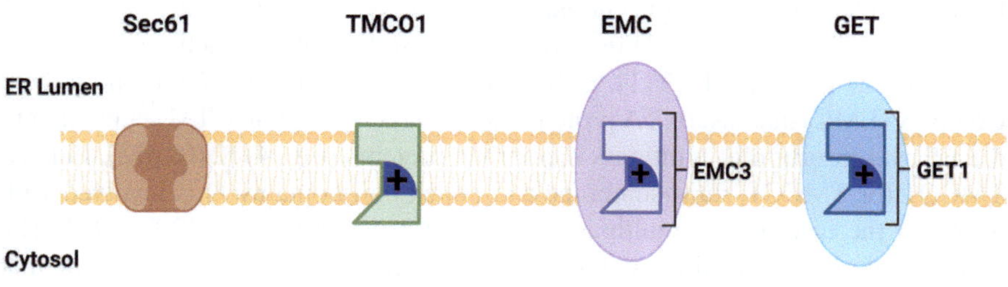

Fig. 4 Conserved translocases/insertases in the ER membrane. The principal machinery employed for translocation and insertion of proteins into the ER is the Sec61 complex. The GET insertase and the EMC complex play a role in the membrane insertion of tail anchored proteins. TMCO1 and EMC play a prominent role in the biogenesis of multispanning membrane proteins. Get1, TMCO1, and EMC3 are evolutionarily related to YidC and belong to the Oxa1 superfamily. (Created with BioRender.com)

complex [194–196]. Sec71 and Sec72 are important for binding exported proteins associated with the cytosolic chaperone, Hsp70 [197]. Sec63 activates the BiP ATPase residing in the ER lumen, allowing it to bind the polypeptide chain after the chain has moved through the Sec channel. The bound BiP prevents passive back sliding of the polypeptide into the cytoplasm. BiP binds the translocated substrates in the ER lumen and thereby functions as a molecular ratchet [196]. This is in contrast with the bacterial Sec system where the ATPase SecA functions to drive translocation through the channel starting with SecA on the cytosolic side of the membrane.

How does the post-translational protein translocation machinery recognize precursors and facilitate translocation? Structures of the Sec61 supercomplex [198, 199] revealed that a cytosolic domain in Sec63 blocks ribosome binding to the channel. The Sec71 and Sec72 subunits bind the translated precursor protein bound to the cytosolic Hsp70 chaperone. Sec63 is also predicted to be involved in the gating of the channel [198, 199]. It is positioned such that it supports the opening of the lateral gate and the central pore [199]. Furthermore, the Sec63 J domain is positioned such that BiP could bind the emerging polypeptide as it enters the lumenal side of the Sec channel [199].

The conserved signal peptidase complex (SPC) cleaves off signal peptides from a large number of preproteins during their translocation into the ER. Specifically, the Sec11a catalytic subunit of the ER SPC is homologous to the archaeal signal peptidase and weakly homologous to the bacterial signal peptidase 1 (also called leader peptidase) [200]. However, the ER SPC is more complex and contains five subunits: Spc12, Spc25, Spc22/23, Spc18 (Sec11a), and Spc21 (Sec11c) [201]. Interestingly, while the ER SPC and archaeal signal peptidase are serine proteases that employ a Ser-His-Asp motif, the bacterial signal peptidase uses a Ser-Lys dyad in catalysis [202–206].

The archaeal YidC homolog Ylp1 has very low sequence homology to the bacterial YidC members, but is similar to the homologs found in the ER. This shows that the archaeal gene was inherited in the ER system of the eukaryotic cell [207, 208]. Ylp1 and the ER homologs possess only 3 TM segments instead of the 5 TM segment core seen with bacterial YidC family members [72, 209].

The new ER family members (part of the Oxa1/YidC superfamily) are GET1, EMC3, and TMCO1 [207, 210]. In each case, they are subunits of a larger complex. GET1 (*G*uided *E*ntry of *T*ail-anchored proteins factor *1*) is a component of the GET complex (GET1/GET2/GET3) involved in the insertion of tail-anchored (TA) proteins into the ER (Fig. 4) [211]. EMC3 is a subunit of the *E*R *M*embrane protein *C*omplex (EMC) that inserts TA proteins, with a less hydrophobic TM segment, into the ER and insertion of multipass membrane proteins (Fig. 4) [210, 212]. Finally, TMCO1 (or TMCC1; *t*rans*m*embrane and *c*oiled-*c*oil domains *1*) is involved with Sec61 in the insertion of multipass membrane proteins (Fig. 4) [207, 212]. The structure of all three of these complexes have been solved recently and show that these YidC homologs (GET3, EMC1 and TMCO1) possess a hydrophilic groove that is open to the membrane and cytosol, like YidC. The groove is formed by 3 TM domains that are related to TM2, TM3, and TM6 in the Gram-negative YidC [211]. This groove provides an opening for hydrophilic regions of the substrate to cross the membrane. In addition, the *E. coli* YidC and the EMC3 protein thin the membrane bilayer by 5–10 Å. [213, 214].

5.2 Mitochondria

Since the origin of eukaryotes over a billion years ago, when an alphaproteobacterium was taken up by an archaeal cell and converted to mitochondria, much of the genome of the endosymbiont was transferred to the host and some was discarded. Therefore, new protein machineries were required that allowed the mitochondria to import more than 1000 different proteins from the cytoplasm [215]. Nowadays, only a fraction of proteins is encoded by the mitochondrial genome.

The translocase in the outer membrane (TOM) functions in the import of proteins into the mitochondria (Fig. 5a) [216–218]. At the inner membrane, there are two translocases that function in transport [218, 219]. The TIM23 is the major inner membrane translocase that transports proteins into the matrix and inserts a subset of membrane proteins into the inner membrane. TIM22 primarily functions to insert carrier proteins, which span the membrane six times, into the inner membrane. In mitochondria, protein traffic is mainly directed inward into the organelle instead of outward from the cytoplasm toward the periplasmic space and outer membrane, as found in Gram-negative bacteria.

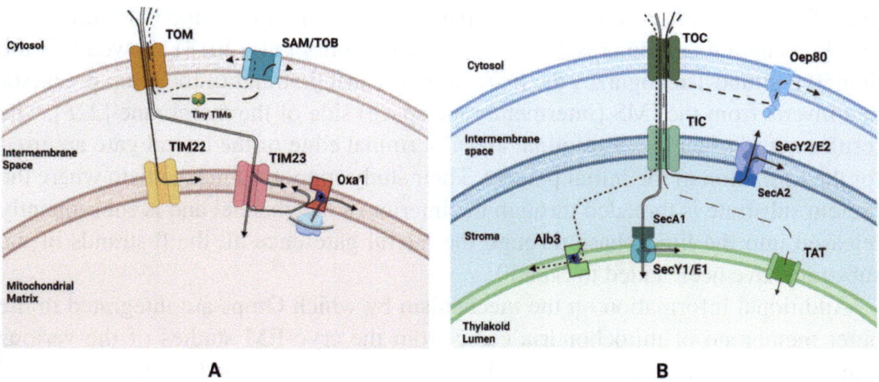

Fig. 5 Conserved machinery involved in membrane insertion and protein export in mitochondria and chloroplasts. (**a**) Proteins are imported into the matrix by the TOM/TIM23 complex (double line). Proteins are imported into mitochondrial intermembrane space by the TOM complex comprised of Tom40, Tom70, Tom20, and Tom22. β-barrel proteins interact with the tiny TIMs and are then inserted and assembled into the outer membrane by the SAM/TOB complex. TIM22 mediates the insertion of carrier proteins into the inner membrane while TIM23 mediates the insertion of a wide variety of inner membrane proteins. A subset of proteins that are imported into the matrix are inserted into the inner membrane by the concerted action of TIM23 and Oxa1. TIM23 is comprised of Tim23, Tim17, Tim44, Tim50, Tim21, and mitochondrial Hsp70. Mitochondrial encoded proteins are inserted into the inner membrane by the Oxa1 insertase. (**b**) The TOC/TIC complexes import proteins from the cytoplasm into the chloroplast stroma. The TOC complex is comprised of Toc75, Toc 159, and Toc34, while the TIC complex consists of Tic20, Tic214, Tic100, and Tic 56, among other subunits. The conserved Oep80 protein is a candidate for insertion of β-barrel proteins into the outer membrane. A subset of proteins is inserted into the chloroplast inner membrane by the SecY2/SecE2/SecA2 machinery after import into the stroma compartment. In the thylakoid membrane, several conserved components (TatABC/SecYE1/Alb3) are employed for protein insertion, and protein translocation into the thylakoid lumen. (Created with BioRender.com)

There are also some of the bacteria-like transport machineries inherited from the prokaryotic ancestor of mitochondria (Bam and YidC) that were widely retained in the existing mitochondria. For example, β-barrel proteins are assembled into the mitochondrial outer membrane by the sorting and assembly machinery (SAM/TOB) complex (Fig. 5a) [219–221]. Sam50 (Tob55) of this complex is homologous to the bacterial BamA that mediates the same function in Gram-negative bacteria (insertion of β-barrel proteins into the outer membrane) [219, 222, 223]. Sam50, like the bacterial homolog BamA, possesses a 16-stranded β-barrel and contains one POTRA domain localized within the mitochondrial intermembrane space compared to the five POTRA domains of BamA [164]. Surprisingly, the other subunits, Sam35 (Tob38), and Sam37 (Tob37), are located on the cytoplasmic side of the mitochondrial outer membrane. In bacteria, the other Bam subunits are localized on the periplasmic side of the outer membrane [224]. While Sam50 and Sam35 are essential for cell viability and β-barrel assembly, Sam37 is not [224].

Most nuclear-encoded Omps of mitochondria are first imported from the cytosol across the outer membrane of mitochondria by the TOM complex (Fig. 5a) [219, 220, 225]. In the intermembrane space, they interact with the tiny TIMs prior to engaging Sam50 of the Sam complex [226]. The lateral β-1/16 seam of Sam50 comprised of β-strands 1 and 16 in the barrel structure of the Sam50 from the mitochondria of *Myceliophthora thermophila* opens to allow binding of the Omp precursor [227]. Using a disulfide crosslinking approach, Hohr et al. (2018) showed that the Sam50 β-strand 1 recognizes the β-signal in the last β-strand of the Omp precursor that inserts from the IMS (intermembrane space) side of the membrane [228]. The results suggest that the N-terminal and C-terminal edge of the lateral gate are used for the formation of the initial β-sheet. Their study supports a mechanism where the protein substrate is threaded through the interior of the channel and is subsequently released into the lipid phase through the lateral gate once all the β-strands of the substrate have been added to Sam50.

Additional information on the mechanism by which Omps are integrated in the outer membrane of mitochondria came from the cryo-EM studies of the various dimeric complexes of Sam50, including one in which the second Sam50 is replaced with the closed β-barrel Mdm10 [229]. Mdm10 is a component of the ER mitochondria encounter structure (ERMES), which tethers the ER to the mitochondria [230]. Also, a substrate-loaded Sam50 complex that lacked a second Sam50 or Mdm10 was isolated. Based on these findings, a β-barrel switching mechanism model was proposed for how Sam50 integrates Omps into the outer membrane [229]. In this model, only one of the Sam50 protomers forms new β-barrel proteins while the second β-barrel protein is a placeholder. As the new Omp is threaded through the Sam50 with the open lateral gate, the growing substrate displaces the second placeholder subunit. A structure of a fully loaded Tom40 substrate bound to a single Sam complex supports this β-barrel switching model [231].

In addition to Sam50/BamA, homologs of the bacterial YidC play prominent roles in membrane insertion in mitochondria in eukaryotic cells. Indeed, the first discovery of the YidC/Oxa1/Alb family was made by scientists studying mitochondrial membrane biogenesis [232, 233]. The YidC homolog, Oxa1, can bind mitochondrial ribosomes and thereby mediate the cotranslational insertion of mitochondrial-encoded proteins from the matrix into the inner membrane (Fig. 5a) [234, 235] and also plays a role for some mitochondrial inner membrane proteins that are imported from the cytosol into the matrix and then exported into the inner membrane. This includes ABC transporters and several metabolite carriers [236–240]. The combined work shows that Oxa1 plays a general role for the inner membrane biogenesis in mitochondria. A second homolog Oxa2 (Cox18) is present in the inner membrane of mitochondria and is required for the translocation of the large C-terminal domain of the Cox2 inner membrane protein from the matrix [241].

In existing mitochondria, a Tat-like pathway is found only occasionally. For example, Tat components have been found in the mitochondrial genome of eukaryotes ranging from jakobid protists, plants, and some early animals, such as sponges [242–244]. Most of the organisms that retain a Tat component encode TatC by the mitochondrial genome, while only some mitochondrial genomes from jakobids, ochrophytes, and cryptophytes also encode TatA. Interestingly, the mitochondrial encoded TatAC from a jakobid lineage can functionally complement the Tat translocase in *E. coli* and transport a Tat substrate [245]. In *Arabidopsis thaliana*, it was shown that there is a TatC-like protein encoded by the mitochondrial genome and a nuclear encoded TatB-like subunit is localized to the mitochondrial inner membrane [243]. This TatBC pathway is required to translocate the folded Fe-S protein Rip1 across the inner membrane to the intermembrane space, and is required for Complex III biogenesis [246]. This study showed for the first time that a Tat pathway is operational in plant mitochondria.

The SecY translocase is rarely present in mitochondria [242, 247]. It is found in the Jakobid protist that encodes over 60 proteins in its mitochondrial genome [247–249]. The jakobid *Reclinomonas americana* mitochondrial genome also encodes a Cox11 protein that possesses a Sec-type signal peptide at its N-terminus. This assembly factor is a candidate SecY substrate that could be exported from the mitochondrial matrix to the mitochondrial intermembrane space. Export of Cox11 by SecY is predicted to be important for the assembly of the cytochrome oxidase [250]. It is puzzling, however, that no SecE or SecG gene has been observed in the mitochondrial genome, nor have they been detected to be encoded in the nuclear genome. Additionally, one would also expect that a SecA protein would be present in the eukaryote to energize Sec-dependent export.

Remarkably, some closely unrelated eukaryotes have homologs of core components of the bacterial type 2 secretion system (TTSS) [251]. In Gram-negative bacteria, the double-membrane spanning TTSS secretes periplasmically-located cargo across the outer membrane in a folded state. The TTSS is composed of 12–15 different types of proteins [252, 253]. While it is not clear whether the hypothetical mitochondrial T2SS is functional to export proteins out of mitochondria from the intermembrane space into the cytosol, the continued presence of key core components suggests it may be playing a valuable role in the export process.

Lastly, a bacterial-like signal peptidase is found in mitochondria, which cleaves signal peptides from mitochondrial precursor proteins. Again, this supports the idea that mitochondria arose from an endosymbiotic bacterium. Imp1 and Imp2, which are involved in precursor processing, are highly homologous to the bacterial signal peptidase and are Ser-Lys serine proteases [200, 254]. Imp1/Imp2 reside in the inner membrane with their catalytic domain protruding into the IMS, analogous to the inner-membrane bacterial signal peptidase 1 with the active site in the periplasmic space.

5.3 Chloroplasts

Chloroplasts are similar to mitochondria in that a vast majority of the genes, following the initial formation of ancestral endosymbiont, were relocated to the host nucleus and the existing chloroplast genomes encode only a small percentage of the proteins. Therefore, they have to import thousands of proteins into the chloroplasts. Translocases in the outer chloroplast membrane (TOC) and translocases in the inner chloroplast membrane (TIC) evolved to take on this function (Fig. 5b) [255–257]. Recently, a huge advance in this area was the determination of the structure of the TOC–TIC (*t*ranslocon on the *o*uter/*i*nner *c*hloroplast membrane) supercomplex [258, 259].

The majority of the imported nuclear-encoded chloroplast proteins are destined to the stroma compartment [260]. Such proteins are synthesized with presequences containing a transit peptide that guide them across the outer and inner membranes of the chloroplast into the stroma where the presequence is proteolytically removed by a stroma-processing peptidase (SPP) [261]. Nuclear-encoded chloroplast proteins that are transported into the thylakoid lumen of the chloroplasts contain presequences with a bipartitite structure [262]. The amino-terminal stroma-targeting transit peptide is used to guide such proteins across the chloroplast envelope into the stroma where it is cleaved by SPP. The second region that directs proteins across the thylakoid membrane, are of two types; a Tat signal peptide containing twin arginines that targets the protein to the Tat machinery, or a Sec signal peptide that directs proteins to the Sec machinery [263]. In both cases, the signal peptide is removed by a thylakoid processing peptidase (TPP), homologous to the bacterial type 1 signal peptidase [200].

Like in mitochondria, there are Omp85 homologs similar to the bacterial BamA (Toc75 and Oep80) in the outer membrane of chloroplasts (Fig. 5b). Toc75 is a component of the TOC protein import machinery, and forms the channel in the outer membrane [258]. While Toc75 functions in the translocation of chloroplast proteins across the outer membrane, Oep80 is a candidate for inserting β-barrel proteins into the outer membrane (Fig. 5b) [264–266]. Both Toc75 and Oep80 possess a 16-stranded β-barrel and 3 POTRA domains located in the IMS [267–269]. The POTRA domain from the *Arabidopsis thaliana* Toc75 shows a chaperone-like function and promotes chloroplast protein import [270]. The Toc75 was shown to have a contact site with inner membrane, Tic22, which is a component of the inner membrane protein import machinery [265, 266].

Interestingly, green plants contain a bacterial-like TamB homolog (Dek5), which is required for envelope biogenesis [271]. It may in fact work with the Omp85 family member, Oep80, which is likely involved in the assembly of β-barrel proteins in the chloroplast outer membrane; this will require further experiments to prove. In many Gram-negative bacteria, TamB functions together with the BamA-related TamA protein (*t*ranslocation and *a*ssembly *m*odule subunit *A*) to insert some β-barrel proteins into the membrane. TamB is an inner membrane protein with a large periplasmic domain that spans the entire periplasm to promote assembly of certain outer membrane proteins [272–276]. While TamA homologs are restricted to proteobacteria and a few other bacteria, TamB homologs are widely distributed in Gram-negative bacteria, including cyanobacteria [277].

SecY and SecE homologs are present in the chloroplast inner membrane [278–280]. The Sec translocation system is comprised of SecY2 (SCY2) and SecE2, respectively, and works with a stroma localized SecA2 to insert candidate proteins into the inner envelope membrane (Fig. 5b). Tic40 and FtsH12 have been identified as substrates [279]. Note that these proteins are nuclear-encoded and imported first into the stroma via TIC/TOC prior to engaging the SecY2/SecE2/SecA2 translocation machinery for inner membrane insertion.

In addition, prokaryotic-like translocases play important roles in the chloroplast protein transport pathways for insertion into or across the thylakoid membrane. Chloroplasts contain a thylakoid SecY1 (SecY homolog) and SecE1, and a stroma localized SecA1 (Fig. 5b). The SecY1/SecE1 complex [281] is energized by SecA1 [282, 283] to transport Sec-dependent cargo from the stroma to the thylakoid lumen [284]. Translocation requires ATP hydrolysis and is stimulated by the proton motive force.

Three Tat translocase components are present: TatA (Tha4), TatB (Hcf106), and TatC (Fig. 5b) [285–289]. Like the bacterial Tat system, proteins exported by this pathway are synthesized with a signal peptide with twin arginines [290–292] and can be transported in a folded state [293, 294]. Export of Tat proteins into the thylakoid lumen requires a pH gradient, leading some scientists originally studying this pathway to call it the delta pH pathway [291].

There are two YidC homologs in the thylakoid membrane: Alb3 and Alb4 (Fig. 5b; Alb4 not shown). Alb3 plays an important role for the post-translational insertion of the *l*ight-*h*arvesting *c*hlorophyll-binding *p*rotein (LHCPs) into the thylakoid membrane [295]. LHCP is a nuclear-encoded protein and is imported first into the stroma compartment. Interestingly, this post-translational insertion into the thylakoid membrane also requires SRP targeting factors. LHCP imported into the stroma forms a complex with the cpSRP43/cpSRP54 [297]. After interaction with the SRP receptor at the thylakoid membrane, the complex is transferred to Alb3 for membrane insertion [297]. In addition, Alb3 cooperates with the SecY1/SecE1 system, and may play a role in the insertion/folding of cpSec1 inserted substrates [298]. Indeed, an intermediate in the membrane insertion process of the photosystem II protein D1 was found in close proximity to the cpSecYE1/Alb3 machinery in the thylakoid membrane [299]. The Alb4 homolog has been implicated in the assembly of the chloroplast F1Fo ATP synthase [300].

6 Conclusion

In bacteria, remarkable advances have been made in our understanding of the biomachineries that insert proteins into the inner membrane (Sec, YidC), transport proteins across the inner membrane (Sec, Tat), integrate β-barrel proteins into the outer membrane (Bam), and sort lipoproteins between the inner and outer membrane (Lol). So many impressive structures of these machines, especially during the last decade by cryo-EM, have provided a clearer picture of the molecular events that occur in the insertion/translocation/sorting processes.

In this review, we also highlighted the biomachineries that found their way into eukaryotic cells through evolution and are now playing important roles in the ER, mitochondria, and chloroplasts. Bacterial-like homologs are present in mitochondria (Sam50, Oxa1, and TatABC), and chloroplasts (Oep80, SecYE2, SecYE1, Alb3, Alb4, TatABC), which function for transport across and into certain membranes. Additionally, ancient archaeal homologs are present in the ER (Sec61αβγ, and Get1, Emc3, and TmcO1) and they play crucial roles in membrane protein transport/insertion. The future in the transport field will be exciting as we discover more about how cells adapted the bacterial/archaeal protein transport mechanisms during the course of evolution.

Acknowledgments This work was supported by a National Science Foundation Grant MCB-1814936 to R.E.D.

References

1. Nikaido H (1988) Structure and functions of the cell envelope of gram-negative bacteria. Rev Infect Dis 10(Suppl 2):S279–S281
2. Shanmugam SK, Dalbey RE (2019) The conserved role of YidC in membrane protein biogenesis. Microbiol Spectr 7(1). https://doi.org/10.1128/microbiolspec.psib-0014-2018
3. Oswald J, Njenga R, Natriashvili A, Sarmah P, Koch HG (2021) The dynamic SecYEG translocon. Front Mol Biosci 8:664241. https://doi.org/10.3389/fmolb.2021.664241
4. Tsirigotaki A, De Geyter J, Sostaric N, Economou A, Karamanou S (2017) Protein export through the bacterial Sec pathway. Nat Rev Microbiol 15(1):21–36. https://doi.org/10.1038/nrmicro.2016.1
5. Cymer F, von Heijne G, White SH (2015) Mechanisms of integral membrane protein insertion and folding. J Mol Biol 427(5):999–1022. https://doi.org/10.1016/j.jmb.2014.09.014
6. Hegde RS, Keenan RJ (2021) The mechanisms of integral membrane protein biogenesis. Nat Rev Mol Cell Biol 23:107. https://doi.org/10.1038/s41580-021-00413-2
7. Palmer T, Stansfeld PJ (2020) Targeting of proteins to the twin-arginine translocation pathway. Mol Microbiol 113(5):861–871. https://doi.org/10.1111/mmi.14461
8. Okuda S, Tokuda H (2011) Lipoprotein sorting in bacteria. Annu Rev Microbiol 65:239–259. https://doi.org/10.1146/annurev-micro-090110-102859
9. Zuckert WR (2014) Secretion of bacterial lipoproteins: through the cytoplasmic membrane, the periplasm and beyond. Biochim Biophys Acta 1843(8):1509–1516. https://doi.org/10.1016/j.bbamcr.2014.04.022

10. Konovalova A, Kahne DE, Silhavy TJ (2017) Outer membrane biogenesis. Annu Rev Microbiol 71:539–556. https://doi.org/10.1146/annurev-micro-090816-093754
11. Noinaj N, Gumbart JC, Buchanan SK (2017) The beta-barrel assembly machinery in motion. Nat Rev Microbiol 15(4):197–204. https://doi.org/10.1038/nrmicro.2016.191
12. Tomasek D, Kahne D (2021) The assembly of beta-barrel outer membrane proteins. Curr Opin Microbiol 60:16–23. https://doi.org/10.1016/j.mib.2021.01.009
13. Crane JM, Randall LL (2017) The Sec system: protein export in *Escherichia coli*. EcoSal Plus 7(2):10.1128/ecosalplus.ESP-0002-2017. https://doi.org/10.1128/ecosalplus.esp-0002-2017
14. Komarudin AG, Driessen AJM (2019) SecA-mediated protein translocation through the SecYEG channel. Microbiol Spectr 7(4). https://doi.org/10.1128/microbiolspec.psib-0028-2019
15. Corey RA, Allen WJ, Collinson I (2016) Protein translocation: what's the problem? Biochem Soc Trans 44(3):753–759. https://doi.org/10.1042/BST20160047
16. Frain KM, Robinson C, van Dijl JM (2019) Transport of folded proteins by the Tat system. Protein J 38(4):377–388. https://doi.org/10.1007/s10930-019-09859-y
17. Berks BC (2015) The twin-arginine protein translocation pathway. Annu Rev Biochem 84:843–864. https://doi.org/10.1146/annurev-biochem-060614-034251
18. Saraogi I, Shan SO (2014) Co-translational protein targeting to the bacterial membrane. Biochim Biophys Acta 1843(8):1433–1441. https://doi.org/10.1016/j.bbamcr.2013.10.013
19. Steinberg R, Knupffer L, Origi A, Asti R, Koch HG (2018) Co-translational protein targeting in bacteria. FEMS Microbiol Lett 365(11). https://doi.org/10.1093/femsle/fny095
20. Miller JD, Bernstein HD, Walter P (1994) Interaction of E. coli Ffh/4.5S ribonucleoprotein and FtsY mimics that of mammalian signal recognition particle and its receptor. Nature 367(6464):657–659. https://doi.org/10.1038/367657a0
21. Lee HC, Bernstein HD (2001) The targeting pathway of *Escherichia coli* presecretory and integral membrane proteins is specified by the hydrophobicity of the targeting signal. Proc Natl Acad Sci U S A 98(6):3471–6. https://doi.org/10.1073/pnas.051484198
22. Parlitz R, Eitan A, Stjepanovic G, Bahari L, Bange G, Bibi E, Sinning I (2007) *Escherichia coli* signal recognition particle receptor FtsY contains an essential and autonomous membrane-binding amphipathic helix. J Biol Chem 282(44):32176–32184. https://doi.org/10.1074/jbc.M705430200
23. Pross E, Soussoula L, Seitl I, Lupo D, Kuhn A (2016) Membrane targeting and insertion of the C-Tail Protein SciP. J Mol Biol 428(20):4218–4227. https://doi.org/10.1016/j.jmb.2016.09.001
24. de Gier JW, Scotti PA, Saaf A, Valent QA, Kuhn A, Luirink J, von Heijne G (1998) Differential use of the signal recognition particle translocase targeting pathway for inner membrane protein assembly in *Escherichia coli*. Proc Natl Acad Sci U S A 95(25):14646–14651. https://doi.org/10.1073/pnas.95.25.14646
25. Wang S, Yang CI, Shan SO (2017) SecA mediates cotranslational targeting and translocation of an inner membrane protein. J Cell Biol 216(11):3639–3653. https://doi.org/10.1083/jcb.201704036
26. Rawat S, Zhu L, Lindner E, Dalbey RE, White SH (2015) SecA drives transmembrane insertion of RodZ, an unusual single-span membrane protein. J Mol Biol 427(5):1023–1037. https://doi.org/10.1016/j.jmb.2014.05.005
27. Saparov SM, Erlandson K, Cannon K, Schaletzky J, Schulman S, Rapoport TA, Pohl P (2007) Determining the conductance of the SecY protein translocation channel for small molecules. Mol Cell 26(4):501–509. https://doi.org/10.1016/j.molcel.2007.03.022
28. Duong F, Wickner W (1997) Distinct catalytic roles of the SecYE, SecG and SecDFyajC subunits of preprotein translocase holoenzyme. EMBO J 16(10):2756–2768. https://doi.org/10.1093/emboj/16.10.2756
29. Scotti PA, Urbanus ML, Brunner J, de Gier JW, von Heijne G (2000) YidC, the *Escherichia coli* homologue of mitochondrial Oxa1p, is a component of the Sec translocase. EMBO J 19:542. https://doi.org/10.1093/emboj/19.4.542

30. Arkowitz RA, Wickner W (1994) SecD and SecF are required for the proton electrochemical gradient stimulation of preprotein translocation. EMBO J 13(4):954–963. https://doi.org/10.1002/j.1460-207
31. Cranford-Smith T, Huber D (2018) The way is the goal: how SecA transports proteins across the cytoplasmic membrane in bacteria. FEMS Microbiol Lett 365(11):fny093. https://doi.org/10.1093/femsle/fny093
32. Chatzi KE, Sardis MF, Economou A, Karamanou S (2014) SecA-mediated targeting and translocation of secretory proteins. Biochim Biophys Acta 1843(8):1466–1474. https://doi.org/10.1016/j.bbamcr.2014.02.014
33. Collinson I, Corey RA, Allen WJ (2015) Channel crossing: how are proteins shipped across the bacterial plasma membrane? Philos Trans R Soc Lond B Biol Sci 370(1679):20150025. https://doi.org/10.1098/rstb.2015.0025
34. Schulze RJ, Komar J, Botte M, Allen WJ, Whitehouse S, Gold VA, Lycklama ANJA, Huard K, Berger I, Schaffitzel C, Collinson I (2014) Membrane protein insertion and proton-motive-force-dependent secretion through the bacterial holo-translocon SecYEG-SecDF-YajC-YidC. Proc Natl Acad Sci U S A 111(13):4844–4849. https://doi.org/10.1073/pnas.1315901111
35. van den Berg B, Clemons WM, Collinson I, Modis Y, Hartmann E (2004) X-ray structure of a protein-conducting channel. Nature 427:36. https://doi.org/10.1038/nature02218
36. Clemons WM Jr, Menetret JF, Akey CW, Rapoport TA (2004) Structural insight into the protein translocation channel. Curr Opin Struct Biol 14(4):390–396. https://doi.org/10.1016/j.sbi.2004.07.006
37. Harris CR, Silhavy TJ (1999) Mapping an interface of SecY (PrlA) and SecE (PrlG) by using synthetic phenotypes and in vivo cross-linking. J Bacteriol 181(11):3438–3444. https://doi.org/10.1128/jb.181.11.3438-3444.1999
38. Nicolaus F, Metola A, Mermans D, Liljenstrom A, Krc A, Abdullahi SM, Zimmer M, Miller Iii TF, von Heijne G (2021) Residue-by-residue analysis of cotranslational membrane protein integration in vivo. Elife 10:e64302. https://doi.org/10.7554/eLife.64302
39. Skach WR, Lingappa VR (1993) Amino-terminal assembly of human P-glycoprotein at the endoplasmic reticulum is directed by cooperative actions of two internal sequences. J Biol Chem 268(31):23552–23561. https://doi.org/10.1016/S0021-9258(19)49498-2
40. Heinrich SU, Rapoport TA (2003) Cooperation of transmembrane segments during the integration of a double-spanning protein into the ER membrane. EMBO J 22(14):3654–3663. https://doi.org/10.1093/emboj/cdg346
41. Kuhn A, Stuart R, Henry R, Dalbey RE (2003) The Alb3/Oxa1/YidC protein family: membrane-localized chaperones facilitating membrane protein insertion? Trends Cell Biol 13(10):510–516. https://doi.org/10.1016/j.tcb.2003.08.005
42. Dalbey RE, Kuhn A (2004) YidC family members are involved in the membrane insertion, lateral integration, folding, and assembly of membrane proteins. J Cell Biol 166:769. https://doi.org/10.1083/jcb.200405161
43. Kol S, Nouwen N, Driessen AJ (2008) Mechanisms of YidC-mediated insertion and assembly of multimeric membrane protein complexes. J Biol Chem 283(46):31269. https://doi.org/10.1074/jbc.R800029200
44. Hennon SW, Soman R, Zhu L, Dalbey RE (2015) YidC/Alb3/Oxa1 Family of Insertases. J Biol Chem 290(24):14866–14874. https://doi.org/10.1074/jbc.R115.638171
45. Zhu L, Kaback HR, Dalbey RE (2013) YidC protein, a molecular chaperone for LacY protein folding via the SecYEG protein machinery. J Biol Chem 288(39):28180–28194. https://doi.org/10.1074/jbc.M113.491613
46. Nagamori S, Smirnova IN, Kaback HR (2004) Role of YidC in folding of polytopic membrane proteins. J Cell Biol 165(1):53–62. https://doi.org/10.1083/jcb.200402067
47. Beck K, Eisner G, Trescher D, Dalbey RE, Brunner J, Muller M (2001) YidC, an assembly site for polytopic *Escherichia coli* membrane proteins located in immediate proximity to the SecYE translocon and lipids. EMBO Rep 2(8):709–714. https://doi.org/10.1093/embo-reports/kve154

48. Wagner S, Pop O, Haan GJ, Baars L, Koningstein G, Klepsch MM, Genevaux P, Luirink J, de Gier JW (2008) Biogenesis of MalF and the MalFGK(2) maltose transport complex in *Escherichia coli* requires YidC. J Biol Chem 283(26):17881–17890. https://doi.org/10.1074/jbc.M801481200
49. Yi L, Jiang F, Chen M, Cain B, Bolhuis A, Dalbey RE (2003) YidC is strictly required for membrane insertion of subunits a and c of the F(1)F(0)ATP synthase and SecE of the SecYEG translocase. Biochemistry 42(35):10537–10544. https://doi.org/10.1021/bi034309h
50. Yi L, Celebi N, Chen M, Dalbey RE (2004) Sec/SRP requirements and energetics of membrane insertion of subunits a, b, and c of the *Escherichia coli* F1F0 ATP synthase. J Biol Chem 279(38):39260–39267. https://doi.org/10.1074/jbc.M405490200
51. Celebi N, Yi L, Facey SJ, Kuhn A, Dalbey RE (2006) Membrane biogenesis of subunit II of cytochrome bo oxidase: contrasting requirements for insertion of N-terminal and C-terminal domains. J Mol Biol 357(5):1428–1436. https://doi.org/10.1016/j.jmb.2006.01.030
52. Zhu L, Klenner C, Kuhn A, Dalbey RE (2012) Both YidC and SecYEG are required for translocation of the periplasmic loops 1 and 2 of the multispanning membrane protein TatC. J Mol Biol 424(5):354–367. https://doi.org/10.1016/j.jmb.2012.09.026
53. Welte T, Kudva R, Kuhn P, Sturm L, Braig D, Muller M, Warscheid B, Drepper F, Koch HG (2012) Promiscuous targeting of polytopic membrane proteins to SecYEG or YidC by the *Escherichia coli* signal recognition particle. Mol Biol Cell 23(3):464–479. https://doi.org/10.1091/mbc.e11-07-0590
54. Samuelson JC, Chen M, Jiang F, Moller I, Wiedmann M, Kuhn A, Phillips GJ, Dalbey RE (2000) YidC mediates membrane protein insertion in bacteria. Nature 406(6796):637–641
55. Chen M, Samuelson JC, Jiang F, Muller M, Kuhn A, Dalbey RE (2002) Direct interaction of YidC with the Sec-independent Pf3 coat protein during its membrane protein insertion. J Biol Chem 277(10):7670–7675. https://doi.org/10.1074/jbc.M110644200
56. Serek J, Bauer-Manz G, Struhalla G, van den Berg L, Kiefer D (2004) *Escherichia coli* YidC is a membrane insertase for Sec-independent proteins. EMBO J 23:294. https://doi.org/10.1038/sj.emboj.76
57. Van Der Laan M, Bechtluft P, Kol S, Nouwen N, Driessen AJ (2004) F1F0 ATP synthase subunit c is a substrate of the novel YidC pathway for membrane protein biogenesis. J Cell Biol 165(2):213–222. https://doi.org/10.1083/jcb.200402100
58. van Bloois E, Jan Haan G, de Gier JW, Oudega B, Luirink J (2004) F(1)F(0) ATP synthase subunit c is targeted by the SRP to YidC in the *E. coli* inner membrane. FEBS Lett 576(1–2):97–100. https://doi.org/10.1016/j.febslet.2004.08.069
59. Facey SJ, Neugebauer SA, Krauss S, Kuhn A (2007) The mechanosensitive channel protein MscL is targeted by the SRP to the novel YidC membrane insertion pathway of *Escherichia coli*. J Mol Biol 365(4):995–1004. https://doi.org/10.1016/j.jmb.2006.10.083
60. Aschtgen MS, Zoued A, Lloubes R, Journet L, Cascales E (2012) The C-tail anchored TssL subunit, an essential protein of the enteroaggregative *Escherichia coli* Sci-1 Type VI secretion system, is inserted by YidC. Microbiologyopen 1(1):71–82. https://doi.org/10.1002/mbo3.9
61. Zhu L, Wasey A, White SH, Dalbey RE (2013) Charge composition features of model single-span membrane proteins that determine selection of YidC and SecYEG translocase pathways in *Escherichia coli*. J Biol Chem 288(11):7704–7716. https://doi.org/10.1074/jbc.M112.429431
62. Soman R, Yuan J, Kuhn A, Dalbey RE (2014) Polarity and charge of the periplasmic loop determine the YidC and sec translocase requirement for the M13 procoat lep protein. J Biol Chem 289(2):1023–1032. https://doi.org/10.1074/jbc.M113.522250
63. Shanmugam SK, Backes N, Chen Y, Belardo A, Phillips GJ, Dalbey RE (2019) New Insights into amino-terminal translocation as revealed by the use of YidC and Sec depletion strains. J Mol Biol 431(5):1025–1037. https://doi.org/10.1016/j.jmb.2019.01.006
64. Hariharan B, Pross E, Soman R, Kaushik S, Kuhn A, Dalbey RE (2021) Polarity/charge as a determinant of translocase requirements for membrane protein insertion. Biochim Biophys Acta Biomembr 1863(2):183502. https://doi.org/10.1016/j.bbamem.2020.183502

65. Kumazaki K, Chiba S, Takemoto M, Furukawa A, Nishiyama K, Sugano Y, Mori T, Dohmae N, Hirata K, Nakada-Nakura Y, Maturana AD, Tanaka Y, Mori H, Sugita Y, Arisaka F, Ito K, Ishitani R, Tsukazaki T, Nureki O (2014) Structural basis of Sec-independent membrane protein insertion by YidC. Nature 509(7501):516–520. https://doi.org/10.1038/nature13167
66. Kumazaki K, Kishimoto T, Furukawa A, Mori H, Tanaka Y, Dohmae N, Ishitani R, Tsukazaki T, Nureki O (2014) Crystal structure of *Escherichia coli* YidC, a membrane protein chaperone and insertase. Sci Rep 4:7299. https://doi.org/10.1038/srep07299
67. Xin Y, Zhao Y, Zheng J, Zhou H, Zhang XC, Tian C, Huang Y (2018) Structure of YidC from Thermotoga maritima and its implications for YidC-mediated membrane protein insertion. FASEB J 32(5):2411–2421. https://doi.org/10.1096/fj.201700893RR
68. Tanaka Y, Izumioka A, Abdul Hamid A, Fujii A, Haruyama T, Furukawa A, Tsukazaki T (2018) 2.8-A crystal structure of *Escherichia coli* YidC revealing all core regions, including flexible C2 loop. Biochem Biophys Res Commun 505(1):141–145. https://doi.org/10.1016/j.bbrc.2018.09.043
69. Shimokawa-Chiba N, Kumazaki K, Tsukazaki T, Nureki O, Ito K, Chiba S (2015) Hydrophilic microenvironment required for the channel-independent insertase function of YidC protein. Proc Natl Acad Sci U S A 112(16):5063–5068. https://doi.org/10.1073/pnas.1423817112
70. He H, Kuhn A, Dalbey RE (2020) Tracking the Stepwise Movement of a Membrane-inserting Protein In Vivo. J Mol Biol 432(2):484–496. https://doi.org/10.1016/j.jmb.2019.10.010
71. Dalbey RE, Kuhn A (2014) How YidC inserts and folds proteins across a membrane. Nat Struct Mol Biol 21(5):435–436. https://doi.org/10.1038/nsmb.2823
72. Laskowski PR, Pluhackova K, Haase M, Lang BM, Nagler G, Kuhn A, Muller DJ (2021) Monitoring the binding and insertion of a single transmembrane protein by an insertase. Nat Commun 12(1):7082. https://doi.org/10.1038/s41467-021-27315-3
73. Petriman NA, Jauss B, Hufnagel A, Franz L, Sachelaru I, Drepper F, Warscheid B, Koch HG (2018) The interaction network of the YidC insertase with the SecYEG translocon, SRP and the SRP receptor FtsY. Sci Rep 8(1):578. https://doi.org/10.1038/s41598-017-19019-w
74. Natale P, Bruser T, Driessen AJ (2008) Sec- and Tat-mediated protein secretion across the bacterial cytoplasmic membrane-distinct translocases and mechanisms. Biochim Biophys Acta 1778(9):1735–1756. https://doi.org/10.1016/j.bbamem.2007.07.015
75. Palmer T, Sargent F, Berks BC (2005) Export of complex cofactor-containing proteins by the bacterial Tat pathway. Trends Microbiol 13(4):175–180. https://doi.org/10.1016/j.tim.2005.02.002
76. Robinson C, Bolhuis A (2004) Tat-dependent protein targeting in prokaryotes and chloroplasts. Biochim Biophys Acta 1694(1–3):135–147. https://doi.org/10.1016/j.bbamcr.2004.03.010
77. Berks BC (1996) A common export pathway for proteins binding complex redox cofactors? Mol Microbiol 22(3):393–404. https://doi.org/10.1046/j.1365-2958.1996.00114.x
78. Hatzixanthis K, Palmer T, Sargent F (2003) A subset of bacterial inner membrane proteins integrated by the twin-arginine translocase. Mol Microbiol 49:1377. https://doi.org/10.1046/j.1365-2958.2003.03642.x
79. Bogsch EG, Sargent F, Stanley NR, Berks BC, Robinson C, Palmer T (1998) An essential component of a novel bacterial protein export system with homologues in plastids and mitochondria. J Biol Chem 273(29):18003–18006. https://doi.org/10.1074/jbc.273.29.18003
80. Wu LF, Ize B, Chanal A, Quentin Y, Fichant G (2000) Bacterial twin-arginine signal peptide-dependent protein translocation pathway: evolution and mechanism. J Mol Microbiol Biotechnol 2(2):179–189
81. Behrendt J, Standar K, Lindenstrauss U, Bruser T (2004) Topological studies on the twin-arginine translocase component TatC. FEMS Microbiol Lett 234(2):303–308. https://doi.org/10.1111/j.1574-6968.2004.tb09548.x
82. Ki JJ, Kawarasaki Y, Gam J, Harvey BR, Iverson BL, Georgiou G (2004) A periplasmic fluorescent reporter protein and its application in high-throughput membrane protein topology analysis. J Mol Biol 341(4):901–909. https://doi.org/10.1016/j.jmb.2004.05.078
83. Rollauer SE, Tarry MJ, Graham JE, Jaaskelainen M, Jager F, Johnson S, Krehenbrink M, Liu SM, Lukey MJ, Marcoux J, McDowell MA, Rodriguez F, Roversi P, Stansfeld PJ, Robinson CV, Sansom MS, Palmer T, Hogbom M, Berks BC, Lea SM (2012) Structure of the TatC

core of the twin-arginine protein transport system. Nature 492(7428):210–214. https://doi.org/10.1038/nature11683

84. Ramasamy S, Abrol R, Suloway CJ, Clemons WM Jr (2013) The glove-like structure of the conserved membrane protein TatC provides insight into signal sequence recognition in twin-arginine translocation. Structure 21(5):777–788. https://doi.org/10.1016/j.str.2013.03.004
85. Matos CF, Di Cola A, Robinson C (2009) TatD is a central component of a Tat translocon-initiated quality control system for exported FeS proteins in *Escherichia coli*. EMBO Rep 10(5):474–479. https://doi.org/10.1038/embor.2009.34
86. Alcock F, Stansfeld PJ, Basit H, Habersetzer J, Baker MA, Palmer T, Wallace MI, Berks BC (2016) Assembling the Tat protein translocase. Elife 5:e20718. https://doi.org/10.7554/eLife.20718
87. Habersetzer J, Moore K, Cherry J, Buchanan G, Stansfeld PJ, Palmer T (2017) Substrate-triggered position switching of TatA and TatB during Tat transport in Escherichia coli. Open Biol 7(8):170091. https://doi.org/10.1098/rsob.170091
88. Dabney-Smith C, Mori H, Cline K (2006) Oligomers of Tha4 organize at the thylakoid Tat translocase during protein transport. J Biol Chem 281(9):5476–5483. https://doi.org/10.1074/jbc.M5124532
89. Alcock F, Baker MA, Greene NP, Palmer T, Wallace MI, Berks BC (2013) Live cell imaging shows reversible assembly of the TatA component of the twin-arginine protein transport system. Proc Natl Acad Sci U S A 110(38):E3650–E3659. https://doi.org/10.1073/pnas.1306738110
90. Rose P, Frobel J, Graumann PL, Muller M (2013) Substrate-dependent assembly of the Tat translocase as observed in live Escherichia coli cells. PloS One 8(8):e69488. https://doi.org/10.1371/journal.pone.0069488
91. Luke I, Handford JI, Palmer T, Sargent F (2009) Proteolytic processing of *Escherichia coli* twin-arginine signal peptides by LepB. Arch Microbiol 191(12):919–925. https://doi.org/10.1007/s00203-009-0516-5
92. Alami M, Luke I, Deitermann S, Eisner G, Koch HG, Brunner J, Muller M (2003) Differential interactions between a twin-arginine signal peptide and its translocase in *Escherichia coli*. Mol Cell 12(4):937–946. https://doi.org/10.1016/S1097-2765(03)00398-8
93. Cline K, Mori H (2001) Thylakoid DeltapH-dependent precursor proteins bind to a cpTatC-Hcf106 complex before Tha4-dependent transport. J Cell Biol 154(4):719–729. https://doi.org/10.1083/jcb.200105149
94. Dabney-Smith C, Mori H, Cline K (2003) Requirement of a Tha4-conserved transmembrane glutamate in thylakoid Tat translocase assembly revealed by biochemical complementation. J Biol Chem 278(44):43027–43033. https://doi.org/10.1074/jbc.M307923200
95. Blummel AS, Haag LA, Eimer E, Muller M, Frobel J (2015) Initial assembly steps of a translocase for folded proteins. Nat Commun 6:7234. https://doi.org/10.1038/ncomms8234
96. Zoufaly S, Frobel J, Rose P, Flecken T, Maurer C, Moser M, Muller M (2012) Mapping precursor-binding site on TatC subunit of twin arginine-specific protein translocase by site-specific photo cross-linking. J Biol Chem 287(16):13430–13441. https://doi.org/10.1074/jbc.M112.343798
97. Cline K (2015) Mechanistic aspects of folded protein transport by the Twin Arginine Translocase (Tat). J Biol Chem 290(27):16530–16538. https://doi.org/10.1074/jbc.R114.626820
98. Alcock F, Berks BC (2022) New insights into the Tat protein transport cycle from characterizing the assembled Tat translocon. Mol Microbiol 118(6):637–651. https://doi.org/10.1111/mmi.14984
99. von Heijne G, Abrahmsen L (1989) Species-specific variation in signal peptide design. Implications for protein secretion in foreign hosts. FEBS Lett 244(2):439–446. https://doi.org/10.1016/0014-5793(89)80579-4
100. New CP, Ma Q, Dabney-Smith C (2018) Routing of thylakoid lumen proteins by the chloroplast twin arginine transport pathway. Photosynth Res 138(3):289–301. https://doi.org/10.1007/s11120-018-0567-z

101. Sala A, Bordes P, Genevaux P (2014) Multitasking SecB chaperones in bacteria. Front Microbiol 5:666. https://doi.org/10.3389/fmicb.2014.00666
102. Hartl FU, Lecker S, Schiebel E, Hendrick JP, Wickner W (1990) The binding cascade of SecB to SecA to SecY/E mediates preprotein targeting to the *E. coli* plasma membrane. Cell 63(2):269–279. https://doi.org/10.1016/0092-8674(90)90160-G
103. Huber D, Rajagopalan N, Preissler S, Rocco MA, Merz F, Kramer G, Bukau B (2011) SecA interacts with ribosomes in order to facilitate posttranslational translocation in bacteria. Mol Cell 41(3):343–353. https://doi.org/10.1016/j.molcel.2010.12.028
104. De Geyter J, Portaliou AG, Srinivasu B, Krishnamurthy S, Economou A, Karamanou S (2020) Trigger factor is a bona fide secretory pathway chaperone that interacts with SecB and the translocase. EMBO Rep 21(6):e49054. https://doi.org/10.15252/embr.201949054
105. Palmer T, Berks BC (2012) The twin-arginine translocation (Tat) protein export pathway. Nat Rev Microbiol 10(7):483–496. https://doi.org/10.1038/nrmicro2814
106. Turner RJ, Papish AL, Sargent F (2004) Sequence analysis of bacterial redox enzyme maturation proteins (REMPs). Can J Microbiol 50(4):225–238. https://doi.org/10.1139/w03-117
107. Robinson C, Matos CF, Beck D, Ren C, Lawrence J, Vasisht N, Mendel S (2011) Transport and proofreading of proteins by the twin-arginine translocation (Tat) system in bacteria. Biochim Biophys Acta 1808(3):876–884. https://doi.org/10.1016/j.bbamem.2010.11.023
108. Ilbert M, Mejean V, Giudici-Orticoni MT, Samama JP, Iobbi-Nivol C (2003) Involvement of a mate chaperone (TorD) in the maturation pathway of molybdoenzyme TorA. J Biol Chem 278(31):28787–28792. https://doi.org/10.1074/jbc.M302730200
109. Oresnik IJ, Ladner CL, Turner RJ (2001) Identification of a twin-arginine leader-binding protein. Mol Microbiol 40(2):323–331. https://doi.org/10.1046/j.1365-2958.2001.02391.x
110. Maillard J, Spronk CA, Buchanan G, Lyall V, Richardson DJ, Palmer T, Vuister GW, Sargent F (2007) Structural diversity in twin-arginine signal peptide-binding proteins. Proc Natl Acad Sci U S A 104(40):15641–15646. https://doi.org/10.1073/pnas.0703967104
111. Lill R, Cunningham K, Brundage LA, Ito K, Oliver D, Wickner W (1989) SecA protein hydrolyzes ATP and is an essential component of the protein translocation ATPase of *Escherichia coli*. EMBO J 8(3):961–966. https://doi.org/10.1002/j.1460-2075.1989.tb03458.x
112. Voorhees RM, Hegde RS (2016) Toward a structural understanding of co-translational protein translocation. Curr Opin Cell Biol 41:91–99. https://doi.org/10.1016/j.ceb.2016.04.009
113. Cannon KS, Or E, Clemons WM Jr, Shibata Y, Rapoport TA (2005) Disulfide bridge formation between SecY and a translocating polypeptide localizes the translocation pore to the center of SecY. J Cell Biol 169(2):219–225. https://doi.org/10.1083/jcb.200412019
114. Economou A, Wickner W (1994) SecA promotes preprotein translocation by undergoing ATP-driven cycles of membrane insertion and deinsertion. Cell 78(5):835–843. https://doi.org/10.1016/S0092-8674(94)90582-7
115. Kusters I, Driessen AJ (2011) SecA, a remarkable nanomachine. Cell Mol Life Sci 68(12):2053–2066. https://doi.org/10.1007/s00018-011-0681-y
116. Dalbey RE, Wickner W (1985) Leader peptidase catalyzes the release of exported proteins from the outer surface of the *Escherichia coli* plasma membrane. J Biol Chem 260(29):15925–15931. https://doi.org/10.1016/S0021-9258(17)36347-0
117. Papanikou E, Karamanou S, Economou A (2007) Bacterial protein secretion through the translocase nanomachine. Nat Rev Microbiol 5(11):839–851. https://doi.org/10.1038/nrmicro1771
118. Zimmer J, Nam Y, Rapoport TA (2008) Structure of a complex of the ATPase SecA and the protein-translocation channel. Nature 455(7215):936–943. https://doi.org/10.1038/nature07335
119. Gelis I, Bonvin AM, Keramisanou D, Koukaki M, Gouridis G, Karamanou S, Economou A, Kalodimos CG (2007) Structural Basis for signal-sequence recognition by the translocase motor SecA as determined by NMR. Cell 131(4):756–769. https://doi.org/10.1016/j.cell.2007.09.039
120. Erlandson KJ, Miller SB, Nam Y, Osborne AR, Zimmer J, Rapoport TA (2008) A role for the two-helix finger of the SecA ATPase in protein translocation. Nature 455(7215):984–987. https://doi.org/10.1038/nature07439

121. Bauer BW, Shemesh T, Chen Y, Rapoport TA (2014) A "push and slide" mechanism allows sequence-insensitive translocation of secretory proteins by the SecA ATPase. Cell 157(6):1416–1429. https://doi.org/10.1016/j.cell.2014.03.063
122. Catipovic MA, Bauer BW, Loparo JJ, Rapoport TA (2019) Protein translocation by the SecA ATPase occurs by a power-stroke mechanism. EMBO J 38(9):e101140. https://doi.org/10.15252/embj.2018101140
123. Whitehouse S, Gold VA, Robson A, Allen WJ, Sessions RB, Collinson I (2012) Mobility of the SecA 2-helix-finger is not essential for polypeptide translocation via the SecYEG complex. J Cell Biol 199(6):919–929. https://doi.org/10.1042/BJ20121314
124. Allen WJ, Corey RA, Oatley P, Sessions RB, Baldwin SA, Radford SE, Tuma R, Collinson I (2016) Two-way communication between SecY and SecA suggests a Brownian ratchet mechanism for protein translocation. Elife 5:e15598. https://doi.org/10.7554/eLife.15598
125. Krishnamurthy S, Sardis MF, Eleftheriadis N, Chatzi KE, Smit JH, Karathanou K, Gouridis G, Portaliou AG, Bondar AN, Karamanou S, Economou A (2022) Preproteins couple the intrinsic dynamics of SecA to its ATPase cycle to translocate via a catch and release mechanism. Cell Rep 38(6):110346. https://doi.org/10.1016/j.celrep.2022.110346
126. Dong L, Yang S, Chen J, Wu X, Sun D, Song C, Li L (2023) Structural basis of SecA-mediated protein translocation. Proc Natl Acad Sci U S A 120(2):e2208070120
127. Tsukazaki T, Mori H, Echizen Y, Ishitani R, Fukai S, Tanaka T, Perederina A, Vassylyev DG, Kohno T, Maturana AD, Ito K, Nureki O (2011) Structure and function of a membrane component SecDF that enhances protein export. Nature 474(7350):235–238
128. Tsukazaki T (2018) Structure-based working model of SecDF, a proton-driven bacterial protein translocation factor. FEMS Microbiol Lett 365(12):fny112. https://doi.org/10.1093/femsle/fny112
129. Gohlke U, Pullan L, McDevitt CA, Porcelli I, de Leeuw E, Palmer T, Saibil HR, Berks BC (2005) The TatA component of the twin-arginine protein transport system forms channel complexes of variable diameter. Proc Natl Acad Sci U S A 102(30):10482–10486
130. Walther TH, Gottselig C, Grage SL, Wolf M, Vargiu AV, Klein MJ, Vollmer S, Prock S, Hartmann M, Afonin S, Stockwald E, Heinzmann H, Nolandt OV, Wenzel W, Ruggerone P, Ulrich AS (2013) Folding and self-assembly of the TatA translocation pore based on a charge zipper mechanism. Cell 152(1–2):316–326. https://doi.org/10.1016/j.cell.2012.12.017
131. Bruser T, Sanders C (2003) An alternative model of the twin arginine translocation system. Microbiol Res 158(1):7–17. https://doi.org/10.1078/0944-5013-00176
132. Hou B, Heidrich ES, Mehner-Breitfeld D, Bruser T (2018) The TatA component of the twin-arginine translocation system locally weakens the cytoplasmic membrane of *Escherichia coli* upon protein substrate binding. J Biol Chem 293(20):7592–7605. https://doi.org/10.1074/jbc.RA118.00
133. Asher AH, Theg SM (2021) Electrochromic shift supports the membrane destabilization model of Tat-mediated transport and shows ion leakage during Sec transport. Proc Natl Acad Sci U S A 118(12):e2018122118. https://doi.org/10.1073/pnas.2018122118
134. Hao B, Zhou W, Theg SM (2022) Hydrophobic mismatch is a key factor in protein transport across lipid bilayer membranes via the Tat pathway. J Biol Chem 298(7):101991. https://doi.org/10.1016/j.jbc.2022.101991
135. Mehner-Breitfeld D, Ringel MT, Tichy DA, Endter LJ, Stroh KS, Lunsdorf H, Risselada HJ, Bruser T (2022) TatA and TatB generate a hydrophobic mismatch important for the function and assembly of the Tat translocon in *Escherichia coli*. J Biol Chem 298(9):102236. https://doi.org/10.1016/j.jbc.2022.102236
136. Sankaran K, Wu HC (1994) Lipid modification of bacterial prolipoprotein. Transfer of diacylglyceryl moiety from phosphatidylglycerol. J Biol Chem 269(31):19701–19706. https://doi.org/10.1016/S0021-9258(17)32077-X
137. Tokunaga M, Tokunaga H, Wu HC (1982) Post-translational modification and processing of *Escherichia coli* prolipoprotein in vitro. Proc Natl Acad Sci U S A 79(7):2255–2259. https://doi.org/10.1073/pnas.79.7.2255
138. Gupta SD, Wu HC (1991) Identification and subcellular localization of apolipoprotein N-acyltransferase in *Escherichia coli*. FEMS Microbiol Lett 62(1):37–41

139. Matsuyama S, Tajima T, Tokuda H (1995) A novel periplasmic carrier protein involved in the sorting and transport of *Escherichia coli* lipoproteins destined for the outer membrane. EMBO J 14(14):3365–3372. https://doi.org/10.1002/j.1460-2075.1995.tb07342.x
140. Matsuyama S, Yokota N, Tokuda H (1997) A novel outer membrane lipoprotein, LolB (HemM), involved in the LolA (p20)-dependent localization of lipoproteins to the outer membrane of *Escherichia coli*. EMBO J 16(23):6947–6955. https://doi.org/10.1093/emboj/16.23.6947
141. Yakushi T, Masuda K, Narita S, Matsuyama S, Tokuda H (2000) A new ABC transporter mediating the detachment of lipid-modified proteins from membranes. Nat Cell Biol 2(4):212–218. https://doi.org/10.1038/35008635
142. Okuda S, Tokuda H (2009) Model of mouth-to-mouth transfer of bacterial lipoproteins through inner membrane LolC, periplasmic LolA, and outer membrane LolB. Proc Natl Acad Sci U S A 106(14):5877–5882. https://doi.org/10.1073/pnas.0900896106
143. Yasuda M, Iguchi-Yokoyama A, Matsuyama S, Tokuda H, Narita S (2009) Membrane topology and functional importance of the periplasmic region of ABC transporter LolCDE. Biosci Biotechnol Biochem 73(10):2310–2316. https://doi.org/10.1271/bbb.90451
144. Tang X, Chang S, Zhang K, Luo Q, Zhang Z, Wang T, Qiao W, Wang C, Shen C, Zhang Z, Zhu X, Wei X, Dong C, Zhang X, Dong H (2021) Structural basis for bacterial lipoprotein relocation by the transporter LolCDE. Nat Struct Mol Biol 28(4):347–355. https://doi.org/10.1038/s41594-021-00573-x
145. Sharma S, Zhou R, Wan L, Feng S, Song K, Xu C, Li Y, Liao M (2021) Mechanism of LolCDE as a molecular extruder of bacterial triacylated lipoproteins. Nat Commun 12(1):4687. https://doi.org/10.1038/s41467-021-24965-1
146. Bei W, Luo Q, Shi H, Zhou H, Zhou M, Zhang X, Huang Y (2022) Cryo-EM structures of LolCDE reveal the molecular mechanism of bacterial lipoprotein sorting in *Escherichia coli*. PLoS Biol 20(10):e3001823. https://doi.org/10.1371/journal.pbio.3001823
147. Mizutani M, Mukaiyama K, Xiao J, Mori M, Satou R, Narita S, Okuda S, Tokuda H (2013) Functional differentiation of structurally similar membrane subunits of the ABC transporter LolCDE complex. FEBS Lett 587(1):23–29. https://doi.org/10.1016/j.febslet.2012.11.009
148. Kaplan E, Greene NP, Crow A, Koronakis V (2018) Insights into bacterial lipoprotein trafficking from a structure of LolA bound to the LolC periplasmic domain. Proc Natl Acad Sci U S A 115(31):E7389–E7397. https://doi.org/10.1073/pnas.1806822115
149. Kaplan E, Greene NP, Jepson AE, Koronakis V (2022) Structural basis of lipoprotein recognition by the bacterial Lol trafficking chaperone LolA. Proc Natl Acad Sci U S A 119(36):e2208662119. https://doi.org/10.1073/pnas.2208662119
150. Masuda K, Matsuyama S, Tokuda H (2002) Elucidation of the function of lipoprotein-sorting signals that determine membrane localization. Proc Natl Acad Sci U S A 99(11):7390–7395. https://doi.org/10.1073/pnas.112085599
151. Narita S, Tokuda H (2010) Sorting of bacterial lipoproteins to the outer membrane by the Lol system. Methods Mol Biol 619:117–129. https://doi.org/10.1007/978-1-60327-412-8_7
152. Sklar JG, Wu T, Kahne D, Silhavy TJ (2007) Defining the roles of the periplasmic chaperones SurA, Skp, and DegP in Escherichia coli. Genes Dev 21(19):2473–2484. https://doi.org/10.1101/gad.1581007
153. De Cock H, Schafer U, Potgeter M, Demel R, Muller M, Tommassen J (1999) Affinity of the periplasmic chaperone Skp of *Escherichia coli* for phospholipids, lipopolysaccharides and non-native outer membrane proteins. Role of Skp in the biogenesis of outer membrane protein. Eur J Biochem 259(1–2):96–103. https://doi.org/10.1046/j.1432-1327.1999.00010.x
154. Harms N, Koningstein G, Dontje W, Muller M, Oudega B, Luirink J, de Cock H (2001) The early interaction of the outer membrane protein phoe with the periplasmic chaperone Skp occurs at the cytoplasmic membrane. J Biol Chem 276(22):18804–18811. https://doi.org/10.1074/jbc.M011
155. Volokhina EB, Grijpstra J, Stork M, Schilders I, Tommassen J, Bos MP (2011) Role of the periplasmic chaperones Skp, SurA, and DegQ in outer membrane protein biogenesis in *Neisseria meningitidis*. J Bacteriol 193(7):1612–1621. https://doi.org/10.1128/jb.00532-10

156. Wu T, Malinverni J, Ruiz N, Kim S, Silhavy TJ, Kahne D (2005) Identification of a multicomponent complex required for outer membrane biogenesis in *Escherichia coli*. Cell 121(2):235–245. https://doi.org/10.1016/j.cell.2005.02.015
157. Sklar JG, Wu T, Gronenberg LS, Malinverni JC, Kahne D, Silhavy TJ (2007) Lipoprotein SmpA is a component of the YaeT complex that assembles outer membrane proteins in *Escherichia coli*. Proc Natl Acad Sci U S A 104(15):6400–6405. https://doi.org/10.1073/pnas.0701579104
158. Doerrler WT, Raetz CR (2005) Loss of outer membrane proteins without inhibition of lipid export in an *Escherichia coli* YaeT mutant. J Biol Chem 280(30):27679–27687. https://doi.org/10.1074/jbc.M504796200
159. Werner J, Misra R (2005) YaeT (Omp85) affects the assembly of lipid-dependent and lipid-independent outer membrane proteins of *Escherichia coli*. Mol Microbiol 57(5):1450–1459. https://doi.org/10.1111/j.1365-2958.2005.04775.x
160. Onufryk C, Crouch ML, Fang FC, Gross CA (2005) Characterization of six lipoproteins in the sigmaE regulon. J Bacteriol 187(13):4552–4561. https://doi.org/10.1128/jb.187.13.4552-4561.2005
161. Malinverni JC, Werner J, Kim S, Sklar JG, Kahne D, Misra R, Silhavy TJ (2006) YfiO stabilizes the YaeT complex and is essential for outer membrane protein assembly in *Escherichia coli*. Mol Microbiol 61(1):151–164. https://doi.org/10.1111/j.1365-2958.2006.05211.x
162. Arnold T, Zeth K, Linke D (2010) Omp85 from the thermophilic cyanobacterium *Thermosynechococcus elongatus* differs from proteobacterial Omp85 in structure and domain composition. J Biol Chem 285(23):18003–18015. https://doi.org/10.1074/jbc.M110.112516
163. Sanchez-Pulido L, Devos D, Genevrois S, Vicente M, Valencia A (2003) POTRA: a conserved domain in the FtsQ family and a class of beta-barrel outer membrane proteins. Trends Biochem Sci 28(10):523–526. https://doi.org/10.1016/j.tibs.2003.08.003
164. Simmerman RF, Dave AM, Bruce BD (2014) Structure and function of POTRA domains of Omp85/TPS superfamily. Int Rev Cell Mol Biol 308:1–34
165. Kim KH, Aulakh S, Paetzel M (2011) Crystal structure of beta-barrel assembly machinery BamCD protein complex. J Biol Chem 286(45):39116–39121. https://doi.org/10.1074/jbc.M111.298166
166. Jansen KB, Baker SL, Sousa MC (2015) Crystal structure of BamB bound to a periplasmic domain fragment of BamA, the central component of the beta-barrel assembly machine. J Biol Chem 290(4):2126–2136. https://doi.org/10.1074/jbc.M114.584524
167. Vuong P, Bennion D, Mantei J, Frost D, Misra R (2008) Analysis of YfgL and YaeT interactions through bioinformatics, mutagenesis, and biochemistry. J Bacteriol 190(5):1507–1517. https://doi.org/10.1128/jb.01477-07
168. Lee J, Sutterlin HA, Wzorek JS, Mandler MD, Hagan CL, Grabowicz M, Tomasek D, May MD, Hart EM, Silhavy TJ, Kahne D (2018) Substrate binding to BamD triggers a conformational change in BamA to control membrane insertion. Proc Natl Acad Sci U S A 115(10):2359–2364. https://doi.org/10.1073/pnas.1711727115
169. Lee J, Xue M, Wzorek JS, Wu T, Grabowicz M, Gronenberg LS, Sutterlin HA, Davis RM, Ruiz N, Silhavy TJ, Kahne DE (2016) Characterization of a stalled complex on the beta-barrel assembly machine. Proc Natl Acad Sci U S A 113(31):8717–8722. https://doi.org/10.1073/pnas.1604100113
170. Noinaj N, Kuszak AJ, Gumbart JC, Lukacik P, Chang H, Easley NC, Lithgow T, Buchanan SK (2013) Structural insight into the biogenesis of beta-barrel membrane proteins. Nature 501(7467):385–390. https://doi.org/10.1038/nature12521
171. Noinaj N, Kuszak AJ, Balusek C, Gumbart JC, Buchanan SK (2014) Lateral opening and exit pore formation are required for BamA function. Structure 22(7):1055–1062. https://doi.org/10.1016/j.str.2014.05.008
172. Bakelar J, Buchanan SK, Noinaj N (2016) The structure of the beta-barrel assembly machinery complex. Science 351(6269):180–186. https://doi.org/10.1126/science.aad3460
173. Gu Y, Li H, Dong H, Zeng Y, Zhang Z, Paterson NG, Stansfeld PJ, Wang Z, Zhang Y, Wang W, Dong C (2016) Structural basis of outer membrane protein insertion by the BAM complex. Nature 531(7592):64–69. https://doi.org/10.1038/nature17199

174. Han L, Zheng J, Wang Y, Yang X, Liu Y, Sun C, Cao B, Zhou H, Ni D, Lou J, Zhao Y, Huang Y (2016) Structure of the BAM complex and its implications for biogenesis of outer-membrane proteins. Nat Struct Mol Biol 23(3):192–196. https://doi.org/10.1038/nsmb.3181
175. Iadanza MG, Higgins AJ, Schiffrin B, Calabrese AN, Brockwell DJ, Ashcroft AE, Radford SE, Ranson NA (2016) Lateral opening in the intact beta-barrel assembly machinery captured by cryo-EM. Nat Commun 7:12865. https://doi.org/10.1038/ncomms12865
176. Tomasek D, Rawson S, Lee J, Wzorek JS, Harrison SC, Li Z, Kahne D (2020) Structure of a nascent membrane protein as it folds on the BAM complex. Nature 583(7816):473–478. https://doi.org/10.1038/s41586-020-2370-1
177. Wu R, Bakelar JW, Lundquist K, Zhang Z, Kuo KM, Ryoo D, Pang YT, Sun C, White T, Klose T, Jiang W, Gumbart JC, Noinaj N (2021) Plasticity within the barrel domain of BamA mediates a hybrid-barrel mechanism by BAM. Nat Commun 12(1):7131
178. Doyle MT, Jimah JR, Dowdy T, Ohlemacher SI, Larion M, Hinshaw JE, Bernstein HD (2022) Cryo-EM structures reveal multiple stages of bacterial outer membrane protein folding. Cell 185(7):1143–1156 e13. https://doi.org/10.1016/j.cell.2022.02.016
179. Kim KH, Aulakh S, Paetzel M (2012) The bacterial outer membrane beta-barrel assembly machinery. Protein Sci 21(6):751–768. https://doi.org/10.1002/pro.2069
180. Estrada Mallarino L, Fan E, Odermatt M, Muller M, Lin M, Liang J, Heinzelmann M, Fritsche F, Apell HJ, Welte W (2015) TtOmp85, a beta-barrel assembly protein, functions by barrel augmentation. Biochemistry 54(3):844–852. https://doi.org/10.1021/bi5011305
181. Doyle MT, Bernstein HD (2019) Bacterial outer membrane proteins assemble via asymmetric interactions with the BamA beta-barrel. Nat Commun 10(1):3358. https://doi.org/10.1038/s41467-019-11230-9
182. Noinaj N, Rollauer SE, Buchanan SK (2015) The beta-barrel membrane protein insertase machinery from Gram-negative bacteria. Curr Opin Struct Biol 31:35–42. https://doi.org/10.1016/j.sbi.2015.02.012
183. Gessmann D, Chung YH, Danoff EJ, Plummer AM, Sandlin CW, Zaccai NR, Fleming KG (2014) Outer membrane beta-barrel protein folding is physically controlled by periplasmic lipid head groups and BamA. Proc Natl Acad Sci U S A 111(16):5878–5883. https://doi.org/10.1073/pnas.1322473111
184. White P, Haysom SF, Iadanza MG, Higgins AJ, Machin JM, Whitehouse JM, Horne JE, Schiffrin B, Carpenter-Platt C, Calabrese AN, Storek KM, Rutherford ST, Brockwell DJ, Ranson NA, Radford SE (2021) The role of membrane destabilisation and protein dynamics in BAM catalysed OMP folding. Nat Commun 12(1):4174. https://doi.org/10.1038/s41467-021-24432-x
185. Petru M, Dohnalek V, Fussy Z, Dolezal P (2021) Fates of Sec, Tat, and YidC translocases in mitochondria and other eukaryotic compartments. Mol Biol Evol 38(12):5241–5254
186. Rapoport TA, Jungnickel B, Kutay U (1996) Protein transport across the eukaryotic endoplasmic reticulum and bacterial inner membranes. Annu Rev Biochem 65:271–303. https://doi.org/10.1146/annurev.bi.65.070196.001415
187. Keenan RJ, Freymann DM, Stroud RM, Walter P (2001) The signal recognition particle. Annu Rev Biochem 70:755–775. https://doi.org/10.1146/annurev.biochem.70.1.755
188. Tajima S, Lauffer L, Rath VL, Walter P (1986) The signal recognition particle receptor is a complex that contains two distinct polypeptide chains. J Cell Biol 103(4):1167–1178
189. Akopian D, Shen K, Zhang X, Shan SO (2013) Signal recognition particle: an essential protein-targeting machine. Annu Rev Biochem 82:693–721
190. Gorlich D, Rapoport TA (1993) Protein translocation into proteoliposomes reconstituted from purified components of the endoplasmic reticulum membrane. Cell 75(4):615–630
191. Do H, Falcone D, Lin J, Andrews DW, Johnson AE (1996) The cotranslational integration of membrane proteins into the phospholipid bilayer is a multistep process. Cell 85(3):369–378. https://doi.org/10.1016/S0092-8674(00)81115-0
192. Ng DT, Brown JD, Walter P (1996) Signal sequences specify the targeting route to the endoplasmic reticulum membrane. J Cell Biol 134(2):269–278. https://doi.org/10.1083/jcb.134.2.269

193. Ast T, Cohen G, Schuldiner M (2013) A network of cytosolic factors targets SRP-independent proteins to the endoplasmic reticulum. Cell 152(5):1134–1145. https://doi.org/10.1016/j.cell.2013.02.003
194. Deshaies RJ, Sanders SL, Feldheim DA, Schekman R (1991) Assembly of yeast Sec proteins involved in translocation into the endoplasmic reticulum into a membrane-bound multisubunit complex. Nature 349(6312):806–808. https://doi.org/10.1038/349806a0
195. Panzner S, Dreier L, Hartmann E, Kostka S, Rapoport TA (1995) Posttranslational protein transport in yeast reconstituted with a purified complex of Sec proteins and Kar2p. Cell 81(4):561–570. https://doi.org/10.1016/0092-8674(95)90077-2
196. Matlack KE, Misselwitz B, Plath K, Rapoport TA (1999) BiP acts as a molecular ratchet during posttranslational transport of prepro-alpha factor across the ER membrane. Cell 97(5):553–564. https://doi.org/10.1016/S0092-8674(00)80767-9
197. Deshaies RJ, Koch BD, Werner-Washburne M, Craig EA, Schekman R (1988) A subfamily of stress proteins facilitates translocation of secretory and mitochondrial precursor polypeptides. Nature 332(6167):800–805. https://doi.org/10.1038/332800a0
198. Wu X, Cabanos C, Rapoport TA (2019) Structure of the post-translational protein translocation machinery of the ER membrane. Nature 566(7742):136–139. https://doi.org/10.1038/s41586-018-0856-x
199. Itskanov S, Park E (2019) Structure of the posttranslational Sec protein-translocation channel complex from yeast. Science 363(6422):84–87. https://doi.org/10.1126/sciimmunol.aav6778
200. Paetzel M, Karla A, Strynadka NC, Dalbey RE (2002) Signal peptidases. Chem Rev 102(12):4549–4580. https://doi.org/10.1021/cr010166y
201. Evans EA, Gilmore R, Blobel G (1986) Purification of microsomal signal peptidase as a complex. Proc Natl Acad Sci U S A 83(3):581–585. https://doi.org/10.1073/pnas.83.3.581
202. Paetzel M (2014) Structure and mechanism of *Escherichia coli* type I signal peptidase. Biochim Biophys Acta 1843(8):1497–1508. https://doi.org/10.1016/j.bbamcr.2013.12.003
203. VanValkenburgh C, Chen X, Mullins C, Fang H, Green N (1999) The catalytic mechanism of endoplasmic reticulum signal peptidase appears to be distinct from most eubacterial signal peptidases. J Biol Chem 274(17):11519–11525. https://doi.org/10.1074/jbc.274.17.11519
204. Bardy SL, Ng SY, Carnegie DS, Jarrell KF (2005) Site-directed mutagenesis analysis of amino acids critical for activity of the type I signal peptidase of the archaeon *Methanococcus voltae*. J Bacteriol 187(3):1188–1191. https://doi.org/10.1128/jb.187.3.1188-1191.2005
205. Paetzel M, Dalbey RE, Strynadka NC (1998) Crystal structure of a bacterial signal peptidase in complex with a beta-lactam inhibitor. Nature 396(6707):186–190. https://doi.org/10.1038/24196
206. Liaci AM, Steigenberger B, Telles de Souza PC, Tamara S, Grollers-Mulderij M, Ogrissek P, Marrink SJ, Scheltema RA, Forster F (2021) Structure of the human signal peptidase complex reveals the determinants for signal peptide cleavage. Mol Cell 81(19):3934–3948 e11. https://doi.org/10.1016/j.molcel.2021.07.031
207. Anghel SA, McGilvray PT, Hegde RS, Keenan RJ (2017) Identification of Oxa1 homologs operating in the eukaryotic endoplasmic reticulum. Cell Rep 21(13):3708–3716
208. Chen Y, Dalbey RE (2018) Oxa1 superfamily: new members found in the ER. Trends Biochem Sci 43(3):151–153. https://doi.org/10.1016/j.tibs.2017.12.005
209. Borowska MT, Dominik PK, Anghel SA, Kossiakoff AA, Keenan RJ (2015) A YidC-like protein in the archaeal plasma membrane. Structure 23(9):1715–1724. https://doi.org/10.1016/j.str.2015.06.025
210. Guna A, Volkmar N, Christianson JC, Hegde RS (2018) The ER membrane protein complex is a transmembrane domain insertase. Science 359(6374):470–473. https://doi.org/10.1126/science.aao3099
211. McDowell MA, Heimes M, Sinning I (2021) Structural and molecular mechanisms for membrane protein biogenesis by the Oxa1 superfamily. Nat Struct Mol Biol 28(3):234–239. https://doi.org/10.1038/s41594-021-00567-9
212. McGilvray PT, Anghel SA, Sundaram A, Zhong F, Trnka MJ, Fuller JR, Hu H, Burlingame AL, Keenan RJ (2020) An ER translocon for multi-pass membrane protein biogenesis. Elife 9:e56889. https://doi.org/10.7554/eLife.56889

213. Chen Y, Capponi S, Zhu L, Gellenbeck P, Freites JA, White SH, Dalbey RE (2017) YidC insertase of *Escherichia coli*: water accessibility and membrane shaping. Structure 25(9):1403–1414 e3. https://doi.org/10.1016/j.str.2017.07.008
214. Pleiner T, Tomaleri GP, Januszyk K, Inglis AJ, Hazu M, Voorhees RM (2020) Structural basis for membrane insertion by the human ER membrane protein complex. Science 369(6502):433–436. https://doi.org/10.1126/science.abb5008
215. Meisinger C, Sickmann A, Pfanner N (2008) The mitochondrial proteome: from inventory to function. Cell 134(1):22–24. https://doi.org/10.1016/j.cell.2008.06.043
216. Araiso Y, Imai K, Endo T (2022) Role of the TOM complex in protein import into mitochondria: structural views. Annu Rev Biochem 91:679–703
217. Wiedemann N, Pfanner N (2017) Mitochondrial machineries for protein import and assembly. Annu Rev Biochem 86:685–714. https://doi.org/10.1146/annurev-biochem-060815-014352
218. Neupert W, Herrmann JM (2007) Translocation of proteins into mitochondria. Annu Rev Biochem 76:723–749. https://doi.org/10.1146/annurev.biochem.76.052705.163409
219. Paschen SA, Waizenegger T, Stan T, Preuss M, Cyrklaff M, Hell K, Rapaport D, Neupert W (2003) Evolutionary conservation of biogenesis of beta-barrel membrane proteins. Nature 426(6968):862–866. https://doi.org/10.1038/nature02208
220. Wiedemann N, Kozjak V, Chacinska A, Schonfisch B, Rospert S, Ryan MT, Pfanner N, Meisinger C (2003) Machinery for protein sorting and assembly in the mitochondrial outer membrane. Nature 424(6948):565–571. https://doi.org/10.1038/nature01753
221. Gentle I, Gabriel K, Beech P, Waller R, Lithgow T (2004) The Omp85 family of proteins is essential for outer membrane biogenesis in mitochondria and bacteria. J Cell Biol 164(1):19–24. https://doi.org/10.1083/jcb.200310092
222. Voulhoux R, Bos MP, Geurtsen J, Mols M, Tommassen J (2003) Role of a highly conserved bacterial protein in outer membrane protein assembly. Science 299(5604):262–265
223. Hagan CL, Silhavy TJ, Kahne D (2011) beta-Barrel membrane protein assembly by the Bam complex. Annu Rev Biochem 80:189–210. https://doi.org/10.1146/annurev-biochem-061408-144611
224. Diederichs KA, Buchanan SK, Botos I (2021) Building Better Barrels - beta-barrel Biogenesis and Insertion in Bacteria and Mitochondria. J Mol Biol 433(16):166894. https://doi.org/10.1016/j.jmb.2021.166894
225. Jores T, Klinger A, Gross LE, Kawano S, Flinner N, Duchardt-Ferner E, Wohnert J, Kalbacher H, Endo T, Schleiff E, Rapaport D (2016) Characterization of the targeting signal in mitochondrial beta-barrel proteins. Nat Commun 7:12036. https://doi.org/10.1038/ncomms12036
226. Weinhaupl K, Lindau C, Hessel A, Wang Y, Schutze C, Jores T, Melchionda L, Schonfisch B, Kalbacher H, Bersch B, Rapaport D, Brennich M, Lindorff-Larsen K, Wiedemann N, Schanda P (2018) Structural basis of membrane protein chaperoning through the mitochondrial intermembrane space. Cell 175(5):1365–1379 e25. https://doi.org/10.1016/j.cell.2018.10.039
227. Diederichs KA, Ni X, Rollauer SE, Botos I, Tan X, King MS, Kunji ERS, Jiang J, Buchanan SK (2020) Structural insight into mitochondrial beta-barrel outer membrane protein biogenesis. Nat Commun 11(1):3290. https://doi.org/10.1038/s41467-020-17144-1
228. Hohr AIC, Lindau C, Wirth C, Qiu J, Stroud DA, Kutik S, Guiard B, Hunte C, Becker T, Pfanner N, Wiedemann N (2018) Membrane protein insertion through a mitochondrial beta-barrel gate. Science 359(6373):eaah6834. https://doi.org/10.1126/science.aah6834
229. Takeda H, Tsutsumi A, Nishizawa T, Lindau C, Busto JV, Wenz LS, Ellenrieder L, Imai K, Straub SP, Mossmann W, Qiu J, Yamamori Y, Tomii K, Suzuki J, Murata T, Ogasawara S, Nureki O, Becker T, Pfanner N, Wiedemann N, Kikkawa M, Endo T (2021) Mitochondrial sorting and assembly machinery operates by beta-barrel switching. Nature 590(7844):163–169. https://doi.org/10.1038/s41586-020-03113-7
230. Flinner N, Ellenrieder L, Stiller SB, Becker T, Schleiff E, Mirus O (2013) Mdm10 is an ancient eukaryotic porin co-occurring with the ERMES complex. Biochim Biophys Acta 1833(12):3314–3325. https://doi.org/10.1016/j.bbamcr.2013.10.006
231. Wang Q, Guan Z, Qi L, Zhuang J, Wang C, Hong S, Yan L, Wu Y, Cao X, Cao J, Yan J, Zou T, Liu Z, Zhang D, Yan C, Yin P (2021) Structural insight into the SAM-mediated assem-

bly of the mitochondrial TOM core complex. Science 373(6561):1377–1381. https://doi.org/10.1126/science.abh0704

232. Bonnefoy N, Chalvet F, Hamel P, Slonimski PP, Dujardin G (1994) OXA1, a *Saccharomyces cerevisiae* nuclear gene whose sequence is conserved from prokaryotes to eukaryotes controls cytochrome oxidase biogenesis. J Mol Biol 239(2):201–212
233. Bauer M, Behrens M, Esser K, Michaelis G, Pratje E (1994) PET1402, a nuclear gene required for proteolytic processing of cytochrome oxidase subunit 2 in yeast. Mol Gen Genet 245(3):272–278. https://doi.org/10.1007/BF00290106
234. Hell K, Neupert W, Stuart RA (2001) Oxa1p acts as a general membrane insertion machinery for proteins encoded by mitochondrial DNA. EMBO J 20(6):1281–1288. https://doi.org/10.1093/emboj/20.6.1281
235. Funes S, Kauff F, van der Sluis EO, Ott M, Herrmann JM (2011) Evolution of YidC/Oxa1/Alb3 insertases: three independent gene duplications followed by functional specialization in bacteria, mitochondria and chloroplasts. Biol Chem 392(1-2):13–19. https://doi.org/10.1515/bc.2011.013
236. Hell K, Herrmann J, Pratje E, Neupert W, Stuart RA (1997) Oxa1p mediates the export of the N- and C-termini of pCoxII from the mitochondrial matrix to the intermembrane space. FEBS Lett 418(3):367–370. https://doi.org/10.1016/S0014-5793(97)01412-9
237. Hell K, Herrmann JM, Pratje E, Neupert W, Stuart RA (1998) Oxa1p, an essential component of the N-tail protein export machinery in mitochondria. Proc Natl Acad Sci U S A 95(5):2250–2255. https://doi.org/10.1073/pnas.95.5.2250
238. Bohnert M, Rehling P, Guiard B, Herrmann JM, Pfanner N, van der Laan M (2010) Cooperation of stop-transfer and conservative sorting mechanisms in mitochondrial protein transport. Curr Biol 20(13):1227–1232. https://doi.org/10.1016/j.cub.2010.05.058
239. Hildenbeutel M, Theis M, Geier M, Haferkamp I, Neuhaus HE, Herrmann JM, Ott M (2012) The membrane insertase Oxa1 is required for efficient import of carrier proteins into mitochondria. J Mol Biol 423(4):590–599. https://doi.org/10.1016/j.jmb.2012.07.018
240. Stiller SB, Hopker J, Oeljeklaus S, Schutze C, Schrempp SG, Vent-Schmidt J, Horvath SE, Frazier AE, Gebert N, van der Laan M, Bohnert M, Warscheid B, Pfanner N, Wiedemann N (2016) Mitochondrial OXA translocase plays a major role in biogenesis of inner-membrane proteins. Cell Metab 23(5):901–908. https://doi.org/10.1016/j.cmet.2016.04.005
241. Saracco SA, Fox TD (2002) Cox18p is required for export of the mitochondrially encoded *Saccharomyces cerevisiae* Cox2p C-tail and interacts with Pnt1p and Mss2p in the inner membrane. Mol Biol Cell 13(4):1122–1131. https://doi.org/10.1091/mbc.01-12-0580
242. Burger G, Gray MW, Forget L, Lang BF (2013) Strikingly bacteria-like and gene-rich mitochondrial genomes throughout jakobid protists. Genome Biol Evol 5(2):418–438. https://doi.org/10.1093/gbe/evt008
243. Carrie C, Weissenberger S, Soll J (2016) Plant mitochondria contain the protein translocase subunits TatB and TatC. J Cell Sci 129(20):3935–3947. https://doi.org/10.1093/gbe/evt008
244. Pett W, Lavrov DV (2013) The twin-arginine subunit C in Oscarella: origin, evolution, and potential functional significance. Integr Comp Biol 53(3):495–502. https://doi.org/10.1242/jcs.190975
245. Petru M, Wideman J, Moore K, Alcock F, Palmer T, Dolezal P (2018) Evolution of mitochondrial TAT translocases illustrates the loss of bacterial protein transport machines in mitochondria. BMC Biol 16(1):141. https://doi.org/10.1186/s12915-018-0607-3
246. Schafer K, Kunzler P, Klingl A, Eubel H, Carrie C (2020) The plant mitochondrial TAT pathway is essential for complex III biogenesis. Curr Biol 30(11):2206. https://doi.org/10.1016/j.cub.2020.01.001
247. Lang BF, Burger G, O'Kelly CJ, Cedergren R, Golding GB, Lemieux C, Sankoff D, Turmel M, Gray MW (1997) An ancestral mitochondrial DNA resembling a eubacterial genome in miniature. Nature 387(6632):493–497. https://doi.org/10.1038/387493a0
248. Edgcomb VP, Roger AJ, Simpson AG, Kysela DT, Sogin ML (2001) Evolutionary relationships among "jakobid" flagellates as indicated by alpha- and beta-tubulin phylogenies. Mol Biol Evol 18(4):514–522. https://doi.org/10.1093/oxfordjournals.molbev.a003830

249. Archibald JM, O'Kelly CJ, Doolittle WF (2002) The chaperonin genes of jakobid and jakobid-like flagellates: implications for eukaryotic evolution. Mol Biol Evol 19(4):422–431. https://doi.org/10.1091/mbc.8.5.923
250. Tong J, Dolezal P, Selkrig J, Crawford S, Simpson AG, Noinaj N, Buchanan SK, Gabriel K, Lithgow T (2011) Ancestral and derived protein import pathways in the mitochondrion of *Reclinomonas americana*. Mol Biol Evol 28(5):1581–1591. https://doi.org/10.1093/molbev/msq305
251. Horvathova L, Zarsky V, Panek T, Derelle R, Pyrih J, Motyckova A, Klapstova V, Vinopalova M, Markova L, Voleman L, Klimes V, Petru M, Vaitova Z, Cepicka I, Hryzakova K, Harant K, Gray MW, Chami M, Guilvout I, Francetic O, Franz Lang B, Vlcek C, Tsaousis AD, Elias M, Dolezal P (2021) Analysis of diverse eukaryotes suggests the existence of an ancestral mitochondrial apparatus derived from the bacterial type II secretion system. Nat Commun 12(1):2947. https://doi.org/10.1038/s41467-021-23046-7
252. Korotkov KV, Sandkvist M (2019) Architecture, function, and substrates of the type II secretion system. EcoSal Plus 8(2):10.1128/ecosalplus.ESP-0034-2018. https://doi.org/10.1128/ecosalplus.esp-0034-2018
253. Thomassin JL, Santos Moreno J, Guilvout I, Tran Van Nhieu G, Francetic O (2017) The trans-envelope architecture and function of the type 2 secretion system: new insights raising new questions. Mol Microbiol 105(2):211–226. https://doi.org/10.1111/mmi.13704
254. Nunnari J, Fox TD, Walter P (1993) A mitochondrial protease with two catalytic subunits of nonoverlapping specificities. Science 262(5142):1997–2004. https://doi.org/10.1126/science.8266095
255. Richardson LGL, Schnell DJ (2020) Origins, function, and regulation of the TOC-TIC general protein import machinery of plastids. J Exp Bot 71(4):1226–1238. https://doi.org/10.1093/jxb/erz517
256. Soll J, Schleiff E (2004) Protein import into chloroplasts. Nat Rev Mol Cell Biol 5(3):198–208
257. Shi LX, Theg SM (2013) The chloroplast protein import system: from algae to trees. Biochim Biophys Acta 1833(2):314–331. https://doi.org/10.1016/j.bbamcr.2012.10.002
258. Liu H, Li A, Rochaix JD, Liu Z (2023) Architecture of chloroplast TOC-TIC translocon supercomplex. Nature 615(7951):349–357. https://doi.org/10.1038/s41586-023-05744-y
259. Jin Z, Wan L, Zhang Y, Li X, Cao Y, Liu H, Fan S, Cao D, Wang Z, Li X, Pan J, Dong MQ, Wu J, Yan Z (2022) Structure of a TOC-TIC supercomplex spanning two chloroplast envelope membranes. Cell 185(25):4788–4800 e13. https://doi.org/10.1016/j.cell.2022.10.030
260. Peltier JB, Friso G, Kalume DE, Roepstorff P, Nilsson F, Adamska I, van Wijk KJ (2000) Proteomics of the chloroplast: systematic identification and targeting analysis of lumenal and peripheral thylakoid proteins. Plant Cell 12(3):319–341. https://doi.org/10.1105/tpc.12.3.319
261. Bruce BD (2000) Chloroplast transit peptides: structure, function and evolution. Trends Cell Biol 10(10):440–447. https://doi.org/10.1016/S0962-8924(00)01833-X
262. Dalbey RE, Robinson C (1999) Protein translocation into and across the bacterial plasma membrane and the plant thylakoid membrane. Trends Biochem Sci 24(1):17–22. https://doi.org/10.1016/S0968-0004(98)01333-4
263. Aldridge C, Cain P, Robinson C (2009) Protein transport in organelles: protein transport into and across the thylakoid membrane. FEBS J 276(5):1177–1186. https://doi.org/10.1111/j.1742-4658.2009.06875.x
264. Eckart K, Eichacker L, Sohrt K, Schleiff E, Heins L, Soll J (2002) A Toc75-like protein import channel is abundant in chloroplasts. EMBO Rep 3(6):557–562. https://doi.org/10.1093/embo-reports/kvf110
265. Inoue K, Potter D (2004) The chloroplastic protein translocation channel Toc75 and its paralog OEP80 represent two distinct protein families and are targeted to the chloroplastic outer envelope by different mechanisms. Plant J 39(3):354–365
266. Day PM, Potter D, Inoue K (2014) Evolution and targeting of Omp85 homologs in the chloroplast outer envelope membrane. Front Plant Sci 5:535. https://doi.org/10.3389/fpls.2014.00535

267. Paila YD, Richardson LG, Inoue H, Parks ES, McMahon J, Inoue K, Schnell DJ (2016) Multi-functional roles for the polypeptide transport associated domains of Toc75 in chloroplast protein import. Elife 5:e12631. https://doi.org/10.7554/eLife.12631
268. Chen YL, Chen LJ, Li HM (2016) Polypeptide transport-associated domains of the Toc75 channel protein are located in the intermembrane space of chloroplasts. Plant Physiol 172(1):235–243. https://doi.org/10.1104/pp.16.00919
269. Day PM, Inoue K, Theg SM (2019) Chloroplast outer membrane beta-Barrel proteins use components of the general import apparatus. Plant Cell 31(8):1845–1855
270. O'Neil PK, Richardson LGL, Paila YD, Piszczek G, Chakravarthy S, Noinaj N, Schnell D (2017) The POTRA domains of Toc75 exhibit chaperone-like function to facilitate import into chloroplasts. Proc Natl Acad Sci U S A 114(24):E4868–E4876. https://doi.org/10.1073/pnas.16211791
271. Zhang J, Wu S, Boehlein SK, McCarty DR, Song G, Walley JW, Myers A, Settles AM (2019) Maize defective kernel5 is a bacterial TamB homologue required for chloroplast envelope biogenesis. J Cell Biol 218(8):2638–2658. https://doi.org/10.1083/jcb.201807166
272. Heinz E, Stubenrauch CJ, Grinter R, Croft NP, Purcell AW, Strugnell RA, Dougan G, Lithgow T (2016) Conserved features in the structure, mechanism, and biogenesis of the inverse autotransporter protein family. Genome Biol Evol 8(6):1690–1705. https://doi.org/10.1093/gbe/evw112
273. Selkrig J, Belousoff MJ, Headey SJ, Heinz E, Shiota T, Shen HH, Beckham SA, Bamert RS, Phan MD, Schembri MA, Wilce MC, Scanlon MJ, Strugnell RA, Lithgow T (2015) Conserved features in TamA enable interaction with TamB to drive the activity of the translocation and assembly module. Sci Rep 5:12905. https://doi.org/10.1038/srep12905
274. Selkrig J, Mosbahi K, Webb CT, Belousoff MJ, Perry AJ, Wells TJ, Morris F, Leyton DL, Totsika M, Phan MD, Celik N, Kelly M, Oates C, Hartland EL, Robins-Browne RM, Ramarathinam SH, Purcell AW, Schembri MA, Strugnell RA, Henderson IR, Walker D, Lithgow T (2012) Discovery of an archetypal protein transport system in bacterial outer membranes. Nat Struct Mol Biol 19(5):506–510, S1. https://doi.org/10.1038/nsmb.2261
275. Shen HH, Leyton DL, Shiota T, Belousoff MJ, Noinaj N, Lu J, Holt SA, Tan K, Selkrig J, Webb CT, Buchanan SK, Martin LL, Lithgow T (2014) Reconstitution of a nanomachine driving the assembly of proteins into bacterial outer membranes. Nat Commun 5:5078. https://doi.org/10.1038/ncomms6078
276. Stubenrauch C, Belousoff MJ, Hay ID, Shen HH, Lillington J, Tuck KL, Peters KM, Phan MD, Lo AW, Schembri MA, Strugnell RA, Waksman G, Lithgow T (2016) Effective assembly of fimbriae in *Escherichia coli* depends on the translocation assembly module nanomachine. Nat Microbiol 1(7):16064. https://doi.org/10.1038/nmicrobiol.2016.64
277. Heinz E, Selkrig J, Belousoff MJ, Lithgow T (2015) Evolution of the Translocation and Assembly Module (TAM). Genome Biol Evol 7(6):1628–1643. https://doi.org/10.1093/gbe/evv097
278. Li Y, Singhal R, Taylor IW, McMinn PH, Chua XY, Cline K, Fernandez DE (2015) The Sec2 translocase of the chloroplast inner envelope contains a unique and dedicated SECE2 component. Plant J 84(4):647–658. https://doi.org/10.1111/tpj.13028
279. Li Y, Martin JR, Aldama GA, Fernandez DE, Cline K (2017) Identification of putative substrates of SEC2, a chloroplast inner envelope translocase. Plant Physiol 173(4):2121–2137
280. Fernandez DE (2018) Two paths diverged in the stroma: targeting to dual SEC translocase systems in chloroplasts. Photosynth Res 138(3):277–287. https://doi.org/10.1007/s11120-018-0541-9
281. Schuenemann D, Amin P, Hartmann E, Hoffman NE (1999) Chloroplast SecY is complexed to SecE and involved in the translocation of the 33-kDa but not the 23-kDa subunit of the oxygen-evolving complex. J Biol Chem 274(17):12177–12182. https://doi.org/10.1074/jbc.274.17.12177
282. Nakai M, Goto A, Nohara T, Sugita D, Endo T (1994) Identification of the SecA protein homolog in pea chloroplasts and its possible involvement in thylakoidal protein transport. J Biol Chem 269(50):31338–31341. https://doi.org/10.1016/S0021-9258(18)31698-3

283. Yuan J, Henry R, McCaffery M, Cline K (1994) SecA homolog in protein transport within chloroplasts: evidence for endosymbiont-derived sorting. Science 266(5186):796–798. https://doi.org/10.1126/science.7973633
284. Celedon JM, Cline K (2013) Intra-plastid protein trafficking: how plant cells adapted prokaryotic mechanisms to the eukaryotic condition. Biochim Biophys Acta 1833(2):341–351. https://doi.org/10.1016/j.bbamcr.2012.06.028
285. Voelker R, Barkan A (1995) Two nuclear mutations disrupt distinct pathways for targeting proteins to the chloroplast thylakoid. EMBO J 14(16):3905–3914
286. Settles AM, Yonetani A, Baron A, Bush DR, Cline K, Martienssen R (1997) Sec-independent protein translocation by the maize Hcf106 protein. Science 278(5342):1467–1470. https://doi.org/10.1126/science.278.5342.1467
287. Walker MB, Roy LM, Coleman E, Voelker R, Barkan A (1999) The maize tha4 gene functions in sec-independent protein transport in chloroplasts and is related to hcf106, tatA, and tatB. J Cell Biol 147(2):267–276. https://doi.org/10.1083/jcb.147.2.267
288. Mori H, Summer EJ, Cline K (2001) Chloroplast TatC plays a direct role in thylakoid (Delta) pH-dependent protein transport. FEBS Lett 501(1):65–68
289. Motohashi R, Nagata N, Ito T, Takahashi S, Hobo T, Yoshida S, Shinozaki K (2001) An essential role of a TatC homologue of a Delta pH- dependent protein transporter in thylakoid membrane formation during chloroplast development in *Arabidopsis thaliana*. Proc Natl Acad Sci U S A 98(18):10499–10504. https://doi.org/10.1073/pnas.181304598
290. Berks BC, Sargent F, Palmer T (2000) The Tat protein export pathway. Mol Microbiol 35(2):260–274. https://doi.org/10.1046/j.1365-2958.2000.01719.x
291. Chaddock AM, Mant A, Karnauchov I, Brink S, Herrmann RG, Klosgen RB, Robinson C (1995) A new type of signal peptide: central role of a twin-arginine motif in transfer signals for the delta pH-dependent thylakoidal protein translocase. EMBO J 14(12):2715–2722. https://doi.org/10.1002/j.1460-2075.1995.tb07272.x
292. Henry R, Carrigan M, McCaffrey M, Ma X, Cline K (1997) Targeting determinants and proposed evolutionary basis for the Sec and the Delta pH protein transport systems in chloroplast thylakoid membranes. J Cell Biol 136(4):823–832. https://doi.org/10.1083/jcb.136.4.823
293. Creighton AM, Hulford A, Mant A, Robinson D, Robinson C (1995) A monomeric, tightly folded stromal intermediate on the delta pH-dependent thylakoidal protein transport pathway. J Biol Chem 270(4):1663–1669. https://doi.org/10.1074/jbc.270.4.1663
294. Clark SA, Theg SM (1997) A folded protein can be transported across the chloroplast envelope and thylakoid membranes. Mol Biol Cell 8(5):923–934. https://doi.org/10.1091/mbc.8.5.923
295. Moore M, Harrison MS, Peterson EC, Henry R (2000) Chloroplast Oxa1p homolog albino3 is required for post-translational integration of the light harvesting chlorophyll-binding protein into thylakoid membranes. J Biol Chem 275(3):1529–1532. https://doi.org/10.1074/jbc.275.3.1529
296. Schuenemann D, Gupta S, Persello-Cartieaux F, Klimyuk VV, Jones JDG, Nussaume L, Hoffman NE (1998) A novel signal recognition particle targets light-harvesting proteins to the thylakoid membranes. Proc Natl Acad Sci U S A 95(17):10312–10316. https://doi.org/10.1073/pnas.95.17.10312
297. Falk S, Sinning I (2010) The C terminus of Alb3 interacts with the chromodomains 2 and 3 of cpSRP43. J Biol Chem 285(53):le25-6; author reply le26-8. https://doi.org/10.1074/jbc.L110.160093
298. Klostermann E, Helling IDG, Carde JP, Schuenemann D (2002) The thylakoid membrane protein ALB3 associates with the cpSecY- translocase in *Arabidopsis thaliana*. Biochem J 368(Pt 3):777–781. https://doi.org/10.1042/bj20021291
299. Walter B, Hristou A, Nowaczyk MM, Schunemann D (2015) In vitro reconstitution of co-translational D1 insertion reveals a role of the cpSec-Alb3 translocase and Vipp1 in photosystem II biogenesis. Biochem J 468(2):315–324. https://doi.org/10.1042/BJ20141425
300. Benz M, Bals T, Gugel IL, Piotrowski M, Kuhn A, Schunemann D, Soll J, Ankele E (2009) Alb4 of *Arabidopsis* promotes assembly and stabilization of a non chlorophyll-binding photosynthetic complex, the CF1CF0-ATP synthase. Mol Plant 2(6):1410–1424. https://doi.org/10.1093/mp/ssp095

Photosynthetic Proteins in Cyanobacteria: from Translocation to Assembly of Photosynthetic Complexes

Julie A. Z. Zedler

Abstract Cyanobacteria are ancient prokaryotes known for their ability to perform oxygenic photosynthesis. The photoautotrophic lifestyle of these organisms requires the efficient synthesis, translocation and assembly of photosynthetic proteins into supramolecular photosynthetic complexes which are typically localised in the thylakoid membranes. Co- and posttranslational translocation of proteins across and into cyanobacterial membranes is carried out by multiple protein translocation systems. These systems, which include the general secretory pathway, the twin-arginine translocation pathway and the signal recognition pathway, are highly conserved across prokaryotes and organelles of endosymbiotic origin. However, cyanobacteria are unique in the requirement to target proteins to and across multiple membranes (cytoplasmic and thylakoid) with a single set of machinery for each translocation pathway. Therefore, one of the current challenges is to understand how photosynthetic proteins and cofactors are synthesized, targeted, and assembled into supramolecular photosynthetic complexes. Recent studies have provided an unprecedented view into thylakoid biogenesis and organisation of photosynthetic complexes in cyanobacteria; however, open questions regarding targeting mechanisms and the existence of a physical connection between membrane systems remain. This chapter will summarize the existing knowledge on protein trafficking in cyanobacteria with a focus on photosynthetic proteins. This information will then be contextualised to discuss similarities and differences between protein targeting in cyanobacteria, other prokaryotes and chloroplast thylakoid membranes.

Keywords Cyanobacteria · Protein targeting · General secretory pathway · Leader peptidase · Membrane biogenesis · Photosystem assembly · Protein sorting · Signal recognition particle · Twin-arginine translocation pathway · Thylapse

---

J. A. Z. Zedler (✉)
Synthetic Biology of Photosynthetic Organisms, Matthias Schleiden Institute for Genetics, Bioinformatics and Molecular Botany, Friedrich Schiller University Jena, Jena, Germany
e-mail: julie.zedler@uni-jena.de

S. D. Schwartzbach et al. (eds.), *Endosymbiotic Organelle Acquisition*,
https://doi.org/10.1007/978-3-031-57446-7_11

1 Introduction

Cyanobacteria are the only known prokaryotes performing oxygenic photosynthesis. With their significant contribution towards carbon fixation and other biogeochemical cycles, the importance of their ecological role is unquestionable. Due to their status as the "progenitors of plastids" [1–3], cyanobacteria have served as an attractive model for generations of photosynthesis researchers.

The thylakoid membrane (TM) is the major site of both the light reactions of photosynthesis, as well as the respiratory electron transport chain in cyanobacteria [4]. These processes are highly dynamic and require the efficient adjustment of light harvesting, photosynthetic electron flow, and respiration to ensure a balanced redox state in the cell. At the core of these processes are multimeric, co-factor-containing, protein complexes that require precise assembly, maintenance and repair. The question arises, how are these complexes assembled and targeted to their final destinations? Generally, protein targeting in cyanobacteria is fulfilled by highly conserved prokaryotic translocation systems. However, the presence of multiple membrane systems, and the common ancestry with chloroplasts, leads to a somewhat unique situation in cyanobacteria. Are the underlying molecular mechanisms more similar to other prokaryotes or to chloroplasts? This book chapter will provide an overview of membrane organisation interwoven with the current state of the art regarding protein targeting in cyanobacteria to give a holistic view into thylakoid biogenesis and organisation of photosynthetic complexes in cyanobacteria.

2 Ultrastructure and Organisation of Membranes in Cyanobacteria

Cyanobacteria typically contain three distinct membranes: the outer membrane (OM), the plasma membrane (PM) and the TM. The only known exception is the basal cyanobacterial genus *Gloeobacter,* where the TM is absent [5, 6]. In *Gloeobacter violaceus*, respiratory or photosynthetic complexes with distinct proteomic compositions are found in highly specialised regions of the PM [7]. Thylakoid membranes are highly specialized, densely packed membranes where the photosynthetic complexes are located. Here, the electron transport chain of oxygenic photosynthesis takes place, as well as a large proportion of the respiratory electron transport chain [4, 8]. The TM encloses a continuous space, the lumen, which is thought to be composed of individual, connected membrane sacs [9, 10]. The pattern of thylakoid arrangements in cyanobacteria is highly diverse with various types described ranging from parietal, radial, parallel to fascicular or also irregular [11]. The evolutionary patterns of thylakoid arrangement in cyanobacteria are complex. This is thought to be partly due to repeated convergent evolution, which limits its value as a taxonomical trait [11]. While higher plant chloroplasts contain densely packed grana stacks [12], these are not present in cyanobacterial thylakoid membranes [13, 14]. In cyanobacteria, soluble light-harvesting complexes, known as phycobilisomes, are attached to the photosystems (mainly to photosystem II (PSII)) [15] requiring a looser TM arrangement. The overall lipid composition of cyanobacterial membranes is similar to chloroplast thylakoid membranes [16] but can vary among species and within regions of the TM harbouring different photosynthetic complexes [17]. The mobility of the individual protein complexes varies [18], leading to what is known as lateral heterogeneity of the TM [8, 19–25]. Consequently, subregions of specialised functions appear, for example, for photosystem (PS) repair [26] or protein biosynthesis [9]. In addition, mobility can be influenced by changes in environmental factors that require fast and efficient regulation of photosynthesis. For instance, higher mobility of complexes was found in the dark compared to light in *Synechocystis* sp. PCC 6803 (hereafter *Synechocystis*) [27] and red light can trigger major rearrangements of photosynthetic complexes in *Synechococcus elongatus* PCC 7942 (hereafter *S. elongatus*) [25, 28].

2.1 PM and TM: Interplay of Membranes?

Whether the PM and TM are connected is a long-standing controversial issue. This is, in part due to similarities in the composition of the PM and the TM, which makes the biochemical analysis and isolation of clean membrane fractions difficult [29–31]. A range of studies have looked at subcellular proteomes [20, 31–38], but reports regarding the organisation of PM and TM are mainly based on ultrastructural analysis. Earlier studies reported connections between PM and TM in *S. elongatus* [39] and *Synechococcus* sp. PCC 7002 [40, 41]. Later studies made similar observations in *Synechocystis*, but these were called regions of convergence [42]. However, an ultrastructural study in *Synechocystis* presented conflicting evidence, suggesting neither a physical connection between PM and TM nor evidence of vesicular transport [43]. A study based on fluorescent dye staining of membranes in *Synechocystis* also concluded that TM and PM seem to be separate entities in cyanobacteria, but suggested that some type of active transport system between the two membranes must exist [44]. Further evidence using state-of-the-art cryo-electron tomography showed no direct connections between TM and PM in *Synechocystis* [9]. However, the authors did not rule out that specialised converging regions of the TM, so-called thylapses, might fuse with the PM under certain environmental conditions [9]. Similar findings were reported in *S. elongatus* [45]. A recently identified protein, AncM (Slr2070), has been suggested to form transient connections between PM and TM based on protein bridges in *Synechocystis* (see Sect. 2.2 for more details) [46] and is a candidate to be involved in transport between these membranes.

2.2 Thylakoid Membrane Biogenesis

Advances have been made in the field; however, to date, TM biogenesis is still poorly understood. For detailed accounts of membrane dynamics and composition, see recent reviews [8, 19, 24]. In brief, TM biogenesis occurs most likely through some form of lipid trafficking [24], for instance by membrane fusion or vesicular processes similar to those known from plastids [47]. A functional TM requires the insertion of lipids and proteins, as well as the formation of functional protein complexes with co-factors (e.g., pigments). Thus, an essential element for TM biogenesis is the cyanobacterial homologue of YidC/Alb3/Oxa1 required for membrane insertion of proteins [48, 49]. More details regarding membrane insertion of proteins via this pathway will be described in Sect. 5.

Regions of the TM that converge close to the PM were first defined as "thylakoid centres" (TCs) in cyanobacteria in 1982 [50]. These TCs seem to be crucial for TM biogenesis and most likely also for the initial steps of PSII assembly (see Sect. 5). To date, most of what we know regarding these subcellular structures is based on the model organism *Synechocystis.* TCs co-localize with the tetratricopeptide repeat protein PratA [51, 52], which is involved in manganese uptake and delivery to PSII [52]. A recent study refined our understanding of these converging regions between PM and TM in *Synechocystis* and defined them as thylapses [9]. These structures are characterized by a distinct lack of phycobilisomes and a high ribosome content, which is consistent with a function in membrane (and PSII) biogenesis [9].

Several proteins are known to be involved in membrane biogenesis although their exact molecular function is often unclear. A famous example of this is the highly conserved vesicle-inducing protein in plastids 1 (VIPP1 or IM30). This protein is a key component involved in TM biogenesis [53]. A cyanobacterial homologue of this protein exists and its partial deletion in *Synechocystis* (Δ*sll0617*) results in malformation of the TM [54]. Vipp1 seems to be important for membrane protection and stabilisation [55], but how it is exactly involved in membrane biogenesis remains uncertain [54, 56]. Furthermore, Vipp1 was also proposed as a candidate for connecting PM and TM [55], for example, by membrane fusion [8, 19], although a recent cryo-ET study in *Synechocystis* did not find evidence to support this [9]. Cryo-EM structure elucidation of Vipp1 from *Synechocystis* showed that the protein forms oligomeric rings and might play an essential role in lipid transport at the converging biogenesis membranes [57].

Another small, membrane-shaping protein is CurT. This protein is known to be needed for thylakoid bending in grana stacks of chloroplasts [58] and was also identified in *Synechocystis* (Slr0483) [14]. A deletion of CurT is viable but results in distorted membrane architecture and growth deficits in agreement with the general role of this protein family in membrane curvature [14]. CurT was suggested to be involved in membrane biogenesis and seems to be important for tubular membrane formation at the thylapse contact site with the PM [9]. Follow-up work on a CurT knockout mutant of *Synechocystis* [14] identified a further protein important for TM maintenance, the aforementioned AncM [46]. Maintenance of a functional TM in

Synechocystis seems to be dependent on the balance of the antagonistically acting proteins CurT and AncM [46]. Further proteins and factors known to be involved in membrane maintenance and biogenesis include the SPFH-domain protein Slr1768 in *Synechocystis*. This member of the band 7 superfamily seems to be tightly PM-associated [59] and involved in TM maintenance under high light [60]. The deletion of *slr1768* can lead to the lack of the TM and cell death [60]. Additionally, hydrocarbons may play a role in TM shaping (curvature) and influence membrane flexibility [61]. Taken together, these studies have provided further stepping stones towards the elucidation of the interplay of TM biogenesis factors and the different membrane systems in cyanobacteria.

Despite the number of factors known to play a key role in TM biogenesis, this knowledge is almost exclusively derived from the model organism *Synechocystis*. Here, major advances in the characterisation of the TC (more specifically the thylapse) ultrastructure and molecular composition have emerged. In *S. elongatus*, where the TM is organised in regular concentric cylinders, evidence for thylapse-like structures was found [45]. It is, however, uncertain to what extent these findings are transferable to other clades of the phylum of cyanobacteria. Some factors such as YidC/Alb3/Oxa1 and Vipp1 are highly conserved beyond cyanobacteria [54, 62, 63], but the phylogeny of other important TM biogenesis factors is more complex. For instance, homologues of CurT and AncM were neither identified in basal cyanobacterial clades nor in the widespread marine picocyanobacterial genus *Prochlorococcus* [14, 46]. This raises the question of how universal the involvement of these proteins in cyanobacterial TM biogenesis is [11]. The lack of these homologues in the basal genus *Gloeobacter* is consistent with the general lack of a TM in this phylogenetic group. A correlation of TM complexity and the presence of CurT/AncM was proposed [46]. Interestingly, no AncM homologue was identified in the Archaeplastida and the AncM *Synechocystis* homologue differed substantially from other cyanobacterial homologues in its N-terminal extension. This could hint at the thylapse being a rather specialized scenario in *Synechocystis*. Other cyanobacteria containing CurT/AncM homologues might have evolved different molecular mechanisms of TM biogenesis and maintenance with or without the formation of TCs.

3 Assembly and Maintenance of Photosynthetic Complexes

The assembly of photosynthetic complexes is tightly associated with TM biogenesis (see Sect. 2.2). Although our understanding of the molecular composition and mechanisms of photosynthetic complexes has substantially progressed over decades of research, many open questions remain. The core complexes involved in the photosynthetic electron transport chain are PSII, photosystem I (PSI), and the cytochrome b_6f complex. In cyanobacteria, PSI is the most abundant photosynthetic complex. Approximately 100,000 copies were found in a cell of *Synechocystis,* while PSII is around two to five times less abundant followed by a two to 20 times lower abundance of the cytochrome b_6f complex [64].

A functional photosynthetic electron transport chain requires the initial assembly of the individual protein subunits together with their cofactors as well as complex maintenance and repair mechanisms. This is an enormous undertaking, considering the final composition of functional photosystems. For instance, PSI from *S. elongatus* forms trimers of which each monomer comprises nine α-helical proteins, three stromal subunits and 127 co-factors, including 96 chlorophyll *a* and 22 carotenoid pigments [65]. The individual subunits of these complexes also need to be efficiently targeted and inserted into the TM. A complex network of further assembly factors and spatiotemporal coordination is needed in PS maturation. These processes also have to be highly dynamic for acclimation of the cell to environmental changes [25, 45, 66]. Early steps of PS assembly were initially described to occur in the PM in *Synechocystis* [67, 68] and thought to be transported to the TM via lateral fusions [68]. More recent studies provide conflicting evidence, suggesting that PS assembly takes place in the TCs (thylapses) [9, 52, 69] or potentially also in the TM [70].

Despite its abundance and pivotal role, the organisation of PSI in the cyanobacterial TM is still relatively poorly understood [22]. PSI was recently found to be tightly packed in macromolecular arrays in the TM [22] and its biogenesis seems to be relatively homogenous across newly synthesised sections and the existing TM *in S. elongatus* [45]. Many PSI assembly factors have been described and are important at different stages of the maturation process (a recent review and references therein provide further information [71]). For example, PsaD is involved in insertion of several PSI subunits [72] and Ycf37 seems to be stabilising PSI intermediates from proteolytic degradation processes [73]. The assembly of PSII occurs as a step-wise, modular process [74] involving a plethora of different assembly factors. Recent reviews provide an in-depth view of PSII assembly beyond what can be mentioned in this book chapter [75, 76]. In brief, complex assembly is initiated by heterodimerisation of D1 (PsbA) and D2 (PsbD). The membrane insertase YidC (Alb3/Oxa1) is essential for membrane insertion of pre-D1 [48] with a C-terminal extension, which is cleaved as the maturation of PSII proceeds by CptA in a two-step process [77] with the assistance of Ycf48 [78]. Loading with manganese takes place in early PSII assembly at the thylapse (earlier referred to as PratA-defined membrane, TC) [52]. HliD is needed for recruitment of the chlorophyll synthase for direct delivery of chlorophyll pigments to nascent PSII apoproteins together with YidC and Ycf39 [79, 80]. During further complex maturation of PSII, core antenna proteins are recruited to the complex: first CP47 (RC47 stage), followed by attachment of CP43. In later stages, the oxygen-evolving complex (OEC) is assembled. PSII maturation proceeds with the insertion of the Mn_4Ca_5O cluster which activates water splitting. The last step towards fully assembled PSII is the dimerisation and attachment of the phycobilisome complexes [75].

Inevitably, many types of reactive oxygen species (ROS) are generated through the light reactions and, thus, efficient detoxification and PS repair mechanisms are needed. While PSI is relatively stable, PSII has a much higher turnover and requires extensive maintenance and repair. D1 has the highest turnover within PSII and, thus, specific mechanisms to exchange this particular protein must exist although the exact mechanisms remain unknown (reviewed in [76]). The repair of PSII seems to be spatially separated from de novo assembly [81] and PSII needs to be partially disassembled to replace D1 [81]. It has been suggested that the economy of costly PSII repair might be improved by recycling of intact CP43/47 core antenna complexes detached from the PSII reaction centre (RC) [82]. However, a later study did not find evidence for the formation of these CP43/47 core antenna complexes lacking the RC proteins [83]. PSII repair seems to be less localised at the TCs than PSII assembly and distributed more evenly across the TM [70]. Repair mechanisms for D1 exchange require a range of proteases (reviewed in [74]). The family of FtsH proteases is involved in D1 degradation [84], and localisation of these proteases in *Synechocystis* suggests specific, localized PSII repair zones [26].

Overall, our understanding of spatiotemporal patterns of the life cycle of photosystems in cyanobacteria has substantially improved over the years, but many questions remain to be answered. Prerequisites for the assembly and maintenance of all photosynthetic complexes are efficient protein translocation and sorting mechanisms. In the case of photosystems, the coordination of pigment and protein biosynthesis is also crucial to avoid the generation of ROS [80, 85]. The following sections will summarise what we know (and don't know) regarding these first important steps of protein targeting in cyanobacteria.

4 Translocation and Targeting of Proteins in Cyanobacteria

The presence of three distinct membrane systems comes with the prerequisite for active translocation mechanisms into and across these membranes. The membrane architecture of these organisms results in a unique variety of potential protein destinations amongst prokaryotes. The following possible routes of protein targeting exist in most cyanobacteria: translocation across the TM (protein import), across the PM (protein export), or membrane insertion into one of the membranes (PM/TM/OM). Furthermore, proteins can be secreted to the extracellular medium across PM and OM in a one- or two-step process. A two-step secretion route requires a first step of protein export to the periplasm, followed by a translocation event across the OM, while one-step secretion systems span both membranes (PM and OM) [86]. This book chapter will focus on these mechanisms with special relevance to photosynthetic proteins and, thus, protein import and membrane insertion routes. For details of common secretion mechanisms and systems in cyanobacteria, the reader is directed to a recent review [86]. In any event of protein translocation in cyanobacteria, first, protein biosynthesis will take place at the cytosolic ribosome, followed by either co-translational or post-translational translocation of the protein. Translocation of a protein cargo can take place in an unfolded or pre- to fully folded state, depending on the translocation route of the protein. Therefore, the translocation requirements in cyanobacteria are distinct from plastids where the bulk of proteins are nucleus-encoded and, thus, need to typically first be imported through the

TOC/TIC complex in an unfolded state [87] and then fold following translocation with various quality control mechanisms in place [88].

The translocation pathways present in cyanobacteria are highly conserved across all prokaryotes and chloroplasts. The General Secretory (Sec) pathway and the twin-arginine translocation (Tat) pathway carry out protein import and export (Fig. 1). The main route for PM/TM insertion is signal-recognition particle (SRP) dependent. How protein insertion into the OM occurs in cyanobacteria is, to date, essentially unknown [89–91]. In bacteria, these processes usually require the cooperation of the translocation and assembly module (TAM) and the β-barrel assembly machinery (BAM) [92]. Interestingly, cyanobacteria contain a single set of genes for all main translocation pathways [93]. Therefore, the translocation complexes themselves must be operating in both TM and PM and cannot ensure membrane-specific targeting. So how are proteins targeted to a specific subcellular destination and efficiently sorted? This is a long-standing question [94, 95] and, to date, our understanding of the underlying molecular mechanisms is still unsatisfactory. Various potential scenarios of pre- and post-translocational sorting models have been discussed over the years [30, 86, 93, 96–99]. In a recent study, the mRNA of *psaA* and *psbA* were detected by RNA-fluorescent in-situ hybridisation

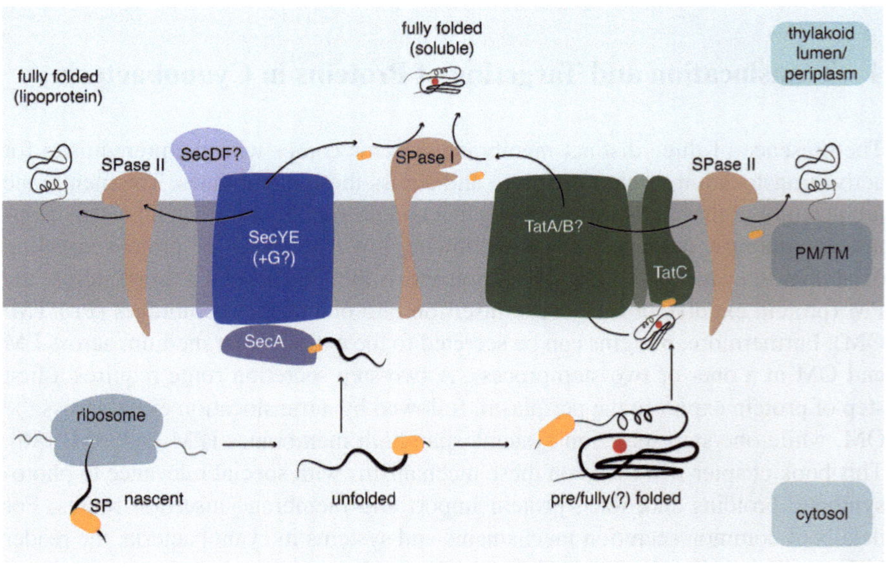

Fig. 1 Overview of protein import/export machinery in cyanobacteria. Proteins are translated at the ribosome in the cytosol and subsequently translocated across the plasma membrane (PM) for protein export to the periplasm or across the thylakoid membrane (TM) for protein import into the thylakoid lumen. A specific signal peptide (SP) sequence targets the protein to the Sec or Tat pathway for translocation. The exact composition of these translocons is not known. Signal (leader) peptidases (SPase) cleave type I Sec/Tat SPs (SPaseI) or type II (SPaseII, Sec and Tat lipoproteins) upon translocation. The specialised case of the type IV SPase for pre-pilin maturation is not shown in this overview. Energetic requirements for different translocation routes are not shown

concentrated around the TM surface in *Synechocystis* and *S. elongatus* [70]. This study suggested that protein targeting in cyanobacteria might be mRNA dependent and involve mRNA-binding proteins that essentially regulate access to the TM [70]. These observations would also be consistent with a potential default export route of translocation [100], that is, "out of the cell", with access to the TM restricted to a subset of proteins regulated by TM-associating mRNA binding proteins. In the following, the general characteristics of import/export and PM/TM membrane insertion in cyanobacteria will be outlined.

4.1 N-Terminal Signal Peptides Determine the Mode of Import/Export

A protein cargo is usually directed to the Sec or Tat translocation route via an N-terminal signal peptide (SP), which is cleaved upon translocation by a leader peptidase (see Sect. 4.4). Sec and Tat SPs share an overall tri-partite structure: (I) a n-region containing basic amino acid (AA) residues, (II) a hydrophobic core (h-region) and (III) a polar c-region at the C-terminus of the SP including a signal (leader) peptidase cleavage site [101]. In the case of classic Tat and Sec substrates, the type I signal peptidase cleavage site is A-x-A, while Tat/Sec lipoproteins have a conserved, lipidated cysteine residue at position 1 of the mature protein preceded by a L-A/S/T-G/A type II consensus motif [101]. Nevertheless, there are crucial differences between Tat and Sec SPs. The n-region of Tat SPs is generally longer than that of Sec SPs with an S-R-R-x-F-L-K consensus sequence containing the eponymous twin arginine residues [102]. Additionally, the h-region of Tat SPs is less hydrophobic compared to Sec SPs [103] and the c-region tends to be more polar serving as a Sec-avoidance motif [104]. While the primary amino acid sequence is highly variable amongst Tat/Sec SPs, the overall structure of SPs is highly conserved across substrates and organisms. For instance, the Trimethylamine-N-oxide reductase 1 (TorA) Tat SP from *E. coli* was recognised as an export signal in *Synechocystis* PCC 6803 [100]. A Sec SP of a secreted lytic polysaccharide monooxygenase from the gram-positive bacterium *Thermobifida fusca* enabled secretion in *Synechococcus elongatus* UTEX 2973 [105] and *Synechocystis* [106].

Finally, pilins that form the type IV pilus (T4P) filament contain a specialised, so-called pre-pilin, SP. This signal sequence is recognised by SRP and directs the protein to the Sec pathway for export followed by SP cleavage through a type IV signal peptidase [107]. The SP itself contains basic amino acids with a conserved C-terminal G residue (−1 cleavage position) followed by the AA residues F-T-L-I-E in the mature pilin sequence [108]. Type IV pili also occur in cyanobacteria (recently reviewed in [86, 109, 110]).

It should be mentioned that, in both Tat and Sec-dependent translocation events, not all SPs are cleaved upon translocation. This applies predominantly to membrane proteins where the signal sequence is part of a transmembrane α-helix (reviewed in

[108, 111], see Sect. 5). One example from cyanobacteria is the Rieske iron–sulphur (hereafter Rieske) protein of the cytochrome b_6f complex. The major Rieske protein in *Synechocystis*, PetC1 (Sll1316) [112], contains a Tat SP that is experimentally verified [97]. The Cryo-EM structure of the cytochrome b_6f complex from *Synechocystis* shows that this PetC1-Tat SP is part of the membrane anchor in the mature protein [113] as previously shown for chloroplasts [114]. In addition, a proteomic study found a subset of soluble proteins in the thylakoid lumen of *Synechocystis* that seemingly lacked SPs, suggesting the existence of alternative transport routes independent of a SP [115]. A process referred to as "Tat hitchhiking" is described in the literature for the large subunit of Hyd2 in *E. coli* [116]. This refers to a cargo protein (the small subunit, HybO) being translocated across a membrane by complex formation with a bona fide Tat substrate (a large subunit, HybC) [116]. Whether similar mechanisms could also exist in cyanobacteria or if there are other, yet-to-be-described, mechanisms is at this stage not known. For instance, less established transport pathways for folded proteins are coming to light in prokaryotes and eukaryotes (reviewed in [117]), which might also be applicable to cyanobacteria.

4.2 *The Sec Pathway Is Responsible for Bulk Protein Translocation*

The Sec pathway is universally conserved amongst all domains of life and operates in bacteria, archaea [118], as well as in the endoplasmic reticulum and plastids in eukaryotic organisms. Via the Sec pathway, proteins are translocated across membranes in an unfolded state driven by ATP hydrolysis. Usually, the Sec route fulfils the "bulk job" of protein translocation in bacteria and this also seems to apply to cyanobacteria [35] (Fig. 1). The Sec pathway is also known to assist in membrane insertion (see Sect. 5).

The Sec translocation pore of bacteria, as present in *E. coli,* is typically formed by SecY, SecE, and SecG. The soluble ATPase SecA is responsible for Sec substrate recognition by binding to the Sec SP post-translation in the cytosol and recruitment to the membrane. Further elements of the *E. coli* Sec pathway are the cytosolic chaperone, SecB, and the non-essential SecDF complex, that associates with the SecYEG translocation pore [119]. Recent reviews describe the mechanistic details of what we know about Sec translocation in further detail [120–122].

While mechanistically equivalent to bacteria, the Sec system in chloroplasts seems to operate with a "minimal" Sec translocon (cpSecY + cpSecE) and cpSecA without SecG, SecB, or SecD/F homologues [123]. Additionally, two functionally distinct Sec translocases are known to operate in chloroplasts with membrane specificity (TM and chloroplast inner envelope membrane) [124–126]. On the other hand, cyanobacteria encode only one set of Sec genes namely SecY (*Synechocystis*: Sll1814), SecE (*Synechocystis*: Ssl3335), SecG (*Synechocystis*: Ssr3307), and SecA

(*Synechocystis*: Sll0616) [86]. Furthermore, in most cyanobacteria, two SecD/F homologues seem to be present (*Synechocystis*: Slr0774, Slr0775) and no SecB homologoues were found [86] (Fig. 1). In the literature, some experimental evidence of the Sec pathway from cyanobacteria can be found: SecA [127] and SecY [95] homologues were experimentally identified in *S. elongatus* associated with both TM and PM and proteomic studies have identified Sec subunits in both TM and PM of *Synechocystis* [31]. SecG has not been found experimentally in cyanobacteria and might therefore be non-essential [86] as known from *E. coli* [128, 129]. Typical cyanobacterial Sec substrates of the photosynthetic machinery include the Sec lipoprotein PsbO (subunit of the oxygen-evolving complex), PetA (cytochrome f of the cytochrome b_6f complex), the soluble electron carrier plastocyanin, and PsaF (PSI reaction centre). Overall, the Sec system of cyanobacteria seems to operate with dual membrane localisation in a similar fashion as in *E. coli,* where the SecB chaperone function is fulfilled by other chaperones [98]. Nevertheless, only experimental evidence can validate this further.

4.3 The Tat Pathway Is Essential for Translocation of Folded Proteins

The Tat pathway is highly conserved in bacteria, archaea, and chloroplasts, where it was originally discovered [130, 131]. Plant mitochondria have retained a Tat pathway [132, 133], but across other mitochondrial lineages, it has been lost to various degrees [134]. Tat depends on the presence of a pH gradient across a membrane [131] and, in contrast to the Sec pathway, transports proteins in a folded state. This is particularly important for co-factor-containing proteins (e.g., iron–sulphur cluster-containing proteins, metal-binding redox enzymes) that require co-factor assembly in the cytosol prior to translocation [99]. Additionally, the Tat system is known for its ability to proofread translocation substrates in bacteria (reviewed in [135, 136]). Nevertheless, this proof-reading seems to be less stringent in chloroplasts [137] and it is not known if or to what extent the cyanobacterial Tat system has proof-reading capacities.

Generally, there are two main types of Tat systems known: First, the TatABC, as found in *E. coli* and chloroplasts. The chloroplastic Tat (cpTat) system operates with the homologues Tha4 (TatA), Hcf106 (TatB) and cpTatC (TatC). Second, a minimal-type Tat system only consisting of TatA and TatC (e.g., *Bacillus subtilis*). It is not clear what type of Tat system operates in cyanobacteria and it is possible that this could vary among cyanobacterial clades (Fig. 1) [86]. For example, *Synechocystis* has two TatA/B homologues (Slr1046/Ssl2823) and one Tat C protein (Sll0194), in contrast to *S. elongatus* where only one TatA/B (Synpcc7942_0226) and one TatC homologue (Synpcc7942_1578) is present. TatA and TatB homologues are overall structurally similar across the phylogenetic tree [138] and are thought to be derived from early events of gene duplication. The TatA/B homologues from *Synechocystis* were shown to both functionally complement knockout mutants of TatB as well as TatAE in *E. coli* [97]. However, the composition of active Tat translocons in cyanobacteria is not known, and thus, no further attribution of function is possible at this stage. Similar to the Sec pathway, it has been experimentally verified that the Tat system operates in PM and TM [97]. Nevertheless, based on proteomic surveys in *Synechocystis,* it might be more abundant in the TM [31, 43].

Independent of the exact composition of the Tat machinery, a Tat-dependent translocation process is initiated by SP recognition through TatC in a Tat(B)C–TatA resting state membrane complex. In the presence of a proton motive force, this leads to activation of the Tat translocase and initiation of translocation through the recruitment of further TatA protomers and formation of the TatA translocation pore [139]. Several recent reviews give detailed insights concerning mechanisms of the Tat pathway [101, 123, 136, 139–141].

In chloroplasts, approximately half of the proteins present in the thylakoid lumen are predicted to be Tat substrates, despite not all of them requiring cofactors [141]. The cpTat translocase has also been shown to be capable of simultaneously transporting up to eight substrates [142]. This emphasises the major role of the Tat system in chloroplast protein trafficking. In cyanobacteria, the predicted number of Tat substrates suggests a more specialised subset of cargo proteins [143], although the total number seems to vary drastically between different cyanobacteria [99]. Generally, the numbers align with a majority of prokaryotes where, for example, in *E. coli*, around 30 Tat substrates exist [101, 144]. There are, however, exceptions to the rule. For example, *Streptomyces coelicolor* [145] and halophilic archaea use Tat as the main protein export route [146]. Despite not being the main import/export route, the Tat system is most likely essential in all cyanobacteria. The main reason being the Tat-dependent translocation of Rieske proteins. The PetC Rieske protein is an essential subunit of the cytochrome b_6f complex and a confirmed Tat substrate in cyanobacteria [97]. A further classic example of Tat substrates operating in cyanobacteria are members of the FutA/IdiA family which are involved in iron uptake [147] and PSII protection during iron stress [148, 149]. Finally, some metal-cofactor-containing enzymes, such as certain carbonic anhydrases, have also been confirmed as Tat substrates in cyanobacteria [150, 151].

In summary, the experimental validation of both the cyanobacterial Tat translocon and its substrates is still limited to a handful of studies. Even fundamental questions such as the composition of the translocation pore remain unanswered. The cyanobacterial Tat system might, therefore, still withhold various surprises, especially in non-model cyanobacteria from more extreme environmental habitats.

4.4 Leader Peptidases Cleave Cargo upon Translocation

When a protein substrate is translocated across or into the membrane via the Sec or Tat pathway, SPs are usually cleaved through membrane-associated type I leader (or signal) peptidases [152] (Fig. 1). Two of these proteases were identified in the PM and TM in *Phormidium laminosum* [94] as type 1 leader peptidases [153]. A genomic survey of *Synechocystis* identified three homologues of leader peptidases: the type I leader peptidases LepB1 (Sll0716), LepB2 (Slr1377) and the type II leader peptidase LspA (Slr1366) [154, 155]. LepB2 is essential in *Synechocystis,* while a LepB1 knockout mutant could be propagated in heterotrophic growth conditions, suggesting a function in photoautotrophic growth [156] and was found to be essential for PsaF maturation [157]. Furthermore, the specialised aspartic peptidase PilD (HofD) [154] is responsible for the maturation of pilins by cleavage and N-terminal methylation. In *Synechocystis,* PilD (Slr1120) is an essential component of the T4P [158]. PilD homologues in *Synechocystis* and *S. elongatus* were capable of processing pre-pilin SPs of heterologous substrates resulting in the secretion of the mature proteins [159, 160]. A PilD knockout (Δ*slr1120*) was found to have a wider impact on protein translocation including impaired PS II biogenesis and Sec-dependent translocation [161]. Pilins are some of the most abundantly secreted proteins in cyanobacteria [159, 162]. Thus, it is not surprising that disruption of pilin processing could cause a backlog and blockage of active Sec translocons and, thereby, have far-reaching consequences throughout the cell.

5 Membrane Insertion of Proteins Is Usually Directed by SRP

In bacteria, several conventional routes of membrane insertion of proteins are known. Most of these are SRP-dependent pathways, that require either the Sec translocon (with or without SecA), the membrane insertase YidC or a combination of both for membrane insertion. Membrane insertion by YidC alone seems to be reserved for smaller proteins and occurs post-translational while all SRP-dependent routes are co-translational [163] (Fig. 2). The SRP pathway is functionally highly conserved and seems to be essential in most organisms with very few known exceptions (e.g., *Saccharomyces cerevisiae* [164]).

Generally, the first step of SRP-dependent membrane targeting is the binding of SRP to a hydrophobic region of the nascent polypeptide chain. The SRP–ribosome–nascent chain complex is then recognised by the membrane-bound FtsY receptor leading to the transfer of the complex to the Sec translocon and/or the YidC membrane insertase where protein translation progresses. SRP-dependent membrane insertion requires GTP hydrolysis upon successful delivery of the nascent polypeptide to Sec/YidC, which is then responsible for membrane insertion, folding and release into the lipid bilayer [163]. In *E. coli*, the SRP complex consists of the protein Ffh (SRP54 homologue) and a 4.5S RNA component. SRP-dependent membrane insertion also occurs in the chloroplast, however, here the situation is somewhat unique. The homologues of the SRP pathway in chloroplasts are cpSRP54 (SRP54 homologue), cpSRP43 (SRP homologue only found in Viridiplantae [165]), cpFtsY (FtsY), the chloroplast Sec translocon (see Sect. 4.2) and Alb3 (YidC). No

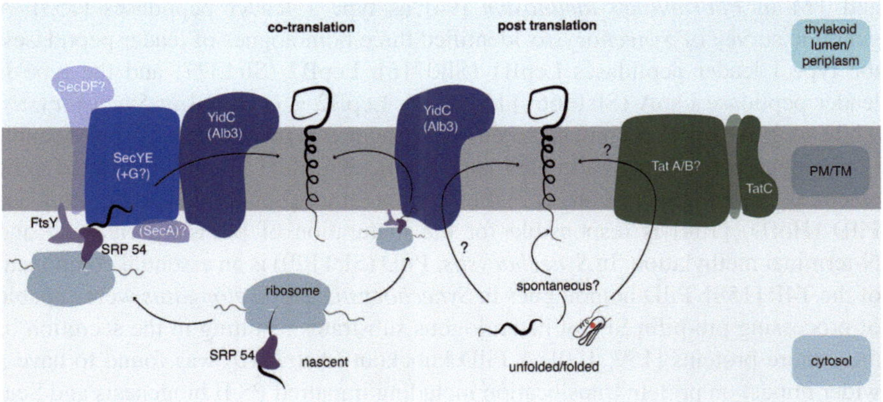

Fig. 2 Overview of potential membrane protein insertion routes in cyanobacteria. The main route is co-translational. Co-translational membrane insertion is dependent on an SRP-complex formation with nascent polypeptides. This complex then binds to FtsY, that facilitates the interaction with either YidC alone or with a complex of the Sec translocon and YidC for membrane insertion of a target protein. Other post-translational pathways might exist. This can occur either via YidC, through a "spontaneous" insertion mechanism, or potentially via the Tat pathway, although it should be stressed that none of these mechanisms are experimentally verified

RNA component was found associated with the cpSRP complex [166]. The cpSRP pathway operates in what appears to be two different modes: post- and co-translational. The post-translational cpSRP system is responsible for delivery of imported light harvesting antenna to the TM (reviewed in [167]) and requires cpSRP54 and cpSRP43. Co-translational cpSRP targeting seems to be an integral part of maintaining photosynthetic complexes where only cpSRP54, not cpSRP43, is involved [168]. In contrast to bacterial co-translational SRP-dependent membrane insertion, cpSRP54 binds to the uL4 domain of the ribosome inside the peptide tunnel instead of at the ribosome exit tunnel [168]. For instance, the plastid-encoded PsbA protein seems to be co-translationally TM inserted under complex formation with cpFtsY, Alb3, cpSecY and Vipp1, and a transient presence of cpSRP54 [169]. The cpSRP43 subunit also seems to have other chaperone functions, for instance, in tetrapyrrole biosynthesis [170].

No direct evidence of how SRP-dependent membrane targeting operates is known from cyanobacteria and it was previously mentioned that it could be a mix of bacterial and chloroplastic mechanisms [93]. Homologues of YidC, Ffh and FtsY are present in cyanobacteria (Fig. 2). In early studies, immunogold labelling showed that YidC in *Synechocystis* (Slr1471) was found to mainly localise to the PM [171] and a proteomic study showed that the protein operates also in the TM [31]. YidC is essential for thylakoid biogenesis in *Synechocystis* and might play a role in cell division [48] and for D1 assembly into PSII [49]. A single Ffh homologue with a GTP-binding site was characterised in cyanobacteria but no evidence of a 4.5S RNA was found [172]. A recent gradient profiling by sequencing approach in *Synechocystis* found the Ffh homologue (Slr1531) clustering with bulk cellular mRNAs [173]. It is most likely that similar mechanisms of membrane insertion operate in cyanobacteria as in other bacteria and the major route is SRP-dependent and occurs co-translationally. One example is the co-purification of the chlorophyll synthase ChlG in *Synechocystis* in a complex with YidC and further assembly factors [80]. It is well possible, that the cyanobacterial SRP system might operate similarly as the co-translational SRP pathway of chloroplasts, especially for the insertion of photosynthetic membrane proteins. Overall, the cyanobacterial SRP could represent an intermediate between bacterial and plastidial solutions, however, experimental evidence in favour or against this hypothesis is yet to be given.

5.1 *Other Membrane Insertion Routes Might Exist in Cyanobacteria*

In bacteria, some membrane proteins can be inserted post-translationally by YidC alone or even integrate into the membrane autonomously (Fig. 2). However, this scenario seems to be only a minor "back-up" route if no vacant translocon is present (reviewed in [163]). Spontaneous/autonomous membrane insertion was suggested to occur also in cyanobacteria [93] as also postulated for chloroplasts [174–176]. It is, however, unclear if this represents a relevant mode of translocation in cyanobacteria. Membrane insertion is not only possible via Sec but can also be Tat-dependent as evident by the Rieske protein of the cytochrome b_6f complex (see Sect. 4.1). This implies that the cargo protein needs to be folded first and then transferred into the membrane [163]. How this can be achieved is unclear. It is possible that this process might require dedicated chaperones or cooperation with other membrane-bound complexes. These putative additional factors and interactions are yet to be identified in cyanobacteria.

6 Future Directions

It is evident that our molecular understanding of protein targeting pathways and mechanisms in cyanobacteria is still far behind that of heterotrophic bacteria and chloroplasts. Based on our current understanding, the overall mechanisms are conserved but tend to present individual twists of a general pattern. This is likely to be true even amongst cyanobacteria. Existing experimental evidence of protein translocation usually derives from a few model organisms (mainly *Synechocystis*) and, in all likelihood, is not representative of the entire phylum. To move forward, there are some areas that would benefit from an increased focus. For example, Sec and Tat translocation processes are strongly influenced by the lipid composition of the membranes [177–180]. This lipid composition can vary not only among organisms but potentially also between TM and PM subdomains. Therefore, a better understanding of the lipid and membrane environment in cyanobacteria might go hand in hand with understanding protein trafficking. A large effort to map the proteome of *Synechocystis* [31, 115, 181] as well as other model unicellular cyanobacteria [29] has been undertaken in recent years. Improving these techniques and applying them to other, understudied, members of the phylum may provide insights into the subcellular localisation of key proteins and, ultimately, contribute to elucidating the molecular intricacies of cyanobacterial protein targeting. State-of-the-art technologies such as cryo-electron tomography and advances in high-resolution mass spectrometry and novel proteomics protocols [29] have already advanced our understanding of subcellular organisation and membrane dynamics in recent years. Now that such methods are becoming commonplace, profound insights into molecular dynamics of the cyanobacterial cell beyond the common model organisms is imaginable. Nevertheless, major limitations remain for this field and despite significant advances, some questions may only be answered when the dynamics of these processes can be captured in vivo and in real time.

Acknowledgements I want to thank Dr. David A. Russo for proofreading and discussions regarding the structure of this chapter and the editors of this book for respecting and accommodating my maternity leave.

References

1. Gould SB, Waller RF, McFadden GI (2008) Plastid evolution. Annu Rev Plant Biol 59:491–517. https://doi.org/10.1146/annurev.arplant.59.032607.092915
2. Archibald JM (2015) Endosymbiosis and eukaryotic cell evolution. Curr Biol 25:R911–R921. https://doi.org/10.1016/j.cub.2015.07.055
3. de Vries J, Archibald JM (2017) Endosymbiosis: did plastids evolve from a freshwater cyanobacterium? Curr Biol 27:R103–R105. https://doi.org/10.1016/j.cub.2016.12.006
4. Mullineaux CW (2014) Co-existence of photosynthetic and respiratory activities in cyanobacterial thylakoid membranes. Biochim Biophys Acta 1837:503–511. https://doi.org/10.1016/j.bbabio.201
5. Mareš J, Hrouzek P, Kaňa R, Ventura S, Strunecký O, Komárek J (2013) The primitive thylakoid-less cyanobacterium *Gloeobacter* is a common Rock-Dwelling organism. PloS One 8:e66323. https://doi.org/10.1371/journal.pone.0066323
6. Rippka R, Waterbury J, Cohen-Bazire G (1974) A cyanobacterium which lacks thylakoids. Arch Microbiol 100:419–436. https://doi.org/10.1007/BF00446333
7. Rexroth S, Mullineaux CW, Ellinger D, Sendtko E, Rögner M, Koenig F (2011) The plasma membrane of the cyanobacterium *Gloeobacter violaceus* contains segregated bioenergetic domains. Plant Cell 23:2379–2390. https://doi.org/10.1105/tpc.111.085779
8. Mullineaux CW, Liu L-N (2020) Membrane dynamics in phototrophic bacteria. Annu Rev Microbiol 74:633–654. https://doi.org/10.1146/annurev-micro-020518-120134
9. Rast A, Schaffer M, Albert S, Wan W, Pfeffer S, Beck F, Plitzko JM, Nickelsen J, Engel BD (2019) Biogenic regions of cyanobacterial thylakoids form contact sites with the plasma membrane. Nat Plants 5:436–446. https://doi.org/10.1038/s41477-019-0399-7
10. Nevo R, Charuvi D, Shimoni E, Schwarz R, Kaplan A, Ohad I, Reich Z (2007) Thylakoid membrane perforations and connectivity enable intracellular traffic in cyanobacteria. EMBO J 26:1467–1473. https://doi.org/10.1038/sj.emboj.7601594

11. Mareš J, Strunecký O, Bučinská L, Wiedermannová J (2019) Evolutionary patterns of thylakoid architecture in cyanobacteria. Front Microbiol 10:277. https://doi.org/10.3389/fmicb.2019.00277
12. Pribil M, Labs M, Leister D (2014) Structure and dynamics of thylakoids in land plants. J Exp Bot 65:1955–1972. https://doi.org/10.1093/jxb/eru090
13. Gu L, Grodzinski B, Han J, Marie T, Zhang Y-J, Song YC, Sun Y (2022) Granal thylakoid structure and function: explaining an enduring mystery of higher plants. New Phytol 236:319–329. https://doi.org/10.1111/nph.18371
14. Heinz S, Rast A, Shao L et al (2016) Thylakoid membrane architecture in *Synechocystis* depends on CurT, a homolog of the granal CURVATURE THYLAKOID1 proteins. Plant Cell 28:2238–2260. https://doi.org/10.1105/tpc.16.00491
15. Watanabe M, Ikeuchi M (2013) Phycobilisome: architecture of a light-harvesting supercomplex. Photosynth Res 116:265–276. https://doi.org/10.1007/s11120-013-9905-3
16. Kobayashi K, Yoshihara A, Kubota-Kawai H (2023) Evolutionary implications from lipids in membrane bilayers and photosynthetic complexes in cyanobacteria and chloroplasts. J Biochem (Tokyo) 174:399. https://doi.org/10.1093/jb/mvad058
17. Yoshihara A, Kobayashi K (2022) Lipids in photosynthetic protein complexes in the thylakoid membrane of plants, algae, and cyanobacteria. J Exp Bot 73:2735. https://doi.org/10.1093/jxb/erac017
18. Mullineaux CW (2008) Factors controlling the mobility of photosynthetic proteins. Photochem Photobiol 84:1310–1316. https://doi.org/10.1111/j.1751-1097.2008.00420.x
19. Siebenaller C, Schneider D (2023) Cyanobacterial membrane dynamics in the light of eukaryotic principles. Biosci Rep 43:BSR20221269. https://doi.org/10.1042/BSR20221269
20. Agarwal R, Matros A, Melzer M, Mock H-P, Sainis JK (2010) Heterogeneity in thylakoid membrane proteome of *Synechocystis* 6803. J Proteomics 73:976–991. https://doi.org/10.1016/j.jprot.2009
21. Liu L-N (2016) Distribution and dynamics of electron transport complexes in cyanobacterial thylakoid membranes. Biochim Biophys Acta 1857:256–265. https://doi.org/10.1016/j.bbabio.2015.11.010
22. MacGregor-Chatwin C, Sener M, Barnett SFH, Hitchcock A, Barnhart-Dailey MC, Maghlaoui K, Barber J, Timlin JA, Schulten K, Hunter CN (2017) Lateral segregation of photosystem I in cyanobacterial thylakoids. Plant Cell 29:1119–1136. https://doi.org/10.1105/tpc.17.00071
23. Folea IM, Zhang P, Aro E-M, Boekema EJ (2008) Domain organization of photosystem II in membranes of the cyanobacterium *Synechocystis* PCC6803 investigated by electron microscopy. FEBS Lett 582:1749–1754. https://doi.org/10.1016/j.febslet.2008.04.044
24. Rast A, Heinz S, Nickelsen J (2015) Biogenesis of thylakoid membranes. Biochim Biophys Acta 1847:821–830. https://doi.org/10.1016/j.bbabio.2015.01.007
25. Casella S, Huang F, Mason D, Zhao G-Y, Johnson GN, Mullineaux CW, Liu L-N (2017) Dissecting the native architecture and dynamics of cyanobacterial photosynthetic machinery. Mol Plant 10:1434–1448. https://doi.org/10.1016/j.molp.2017.09.019
26. Sacharz J, Bryan SJ, Yu J, Burroughs NJ, Spence EM, Nixon PJ, Mullineaux CW (2015) Sub-cellular location of FtsH proteases in the cyanobacterium *Synechocystis* sp. PCC 6803 suggests localised PSII repair zones in the thylakoid membranes. Mol Microbiol 96:448–462. https://doi.org/10.1111/mmi.1294
27. Stingaciu L-R, O'Neill HM, Liberton M, Pakrasi HB, Urban VS (2019) Influence of chemically disrupted photosynthesis on cyanobacterial thylakoid dynamics in *Synechocystis* sp. PCC 6803. Sci Rep 9:5711. https://doi.org/10.1038/s41598-019-42024-0
28. Sarcina M, Bouzovitis N, Mullineaux CW (2006) Mobilization of photosystem II induced by intense red light in the cyanobacterium *Synechococcus* sp PCC7942. Plant Cell 18:457–464. https://doi.org/10.1105/tpc.105.035808
29. Dahlgren KK, Gates C, Lee T, Cameron JC (2021) Proximity-based proteomics reveals the thylakoid lumen proteome in the cyanobacterium *Synechococcus* sp. PCC 7002. Photosynth Res 147:177–195. https://doi.org/10.1007/s11120-020-00806-y

30. Pisareva T, Kwon J, Oh J, Kim S, Ge C, Wieslander Å, Choi J-S, Norling B (2011) Model for membrane organization and protein sorting in the Cyanobacterium *Synechocystis* sp. PCC 6803 inferred from proteomics and multivariate sequence analyses. J Proteome Res 10:3617–3631. https://doi.org/10.1021/pr200268r
31. Baers LL, Breckels LM, Mills LA, Gatto L, Deery MJ, Stevens TJ, Howe CJ, Lilley KS, Lea-Smith DJ (2019) Proteome mapping of a cyanobacterium reveals distinct compartment organization and cell-dispersed metabolism. Plant Physiol 181:1721–1738. https://doi.org/10.1104/pp.19.00897
32. Liberton M, Saha R, Jacobs JM, Nguyen AY, Gritsenko MA, Smith RD, Koppenaal DW, Pakrasi HB (2016) Global proteomic analysis reveals an exclusive role of thylakoid membranes in bioenergetics of a model cyanobacterium. Mol Cell Proteomics 15:2021–2032. https://doi.org/10.1074/mcp.M115.057240
33. Huang F, Parmryd I, Nilsson F, Persson AL, Pakrasi HB, Andersson B, Norling B (2002) Proteomics of *Synechocystis* sp. Strain PCC 6803: identification of plasma membrane proteins. Mol Cell Proteomics 1:956–966. https://doi.org/10.1074/mcp.M200043-MCP200
34. Srivastava R, Pisareva T, Norling B (2005) Proteomic studies of the thylakoid membrane of *Synechocystis* sp. PCC 6803. Proteomics 5:4905–4916. https://doi.org/10.1002/pmic.200500111
35. Fulda S, Huang F, Nilsson F, Hagemann M, Norling B (2000) Proteomics of *Synechocystis* sp. strain PCC 6803. Eur J Biochem 267:5900–5907. https://doi.org/10.1046/j.1432-1327.2000.01642.x
36. Wang Y, Xu W, Chitnis PR (2009) Identification and bioinformatic analysis of the membrane proteins of *Synechocystis* sp. PCC 6803. Proteome Sci 7:11. https://doi.org/10.1186/1477-5956-7-11
37. Huang F, Hedman E, Funk C, Kieselbach T, Schröder WP, Norling B (2004) Isolation of outer membrane of *Synechocystis* sp. PCC 6803 and its proteomic characterization. Mol Cell Proteomics 3:586–595. https://doi.org/10.1074/mcp.M300137-MCP200
38. Herranen M, Battchikova N, Zhang P, Graf A, Sirpiö S, Paakkarinen V, Aro E-M (2004) Towards functional proteomics of membrane protein complexes in *Synechocystis* sp. PCC 6803. Plant Physiol 134:470–481. https://doi.org/10.1104/pp.103.032326
39. Allen MM (1968) Photosynthetic membrane system in *Anacystis nidulans*. J Bacteriol 96:836–841. https://doi.org/10.1128/jb.96.3.836-841.1968
40. Nierzwicki-Bauer SA, Balkwill DL, Stevens SE (1983) Three-dimensional ultrastructure of a unicellular cyanobacterium. J Cell Biol 97:713–722. https://doi.org/10.1083/jcb.97.3.713
41. Nierzwicki-Bauer SA, Balkwill DL, Stevens SE Jr (1984) The use of high-voltage electron microscopy and semi-thick sections for examination of cyanobacterial thylakoid membrane arrangements. J Microsc 133:55–60. https://doi.org/10.1111/j.1365-2818.1984.tb00462.x
42. van de Meene AML, Hohmann-Marriott MF, Vermaas WFJ, Roberson RW (2006) The three-dimensional structure of the cyanobacterium *Synechocystis* sp. PCC 6803. Arch Microbiol 184:259–270. https://doi.org/10.1007/s00203-005-0027-y
43. Liberton M, Howard Berg R, Heuser J, Roth R, Pakrasi HB (2006) Ultrastructure of the membrane systems in the unicellular cyanobacterium *Synechocystis* sp. strain PCC 6803. Protoplasma 227:129–138. https://doi.org/10.1007/s00709-006-0145-7
44. Schneider D, Fuhrmann E, Scholz I, Hess WR, Graumann PL (2007) Fluorescence staining of live cyanobacterial cells suggest non-stringent chromosome segregation and absence of a connection between cytoplasmic and thylakoid membranes. BMC Cell Biol 8:39. https://doi.org/10.1186/1471-2121-8-39
45. Huokko T, Ni T, Dykes GF et al (2021) Probing the biogenesis pathway and dynamics of thylakoid membranes. Nat Commun 12:3475. https://doi.org/10.1038/s41467-021-23680-1
46. Ostermeier M, Heinz S, Hamm J, Zabret J, Rast A, Klingl A, Nowaczyk MM, Nickelsen J (2021) Thylakoid attachment to the plasma membrane in *Synechocystis* sp. PCC 6803 requires the AncM protein. Plant Cell 34:655–678. https://doi.org/10.1093/plcell/koab253

47. Heidrich J, Thurotte A, Schneider D (2017) Specific interaction of IM30/Vipp1 with cyanobacterial and chloroplast membranes results in membrane remodeling and eventually in membrane fusion. Biochim Biophys Acta 1859:537–549. https://doi.org/10.1016/j.bbamem.2016.09.025
48. Spence E, Bailey S, Nenninger A, Møller SG, Robinson C (2004) A homolog of Albino3/OxaI is essential for thylakoid biogenesis in the cyanobacterium *Synechocystis* sp. PCC6803. J Biol Chem 279:55792–55800. https://doi.org/10.1074/jbc.M411041200
49. Ossenbühl F, Inaba-Sulpice M, Meurer J, Soll J, Eichacker LA (2006) The *Synechocystis* sp PCC 6803 Oxa1 homolog is essential for membrane integration of reaction center Precursor Protein pD1. Plant Cell 18:2236–2246. https://doi.org/10.1105/tpc.106.043646
50. Kunkel DD (1982) Thylakoid centers: structures associated with the cyanobacterial photosynthetic membrane system. Arch Microbiol 133:97–99. https://doi.org/10.1007/BF00413518
51. Schottkowski M, Gkalympoudis S, Tzekova N, Stelljes C, Schünemann D, Ankele E, Nickelsen J (2009) Interaction of the periplasmic PratA factor and the PsbA (D1) protein during biogenesis of photosystem II in *Synechocystis* sp. PCC 6803. J Biol Chem 284:1813–1819. https://doi.org/10.1074/jbc.M806116200
52. Stengel A, Gügel IL, Hilger D, Rengstl B, Jung H, Nickelsen J (2012) Initial steps of photosystem II de Novo assembly and preloading with manganese take place in biogenesis centers in *Synechocystis*. Plant Cell 24:660–675. https://doi.org/10.1105/tpc.111.093914
53. Kroll D, Meierhoff K, Bechtold N, Kinoshita M, Westphal S, Vothknecht UC, Soll J, Westhoff P (2001) *VIPP1*, a nuclear gene of *Arabidopsis thaliana* essential for thylakoid membrane formation. Proc Natl Acad Sci 98:4238–4242. https://doi.org/10.1073/pnas.061500998
54. Westphal S, Heins L, Soll J, Vothknecht UC (2001) Vipp1 deletion mutant of *Synechocystis*: a connection between bacterial phage shock and thylakoid biogenesis? Proc Natl Acad Sci 98:4243–4248. https://doi.org/10.1073/pnas.061500998
55. Junglas B, Schneider D (2018) What is Vipp1 good for? Mol Microbiol 108:1–5. 108:1–5. https://doi.org/10.1111/mmi.13924
56. Bryan SJ, Burroughs NJ, Shevela D et al (2014) Localisation and interactions of the Vipp1 protein in cyanobacteria. Mol Microbiol 94:1179–1195. https://doi.org/10.1111/mmi.12826
57. Gupta TK, Klumpe S, Gries K et al (2021) Structural basis for VIPP1 oligomerization and maintenance of thylakoid membrane integrity. Cell 184:3643–3659.e23. https://doi.org/10.1016/j.cell.2021
58. Armbruster U, Labs M, Pribil M et al (2013) Arabidopsis CURVATURE THYLAKOID1 proteins modify thylakoid architecture by inducing membrane curvature. Plant Cell 25:2661–2678. https://doi.org/10.1105/tpc.113.113118
59. Boehm M, Nield J, Zhang P, Aro E-M, Komenda J, Nixon PJ (2009) Structural and mutational analysis of band 7 proteins in the Cyanobacterium *Synechocystis* sp. Strain PCC 6803. J Bacteriol 191:6425–6435. https://doi.org/10.1128/jb.00644-09
60. Bryan SJ, Burroughs NJ, Evered C et al (2011) Loss of the SPHF Homologue Slr1768 Leads to a Catastrophic Failure in the Maintenance of Thylakoid Membranes in *Synechocystis* sp. PCC 6803. PloS One 6:e19625. https://doi.org/10.1371/journal.pone.0019
61. Lea-Smith DJ, Ortiz-Suarez ML, Lenn T et al (2016) Hydrocarbons are essential for optimal cell size, division, and growth of cyanobacteria. Plant Physiol 172:1928–1940. https://doi.org/10.1104/pp.1
62. Kuhn A, Stuart R, Henry R, Dalbey RE (2003) The Alb3/Oxa1/YidC protein family: membrane-localized chaperones facilitating membrane protein insertion? Trends Cell Biol 13:510–516. https://doi.org/10.1016/j.tcb.2003.08.005
63. Saller MJ, Wu ZC, de Keyzer J, Driessen AJM (2012) The YidC/Oxa1/Alb3 protein family: common principles and distinct features. Biol Chem 393:1279–1290. https://doi.org/10.1515/hsz-2012-019
64. Jackson PJ, Hitchcock A, Brindley AA, Dickman MJ, Hunter CN (2023) Absolute quantification of cellular levels of photosynthesis-related proteins in *Synechocystis* sp. PCC 6803. Photosynth Res 155:219–245. https://doi.org/10.1007/s11120-022-00990-z

65. Jordan P, Fromme P, Witt HT, Klukas O, Saenger W, Krauß N (2001) Three-dimensional structure of cyanobacterial photosystem I at 2.5 Å resolution. Nature 411:909–917.
66. Zhao L-S, Li C-Y, Chen X-L, Wang Q, Zhang Y-Z, Liu L-N (2022) Native architecture and acclimation of photosynthetic membranes in a fast-growing cyanobacterium. Plant Physiol Kiac 190:1883. https://doi.org/10.1093/plphys/kiac372
67. Klinkert B, Ossenbühl F, Sikorski M, Berry S, Eichacker L, Nickelsen J (2004) PratA, a periplasmic Tetratricopeptide repeat protein involved in biogenesis of photosystem II in *Synechocystis* sp. PCC 6803. J Biol Chem 279:44639–44644. https://doi.org/10.1074/jbc.M405393200
68. Zak E, Norling B, Maitra R, Huang F, Andersson B, Pakrasi HB (2001) The initial steps of biogenesis of cyanobacterial photosystems occur in plasma membranes. Proc Natl Acad Sci 98:13443–13448. https://doi.org/10.1073/pnas.241503898
69. Selão TT, Zhang L, Knoppová J, Komenda J, Norling B (2016) Photosystem II assembly steps take place in the thylakoid membrane of the cyanobacterium *Synechocystis* sp. PCC6803. Plant Cell Physiol 57:95–104. https://doi.org/10.1093/pcp/pcv178
70. Mahbub M, Hemm L, Yang Y et al (2020) mRNA localization, reaction centre biogenesis and thylakoid membrane targeting in cyanobacteria. Nat Plants 6:1179–1191. https://doi.org/10.1038/s4147
71. Chen M, Liu X, He Y, Li N, He J, Zhang Y (2022) Diversity among cyanobacterial photosystem I oligomers. Front Microbiol 12:781826. https://doi.org/10.3389/fmicb.2021.781826
72. Xu W, Tang H, Wang Y, Chitnis PR (2001) Proteins of the cyanobacterial photosystem I. Biochim Biophys Acta 1507:32–40. https://doi.org/10.1016/S0005-2728(01)00208-0
73. Dühring U, Ossenbühl F, Wilde A (2007) Late assembly steps and dynamics of the cyanobacterial photosystem. J Biol Chem 282:10915–10921. https://doi.org/10.1074/jbc.M609206200
74. Nixon PJ, Michoux F, Yu J, Boehm M, Komenda J (2010) Recent advances in understanding the assembly and repair of photosystem II. Ann Bot 106:1–16. https://doi.org/10.1093/aob/mcq059
75. Heinz S, Liauw P, Nickelsen J, Nowaczyk M (2016) Analysis of photosystem II biogenesis in cyanobacteria. Biochim Biophys Acta 1857:274–287. https://doi.org/10.1016/j.bbabio.2015.11.007
76. Johnson VM, Pakrasi HB (2022) Advances in the understanding of the lifecycle of photosystem II. Microorganisms 10:836. https://doi.org/10.3390/microorganisms10050836
77. Komenda J, Kuviková S, Granvogl B, Eichacker LA, Diner BA, Nixon PJ (2007) Cleavage after residue Ala352 in the C-terminal extension is an early step in the maturation of the D1 subunit of Photosystem II in *Synechocystis* PCC 6803. Biochim Biophys Acta 1767:829–837. https://doi.org/10.1016/j.bbabio.2007.01
78. Komenda J, Nickelsen J, Tichý M, Prásil O, Eichacker LA, Nixon PJ (2008) The cyanobacterial homologue of HCF136/YCF48 is a component of an early photosystem II assembly complex and is important for both the efficient assembly and repair of photosystem II in *Synechocystis* sp. PCC 6803. J Biol Chem 283:22390–22399. https://doi.org/10.1074/jbc.M801917200
79. Knoppová J, Sobotka R, Tichy M, Yu J, Konik P, Halada P, Nixon PJ, Komenda J (2014) Discovery of a chlorophyll binding protein complex involved in the early steps of photosystem II assembly in *Synechocystis*. Plant Cell 26:1200–1212. https://doi.org/10.1105/tpc.114.123919
80. Chidgey JW, Linhartová M, Komenda J, Jackson PJ, Dickman MJ, Canniffe DP, Koník P, Pilný J, Hunter CN, Sobotka R (2014) A cyanobacterial chlorophyll Synthase-HliD complex associates with the Ycf39 protein and the YidC/Alb3 insertase. Plant Cell 26:1267–1279. https://doi.org/10.1105/tpc.114.12449
81. Nickelsen J, Rengstl B (2013) Photosystem II assembly: from cyanobacteria to plants. Annu Rev Plant Biol 64:609–635. https://doi.org/10.1146/annurev-arplant-050312-120124
82. Weisz DA, Johnson VM, Niedzwiedzki DM, Shinn MK, Liu H, Klitzke CF, Gross ML, Blankenship RE, Lohman TM, Pakrasi HB (2019) A novel chlorophyll protein complex

in the repair cycle of photosystem II. Proc Natl Acad Sci 116:21907–21913. https://doi.org/10.1073/pnas.1909644116
83. Bečková M, Sobotka R, Komenda J (2022) Photosystem II antenna modules CP43 and CP47 do not form a stable 'no reaction centre complex' in the cyanobacterium *Synechocystis* sp. PCC 6803. Photosynth Res 152:363–371. https://doi.org/10.1007/s11120-022-00896-w
84. Bailey S, Silva P, Nixon P, Mullineaux C, Robinson C, Mann N (2001) Auxiliary functions in photosynthesis: the role of the FtsH protease. Biochem Soc Trans 29:455–459. https://doi.org/10.1042/bst0290455
85. Apel K, Hirt H (2004) Reactive oxygen species: metabolism, oxidative stress, and signal transduction. Annu Rev Plant Biol 55:373–399. https://doi.org/10.1146/annurev.arplant.55.031903.141701
86. Russo DA, Zedler JAZ (2021) Genomic insights into cyanobacterial protein translocation systems. Biol Chem 402:39–54. https://doi.org/10.1515/hsz-2020-0247
87. Ruprecht M, Bionda T, Sato T, Sommer MS, Endo T, Schleiff E (2010) On the impact of precursor unfolding during protein import into chloroplasts. Mol Plant 3:499–508. https://doi.org/10.1093/mp/ssp116
88. Xing J, Pan J, Yi H et al (2022) The plastid-encoded protein Orf2971 is required for protein translocation and chloroplast quality control. Plant Cell 34:3383–3399. https://doi.org/10.1093/plcell/koac1
89. Arnold T, Zeth K, Linke D (2010) Omp85 from the Thermophilic Cyanobacterium *Thermosynechococcus elongatus* differs from Proteobacterial Omp85 in structure and domain composition. J Biol Chem 285:18003–18015. https://doi.org/10.1074/jbc.M110.112516
90. Koenig P, Mirus O, Haarmann R, Sommer MS, Sinning I, Schleiff E, Tews I (2010) Conserved Properties of Polypeptide Transport-associated (POTRA) domains derived from cyanobacterial Omp85. J Biol Chem 285:18016–18024. https://doi.org/10.1074/jbc.M110.112649
91. Schleiff E, Soll J (2005) Membrane protein insertion: mixing eukaryotic and prokaryotic concepts. EMBO Rep 6:1023–1027. https://doi.org/10.1038/sj.embor.7400563
92. Ranava D, Caumont-Sarcos A, Albenne C, Ieva R (2018) Bacterial machineries for the assembly of membrane-embedded β-barrel proteins. FEMS Microbiol Lett 365:fny087. https://doi.org/10.1093/femsle/fny087
93. Frain KM, Gangl D, Jones A, Zedler JAZ, Robinson C (2016) Protein translocation and thylakoid biogenesis in cyanobacteria. Biochim Biophys Acta 1857:266–273 https://doi.org/10.1016/j.bbabio.2015.08
94. Barbrook AC, Packer JCL, Howe CJ (1993) Components of the protein translocation machinery in the thermophilic cyanobacterium *Phormidium laminosum*. Biochem Biophys Res Commun 197:874–877. https://doi.org/10.1006/bbrc.1993.2560
95. Nakai M, Sugita D, Omata T, Endo T (1993) Sec-Y protein is localized in both the cytoplasmic and thylakoid membranes in the cyanobacterium *Synechococcus* PCC7942. Biochem Biophys Res Commun 193:228–234. https://doi.org/10.1006/bbrc.1993.1613
96. Howe CJ, Barbrook AC, Packer JCL (1996) Protein targeting and translocation in cyanobacterial membrane biogenesis. Biochem Soc Trans 24:750–753. https://doi.org/10.1042/bst0240750
97. Aldridge C, Spence E, Kirkilionis MA, Frigerio L, Robinson C (2008) Tat-dependent targeting of Rieske iron-sulphur proteins to both the plasma and thylakoid membranes in the cyanobacterium *Synechocystis* PCC6803. Mol Microbiol 70:140–150. https://doi.org/10.1111/j.1365-2958.2008.06401.x
98. Schneider D (2014) Protein targeting, transport and translocation in cyanobacteria. In: Flores E, Herrero A (eds) Cell Biol. Cyanobacteria. Caister Academic Press, Seville, pp 121–147.
99. Barnett JP, Robinson C, Scanlan DJ, Blindauer CA (2011) The Tat protein export pathway and its role in cyanobacterial metalloprotein biosynthesis. FEMS Microbiol Lett 325:1–9. https://doi.org/10.1111/j.1574-6968.2011.02391.x
100. Spence E, Sarcina M, Ray N, Moller SG, Mullineaux CW, Robinson C (2003) Membrane-specific targeting of green fluorescent protein by the Tat pathway in the cyanobacterium *Synechocystis* PCC6803. Mol Microbiol 48:1481–1489. https://doi.org/10.1046/j.1365-2958.2003.03519.x

101. Palmer T, Berks BC (2012) The twin-arginine translocation (Tat) protein export pathway. Nat Rev Microbiol 10:483–496. https://doi.org/10.1038/nrmicro2814
102. Freudl R (2018) Signal peptides for recombinant protein secretion in bacterial expression systems. Microb Cell Fact 17:52. https://doi.org/10.1186/s12934-018-0901-3
103. Cristóbal S, de Gier JW, Nielsen H, von Heijne G (1999) Competition between Sec- and TAT-dependent protein translocation in *Escherichia coli*. EMBO J 18:2982–2990
104. Bogsch E, Brink S, Robinson C (1997) Pathway specificity for a delta pH-dependent precursor thylakoid lumen protein is governed by a "Sec-avoidance" motif in the transfer peptide and a "Sec-incompatible" mature protein. EMBO J 16:3851–3859. https://doi.org/10.1093/emboj/16.13.3851
105. Russo DA, Zedler JAZ, Wittmann DN, Möllers B, Singh RK, Batth TS, van Oort B, Olsen JV, Bjerrum MJ, Jensen PE (2019) Expression and secretion of a lytic polysaccharide monooxygenase by a fast-growing cyanobacterium. Biotechnol Biofuels 12:74. https://doi.org/10.1186/s13068-019-1416-9
106. Russo DA, Zedler JAZ, Conradi FD, Schuergers N, Jensen PE, Mullineaux CW, Wilde A, Pohnert G (2021) Development of a highly sensitive Luciferase-based reporter system to study two-step protein secretion in cyanobacteria. J Bacteriol 204:e00504–e00521. https://doi.org/10.1128/jb.00504-21
107. Dalbey RE, Kuhn A (2012) Protein Traffic in Gram-negative bacteria – how exported and secreted proteins find their way. FEMS Microbiol Rev 36:1023–1045. https://doi.org/10.1111/j.1574-6976.2
108. Kaushik S, He H, Dalbey RE (2022) Bacterial signal peptides- navigating the journey of proteins. Front Physiol 13:933153. https://doi.org/10.3389/fphys.2022.933153
109. Conradi FD, Mullineaux CW, Wilde A (2020) The role of the cyanobacterial type IV Pilus machinery in finding and maintaining a favourable environment. Life 10:252. https://doi.org/10.3390/life10110252
110. Chen Z, Li X, Tan X, Zhang Y, Wang B (2020) Recent advances in biological functions of thick Pili in the Cyanobacterium *Synechocystis* sp. PCC 6803. Front Plant Sci 11:241. https://doi.org/10.3389/fpls.2020.00241
111. Fernandez DE (2018) Two paths diverged in the stroma: targeting to dual SEC translocase systems in chloroplasts. Photosynth Res 138:277–287. https://doi.org/10.1007/s11120-018-0541-9
112. Schneider D, Berry S, Volkmer T, Seidler A, Rögner M (2004) PetC1 is the major rieske iron-sulfur protein in the cytochrome b_6f complex of *Synechocystis* sp. PCC 6803. J Biol Chem 279:39383–39388. https://doi.org/10.1074/jbc.M406288200
113. Proctor MS, Malone LA, Farmer DA et al (2022) Cryo-EM structures of the *Synechocystis* sp. PCC 6803 cytochrome b_6f complex with and without the regulatory PetP subunit. Biochem J 479:1487–1503. https://doi.org/10.1042/BCJ20220124
114. Karnauchov I, Herrmann RG, Klösgen RB (1997) Transmembrane topology of the Rieske Fe/S protein of the cytochrome b_6/f complex from spinach chloroplasts. FEBS Lett 408:206–210. https://doi.org/10.1016/S0014-5793(97)00427-4
115. Xiao Z, Huang C, Ge H et al (2022) Proximity labeling facilitates defining the proteome neighborhood of photosystem II oxygen evolution complex in a model cyanobacterium. Mol Cell Proteomics MCP 21:100440. https://doi.org/10.1016/j.mcpro.2022.100440
116. Rodrigue A, Chanal A, Beck K, Müller M, Wu LF (1999) Co-translocation of a periplasmic enzyme complex by a hitchhiker mechanism through the bacterial tat pathway. J Biol Chem 274:13223–13228 https://doi.org/10.1074/jbc.274.19.13223
117. Pei D, Dalbey RE (2022) Membrane translocation of folded proteins. J Biol Chem 298:102107. https://doi.org/10.1016/j.jbc.2022.102107
118. Pohlschröder M, Giménez MI, Jarrell KF (2005) Protein transport in Archaea: sec and twin arginine translocation pathways. Curr Opin Microbiol 8:713–719. https://doi.org/10.1016/j.mib.2005.10.006
119. Driessen AJM, Nouwen N (2008) Protein translocation across the bacterial cytoplasmic membrane. Annu Rev Biochem 77:643–667. https://doi.org/10.1146/annurev.biochem.77.061606.160747

120. Denks K, Vogt A, Sachelaru I, Petriman N-A, Kudva R, Koch H-G (2014) The Sec translocon mediated protein transport in prokaryotes and eukaryotes. Mol Membr Biol 31:58–84. https://doi.org/10.3109/09687688.2014.907455
121. Tsirigotaki A, Geyter JD, Šoštaric N, Economou A, Karamanou S (2017) Protein export through the bacterial Sec pathway. Nat Rev Microbiol 15:21–36. https://doi.org/10.1038/nrmicro.2016.161
122. Rapoport TA, Li L, Park E (2017) Structural and mechanistic insights into protein translocation. Annu Rev Cell Dev Biol 33:369–390. https://doi.org/10.1146/annurev-cellbio-100616-060439
123. Albiniak AM, Baglieri J, Robinson C (2012) Targeting of lumenal proteins across the thylakoid membrane. J Exp Bot 63:1689–1698. https://doi.org/10.1093/jxb/err444
124. Skalitzky CA, Martin JR, Harwood JH, Beirne JJ, Adamczyk BJ, Heck GR, Cline K, Fernandez DE (2011) Plastids contain a second sec translocase system with essential functions. Plant Physiol 155:354–369. https://doi.org/10.1104/pp.110.166546
125. Singhal R, Fernandez DE (2017) Sorting of SEC translocase SCY components to different membranes in chloroplasts. J Exp Bot 68:5029–5043. https://doi.org/10.1093/jxb/erx318
126. Anderson SA, Singhal R, Fernandez DE (2019) Membrane-specific targeting of Tail-Anchored proteins SECE1 and SECE2 within chloroplasts. Front Plant Sci 10:1401. https://doi.org/10.3389/fpls.2019.01401
127. Nakai M, Nohara T, Sugita D, Endo T (1994) Identification and characterization of the SecA protein homologue in the cyanobacterium *Synechococcus* PCC7942. Biochem Biophys Res Commun 200:844–851. https://doi.org/10.1006/bbrc.1994.1528
128. Nishiyama K, Hanada M, Tokuda H (1994) Disruption of the gene encoding p12 (SecG) reveals the direct involvement and important function of SecG in the protein translocation of *Escherichia coli* at low temperature. EMBO J 13:3272–3277. https://doi.org/10.1002/j.1460-2075.1994.tb06628.x
129. Flower AM, Hines LL, Pfennig PL (2000) SecG is an auxiliary component of the protein export apparatus of *Escherichia coli.* Mol Gen Genet MGG 263:131–136. https://doi.org/10.1007/s0043800
130. Mould RM, Robinson C (1991) A proton gradient is required for the transport of two lumenal oxygen-evolving proteins across the thylakoid membrane. J Biol Chem 266:12189–12193. https://doi.org/10.1016/S0021-9258(18)98879-4
131. Cline K, Ettinger WF, Theg SM (1992) Protein-specific energy requirements for protein transport across or into thylakoid membranes. Two lumenal proteins are transported in the absence of ATP. J Biol Chem 267:2688–2696. https://doi.org/10.1016/S0021-9258(18)45935-2
132. Schäfer K, Künzler P, Schneider K, Klingl A, Eubel H, Carrie C (2020) The plant mitochondrial TAT pathway is essential for complex III biogenesis. Curr Biol 30:840–853.e5. https://doi.org/10.1016/j.cub.2020.01.001
133. Cline SG, Gabilly ST, Subrahmanian N, Hamel PP (2016) Cofactor assembly of Cytochrome bc1-b6f complexes. In: Cramer WA, Kallas T (eds) Cytochrome complexes Evol. Struct. Energy Transduct. Signal. Springer Netherlands, Dordrecht, pp 501–525. https://doi.org/10.1007/978-94-017-7481-9_26
134. Petrů M, Wideman J, Moore K, Alcock F, Palmer T, Doležal P (2018) Evolution of mitochondrial TAT translocases illustrates the loss of bacterial protein transport machines in mitochondria. BMC Biol 16:141. https://doi.org/10.1093/molbev/msab090
135. Robinson C, Matos CFRO, Beck D, Ren C, Lawrence J, Vasisht N, Mendel S (2011) Transport and proofreading of proteins by the twin-arginine translocation (Tat) system in bacteria. Biochim Biophys Acta 1808:876–884. https://doi.org/10.1016/j.bbamem.2010.11.023
136. Frain KM, Robinson C, van Dijl JM (2019) Transport of folded proteins by the tat system. Protein J 38:377–388. https://doi.org/10.1007/s10930-019-09859-y
137. Hynds PJ, Robinson D, Robinson C (1998) The sec-independent twin-arginine translocation system can transport both tightly folded and malfolded proteins across the thylakoid membrane. J Biol Chem 273:34868–34874. https://doi.org/10.1074/jbc.273.52.34868

138. Alcock F, Stansfeld PJ, Basit H, Habersetzer J, Baker MA, Palmer T, Wallace MI, Berks BC (2016) Assembling the Tat protein translocase. eLife 5:e20718. https://doi.org/10.7554/eLife.20718
139. Palmer T, Stansfeld PJ (2020) Targeting of proteins to the twin-arginine translocation pathway. Mol Microbiol 113:861–871. https://doi.org/10.1111/mmi.14461
140. Cline K (2015) Mechanistic aspects of folded protein transport by the Twin Arginine Translocase (Tat). J Biol Chem 290:16530–16538. https://doi.org/10.1074/jbc.R114.626820
141. New CP, Ma Q, Dabney-Smith C (2018) Routing of thylakoid lumen proteins by the chloroplast twin arginine transport pathway. Photosynth Res 138:289–301. https://doi.org/10.1007/s11120-018-05
142. Dabney-Smith C, Mori H, Cline K (2006) Oligomers of Tha4 organize at the thylakoid Tat translocase during protein transport. J Biol Chem 281:5476–5483. https://doi.org/10.1074/jbc.M512453200
143. Dilks K, Rose RW, Hartmann E, Pohlschröder M (2003) Prokaryotic utilization of the twin-arginine translocation pathway: a genomic survey. J Bacteriol 185:1478–1483. https://doi.org/10.1128/JB.185.4.1478-1483.2003
144. Palmer T, Sargent F, Berks BC (2010) The Tat protein export pathway. EcoSal Plus 4. https://doi.org/10.1128/ecosalplus.4.3.2
145. Widdick DA, Dilks K, Chandra G, Bottrill A, Naldrett M, Pohlschröder M, Palmer T (2006) The twin-arginine translocation pathway is a major route of protein export in *Streptomyces coelicolor*. Proc Natl Acad Sci 103:17927–17932. https://doi.org/10.1073/pnas.0607025103
146. Thomas JR, Bolhuis A (2006) The *tatC* gene cluster is essential for viability in halophilic archaea. FEMS Microbiol Lett 256:44–49. https://doi.org/10.1111/j.1574-6968.2006.00107.x
147. Polyviou D, Machelett MM, Hitchcock A, Baylay AJ, MacMillan F, Moore CM, Bibby TS, Tews I (2018) Structural and functional characterization of IdiA/FutA (Tery_3377), an iron-binding protein from the ocean diazotroph *Trichodesmium erythraeum*. J Biol Chem 293:18099–18109. https://doi.org/10.1074/jbc.RA118.001929
148. Exss-Sonne P, Tölle J, Bader KP, Pistorius EK, Michel K-P (2000) The IdiA protein of *Synechococcus* sp. PCC 7942 functions in protecting the acceptor side of Photosystem II under oxidative stress. Photosynth Res 63:145–157. https://doi.org/10.1023/A:1006322925324
149. Michel K-P, Pistorius EK (2004) Adaptation of the photosynthetic electron transport chain in cyanobacteria to iron deficiency: the function of IdiA and IsiA. Physiol Plant 120:36–50. https://doi.org/10.1111/j.0031-9317.2004.0229.x
150. Kupriyanova EV, Sinetova MA, Bedbenov VS, Pronina NA, Los DA (2018) Putative extracellular α-class carbonic anhydrase, EcaA, of *Synechococcus elongatus* PCC 7942 is an active enzyme: a sequel to an old story. Microbiology 164:576–586. https://doi.org/10.1099/mic.0.000634
151. Kupriyanova EV, Sinetova MA, Leusenko AV, Voronkov AS, Los DA (2022) A leader peptide of the extracellular cyanobacterial carbonic anhydrase ensures the efficient secretion of recombinant proteins in *Escherichia coli*. J Biotechnol 344:11–23. https://doi.org/10.1016/j.jbiotec.2021.12.006
152. Auclair SM, Bhanu MK, Kendall DA (2012) Signal peptidase I: cleaving the way to mature proteins. Protein Sci Publ Protein Soc 21:13–25. https://doi.org/10.1002/pro.757
153. Barbrook AC, Packer JCL, Howe CJ (1996) Inhibition by penem of processing peptidases from cyanobacteria and chloroplast thylakoids. FEBS Lett 398:198–200. https://doi.org/10.1016/S0014-579
154. Sokolenko A, Pojidaeva E, Zinchenko V, Panichkin V, Glaser VM, Herrmann RG, Shestakov SV (2002) The gene complement for proteolysis in the cyanobacterium *Synechocystis* sp. PCC 6803 and *Arabidopsis thaliana* chloroplasts. Curr Genet 41:291–310. https://doi.org/10.1007/s00294-002-0309-8
155. Mrázek J, Bhaya D, Grossman AR, Karlin S (2001) Highly expressed and alien genes of the *Synechocystis* genome. Nucleic Acids Res 29:1590–1601. https://doi.org/10.1093/nar/29.7.1590

156. Zhbanko M, Zinchenko V, Gutensohn M, Schierhorn A, Klösgen RB (2005) Inactivation of a predicted leader peptidase prevents photoautotrophic growth of *Synechocystis* sp. Strain PCC 6803. J Bacteriol 187:3071–3078. https://doi.org/10.1128/jb.187.9.3071-3078.2005
157. Zhang L, Selão TT, Pisareva T, Qian J, Sze SK, Carlberg I, Norling B (2013) Deletion of *Synechocystis* sp. PCC 6803 leader peptidase LepB1 affects photosynthetic complexes and respiration. Mol Cell Proteomics 12:1192–1203. https://doi.org/10.1074/mcp.M112.022145
158. Bhaya D, Bianco NR, Bryant D, Grossman A (2000) Type IV pilus biogenesis and motility in the cyanobacterium *Synechocystis* sp. PCC6803. Mol Microbiol 37:941–951. https://doi.org/10.1046/j.1365-2958.2000.02068.x
159. Sergeyenko TV, Los DA (2003) Cyanobacterial leader peptides for protein secretion. FEMS Microbiol Lett 218:351–357. https://doi.org/10.1016/S0378-1097(02)01197-7
160. Chen P-H, Liu H-L, Chen Y-J, Cheng Y-H, Lin W-L, Yeh C-H, Chang C-H (2012) Enhancing CO_2 bio-mitigation by genetic engineering of cyanobacteria. Energ Environ Sci 5:8318–8327. https://doi.org/10.1039/C2EE21124F
161. Linhartová M, Bučinská L, Halada P, Ječmen T, Šetlík J, Komenda J, Sobotka R (2014) Accumulation of the Type IV prepilin triggers degradation of SecY and YidC and inhibits synthesis of Photosystem II proteins in the cyanobacterium *Synechocystis* PCC 6803. Mol Microbiol 93:1207–1223. https://doi.org/10.1111/mmi.12730
162. Oliveira P, Martins NM, Santos M, Pinto F, Büttel Z, Couto NAS, Wright PC, Tamagnini P (2016) The versatile TolC-like Slr1270 in the cyanobacterium *Synechocystis* sp. PCC 6803. Environ Microbiol 18:486–502. https://doi.org/10.1111/1462-2920.13172
163. Kuhn A, Koch H-G, Dalbey RE (2017) Targeting and insertion of membrane proteins. EcoSal Plus 7. https://doi.org/10.1128/ecosalplus.ESP-0012-2016
164. Hann BC, Walter P (1991) The signal recognition particle in *S. cerevisiae*. Cell 67:131–144. https://doi.org/10.1016/0092-8674(91)90577-L
165. Ziehe D, Dünschede B, Schünemann D (2017) From bacteria to chloroplasts: evolution of the chloroplast SRP system. Biol Chem 398:653–661. https://doi.org/10.1515/hsz-2016-0292
166. Li X, Henry R, Yuan J, Cline K, Hoffman NE (1995) A chloroplast homologue of the signal recognition particle subunit SRP54 is involved in the posttranslational integration of a protein into thylakoid membranes. Proc Natl Acad Sci 92:3789–3793. https://doi.org/10.1073/pnas.92.9.3789
167. Ziehe D, Dünschede B, Schünemann D (2018) Molecular mechanism of SRP-dependent light-harvesting protein transport to the thylakoid membrane in plants. Photosynth Res 138:303–313. https://doi.org/10.1007/s11120-018-0544-6
168. Hristou A, Gerlach I, Stolle DS, Neumann J, Bischoff A, Dünschede B, Nowaczyk MM, Zoschke R, Schünemann D (2019) Ribosome-associated chloroplast SRP54 enables efficient cotranslational membrane insertion of key photosynthetic proteins. Plant Cell 31:2734–2750. https://doi.org/10.1105/tpc.19.00169
169. Walter B, Hristou A, Nowaczyk MM, Schünemann D (2015) *In vitro* reconstitution of co-translational D1 insertion reveals a role of the cpSec–Alb3 translocase and Vipp1 in photosystem II biogenesis. Biochem J 468:315–324. https://doi.org/10.1042/BJ20141425
170. Ji S, Siegel A, Shan S, Grimm B, Wang P (2021) Chloroplast SRP43 autonomously protects chlorophyll biosynthesis proteins against heat shock. Nat Plants 7:1420–1432. https://doi.org/10.1038/s4147
171. Gathmann S, Rupprecht E, Kahmann U, Schneider D (2008) A conserved structure and function of the YidC homologous protein Slr1471 from *Synechocystis* sp. PCC 6803. J Microbiol Biotechnol 18:1090–1094
172. Packer JCL, Howe CJ (1996) The cyanobacterial genome contains a single copy of the *ffh* gene encoding a homologue of the 54 kDa subunit of signal recognition particle. Plant Mol Biol 30:659–665. https://doi.org/10.1007/BF00049341
173. Riediger M, Spät P, Bilger R, Voigt K, Maček B, Hess WR (2021) Analysis of a photosynthetic cyanobacterium rich in internal membrane systems via gradient profiling by sequencing (Grad-seq). Plant Cell 33:248–269. https://doi.org/10.1093/plcell/koaa017

174. Schleiff E, Klösgen RB (2001) Without a little help from 'my' friends: direct insertion of proteins into chloroplast membranes? Biochim Biophys Acta 1541:22–33. https://doi.org/10.1016/S0167-48
175. Aldridge C, Cain P, Robinson C (2009) Protein transport in organelles: protein transport into and across the thylakoid membrane: protein transport across thylakoid membranes. FEBS J 276:1177–1186. https://doi.org/10.1111/j.1742-4658.2009.06875.x
176. Woolhead CA, Thompson SJ, Moore M, Tissier C, Mant A, Rodgers A, Henry R, Robinson C (2001) Distinct Albino3-dependent and -independent Pathways for Thylakoid Membrane Protein Insertion. J Biol Chem 276:40841–40846. https://doi.org/10.1074/jbc.M106523200
177. Hao B, Zhou W, Theg SM (2022) Hydrophobic mismatch is a key factor in protein transport across lipid bilayer membranes via the Tat pathway. J Biol Chem 298:101991. https://doi.org/10.1016/j.jbc.2022.101991
178. Rathmann C, Schlösser AS, Schiller J, Bogdanov M, Brüser T (2017) Tat transport in *Escherichia coli* requires zwitterionic phosphatidylethanolamine but no specific negatively charged phospholipid. FEBS Lett 591:2848–2858. https://doi.org/10.1002/1873-3468.12794
179. Koch S, de Wit JG, Vos I, Birkner JP, Gordiichuk P, Herrmann A, van Oijen AM, Driessen AJM (2016) Lipids activate SecA for high affinity binding to the SecYEG complex. J Biol Chem 291:22534–22543. https://doi.org/10.1074/jbc.M116.743831
180. Prabudiansyah I, Kusters I, Caforio A, Driessen AJM (2015) Characterization of the annular lipid shell of the Sec translocon. Biochim Biophys Acta 1848:2050–2056. https://doi.org/10.1016/j.bbamem
181. Wang J, Huang X, Ge H et al (2022) The quantitative proteome atlas of a model cyanobacterium. J Genet Genomics 49:96–108. https://doi.org/10.1016/j.jgg.2021.09.007

Part III
Organelle Integration Within the Eukaryotic Host Cell

What Has *Paulinella* Taught us About Endosymbiont Metabolic Integration?

Victoria Calatrava, Timothy G. Stephens, Arthur R. Grossman, and Debashish Bhattacharya

Abstract Primary endosymbiosis between a cyanobacterium and a protist marked the origin of photosynthetic eukaryotes, profoundly impacting the history of life on our planet. This event occurred 1.6–2.1 Bya and gave rise to the putative single common ancestor of the Archaeplastida (i.e., glaucophytes, red algae, and green algae plus land plants). A second independent primary endosymbiotic event took place more recently (~120 Mya) and resulted in photosynthetic *Paulinella* species that harbor primary plastids known as chromatophores. Whereas there are multiple examples of plastid secondary and tertiary endosymbiosis, fueling massive diversification of photosynthetic eukaryotes on Earth, instances of primary plastid origin are fleetingly rare. This review summarizes what has been learned about organellogenesis using the nascent *Paulinella* system, emphasizing mechanisms and constraints with respect to host–endosymbiont metabolic integration. Similarities and differences between the nascent *Paulinella* symbiosis with the more ancient Archaeplastida system are described, and possible convergent outcomes of these systems are discussed.

Keywords *Paulinella* · Endosymbiosis · Primary endosymbiosis · Photosynthesis · Plastids · Chromatophore · Organellogenesis

---

Victoria Calatrava and Timothy G. Stephens are co-first authors.

V. Calatrava (✉) · A. R. Grossman
The Carnegie Institution for Science, Department of Plant Biology, Stanford, CA, USA
e-mail: vcalatrava@carnegiescience.edu; agrossman@carnegiescience.edu

T. G. Stephens (✉) · D. Bhattacharya
Rutgers, The State University of New Jersey, Department of Biochemistry and Microbiology, New Brunswick, NJ, USA
e-mail: ts942@sebs.rutgers.edu; dbhattac@rutgers.edu

S. D. Schwartzbach et al. (eds.), *Endosymbiotic Organelle Acquisition*,
https://doi.org/10.1007/978-3-031-57446-7_12

1 Introduction

1.1 Brief History of Primary Plastid Origin

Around 1.6–2.1 Bya [1], a β-cyanobacterium was engulfed and retained by a eukaryotic heterotroph. The captured cell evolved into a plastid in the putative common ancestor of all Archaeplastida (i.e., glaucophytes, red algae, and green algae plus land plants) [2]. Following this primary endosymbiosis, some members of the Archaeplastida were engulfed by different eukaryotic heterotrophs, which became photosynthetic via complex (e.g., secondary or tertiary) endosymbiosis. These events enabled oxygenic photosynthesis to diversify and spread to distantly related eukaryotes in the tree of life. This process profoundly impacted the history of life on our planet by generating a vast array of primary producers that support different aquatic and terrestrial ecosystems.

1.2 The Paulinella *System*

The only other known example of primary endosymbiosis involving a cyanobacterium is represented by the genus *Paulinella*, which harbors primary plastids designated as chromatophores. The *Paulinella* genus contains a handful of species that have been isolated from both marine and freshwater low irradiance habitats. This lineage was first discovered in 1894 by Robert Lauterborn [3, 4] and is a member of the Cercozoa (Rhizaria) that originally maintained a phagotrophic (heterotrophic) lifestyle. Photoautotrophic *Paulinella* species (green *Paulinella*) are monophyletic and evolved ~120 Mya, when the protist ancestor of this lineage engulfed and retained a α-cyanobacterium related to *Synechococcus* spp. [5–10] (Fig. 1).

Photosynthetic *Paulinella* species accumulate toxic reactive oxygen species (ROS) during growth in the light, a consequence of photosynthetic electron transport and elevated intracellular redox conditions which are exacerbated under moderate and high light intensities. These species are sensitive even to relatively low light intensities when compared to the free-living cyanobacterial relatives of their chromatophores [11–13]. This difference likely reflects the nascent, suboptimal metabolic integration of the partner organisms that evolved into the green *Paulinella* lineage.

The evolution of efficient host–endosymbiont metabolic integration requires gene transfer from the endosymbiont genome to the host genome, gene loss and modification, genomic rearrangements, the tailoring of gene expression, and the development of mechanisms to facilitate the entry of cytoplasmically synthesized proteins into the nascent organelle. New genetic resources may be acquired via endosymbiotic gene transfer (EGT) and (non-endosymbiotic) horizontal gene transfer (HGT), and genetic rearrangements within the host genome. The genetic mechanisms and evolutionary forces driving these reorganizations are obscured in the

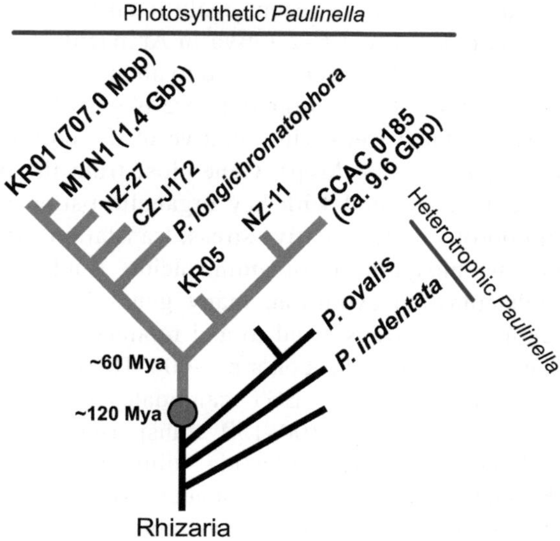

Fig. 1 Schematic tree of the *Paulinella* lineage based on Lhee et al. [11]. The heterotrophic species are shown with thin black lines. These species are sister to the photosynthetic lineage, shown with thicker grey lines. The primary plastid endosymbiosis occurred ~120 million years ago (Mya) and the deepest split in the tree of photosynthetic *Paulinella* species occurred ~60 Mya [11]. Genome size is shown, when known, next to each isolate

ancient origins of algal and plant plastids. Therefore, and importantly, this recent primary endosymbiosis represents a unique model for investigating early events in the evolution of a primary plastid (recently reviewed in [9]).

1.3 Comparison of Paulinella *and Archaeplastida*

During plastid evolution, the genomes of the cyanobacterial endosymbiont and the host become dramatically reorganized to tailor the expression and regulatory features of genes and to integrate the metabolisms of the partner organisms. These processes allow for the establishment of a functional photosynthetic organelle and consequent loss of the capacity for phagotrophic feeding by the newly evolved photoautotrophic host cell. Canonical features among Archaeplastida include massive reduction of plastid genome size, which can be compensated for by existing eukaryotic genes present on the host genome and those derived via HGT and EGT (including genes involved in photosynthesis). Regardless of the origin of the nuclear genes integral to the endosymbiotic process and organellogenesis, these transferred genes, many of which are prokaryotic, must become active and properly regulated in the recipient eukaryotic host, which likely poses a critical hurdle for primary endosymbiosis [14]. Moreover, proteins encoded by nuclear genes that function within the endosymbiont must be translocated from the host cytoplasm into the endosymbiont. Thus, the host must evolve a mechanism for importing proteins into the endosymbiont.

Whereas analogous innovations are occurring in photosynthetic *Paulinella* species, they are not nearly as developed as in the Archaeplastida, likely because of the more recent evolutionary origin of the chromatophore (~120 Mya compared to 1.6–2.1 Bya in Archaeplastida). The *Paulinella* chromatophore genome has been reduced to about one-third of its original size (i.e., the genome size of the chromatophore is ~1 Mbp, whereas the genome size of a closely related free-living relative of the chromatophore, *Synechococcus* sp. WH 5701, is 3.28 Mbp). Genes lost from the plastid genome encode many proteins associated with key organelle pathways, including photosynthesis, photoprotection, oxidative stress, translation, transcription, DNA replication, and the biosynthesis of amino acids, lipids, starch, and pigments. During Archaeplastida evolution, many genes lost from the endosymbiont/plastid genome were transferred to and retained in the nuclear genome of the host. The Archaeplastida contain ~600–1000 EGT candidate genes, whereas *Paulinella* has <100 EGT candidates (see below) [15–18]. Furthermore, whereas there are 100–150 transporters involved in plastid metabolite exchange in the Archaeplastida, solute transporters are nearly absent from the chromatophore. Metabolic connectivity between the host and chromatophore appears to rely on small transmembrane proteins that alter membrane permeability [19] but may not be very specific for individual metabolites. In the Archaeplastida, N-terminal targeting sequences (transit peptides) have evolved to traffic cytoplasmic proteins into the plastid [20]. *Paulinella* has evolved analogous N-terminal sequences, known as chromatophore targeting peptides (crTPs). These crTPs, which are associated with the targeting of 'long', soluble *Paulinella* proteins, can also route engineered proteins into tobacco chloroplasts [21]. However, the precise mechanisms of protein transport have not been analyzed in *Paulinella* and it is not clear whether a specific transport apparatus is present on the chromatophore envelope. Additional gene transfers from the endosymbiont to the *Paulinella* nuclear genome, followed by retention and optimization of expression of these genes, may allow further reduction of the chromatophore genome and help stabilize interactions between the host and the endosymbiont. Continued integration of chromatophore genes into the *Paulinella* nuclear genome and subsequent tailoring of these newly acquired nuclear genes, may confer this green protist with the capacity for elevated growth rates and/or the ability to colonize environments with higher intensity illumination. Unlike the Archaeplastida, photosynthetic *Paulinella* came into existence in a crowded world dominated by diatoms, dinoflagellates, and a variety of picoeukaryotes. Furthermore, there are significant differences between current abiotic environmental conditions and those that existed 1.6–2.1 Bya when cyanobacteria were dominant. Whereas it is tempting to assume that *Paulinella* will follow a similar evolutionary trajectory as the Archaeplastida, evolving to colonize more "standard" photoautotrophic niches, this is not assured because of differences in current biotic and abiotic conditions relative to those associated with the establishment of the Archaeplastida plastid.

1.4 Chassis and Engine Hypothesis

Whereas multiple examples of endosymbiosis involving algae with primary and secondary plastids have been reported, the establishment of primary plastid endosymbiosis is very rare, suggesting that there are serious constraints associated with integrating the genetic and metabolic processes of the cyanobacterial endosymbiont and its eukaryotic host [7]. The independent origin (with respect to the Archaeplastida) and the early stage of primary plastid evolution in *Paulinella*, make this system valuable for elucidating early events in the integration process and has allowed for the generation of multiple models reviewed here, including the "chassis and engine" model for plastid evolution.

Genome rearrangements and adaptations have occurred in *Paulinella* to accommodate the integration of the endosymbiont, yet it appears to be a 'work-in-progress' that may be improved with time and if conditions allow. The 'chassis and engine' hypothesis uses an analogy to highlight major issues associated with the evolution of a primary plastid. In the *Paulinella* system, the 'engine' is the chromatophore that powers the 'vehicle'–the photosynthetic *Paulinella* holobiont. This engine relies on inorganic nutrients and other resources that the 'chassis'–the host cell–must supply, such as sunlight, water, proteins, and nutrients including inorganic carbon (Ci). The engine produces the energy for the system (in the form of ATP and fixed carbon), inevitably generating excess 'heat' (harmful reactive oxygen species) that must be managed. In this analogy, the radiator (which is initially part of the engine but becomes part of the chassis over time) maintains mechanisms that can prevent or limit photodamage (i.e., non-photochemical quenching and water-to-water cycles) and enzymes that detoxify reactive oxygen species (e.g., superoxide dismutase, glutathione peroxidases, and catalases). The chassis also maintains control over the engine 'speed' by modulating chloroplast metabolism (i.e., production and utilization of ATP and reductant/fixed carbon production), which powers vehicular activity/motion with photosynthate.

In the nascent plastids of *Paulinella*, the engine/endosymbiont is initially equipped to fully operate (i.e., perform photosynthesis efficiently under high light intensity), whereas the chassis/host must experience multiple alterations to control the engine and avoid uncontrolled power generation that can damage the vehicle (e.g., photoinhibition), ensuring smooth operation of the vehicle, with infrequent breakdowns. In *Paulinella,* the host cell has extensively remodeled its genome and metabolism to accommodate the endosymbiont, including the evolution of a mechanism for importing nuclear-encoded proteins into the chromatophore. These developments precipitated the loss of vital genes in the genome of the endosymbiont and the phagotrophic capacity of the host, to the point at which both microbes strictly rely on each other. Likely because of the inability of the host to optimally utilize resources from the endosymbiont and to assist in controlling the 'speed' at which the 'engine' is running, the habitat of *Paulinella* is restricted to low irradiance environments (Fig. 2), which prevents the engine from 'overheating'. Based on further chromatophore genome reduction, as observed for plastid genomes of the

Archaeplastida, it seems likely that the next steps in the integration of the *Paulinella* association would involve further remodeling of the host to better control the endosymbiont through the establishment of both anterograde and retrograde signaling (see Sect. 2.6 for additional details). This may allow additional reduction of the chromatophore genome, potentially attaining a similar plastid genome size as found in the Archaeplastida. Alternatively, the endosymbiont may not be able to escape its current low light conditions due to an inability to compete effectively with other algae. In this case, there may be minimal additional reduction of the endosymbiont genome, or the host might lose photosynthetic activity altogether, as has occurred in many other algal lineages [22].

2 Integration of Host and Chromatophore Genomes

2.1 Overview of Paulinella *Genome Biology*

The only high-quality nuclear genome assembly available for photosynthetic *Paulinella* is from *P. micropora* KR01. This genome is 707 Mbp and contains ~32,361 predicted genes. More fragmented genome data exist for the sister taxon of *P. micropora* KR01, *P. micropora* MYN1, whose assembly totals 1.35 Gbp with 36,763 predicted genes [23]. The more diverged *P. chromatophora* CCAC 0185, has a massive genome, estimated to be ~9.6 Gbp with 78,140 putative genes. The latter value is based on a cDNA assembly and is likely an over-estimate of the size of the gene inventory [16]. Therefore, photosynthetic *Paulinella* species are highly diverged with respect to genome size, with all known species having very large genomes. In contrast, the chromatophore genome in these taxa is of uniform size (~1 Mbp) with 860–878 protein coding genes [15, 24]. These organelle genomes have undergone significant reduction when compared to their free-living α-cyanobacterial (likely *Synechococcus* sp.) ancestor. This process of genome

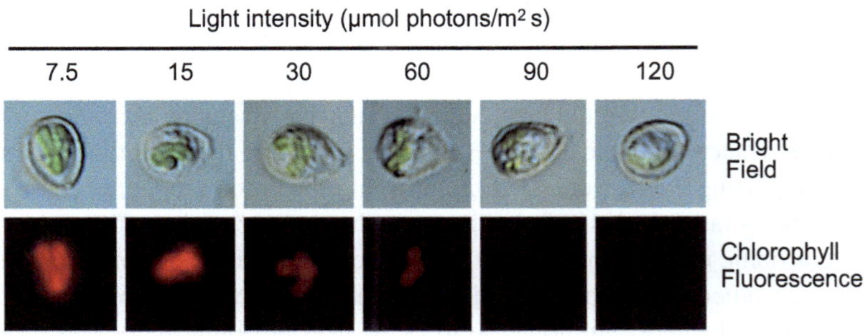

Fig. 2 Light sensitivity of *Paulinella micropora* KR01. The cells were grown for 3 weeks under the light intensities (7.5–120 μmol photons m^2s) shown at the top of each pair of images. Representative cells are shown for each treatment using bright field (top) and fluorescence (bottom) microscopy; the latter shows red chlorophyll fluorescence

reduction is driven by Muller's ratchet acting on the captured endosymbiont, which is no longer able to undergo recombination with other free-living cyanobacterial cells [25–27]. Massive gene loss is both a challenge and an opportunity for the host cell to gain control (via anterograde signaling) over the fate and function of the endosymbiont and is therefore an important focus of research on primary endosymbiosis. Although stabilized in size since the deepest split in the photosynthetic *Paulinella* lineage, between *P. micropora* and *P. chromatophora* (~60 Mya [10] [Fig. 1]), the chromatophore genome is likely destined for additional size and gene content reduction as EGT to the nuclear genome continues. Naturally, the tempo and extent of this process is not known. Nonetheless, in their present state, the *Paulinella* chromatophore genomes have only undergone moderate levels of genome reduction compared to algal and plant lineages, with only ~20 unique gene families, or less, having been relocated to the *Paulinella* nuclear genome via EGT [15–18]. In comparison, ~600–1000 genes were relocated to the nuclear genome of algae and plants via EGT, and most Archaeplastida plastid genomes are only 80–200 kbp [8, 28], although some red algal plastid genomes can encode as many as ~250 genes [29].

2.2 *Metabolic Integration in the Face of Chromatophore Gene Loss*

Characterization of the chromatophore genome has demonstrated the loss of essential genes for survival of the endosymbiont, such as those involved in the TCA cycle, metabolite transport, enzyme cofactor biosynthesis, and amino acid and nucleotide metabolism pathways [9, 16, 18, 21, 30]. For this union to survive, the host amoeba needs to 'rescue' these essential, lost endosymbiont functions. Specifically, missing proteins and metabolites need to be supplied, which beyond rescuing chromatophore functions, places the compartment under host control. Here, we discuss how this remarkable process of metabolic integration develops, its major targets, and strategies used by the host cell to achieve control over the nascent photosynthetic organelle.

Previous work has shown that most genes whose proteins rescue chromatophore functions are host derived, with a limited number originating via EGT or HGT [18, 21]. The most direct way for the host to provide needed genetic resources is through duplication of nuclear loci, with the novel gene family members free to evolve chromatophore targeting peptides (crTPs) and therefore, perform their function in the nascent organelle. Given that longer chromatophore-destined proteins (so-called 'long proteins'; >268 amino acids [aa]) contain this novel, distinctive sorting signal of ~200 aa, it is relatively straightforward using in silico tools to identify the collection of nuclear encoded, chromatophore targeted proteins in *Paulinella*. These crTP-containing proteins of host provenance are obvious candidates for integrating host and endosymbiont metabolisms [18, 21]. Comparative analyses reveals that there are ~208 long proteins shared between *P. micropora* KR01 and *P. chromatophora* that are candidates for import into the chromatophore [18]. These define the ancestral set of nuclear-encoded chromatophore-targeted proteins in the photosynthetic (green) *Paulinella* lineage that had its basal split ~60 Mya (Fig. 1). Of these crTP-containing proteins, two thirds have annotated functions, with the largest category (58 proteins) being 'genetic information processing', raising the possibility that members of this protein subset were critical for the host to gain control of endosymbiont biology [18].

Gabr and co-authors used this information and gene inventory data from the chromatophore genome to study metabolic integration in the photosynthetic *Paulinella* lineage [31]. They classified genes putatively involved in metabolic integration into two classes: genes novel to the *Paulinella* lineage (i.e., endosymbiont encoded or originating via EGT/HGT) and those that are redundant (i.e., with homologs existing in both the host and endosymbiont genomes). These two gene classes are expected to have divergent evolutionary trajectories in the novel algal lineage. Specifically, the former gene set may be largely immune to loss due to its irreplaceable role in organelle maintenance and functions (e.g., photosynthesis), whereas the latter offers the host the opportunity to 'divide and conquer' the endosymbiont by controlling specific steps in different, redundant pathways.

In *P. chromatophora*, the majority of genes involved in DNA replication such as DNA helicase (*dnaB*), the single-strand binding protein SSB, DNA primase (*dnaG*), gyrase (*gyrA* and *gyrB*), topoisomerase (*topA*), the replication initiation protein DnaA, and DNA polymerase III subunits [30] are chromatophore-encoded. In contrast, as mentioned above, genes involved in 'genetic information processing' (e.g., encoding DNA polymerase I-like, *polA*, and NAD-dependent DNA ligase) are present on the nuclear genome of *P. micropora* KR01, with many of the encoded proteins having a crTP [18]. So, what do these observations reveal about host–endosymbiont metabolic integration? We will review results presented in Gabr et al. [31] from two areas of investigation, control of chromatophore DNA replication and nucleotide biosynthesis.

2.3 Chromatophore Localization of Nucleotide Biosynthesis

Current evidence suggests that the host controls de novo chromatophore nucleotide biosynthesis because key enzymes required for the initial (i.e., generation of inosine monophosphate [IMP] and uridine monophosphate [UMP]) and final (i.e., generation of guanosine, adenine, and thymine) steps of this pathway are nuclear encoded and contain a *Paulinella* crTP (Fig. 3). This situation presumably allows the host to regulate the generation of precursor molecules required for the function of this pathway in the endosymbiont. It is surprising that the enzyme responsible for the bidirectional conversion of uracil from UMP (part of the salvage pathway) is chromatophore encoded, whereas all other salvage and degradation pathway enzymes are encoded by nuclear loci but lack a crTP encoding sequence [13]. It was also discovered that some enzymes catalyzing key reactions, such as the conversion

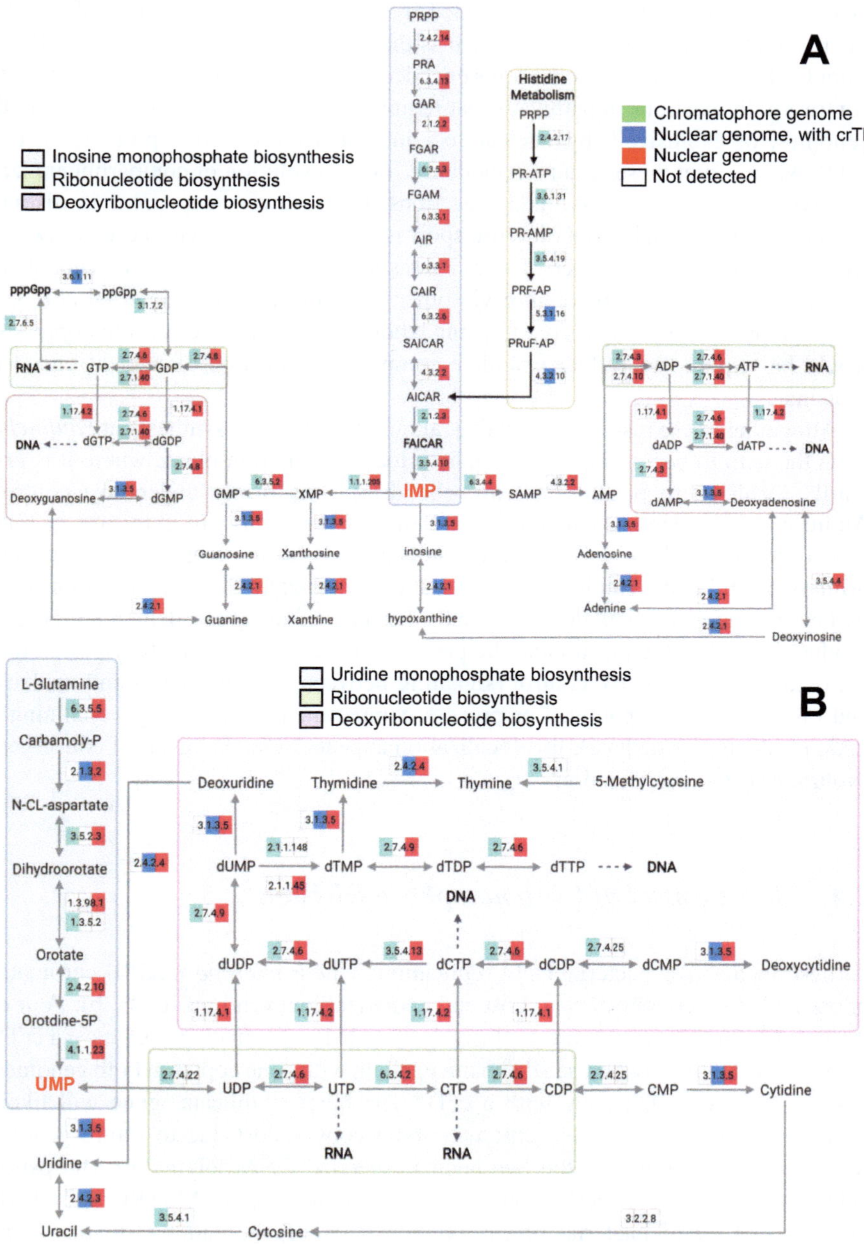

Fig. 3 Reconstruction of the putative purine (**a**) and pyrimidine (**b**) metabolism pathways in *P. micropora* KR01. The colored boxes show proteins whose genes are chromatophore-encoded (green), nuclear-encoded with a crTP (blue), nuclear-encoded without a crTP (red), or not detected in our analysis (uncolored). A legend is given at the top of each pathway

of UDP to dUDP (Fig. 3b), are encoded by nuclear genes that lack a crTP [31]. There are also some key reactions, such as the bidirectional conversion of UMP to/from UDP, that are limited to proteins encoded by chromatophore genes. These data contrast with reactions in pathways that require enzymes encoded by genes in both genomes, or exclusively by nuclear loci in which the encoded proteins have a crTP. We interpret these results as evidence for the evolution of compartmentalization whereby specific biosynthetic reactions in a pathway may occur in distinct cellular locations, and metabolite transport is needed to complete these pathways. However, because the absence of a crTP does not mean that the protein is unable to be routed into the chromatophore via other mechanisms (or an internal crTP), a thorough proteomic analyses (contrasting whole cells and isolated chromatophores) would be required to further elucidate compartmentalization of nuclear encoded proteins.

Altogether, these results suggest that, although nucleotide synthesis in *Paulinella* is on the path to becoming chromatophore localized (like in plants where it is primarily localized to plastids and mitochondria), it is firmly under host control. Additionally, nucleotide catabolism and salvage reactions are cytosolic in both plants and *Paulinella* [32, 33]. These results are consistent across two independent primary plastid endosymbiosis and suggest that localization of some reactions may be better suited to organelles, potentially because of energetic demands (the end product is used in the plastid and the precursor doesn't have to be transported into the organelle), interactive control (signaling between the chloroplast and nucleus) and other benefits of pathway or process compartmentalization (e.g., confining a toxic product/intermediate); this localization appears to be the result of convergent evolution by natural selection.

2.4 Host Control of Chromatophore Division

As described above, bacterial DNA replication in the *P. micropora* KR01 chromatophore includes proteins of both host and endosymbiont origin (Fig. 4a, b). Four of the proteins in this process are exclusively encoded in the nucleus and have a crTP, whereas the DNA polymerase III subunit epsilon, with gene copies in both genomes, has a nuclear-encoded copy with a crTP. The latter 'duplicate' gene will likely become single copy once the chromatophore copy is lost due to future genome reduction. Interestingly, of the two nuclear encoded DNA polymerase III epsilon subunits in *P. micropora* KR01, one is mitochondrial-targeted (MSTRG.8071.1.p1) and most likely of bacterial origin, whereas the crTP-containing subunit (gene MSTRG.13717.1.p1) may also be of bacterial origin, but this is less clear (Fig. 4c) because it contains significant stretches of serine residues that are not present in the mitochondrial-targeted copy [31]. The long, predicted sequence between the crTP and the mature protein and the regions of repeated serine residues within the protein may be signals for crTP cleavage. Oberleitner et al. [19] identified a high frequency of serine residues around the predicted crTP cleavage sites, which suggests that

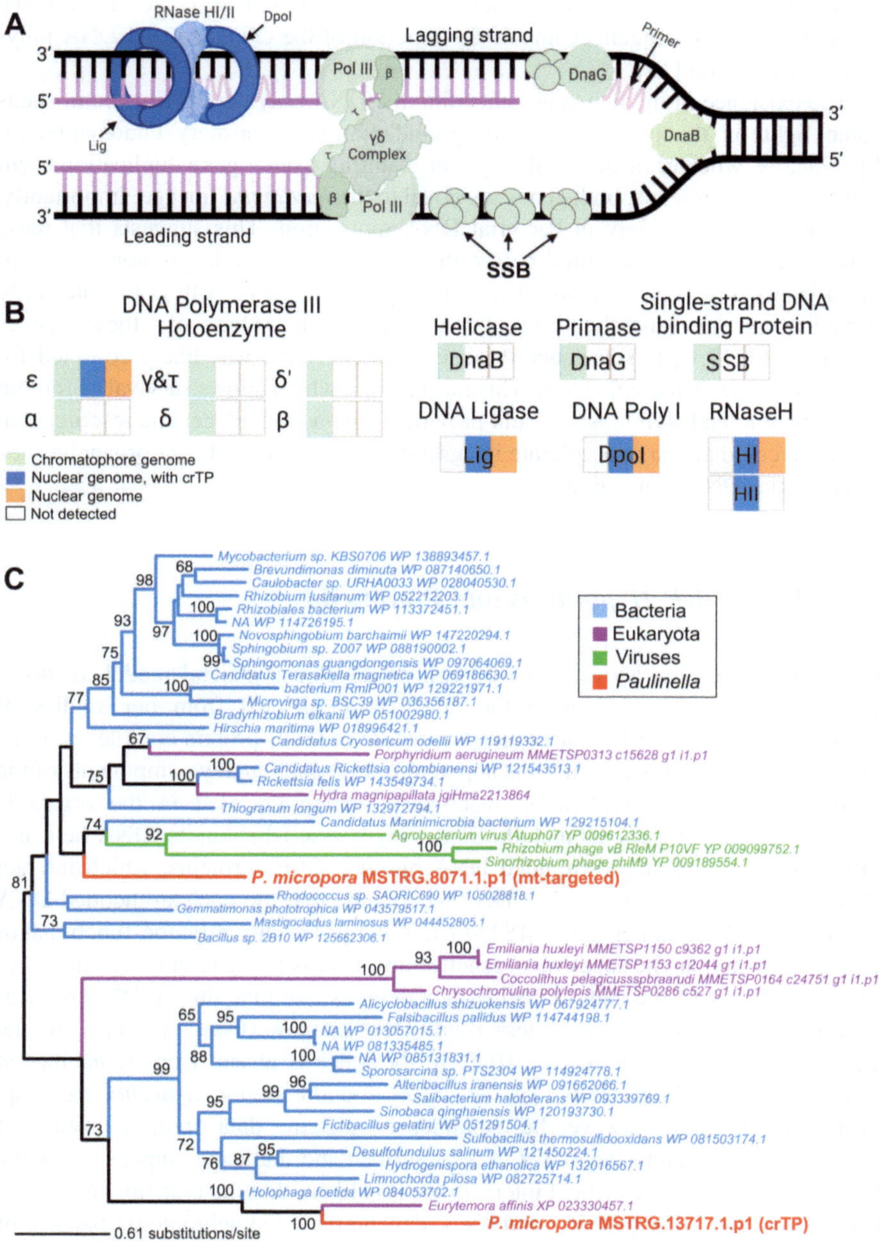

Fig. 4 (**a**) Reconstruction of the bacterial DNA replication pathway in the *Paulinella micropora* KR01 chromatophore. DNA replication complex on a segment of DNA showing proteins of the complex that are nuclear-encoded with a crTP (blue) and chromatophore-encoded (light green). (**b**) Colored boxes showing proteins that are chromatophore-encoded (green), nuclear-encoded with a crTP (blue), or nuclear-encoded without a crTP (orange). (**c**) IQ-TREE phylogeny of DNA polymerase III subunit epsilon proteins showing the positions of *P. micropora* KR01 genes within a broadly sampled dataset. Branch support is based on 2000 ultrafast bootstrap replicates (only values >60% are shown). The legend identifies taxonomic origins of sequences

these repeats may have evolved to facilitate crTP function. It is, however, unknown if these additional sequences impair the function of the crTP-targeted DNA polymerase III subunit [19].

The existence of the two genes encoding the DNA polymerase III subunit sheds some light on the potential role of pre-adaptation to primary endosymbiosis. Specifically, whether of bacterial origin or resulting from a gene duplication event followed by divergence, both genes are novel to the *Paulinella* lineage. Importantly, they function exclusively in bacterial DNA replication. This suggests that these genes may have been acquired before the cyanobacterial endosymbiont was captured by its host (i.e., they are absent in most eukaryotes), with both potentially participating in mitochondrial DNA replication. One of these genes (MSTRG.13717.1.p1) subsequently acquired a crTP and was likely co-opted for chromatophore DNA replication. This mechanism, which allows survival of foreign genes prior to endosymbiosis, would provide a "pre-adaptive" genetic resource that selection could act on to accelerate integration and host control over organelle (i.e., prokaryotic) DNA replication.

2.5 *The CORR Hypothesis and* Paulinella

Given the discussion above, can we begin to develop a general model for host–endosymbiont integration using the information gleaned from our studies of *Paulinella*? We can start by dissecting the nature of multi-protein complexes in the chromatophore. In this regard, as is apparent in Fig. 4a, b, many components of the chromatophore DNA replication machinery remain encoded in the organelle genome. These components include DNA polymerase III subunits, DNA helicase, DNA primase, and single-stranded DNA binding (SSB) proteins, which interact during DNA replication [34, 35]. Here we assume that the nuclear-encoded DNA polymerase III subunit epsilon (MSTRG.13717.1.p1) is currently non-functional or is less efficient than the ancestral organelle copy, necessitating retention of the chromatophore copy of the gene. Existing data demonstrate that, during DNA replication in bacteria, SSB proteins interact with DNA primase (DnaG) proteins and the chí subunit of DNA polymerase III [35]; the latter is absent from *P. micropora* KR01 and in a close relative of the chromatophore donor, *Synechococcus* sp. WH5701. This may however be explained by missing data or the existence of another protein subunit that is yet to be described and that can compensate for the loss of chí. Given these critical interactions, it is likely that the assembly and function of this, and other protein complexes, depend on the stoichiometric balance of subunits, that is, they require strict regulatory control [36, 37]. For this reason, we speculate that the only genes that have been successfully transferred to the nuclear genome of *P. micropora* KR01 (and the encoded protein products imported into the organelle) are the more peripheral components of the DNA replication machinery. These are DNA ligase, DNA polymerase I, and ribonuclease HI and HII proteins, which do not engage in strong protein–protein interactions with other components of the DNA replication machinery. In other words, genes encoding the strongly interacting DNA polymerase III subunits that potentially initiate assembly of the complex (DnaB, DnaG, and SSB proteins) cannot be transferred to the nuclear genome because prior to transfer their expression needs to be strictly controlled/coordinated, which may not yet have evolved in photosynthetic *Paulinella*: see discussion below that describes retrograde signaling in these species [31].

We consider this observation to be especially important because it dovetails with the existing "co-localization for redox regulation" (CORR) hypothesis [38]. The CORR hypothesis posits that some genes (e.g., photosystem and ATP-synthase subunits) must remain in the organelle genome to ensure that their expression is controlled by the redox state of their products. *Paulinella* offers an ancient forerunner of the CORR, whereby it is not the redox state, but rather, the stoichiometric balance of proteins involved in DNA replication that maintains genes encoding interacting proteins in the organelle genome. In fact, the energy producing machineries are still more-or-less completely encoded on the chromatophore genome (in most cases, only minor subunits–not required for the function of the photosynthetic apparatus–are nuclear encoded), therefore an analog of the CORR is acting on a variety of other pathways impacted by gene loss. Once a mechanism to precisely control the stoichiometry of nuclear-encoded proteins that function in the chromatophore is attained in *Paulinella*, the transfer of all DNA replication machinery genes from the chromatophore to the nuclear genome may be facilitated.

In plastids of algae and land plants, the photosynthetic apparatus uses oligomeric protein complexes comprised of protein subunits encoded by genes located in both plastid and nuclear genomes. To ensure the required stoichiometry for the functional assembly of the constituent subunits, the expression of both genetic compartments must be coordinated. An assembly-dependent regulatory process referred to as the CES process (for *C*ontrol by *E*pistasy of *S*ynthesis) controls the synthesis rate of plastid-encoded subunits—CES subunits—from a protein complex when its assembly is compromised because certain core subunits are missing [39–41]. The components of PSI, PSII, cytochrome b_6f, and ATP synthase contain (chloroplast encoded) CES subunits whose translation is downregulated when the assembly of their respective complex is compromised [42]. Whether a CES process has evolved in photosynthetic *Paulinella* is unknown. The potential evolution of CES regulation in the chromatophore may favor the 'decentralization' of protein synthesis of certain complexes that function in the chromatophore, such as those of the DNA replication machinery or complexes involved in photosynthetic electron transport, by passing on the synthesis of some subunits to the host and synchronizing their accumulation across compartments. This regulatory process could facilitate the transfer of chromatophore genes encoding subunits of protein complexes to the nuclear genome in a stepwise manner, as opposed to an "all or nothing" process.

A final point regarding integration of the host and endosymbiont genome is the difference between a gene product being targeted to the chromatophore (presumably after acquiring a crTP) and the regulation of its expression by the host; the former would almost surely precede the latter. In *P. micropora* KR01, two of the nuclear genes encoding a crTP follow a diurnal cycle (DNA polymerase III subunit

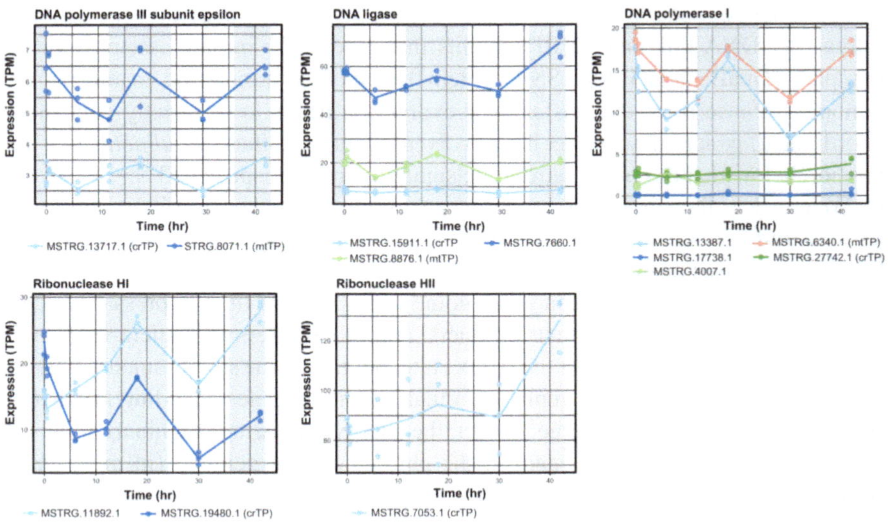

Fig. 5 Expression of nuclear-encoded bacterial DNA replication pathway genes over a 48 h diurnal cycle under ambient light. Dark periods are shown with gray and light periods with white boxes. Gene expression values are in transcripts per million (TPM). Each point represents a replicate sample (n = 3 per timepoint), lines connect the mean expression values of each timepoint. Gene names and the presence of a crTP, if predicted, are shown at the bottom of each panel

epsilon, Ribonuclease HI), whereas the other three also contain a crTP but exhibit no diurnal control (Fig. 5). In contrast, all mitochondrial targeted proteins (i.e., they contain a mtTP) are under strong diurnal control, with peak expression occurring during the middle of the dark period when cell division presumably occurs. It is likely that the chromatophore-targeted proteins will one day exhibit diurnal expression similar to their polypeptide counterparts that function in the mitochondrion, which might optimize chromatophore DNA replication and cell division. This result highlights the "all or nothing" idea because in the absence of a coordinated regulation of these genes, selection would invariably favor keeping the core subunit of DNA replication in the same location (i.e., the chromatophore genome).

2.6 Retrograde Signaling in Nuclear Gene Regulation

An important evolutionary step in organellogenesis is the origin of retrograde signaling, which is the transmission of signals from the organelle to the nucleus, allowing the host to respond to and modulate (through anterograde signaling) physiological/metabolic processes occurring in the organelle. This coordination is a landmark feature of eukaryotes, but has it evolved in the photosynthetic *Paulinella* lineage? The Archaeplastida have complex signaling pathways because the vast majority (>90%) of plastid functions are encoded by nuclear-encoded proteins that must be sorted efficiently and quantitatively to maintain plastid functions [43]. Retrograde signaling enables the metabolic state of the organelle to directly impact expression of nuclear-encoded genes [44]. In contrast, anterograde signaling enables regulation of organelle gene expression and protein activities in response to signals sensed in the cell nucleus that reflect the cellular metabolic state [45]. As an example, plant retrograde signaling impacts chloroplast biogenesis [46], responses to excess irradiation [47], and redox stress [48]. In photosynthetic *Paulinella*, the limited information we have acquired thus far does not support the presence of retrograde signaling. Analysis of differential gene expression in cultures of *P. chromatophora* exposed to high light, when compared to low light treatment, did not yield evidence for light-induced transcriptional regulation for a target set of chromatophore-encoded genes, and for most EGT-derived PSII and PSI genes in the nuclear genome [12]. This is a preliminary result, but surprising, given that the light regulation of nuclear genes that encode plastid-destined proteins associated with various plastid processes is a common feature of algae and plants [49]. We speculate that the constitutive expression of chromatophore-encoded genes is the default setting for this (and perhaps all) nascent organelles, and that natural selection will ultimately lead to the evolution of nuclear regulation of chromatophore protein abundance and activity; for chloroplasts, the synthesis of many nuclear encoded RNA-binding factors regulates the stability/translation of plastid encoded RNAs. This will increase the efficiency of metabolic/physiological processes and the biogenesis of macromolecular complexes within the chromatophore and ensure coordination of processes among cellular compartments. This process is already apparent with respect to chromatophore DNA replication, for which a subset of nuclear genes encoding chromatophore-destined proteins show a diurnal pattern of gene expression. A significant expansion of this area of research would afford a more complete understanding of the integration of nuclear-organelle metabolism.

3 Integration of Metabolisms

3.1 Gene Distribution Between Nuclear and Chromatophore Genomes

As mentioned above, much of the chromatophore function is satisfied by proteins encoded by nuclear genes. Some of these genes are of endosymbiont origin, including those involved in photosynthesis and photoprotection, whereas others were acquired by the host from different microbes, including genes involved in peptidoglycan biosynthesis. Most of the nuclear-encoded, chromatophore-localized proteins, however, are related to the ancestral host and were retailored to complement endosymbiont functions [18]. Their different phylogenetic origins have led to the establishment of mosaic metabolic pathways that reflect complex genomic rearrangements that are enabling metabolic integration of host and endosymbiont.

The analysis of the chromatophore proteome of *P. chromatophora* combined with in silico predictions indicated that almost 450 nuclear-encoded proteins are targeted to the chromatophore, most of which (~90%) are of eukaryotic (host) origin. Endosymbiont- and other bacteria-derived genes together encode <10% of the imported proteins [21]. Similar proportions were estimated for the sister lineage *P. micropora* based on genomic analysis [18]. Hence, EGT/HGT contributed relatively few genes to the current integration state of photosynthetic *Paulinella* species; most contributions to chromatophore processes appear to be derived from the host and rely on the retailoring of eukaryotic genes present in the host genome. Nevertheless, all EGT, HGT, and host-derived genes evolved in the nuclear genome to fill gaps in biosynthetic pathways vital for endosymbiont function and that foster metabolic integration necessary for a photosynthetic lifestyle.

3.2 Nutrient Transport Function

Some biochemical pathways, such as those involved in the biosynthesis of purines or some amino acids (Met, Ser and Gly), were lost from the chromatophore genome and must be metabolically compensated for by the host; these amino acids are expected to be transported into the chromatophore [16]. However, the genes for most solute uptake systems (~70 in free-living *Synechococcus* spp.) have been lost from the chromatophore genome including those involved in the transport of ammonium, nitrate, urea, amino acids, and sulfate [19, 21, 30, 50]. The remaining transporters encoded on the chromatophore genome are limited to the acquisition of inorganic ions (e.g., K^+, Na^+, Fe^{2+}, Mg^{2+}, PO^{2-}, HCO^-). Moreover, no putative transporters, or proteins with multiple transmembrane domains encoded by nuclear genes, have been identified in the chromatophore of *P. chromatophora*. Thus, although a mechanism to transport large soluble proteins from the cytoplasm to the chromatophore (with the addition of a crTP) has evolved in *Paulinella*, it seems that specialized systems to assist the more complicated task of targeting multi-spanning transmembrane proteins to the chromatophore is yet to evolve [19, 21]. Instead, *Paulinella* seems to target proteins with a single transmembrane domain and antimicrobial peptide-like orphan proteins to the chromatophore that could modulate membrane permeability and allow metabolite exchange between the host cytoplasm and the chromatophore [19].

3.3 Regulation of Photosynthesis

The acquisition of oxygenic photosynthesis by *Paulinella* provided a clear selective advantage over its phagotrophic ancestor. These cells no longer had to locate, capture, and digest other microbes to acquire fixed carbon. However, the absorption of light by oxygenic photosynthetic organisms can generate reactive oxygen species

(ROS) which can cause oxidative damage to cells if this process is not tightly controlled. Photosynthetic organisms have developed a variety of mechanisms to avoid ROS accumulation, which can cause inhibition of the photosynthetic machinery, protein denaturation, and lipid peroxidation that in turn would impair multiple cell processes and potentially result in death. Elevated ROS levels can be generated and accumulate when the light energy absorbed by cells exceeds the maximum photosynthetic capacity or the ability of the cells to use the photosynthate. To manage cellular redox and reduce the accumulation of ROS, oxygenic phototrophs have evolved multiple mechanisms, including the production of enzymes such as superoxide dismutase (SOD) and glutathione-S-transferase, which can rapidly eliminate specific ROS, as well as specific proteins that dissipate excess absorbed light energy and release it as heat (non-photochemical quenching [NPQ]). Exposure of *Paulinella* to relatively low levels of light (>60 μmol photons m^{-2} s^{-1}) can induce cell bleaching (Fig. 2) and impair growth [12]. These findings suggest that ROS accumulation in *Paulinella* is a consequence of photosynthetic activity, and that the amoeba is unable to optimally manage redox and the associated ROS accumulation. Genomic and biophysical data support the idea that the activities of the photosynthetic complexes are normal, but that hyperaccumulation of ROS occurs even when the cells are exposed to moderate light intensities: i.e., PSII is fully active in this amoeba but highly susceptible to photodamage [13], potentially because the products of photosynthesis are not rapidly distributed to the different compartments of the cell (suboptimal integration of the metabolic activities in the chromatophore and amoeba host). *Paulinella* has evolved a thioredoxin system for redox regulation, like that present in algae and plants, through integration of gene/protein functions derived from diverse sources, including the endosymbiont, the host, bacteria, and red algae [51]. Other genes involved in photoprotection have duplicated independently (i.e., via convergence) in the genomes of *P. chromatophora* and *P. micropora* [12, 18]. Moreover, after the development of the symbiotic association, some photoprotective genes acquired high-light responsive transcriptional regulation, which strongly suggests that oxidative stress generated during photosynthesis is a primary driver for the evolution of this system.

3.4 Relocation of Photosynthesis Genes

In Paulinella, genes on the chromatophore genome involved in photosynthesis (e.g., *psaA*, *psbA*, and *cpcC*) show weak transcriptional regulation when compared to free-living cyanobacteria [12, 52], which is also observed in Archaeplastida plastids [53]. A few *Paulinella* genes that are involved in photosynthesis and photoprotection were transferred to the nuclear genome, including those encoding PSI subunits and high light inducible proteins (HLIPs). Whereas some of these genes are also present on the chromatophore genome (e.g., *psaI* and *csos4A* genes in *P. chromatophora*), others have been lost (e.g., *psaE* and *hli* genes), which suggests that they became optimally functional in the host nuclear genome. Whereas the nuclear copies of *PSAI* and *PSAE* are not significantly induced in *P. chromatophora* by high light, as they are in free-living cyanobacteria, one copy of each of these genes is strongly induced by high light in *P. micropora* KR01, which suggests that a gain of transcriptional control of EGT genes by the host is ongoing. In contrast, nuclear copies of the *HLI* genes, which are absent in the chromatophore genome of both sister *Paulinella* lineages, can be strongly induced by high light in both species. The occurrence of EGT genes related to photoprotection, and the gain of transcriptional control of these genes in the nuclear genome of the host, supports the hypothesis that photooxidative stress is likely a force driving the integration of host and endosymbiont metabolic processes.

4 The *HLI* Gene Family

4.1 EGT Origin of HLI *Genes*

Among the genes in the nuclear genome that are of putative EGT-origin, >80% encode HLI proteins (>50 genes in *P. micropora* KR01). The *HLI* genes, and their relatives in eukaryotes, the OHPs (One-Helix proteins), are involved in photoprotection [54, 55]. In cyanobacteria, the HLI proteins may regulate chlorophyll biosynthesis, transiently bind chlorophyll or chlorophyll intermediates, participate in assembly of photosynthetic complexes, elicit NPQ activity, and function in quenching singlet oxygen, which makes them crucial to survival under high light stress [55, 56].

The *HLI* gene family in *Paulinella* is polyphyletic with at least three clades, including members of the *Synechococcus* sp. WH5701 lineage which is closely related to the chromatophore donor. Our findings suggest that multiple *hli* genes (≥3 genes) were transferred to the nucleus before the separation of *P. micropora* and *P. chromatophora* lineages. In the nucleus, some of these genes were retrocopied and rearranged in pairs. After divergence of the *HLI* genes in the two *Paulinella* species, these gene pairs underwent extensive duplications in both of the *Paulinella* species, generating expanded gene families that arose independently through convergent evolution (Fig. 6a).

4.2 Retrotransposition and Functional Diversification of HLI *Genes*

Genomic and transcriptomic analyses in *P. micropora* suggest that most of the *HLI* genes are retrocopies; products of the RNA-based retrotransposition gene duplication mechanism [57]. Retrocopied genes generally acquire new regulatory elements from the region of the genome in which they are inserted, which allows them to gain

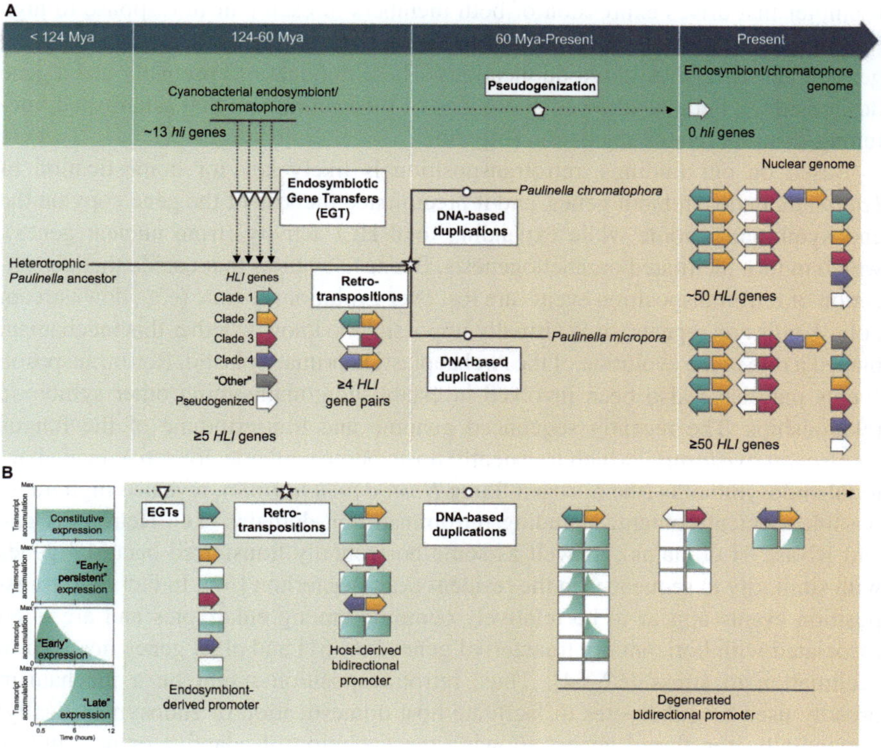

Fig. 6 Hypothetical model for the evolution of the *HLI* gene family in *Paulinella*. (**a**) At the early stages of primary endosymbiosis, before *P. chromatophora* and *P. micropora* diverged, different *HLI* genes were transferred via independent endosymbiotic gene transfer (EGT) events from the cyanobacterial endosymbiont/chromatophore to the host nuclear genome. These genes were duplicated through retrotransposition, forming at least three types of head-to-head gene pairs with different *HLI* gene compositions. After *P. chromatophora* and *P. micropora* KR01 diverged from each other, the *HLI* gene pairs were independently duplicated in both species (via a DNA-mediated mechanism), resulting in expansion of this gene family. Genes were also lost after transfer to the nuclear genome due to pseudogenization, likely because they were not effectively expressed and/or not functionally optimal. (**b**) The ancestral EGT-derived *HLI* genes were most likely associated with their original promoter and not efficiently regulated by stress in the host's nuclear genome. Different *HLI* paired retrocopies may have associated with or acquired a bidirectional promoter that is induced by stress, leading to an early-persistent expression pattern through retrotransposition. Extensive duplications of the *HLI* pairs and subsequent degeneration of their ancestral promoters could have resulted in the evolution of two new expression patterns that are complementary, partitioning the early-persistent pattern into the more specialized early and late expression patterns

new modes of transcriptional regulation [58]. In *Paulinella*, retrotransposition of *HLI*, and likely the nuclear genes encoding the PSI subunits, PSAE and PSAI, allowed the host to gain transcriptional control in response to high light stress. Likely as a consequence of this process, retrocopied *HLI* genes became arranged in pairs that contain phylogenetically distinct *HLI* genes oriented in a head-to-head configuration. These gene pairs are under the control of a putative bidirectional promoter that drives expression of both members of each pair in response to high light stress. After being retrocopied, the *HLI* gene pairs were further expanded in the genome through DNA-based duplications. These duplications probably had a gene dosage effect, but also allowed alterations of their promoters that established fine-tuning of stress related regulation (Fig. 6b).

Based on our findings, retrotransposition is likely key for domestication of *Paulinella* endosymbiont genes, and it precipitated the loss of the gene copy on the endosymbiont genome while expanding total HLI activity (from nuclear genes), which in turn facilitated organellogenesis. Because features that enable the identification of retrotransposition events are lost over evolutionary time (e.g., downstream poly-A tails and repeats), it is virtually impossible to know whether this mechanism played a role in the evolution of the Archaeplastida primary plastid. Retrotransposon events may have also been involved in establishing/maintaining other symbiotic relationships. The recently sequenced genome and transcriptome of the fungus *Geosiphon pyriformis,* which transiently hosts photosynthetically active cyanobacterial endosymbionts (the nitrogen fixing *Nostoc punctiforme*), revealed high transposable element content, including retrotransposon domains (i.e., Retrotrans gag and RNase_H domains), as well as some horizontally transferred bacterial genes with similarity to sequences in the resident cyanobacterium [59]. In fact, retrotransposition events appear to be relatively common among eukaryotes and are often associated with horizontally transferred genes [60, 61] and algal genes involved in acclimation to stress [62–64]. Thus, retrotransposition might be a mechanism broadly used by eukaryotes to facilitate host domestication of endosymbiotically/horizontally transferred genes, allowing the evolution of adaptive or new fitness functions that emerge under “extreme” conditions [57].

4.3 A General Model for Optimization of Transcriptional Control of EGT-Derived Genes in the Nucleus

Many of the EGT-derived genes present on the *Paulinella* nuclear genome have become functional, and thereby allow loss of the endosymbiont gene copy. The latter is likely key for the integration of endosymbiont-specific metabolic function in the host. We have observed duplications, including retrotransposition of EGT genes in the *Paulinella* host nuclear genome, which likely plays an important role in gene domestication as previously mentioned. Based on our results and those of others [14, 57], we proposed a general model for the origin and optimization of EGT-derived genes in the host nuclear genome (Fig. 7) [57]. In the process of

Fig. 7 (continued) improve transcription of the EGT-derived gene retrocopies and be further optimized over time. Additional DNA-based duplications allow further optimization of transcription and subfunctionalization of some of the retrocopies, while the suboptimally regulated copies become pseudogenized (5). The increased gene dosage and optimization of transcriptional control of the endosymbiont-derived genes in the host nucleus relaxes selective pressure on the endosymbiont copies, resulting in their pseudogenization and loss over time (6). Consequently, endosymbiont genes become subjected to host control in the nucleus while the endosymbiont genome is reduced. Abbreviations: mRNA, messenger RNA

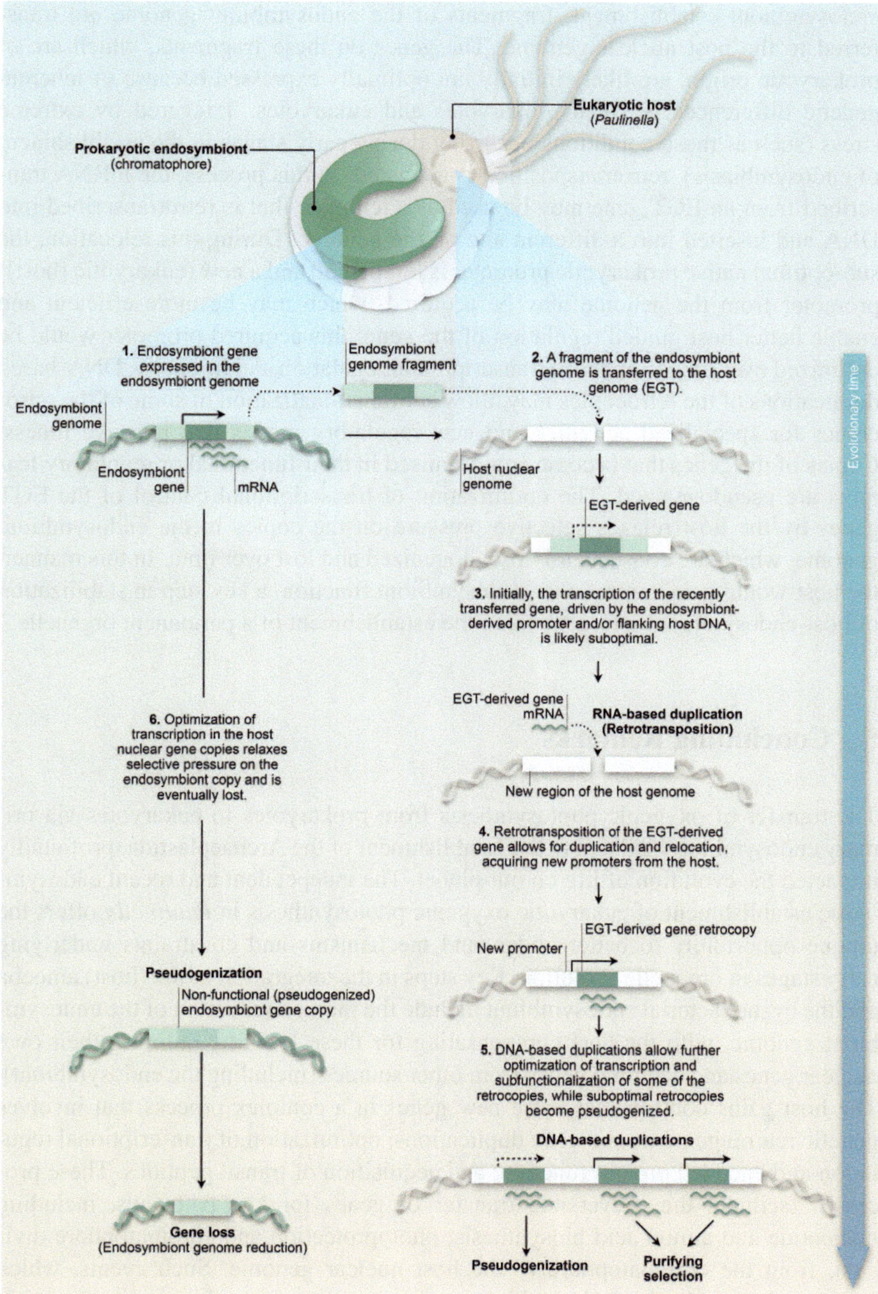

Fig. 7 Hypothetical role of retrotransposition in domestication of prokaryotic genes in eukaryotic genomes based on the *Paulinella* model. Genome fragments from the endosymbiont containing genes (1) are transferred from the endosymbiont into the host nuclear genome (2). The recently transferred endosymbiont genes of prokaryotic origin are likely not optimally expressed in the eukaryotic genome (3). Retrotransposition events result in duplication and relocation of the genes in the nuclear genome that allow acquisition of new promoters (4). These new promoters may

endosymbiont establishment, fragments of the endosymbiont genome are transferred to the host nuclear genome. The genes on these fragments, which are of prokaryotic origin, are likely initially not optimally expressed because of inherent genetic differences between prokaryotes and eukaryotes. Triggered by extreme stress (such as those conditions occurring during early stages in the establishment of endosymbiosis), retrotransposition is activated. In this process, the mRNA transcribed from an EGT gene may be used as a template that is retrotranscribed into DNA and inserted into a different site on the genome. During this relocation, the sub-optimal native prokaryotic promoter is left behind and a new (eukaryotic [host]) promoter from the genome may be acquired, which may be more efficient and enable better host guided regulation of the gene; this acquired promoter would be optimized over time to enhance transcriptional regulation. Additionally, DNA-based duplications of the retrocopies may allow subfunctionalization of some of the retrocopies for specialized activities and new regulatory modes that improve fitness. Copies of the genes that become compromised in their functional or regulatory features are pseudogenized. The optimization of transcriptional control of the EGT genes by the host relaxes selective pressure on the copies in the endosymbiont genome, which are consequently pseudogenized and lost over time. In this manner, the host would gain control over endosymbiont function, a key step in stabilization of host-endosymbiont interaction and the establishment of a permanent organelle.

5 Concluding Remarks

The transfer of oxygenic photosynthesis from prokaryotes to eukaryotes via primary endosymbiosis that led to the establishment of the Archaeplastida, profoundly impacted the evolution of life on our planet. The independent and recent endosymbiotic establishment of eukaryotic oxygenic photosynthesis in *Paulinella* offers the unique opportunity to better understand mechanisms and constraints underlying early stages in organelle evolution. Key steps in the integration of the (host) amoeba and the cyanobacterial endosymbiont include the marked reduction of the endosymbiont genome, with the host compensating for these loses by tailoring their own nuclear genes and acquired genes from other sources (including the endosymbiont). The host gains control over these new genes in a complex process that involves genetic rearrangements and gene duplications, optimization of transcriptional regulation and encoded protein function, and acquisition of transit peptides. These processes facilitate the irreversible transfer of genes for key pathways, including nucleotide and amino acid biosynthesis, photoprotection and chromatophore division, from the chromatophore to the host nuclear genome. Such events, which appear to largely be driven by oxidative stress resulting from photosynthetic activities, are allowing for domestication of the *Paulinella* endosymbiont and the establishment of a primary photosynthetic organelle (chromatophore). Future evolutionary remodeling of *Paulinella* is likely to include mechanisms that improve metabolic connectivity and the more efficient import of proteins into the chromatophore.

Ultimately, continued domestication would improve metabolic coupling and stress tolerance of the "partnering systems", which would increase the light tolerance of *Paulinella* cells, their growth rate, and potentially enable them to colonize other ecological niches.

Acknowledgments All authors were supported by a grant from the National Aeronautics and Space Administration (80NSSC19K0462) awarded to D.B. and A.R.G. D.B. was also supported by a NIFA-USDA Hatch grant (NJ01180). A.R.G. and V.C. were also supported by the Carnegie Institution for Science.

References

1. Sánchez-Baracaldo P, Raven JA, Pisani D, Knoll AH (2017) Early photosynthetic eukaryotes inhabited low-salinity habitats. Proc Natl Acad Sci U S A 114:E7737–E7745. https://doi.org/10.1073/pnas.1620089114
2. Yoon HS, Hackett JD, Ciniglia C, Pinto G, Bhattacharya D (2004) A molecular timeline for the origin of photosynthetic eukaryotes. Mol Biol Evol 21:809–818. https://doi.org/10.1093/molbev/msh07
3. Lauterborn R, Ii P (1895) *Paulinella chromatophora* nov. gen., nov. spec., ein beschalter Rhizopode des Siisswassers mit blaugrünen chromatophorenartigen Einschlussen. Zeitschrift für Wiss Zoologische
4. Melkonian M, Mollenhauer D (2005) Robert Lauterborn (1869–1952) and his *Paulinella chromatophora*. Protist 156:253–262. https://doi.org/10.1016/j.protis.2005.06.001
5. Marin B, Nowack ECM, Melkonian M (2005) A plastid in the making: evidence for a second primary endosymbiosis. Protist 156:425–432. https://doi.org/10.1016/j.protis.2005.09.001
6. Bhattacharya D, Price DC, Yoon HS, Yang EC, Poulton NJ, Andersen RA et al (2012) Single cell genome analysis supports a link between phagotrophy and primary plastid endosymbiosis. Sci Rep 2:356. https://doi.org/10.1038/srep00356
7. Stephens TG, Gabr A, Calatrava V, Grossman AR, Bhattacharya D (2021) Why is primary endosymbiosis so rare? New Phytol 231:1693–1699. https://doi.org/10.1111/nph.17478
8. Nowack ECM, Weber APM (2018) Genomics-informed insights into endosymbiotic organelle evolution in photosynthetic eukaryotes. Annu Rev Plant Biol 69:51–84. https://doi.org/10.1146/annurev-arplant-042817-040209
9. Gabr A, Grossman AR, Bhattacharya D (2020) Paulinella, a model for understanding plastid primary endosymbiosis. J Phycol 56:837–843. https://doi.org/10.1111/jpy.13003
10. Lhee D, Ha J-S, Kim S, Park MG, Bhattacharya D, Yoon HS (2019) Evolutionary dynamics of the chromatophore genome in three photosynthetic *Paulinella* species. Sci Rep 9:2560. https://doi.org/10.1038/s41598-019-38621-8
11. Lhee D, Yang EC, Kim JI, Nakayama T, Zuccarello G, Andersen RA et al (2017) Diversity of the photosynthetic *Paulinella* species, with the description of Paulinella micropora sp. nov. and the Chromatophore genome sequence for strain KR01. Protist 168:155–170. https://doi.org/10.1016/j.protis.2017.01.003
12. Zhang R, Nowack ECM, Price DC, Bhattacharya D, Grossman AR (2017) Impact of light intensity and quality on chromatophore and nuclear gene expression in *Paulinella chromatophora*, an amoeba with nascent photosynthetic organelles. Plant J 90:221–234. https://doi.org/10.1111/tpj.13488
13. Gabr A, Zournas A, Stephens TG, Dismukes GC, Bhattacharya D (2022) Evidence for a robust photosystem II in the photosynthetic amoeba *Paulinella*. New Phytol 234:934–945. https://doi.org/10.1111/nph.18052
14. Husnik F, McCutcheon JP (2018) Functional horizontal gene transfer from bacteria to eukaryotes. Nat Rev Microbiol 16:67–79. https://doi.org/10.1038/nrmicro.2017.137

15. Nowack ECM, Vogel H, Groth M, Grossman AR, Melkonian M, Glöckner G (2011) Endosymbiotic gene transfer and transcriptional regulation of transferred genes in *Paulinella chromatophora*. Mol Biol Evol 28:407–422. https://doi.org/10.1093/molbev/msq209
16. Nowack ECM, Price DC, Bhattacharya D, Singer A, Melkonian M, Grossman AR (2016) Gene transfers from diverse bacteria compensate for reductive genome evolution in the chromatophore of *Paulinella chromatophora*. Proc Natl Acad Sci U S A 113:12214–12219. https://doi.org/10.1073/pnas.160801611
17. Nowack ECM (2014) *Paulinella chromatophora* – rethinking the transition from endosymbiont to organelle. Acta Soc Bot Pol Pol Tow Bot 83:387–397. https://doi.org/10.5586/asbp.2014.049
18. Lhee D, Lee J, Ettahi K, Cho CH, Ha J-S, Chan Y-F et al (2021) Amoeba genome reveals dominant host contribution to plastid endosymbiosis. Mol Biol Evol 38:344–357. https://doi.org/10.1093/molbev/msaa206
19. Oberleitner L, Poschmann G, Macorano L, Schott-Verdugo S, Gohlke H, Stühler K et al (2020) The puzzle of metabolite exchange and identification of Putative Octotrico Peptide repeat expression regulators in the Nascent Photosynthetic Organelles of *Paulinella chromatophora*. Front Microbiol 11:607182. https://doi.org/10.3389/fmicb.2020.607182
20. Knopp M, Garg SG, Handrich M, Gould SB (2020) Major changes in plastid protein import and the origin of the Chloroplastida. iScience 23:100896. https://doi.org/10.1016/j.isci.2020.100896
21. Singer A, Poschmann G, Mühlich C, Valadez-Cano C, Hänsch S, Hüren V et al (2017) Massive protein import into the early-evolutionary-stage photosynthetic organelle of the Amoeba *Paulinella chromatophora*. Curr Biol 27:2763–2773.e5. https://doi.org/10.1016/j.cub.2017.08.010
22. Dorrell RG, Azuma T, Nomura M, Audren de Kerdrel G, Paoli L, Yang S et al (2019) Principles of plastid reductive evolution illuminated by nonphotosynthetic chrysophytes. Proc Natl Acad Sci U S A 116:6914–6923. https://doi.org/10.1073/pnas.1819976116
23. Matsuo M, Katahata A, Tachikawa M, Minakuchi Y, Noguchi H, Toyoda A et al (2019) Large DNA virus promoted the endosymbiotic evolution to make a photosynthetic eukaryote. bioRxiv:809541. https://doi.org/10.1101/809541
24. Reyes-Prieto A, Yoon HS, Moustafa A, Yang EC, Andersen RA, Boo SM et al (2010) Differential gene retention in plastids of common recent origin. Mol Biol Evol 27:1530–1537. https://doi.org/10.1093/molbev/msq032
25. Martin W, Herrmann RG (1998) Gene transfer from organelles to the nucleus: how much, what happens, and why? Plant Physiol 118:9–17. https://doi.org/10.1104/pp.118.1.9
26. Muller HJ (1964) The relation of recombination to mutational advance. Mutat Res 106:2–9. https://doi.org/10.1016/0027-5107(64)90047-8
27. Muller HJ (1932) Some genetic aspects of sex. Am Nat 66:118–138. https://doi.org/10.1086/280418
28. Ponce-Toledo RI, López-García P, Moreira D (2019) Horizontal and endosymbiotic gene transfer in early plastid evolution. New Phytol 224:618–624. https://doi.org/10.1111/nph.15965
29. Green BR (2011) Chloroplast genomes of photosynthetic eukaryotes. Plant J 66:34–44. https://doi.org/10.1111/j.1365-313X.2011.04541.x
30. Nowack ECM, Melkonian M, Glöckner G (2008) Chromatophore genome sequence of *Paulinella* sheds light on acquisition of photosynthesis by eukaryotes. Curr Biol 18:410–418. https://doi.org/10.1016/j.cub.2008.02.051
31. Gabr A, Stephens TG, Bhattacharya D (2022) Loss of key endosymbiont genes may facilitate early host control of the chromatophore in *Paulinella*. iScience 25:104974. https://doi.org/10.1016/j.isci.2022.104974
32. Witte C-P, Herde M (2020) Nucleotide metabolism in plants. Plant Physiol 182:63–78. https://doi.org/10.1104/pp.19.00955
33. Zrenner R, Stitt M, Sonnewald U, Boldt R (2006) Pyrimidine and purine biosynthesis and degradation in plants. Annu Rev Plant Biol 57:805–836. https://doi.org/10.1146/annurev.arplant.57.032905

34. Antony E, Lohman TM (2019) Dynamics of E. coli single stranded DNA binding (SSB) protein-DNA complexes. Semin Cell Dev Biol 86:102–111. https://doi.org/10.1016/j.semcdb.2018.03.017
35. Shereda RD, Kozlov AG, Lohman TM, Cox MM, Keck JL (2008) SSB as an organizer/mobilizer of genome maintenance complexes. Crit Rev Biochem Mol Biol 43:289–318. https://doi.org/10.1080/10409230802341296
36. Birchler JA, Veitia RA (2012) Gene balance hypothesis: connecting issues of dosage sensitivity across biological disciplines. Proc Natl Acad Sci U S A 109:14746–14753. https://doi.org/10.1073/pnas.1207726109
37. Wang Y, Huan Q, Chu X, Li K, Qian W (2020) Single-cell transcriptome analyses recapitulate the cellular and developmental responses to abiotic stresses in rice. bioRxiv:2020.01.30.926329. https://doi.org/10.1101/2020.01.30.926329
38. Allen JF (1993) Control of gene expression by redox potential and the requirement for chloroplast and mitochondrial genomes. J Theor Biol 165:609–631. https://doi.org/10.1006/jtbi.1993.1210
39. Wollman F-A, Minai L, Nechushtai R (1999) The biogenesis and assembly of photosynthetic proteins in thylakoid membranes. Biochim Biophys Acta Bioenerg 1411:21–85. https://doi.org/10.1016/S0005-2728(99)00043-2
40. Drapier D, Rimbault B, Vallon O, Wollman F-A, Choquet Y (2007) Intertwined translational regulations set uneven stoichiometry of chloroplast ATP synthase subunits. EMBO J 26:3581–3591. https://doi.org/10.1038/sj.emboj.7601802
41. Wietrzynski W, Traverso E, Wollman F-A, Wostrikoff K (2021) The state of oligomerization of Rubisco controls the rate of synthesis of the Rubisco large subunit in *Chlamydomonas reinhardtii*. Plant Cell 33:1706–1727. https://doi.org/10.1093/plcell/koab061
42. Grossman AR, Wollman FA (eds) (2023) The *Chlamydomonas* sourcebook, 3rd edn. Elsevier/Academic. Available: https://www.elsevier.com/books/the-chlamydomonas-sourcebook/grossman/978-0-12-821430-5
43. Jarvis P, López-Juez E (2013) Biogenesis and homeostasis of chloroplasts and other plastids. Nat Rev Mol Cell Biol 14:787–802. https://doi.org/10.1038/nrm3702
44. Nott A, Jung H-S, Koussevitzky S, Chory J (2006) Plastid-to-nucleus retrograde signaling. Annu Rev Plant Biol 57:739–759. https://doi.org/10.1146/annurev.arplant.57.032905.105310
45. Woodson JD, Chory J (2008) Coordination of gene expression between organellar and nuclear genomes. Nat Rev Genet 9:383–395. https://doi.org/10.1038/nrg2348
46. Terry MJ, Smith AG (2013) A model for tetrapyrrole synthesis as the primary mechanism for plastid-to-nucleus signaling during chloroplast biogenesis. Front Plant Sci 4:14. https://doi.org/10.3389/fpls.2013.00014
47. Stavillo GM, Crisp PA, Pornsiriwong W (2011) Evidence for a SAL1-PAP chloroplast retrograde pathway that functions in drought and high ligh. Plant Cell 23(11):3992. https://doi.org/10.1105/tpc.111.091033
48. Pfalz J, Liebers M, Hirth M, Grübler B, Holtzegel U, Schröter Y et al (2012) Environmental control of plant nuclear gene expression by chloroplast redox signals. Front Plant Sci 3:257. https://doi.org/10.3389/fpls.2012.00257
49. Leister D (2012) Retrograde signaling in plants: from simple to complex scenarios. Front Plant Sci 3:135. https://doi.org/10.3389/fpls.2012.00135
50. Valadez-Cano C, Olivares-Hernández R, Resendis-Antonio O, DeLuna A, Delaye L (2017) Natural selection drove metabolic specialization of the chromatophore in *Paulinella chromatophora*. BMC Evol Biol:17. https://doi.org/10.1186/s12862-017-0947-6
51. Lhee D, Bhattacharya D, Yoon HS (2021) Independent evolution of the thioredoxin system in photosynthetic *Paulinella* species. Curr Biol 31:R328–R329. https://doi.org/10.1016/j.cub.2021.02.029
52. Tan X, Hou S, Song K, Georg J, Klähn S, Lu X et al (2018) The primary transcriptome of the fast-growing cyanobacterium *Synechococcus elongatus* UTEX 2973. Biotechnol Biofuels 11:218. https://doi.org/10.1186/s13068-018-1215-8

53. Mettler T, Mühlhaus T, Hemme D, Schöttler M-A, Rupprecht J, Idoine A et al (2014) Systems analysis of the response of photosynthesis, metabolism, and growth to an increase in irradiance in the photosynthetic model organism *Chlamydomonas reinhardtii*. Plant Cell 26:2310–2350. https://doi.org/10.1105/tpc.114.124537
54. Beck J, Lohscheider JN, Albert S, Andersson U, Mendgen KW, Rojas-Stütz MC et al (2017) Small one-Helix Proteins are essential for photosynthesis in *Arabidopsis*. Front Plant Sci 8:7. https://doi.org/10.3389/fpls.2017.00007
55. Komenda J, Sobotka R (2016) Cyanobacterial high-light-inducible proteins—protectors of chlorophyll--protein synthesis and assembly. Biochimica et Biophysica Acta (BBA)-Bioenergetics 1857:288–295. https://doi.org/10.1016/j.bbabio.2015.08.011
56. He Q, Dolganov N, Bjorkman O, Grossman AR (2001) The high light-inducible polypeptides in *Synechocystis* PCC6803. Expression and function in high light. J Biol Chem 276:306–314. https://doi.org/10.1074/jbc.M008686200
57. Calatrava V, Stephens TG, Gabr A, Bhaya D, Bhattacharya D, Grossman AR (2022) Retrotransposition facilitated the establishment of a primary plastid in the thecate amoeba *Paulinella*. Proc Natl Acad Sci U S A 119:e2121241119. https://doi.org/10.1073/pnas.2121241119
58. Casola C, Betrán E (2017) The genomic impact of gene retrocopies: what have we learned from comparative genomics, population genomics, and transcriptomic analyses? Genome Biol Evol 9:1351–1373. https://doi.org/10.1093/gbe/evx081
59. Malar CM, Krüger M, Krüger C, Wang Y, Stajich JE, Keller J et al (2021) The genome of *Geosiphon pyriformis* reveals ancestral traits linked to the emergence of the arbuscular mycorrhizal symbiosis. Curr Biol 31:1570–1577.e4. https://doi.org/10.1016/j.cub.2021.01.058
60. Pauchet Y, Heckel DG (2013) The genome of the mustard leaf beetle encodes two active xylanases originally acquired from bacteria through horizontal gene transfer. Proc Biol Sci 280:20131021. https://doi.org/10.1098/rspb.2013.1021
61. McNulty SN, Foster JM, Mitreva M, Dunning Hotopp JC, Martin J, Fischer K et al (2010) Endosymbiont DNA in endobacteria-free filarial nematodes indicates ancient horizontal genetic transfer. PloS One 5:e11029. https://doi.org/10.1371/journal.pone.0011029
62. Zhang X, Cvetkovska M, Morgan-Kiss R, Hüner NPA, Smith DR (2021) Draft genome sequence of the Antarctic green alga *Chlamydomonas* sp. UWO241. iScience 24:102084. https://doi.org/10.1016/j.isci.2021.102084
63. Song B, Morse D, Song Y, Fu Y, Lin X, Wang W et al (2017) Comparative genomics reveals two major bouts of gene retroposition coinciding with crucial periods of *Symbiodinium* evolution. Genome Biol Evol 9:2037–2047. https://doi.org/10.1093/gbe/evx144
64. Lin S, Cheng S, Song B, Zhong X, Lin X, Li W et al (2015) The *Symbiodinium kawagutii* genome illuminates dinoflagellate gene expression and coral symbiosis. Science:691–694. https://doi.org/10.1126/science.aad0408

Translocation of Proteins into Primary Plastids

Venkatasalam Shanmugabalaji and Felix Kessler

Abstract Most chloroplast proteins are encoded in the nucleus and must be imported into plastids. They are synthesized as preproteins in the cytoplasm and have N-terminal transit peptides that enable their translocation across the dual plastid membrane. This requires the translocons at the chloroplast's outer (TOC) and inner membranes (TIC). The transit peptide first engages the TOC import receptors, then inserts across the outer membrane protein translocating channel followed by full translocation across the inner membrane with the help of ATP-driven motors. Once the preprotein reaches the chloroplast stroma, stromal processing peptidase cleaves the transit peptide, allowing protein maturation to progress. In *Arabidopsis*, a family of TOC receptors exists. In varying compositions, they may mediate the import of specific proteome subsets into different plastid types. The remodeling of TOC -complexes is mediated by the ubiquitin–proteasome system in a process termed CHLORAD. This chapter will discuss the latest advances in the field:

- Composition of TOC and TIC and the debate surrounding the composition of TIC
- Energy requirements of translocation and proposed motors of translocation
- Quality control in translocation by the ubiquitin–proteasome system

Keywords Chloroplast · Plastid · TOC-TIC · CHLORAD · Translocation · ATP · Transit peptide · Preprotein · Ubiquitin · Sp1 · DELLA

1 Introduction

Plastids originated from an endosymbiotic relationship between a prokaryotic photosynthetic bacterium and a eukaryotic cell around 1.5 billion years ago. During coevolution, only around a hundred genes were retained in the plastid genome,

V. Shanmugabalaji (✉) · F. Kessler (✉)
Laboratoire de Physiologie Végétale, Université de Neuchâtel, Neuchâtel, Switzerland
e-mail: shanmugabalaji.venkatasalam@unine.ch; felix.kessler@unine.ch

S. D. Schwartzbach et al. (eds.), *Endosymbiotic Organelle Acquisition*,
https://doi.org/10.1007/978-3-031-57446-7_13

whereas many genes from the endosymbiont were lost or transferred to the host cell nucleus [1]. Thus, a protein import machinery is needed to transport nuclear-encoded chloroplast proteins back to the chloroplasts, making chloroplasts semiautonomous. The translocon complexes at the outer (TOC) and inner (TIC) chloroplast envelope membranes play a crucial role in this process, ensuring the precise import of proteins into the chloroplast [2]. Chloroplasts are part of the plastid organelle family and are responsible for global carbon fixation and therefore primarily known for photosynthesis but also various other metabolic processes [3]. Different types of plastids exists, including chloroplasts (leaf), amyloplasts (tubers, roots), and chromoplasts (flowers, fruits), and can change their types in response to developmental or environmental signals. For example, chloroplasts can transform into chromoplasts during fruit ripening [4–6].

Proteins destined for chloroplasts (preproteins) have cleavable targeting sequences (transit peptides; TP) at their N-terminal. Preproteins with N-terminal transit peptides are guided through the cytosol, escorted by a chaperone complex [7]. At the outer envelope of the chloroplast, the transit peptide interacts with the TOC complex, specifically the TOC159 and TOC34 GTP-binding receptors [8]. The preprotein is inserted across the outer membrane through a large hybrid protein-conducting channel involving TOC159 and TOC75, a β-barrel protein of the OMP85 family [9, 10]. The transit peptide interacts with the components of the TIC complex, such as TIC22 and TIC20, at first without traversing the inner membrane. The preprotein crosses the TIC20 inner membrane protein-conducting channel and enters the chloroplast stroma with the assistance of ATP-dependent motor components [11]. After import, stromal processing peptidase (SPP) cleaves the transit peptide, allowing the protein to fold, assemble, and perform its function in the chloroplast stroma, inner envelope membrane, or be further directed to the thylakoids [11–13].

The chloroplast has evolved sophisticated protein turnover systems, including intrinsic proteases and the nucleocytosolic ubiquitin–proteasome system (UPS), to regulate the chloroplast proteome. These protein turnover systems can change significantly due to developmental stages and environmental signals as part of elaborate protein homeostasis systems, also known as proteostasis systems. Chloroplast proteostasis involves synthesizing and folding chloroplast-encoded proteins, import of nucleus-encoded chloroplast proteins, assembly of protein complexes and elimination of damaged or unwanted proteins [14]. The nucleocytosolic UPS regulates chloroplast outer envelope membrane (OEM) proteins via CHLORAD and/or the TOC159-DELLA pathway, thereby controlling the protein import [15–17]. This involves the ubiquitin-tagging of TOC components and their subsequent degradation by the proteasome, thus playing a crucial role in regulating the chloroplast proteome. This system assists the target-specific turnover of preprotein receptors, thereby controlling the balance and quality of proteins within the chloroplast.

2 Composition of TOC and TIC Complexes and Chloroplast Protein Import

The TOC–TIC complexes are responsible for the import of proteins into the chloroplasts. These complexes work together to transport proteins from the cytosol into the different compartments of the chloroplast. In the early 1990s, studies on isolated pea chloroplasts identified several key components of the TOC–TIC complexes. These components were initially named Import intermediate Associated Proteins (IAP) or Outer Envelope Proteins (OEP) due to their association with preproteins "in transit" or the outer envelope membrane, respectively. IAP/OEP34 was renamed Toc34. IAP/OEP75 was renamed Toc75. IAP/OEP86 was renamed Toc159. Inner Envelope Protein IAP100/IEP110 was renamed TIC110 [18–21].

2.1 Translocon at the Outer Chloroplast Membrane

The TOC complex serves as a gateway for the import of nucleus-encoded proteins from the cytoplasm into the chloroplast. During protein translocation into chloroplasts, protein recognition is one of the crucial steps. Precursor proteins contain transit peptides, which are sequences that direct them to the appropriate destination within the chloroplast. Two GTPase domain-containing proteins, TOC159 and TOC34, are involved in protein recognition [22]. These proteins function as transit peptide receptors. TOC159 and TOC34 interact with the channel protein TOC75 to form the fully functional TOC complex. These three proteins, TOC159, TOC34, and TOC75, constitute the core components of the TOC complex (Fig. 1; Table 1).

TOC159

TOC159 is one of the key components of the translocon at the outer chloroplast membrane, which is responsible for recognizing and translocating proteins from the cytoplasm into the chloroplast. TOC159 acts as a receptor recognizing the specific targeting signals of pre-proteins meant for chloroplast import. The TOC159 mutant *ppi2* displays an albino phenotype, revealing a critical component of chloroplast import machinery [23]. The homologs of TOC159 mutants in Arabidopsis, AtToc90, AtToc120, and AtToc132 single knockout mutants do not show apparent phenotypes. However, the double knockout mutant of AtToc120 and AtToc132 also has an albino phenotype [24]. TOC159 and its three homologs (atTOC120, atTOC132, and atTOC90) share a typical structural organization.

They possess three main domains:

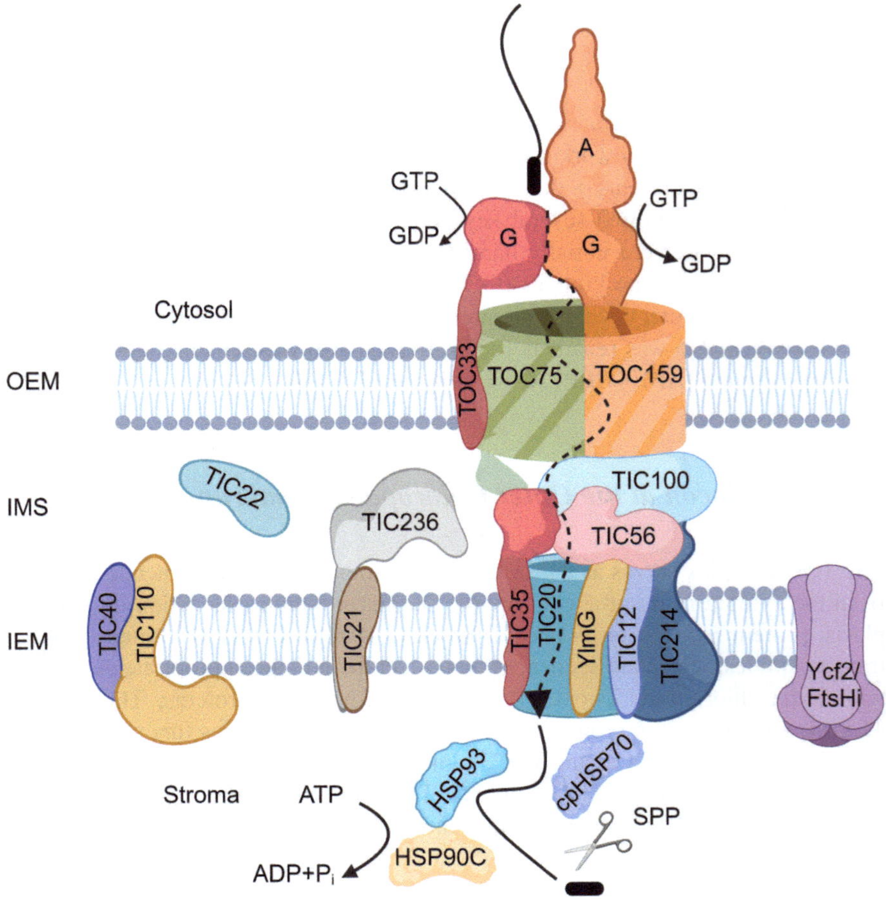

Fig. 1 Chloroplast protein translocation structure and pathway. The TOC complex in the chloroplast outer envelope membrane (OEM) consists of transit peptide receptors (TOC159 and TOC33) and the channel-forming protein TOC75. These receptors have cytosolic GTPase domains, crucial for recognizing transit peptides and initiating the import process. The TIC complex, located in the inner envelope membrane (IEM) of chloroplasts, is responsible for translocating proteins from the TOC complex into the stroma of chloroplasts. Different models exist for the TIC translocon. One model involves a 1-MDa TIC complex incorporating plastome-encoded Tic214 (or Ycf1) and nucleus-encoded subunits (TIC100, TIC56, TIC35, YlmG, TIC20, and TIC12). TIC110 and Tic40 are proposed to recruit and regulate various stromal chaperones (cpHsp70, Hsp90C, and Hsp93), aiding protein import and folding through ATP hydrolysis. TIC22 is located in the inner membrane space (IMS), and TIC236 is between IEM and IMS. In addition, TIC21, located at IEM, is associated with TIC236 (not a part of the 1-MDa TIC complex). Alternatively, a 2-MDa motor complex is proposed, comprising plastome-encoded Ycf2, and five FtsH-like ATPase subunits. The transit peptide is cleaved by stromal peptide protease (SPP)

- N-terminal acidic (A-) domain: Located at the N-terminus, this domain is acidic and predicted to be intrinsically disordered. The TOC159A domain might play a role in protein–protein interactions or signal recognition [25].

Table 1 TOC-TIC components in *C. reinhardtii* and *A. thaliana*

	Protein	*C.reinhardtii*	*A. thaliana*
TOC-TIC components elucidated by Cryo-EM			
Conserved TOC components			
	TOC120	Cre17.g707500	At3g16620
	TOC90	Cre17.g734300	At5g20300
	TOC75-III	Cre03.g175200	At3g46740
	TOC34	Cre06.g252200	At1g02280
Green algae specific TOC components			
	TOC10	Cre02.g142246	X
	TOC39	Cre17.g722750	X
	TOC52	Cre12.g532100	X
Conserved TIC components			
	TIC20-1	Cre08.g379650	At1g04940
	TIC12	Cre09.g402100	At1g80920
	TIC35	Cre03.g164700	At2g43630
	YImG	Cre02.g080250	At3g07430
	TIG214	TIC214	ATCG01130
	TIC100	Cre06g300550	AT5G22640
	TIC56	Cre17.g727100	AT5G01590
Green algae specific TOC components			
	TIC13	Cre12.g527550	X
Other TOC-TIC components			
Plant specific TOC components			
	TOC159	X	At4g02510
	TOC132	X	At2g16640
	TOC33	x	At5g05000
Conserved TIC components			
	TIC20-II	Cre01.g039150	At2g47840
	TIC20-IV	Cre04.g225050	At4g03320
Plant specific TIC components			
	TIC20-V	X	At5g55710
Associated TIC components			
	TIC21	Cre01.g039450	At2g15290
	Tic22	Cre14.g625750	At4g33350
	TIC40	Cre12.g508000	At5g16620
	TIC110	Cre10.g452450	At1g06950
	TIC236	Cre05.g243150	AT2G25660

- Central GTPase (G-) domain: This domain is involved in GTP binding and hydrolysis, has receptor function by recognizing and interacting with preproteins [26].
- C-terminal membrane-anchoring (M-) domain: This domain is located at the C-terminus and has been shown to form a ß-barrel structure along with TOC75 to make a hybrid channel structure at the outer chloroplast membrane [9, 10].

Once TOC159 recognizes the transit peptide, it guides the protein through the TOC complex and facilitates translocation across the outer chloroplast membrane.

Toc159 is of eukaryotic origin and does not appear to have homologs in cyanobacteria. Hydrophobic cluster analysis suggests that the G-, M-, and A-domains of Toc159 originated from the tandem duplication of a GTP-binding domain. A survey of 31 Viridiplantae genomes reveals that most plant species, except for some exceptions like Volvox carteri, contain more than one Toc159 homolog. These homologs can vary in sequence and function [27].

TOC34

TOC34 and TOC159 are homologous GTP-binding proteins located at the chloroplast surface. TOC34 was first identified in pea as a GTP-binding protein. This GTP-binding activity is vital for its preprotein recognition function. TOC34 has a GTP-binding domain (G-domain) and a short C-terminal transmembrane domain forming an α-helix. The GTP binding is critical for preprotein recognition, and the GTPase (G-) domain of TOC34 interacts with other components of the TOC complex [28]. The mechanism by which GTP hydrolysis facilitates insertion is not fully understood, but it has been suggested that the G-domain of TOC34 may play a role in its own targeting and/or assembly into TOC complexes [27]. TOC34 interacts directly with TOC75 and GTPase activity could facilitate the insertion of TOC34 and/or stabilize its interaction with Toc75. This interaction is thought to be highly stable. TOC34 has also been shown to participate in the insertion of TOC159 into the TOC complex [29].

In Arabidopsis, two homologs (AtToc33 and AtToc34) exist. AtToc33 is primarily expressed in rapidly growing photosynthetic tissues, while AtToc34 is expressed at low levels throughout development. Knockout mutants of AtToc33 (*ppi1*) in Arabidopsis exhibit a yellow-green, non-lethal phenotype due to a reduced accumulation of photosynthetic proteins compared to non-photosynthetic proteins [30]. Knockout mutants of AtToc34 (*ppi3*) do not show a strong phenotype apart from retarded root growth. Attempts to isolate a homozygous double mutant of AtToc33 and AtToc34 failed because it had an embryo-lethal phenotype. This suggests that AtToc33 and AtToc34 have overlapping functions and can complement each other in protein import processes [31].

Similar to Toc159, Toc34 is not found in cyanobacterial genomes, suggesting that it may have a eukaryotic origin. This supports the idea that the chloroplast protein import machinery, at least in part, has evolved in eukaryotic organisms [27].

TOC75

TOC75 is a key component of the TOC complex, which plays a central role in the import of proteins into the chloroplast. TOC75 forms a β-barrel with 16 β-strands. This protein was first identified in peas and later found to have homologs in cyanobacteria and non-photosynthetic gram-negative bacteria, indicating that it, as well as the outer chloroplast envelope, is of prokaryotic origin. TOC75 belongs to the Omp85 superfamily, which includes proteins found in gram-negative bacteria (e.g., BamA), mitochondria (e.g., SAM50), and plastids (e.g., TOC75) [32]. Members of this family share two conserved structural features: a polypeptide transport-associated (POTRA) domain at the N-terminus and a β-barrel of 16 β-strands at the C-terminus [33]. Specifically, it fulfills the criteria for being a protein-conducting channel at the outer chloroplast membrane, which implies that it plays a role in transporting proteins across the outer membrane of chloroplasts. The POTRA domain in TOC75 is known to be involved in preprotein recognition and has chaperone-like activity. They help guide incoming preproteins across the intermembrane space, facilitating their import into the chloroplast [34–37].

In Arabidopsis, there are three isoforms of Toc75. AtToc75-III, which forms the protein import channel, is a crucial gene, and mutants lacking this isoform are arrested at the two-cell stage of embryogenesis. Toc75-III has three POTRA domains, each appearing essential for its function. AtToc75-IV is expressed at lower levels and non-essential. AtToc75-I is a pseudogene [38]. Toc75 is believed to have evolved from the cynobacterial Omp85 [39]. Toc75 is present in the green lineage, while Rhodophytes and Glaucophytes harbor a Omp85 homolog [27].

2.2 *Translocon at the Inner Chloroplast Membrane*

The TOC complex is just one part of the chloroplast protein import machinery. Once proteins pass through the TOC complex and enter the intermembrane space, they encounter the translocon at the inner chloroplast membrane (TIC complex), which helps them move across the inner membrane into the stroma or further on to the thylakoid membranes or other chloroplast compartments. Early studies identified TIC110 and TIC40 as TIC components [40, 41]. However, the more recent discovery of the 1-MDa TIC complex challenges previously proposed models for the composition and function of the TIC machinery. Tic214, Tic100, Tic56, Tic20, and Tic12 assemble a 1-MDa complex in green algae and plants [42] (Fig. 1; Table 1). Grasses may have alternative components. Any cooperation between the distinct 1-MDa and TIC110-based systems during protein import has remained elusive.

TIC20

TIC20 was originally named IAP21 or TIC21 when it was first discovered by chemical crosslinking of importing precursor proteins in pea chloroplasts [37]. TIC20 is a critical complex component forming the translocon channel at the chloroplast's inner membrane. It contains four membrane-spanning α-helices, suggesting it creates a channel for proteins to pass through the chloroplast envelope membranes.

Both the N- and C-termini of TIC20 are exposed to the stroma [11]. *Arabidopsis thaliana* has four Tic20 isoforms: AtTic20-I, -II, -IV, and -V. These isoforms are localized in the chloroplast envelope, with varying functions. Knockdown of Tic20 levels in Arabidopsis exhibits pale leaves, reduced chloroplast protein levels, and defects in protein import. In addition, reconstituted liposomes containing purified TIC20 have demonstrated cation-selective channel activity, further supporting the idea that TIC20 functions as a protein channel [43, 44]. TIC20 within the 1-MDa complex may form a translocon channel with a relatively large pore size, up to 30–35 Å, allowing folded proteins pass through [45, 46]. The more recent structural studies shows that the new TIC component YlmG (well conserved in plant and algae) together with TIC20 may constitute a putative translocation pathway with a lateral opening in the inner envelope membrane, suggesting their role in protein translocation as well as insertion into the inner membrane [9, 10]. However, the current structural models are unable to fully elucidate the import mechanisms.

Tic20 is one of the TIC proteins of prokaryotic origin. Tic20 genes are found in various organisms, including green algae and flowering plants, and they can be divided into two subfamilies or groups. Two evolutionary branches have developed since then, with Tic20 genes in group 1 evolving faster than those in group 2. Tic20-I and Tic20-IV in group 1 are essential for chloroplast development and protein import in *A. thaliana* [43]. Mutants lacking these proteins exhibit albino phenotypes, are seedling-lethal, lack thylakoid membranes, and fail to accumulate photosynthetic proteins. Group 2 Tic20 members (Tic20-II and Tic20-V) are expressed at relatively high levels during development, similar to Tic20-I. However, unlike group 1 Tic20 proteins, group 2 Tic20s are not essential for chloroplast development and do not have redundant roles in protein import [47].

TIC214

TIC214 is notable for being the only chloroplast-encoded translocon subunit. It is anchored in the inner envelope membrane by a C-terminal membrane domain. TIC214 is connected to the TOC channel via its large intermembrane space (IMS) domain. The IMS domains of TIC214 intricately fold to form a large complex with TIC100, TIC56, TIC35, and TOC52 [9, 10]. Conditional repression of Tic214 in *C. reinhardtii* (a green alga) impairs chloroplast protein import and has broader effects on chloroplast ribosome biogenesis, protein folding, and stress response [48]. Tic214 appears to have evolved within the chloroplast rather than the nuclear genome. This occurred before Chlorophyta (green algae) divergence from Streptophyta (land plants and their green algal relatives). Tic20, Tic56, Tic100, and Tic214 were established when land plants (Embryophyta) evolved. Most seed plants, including eudicots and monocots, possess these TIC components, highlighting their importance. However, a unique class of monocots, including grasses like rice, maize, and wheat, lacks Tic56, Tic100, and Tic214 in their nuclear and chloroplast genomes. This suggests the existence of a distinct TIC complex in these grasses that still include Tic20 [49].

TIC56

TIC56 is one of the subunits of the 1-MDa TIC complex. TIC56 likely plays a structural role within the TIC complex, helping to stabilize the translocon channel that allows proteins to cross the inner envelope membrane. In Arabidopsis, *tic56* mutant has defects in ribosome assembly and plastid translation [50, 51].

TIC100

TIC100 is one of the subunits of the 1-MDa TIC complex, helping to stabilize the translocon channel and facilitate the translocation of proteins across the inner envelope membrane. In addition, loss of Tic100 changes RNA editing and may alter retrograde signaling [52].

TIC12

TIC12 is part of the TIC core, potentially located on the inside wall of the channel. TIC12 interacts with the transit peptide of preprotein along with TIC20 [53].

TIC110

TIC110 is an integral membrane protein with two transmembrane helices at its N-terminus, anchoring it at the inner envelope membrane of the chloroplast. The majority of the protein extends into the stroma [40, 54]. The TIC110 stromal portion (greater than 90 kDa) is thought to play a role in recruiting stromal chaperones to import precursor proteins [40]. It is also suggested to form a docking site in the stroma for precursor proteins emerging from the Tic complex during the late stages of translocation [55]. TIC110 is essential for plastid biogenesis and plant viability. Down-regulation of atTIC110 expression leads to reduced plastid protein levels and impaired translocation of precursors across the inner envelope membrane. Homozygous TIC110 mutants are embryo-lethal, while heterozygous mutants display chlorotic phenotypes and defective protein import [56]. TIC110 is believed to have originated in eukaryotes. It is conserved not only in higher plants but also in a varied range of organisms, including red algae and glaucophytes. TIC110 is also present in different plastid types [27].

Components like TIC236 and TIC22, proposed to connect the TOC and TIC complexes in plants, were missing in the recent structural studies in algae [9, 10, 57, 58]. This suggests potential fundamental differences in the import machinery of these evolutionary groups or transient interactions specific to assisting complex assembly.

3 Energy Requirements of Translocation and Proposed Motors of Translocation

The energy requirements for protein translocation into chloroplasts are satisfied by the hydrolysis of ATP and GTP. These high energy molecules are essential at various stages of the translocation process, including receptor binding, unfolding, translocon function, motor complex activity, and chaperone-driven import [22, 59].

3.1 Energy Requirement for Preprotein Transport

Initial experimentation showed that the import of preproteins into chloroplasts was energy-dependent, and the energy source was primarily ATP. Dark-adapted chloroplasts had compromised import, which could be rescued by adding ATP, highlighting its role as the primary energy source [60, 61]. It was observed that different steps of chloroplast protein import required distinct concentrations of ATP. Low ATP concentrations (50–100 μM) were sufficient for preprotein binding to the chloroplast's surface. Whereas high ATP concentrations (1 mM or more) were necessary to complete protein translocation across the chloroplast envelope. The research exploited these energetics findings to generate preprotein translocation intermediates arrested at either of the two stages and permitting to isolate the intermediate-associated proteins (IAP; see Sect. 3.2). Subsequent studies demonstrated that ATP hydrolysis occurred inside chloroplasts and was essential for driving the import of preproteins. This finding indicated that ATP was needed for initial binding and preprotein translocation across the chloroplast envelope [62, 63].

GTP hydrolysis by the receptors plays a crucial role in controlling the initial stages of protein import. Protein binding and crosslinking studies show that GTP hydrolysis is checkpoint to transition from the transit peptide's interaction with TOC75 and TIC20 to preprotein engagement with the ATP-dependent import motor [22, 64].

3.2 Early and Late Translocation Intermediates

Experiments employing low and high concentrations of ATP revealed two distinct translocation intermediates. These intermediates were created using a recombinant preprotein fused to IgG-binding domains called pS-ProtA. When incubated at low ATP concentrations, pS-ProtA remained bound to the chloroplasts, defining the "early translocation intermediate." At high ATP concentrations, pS-ProtA was fully imported but could be arrested by chilling while inserted across both the inner and outer membrane, defining the "late translocation intermediate." In the late translocation intermediate, the N-terminal transit peptide was cleaved while the C-terminus remained accessible to exogenously added protease. Thus, clearly the mature S-ProtA of the late translocation intermediate spanned the entire envelope. The "early" translocation intermediate pS-ProtA was associated with three bands visible on an SDS-PAGE gel. These proteins were molecularly cloned and sequenced, leading to the identification of TOC159, TOC75, and TOC34, which form the core of the TOC complex involved in chloroplast protein import. The "late" translocation intermediate pS-ProtA and S-ProtA were associated with two additional bands. One of these bands was identified as TIC110, while the second band (formerly IAP36) remains unidentified and its role in chloroplast protein import unknown [18, 19, 65].

3.3 Motor Systems for Protein Translocation

For many years, it was proposed that stromal chaperones, including cpHsp70, Hsp90C, and Hsp93, constitute the sole import motor system [66–68]. These chaperones were shown to be involved after recruitment by components of the inner envelope membrane (IEM), including TIC110 and TIC40. Recently, a large AAA-ATPase complex consisting of chloroplast-encoded Ycf2, five FtsH-like proteins (FtsHi1/2/4/5, and FtsH12), and a NAD-malate dehydrogenase subunit has been identified and proposed to fulfill motor functions in the import process [69]. Another study, however, demonstrated that FtsHi1 and FtsHi2 ATPase domains reside in the inner membrane space, and would therefore be unable to function as motor proteins [70]. The possible co-existence of multiple motor systems suggests that different systems may have different substrate specificities and properties. Chloroplasts may have evolved to utilize multiple motor systems to meet various requirements of the organelle. Notably, neither of these proposed motor systems has been observed in the currently available Cryo-EM (Cryo-electron microscopy) structures [9, 10], which adds to the complexity and uncertainty surrounding their roles.

4 Quality Control in Translocation by the Ubiquitin–Proteasome System (UPS)

The relative abundance of translocon isoforms dependent on transit peptide receptors TOC132-TOC34 (non-photosythetic) vs TOC159-TOC33 (photosynthetic) is crucial for balancing the import of different classes of preproteins. This balance is required for proteome remodeling during transitions between different plastid types or in response to physiological changes. There are two known pathways, chloroplast-associated protein degradation (CHLORAD) and the

TOC159-DELLA pathway, involved in the regulation of TOC complexes by the UPS [16, 17, 71] (Fig. 2).

4.1 Chloroplast-Associated Protein Degradation (CHLORAD)

The pathway involving the degradation of TOC complexes by the UPS, including SP1, SP2, and CDC48, has been named the CHLORAD pathway [17, 71]. This pathway is crucial for the turnover and regulation of TOC proteins during chloroplast development and function.

SP1-Mediated TOC Protein Ubiquitination

A forward-genetic screen identified a RING-type E3 ligase called "suppressor of ppi1 locus 1"(SP1). The ppi1 mutant, which lacks the Toc33 protein, exhibits a pale-yellow phenotype. SP1 is located in the Outer Envelope Membrane (OEM) of the chloroplast and possesses domains that facilitate target recognition, including two transmembrane domains separated by an Inter Membrane Space (IMS) domain and a cytosolic RING domain. SP1 directly interacts with all TOC proteins and mediates their ubiquitination and subsequent degradation. When SP1 is lost, there is an increase in the abundance of TOC proteins. When combined with the *ppi1* mutation, this increase in TOC protein levels leads to enhanced greening (suppression of *ppi1* phenotype). This finding indicates that the degradation of TOC proteins is essential for regulating the import machinery during developmental processes [17].

TOC Retrotranslocation

Polyubiquitinated TOC proteins at the OEM need to overcome membrane-association to encounter the cytosolic 26S proteasome (26SP). Degradation of proteins in other cellular membranes involves retrotranslocation across the membrane before proteasomal degradation. SP2 "suppressor of *ppi1* locus 2" (a Omp85 family protein), assists TOC retrotranslocation. SP2 is characterized as a β-barrel membrane protein that is proposed to provide the retrotranslocation channel in the OEM. SP2, like SP1, physically interacts with TOC proteins. A cytosolic factor provides the motive force for the extraction of TOC proteins through SP2 called "cell division cycle 48" (CDC48). CDC48 is an ATPase that undergoes conformational changes induced by ATP hydrolysis to create the motive force for the extraction and denaturation of bound substrates and also function in other cellular processes and retrotranslocation systems [71].

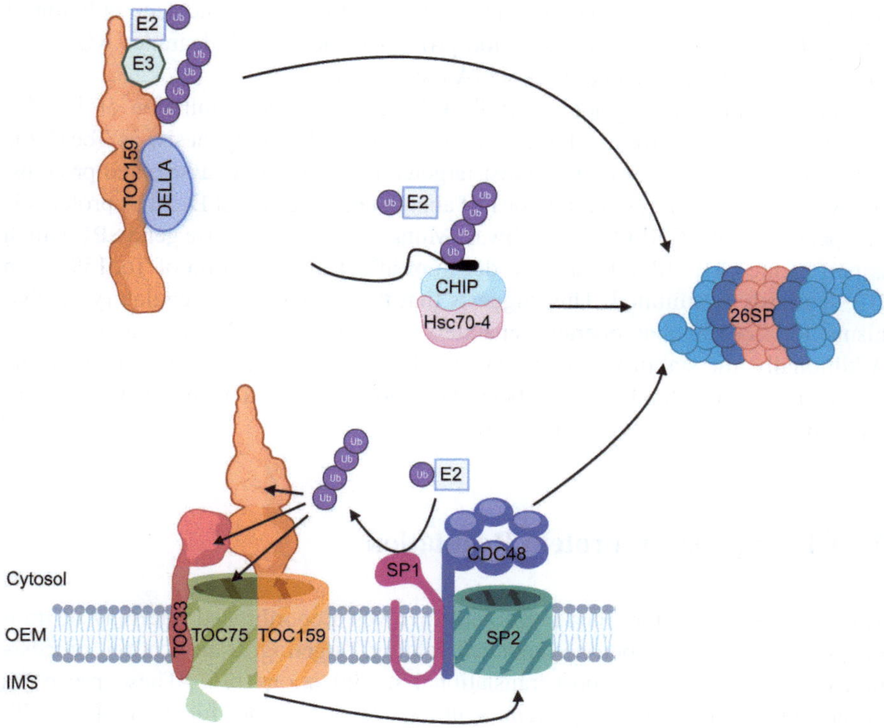

Fig. 2 Regulation of chloroplast protein import pathway by UPS. The unassembled preprotein receptor TOC159 under low gibberellic acid (GA) conditions (unfavorable condition) interacts with the DELLA RGL2, triggering degradation of TOC159 by the UPS while further TOC components (TOC33, TOC75) are targeted by CHLORAD. This represses precocious chloroplast biogenesis, particularly before germination. The UPS targets unimported preproteins in the cytosol to prevent their accumulation and the formation of cytotoxic aggregates. This process is mediated by the chaperone Hsc70-4 and the E3 ligase CHIP. By tagging these preproteins with ubiquitin, the UPS marks them for degradation, ensuring the cytosol is not overloaded with unassembled proteins. The CHLORAD system can target the TOC components by the E3 ligase SP1. Ubiquitinated proteins are retrotranslocated from the membrane via the channel protein SP2, driven by the cytosolic AAA+ ATPase CDC48. This process allows CHLORAD to regulate chloroplast protein import and, consequently, the composition of the chloroplast proteome. In all cases, the final degradation of ubiquitinated proteins is mediated by the nucleocytosolic 26S proteasome (26SP)

4.2 *TOC159-DELLA Pathway*

Gibberellic acid (GA) is a phytohormone that plays a crucial role in regulating the abundance of the Toc159 protein during seed germination. Under conditions of low GA concentration, the DELLA protein RGA-LIKE2 (RGL2), which acts as a negative regulator of GA signaling [72], targets Toc159 for degradation via the ubiquitin–proteasome system (UPS). This leads to a reduction in Toc159 levels. Any photosynthetic preproteins that accumulate in the cytoplasm are also targeted for UPS-mediated degradation. This suppression of chloroplast development in the dark prevents premature greening and photooxidation due to the accumulation of chloroplast precursors. As germination progresses and SLY1 degrades RGL2 via the UPS system the concentration of GA increases [73].

Consequently, the degradation of RGL2 leads to the accumulation of Toc159. This increase in the Toc159 levels promotes chloroplast biogenesis, as Toc159 is essential for the import of chloroplast-targeted photosynthesis-associated proteins. It's worth noting that regulating Toc159 abundance by GA and DELLA proteins is independent of the CHLORAD pathway. Mutation of the E3 ligase gene SP1, which is part of the CHLORAD pathway, does not affect the reduction of Toc159 when GA synthesis is inhibited. This suggests that there are distinct regulatory mechanisms for Toc159 that operate separately from the CHLORAD pathway [16]. Additionally, the stability of components of the TOC complex (Translocon at the Outer Chloroplast Membrane) has been shown to be regulated by the small ubiquitin-like modifier (SUMO) system [74, 75].

5 Chloroplast Preprotein Regulation

The regulation of chloroplast preprotein involves a coordinated interplay of transcriptional control, chaperone-mediated folding, targeting, translocation, processing, quality control, and post-translational modifications [76]. These processes ensure proper chloroplast protein synthesis, import, and function. The UPS (ubiquitin-proteasome system) plays a significant role in regulating chloroplast preproteins at various stages.

5.1 UPS-Dependent Preprotein Regulation

Initial evidence for the UPS regulation of chloroplast proteins came from observations of cytosolic E3 ubiquitin ligase activity targeting pre-proteins. The UPS pathway monitors and prevents the premature accumulation of photosynthetic preproteins in the cytoplasm. This regulatory system is crucial in preventing these preproteins from accumulating to toxic levels in the cytoplasm. This UPS pathway involves specific components, including the Hsp70 isoform known as Hsc70-4 and the C-terminus of Hsc70-interacting protein (CHIP) cytosolic E3 ubiquitin ligase. Hsc70-4 selectively recognizes cytosolic preprotein based on a specific motif present within the transit peptides of these preproteins. Once recruited to CHIP, the pre-proteins undergo ubiquitination and are degraded by the 26S proteasome [77]. An Arabidopsis homolog of BCL2-associated athanogene 1 (BAG1) was identified as a cofactor in Hsc70-mediated preprotein degradation. BAG1 likely plays a role in facilitating the degradation of preproteins in coordination with Hsc70. Another cytosolic E3 ligase called stress-associated protein5 (SAP5) was identified in wheat. SAP5 interacts with the preprotein of Hsp90C, leading to its degradation. This further illustrates the diversity of E3 ligases involved in the UPS-mediated regulation of preproteins [78]. Mild impairment of the proteasome, resulting in an enlarged pool of cytosolic preprotein precursors, leads to increased levels of functional photosynthetic complexes and improved photosynthetic performance. This suggests that the precise regulation of preprotein levels by the UPS is essential for optimizing photosynthesis and chloroplast function [79] . In Arabidopsis, there are two redundant Glk transcription factors, namely Glk1 and Glk2. These factors play a role in promoting the expression of genes encoding photosynthetic proteins, which, in turn, contribute to chloroplast biogenesis. Interestingly, Glk1 is also subject to posttranslational regulation by the UPS [80].

6 Conclusion

The research community has significantly advanced in comprehending chloroplast protein import and routing processes but knowledge gaps remain or have been newly identified. Recent cryo-EM structural studies have provided detailed insights into the arrangement of TOC–TIC components and the likely path of preproteins and their transit peptides across the chloroplast envelope. The cytosolic GTPase domains of TOC34 and TOC159, crucial for transit peptide recognition, were not fully revealed in recent cryo-EM structures. These domains remain important targets for future structural research [9, 10]. However, understanding how transit peptides belonging to different classes of preproteins (e.g., photosynthesis-associated vs. non-photosynthetic housekeeping) are recognized and distinguished is still a fundamental gap in knowledge [81]. In addition, the exact mechanisms by which cytosolic factors target preproteins to the chloroplast are still unclear [82]. Further research is needed to elucidate the intricate processes involved in this targeting mechanism. While a well-accepted model for TOC composition exists and is underscored by cryo-EM structures, a consensus regarding the components and functioning of the TIC apparatus is lacking. Understanding the composition and regulation of TIC is crucial for a comprehensive understanding of chloroplast protein import. Proteolysis, including processes regulated by mechanisms like the nucleocytosolic UPS and CHLORAD, is vital for maintaining the balance of proteins within chloroplasts. Although CHLORAD and the nucleocytosolic UPS have been identified as major regulators of chloroplast protein homeostasis, there is a high likelihood that additional components and regulatory pathways remain undiscovered [14]. Ongoing mechanistic investigations are crucial for identifying these components and understanding their roles in protein import and degradation.

Acknowledgments This work was supported by grants from the Swiss National Science Foundation (31003A_156998 and 31003A _176191) and by the University of Neuchâtel. BioRender.com for the images.

References

1. Chan CX et al (2011) Plastid origin and evolution: new models provide insights into old problems. Plant Physiol 155(4):1552–1560. https://doi.org/10.1104/pp.111.173500
2. Richardson LG et al (2014) Targeting and assembly of components of the TOC protein import complex at the chloroplast outer envelope membrane. Front Plant Sci 5:269. https://doi.org/10.3389/fpls.2014.00269
3. Jarvis P, López-Juez E (2013) Biogenesis and homeostasis of chloroplasts and other plastids. Nat Rev Mol Cell Biol 14(12):787–802. https://doi.org/10.1038/nrm3702
4. Egea I et al (2010) Chromoplast differentiation: current status and perspectives. Plant Cell Physiol 51(10):1601–1611. https://doi.org/10.1093/pcp/pcq136
5. Choi H, Yi T, Ha SH (2021) Diversity of plastid types and their interconversions. Front Plant Sci 12:692024. https://doi.org/10.3389/2Ffpls.2021.692024
6. Bathgate B et al (1985) Plastid changes during the conversion of chloroplasts to chromoplasts in ripening tomatoes. Planta 165(2):197–204. https://doi.org/10.1007/bf00395042
7. Bruce BD (2000) Chloroplast transit peptides: structure, function and evolution. Trends Cell Biol 10(10):440–447. https://doi.org/10.1016/s0962-8924(00)01833-x
8. Bruce BD (2001) The paradox of plastid transit peptides: conservation of function despite divergence in primary structure. Biochim Biophys Acta 1541(1-2):2–21. https://doi.org/10.1016/s0167-4889(01)00149-5
9. Jin Z et al (2022) Structure of a TOC-TIC supercomplex spanning two chloroplast envelope membranes. Cell 185(25):4788–4800.e13. https://doi.org/10.1016/j.cell.2022.10.030
10. Liu H et al (2023) Architecture of chloroplast TOC-TIC translocon supercomplex. Nature 615(7951):349–357. https://doi.org/10.1038/s41586-023-05744-y
11. Kouranov A et al (1998) Tic20 and Tic22 are new components of the protein import apparatus at the chloroplast inner envelope membrane. J Cell Biol 143(4):991–1002. https://doi.org/10.1083/jcb.143.4.991
12. Ballabani G et al (2023) The journey of preproteins across the chloroplast membrane systems. Front Physiol 14:1213866. https://doi.org/10.3389/fphys.2023.1213866
13. Richter S, Lamppa GK (2002) Determinants for removal and degradation of transit peptides of chloroplast precursor proteins. J Biol Chem 277(46):43888–43894. https://doi.org/10.1074/jbc.M206020200
14. Sun Y, Jarvis RP (2023) Chloroplast proteostasis: import, sorting, ubiquitination, and proteolysis. Annu Rev Plant Biol 74:259–283. https://doi.org/10.1146/annurev-arplant-070122-032532
15. Nellaepalli S, Lau AS, Jarvis RP (2023) Chloroplast protein translocation pathways and ubiquitin-dependent regulation at a glance. J Cell Sci 136(18):jcs241125. https://doi.org/10.1242/jcs.241125
16. Shanmugabalaji V et al (2018) Chloroplast biogenesis controlled by DELLA-TOC159 interaction in early plant development. Curr Biol 28(16):2616–2623.e5. https://doi.org/10.1016/j.cub.2018.06.006
17. Ling Q et al (2012) Chloroplast biogenesis is regulated by direct action of the ubiquitin-proteasome system. Science 338(6107):655–659. https://doi.org/10.1126/science.1225053
18. Schnell DJ, Kessler F, Blobel G (1994) Isolation of components of the chloroplast protein import machinery. Science 266(5187):1007–1012. https://doi.org/10.1126/science.7973649
19. Kessler F et al (1994) Identification of two GTP-binding proteins in the chloroplast protein import machinery. Science 266(5187):1035–1039. https://doi.org/10.1126/science.7973656
20. Hirsch S et al (1994) A receptor component of the chloroplast protein translocation machinery. Science 266(5193):1989–1992. https://doi.org/10.1126/science.7801125
21. Perry SE, Keegstra K (1994) Envelope membrane proteins that interact with chloroplastic precursor proteins. Plant Cell 6(1):93–105. https://doi.org/10.1105/tpc.6.1.93
22. Schnell DJ (2019) The TOC GTPase receptors: regulators of the fidelity, specificity and substrate profiles of the general protein import machinery of chloroplasts. Protein J 38(3):343–350. https://doi.org/10.1007/s10930-019-09846-3

23. Bauer J et al (2000) The major protein import receptor of plastids is essential for chloroplast biogenesis. Nature 403(6766):203–207. https://doi.org/10.1038/35003214
24. Kubis S et al (2004) Functional specialization amongst the Arabidopsis Toc159 family of chloroplast protein import receptors. Plant Cell 16(8):2059–2077. https://doi.org/10.1105/2Ftpc.104.023309
25. Agne B et al (2010) The acidic A-domain of Arabidopsis TOC159 occurs as a hyperphosphorylated protein. Plant Physiol 153(3):1016–1030. https://doi.org/10.1104/pp.110.158048
26. Wang F et al (2008) The role of GTP binding and hydrolysis at the atToc159 preprotein receptor during protein import into chloroplasts. J Cell Biol 183(1):87–99. https://doi.org/10.1083/2Fjcb.200803034
27. Shi LX, Theg SM (2013) The chloroplast protein import system: from algae to trees. Biochim Biophys Acta 1833(2):314–331. https://doi.org/10.1016/j.bbamcr.2012.10.002
28. Sun YJ et al (2002) Crystal structure of pea Toc34, a novel GTPase of the chloroplast protein translocon. Nat Struct Biol 9(2):95–100. https://doi.org/10.1038/nsb744
29. Wallas TR et al (2003) The roles of toc34 and toc75 in targeting the toc159 preprotein receptor to chloroplasts. J Biol Chem 278(45):44289–44297. https://doi.org/10.1074/jbc.M307873200
30. Jarvis P et al (1998) An *Arabidopsis* mutant defective in the plastid general protein import apparatus. Science 282(5386):100–103. https://doi.org/10.1126/science.282.5386.100
31. Constan D et al (2004) An outer envelope membrane component of the plastid protein import apparatus plays an essential role in *Arabidopsis*. Plant J 38(1):93–106. https://doi.org/10.1111/j.1365-313x.2004.02024.x
32. Hsu SC, Inoue K (2009) Two evolutionarily conserved essential beta-barrel proteins in the chloroplast outer envelope membrane. Biosci Trends 3(5):168–178
33. Simmerman RF, Dave AM, Bruce BD (2014) Structure and function of POTRA domains of Omp85/TPS superfamily. Int Rev Cell Mol Biol 308:1–34. https://doi.org/10.1016/b978-0-12-800097-7.00001-4
34. Sánchez-Pulido L et al (2003) POTRA: a conserved domain in the FtsQ family and a class of beta-barrel outer membrane proteins. Trends Biochem Sci 28(10):523–526. https://doi.org/10.1016/j.tibs.2003.08.003
35. O'Neil PK et al (2017) The POTRA domains of Toc75 exhibit chaperone-like function to facilitate import into chloroplasts. Proc Natl Acad Sci U S A 114(24):E4868–e4876. https://doi.org/10.1073/pnas.1621179114
36. Paila YD et al (2016) Multi-functional roles for the polypeptide transport associated domains of Toc75 in chloroplast protein import. Elife 5:e12631. https://doi.org/10.7554/2FeLife.12631
37. Kouranov A, Schnell DJ (1997) Analysis of the interactions of preproteins with the import machinery over the course of protein import into chloroplasts. J Cell Biol 139(7):1677–1685. https://doi.org/10.1083/2Fjcb.139.7.1677
38. Baldwin A et al (2005) A molecular-genetic study of the Arabidopsis Toc75 gene family. Plant Physiol 138(2):715–733. https://doi.org/10.1104/pp.105.063289
39. Reumann S, Davila-Aponte J, Keegstra K (1999) The evolutionary origin of the protein-translocating channel of chloroplastic envelope membranes: identification of a cyanobacterial homolog. Proc Natl Acad Sci U S A 96(2):784–789. https://doi.org/10.1073/pnas.96.2.784
40. Jackson DT, Froehlich JE, Keegstra K (1998) The hydrophilic domain of Tic110, an inner envelope membrane component of the chloroplastic protein translocation apparatus, faces the stromal compartment. J Biol Chem 273(26):16583–16588. https://doi.org/10.1074/jbc.273.26.16583
41. Stahl T et al (1999) Tic40, a new "old" subunit of the chloroplast protein import translocon. J Biol Chem 274(52):37467–37472. https://doi.org/10.1074/jbc.274.52.37467
42. Kikuchi S et al (2013) Uncovering the protein translocon at the chloroplast inner envelope membrane. Science 339(6119):571–574. https://doi.org/10.1126/science.1229262
43. Hirabayashi Y et al (2011) In vivo studies on the roles of two closely related Arabidopsis Tic20 proteins, AtTic20-I and AtTic20-IV. Plant Cell Physiol 52(3):469–478. https://doi.org/10.1093/pcp/pcr010

44. Kovács-Bogdán E et al (2011) Tic20 forms a channel independent of Tic110 in chloroplasts. BMC Plant Biol 11:133. https://doi.org/10.1186/1471-2229-11-133
45. Kikuchi S et al (2009) A 1-megadalton translocation complex containing Tic20 and Tic21 mediates chloroplast protein import at the inner envelope membrane. Plant Cell 21(6):1781–1797. https://doi.org/10.1105/2Ftpc.108.063552
46. Ganesan I et al (2018) Evaluating the functional pore size of chloroplast TOC and TIC protein translocons: import of folded proteins. Plant Cell 30(9):2161–2173. https://doi.org/10.1105/tpc.18.00427
47. Kasmati AR et al (2011) Molecular and genetic analyses of Tic20 homologues in *Arabidopsis thaliana* chloroplasts. Plant J 66(5):877–889. https://doi.org/10.1111/j.1365-313X.2011.04551.x
48. Ramundo S et al (2020) Coexpressed subunits of dual genetic origin define a conserved supercomplex mediating essential protein import into chloroplasts. Proc Natl Acad Sci U S A 117(51):32739–32749. https://doi.org/10.1073/pnas.2014294117
49. de Vries J et al (2015) YCF1: a Green TIC? Plant Cell 27(7):1827–1833. https://doi.org/10.1105/tpc.114.135541
50. Köhler D et al (2015) Characterization of chloroplast protein import without Tic56, a component of the 1-megadalton translocon at the inner envelope membrane of chloroplasts. Plant Physiol 167(3):972–990. https://doi.org/10.1104/pp.114.255562
51. Köhler D et al (2016) Importance of translocon subunit Tic56 for rRNA processing and chloroplast ribosome assembly. Plant Physiol 172(4):2429–2444. https://doi.org/10.1104/2Fpp.16.01393
52. Loudya N et al (2022) Mutations in the chloroplast inner envelope protein TIC100 impair and repair chloroplast protein import and impact retrograde signaling. Plant Cell 34(8):3028–3046. https://doi.org/10.1093/plcell/koac153
53. Zhao X, Higa T, Nakai M (2022) Tic12, a 12-kDa essential component of the translocon at the inner envelope membrane of chloroplasts in *Arabidopsis*. Plant Cell 34(11):4569–4582. https://doi.org/10.1093/plcell/koac240
54. Kessler F, Blobel G (1996) Interaction of the protein import and folding machineries of the chloroplast. Proc Natl Acad Sci U S A 93(15):7684–7689. https://doi.org/10.1073/pnas.93.15.7684
55. Inaba T et al (2003) atTic110 functions as a scaffold for coordinating the stromal events of protein import into chloroplasts. J Biol Chem 278(40):38617–38627. https://doi.org/10.1074/jbc.m306367200
56. Inaba T et al (2005) Arabidopsis tic110 is essential for the assembly and function of the protein import machinery of plastids. Plant Cell 17(5):1482–1496. https://doi.org/10.1105/tpc.105.030700
57. Chen YL et al (2018) TIC236 links the outer and inner membrane translocons of the chloroplast. Nature 564(7734):125–129. https://doi.org/10.1038/s41586-018-0713-y
58. Kouranov A, Wang H, Schnell DJ (1999) Tic22 is targeted to the intermembrane space of chloroplasts by a novel pathway. J Biol Chem 274(35):25181–25186. https://doi.org/10.1074/jbc.274.35.25181
59. Ruprecht M et al (2010) On the impact of precursor unfolding during protein import into chloroplasts. Mol Plant 3(3):499–508. https://doi.org/10.1093/mp/ssp116
60. Grossman AR et al (1980) Post-translational uptake of cytoplasmically synthesized proteins by intact chloroplasts *in vitro*. Ann N Y Acad Sci 343:266–274. https://doi.org/10.1111/j.1749-6632.1980.tb47257.x
61. Cline K et al (1985) Precursors to two nuclear-encoded chloroplast proteins bind to the outer envelope membrane before being imported into chloroplasts. J Biol Chem 260(6):3691–3696
62. Olsen LJ et al (1989) ATP is required for the binding of precursor proteins to chloroplasts. J Biol Chem 264(12):6724–6729
63. Olsen LJ, Keegstra K (1992) The binding of precursor proteins to chloroplasts requires nucleoside triphosphates in the intermembrane space. J Biol Chem 267(1):433–439. https://doi.org/10.1016/S0021-9258(18)48513-4
64. Richardson LGL et al (2018) Molecular topology of the transit peptide during chloroplast protein import. Plant Cell 30(8):1789–1806. https://doi.org/10.1105/tpc.18.00172

65. Schnell DJ, Blobel G (1993) Identification of intermediates in the pathway of protein import into chloroplasts and their localization to envelope contact sites. J Cell Biol 120(1):103–115
66. Su PH, Li HM (2010) Stromal Hsp70 is important for protein translocation into pea and *Arabidopsis* chloroplasts. Plant Cell 22(5):1516–1531. https://doi.org/10.1083/jcb.120.1.103
67. Inoue H, Li M, Schnell DJ (2013) An essential role for chloroplast heat shock protein 90 (Hsp90C) in protein import into chloroplasts. Proc Natl Acad Sci U S A 110(8):3173–3178. https://doi.org/10.1073/pnas.1219229110
68. Huang PK et al (2016) Chloroplast Hsp93 directly binds to transit peptides at an early stage of the preprotein import process. Plant Physiol 170(2):857–866. https://doi.org/10.1104/pp.15.01830
69. Kikuchi S et al (2018) A Ycf2-FtsHi Heteromeric AAA-ATPase complex is required for chloroplast protein import. Plant Cell 30(11):2677–2703. https://doi.org/10.1105/tpc.18.00357
70. Chang CY, Chen LJ, Li HM (2023) Chloroplast import motor subunits FtsHi1 and FtsHi2 are located on opposite sides of the inner envelope membrane. Proc Natl Acad Sci U S A 120(37):e2307747120. https://doi.org/10.1073/pnas.2307747120
71. Ling Q et al (2019) Ubiquitin-dependent chloroplast-associated protein degradation in plants. Science 363(6429):eaav4467. https://doi.org/10.1105/tpc.18.00357
72. Piskurewicz U et al (2008) The gibberellic acid signaling repressor RGL2 inhibits Arabidopsis seed germination by stimulating abscisic acid synthesis and ABI5 activity. Plant Cell 20(10):2729–2745. https://doi.org/10.1105/tpc.108.061515
73. Ariizumi T, Lawrence PK, Steber CM (2011) The role of two f-box proteins, SLEEPY1 and SNEEZY, in Arabidopsis gibberellin signaling. Plant Physiol 155(2):765–775. https://doi.org/10.1104/pp.110.166272
74. Accossato S, Kessler F, Shanmugabalaji V (2020) SUMOylation contributes to proteostasis of the chloroplast protein import receptor TOC159 during early development. Elife 9:e60968. https://doi.org/10.7554/elife.60968
75. Watson SJ et al (2021) Crosstalk between the chloroplast protein import and SUMO systems revealed through genetic and molecular investigation in *Arabidopsis*. Elife 10:e60960. https://doi.org/10.7554/elife.60960
76. Gao LL et al (2023) Chloroplast proteostasis: a story of birth, life, and death. Plant Commun 4(1):100424. https://doi.org/10.1016/j.xplc.2022.100424
77. Lee S et al (2009) Heat shock protein cognate 70-4 and an E3 ubiquitin ligase, CHIP, mediate plastid-destined precursor degradation through the ubiquitin-26S proteasome system in *Arabidopsis*. Plant Cell 21(12):3984–4001. https://doi.org/10.1105/tpc.109.071548
78. Zhang N et al (2019) Identification of HSP90C as a substrate of E3 ligase TaSAP5 through ubiquitylome profiling. Plant Sci 287:110170. https://doi.org/10.1016/j.plantsci.2019.110170
79. Grimmer J et al (2020) Mild proteasomal stress improves photosynthetic performance in *Arabidopsis* chloroplasts. Nat Commun 11(1):1662. https://doi.org/10.1038/s41467-020-15539-8
80. Tokumaru M et al (2017) Ubiquitin-Proteasome dependent regulation of the GOLDEN2-LIKE 1 transcription factor in response to plastid signals. Plant Physiol 173(1):524–535. https://doi.org/10.1104/pp.16.01546
81. Chu CC, Swamy K, Li HM (2020) Tissue-specific regulation of plastid protein import via Transit-Peptide Motifs. Plant Cell 32(4):1204–1217. https://doi.org/10.1105/tpc.19.00702
82. Lee DW, Jung C, Hwang I (2013) Cytosolic events involved in chloroplast protein targeting. Biochim Biophys Acta 1833(2):245–252. https://doi.org/10.1016/j.bbamcr.2012.03.006

Translocation of Proteins into Complex Plastids with Three Envelope Membranes

Dion G. Durnford and Steven D. Schwartzbach

Abstract The unrelated euglenophytes and peridinin dinoflagellates have secondary plastids with three envelope membranes. All other secondary plastids have four envelope membranes including Apicomplexans, which are close relatives of dinoflagellates. The endosymbiont that gave rise to the plastid in *Euglena* was a green alga while the peridinin dinoflagellate endosymbiont was an alga with a red alga-derived plastid. Despite this, the euglenophytes and dinoflagellates share a number of unique features that are not present in organisms with four envelope membrane complex plastids. Proteins are targeted to all complex plastids regardless of the number of envelope membranes by a bipartite presequence composed of a signal peptide domain targeting the precursor to an ER-like compartment and a chloroplast targeting transit peptide domain. Unique to some *Euglena* and dinoflagellate bipartite targeting sequences is a hydrophobic domain, the stop-transfer membrane anchor sequence, which stops translocation into the ER lumen anchoring the protein in the ER membrane with the N-terminal transit peptide in the ER lumen and most of the protein in the cytoplasm.

In vivo and in vitro studies have shown that *Euglena* and dinoflagellate plastid proteins are transported in vesicles from the ER to the Golgi apparatus to the plastid. Upon vesicle fusion with the outer envelope membrane, proteins with a presequence containing a stop-transfer membrane anchor sequence become integral envelope membrane proteins, while those without a stop-transfer membrane anchor sequence are soluble proteins in the intermembrane space between the outer and middle envelope membrane. During protein transit through the ER and Golgi apparatus proteins fold, they can be glycosylated, and disulfide bonds are formed. Import of these post-translationally modified proteins through the middle and inner envelope membrane requires translocation machinery that can transport bulky glycosylated or folded

D. G. Durnford
Department of Biology, University of New Brunswick, Fredericton, NB, Canada
e-mail: durnford@unb.ca

S. D. Schwartzbach (✉)
Department of Biological Sciences, University of Memphis, Memphis, TN, USA
e-mail: sdschwrt@memphis.edu

S. D. Schwartzbach et al. (eds.), *Endosymbiotic Organelle Acquisition*,
https://doi.org/10.1007/978-3-031-57446-7_14

proteins, remove integral membrane proteins from the outer envelope membrane, and translocate them and intermembrane soluble proteins across the middle and inner envelope membranes into the stroma. It is unlikely that the typical translocons at the outer envelope membrane of chloroplasts (TOC) and at the inner envelope membrane of chloroplasts (TIC) can perform these functions. Genome, transcriptome, and proteome studies have not found homologues of these translocons in *Euglena* plastids. In this review, we will discuss the similarities between the unrelated euglenophyte and peridinin dinoflagellate plastid protein presequences, plastid protein import pathways, and suggest selection pressures that may have led to modifications of the plastid glycoprotein import pathway rather than the conventional TOC/TIC pathway for translocation of proteins into their complex plastids.

Keywords Chloroplast precursor proteins · Chloroplast protein import · Complex plastids · Dinoflagellates · Endosymbiosis · *Euglena* · Glycoprotein · Import apparatus · Plastid evolution · Precursor protein · Protein translocation · Signal peptide · Stop-transfer sequence · Targeting signal · Transit peptide

1 Introduction

The challenge of targeting proteins to complex plastids was recognized when microscopy studies found that plastids in a variety of algal groups [1–5] had three or four envelop envelope membranes rather than the two membranes of higher plants and red and green algae. The additional membranes clearly posed a challenge for protein import but also provided a "fossil record" as to the origin of the plastid [2, 3, 6]. These extra membranes were proposed to be residual membranes from an endosymbiotic event and were vital evidence for developing the now generally accepted endosymbiotic origin of simple plastids. These simple plastids were derived from a "primary" endosymbiotic event with a prokaryote, such as a photosynthetic cyanobacteria-like endosymbiont, while the more complex plastids with three or four envelope membranes originated from a "secondary" endosymbiotic event with a photosynthetic eukaryote [7, 8]. The acquisition of photosynthesis through eukaryote–eukaryote endosymbiosis has created a complex pattern of plastid ultrastructure and, with it, unique protein transport barriers.

Of the organisms with complex plastids, *Euglena* and the dinoflagellates are unique in having three plastid envelope membranes rather than four as with the stramenopiles, cryptomonads, haptophytes, and photosynthetic members of the Rhizaria [4, 8]. *Euglena* and dinoflagellates are unrelated, indicating that these evolutionary endosymbiotic events happened independently. The *Euglena* plastid arose through secondary endosymbiosis between a eukaryotic host and a green algal endosymbiont, while the dinoflagellate plastid arose through secondary endosymbiosis between a eukaryotic host and a red algal endosymbiont [9, 10]. In the case of the peridinin-dinoflagellates, it is particularly surprising considering that the apicoplastida, which are closely related to the dinoflagellates [11–15], have four plastid envelope membranes rather than three. This suggests that the apicoplast plastid with four envelope membranes may represent the ancestral state [13].

The outermost of the three envelope membranes lacks bound ribosomes and it is generally hypothesized to have evolved from the phagosomal membrane that surrounded the eukaryotic photosynthetic endosymbiont, during phagocytic uptake. Although plastids can be isolated from *Euglena*, it is difficult to isolate each of the three plastid envelope membranes. A detailed analysis of the envelope membranes of *Euglena* has been performed, and based on lipid composition, the middle membrane has been identified as the outer of the two envelope membranes of simple chloroplasts [16]. This contrasts with the complex plastids that have a total of four envelope membranes where the outermost membrane is usually studded with ribosomes, implying that this membrane is part of the host's endomembrane system and directly derived from the endoplasmic reticulum (ER). It has been named the chloroplast ER (CER) [4]. The discovery of the CER provided the first clue that nucleus-encoded chloroplast localized proteins are co-translationally imported into complex plastids with four envelope membranes.

Targeting proteins and translocating them into an organelle requires specific targeting information that resides within the protein to be targeted. For many organelles, like the chloroplast, mitochondria, and the ER, this information is found in an amino-terminal (N-terminal) presequence that is proteolytically removed within the organelle, yielding the mature protein. Nucleus encoded, cytoplasmically synthesized proteins that are post-translationally imported into the simple, primary plastids of red algae, green algae, glaucophytes, and plants have N-terminal presequences called transit peptides that are necessary and sufficient for targeting to the organelle [17]. The transit peptide is usually removed by a specific protease once the protein enters the plastid.

One of the more startling discoveries with respect to *Euglena* biology was the detection of the light-harvesting, chlorophyll a/b binding protein of Photosystem II (LHCPII) in the Golgi apparatus by immunoelectron microscopy [18–21]. Equally surprising was the detection by immunoelectron microscopy of ribulose bisphosphate carboxylase oxygenase (Rbc) and peridinin-chlorophyll a-protein (PCP) in the dinoflagellate Golgi [22]. These results pointed to a unique plastid-targeting mechanism for nucleus-encoded proteins and an unexpected solution to traversing the additional plastid membrane. This Golgi-directed mechanism was supported by sequencing of *Euglena* [23–32] and dinoflagellate [22, 33–35] genes encoding plastid-targeted proteins. These proteins had a bipartite presequence composed of an N-terminal domain having characteristics of a signal peptide that mediates co-translational import into the ER and a second domain having characteristics of a transit peptide that mediates import into chloroplasts.

The absence of ribosomes on the outer plastid envelope membrane and the presence of an N-terminal presequence having properties of an ER targeting signal peptide on nuclear encoded plastid localized proteins provided compelling evidence that plastid-targeted proteins are first co-translationally imported into the ER in both

Euglena and dinoflagellates. Immunogold localization of nucleus-encoded plastid proteins to the Golgi apparatus in both groups also suggests that there must be vesicular transport from the ER to the Golgi apparatus, and from the Golgi apparatus to the chloroplast for subsequent translocation through the three-plastid envelope membranes into the plastid. The purpose of this review is to present the presequence structural features, the in vivo and in vitro experimental evidence demonstrating vesicular transport of plastid proteins from their site of synthesis in the ER to the *Euglena* and dinoflagellate plastid. A model for presequence evolution and protein import into *Euglena and* dinoflagellate complex plastids based on this experimental evidence will also be presented.

This review focuses primarily on work from *Euglena gracilis* and the typical peridinin-containing dinoflagellates that acquired a plastid ultimately from a red alga, either secondarily [11] or in a more complex tertiary endosymbiosis [36, 37]. The dinoflagellates, however, seemed to have a high propensity for acquiring endosymbionts and, over time, new plastids [38, 39]. These serial endosymbiotic events essentially can replace an existing plastid, perhaps with a non-photosynthetic intermediate. Examples of this are found with the green dinoflagellates, like *Lepidodinium chlorophorum* that acquired a plastid through endosymbiosis with a green alga [40], likely replacing the one of red-algal origin that is typical of the peridinin-dinoflagellates [41]. In this example, the plastid has four surrounding membranes, not three [42]. For these dinoflagellates, you would predict targeting mechanisms more related to those of the stramenopiles that also have 4 membranes and a CER. Initial work has indicated that the precursor proteins targeted to the plastid of *Lepidodinuim* do have a N-terminal presequence with a signal peptide followed by a transit peptide [43], as you would predict.

2 An Unusual Case of Convergent Evolution

The similarities in plastid ultrastructure, the plastid-targeting presequences, and the chloroplast protein localization mechanism that routes precursors through the Golgi apparatus of dinoflagellates and *Euglena* are intriguing, as we will discuss. Unexpectedly, however, dinoflagellates and Euglenids also share a variety of additional molecular features that differ in most other organisms [12]. One such similarity is the presence of a large number of polycistronic mRNAs in both groups, though the prevalence in each is variable [12]. This is particularly obvious by the large number of plastid-targeted proteins in *Euglena* [23, 44, 45] and dinoflagellates [46–48] that are encoded by polycistronic mRNAs that encode polyproteins with multiple proteins separated by a short cleavable linker. Once imported into the plastid, the polyprotein is processed by a stromal protease at the linkers, releasing the individual protein units [49]. While there are rare examples of polyproteins in eukaryotes, one example being the polyubiquitin gene [50], plastid polyprotein precursors have not been found in other organisms.

There are also uncanny similarities in the chromosome structure and mRNA processing. *Euglena* and dinoflagellates both have permanently condensed chromosomes [12, 51] and cytoplasmic mRNAs with an unusual 5′ spliced leader that is added to all mRNAs by trans splicing [12, 52, 53]. Perhaps related to the odd chromosome structure and spliced leaders, gene expression is typically controlled at the translational rather than the transcriptional level in *Euglena* [54–57] and dinoflagellates [58, 59].

Despite these shared features, phylogenetic analyses indicate that dinoflagellates and *Euglena* are unrelated and do not appear to share a common ancestor [60]. The uncanny number of molecular similarities between these unrelated groups, suggest that there are limits in the solutions to particular molecular challenges [12]. Thus, these similarities must be the result of convergent evolution and the similar targeting mechanism possibly reflecting similar selection pressures related to the presence of three plastid envelope membranes [22].

3 In Vivo Discovery of the Endomembrane Trafficking Pathway to *Euglena* and Dinoflagellate Plastids

Immunoelectron microscopy provided the first indication that chloroplast proteins synthesized in the cytoplasm were not directly imported into the triple envelope complex chloroplasts of *Euglena.* In light-dark (14:10) division synchronized *Euglena*, rapid pLHCPII synthesis begins 6–8 h after the start of the light period, peaking at 10 h and then declining during the dark period, being undetectable during the last six hours of darkness [61]. Immunogold electron microscopy of synchronously dividing cells at 10 h into the light period of a 14:10 light dark cycle localized LHCPII to the chloroplast thylakoids and the Golgi apparatus (Fig. 1a). A three-dimensional reconstruction from serial sections of synchronous cells immunogold labeled with antibody to LHCPII at 10 h into the light period (Fig. 1b yellow dots), a time of rapid pLHCPII synthesis, localizes LHCPII only to the Golgi apparatus (Fig. 1b brown structure) and chloroplast (Fig. 1b green structure [62]. LHCPII was absent from the cytoplasm (Fig. 1b black space) and the nucleus (Fig. 1b blue dots). LHCPII is not immunolocalized to the Golgi apparatus during the latter phase of the dark period and the start of the light period when pLHCPII synthesis is low [19, 21].

Studies of pLHCPII synthesis rates and immunoelectron microscopy studies of LHCPII in dark-grown, resting *Euglena* exposed to light provided additional evidence that pLHCPII is transported from the Golgi apparatus to the *Euglena* plastid. Exposure of dark grown, resting *Euglena* to light slowly increases the rate of pLHCPII synthesis 50- to 100-fold with the maximum rate being attained 12–24 hours after light exposure [21, 63]. Immunoelectron microscopy localized LHCPII during the early hours of light exposure to the Golgi apparatus with little immunogold present in the developing thylakoids [20, 21]. Immunoelectron microscopy with LHCPII antibody found that both the Golgi and thylakoids were immunogold labeled in cells

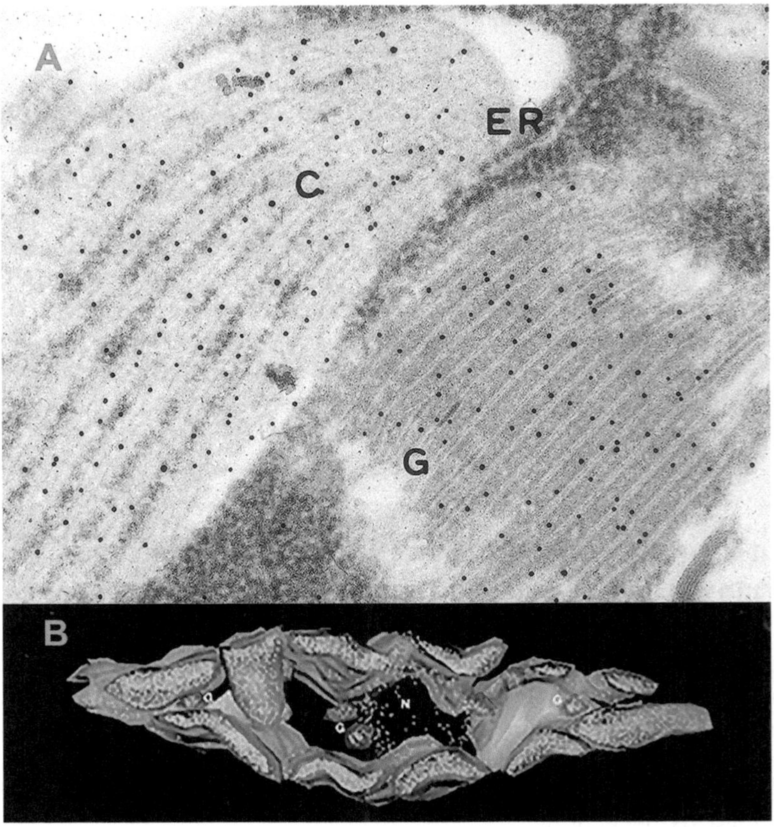

Fig. 1 Immunogold localization of LHCPII in synchronously dividing *Euglena*. (**a**) A section of a synchronously dividing cell at 10 h of a 14:10 light dark cycle immunogold labeled with antibody to LHCPII. Immunogold is localized in the chloroplast (*C*) and Golgi apparatus (*G*) but not in the endoplasmic reticulum (*ER*) (Modified from Osafune et al. [19]) (Reproduced from Durnford and Schwartzbach [108]). (**b**) A computer generated three dimensional model constructed from serial sections of a synchronously dividing cell at 10 h of a 14:10 light dark cycle immunogold labeled with antibody to LHCPII. Immunogold labeled LHCPII (yellow dots) is localized in the chloroplasts (green) and Golgi apparatus (brown) but not in the nucleus (blue dots) or the cytoplasm. (Modified from Schwartzbach et al. [62]) (Reproduced from Durnford and Schwartzbach [108])

exposed to light for 48 h, a time of active pLHCPII synthesis. However, only the thylakoids were immunolabeled at 96 h when the rate of pLHCPII synthesis is low [20, 21]. Exposure of dark grown *Euglena* to a low-light intensity does not induce thylakoid formation and LHCPII synthesis [64]. Thylakoid formation and pLHCPII synthesis are induced upon transfer to high light. Immunoelectron microscopy did not detect LHCPII in cells exposed to a low-light intensity. In cells transferred from low-light intensity to high-light intensity, a treatment inducing LHCPII synthesis and thylakoid formation, immunoelectron microscopy localized LHCPII to both the Golgi apparatus and developing thylakoids [18, 21]. Taken together, the immunolocalization of LHCPII to the Golgi apparatus only when cells are actively synthesizing LHCP and prior to thylakoid immunogold labeling suggests that pLHCPII is transported into the Golgi apparatus prior to chloroplast localization.

In light-dark synchronized cells of the dinoflagellate *Gonyaulax*, the rate of synthesis of ribulose bisphosphate carboxylase-oxygenase (Rbc,), a stromal protein, and PCP, a thylakoid protein, is high at the start of the light phase [22]. In *Gonyaulax*, both Rbc and PCP were immunolocalized to the Golgi apparatus and chloroplast at the time of rapid protein synthesis [22], similar to the observations of LHCP localization in *Euglena*. At a time when Rbc and PCP synthesis is low, however, these proteins were immunolocalized to the chloroplast but not to the Golgi indicating that the Golgi was a transient localization for newly synthesized plastid-targeted proteins. These results suggest that in dinoflagellates just as in *Euglena*, proteins are transported from the Golgi apparatus to the chloroplast.

The precursors for most of the *Euglena* light-harvesting complexes (LHCs) are polyproteins composed of multiple LHC subunits joined by a conserved decapeptide linker [25, 44, 45, 65, 66]. They are encoded by high molecular weight mRNAs. Immunoprecipitation of cell free translation products and total cellular protein from pulse labeled cells with antibody to *Euglena* LHCPII identified four high molecular weight pLHCPII polyproteins [63, 67, 68]. The protease that cleaves the decapeptide linker releasing individual mature LHCPIIs from the pLHCPII polyprotein is localized in the chloroplast [49]. Pulse-chase immunoprecipitation experiments have shown that the half-life for conversion of the pLHCPII polyproteins to chloroplast-localized, mature LHCPII is approximately 20 minutes [67]. pLHCPII mRNA is associated predominantly with membrane bound polysomes [54], suggesting the pLHCPII polyproteins are translocated into the ER prior to chloroplast localization. It appears that approximately 20 minutes is required to transport the pLHCPII polyprotein from its site of synthesis on ER associated polysomes to the chloroplast, where it is proteolytically processed into mature LHCPIIs.

Pulse-chase intracellular localization experiments were performed to determine directly if pLHCPII polyproteins are transported from their site of synthesis on ER bound polysomes, to the Golgi apparatus, and then to the chloroplast where the polyprotein is proteolytically processed to mature LHCPII. When dark grown, resting *Euglena* were exposed to light for 24 h to induce LHCPII synthesis and pulse labeled for 10 min with ^{35}S-sulfate, the only proteins immunoprecipitated with antibody to LHCPII were the high molecular weight pLHCPII polyproteins [63, 67, 68]. The pLHCPII polyproteins were found in the ER and Golgi fractions on sucrose gradients [68] but could not be detected in chloroplast fractions, ruling out a direct co-translational import into the chloroplast. During a 20- and 40-minute chase, the fraction of pLHCPII polyproteins in the ER decreased and the amount in the Golgi increased. Mature LHCPII only accumulated in the chloroplast and thylakoid fraction, never in the ER and Golgi fractions during the 20- and 40-minute chase [68]. The time-dependent decrease in ER and Golgi-localized pLHCPII polyproteins coincident with the accumulation of mature LHCPII in the chloroplast and thylakoids indicates that pLHCPII polyproteins are synthesized in the ER, transported to the Golgi apparatus and then from the Golgi apparatus to the chloroplast.

The precursor to the small subunit of *Euglena* ribulose bisphosphate carboxylase-oxygenase (pRbcS), a soluble stromal protein, is also a polyprotein composed of 8 RbcS units covalently joined by a decapeptide linker [23]. The half-life for conversion of pRbcS polyproteins to mature RbcS is 12 minutes [69, 70], indicating that pRbcS transport from its site of synthesis to the chloroplast, where the polyprotein is processed, is approximately twice as fast as the transport of pLHCPII to the chloroplast. Pulse-chase immunoprecipitation studies found that, as seen for pLHCPII polyproteins, the pRbcS polyproteins were first detected in the ER fractions, then in the Golgi fractions, and are never detected in the chloroplast fractions [70]. Concomitant with a decrease in Golgi localized pRbcS is an increase in chloroplast localized mature RbcS, indicating that as found for pLHCPII, a thylakoid protein, pRbcS, a stromal protein, is transported from the ER to the Golgi apparatus and then to the chloroplast where the polyprotein is proteolytically converted into mature RbcS.

An N-terminal signal peptide initiates co-translational translocation across the ER membrane into the lumen [71]. Hydrophobic domains within the presequence or the mature protein can stop translocation, anchoring the protein in the ER membrane. The pLHCPII and pRbcS N-terminal Class I presequences, as discussed below, contain two hydrophobic regions within the presequence. The first is part of the signal peptide domain that targets the emerging protein to the ER and which is removed by signal peptidase in the ER lumen. The second hydrophobic region follows a transit peptide domain, which remains until the precursor polyprotein is processed in the chloroplast [23, 26, 72, 73]. If this second hydrophobic domain within the presequence functions as a stop-transfer, membrane anchor sequence, pLHCPII and pRbcS, would be co-translationally inserted into the ER membrane and transported to the chloroplast in vesicles as an integral membrane protein rather than in the vesicle lumen as a soluble protein.

The ability of different salt washes to remove proteins from membrane fractions can be used to determine if they are peripheral or integral membrane proteins. Peripheral membrane proteins can be stripped from their associated membranes by washing with potassium acetate, leaving behind proteins integrated in the membrane or located in the lumen/stroma of the organelles being washed. Washing with sodium carbonate opens the membrane releasing luminal and stromal proteins but not integral membrane proteins. Washing intact chloroplasts or ER and Golgi membranes with potassium acetate did not remove in vivo synthesized pLHCPII, LHCPII, pRbcS, or RbcS from these membranes [68, 70]. pLHCPII and pRbcS were not removed from ER and Golgi membranes by sodium carbonate, indicating they are integral membrane proteins that were co-translationally inserted into the ER membrane and transferred in vesicles as integral membrane proteins to the Golgi apparatus. Mature LHCPII, an integral thylakoid membrane protein, was also not released from the thylakoid by washing chloroplasts with sodium carbonate [68] but the mature RbcS, a soluble stromal protein, was released [70]. Since the only hydrophobic domain in pRbcS, after cleavage of the presequence signal peptide is the second presequence hydrophobic domain, these results suggest that this domain must function as a stop-transfer sequence anchoring pLHCPII and pRbcS in the ER membrane.

Treatment of an ER and Golgi membrane fraction with a protease that cannot penetrate the membrane, a protease protection assay, can provide information regarding the topology of an integral membrane protein. Trypsin digestion of ER and Golgi membranes containing pLHCPII or pRbcS produced multiple protected fragments [68, 70]. It appears that both pLHCPII, the precursor of an integral thylakoid membrane protein, and pRbcS, the precursor of a soluble stromal protein, are transported from the ER to the Golgi apparatus to the chloroplast as integral membrane proteins. Upon fusion of the transport vesicle with the chloroplast outer membrane, the precursor would not be within the outermost and middle envelope membrane intermembrane space but would be an integral outer envelope membrane protein with multiple membrane spanning domains. Not only would the precursor need to be translocated through the middle and innermost chloroplast envelope membranes, it would also need to be removed from and translocated through the outermost envelope membrane.

4 The Structure of Precursors Targeted to Complex Chloroplasts with Three Envelope Membranes

4.1 The Structure of the Chloroplast Protein Presequence

The additional membrane around the plastids of euglenophytes and dinoflagellates poses a further barrier to import. In *Euglena*, the plastid targeting sequences are a complex, modular construction [72] that was surprisingly similar to the dinoflagellate plastid-targeting presequences [22, 35]. This implied similar strategies for targeting and traversing this additional membrane barrier.

In plants and green algae, plastid-targeted proteins typically have an N-terminal transit peptide that mediates plastid targeting and import. While there is no sequence conservation in transit peptides, they all share some biochemical similarities [17] such as an elevated content of hydroxylated amino acids (Ser, Thr) and an overall net positive charge due to an under-representation of acidic amino acids. The transit peptide may form regions of amphipathic alpha helices in the hydrophobic environment of the membrane [17], which could be important for receptor recognition and engagement with the translocation apparatus (See Venkatasalam and Kessler, chapter "Translocation of Proteins into Primary Plastids", this volume). The transit peptide is usually removed by a stromal processing peptidase as the protein is imported into the chloroplast. However, not all chloroplast-localized proteins in plants have a transit peptide. Proteins targeted to the outer membrane often lack a transit peptide and instead have other features such as a N-terminal signal anchor or C-terminal tail-anchor [74]. So, while the transit peptide is the canonical mode of targeting to plastids, there are variations depending on target destination. For instance, proteins targeted to the thylakoid lumen have a second functional module located immediately after the transit peptide containing additional sequence information required to cross the thylakoid membrane. This extra region is like the hydrophobic region of classic signal peptides. In these thylakoid lumen-targeted proteins, once imported into the stroma, the transit peptide is cleaved exposing the lumen-targeting domain that mediates import into the thylakoid lumen whereby the targeting domain is removed by a thylakoid protease [75, 76].

In both the euglenophytes and dinoflagellates, there are two major classes (I and II) of plastid-targeting presequences (Fig. 2) [35, 72]. In both, the presequences are composed of distinct functional modules that are pieced together to target proteins to the plastid and across the three membranes. The Class I presequences have two hydrophobic domains on either side of a classic transit peptide domain. The first hydrophobic domain is a predicted signal peptide, which is ultimately for targeting to the ER. The second hydrophobic domain follows the transit peptide region and is effectively a stop-transfer sequence, which anchors the

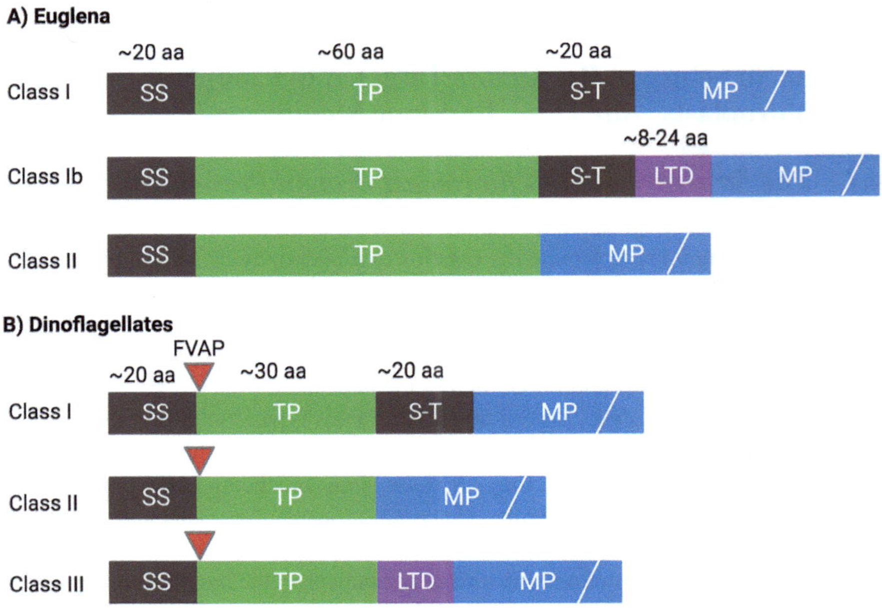

Fig. 2 Main structural features of plastid-targeting presequences from (**a**) *Euglena* and (**b**) peridinin-containing dinoflagellates. Both groups have Class I and Class II presequences that were convergently acquired during endosymbiogenesis. Depending on the group, the different Class of presequences can be modified with the addition of a thylakoid-lumen targeting domain (Class Ib, *Euglena*; Class III, dinoflagellates) for proteins that are targeted to the thylakoid lumen. SS Signal Sequence cleaved in the ER by signal peptidase, TP Transit Peptide cleaved in the stroma by stromal processing peptidase, S-T Stop-Transfer Membrane Anchor, LTD Thylakoid Lumen-Targeting Domain, MP Mature Protein. The location of the FVAP motif (Phenylalanine, Valine, Alanine, Proline) characteristic of *red* algal-derived plastid presequences is shown in the dinoflagellate presequences. (Created with BioRender.com)

protein into the ER membrane (Fig. 2). The Class I-type presequence in *Euglena* was recognized in earlier studies that sequenced a limited number of genes encoding plastid-targeted proteins [23–32]. A subset of Class I presequences in *Euglena*, Class Ib (Fig. 2a) [72], are thylakoid lumen proteins that have an additional hydrophobic domain (Fig. 2a) similar to a signal sequence that is required for translocation across the thylakoid membrane into the lumen. To our knowledge, an equivalent of a Class Ib presequence has not yet been identified in dinoflagellates, but we would predict they exist. Upon presequence cleavage by signal peptidase in the stroma, this domain is at the presequence N-terminus then targeting the protein to the thylakoid lumen.

A second major category of presequences in *Euglena* and dinoflagellates are called Class II. They resemble Class I presequences except they lack the stop-transfer sequence after the transit peptide (Fig. 2) [35, 72]. In dinoflagellates, there is a subset of the Class II presequence where the transit peptide domain is followed by a luminal targeting domain (LTD), which was called a Class III presequence (Fig. 2b). As with the Euglena Class Ib presequence, this LTD was for targeting proteins across the thylakoid lumen once they reach the stroma of the plastid. A Class III-equivalent presequence has not yet been identified in *Euglena*.

The relative proportion of Class II presequences was initially estimated to be low, around 11% [72]. The analysis of a larger number of *Euglena* plastid-targeted proteins identified in a proteomic study suggests that the number of proteins with a Class II presequence is much greater, closer to 44% of the Class I + II total [77]. In a non-photosynthetic *Euglena* (*Euglena longa*), about 30% of the detected plastid-targeted proteins were categorized as having Class II presequences [78]. In the dinoflagellate *Heterocapsa triquerta* [35], there were roughly equal numbers of proteins with Class I and II presequences, agreeing with the larger estimates of the Class II sequences in *Euglena*.

4.2 Domains in the Chloroplast Protein Presequences

The Signal Sequence

In *Euglena*, the signal sequence domain of the plastid targeting presequence is a clear hydrophobic domain that is variable in length (roughly 20 amino acids long, Fig. 2). It is often preceded by a short stretch of amino acids and often possessing a basic amino acid [72]. A similar organization was seen in the dinoflagellate, *Heterocapsa* presequences [35], so presumably these would be functionally equivalent. Biochemical studies confirmed that this domain functions as a signal peptide responsible for initiating translocation into the ER lumen of *Euglena* [26, 68, 73] (see Sect. 5.1) and also of dinoflagellates [22].

The Transit Peptide

In both *Euglena* and dinoflagellates, the signal peptide domain is followed by a region that resembles a chloroplast transit peptide in composition (Fig. 2). In *Euglena*, the transit peptide is roughly 60 amino acids in length, but in the dinoflagellate *Hetercapsa triquerta*, it is about half that length on average. This transit peptide domain is enriched in hydroxylated amino acids (Ser/Thr; ca. 20%), has an elevated alanine content, but is deficient in acidic amino acids (Glu/Asp) compared to the mature protein [35, 72, 77]. At least in *Euglena*, the transit peptide region contains an elevated proline content, upwards of 20% [72, 77], that is higher than the typical proline enrichment of plant transit peptides. In plants, the prolines in the transit peptides were crucial for proper targeting and efficient import of precursor proteins, especially those with transmembrane domains [79]. The authors propose that the prolines help prevent insertion of the precursor in the chloroplast envelope, which may be particularly important for the precursors with stop-transfer sequences as you find in many of the proteins entering plastids with three membranes.

Organisms with three plastid membranes have what looks like a transit peptide to cross the inner-two most plastid envelopes, and it is presumed this region would function in the same fashion as the plant transit peptide, namely interacting with the translocation complex in the chloroplast envelope membranes. Heterologous protein import studies with *Euglena* precursors and plant chloroplasts identified this region as a functional transit peptide [80, 81] (see Sect. 5.1). In dinoflagellates, the transit peptide domain was effective in targeting proteins to a plant plastid, confirming the region's functional capacity to target and import proteins across the chloroplast envelope [22, 34].

While there is little sequence conservation in canonical transit peptides, in the dinoflagellate *Heterocapsa*, there is a conserved FVAP-motif at the start of the transit peptide [35] that is absent in *Euglena* transit peptides [72]. This Phenylalanine (F) is followed by three non-polar, hydrophobic amino acids (VAP) in *Heterocapsa*, likely just after the signal sequence cleavage site. The requirement for phenylalanine in the FVAP motif reflects the different ancestry of the plastid in peridinin-dinoflagellates. Red algae, glaucophytes, and organisms with red-algal-derived plastids (e.g., stramenopiles, cryptophytes, haptophytes etc.) all have a requirement for a conserved F for efficient protein import into the plastid [35, 82]. In diatoms and cryptophytes, removal of the phenylalanine abolished protein import [83]. The ability of diatom plastids to import proteins with transit peptides from a variety of different organisms with red-derived plastids indicates that these complex plastids share a similar import mechanism [82, 84, 85].

The Stop-Transfer Sequence

In the classic Class I presequence in *Euglena* and dinoflagellates, a second hydrophobic sequence follows the transit peptide domain (Fig. 2). This region functions as a membrane anchor stop-transfer sequence that prevents the co-translational import of a protein into the ER, anchoring it in the ER membrane [73]. This sequence is, on average, approximately 20 amino acids in length, though there is a considerable variation [35, 72]. Immediately following the stop-transfer sequence in *Euglena* there are often two or three basic amino acids, a feature commonly seen in stop-transfer sequences in other organisms [86], plus a variable number of additional amino acids N-terminal to the stromal processing peptidase cleavage site.

The Thylakoid Lumen Targeting Domain

Plastid proteins targeted to the thylakoid lumen must pass through yet another membrane after passing through the inner envelope membrane, and these proteins are specifically targeted to the thylakoid via a luminal targeting domain. In plants, the transit peptide is removed upon import into the chloroplast by the stromal processing peptidase leaving a thylakoid lumen-targeting domain at the N-terminus of the protein [75, 76, 87]. These thylakoid lumen-targeting domains have been identified in the presequences of lumen-targeted proteins of *Euglena* (called Class Ib) [72]. The presequence thylakoid lumen targeting domain is an approximately 8–24 amino acid hydrophobic domain, C-terminal to the stop-transfer sequence. In the dinoflagellates *Heterocapsa triquerta* and *Amphidinium carterae*, only a total of five proteins were identified with hydrophobic thylakoid luminal targeting domains and both of these were variants of a Class II presequence (Called Class III, [35]). We predict that as more data becomes available, thylakoid luminal targeting domains will be found associated with both Class I and II presequences in dinoflagellates and *Euglena*.

Unclassified Presequences

In a proteomic study of isolated *Euglena* plastids [77], a significant portion (16%) of predicted, plastid localized proteins had presequences with limited resemblance to Class I or II presequences. There is a region with an amino acid composition characteristic of a transit peptide, but an identifiable signal peptide or stop-transfer sequence was not present. This interesting finding suggests that there is a novel targeting mechanism, but one that requires further investigation. It's always hard to rule out cross contamination when organelles are isolated, but this certainly needs to be explored further.

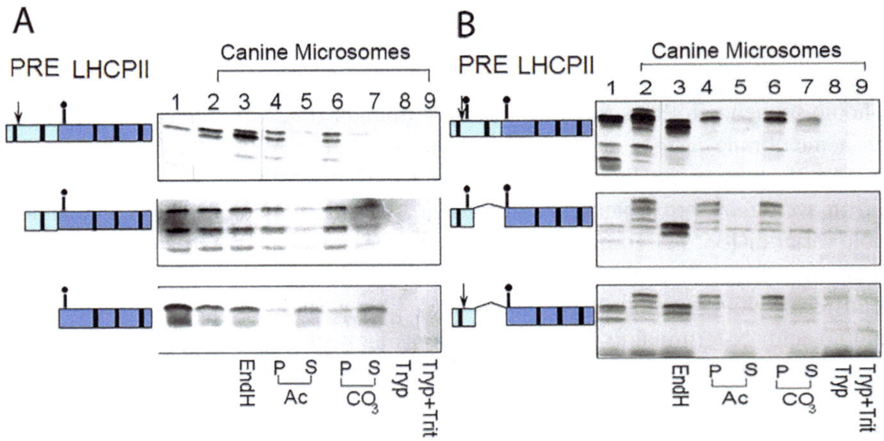

Fig. 3 Microsomal import and glycosylation of pLHCPII and pLHCPII presequence deletion glycosylation reporter constructs. (**a**, **b**) Schematic diagrams of the presequence (light blue) deletion glycosylation reporter constructs indicating the locations of the predicted signal peptidase cleavage site (arrow), hydrophobic domains (black boxes) within the presequence and mature protein (dark blue), and potential glycosylation sites within the presequence (Asp40) and mature LHCPII (Asp148) (a lollipop shape). mRNA transcribed from the indicated constructs was translated in the absence (lane 1) or presence (lanes 2–9) of canine microsomes. After terminating translation, microsomal membranes were incubated with Endo H (EndH), washed with 0.5 M potassium acetate (Ac) or 0.1 M Na_2CO_3, pH 11.5 (CO_3), and separated into a pellet (P) and supernatant (S) fraction or treated with trypsin (Tryp) in the presence and absence of Triton X-100 (Trit). Products were separated by SDS gel electrophoresis and visualized by fluorography. (Modified from Sulli et al. [73]) (Reproduced from Durnford and Schwartzbach [108])

5 Elucidation of Functional Domains in *Euglena* and Dinoflagellate Bipartite Plastid Targeting Presequences

5.1 Euglena *Presequence Functional Domains*

In vitro organelle import experiments have shown that the *Euglena* pLHCPII and pRbcS, Class I presequences contain a signal peptide domain responsible for initiating co-translational translocation of the precursor into the ER [26, 70, 73]. A pLHCPII (Fig. 3a top) [26, 73] and a pRbcS, [70] composed of the complete Class I presequence and one mature protein unit were co-translationally imported in vitro into canine microsomes and inserted into the microsomal membrane oriented with a portion of the imported protein remaining outside the microsomes. The pLHCPII presequence was cleaved and the change in the molecular weight of the imported protein was consistent with the cleavage of the N-terminal presequence at the predicted signal peptidase cleavage site [26, 73]. Although the pRbcS, presequence has a functional signal peptide domain initiating translocation into canine microsomes, the presequence was not cleaved [70] suggesting that if it has a signal peptidase cleavage site, it is not recognized by the canine microsomal signal peptidase. Mature LHCPII (Fig. 3a bottom), mature RbcS, and pea mature LHCPII [26, 70, 73] were not imported into canine microsomes, indicating that the signal peptide is a presequence domain and not a domain within the mature protein.

The sodium carbonate-resistant association of pLHCPII (Fig. 3a top) [26, 73] and pRbcS [70] with the canine microsomal membrane indicates that they have been co-translationally inserted into the microsomal membrane; a process requiring a stop-transfer membrane anchor sequence. A truncated pLHCPII lacking amino acids 1–43, the presumptive presequence N-terminal signal peptide based on the criteria of von Heijne [71], but containing the presequence second hydrophobic domain was also imported in vitro into canine microsomes and inserted into the microsomal membrane oriented with a portion of the imported protein remaining outside the microsomes (Fig. 3a middle) [73]. This shows that the presequence region C-terminal to the signal peptidase cleavage site including the second hydrophobic domain can function as an uncleaved signal peptide initiating translocation of the precursor across the microsomal membrane with the hydrophobic region anchoring the protein in the membrane. The ability of the pLHCPII Class I presequence second hydrophobic domain to function as the hydrophobic domain of a signal peptide when it is at the precursor's N-terminus suggests that the pLHCPII, and by inference the pRbcS, Class I presequence second hydrophobic domain is in fact the stop-transfer membrane anchor domain responsible for co-translational insertion of the precursors into the ER membrane in vivo [68, 70].

Protein glycosylation occurs within the ER lumen as proteins are co-translationally translocated across the ER membrane. Glycosylation increases the apparent molecular weight of a protein and digestion with Endo H removes the glycan, decreasing the apparent molecular weight. The changes in the apparent molecular weight upon glycosylation and deglycosylation allow glycosylation sites to serve as reporters for regions of a protein translocated across the ER membrane into the ER lumen. By studying the import of pLHCPII presequence deletions and glycosylation reporters into canine microsomes, the functional domains of the LHCPII presequence and the topology of the imported pLHCPII were determined [73].

Mature LHCPII contains a potential glycosylation site eight amino acids from the N-terminus. Upon translocation of pLHCPII across the microsomal membrane, the apparent molecular weight of the protein did not change after treatment with endoglycosidase H (EndoH) (Fig. 3a top) [73], indicating that the protein was not glycosylated. Sodium carbonate washes did not release the translocated protein from the microsomal membrane indicating that it is an integral microsomal membrane protein oriented with the presequence in the microsomal lumen and the N-terminal region of mature LHCPII containing the glycosylation site outside the microsomes (Fig. 3a top). A pLHCPII with a glycosylation site inserted into the presequence at amino acid 40 between the predicted signal peptidase cleavage site and the second presequence hydrophobic domain was imported into canine microsomes and inserted into the membrane [73], as evidenced by the failure of sodium carbonate washes to release the protein from the microsomal membrane (Fig. 3b top). The apparent molecular weight of the translocated protein was higher than the molecular weight of the protein synthesized in the absence of microsomes. EndoH digestion decreased the molecular weight of the protein indicating the protein was glycosylated and the molecular weight of the Endo H digested deglycosylated protein was less than that of the protein synthesized in the absence of microsomes (Fig. 3b top), indicative of signal peptide cleavage. Since the glycosylation site in the mature protein C-terminal to the second presequence hydrophobic domain was not glycosylated and the glycosylation site at presequence amino acid 40, N-terminal to the second presequence hydrophobic domain was glycosylated, the second presequence hydrophobic domain must function as the stop-transfer sequence orienting pLHCPII with the N-terminus formed after signal peptidase cleavage in the ER lumen and the mature protein on the cytoplasmic membrane face [73]. The Class I presequence stop transfer hydrophobic domain does in fact function to prevent glycosylation of the mature LHCPII; a process which could lead to formation of a non-functional protein.

Truncated pLHCPIIs containing only the presequence N-terminal 42 amino acids, the predicted presequence signal peptide domain containing (Fig. 3b middle) or lacking (Fig. 3b bottom) a glycosylation site at amino acid 40, were co-translationally transported into canine microsomes definitively identifying the presequence N-terminal domain as a functional signal peptide [73]. Although the presequence lacks the second hydrophobic membrane anchor domain, the protein was not released from the membrane by sodium carbonate washes but it was digested by an exogenous protease. This is indicative of the protein N-terminus being in the microsomal lumen and the C-terminus outside the microsomal membrane. There are three hydrophobic domains in mature LHCPII [45] and one of these domains probably functioned as the membrane-anchor sequence. The apparent molecular weight of both proteins increased upon import but the increase for the protein containing two glycosylation sites was greater than the increase of the protein with a single glycosylation site (Fig. 3b middle bottom). Endo H treatment decreased the apparent molecular weight of both proteins, confirming the translocation of the asparagine eight amino acids from the N-terminus of the mature protein into the microsome, where it was glycosylated. Glycosylation of the asparagine eight amino acids from the N-terminus of the mature protein in the absence but not in the presence of the presequence second hydrophobic domain identifies this domain as the membrane anchor sequence responsible for orienting pLHCPII in the ER membrane with the presequence region between the signal peptidase cleavage site and the second hydrophobic domain within the ER lumen and the remainder of the presequence and mature protein in the cytoplasm.

Euglena plastid proteins are transported from the Golgi apparatus to the chloroplast in vesicles oriented with the N-terminus in the vesicle lumen and the mature protein on the cytoplasmic membrane face [26, 68, 70, 73]. Upon fusion of the transport vesicles with the outermost of the three chloroplast membranes, the precursor would be anchored in the envelope membrane with the N-terminus in the intermembrane space and the remainder of the protein in the cytoplasm. The molecular mechanism of vesicle budding and target membrane fusion are well characterized reactions requiring numerous proteins to form the budding and fusion complexes [88]. N-ethylmaleimide-Sensitive Factor [89] is an ATPase that functions in the dissociation of the soluble NSF attachment protein receptor (SNARE) mediated membrane fusion complex [89] and it is inactivated by N-ethylmaleimide [88]. A number of other proteins binding and hydrolyzing ATP or GTP are also required during the budding and fusion reactions. Multiple GTPases and ATP as a source of energy supplied exogenously or generated internally by photosynthesis is also required for protein import into simple plastids of plants [90].

pLHCPII was transferred from purified *Euglena* Golgi membranes to intact purified *Euglena* chloroplasts in a time-dependent manner [81]. pLHCPII was processed by the stromal processing protease to mature LHCPII and a protein having a molecular weight consistent with a LHCPII polyprotein containing two mature LHCPIIs indicating the pLHCPII was translocated through all three envelope membranes. Golgi to chloroplast transport and chloroplast import required exogenous ATP, intrachloroplast ATP synthesized photosynthetically in the light, and GTP [81]. This differs from import into plant chloroplasts where exogenous or photosynthetically generated ATP supports maximal import rates [91]. This is however consistent with the requirement for ATP in vesicle fusion reactions and in protein translocation through the intermediate and inner envelope membrane. *Euglena* Golgi to chloroplast transport was not inhibited by N-ethylmaleimide [81]. This suggests that the highly conserved SNARE vesicle fusion apparatus used for all fusions on the endocytic and exocytic pathway [88] are not used for fusion of *Euglena* Golgi derived vesicles with the outer plastid envelope membrane.

The bipartite Class I and Class II presequences of *Euglena* chloroplast proteins are composed of an N-terminal signal peptide domain and a domain having characteristics of a transit peptide. In plants and green algae, a transit peptide is responsible for post translational import of proteins into chloroplasts. Upon transport into the ER, the *Euglena* presequence signal peptide is cleaved exposing the transit peptide domain. A *Euglena* pRbcS, containing the complete N-terminal bipartite presequence linked to a mature RbcS was not imported into pea chloroplasts [81]. A *Euglena* pRbcS [81] and the precursor to the 30-kDa subunit of the oxygen-evolving complex (pOEC30) [80] lacking the signal peptide domain but containing the presumed transit peptide and stop-transfer membrane anchor sequence are post translationally imported into pea chloroplasts identifying the domain after the signal sequence as a functional transit peptide. A *Euglena* pRbcS, Class I presequence which lacked the N-terminal signal peptide domain, in effect converting it into a Class II presequence by deletion of the stop-transfer membrane anchor sequence, was also post-translationally imported into pea chloroplasts [81]. Import into pea chloroplasts of a *Euglena* pRbcS, lacking the signal peptide domain and containing only the presequence region N-terminal to the hydrophobic stop-transfer membrane anchor sequence, identifies this region as the functional transit peptide domain. The *Euglena* pRbcS lacking the signal peptide domain (which was imported post translationally in vitro into pea chloroplast) was not post-translationally imported in vitro into *Euglena* chloroplasts [81], indicating that vesicular transport from the ER to the

Golgi apparatus to the chloroplast is the obligate pathway for protein import into *Euglena* chloroplasts.

5.2 Dinoflagellate Presequence Functional Domains

A pPCP-Luciferase (LUC) fusion protein construct with a Class I presequence from the dinoflagellate *Gonyaulax* [22] and 33 kDa oxygen-enhancer I precursor protein (pPsbO) from the dinoflagellate *Heterocapsa* [34] were co-translationally translocated into canine microsomes and the N-terminal signal peptide was cleaved in both cases. The inserted pPCP-LUC fusion protein was protease sensitive, indicating that it was inserted into the microsomal membrane oriented with the majority of the protein remaining outside the microsomes. Since the second presequence hydrophobic domain is the only hydrophobic domain remaining after signal peptide cleavage, it must be the functional stop-transfer membrane anchor sequence.

A *Gonyaulax* pPCP-LUC fusion protein lacking the N-terminal presequence signal peptide domain was inserted into the microsomal membrane [22]. The size of the inserted protein was not changed and the inserted protein was protease sensitive. This indicates that the second presequence hydrophobic region can function as a signal sequence to target proteins to the ER, but it is not recognized by the signal peptidase. The protease sensitivity of the fusion protein also demonstrates that the second hydrophobic region of the presequence acts as a stop-transfer membrane anchor sequence, anchoring the protein in the membrane with the N-terminus in the ER lumen and the majority of the protein in the cytoplasm.

The pPCP-LUC fusion protein lacking the N-terminal signal peptide domain was imported in vivo into potato (*Solanum chacoense*) chloroplasts [22], identifying this region as a plastid-targeting transit peptide. The full length *Heterocapsa* pPsbO [34] and a truncated pPsbO containing the region C-terminal to the signal peptide hydrophobic domain (including the putative transit peptide and the stop-transfer sequence) were imported in vitro into pea chloroplasts. Two proteins smaller than the incubated precursors were found in the intact chloroplasts and in protease-treated chloroplasts. The larger of the two imported proteins was found in the stromal fraction, suggesting that upon import, the stromal processing protease removed the presequence plastid-targeting transit peptide. The smaller protein was found in protease-treated thylakoids, demonstrating that it was imported across the thylakoid membrane, into the lumen, and the remaining thylakoid targeting presequence was cleaved by a thylakoid protease. The in vivo and in vitro experiments with the *Gonyaulax* pPCP-LUC fusion protein and the *Heterocapsa* pPsbO demonstrate that the dinoflagellate presequence containing the second hydrophobic domain is a functional transit peptide mediating import of the precursor into simple chloroplasts. The in vitro studies with the *Gonyaulax* pPCP-LUC fusion protein lacking the N-terminal signal peptide indicate that, as found for *Euglena*, the transit peptide domain containing the second hydrophobic region can also a function as a signal peptide mediating the translocation of the precursor into canine microsome [22].

6 ER to Golgi to Chloroplast Trafficking of Plastid Glycoproteins to Simple Chloroplasts

N-linked glycosylation is a common co-translational modification of proteins. An oligosaccharide attached to dolichol phosphate is initially assembled on the cytoplasmic side of the ER membrane by sequential addition of individual sugars [92]. At some point in the assembly process, the dolichol phosphate oligosaccharide flips to the luminal side of the ER membrane, and assembly of the oligosaccharide is completed by further sequential addition of individual sugars. The completed oligosaccharide is transferred from dolichol phosphate to the asparagine of a protein during the co-translational translocation of the protein into the ER lumen. The protein is folded in the ER, disulfide bonds are formed and the oligosaccharide is further modified prior to vesicular transport to the Golgi apparatus [92]. Glycan modification is completed in the Golgi apparatus and the glycoprotein is transported in vesicles from the Golgi apparatus to its final destination [93]. The presence of the glycan produces a significant increase in the bulk of the folded protein.

Simple and complex chloroplasts contain glycoproteins. Arabidopsis α-carbonic anhydrase (CA) is a chloroplast-localized, disulfide containing glycoprotein whose folding and activity are dependent on the modifications, glycosylation, and disulfide bond formation, occurring in the ER and Golgi apparatus [94]. The protein cannot be directly imported into isolated pea chloroplasts but is imported into microsomes. α-CA has a functional signal peptide targeting the protein to the Arabidopsis protoplast ER, and when Golgi-mediated vesicular transport is inhibited, α-CA is not found in chloroplasts [95], clearly demonstrating protein transport from the ER to the Golgi apparatus to the chloroplast.

Arabidopsis α-CA is just one of a number of chloroplast glycoproteins that have been definitively shown to be transported from the ER to the Golgi apparatus to simple chloroplasts. Rice nucleotide pyrophosphatase/phosphodiesterase (NPP) is a glycosylated chloroplast protein whose localization to the chloroplast is dependent upon a functional ER-Golgi vesicular transport system [96, 97]. Rice chloroplast α-amylase is also a glycoprotein and disruption of the ER to Golgi transport system inhibits its transport to the chloroplast, leaving the protein in the ER [98]. Taken together, the studies of glycoprotein import into simple chloroplasts have demonstrated the existence of a second protein import pathway involving vesicular transport from the ER to the Golgi apparatus to the chloroplast. After fusion of the Golgi transport vesicle with the outer plastid membrane, the protein is in the intermembrane space. Due to protein folding and glycosylation in the ER, the protein is probably too large to utilize the TIC translocation channel, indicating that a yet to be identified translocation system is used to transport proteins across the simple chloroplast inner membrane into the stroma. The identification of an ER-to-Golgi-to-plastid protein transport system in these simple plastids establishes that such protein targeting mechanisms may be an ancestral feature for organelles, which may have been more prevalent during endosymbiogenesis that saw the transition of the endosymbiont to an organelle over time. If so, this established protein-targeting mechanism in complex plastids with three or more membranes could be viewed as an evolutionary step in endosymbiogenesis.

7 The Mysterious Dinoflagellate and *Euglena* Plastid Envelope Translocons

The *Euglena* plastid and the typical peridinin-containing dinoflagellate plastid were acquired through an endosymbiotic association between a eukaryotic host and a eukaryotic green or an alga with a red alga-derived plastid, respectively [8–10, 36, 37, 99]. Just like simple plastids that evolved from a prokaryotic cyanobacterial endosymbiont, the eukaryotic algal endosymbiont ultimately became enslaved by the host becoming fully integrated into the hosts' metabolic and regulatory pathways. The nature of such events, which include massive endosymbiont gene loss and the transfer of genes to the host nucleus, have been discussed extensively [8, 100]. Transfer of endosymbiont genes to the host nucleus resulted in most of the chloroplast proteins being translated in the cytoplasm, necessitating a mechanism to translocate the proteins from the cytoplasm across the chloroplast envelope membranes into the plastid.

The process for translocating proteins from the cytoplasm across the envelope membranes of simple chloroplasts that evolved from a prokaryotic endosymbiont is well characterized [101] (See Venkatasalam and Kessler, chapter "Translocation of Proteins into Primary Plastids", this volume). The multiprotein translocon at the outer envelope membrane of chloroplasts (TOC) and the multiprotein translocon at the inner envelope membrane of chloroplasts (TIC) associate at membrane contact sites, providing a passage for simultaneous translocation of the precursor from the cytoplasm to the stroma. The N-terminal transit peptide on chloroplast precursors is recognized by the TOC complex and the unfolded precursor is transported through the two envelope membranes into the plastid [101] (See Venkatasalam and Kessler, chapter "Translocation of Proteins into Primary Plastids", this volume).

Of the three membranes surrounding the *Euglena* and dinoflagellate plastid, it is generally agreed that the inner two membranes are derived from the chloroplast envelope of the endosymbiont. The lipid composition of the *Euglena* plastid envelope supports this assumption [16]. If the *Euglena* and dinoflagellate middle and inner plastid membranes evolved from the plastid membranes of the eukaryotic endosymbiont, it is reasonable to expect that they retained the TOC/TIC transit peptide-dependent protein import machinery during evolution of the endosymbiont to a complex plastid. Consistent with this expectation is the presence on both *Euglena* and dinoflagellate chloroplast protein presequences of a functional plastid targeting transit peptide domain that initiates translocation of the precursors through both envelope membranes of pea chloroplasts [34, 80, 81] and *Solanum chacoense* chloroplasts [22]. The presence of functional transit peptides on *Euglena* and dinoflagellate presequences provides experimental support, suggesting that import through the middle and inner plastid membranes uses import machinery evolutionarily related to the TOC/TIC import machinery of simple chloroplasts.

Although all *Euglena* and dinoflagellate precursor presequences contain a transit peptide domain that can interact with the TIC/TOC translocation system, there are fundamental differences between the requirements for import of proteins into simple chloroplasts and import into *Euglena* and dinoflagellate complex plastids. Class I presequences have a stop transfer sequence, resulting in them being anchored in the Golgi transport vesicle with a majority of the protein remaining in the cytoplasm [22, 26, 35, 68, 70, 72, 73]. After fusion of the Golgi transport vesicle with the outer chloroplast membrane, proteins with Class I presequences become integral outer envelope membrane proteins. Import of proteins with a Class I presequence requires a mechanism to transfer an integral outer envelope membrane protein from the membrane into a translocation channel and then transfer the protein through the outer membrane. As the precursor contains a functional transit peptide and is unfolded, a reasonable expectation is that the transit peptide would interact with the TOC/TIC import apparatus of the endosymbiont envelope or proteins that evolved from them for translocation of the protein removed from the outer membrane across the middle and inner membranes at contact sites.

Proteins with Class II presequences could be glycosylated, acquire disulfide bonds, and fold during passage through the ER and Golgi, as they are likely translocated into the ER lumen because they lack a stop-transfer domain. If the mature protein has a hydrophobic membrane spanning domain, then these proteins could also be integrated into the ER membrane. After fusion of the Golgi transport vesicle containing these precursor proteins with the outer plastid envelope membrane, the precursors are in the space between the outer and middle envelopes. A glycosylated, folded, and disulfide bond-containing Class II precursors would be too large for transport by the TOC/TIC system. Simple chloroplasts contain an uncharacterized system to target and fuse chloroplast glycoprotein containing Golgi transport vesicles with the outer plastid envelope membrane and translocate the folded glycoprotein across the inner envelope membrane to the stroma [94–98]. If this translocation system was functional in the endosymbiont, it is likely to have evolved into the import system for proteins with Class II presequences, replacing TOC/TIC import components.

Euglena genomic and plastid proteomic studies have identified TIC 32 and TIC 21 in *Euglena* but TOC components appeared to be absent [77, 78, 102]. TIC 110 was identified in the dinoflagellate genome but TOC components, as with *Euglena*, were not identified [33, 34]. Homologues of proteins involved in vesicular transport; derlins, coat complex subunits, several Rab GTPases and SNARE proteins, were identified in the plastid proteome [77]. As pointed out previously [102], the presence of a vesicular transport pathway from the Golgi to the plastid requires a retrograde vesicular transport pathway from the plastid back to the Golgi. Thus, the presence in the plastid of proteins involved in vesicular transport could be indicative of vesicular transport being the mechanism of protein import across the envelope membrane to the stroma or alternatively, indicative of a retrograde vesicular transport system from the plastid envelope back to the Golgi. Contrary to expectations, the preponderance of evidence clearly indicates the transport pathway across the *Euglena* and dinoflagellate envelope did not evolve from the TOC/TIC transport machinery. Further work is required to characterize the enigmatic components of the *Euglena* and dinoflagellate plastid protein import system and determine if they evolved from the simple chloroplast glycoprotein import system or from a yet uncharacterized import pathway.

8 Evolution of the Dinoflagellate and *Euglena* Plastid Targeting and Import System

8.1 *Evolution of the Dinoflagellate and* Euglena *Plastid Protein Presequence*

A probable evolutionary origin of the outermost *Euglena* and Dinoflagellate plastid envelope membrane is the phagosomal membrane that would have surrounded the phagocytosed green and red algal-derived endosymbiont [2]. The middle and inner membrane appear to have evolved from the outer and inner envelope membrane of the endosymbiont based on the middle membrane lipid composition [16]. Primary plastids contain glycoproteins which are transported from the Golgi apparatus to the plastid in vesicles just as proteins are transported from the Golgi apparatus to the phagosome [94–98]. While difficult to prove, an attractive hypothesis in terms of plastid evolution is that the mechanism that already existed for targeting proteins to phagosomes or glycoproteins to simple chloroplasts provided the basis for evolution of a mechanism for targeting nucleus encoded proteins to the *Euglena* and dinoflagellate plastid. Phagosome proteins and glycoproteins are synthesized by ER bound ribosomes and transported in vesicles from the ER to the Golgi apparatus and then to the phagosome or chloroplast. A signal peptide is required for co-translational ER localization [71]. Signals on the proteins interacting with membrane receptors sort proteins into vesicles that contain targeting proteins such as SNAREs ensuring they will dock and fuse with the phagosome or plastid delivering their cargo to the appropriate cellular location [88, 103, 104]. Thus, endosymbiont genes transferred to the host nucleus would have to acquire a presequence with both a signal peptide for targeting to the ER and sorting signals for incorporation into budding vesicles targeted to the endosymbiont within the phagosome. Upon vesicular fusion with and translocation through the outermost *Euglena* and dinoflagellate plastid envelope membrane derived from the phagosomal membrane, a transit peptide-dependent protein import system, possibly also used for glycoprotein import, could be used to transport proteins through the endosymbiont's double envelope membrane. The existing targeting system for delivery of glycoproteins to the simple plastid or proteins to the phagosome and the system for translocating glycoproteins through the outer and inner endosymbiont plastid envelope membranes could thus have been the basis for the evolution of a mechanism to import proteins into the *Euglena* and dinoflagellate plastids.

Based on these assumptions, the *Euglena* and dinoflagellate chloroplast protein presequences are proposed to have evolved through addition of an ER-targeting signal peptide to an endosymbiont derived presequence having a chloroplast targeting transit peptide, the subsequent evolution of additional structural/sequence information for vesicular cargo sorting to the chloroplast and import through the middle and inner envelope membranes. A clue to how this occurred comes from in vitro experimental studies of the functionality of the *Euglena* and dinoflagellate Class I presequence. When the presequence signal peptide was removed, the *Euglena* pLHCPII and the dinoflagellate pPCP-LUC fusion protein containing only the presequence transit peptide domain including the hydrophobic stop-transfer membrane anchor sequence were still imported into canine microsomes (Fig. 3a middle) [22, 73]. The presence of the stop-transfer sequence in the presequence transit peptide domain did not prevent the transit peptide domain-dependent in vitro import of Euglena pRbcS into pea chloroplasts [81], the in vivo import of the *Gonyaulax* pPCP-LUC fusion protein lacking the N-terminal signal peptide domain into *Solanum chacoense* chloroplasts [22] and the in vivo import of a truncated *Heterocapsa* pPsbO containing the region C-terminal to the signal peptide hydrophobic domain including the stop transfer sequence into pea chloroplast [34]. Taken together, these experiments demonstrated that the Class I presequence transit peptide domain is bifunctional, functioning as a transit peptide and, due to the presence of the hydrophobic stop transfer sequence, a signal peptide.

The first step in the evolution of the Class I *Euglena* and dinoflagellate plastid targeting presequence may not have been the addition of a signal peptide domain to the endosymbiont transit peptide presequence but rather the addition of a hydrophobic stop-transfer membrane anchor domain to the existing endosymbiont transit peptide presequence. This would create a bifunctional presequence containing an ER targeting, but uncleaved, signal peptide and a chloroplast targeting transit peptide. Subsequent Class I presequence evolution resulted in acquisition of a distinct N-terminal cleaved signal peptide optimizing ER targeting, an optimized transit peptide for import across the middle and inner plastid envelope membranes, and the addition to the presequence of vesicular protein sorting sequences for incorporation of chloroplast protein precursors into budding vesicles that will fuse with the outermost plastid envelope membrane. The sorting information remains to be identified but is proposed to be within the transit peptide domain. This would be fundamentally different from the vesicular sorting chloroplast import signal which is in the mature glycoprotein targeted to simple chloroplasts [98, 105]. In this scenario, evolution of the Class II presequence simply involved the loss of the stop-transfer hydrophobic region within the transit peptide domain. Precursors would be transported into the ER lumen and could be glycosylated, form disulfide bonds, and fold. If these modifications were required for protein function as found for some glycoproteins transported from the ER to simple chloroplasts [94], the dependence of function on these post translational modifications would provide a selection pressure for the loss of the second hydrophobic domain creating a Class II presequence. It is also possible that some precursors with Class II presequences are imported more efficiently due to obviating the need for precursor removal from and transport through the outer envelope membrane providing a selection pressure for Class II presequences independent of protein post translational modifications. The fact that the entire Class II precursors are within the transport vesicle supports the proposal that the transit peptide presequence domain contains the information for precursor sorting into Golgi to Chloroplast transport vesicles.

The signal peptide and stop-transfer sequence could have been acquired de novo through mutations that extend the reading frame or the insertion of genetic elements from a variety of sources. Alternatively, these could have been acquired from other endomembrane-targeted proteins through events such as unequal crossing over, exon shuffling, or integration of the transferred gene into a gene already possessing an ER-targeting signal peptide. Once effective targeting sequences arose, these could spread to other plastid proteins in a similar fashion. Evidence for the acquisition of chloroplast targeting presequences via crossing over or exon shuffling was obtained from the *Euglena* EST project where the sequence similarity of the presequences of several unrelated chloroplast proteins were unusually high, and in one pair, nearly identical [72]. This suggests that the spread of targeting information between genes was common and an effective way to acquire targeting information.

8.2 *Why Did Multiple Classes of* Euglena *and Dinoflagellate Plastid Targeting Presequences Requiring Two Envelope Translocation Systems Evolve?*

The presence of *Euglena* and dinoflagellate chloroplast-localized proteins with different N-terminal presequences requiring different envelope translocation systems (Fig. 4) suggests that there are selective advantages that maintain the two presequence classes. The stop-transfer sequence in Class I presequences prevents the protein from being translocated into the ER lumen, anchoring the protein in the membrane [68, 70, 73]. The lack of a stop-transfer sequence in the Class II presequences, barring a substantial hydrophobic domain in the mature protein, allows the protein to be translocated through the ER membrane becoming a soluble ER luminal protein (Fig. 4) that could fold, form disulfide bonds, and be glycosylated [92, 93], making it too large to be translocated through the translocon used to translocate the unfolded proteins with a Class I presequence.

Fig. 4 (continued) potentially at a contact site. How this occurs is unknown, but likely involves a series of chaperonins that mediate the folding-unfolding process that allows the protein to be transport competent. Class II proteins dissociate from the membrane localized transit peptide receptor being released into the space between the outer membrane (membrane 3) middle membrane (membrane 2). The transit peptide interacts with a transit peptide receptor on the middle membrane translocator and is imported through unidentified middle and inner membrane translocators at a contact site. The translocators may have evolved from the unidentified translocators in simple chloroplasts that import glycoproteins. Once in the stroma, the Class I and Class II presequences would be removed by a stromal peptidase and the protein targeted to its final destination. (Created with BioRender.com)

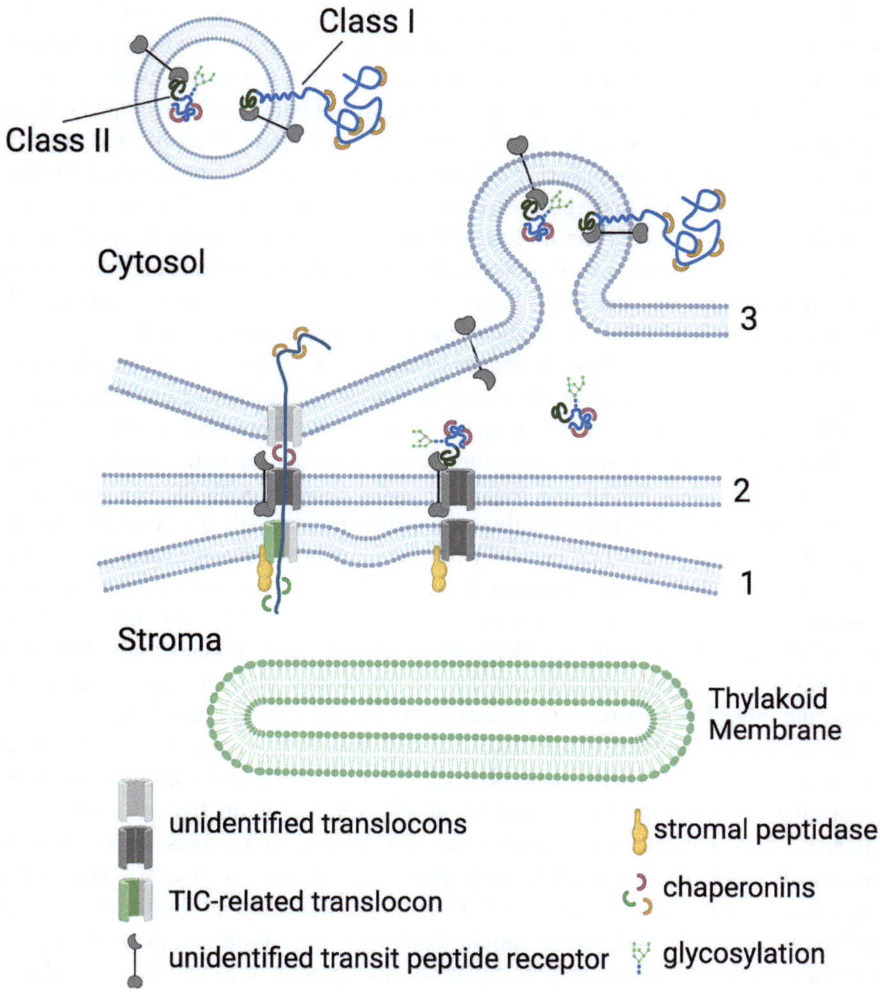

Fig. 4 Proposed model for the localization of nuclear encoded proteins to the plastids of euglenophytes and dinoflagellates. Cytoplasmically synthesized plastid proteins are co-translationally translocated into the ER and targeted to the chloroplast in vesicles that bud from the Golgi Apparatus. The panel shows a vesicle derived from the Golgi, containing proteins with either a Class I or Class II presequence. Within the vesicle, the Class I and Class II presequences interact with the luminal domain of the unidentified transit peptide receptor. The Class I precursors are integral vesicle membrane proteins with the presequence in the vesicle lumen and the majority of the protein in the cytoplasm maintained in an unfolded state by chaperones. Class II presequence-containing proteins are in a folded state entirely inside the vesicle. Upon fusion of the vesicle with the outer membrane of the plastid (membrane 3), proteins with Class I presequences are integral outer membrane proteins. The transit peptide domain in the outer membrane middle membrane (membrane 2) intermembrane space interacts with a transit peptide receptor on the middle membrane translocator. The precursor migrates laterally out of the outer membrane into the unidentified outer membrane translocator, is imported into the stroma through an unidentified middle membrane translocator and the inner membrane translocator that contains some TIC homologous subunits,

The chloroplast precursor proteins with Class I presequences would be transported in vesicles as integral membrane proteins from the ER to the Golgi apparatus and then to the chloroplast (Fig. 4). Upon fusion of the transport vesicle with the chloroplast, the Class I presequence precursors would be integral envelope proteins that must be extracted from the outer envelope membrane and translocated through all three envelope membranes (Fig. 4). As these proteins have not entered the ER lumen, they would not undergo post-translational modifications that increase their size. Translocation across the middle and inner membranes could be mechanistically similar to the transport of proteins into simple chloroplasts using translocons whose proteins include the TIC 32 and TIC 21 homologues found in *Euglena* [77, 78, 102] and the TIC 110 homologue found in dinoflagellates [33, 34].

Precursors with Class II presequences would be soluble proteins within the transport vesicle (Fig. 4). Fusion of the transport vesicle with the outer plastid envelope would release the precursor to the space between the outer and middle envelope membranes effectively translocating the precursor across the outer envelope membrane. Transport into the stroma requires translocation through the two additional envelope membranes in a process that is probably mechanistically simpler than the necessity to extract Class I precursors from the outermost envelope membrane prior to translocation through the three envelope membranes. The identity of the translocons for translocation of Class II precursors through the middle and inner envelope membranes is unknown but it is likely the inner membrane translocon is related to the translocon in simple chloroplasts used to translocate glycoproteins, folded proteins, and other ER modified proteins through the inner plastid membrane.

An important question is why two different targeting presequences and import mechanisms evolved. There are no obvious differences in protein structure or inter-chloroplast localization that could explain why some proteins have Class I presequences, that prevent them entering the ER lumen, and others have Class II presequences, that allow them to enter the lumen of the ER. The ER and Golgi apparatus are the site of a variety of co- and post-translational modifications that include protein N-linked glycosylation, disulfide bond formation, proteolytic pre-protein processing, O-linked glycosylation, and protein folding [92, 93]. These modifications could impart a strong selection pressure for or against entry of specific proteins into the ER. For instance, certain post-translational modifications could impair protein function or prevent protein folding in the absence of specific co-factors/subunits that may be in the chloroplast. For instance, using the GlycoEP prediction program for N-linked glycosylation (http://www.imtech.res.in/raghava/glycoep/), 44% of the *Euglena* proteins with a Class I presequence had a predicted N-linked glycosylation site while only 7% of the proteins with a Class II presequence had a predicted N-linked glycosylation site. While this analysis is flawed by the small sample size and the fact that not all of the proteins in the dataset were full-length (thus missing potential sites), it provides evidence supporting the hypothesis that some Class I presequences containing a stop-transfer membrane anchor sequence evolved to prevent co-translational transport of the precursor into the ER where N-linked glycosylation of plastid proteins and other inactivating modifications could occur. Of course, a Class II presequence could be positively selected for if certain post-translational modifications in the ER lumen are required for proper folding or activity once they reach their chloroplast destination.

One of the most abundant *Euglena* chloroplast proteins, the thylakoid integral membrane protein LHCPII, contains a potential N-linked glycosylation site eight amino acids from the N-terminus in front of the first hydrophobic membrane spanning domain [73]. Glycosylation of LHCPII would clearly interfere with the thylakoid membrane insertion of LHCPII and thus the assembly of the LHC. In vitro studies of LHCPII co-translational translocation into canine microsomes [73] (Fig. 3b) have provided direct evidence that this site would be co-translationally glycosylated during translocation into the ER and the stop-transfer membrane anchor sequence prevents translocation of this site into the ER, preventing its N-linked glycosylation. This finding provides experimental support for the hypothesis that Class I presequences evolved to prevent modifications such as N-linked glycosylation of chloroplast proteins.

Euglena chloroplast proteins with a Class II presequence are co-translationally imported into the ER due to the lack of a stop-transfer membrane anchor sequence in the presequence. If these proteins contain an N-linked glycosylation site they will be glycosylated. *Euglena*, plant, and diatom plastids contain glycoproteins that are glycosylated in the ER (plants and *Euglena*) or CER (diatoms) and glycosylation does not prevent their import into the plastid stroma [94, 95, 106, 107]. The activity of some proteins in simple chloroplasts is dependent upon N-linked glycosylation [94] providing a strong selection pressure for evolution of Class II presequences.

In addition to N-linked glycosylation, there are other post-translational modifications that can occur in the endomembrane system. Any of these modifications could render a protein non-functional providing a selection pressure for maintenance of Class I presequences or the modifications may be required for protein function providing a selection pressure for the evolution of Class II presequences. The point being that there may not be a single biochemical rational for the existence of Class I and Class II presequences but rather a variety of explanations that are specific to the transported protein. For some proteins, there may be no effect of ER modifications on protein function and thus no selection pressure for a Class I or Class II presequence. What if any role post translational endomembrane protein modifications play in protein function and thus a basis for the presence of a Class I or Class II presequence cannot be assessed solely from sequence data but will require biochemical studies of the modified and unmodified protein.

8.3 A Working Model for Localization of Nuclear Encoded Proteins to Euglena *and Dinoflagellate Chloroplasts*

Based on in vivo trafficking studies, in vitro chloroplast import studies, the structure of the chloroplast targeting presequence and the chloroplast proteome, a working model that illustrates a similar solution to a common import challenge has been developed for protein import into *Euglena* and dinoflagellate complex plastids (Fig. 4). When plastid proteins are translated and the N-terminal signal sequence emerges from the ribosome, it interacts with the signal recognition particle, binds to the ER membrane forming a membrane bound ribosome. There, the protein begins to be co-translationally transported into the ER membrane where the signal peptide is cleaved, leaving the remainder of the presequence with the transit peptide at the N-terminus. Proteins with a Class I presequence contain a stop transfer sequence that stops precursor translocation into the ER lumen and inserts the protein into the ER membrane oriented with the presequence in the ER lumen and the majority of the protein in the cytoplasm. Presumably, chaperonins would maintain the cytoplasmic portion of the protein in a transport-competent, unfolded state. Proteins with a Class II presequence that lack the stop-transfer domain are co-translationally translocated into the ER lumen where they fold and are potentially modified by glycosylation or disulphide bond formation. The potential of such modifications to impact function and ability of the protein to fold or assemble into higher order complexes may be a selective force determining whether a protein has a Class I or Class II presequence.

Vesicles bud from the ER membrane transporting the chloroplast precursor proteins to the Golgi apparatus. Proteins with a Class I presequence are transported as integral vesicle membrane proteins with the majority of the chaperone bound unfolded protein in the cytoplasm while proteins with Class II presequences are transported as folded, luminal vesicle proteins. Upon vesicle fusion with the Golgi membrane, the Class I proteins are integral membrane proteins with the presequence in the lumen and the majority of the chaperone bound unfolded protein in the cytoplasm while folded Class II proteins are lumenal ER proteins.

The transit peptide of both Class I and Class II presequences is exposed to the Golgi lumen where it interacts with an unidentified transit peptide receptor sorting the proteins into vesicles that are ultimately targeted to the plastid outer envelope (Fig. 4). Through interaction with chaperones, the cytoplasmic portion of proteins with a class I presequence remains in an unfolded state. Upon vesicle fusion with the outer plastid membrane (membrane 3, Fig. 4), proteins with Class I presequences are integral outer membrane proteins that must migrate from the outer membrane into the outer membrane translocation channel, perhaps mediated by the transit peptide receptor (Fig. 4). Presumably, once the protein is loading into the outer membrane translocon, the translocation across the outer membrane would resume. This could allow recognition and insertion of the plastid protein into the unidentified translocons in the second and first membrane (membranes 2 and 1, Fig.4), possibly at contact sites as seen in simple chloroplasts (Fig. 4). Both *Euglena* [77, 78, 102] and dinoflagellates [33, 34] lack TOC but contain some, but not all, homologues of TIC translocon proteins which may be part of the unidentified inner membrane translocon which functions in a similar fashion to the classic TIC transporter found in primary plastids (Fig. 4).

After vesicle fusion with the outer chloroplast membrane and dissociation from the integral membrane transit peptide receptor, Class II presequence-containing proteins would be in the intermembrane space between the outer membrane

(membrane 3) and middle membrane (membrane 2). The presequence transit peptide would bind the transit peptide receptor on the middle membrane (membrane 2) and be translocated into the stroma through the remaining middle and inner membranes (membrane 1 Fig.4) via unidentified translocons, perhaps at contact sites (Fig. 4).

Proteins with Class II presequences have been in the ER and Golgi lumen and could have substantial post-translational modifications, like glycosylation (illustrated in Fig. 4), disulfide bonds, and be folded, resulting in an increased size. They may be too large to move through TOC/TIC-like membrane transporters which would be suitable for Class I proteins. Simple chloroplasts contain glycoproteins which are imported through yet unidentified translocons composed of unidentified proteins. The Class II transporters may be similar to the unidentified transporters used for glycoprotein import into simple chloroplasts or they may be unique to chloroplasts with three envelope membranes. Not shown in this model is an hypothesis for how any plastid protein that lacks the canonical presequence structure [77] is imported into the plastid, which may represent an entirely different approach to plastid import. Once in the plastid, the stomal peptidase would remove the transit peptide presequence domain and the proteins are targeted to their correct intraplastid location.

9 Conclusions

Simple chloroplasts have two envelope membranes and arose through a singular primary endosymbiotic association between a eukaryotic host and a prokaryotic cyanobacterial ancestor [7, 8]. Complex chloroplasts contain three or four envelope membranes and evolved multiple times from secondary endosymbiosis between a photosynthetic eukaryotic endosymbiont and a eukaryotic host [7, 8]. The transition from an endosymbiont to a chloroplast involved massive horizontal gene transfer from the endosymbiont to the host cell nucleus, necessitating evolution of a mechanism to return the encoded proteins to the endosymbiont as it evolved into a plastid. The plastid envelope membranes presented a barrier to this transfer necessitating the evolution of a mechanism to target the proteins to the envelope and translocate them through the multiple envelope membranes into the stroma.

The problem of returning nuclear encoded proteins to simple chloroplasts was solved by evolution of the chloroplast transit peptide and the TOC/TIC translocons for protein translocation through the outer and inner envelope membranes [17, 101]. In the case of the endosymbionts that evolved into the *Euglena* and dinoflagellate plastids, this targeting system would have been well established. However, during its evolution, the endosymbiont would have been encapsulated by a phagosomal membrane, ultimately forming the outer membrane of the complex chloroplasts with three envelope membranes [2] as all the other cellular components of the endosymbiont were lost. This would leave the chloroplast at least functionally a part of the endomembrane system. Proteins targeted to the phagosome are synthesized on

ER membrane bound ribosomes and transported in vesicles from the ER to the Golgi apparatus to the phagosome. Therefore, as endosymbiont genes were transferred to the nucleus, the first step in targeting the proteins back to the *Euglena* or dinoflagellate plastid through the endomembrane system was likely the addition of a hydrophobic domain (a signal sequence) to the transit peptide presequence used for targeting proteins to the simple chloroplasts of the endosymbiont. This would also require modifications and neofunctionalization of existing translocation proteins to adapt to the unique challenges of targeting through the endomembrane system and traversing the additional membranes of the *Euglena* and dinoflagellate plastids. It is still uncanny that these quite different groups seemed to have stumbled upon similar solutions to the protein import problem. As pointed out previously, it was the number and origin of plastid envelope membranes [22] in euglenophytes and dinoflagellates that was possibly the driving force for the convergent evolution of their similar protein targeting translocation system [22] and many of the other features shared by these organisms [12].

Despite the important progress to date, there is much we don't know about the transport process into these complex plastids and the proteins involved. Identifying the presequence transit peptide receptor(s), the Golgi vesicle sorting sequences, the vesicle targeting sequences, and the membrane translocons will not only be vital to understanding this complex targeting and import mechanism but will allow us to understand how these two distinct lineages converged onto a similar solution for getting nucleus-encoded proteins to their proper location in the plastid. It very well could be as previously suggested, there is indeed a limited number of evolutionary solutions for specific biochemical problems [12].

References

1. Gibbs SP (1970) Comparative ultrastructure of algal chloroplast. Ann N Y Acad Sci 175(2):454–473. https://doi.org/10.1111/j.1749-6632.1970.tb45167.x
2. Gibbs SP (1978) The chloroplasts of *Euglena* may have evolved from symbiotic green algae. Can J Bot 56:2883–2889. https://doi.org/10.1139/b78-345
3. Gibbs SP (1981) The chloroplasts of some algal groups may have evolved from endosymbiotic eukaryotic algae. Ann NY Acad Sci 361:193–207. https://doi.org/10.1111/j.1749-6632.1981.tb54365.x
4. Gibbs SP (1981) The chloroplast endoplasmic reticulum: structure, function and evolutionary significance. Int Rev Cytol 72:49–99. https://doi.org/10.1016/S0074-7696(08)61194-8
5. Lefortran M, Pineau B (1980) Structure and functional-organization of chloroplastic membranes envelope in *Euglena gracilis*. Eur J Cell Biol 22(1):279–279
6. Whatley JM, Whatley FR (1981) Chloroplast evolution. New Phytol 87(2):233–247. https://doi.org/10.1111/j.1469-8137.1981.tb03195.x
7. Cavalier-Smith T (2003) Genomic reduction and evolution of novel genetic membranes and protein-targeting machinery in eukaryote-eukaryote chimaeras (meta-algae). Philos Trans R Soc Lond B Biol Sci 358(1429):109–133. https://doi.org/10.1098/2Frstb.2002.1194
8. Keeling PJ (2013) The number, speed, and impact of plastid endosymbioses in eukaryotic evolution. Annu Rev Plant Biol 64:583–607. https://doi.org/10.1146/annurev-arplant-050312-120144

9. Cavalier-Smith T (1999) Principles of protein and lipid targeting in secondary symbiogenesis: euglenoid, dinoflagellate, and sporozoan plastid origins and the eukaryote family tree. J Eukaryot Microbiol 46(4):347–366. https://doi.org/10.1111/j.1550-7408.1999.tb04614.x
10. Vesteg M, Vacula R, Krajcovic J (2009) On the origin of chloroplasts, import mechanisms of chloroplast-targeted proteins, and loss of photosynthetic ability. Folia Microbiol (Praha) 54(4):303–321. https://doi.org/10.1007/s12223-009-0048-z
11. Janouskovec J et al (2010) A common red algal origin of the apicomplexan, dinoflagellate, and heterokont plastids. Proc Natl Acad Sci U S A 107(24):10949–10954. https://doi.org/10.1073/pnas.1003335107
12. Lukes J, Leander BS, Keeling PJ (2009) Cascades of convergent evolution: the corresponding evolutionary histories of euglenozoans and dinoflagellates. Proc Natl Acad Sci U S A 106(Suppl 1):9963–9970. https://doi.org/10.1073/pnas.0901004106
13. Obornik M et al (2009) Evolution of the apicoplast and its hosts: from heterotrophy to autotrophy and back again. Int J Parasitol 39(1):1–12. https://doi.org/10.1016/j.ijpara.2008.07.010
14. Simpson AG, Stevens JR, Lukes J (2006) The evolution and diversity of kinetoplastid flagellates. Trends Parasitol 22(4):168–174. https://doi.org/10.1016/j.pt.2006.02.006
15. Vesteg M et al (2019) Comparative molecular cell biology of phototrophic euglenids and parasitic trypanosomatids sheds light on the ancestor of Euglenozoa. Biol Rev Camb Philos Soc 94(5):1701–1721. https://doi.org/10.1111/brv.12523
16. Tomeckova L et al (2020) The lipid composition of *Euglena gracilis* middle plastid membrane resembles that of primary plastid envelopes. Plant Physiol 184(4):2052–2063. https://doi.org/10.1104/pp.20.00505
17. Bruce BD (2000) Chloroplast transit peptides: structure, function and evolution. Trends Cell Biol 10(10):440–447. https://doi.org/10.1016/s0962-8924(00)01833-x
18. Osafune T, Schiff JA, Hase E (1990) Immunogold localization of LHCPII apoprotein in the Golgi of *Euglena*. Cell Struct Funct 15(2):99–105. https://doi.org/10.1016/0014-4827(91)90103-2
19. Osafune T, Schiff JA, Hase E (1991) Stage-dependent localization of LHCP II apoprotein in the Golgi of synchronized cells of *Euglena gracilis* by immunogold electron microscopy. Exp Cell Res 193(2):320–330. https://doi.org/10.1016/0014-4827(91)90103-2
20. Osafune T et al (1991) Immunolocalization of LHCPII apoprotein in the Golgi during light-induced chloroplast development in non-dividing *Euglena* cells. J Electron Microsc (Tokyo) 40:41–47
21. Schiff JA et al (1991) Photocontrol and processing of LHCPII apoprotein in *Euglena*: possible role of Golgi and other cytoplasmic sites. J Photchem Photobiol B Biol 11:219–236. https://doi.org/10.1016/1011-1344(91)80262-g
22. Nassoury N, Cappadocia M, Morse D (2003) Plastid ultrastructure defines the protein import pathway in dinoflagellates. J Cell Sci 116(Pt 14):2867–2874. https://doi.org/10.1242/jcs.00517
23. Chan RL et al (1990) Eight small subunits of Euglena ribulose 1–5 bisphosphate carboxylase/oxygenase are translated from a large mRNA as a polyprotein. EMBO J 9(2):333–338. https://doi.org/10.1002/j.1460-2075.1990.tb08115.x
24. Henze K et al (1995) A nuclear gene of eubacterial origin in *Euglena gracilis* reflects cryptic endosymbioses during protist evolution. Proc Natl Acad Sci U S A 92(20):9122–9126. https://doi.org/10.1073/2Fpnas.92.20.9122
25. Houlne G, Schantz R (1987) Molecular analysis of the transcripts encoding the light-harvesting chlorophyll a/b protein in *Euglena gracilis*: unusual size of the mRNA. Curr Genet 12(8):611–616. https://doi.org/10.1007/bf00368064
26. Kishore R, Muchhal U, Schwartzbach SD (1993) The presequence of Euglena LHCPII, a cytoplasmically synthesized chloroplast protein, contains a functional endoplasmic reticulum targeting domain. Proc Natl Acad Sci U S A 90:11845–11849. https://doi.org/10.1073/pnas.90.24.11845
27. Lin Q et al (1994) Isolation and characterization of cDNA clones for chloroplast translational initiation factor-3 from *Euglena gracilis*. J Biol Chem 269(24):16984

28. Plaumann M et al (1997) Multiple recruitment of class-I aldolase to chloroplasts and eubacterial origin of eukaryotic class-II aldolases revealed by cDNAs from *Euglena gracilis*. Curr Genet 31(5):430–438. https://doi.org/10.1007/s002940050226
29. Santillan Torres JL et al (2003) Cytochrome f and subunit IV, two essential components of the photosynthetic bf complex typically encoded in the chloroplast genome, are nucleus-encoded in *Euglena gracilis*. Biochim Biophys Acta 1604(3):180–189. https://doi.org/10.1016/s0005-2728(03)00058-6
30. Sharif AL, Smith AG, Abell C (1989) Isolation and characterisation of a cDNA clone for a chlorophyll synthesis enzyme from *Euglena gracilis*. The chloroplast enzyme hydroxymethylbilane synthase (porphobilinogen deaminase) is synthesised with a very long transit peptide in Euglena. Eur J Biochem 184(2):353–359. https://doi.org/10.1111/j.1432-1033.1989.tb15026.x
31. Shigemori Y et al (1994) The presequence of the precursor to the nucleus-encoded 30 kDa protein of photosystem II in *Euglena gracilis* Z includes two hydrophobic domains. Plant Mol Biol 24(1):209–215. https://doi.org/10.1007/bf00040587
32. Vacula R et al (1999) Nucleus-encoded precursors to thylakoid lumen proteins of *Euglena gracilis* possess tripartite presequences. DNA Res 6(1):45–49. https://doi.org/10.1093/dnares/6.1.45
33. Hehenberger E et al (2014) Evidence for the retention of two evolutionary distinct plastids in dinoflagellates with diatom endosymbionts. Genome Biol Evol 6(9):2321–2334. https://doi.org/10.1093/gbe/evu182
34. Chaal BK, Green BR (2005) Protein import pathways in 'complex' chloroplasts derived from secondary endosymbiosis involving a red algal ancestor. Plant Mol Biol 57(3):333–342. https://doi.org/10.1007/s11103-004-7848-y
35. Patron NJ et al (2005) Complex protein targeting to dinoflagellate plastids. J Mol Biol 348(4):1015–1024. https://doi.org/10.1016/j.jmb.2005.03.030
36. Sibbald SJ, Archibald JM (2020) Genomic Insights into Plastid Evolution. Genome Biol Evol 12(7):978–990. https://doi.org/10.1093/gbe/evaa096
37. Yoon HS et al (2005) Tertiary endosymbiosis driven genome evolution in dinoflagellate algae. Mol Biol Evol 22(5):1299–1308. https://doi.org/10.1093/molbev/msi118
38. Saldarriaga JF et al (2001) Dinoflagellate nuclear SSU rRNA phylogeny suggests multiple plastid losses and replacements. J Mol Evol 53(3):204–213. https://doi.org/10.1007/s002390010210
39. Waller RF, Kořený L (2017) Plastid complexity in Dinoflagellates: a picture of gains, losses, replacements and revisions. Adv Botanic Res 84:105–143. https://doi.org/10.1007/s002390010210
40. Matsumoto T et al (2011) Green-colored plastids in the dinoflagellate genus *Lepidodinium* are of core chlorophyte origin. Protist 162(2):268–276. https://doi.org/10.1016/j.protis.2010.07.001
41. Shalchian-Tabrizi K et al (2006) Combined heat shock protein 90 and ribosomal RNA sequence phylogeny supports multiple replacements of dinoflagellate plastids. J Eukaryot Microbiol 53(3):217–224. https://doi.org/10.1111/j.1550-7408.2006.00098.x
42. Hansen G, Botes L, Salas D (2007) Ultrastructure and large subunit rDNA sequences of Lepidodinium viride reveal a close relationship to *Lepidodinium chlorophorum* comb. nov. (=Gymnodinium chlorophorum). Phycol Res 55:25–41. https://doi.org/10.1111/j.1440-1835.2006.00442.x
43. Minge MA et al (2010) A phylogenetic mosaic plastid proteome and unusual plastid-targeting signals in the green-colored dinoflagellate *Lepidodinium chlorophorum*. BMC Evol Biol 10:191. https://doi.org/10.1186/1471-2148-10-191
44. Koziol AG, Durnford DG (2008) *Euglena* light-harvesting complexes are encoded by multifarious polyprotein mRNAs that evolve in concert. Mol Biol Evol 25(1):92–100. https://doi.org/10.1093/molbev/msm232
45. Muchhal US, Schwartzbach SD (1992) Characterization of a *Euglena* gene encoding a polyprotein precursor to the light-harvesting chlorophyll a/b-binding protein of photosystem II. Plant Mol Biol 18(2):287–299. https://doi.org/10.1007/bf00034956

46. Hiller RG, Wrench PM, Sharples FP (1995) The light-harvesting chlorophyll a-c-binding protein of dinoflagellates: a putative polyprotein. FEBS Lett 363(1–2):175–178. https://doi.org/10.1016/0014-5793(95)00297-m
47. Maruyama S et al (2015) Diversification of the light-harvesting complex gene family via intra- and intergenic duplications in the coral symbiotic alga Symbiodinium. PloS One 10(3):e0119406. https://doi.org/10.1371/2Fjournal.pone.0119406
48. Rowan R et al (1996) Rubisco in marine symbiotic dinoflagellates: form II enzymes in eukaryotic oxygenic phototrophs encoded by a nuclear multigene family. Plant Cell 8(3):539–553. https://doi.org/10.1105/tpc.8.3.539
49. Enomoto T, Sulli C, Schwartzbach SD (1997) A soluble chloroplast protease processes the *Euglena* polyprotein precursor to the light harvesting chlorophyll a/b binding protein of Photosystem II. Plant Cell Physiol 38:743–746
50. Graham RW, Jones D, Candido EP (1989) UbiA, the major polyubiquitin locus in *Caenorhabditis elegans*, has unusual structural features and is constitutively expressed. Mol Cell Biol 9(1):268–277. https://doi.org/10.1074/jbc.270.22.13084
51. Leander BS et al (2017) Euglenida. In: Archibald JM, Simpson AGB, Slamovits CH (eds) Handbook of the protists. Springer International Publishing, Cham, pp 1047–1088
52. McWatters DC, Russell AG (2017) Euglena transcript processing. Adv Exp Med Biol 979:141–158. https://doi.org/10.1007/978-3-319-54910-1_8
53. Zhang H et al (2007) Spliced leader RNA trans-splicing in dinoflagellates. Proc Natl Acad Sci U S A 104(11):4618–4623. https://doi.org/10.1073/pnas.0700258104
54. Kishore R, Schwartzbach SD (1992) Translational control of the synthesis of the *Euglena* light harvesting chlorophyll a/b binding protein of photosystem II. Plant Sci 85:79–89. https://doi.org/10.1016/0168-9452(92)90096-5
55. Monroy AF, McCarthy SA, Schwartzbach SD (1987) Evidence for translational regulation of chloroplast and mitochondrial biogenesis in *Euglena*. Plant Sci 51:61–76. https://doi.org/10.1016/0168-9452(87)90221-4
56. Schwartzbach SD (2017) Photo and nutritional regulation of *Euglena* organelle development. Adv Exp Med Biol 979:159–182. https://doi.org/10.1007/978-3-319-54910-1_9
57. Rikin A, Schwartzbach S (1989) Translational regulation of the synthesis of *Euglena* fumarase by light and ethanol. Plant Physiol 90:63–69. https://doi.org/10.1104/pp.90.1.63
58. Morse D et al (1989) Circadian regulation of bioluminescence in *Gonyaulax* involves translational control. Proc Natl Acad Sci U S A 86(1):172–176. https://doi.org/10.1073/pnas.86.1.172
59. Roy S, Jagus R, Morse D (2018) Translation and translational control in Dinoflagellates. Microorganisms 6(2):30. https://doi.org/10.3390/2Fmicroorganisms6020030
60. Burki F (2014) The eukaryotic tree of life from a global phylogenomic perspective. Cold Spring Harb Perspect Biol 6(5):a016147. https://doi.org/10.1101/cshperspect.a016147
61. Brandt P, von Kessel B (1983) Cooperation of cytoplasmic and plastidial translation in formation of the photosynthetic apparatus and its stage-specific efficiency. Plant Physiol 72(3):616–619. https://doi.org/10.1104/pp.72.3.616
62. Schwartzbach SD, Osafune T, Löffelhardt W (1998) Protein import into cyanelles and complex chloroplasts. Plant Mol Biol 38:247–263
63. Rikin A, Schwartzbach SD (1989) Regulation by light and ethanol of the synthesis of the light harvesting chlorophyll a/b binding protein of photosystem II in Euglena. Planta 178:76–83. https://doi.org/10.1007/bf00392529
64. Spano AJ, Ghaus H, Schiff JA (1987) Chlorophyll-protein complexes and other thylakoid components at the low intensity threshold in euglena chloroplast development. Plant Cell Physiol 28(6):1101–1108. https://doi.org/10.1093/oxfordjournals.pcp.a077390
65. Houlne G, Schantz R (1988) Characterization of cDNA sequences for LHCI apoproteins in *Euglena gracilis*: the mRNA encodes a large precursor containing several consecutive divergent polypeptides. Mol Gen Genet 213(2–3):479–486. https://doi.org/10.1093/oxfordjournals.pcp.a077390

66. Muchhal US, Schwartzbach SD (1994) Characterization of the unique intron-exon junctions of *Euglena* gene(s) encoding the polyprotein precursor to the light-harvesting chlorophyll a/b binding protein of photosystem II. Nucleic Acids Res 22:5737–5744. https://doi.org/10.1093/nar/22.25.5737
67. Rikin A, Schwartzbach SD (1988) Extremely large and slowly processed precursors to the Euglena light harvesting chlorophyll a/b binding proteins of photosystem II. Proc Natl Acad Sci U S A 85:5117–5121. https://doi.org/10.1073/pnas.85.14.5117
68. Sulli C, Schwartzbach SD (1995) The polyprotein precursor to the Euglena light harvesting chlorophyll a/b-binding protein is transported to the Golgi apparatus prior to chloroplast import and polyprotein processing. J Biol Chem 270:13084–13090. https://doi.org/10.1074/jbc.270.22.13084
69. Keller M et al (1991) Post-transcriptional regulation by light of the biosynthesis of Euglena ribulose-1,5-bisphosphate carboxylase/oxygenase small subunit. Plant Mol Biol 17(1):73–82. https://doi.org/10.1016/2Fj.ceb.2018.02.011
70. Sulli C, Schwartzbach SD (1996) A soluble protein is imported into Euglena chloroplasts as a membrane-bound precursor. Plant Cell 8:43–53. https://doi.org/10.1105/2Ftpc.8.1.43
71. von Heijne G (1988) Transcending the impenetrable: how proteins come to terms with membranes. Biochim Biophys Acta 947:307–333. https://doi.org/10.1016/0304-4157(88)90013-5
72. Durnford DG, Gray MW (2006) Analysis of *Euglena gracilis* plastid-targeted proteins reveals different classes of transit sequences. Eukaryot Cell 5(12):2079–2091
73. Sulli C et al (1999) Topology of Euglena chloroplast protein precursors within endoplasmic reticulum to Golgi to chloroplast transport vesicles. J Biol Chem 274:457–463. https://doi.org/10.1128/ec.00222-06
74. Lee J, Kim DH, Hwang I (2014) Specific targeting of proteins to outer envelope membranes of endosymbiotic organelles, chloroplasts, and mitochondria. Front Plant Sci 5:173. https://doi.org/10.3389/fpls.2014.00173
75. Bassham DC et al (1991) Transport of proteins into chloroplasts. Delineation of envelope "transit" and thylakoid "transfer" signals within the pre-sequences of three imported thylakoid lumen proteins. J Biol Chem 266(35):23606–23610
76. Smeekens S et al (1986) The role of the transit peptide in the routing of precursors toward different chloroplast compartments. Cell 46(3):365–375. https://doi.org/10.1016/0092-8674(86)90657-4
77. Novak Vanclova AMG et al (2020) Metabolic quirks and the colourful history of the *Euglena gracilis* secondary plastid. New Phytol 225(4):1578–1592. https://doi.org/10.1111/nph.16237
78. Zahonova K et al (2018) Peculiar features of the plastids of the colourless alga *Euglena longa* and photosynthetic euglenophytes unveiled by transcriptome analyses. Sci Rep 8(1):17012. https://doi.org/10.1038/s41598-018-35389-1
79. Lee DW et al (2018) Prolines in transit peptides are crucial for efficient preprotein translocation into chloroplasts. Plant Physiol 176(1):663–677. https://doi.org/10.1104/pp.17.01553
80. Inagaki J et al (2000) Protein translocation within chloroplast is similar in *Euglena* and higher plants. Biochem Biophys Res Commun 277(2):436–442. https://doi.org/10.1006/bbrc.2000.3702
81. Slavikova S et al (2005) Homologous and heterologous reconstitution of Golgi to chloroplast transport and protein import into the complex chloroplasts of Euglena. J Cell Sci 118(Pt 8):1651–1661. https://doi.org/10.1242/jcs.02277
82. Kilian O, Kroth PG (2005) Identification and characterization of a new conserved motif within the presequence of proteins targeted into complex diatom plastids. Plant J 41(2):175–183. https://doi.org/10.1111/j.1365-313x.2004.02294.x
83. Gruber A et al (2007) Protein targeting into complex diatom plastids: functional characterisation of a specific targeting motif. Plant Mol Biol 64(5):519–530. https://doi.org/10.1007/s11103-007-9171-x
84. Gould SB et al (2006) Protein targeting into the complex plastid of cryptophytes. J Mol Evol 62(6):674–681. https://doi.org/10.1007/s00239-005-0099-y

85. Kroth PG, Schroers Y, Kilian O (2005) The peculiar distribution of class I and class II aldolases in diatoms and in red algae. Curr Genet 48(6):389–400. https://doi.org/10.1007/s00294-005-0033-2
86. Kuroiwa T et al (1991) Systematic analysis of stop-transfer sequence for microsomal membrane. J Biol Chem 266(14):9251–9255
87. Ko K, Cashmore AR (1989) Targeting of proteins to the thylakoid lumen by the bipartite transit peptide of the 33 kd oxygen-evolving protein. EMBO J 8(11):3187–3194. https://doi.org/10.1002/2Fj.1460-2075.1989.tb08477.x
88. Bonifacino JS, Glick BS (2004) The mechanisms of vesicle budding and fusion. Cell 116(2):153–166. https://doi.org/10.1016/s0092-8674(03)01079-1
89. Loewenstein T et al (2010) Textile-compatible substrate electrodes with electrodeposited ZnO--a new pathway to textile-based photovoltaics. ChemPhysChem 11(4):783–788. https://doi.org/10.1002/cphc.200900428
90. Paila YD, Richardson LGL, Schnell DJ (2015) New insights into the mechanism of chloroplast protein import and its integration with protein quality control, organelle biogenesis and development. J Mol Biol 427(5):1038–1060. https://doi.org/10.1016/j.jmb.2014.08.016
91. Theg SM et al (1989) Internal ATP is the only energy requirement for the translocation of precursor proteins across chloroplastic membranes. J Biol Chem 264(12):6730–6736. https://doi.org/10.1016/S0021-9258(18)83490-1
92. Burda P, Aebi M (1999) The dolichol pathway of N-linked glycosylation. Biochim Biophys Acta 1426(2):239–257. https://doi.org/10.1016/s0304-4165(98)00127-5
93. Stanley P (2011) Golgi glycosylation. Cold Spring Harb Perspect Biol 3(4):a005199. https://doi.org/10.1101/2Fcshperspect.a005199
94. Buren S et al (2011) Importance of post-translational modifications for functionality of a chloroplast-localized carbonic anhydrase (CAH1) in *Arabidopsis thaliana*. PloS One 6(6):e21021. https://doi.org/10.1371/journal.pone.0021021
95. Villarejo A et al (2005) Evidence for a protein transported through the secretory pathway en route to the higher plant chloroplast. Nat Cell Biol 7(12):1224–1231. https://doi.org/10.1371/journal.pone.0021021
96. Kaneko K et al (2016) N-glycomic and microscopic subcellular localization analyses of NPP1, 2 and 6 strongly indicate that trans-Golgi compartments participate in the Golgi to plastid traffic of nucleotide Pyrophosphatase/Phosphodiesterases in rice. Plant Cell Physiol 57(8):1610–1628. https://doi.org/10.1093/pcp/pcw089
97. Nanjo Y et al (2006) Rice plastidial N-glycosylated nucleotide pyrophosphatase/phosphodiesterase is transported from the ER-golgi to the chloroplast through the secretory pathway. Plant Cell 18(10):2582–2592. https://doi.org/10.1105/tpc.105.039891
98. Kitajima A et al (2009) The rice alpha-amylase glycoprotein is targeted from the Golgi apparatus through the secretory pathway to the plastids. Plant Cell 21(9):2844–2858. https://doi.org/10.1105/tpc.109.068288
99. Turmel M, Otis C, Lemieux C (2009) The chloroplast genomes of the green algae *Pedinomonas minor*, *Parachlorella kessleri*, and *Oocystis solitaria* reveal a shared ancestry between the Pedinomonadales and Chlorellales. Mol Biol Evol 26(10):2317–2331. https://doi.org/10.1093/molbev/msp138
100. Archibald JM (2015) Endosymbiosis and eukaryotic cell evolution. Curr Biol 25(19):R911–R921. https://doi.org/10.1016/j.cub.2015.07.055
101. Richardson LGL, Schnell DJ (2020) Origins, function, and regulation of the TOC-TIC general protein import machinery of plastids. J Exp Bot 71(4):1226–1238. https://doi.org/10.1093/jxb/erz517
102. Ebenezer TE et al (2019) Transcriptome, proteome and draft genome of *Euglena gracilis*. BMC Biol 17(1):11. https://doi.org/10.1186/s12915-019-0626-8
103. Braulke T, Bonifacino JS (2009) Sorting of lysosomal proteins. Biochim Biophys Acta 1793(4):605–614. https://doi.org/10.1016/j.bbamcr.2008.10.016

104. Neuhaus JM, Rogers JC (1998) Sorting of proteins to vacuoles in plant cells. Plant Mol Biol 38(1–2):127–144. https://doi.org/10.1023/A:1006032627036
105. Kaneko K et al (2011) Differential localizations and functions of rice nucleotide pyrophosphatase/phosphodiesterase isozymes 1 and 3. Plant Biotechnol 28(1):69–76. https://doi.org/10.5511/plantbiotechnology.10.1228a
106. O'Neill EC (2022) Glycosylated proteins in the protozoan alga *Euglena gracilis*: a proteomic approach. FEMS Microbiol Lett 370:fnac120. https://doi.org/10.1093/femsle/fnac120
107. Peschke M et al (2013) Evidence for glycoprotein transport into complex plastids. Proc Natl Acad Sci U S A 110(26):10860–10865. https://doi.org/10.1073/pnas.1301945110
108. Durnford DG, Schwartzbach SD (2017) Protein targeting to the plastid of Euglena. Adv Exp Med Biol 979:183–205. https://doi.org/10.1007/978-3-319-54910-1_10

Translocation of Proteins into Four Membrane-Bound Complex Plastids of Red Algal Origin

Ansgar Gruber and Peter G. Kroth

Abstract Plastids initially evolved via endosymbiotic uptake of cyanobacteria into a non-photosynthetic eukaryote, however, in the majority of algal groups, plastids have been indirectly acquired, via endosymbiotic uptake of photosynthetic eukaryotes into other eukaryotic cells. Several types of these so-called complex plastids exist, differing mainly in the phylogenies of the former endosymbionts, and in the plastid structure, especially in the numbers and topologies of envelope membranes surrounding the plastid. *Cryptophyta*, *Haptophyta,* and *Ochrophyta* (the photosynthetic members of *Stramenopiles*) possess plastids of a similar topology, with four membranes surrounding the stroma, and the outermost membrane being part of the eukaryotic endomembrane system. These additional membranes compared to plant plastids required the establishment of further transport systems of nucleus-encoded proteins into the organelles. In this chapter, we are reviewing the targeting signals and targeting mechanisms of proteins into the complex plastids of these algal groups, and discuss the physiological and evolutionary implications.

Keywords Algae · Organelle · Chloroplast · Endosymbiosis · Protein import · Evolution · Metabolism · Membranes · Endoplasmic reticulum · *Stramenopiles* · Diatom

A. Gruber
Institute of Parasitology, Biology Centre, Czech Academy of Sciences, České Budějovice, Czech Republic

Faculty of Science, University of South Bohemia, České Budějovice, Czech Republic
e-mail: Ansgar.Gruber@paru.cas.cz

P. G. Kroth (✉)
Department of Biology, Universität Konstanz, Konstanz, Germany
e-mail: Peter.Kroth@uni-konstanz.de

S. D. Schwartzbach et al. (eds.), *Endosymbiotic Organelle Acquisition*,
https://doi.org/10.1007/978-3-031-57446-7_15

1 Introduction

The term "algae" comprises aquatic organisms that primarily are photosynthetic and that possess photosystems that are similar both to those of cyanobacteria (from which eukaryotic algae obtained their photosystems) and to those of land plants (which are descendants of green algae). One reason why the concept of "algae" does not cover a phylogenetically consistent group is that several endosymbiotic processes involving different host cells led to the development of these organisms [1–3]. Obvious examples for results of endosymbiotic processes are organelles like plastids and mitochondria that show high similarities to their bacterial ancestors, both on the physiological and genetic levels [4, 5]. Mitochondria evolved early in eukaryote evolution, so early that their appearance might mark the origin of eukaryotes, and it is not clear whether their host ancestor, if investigated today, would be classified as a eukaryote [6]. What appears clear is that *Alphaproteobacteria*, among the extant bacterial groups, are the ones most closely related to the symbiont ancestor of mitochondria, and that the host ancestor was from the group of *Archaea* [6].

Plastids, however, evolved later than mitochondria, and the host of this endosymbiosis must unambiguously have been a eukaryote, while the symbiont was from the group of *Cyanobacteria* [7, 8]. Plastids that can be directly traced back to this initial endosymbiosis are called primary plastids. In evolutionary terms, with the newly acquired photosynthetic abilities, the first eukaryotic algae were able to spread and diversify in the aquatic systems, and in some cases became land-based, living, and reproducing outside of the aquatic systems [9].

In principle, the uptake of cells into other cells occurs frequently in nature. Such endosymbioses may either have a mutualistic symbiotic background, like it has been shown in several cases like in ciliates that take up algae [10], or they may be a form of parasitism of the endosymbionts on the host, like in the bacterial group of *Chlamydiae* [11]. Furthermore, they can be based on phagotrophic behavior of the host organisms, like in sea slugs [12], *Dinoflagellata* [13], or other organisms feeding on algal cells taking up algal cells as food. These latter uptakes for feeding purposes can further be differentiated into either uptake of the whole cells, or uptake of subcellular organelles like plastids (kleptoplasty, see chapter "Dinotoms Illuminate Early Pathways Towards the Stable Acquisition of Photosynthetic Endosymbionts" by Yamada et al.). While endosymbioses are widely spread, only a few of them led to the evolution of organelles [4, 14–16].

Primary endosymbiosis led to three major groups of algae: green algae, red algae, as well as the glaucophytes, together called the *Archaeplastida* [17], but also, clearly independently of these groups, to *Paulinella chromatophora*, an amoeboid member of *Rhizaria* with a so-called chromatophore, which can be seen as a primary plastid of independent origin, compared to all other plastids [15, 18, 19]. In both cases, the *Archaeplastida* and the photosynthetic *Paulinella* species, the primary plastids are monophyletic [20, 21], diversification and speciation of the organisms with the newly acquired organelles therefore appear to be more easily explainable than the *de-novo* acquisition of plastids (see also chapters "The

Evolutionary Origin of Primary Plastids" and "What Has *Paulinella* Taught Us About Endosymbiont Metabolic Integration?").

However, evolution did not stop after establishing primary endosymbiosis. Instead, a number of further combinations of eukaryotic endosymbionts and eukaryotic hosts in so-called secondary or tertiary endosymbioses yielded a number of new algal groups. These algae were different to the algae with primary plastids, because their plastid-ancestor endosymbionts were eukaryotes, and, very importantly, also the eukaryotic host cells are remarkably diverse [1, 22]. The distribution of organisms with such indirectly acquired plastids across the eukaryotic tree of life is wide, but patchy, and for all groups of eukaryotes that contain these plastids, plastid-lacking sister groups can be found. Immediate plastid related features like plastid genomes or pigment contents allow for the clear distinction of plastids that evolved from green- and red algal ancestors. But especially within the red-algal derived plastids, it is difficult to infer if the distribution pattern of plastids within the eukaryotic diversity arose via independent plastid losses after one common acquisition, or via independent acquisitions (in parallel or in series) [2, 7].

The combination of a eukaryotic host cell that possesses a plastid, which again originates from an endosymbiotic eukaryotic alga, and which itself had plastids originating from a prokaryotic cyanobacterium (thus "a cell in a cell in a cell"), strongly increased the genetic, but also the cellular complexity [23–27]. Plastids that evolved via secondary or tertiary endosymbioses are therefore also commonly called complex plastids. One of the hallmark features of such complex plastids is the number of membranes that surround the plastids. Looking at the currently known groups of algae and plants, the following types of plastids can be distinguished: primary plastids with an envelope consisting of two membranes (see chapter "Translocation of Proteins into Primary Plastids"), complex plastids surrounded by three membranes (of green- and of red-algal origins, see chapter "Translocation of Proteins into Complex Plastids with Three Envelope Membranes"), and complex plastids surrounded by four membranes (also of green- and of red-algal origins, with some exceptions in the membrane numbers that we will discuss later). In this chapter, we will focus on four membrane-bound complex plastids of red algal origin.

2 The Role of Endosymbioses

2.1 Endosymbioses and Protein Targeting

All these primary/secondary/tertiary processes have in common that a cell was engulfed by a eukaryotic host cell, thereby losing its genetic and functional independence within an evolutionary time frame. This came with a loss of the majority of the symbiont genes, leading to the much smaller organelle genomes compared to genomes of free-living bacteria [28, 29]. The process of organelle genome reduction could have started immediately with genes that became obsolete with the endosymbiosis. However, not all genes that are missing on organelle genomes got lost. Genetic material was, supposedly passively, transferred from the genome of the respective endosymbiont to the genome of the respective host cell by intracellular gene transfer. This process is called endosymbiotic gene transfer (see also chapter "The Extent, Role, and Timing of Endosymbiotic Gene Transfer in Plastids"). Indeed, in higher plants, it was demonstrated experimentally that, when plants with transgenic chloroplasts carrying a nuclear selectable marker gene were analyzed, the selection marker with a certain rate was found within the nuclear genome, indicating intracellular gene transfer from the plastid to the nuclear genome [30].

At the same time, during the ongoing endosymbiosis, the host evolved the capability of transporting proteins to the symbiont/nascent organelle, which helped in controlling the organelle. Once protein transport systems were established, also genes which originally were encoded on the organelle/symbiont genome, could acquire promoters and targeting signals to allow them being targeted back to the organelle. This would then create redundancy between the two copies of the gene, inevitably resulting in a loss of one of them. This mechanism has been suggested by Doolittle [31], who termed it "gene transfer ratchet." It should be noted that no photosynthetic organism is known to have completely lost the plastid genome, which is likely due to physiological restrictions [32]. The redox reactions involved in photosynthesis (and similarly in respiration in the case of mitochondria) require a tight regulation of enzyme expression, which again requires that the genome is located in the same cellular compartment as the enzyme complexes (this is the "co-location for redox regulation", or CORR hypothesis, first suggested by Allen [33, 34].

2.2 Endosymbioses and Cellular Complexity

As mentioned, algae with complex plastids often can be identified on the ultrastructural level by the presence of additional membranes surrounding the plastids. These additional membranes were actually the tell-sign that led to the hypothesis of their secondary origin [35, 36]. While primary plastids generally have two membranes, often termed inner and outer envelope membranes, complex plastids, as a consequence of eukaryote – eukaryote endosymbiosis, can possess two to four surrounding membranes. The outermost membrane likely originates from the phagotrophic membrane of the host cell, while the second membrane from outside corresponds to the plasma membrane of the eukaryotic endosymbiont [8]. In this point of view, no loss of membranes during the evolution of the complex plastids needs to be postulated. However, it is far from clear if all complex plastids are secondary plastids, or if some of them arose in higher order endosymbioses, thus going through additional reductive steps in which the number of envelope membranes is reduced. Such scenarios might appear non-parsimonious at first sight; however, with respect to parsimony, several aspects and their associated costs need to be considered. One such aspect is the number of postulated plastid gains to explain the occurrence of the plastids across the eukaryotic diversity, and another one, equally important but less

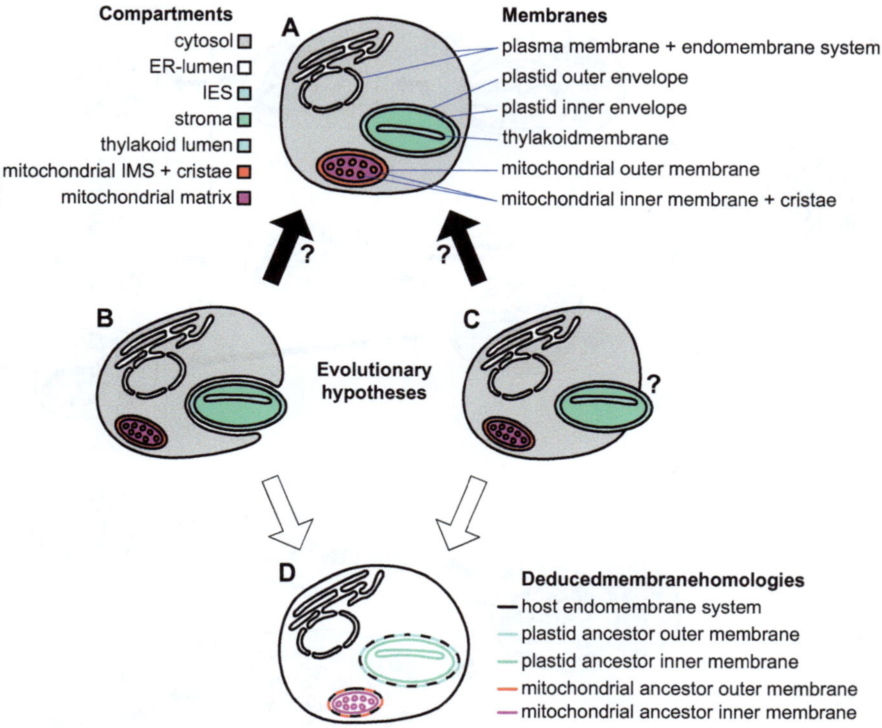

Fig. 1 Origin of cellular compartments and membrane homologies in algae with primary plastids. *A* Observed cell structure of a cell with primary plastids. *B/C*: two evolutionary hypotheses of the origin of primary plastids: phagocytotic uptake, (*B*) or other invasion of the plastid ancestor (*C*). The observed structurea (*A*) can be explained by evolutionary hypotheses (black arrows), and in the case of primary plastids, these hypotheses result in the same deduced membrane homologies, shown in *D* (white arrows). Mitochondrial and plastid outer membranes combine the properties of bacterial outer membranes (presence of beta-barrel proteins) with the properties of eukaryotic membranes (no lipopolysaccharides), and might therefore result from fusions of these membranes, see text for details. Abbreviations: ER endoplasmic reticulum, IES inter-envelope space, IMS intermembrane space

often recognized, is the number of plastid losses that need to be postulated in turn (less plastid gains mean that more plastid losses need to be postulated). This topic has been discussed in larger detail recently [2]. Here, we would like to point out another aspect to be considered with respect to parsimony, which lies in the mechanisms and number of membrane gains and losses that need to be postulated in order to explain the topology of complex plastids as either being secondary, tertiary, or higher order plastids (Figs. 1 and 2).

As mentioned, the assignment of membrane origins appears straightforward when the topology of four membrane-bound complex plastids of red algal origin is analyzed in isolation, and under the assumption that they arose via secondary endosymbiosis. With this frame, the innermost two membranes trace back to the original

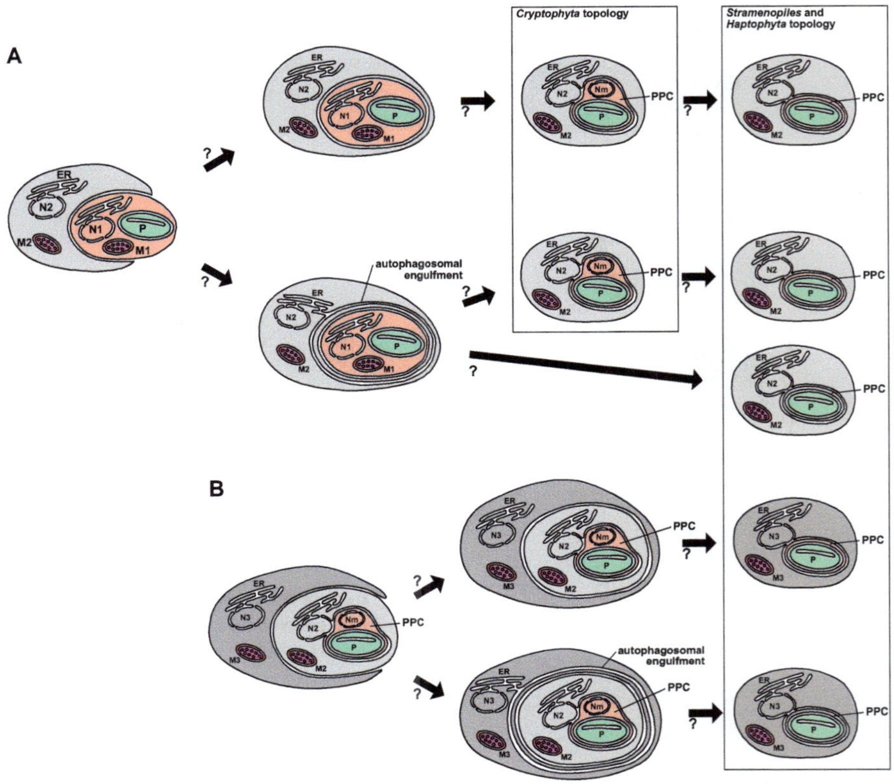

Fig. 2 Origin of cellular compartments and membrane homologies in cells with complex plastids resulting from higher order endosymbioses (**A** secondary, **B** tertiary), depending on the mechanism of membrane reduction during evolution of the complex plastids (autophagosomal engulfment [37–39], or other unspecified and more accidental mechanisms), different membrane and compartment homologies need to be postulated. In case of organisms without nucleomorph (*Stramenopiles, Haptophyta*), the cellular topology can be explained by secondary as well as higher order (tertiary) endosymbiosis (for a discussion of the current hypotheses see [2]), and depending on the mechanisms of cellular reduction, the PPC of *Stramenopiles* and *Haptophyta* may or may not be homologous to the cytosol of the former eukaryotic algal endosymbiont. See text for details. Abbreviations: ER: endoplasmic reticulum, M1: mitochondria of the eukaryotic algal endosymbiont during secondary endosymbiosis, M2: mitochondria of the eukaryotic host during secondary endosymbiosis, M3: mitochondria of the eukaryotic host during tertiary endosymbiosis, N1: nucleus of the eukaryotic algal endosymbiont during secondary endosymbiosis (corresponding to Nm in *Cryptophyta*), N2: nucleus of the eukaryotic algal host during secondary endosymbiosis, N3: nucleus of the eukaryotic host during tertiary endosymbiosis, Nm: nucleomorph, P: plastid, PPC: periplastidial compartment

plastid envelope of the endosymbiont, thus to the cyanobacterial inner and outer membranes, while the first membrane from the outside corresponds to the former phagocytotic vacuole (now fused with the host endoplasmic reticulum and hence also referred to as cER, for chloroplast endoplasmic reticulum), and the second membrane from the outside (referred to as the periplastid membrane) corresponds to the former cytoplasmic membrane of the eukaryotic endosymbiont (Fig. 2). With this clear picture in mind, Cavalier-Smith stated that the membrane topologies found in plastid containing cells are "frozen relics of their chequered symbiogenetic history, not purely functional engineering designs" [40]. This statement however, appears to contain a false dichotomy, because nature proves us again and again, that even though evolution is based on inheritance, the preservation of a heritage is not an evolutionary mechanism; just as much as functional engineering is not an evolutionary mechanism. It is, therefore most useful to look at alternative explanations of the membrane topologies. In the case of four membrane-bound plastids of red algal origin, one such explanation could be that both outer plastid membranes (cER membrane and periplastid membrane) could represent ER membranes, derived from an autophagosomal engulfment of the plastid by the ER, as suggested by Melkonian [37, 38], and more recently discussed by Gould et al. [39] (see also Fig. 2). This scenario is appealing because the original function of autophagosomes is the degradation of their enclosed content [41], which directly provides a hypothetical mechanism how the extra membranes of the nascent organelle could have been lost. And such losses of membranes need to be postulated for a number of reasons: (i) Already during the evolution of primary plastids from *Cyanobacteria*, one membrane must have been lost, if the uptake of the endosymbionts was via phagocytosis [7] (Fig. 1). (ii) Three membrane-bound complex plastids do exist, in the *Euglenophyta* [42, 43] and in the *Dinoflagellata* [44] (see also chapter "Translocation of Proteins into Complex Plastids with Three Envelope Membranes"). (iii) Increasing phylogenomic evidence points to the horizontal transfer of complex plastids in higher order endosymbioses [45], and loss of membranes is a required consequence of these scenarios (Fig. 2).

It is also interesting to note that kleptoplasts are also frequently surrounded by fewer membranes than would be expected from simple addition of the membrane numbers of the organisms involved [13, 46], this however, can also be a result of the uptake of cell content without uptake of the prey plasma membrane, as for example in myzocytosis [13].

Furthermore, there are two biases which influence the number of envelope membranes beyond simple addition, and which could explain the observed numbers of membranes. Firstly, while there is no evolutionary or ecological advantage in maintaining exactly the number of membranes that might appear logical to scientists, the presence or absence of additional membranes around the plastids can have severe physiological effects, with direct consequences for evolution and ecology of the respective organisms [2, 25]. Consequently, the membrane numbers that we see are under evolutionary pressure and therefore may or may not fit the "historically correct" number of membranes. Secondly, the cellular integration of an endosymbiont is closely connected to the evolution of protein targeting to the nascent organelle [4, 8, 14, 15]. Looking at dinotoms, dinoflagellates that contain complex plastids acquired from diatoms [13, 47] and that are surrounded by five membranes (four from the diatom, plus the myzocytosis vacuole), it is striking that the genetic integration of the symbiont is low and that protein targeting from the host (dinoflagellate) nucleus to the acquired diatom plastid apparently does not happen [47]. It is therefore quite possible that the number of four plastid membranes (thylakoid membranes/lumen not counted) is a maximum of biological membranes that can be overcome by targeting systems, and hence represents a functional and structural constraint that can only be overcome by membrane loss, but not by additional protein targeting steps. This could then explain why the currently known and well-integrated complex plastids are always surrounded by four or less membranes.

That said, there still is a strong core in Cavalier-Smith's above-cited statement, and that is the clear distinction between the plastid and mitochondrial membranes (which both are of bacterial origin), and the membranes of the eukaryotic endomembrane system. In the outer membrane of Gram negative bacteria, unlike in any other biological membrane, the outer leaflet contains lipopolysaccharides (LPS), and membrane spanning beta-barrel proteins can be found [48]. While LPS is not present in plastid or mitochondrial outer membranes, membrane spanning beta-barrel proteins are, and they are exclusively found in these three types of membranes (bacterial, mitochondrial and plastid outer membranes). Since the membrane insertion of beta-barrel proteins depends on the existence of pre-existing beta-barrel proteins in the membrane [49], the membranes themselves are carrying a part of the heritable information of the cell, because the genes alone would be useless, in case that a beta-barrel protein containing target membrane would be lost. Cavalier-Smith therefore called these membranes "genetic membranes" [40]. There is a clear consequence of this mechanism in the topology of plastids, and that is, that due to the presence of membrane spanning beta-barrel proteins in the second membrane counted from inside, it can be seen as certain that this membrane, and consequently also the inner most plastid membrane, in complex plastids, trace back to the primary plastid envelope membranes, which in turn trace back to the bacterial membranes of the plastid ancestor (Figs. 1 and 2). Similar to the protein content, also the lipid composition of plastid membranes is similar to the one of cyanobacterial membranes, with galactolipids being the predominant class [50, 51]. Also from the lipid composition of the inner envelope in a variety of organisms with complex plastids, it can be concluded that the two innermost plastid envelope membranes trace back to the cyanobacterial inner and outer membranes [43, 52].

Interestingly, even though the outer membranes of plastids are highly distinct from other eukaryotic membranes in terms of protein and lipid content, there must be some interaction between the eukaryotic endomembrane system and the outer plastid membrane in cells with primary plastids, as can be seen from the – exceptional – import of nucleus-encoded plastid proteins via the secretory pathway [53, 54].

There is another membrane which can be traced with certainty, and that is the outermost membrane, in both scenarios, whether it corresponds to a phagocytotic membrane or to an autophagosomal membrane, it must be host-organism derived (Fig. 2).

3 Ultrastructure of Complex Plastids of Red Algal Origin

One important feature of eukaryotic cells is their compartmentalization, as it allows the spatial separation of metabolic reactions. Therefore, there is a tight connection between physiology and ultrastructure. Some of the ecologically most relevant groups like *Haptophyta*, diatoms, and brown algae posses four membrane-bound complex plastids of red algal origin [27]. This topology requires additional transportation steps for substrates and proteins compared to plastids with two envelope membranes [25, 55]. Although sharing the same number of envelope membranes, the complex plastids in the aforementioned algal groups do not have uniform structures, which is why in the following, we describe some ultrastructural aspects of the individual algal groups.

In various algal groups, the outermost of the four plastid membranes is studded with ribosomes and is continuous with the endoplasmic reticulum, hence the name cER [56]. The narrow space between the outermost and the next membrane is considered to represent the chloroplast ER cisternae. The periplastidic space between the second and the third membrane is considered to be the remnant of the previous cytoplasm of the endosymbiont. This compartment in EM images appears to lack internal structures, indicating that the mitochondria, nucleus, and other structures disappeared during evolution with the exception of *Cryptophyta* and *Chlorachniophyta* that possess a remnant nucleus, the nucleomorph (see below). However, in some cases, vesicular structures have been observed in this compartment [57, 58]. As mentioned before, the two innermost membranes are supposed to be related to the plastid envelope membranes of the endosymbiotic eukaryotic alga. Unlike the outer membranes, these two membranes are highly conserved across the diversity of all plastids.

3.1 *Cryptophyta*

This primarily flagellated group of algae for a long time held a secret that was essential for our understanding of secondary endosymbiosis, the nucleomorph [59, 60], which is found only in *Cryptophyta* and *Chlorarachniophyta* [61]. Nucleomorphs represent vestigial but active nuclei of the endosymbiont, and thus can be considered as models for an intermediate stage to algal groups like diatoms, in which no such nuclear remnants have been detected. It should be noted that in the aforementioned dinotoms, the plastid-providing diatoms also retain their nucleus (and their mitochondria) [13, 47] (see chapter "Dinotoms Illuminate Early Pathways Towards the Stable Acquisition of Photosynthetic Endosymbionts"), however, these nuclei are not considered to be nucleomorphs, due to the low amount of genetic integration with the host cell [47].

One further feature that is also shared with *Chlorarachniophyta* is the presence of starch grains between the two membrane pairs. Photosynthetically produced carbohydrates are stored as starch in the periplastidic space and deposited close to the nucleomorph [62]. As *Cryptophyta* and *Chlorarachniophyta* have obtained their plastids from different endosymbionts in independent endosymbioses, nucleomorphs seem to be a parallel development, indicating convergent evolutionary mechanisms, possibly representing a somehow transitory state. A common explanation why nucleomorphs "still" exist, is, that with only one plastid per cell in some lines, lysis of an organelle (a prerequisite for gene transfer to the nucleus), is lethal for the cell, and that therefore the transfer of nucleomorph or plastid-encoded genes to the nucleus has come to a halt [63]. The nucleomorph in *Guillardia theta* contains three small eukaryotic chromosomes with more than 500 genes [60]. At least some of the respective gene products (about 30 proteins in *Cryptophyta*) have to be transported from the periplastidic space into the plastid stroma [61]. The nucleomorph itself is surrounded by a double membrane with pores, which are similar to those that are found in the nuclear double membrane. Furthermore, microtubules that might be involved in chromosome segregation during organelle division can be found [64]. Within the plastid stroma of *Cryptophyta*, the thylakoids are grouped pairwise and without interconnections [65].

3.2 *Stramenopiles*

The *Stramenopiles* are also called *Heterokontophyta* due to the two different types of flagella, of which one possesses stiff hairs (mastigonemes) that increase the drag of flagellar movement in water [66]. *Stramenopiles* together with *Alveolata* and *Rhizaria* are considered to be clades of the SAR supergroup, which is separate from *Cryptophyta* and *Haptophyta* [45]. The plastids of *Stramenopiles* are usually surrounded by four membranes, and the cER is continuous with the endoplasmic reticulum of the former host cell, and studded with ribosomes. The thylakoids of *Stramenopiles* are usually organized in triple lamellae thylakoid stacks, as well as a girdle lamella, and of the typical four membrane envelope membranes. Recent 3D imaging techniques revealed that there might be close contact sites between mitochondria and plastids in diatoms and a preserved volumetric ratio between plastids and mitochondria within this group [67]. The plastids in most *Stramenopiles* possess a pyrenoid, a structure composed mostly of Rubisco protein, which is mostly central and penetrated by a single thylakoid membrane [68–70].

Among the *Stramenopiles*, also two groups with unusual, and highly interesting plastid ultrastructures are found, the *Synchromophyceae*, and the so far orphan species *Chrysoparadoxa australica*. The *Synchromophyceae* possess several yellowish-green chloroplasts grouped in complexes that do not have a girdle lamella. Most strikingly, members of this group do not have individual envelope membranes including the ER membrane, instead the two to three plastids in *Synchroma grande* share the same outer pair of enveloping membranes that closely follows the outline of the chloroplast complex. This feature so far has not been found in any other organism with complex plastids [71], and might imply that already the ancestor of the plastids of *Stramenopiles* possessed more than one plastid.

Chrysoparadoxa australica was discovered as a benthic unicellular algae, that, based on sequence data, pigmentation, and flagellar morphology, is clearly a member of the *Stramenopiles* [72]. However, unlike any other organism with complex plastids of red algal origin, in this alga, the plastid is surrounded by only two envelope membranes. Furthermore, the outer of these envelope membranes is continuous with the nuclear envelope, and therefore a part of the endomembrane system [72]. The continuity of the outer plastid membrane with the nuclear envelope points to the direction that *C. australica* might be the first known organisms that has lost the original plastid outer envelope, the one that in other organisms contains membrane spanning beta-barrel proteins. Future molecular studies will show if these plastids indeed function without beta-barrel proteins in their outer envelope, and this would give hints to which of the membranes have been lost to account for this unique plastid topology.

3.3 Haptophyta

The *Haptophyta*, also known as *Prymnesiophyta*, are an independent evolutionary group [73] that is characterized by calciferous scales and by a haptonema [74]. This filamentous appendage with an unusual microtubule structure is located between the two flagella. It is supposed to be involved in food capture [75]. The plastids of these algae also originate from red algae. They have four plastid membranes, with the outermost membrane studded with ribosomes. The storage products are related to chrysolaminarin, a ß-glucan that is also found in *Stramenopiles* [76].

3.4 **Dinoflagellata** *(and Other* **Alveolata***)*

Dinoflagellata are a peculiar group of algae, because, while within other algal groups usually similar plastids types can be found, *Dinoflagellata* are more promiscuous, having taken up and replaced plastids several time during evolution [77]. Accordingly, *Dinoflagellata* are monophyletic regarding the host cell, but polyphyletic regarding their plastids. The plastids of *Dinoflagellata* are of the same ancestry as other complex plastids of red algal origin, however, they are difficult to characterize via molecular phylogenies, due to the extreme reduction of their plastid genomes, which are present in so called mini-circles (rather than a full circular mapping plastid genome), which encode one gene per mini-circle [78]. The plastids are surrounded by three membranes, with no permanent continuity between the outer most membrane and the endoplasmic reticulum, or nuclear envelope. Protein targeting to these plastids takes place via the endoplasmic reticulum and Golgi, and this extra step, from the perspective of the protein, also adds an additional membrane to be crossed, rendering the process similar to protein import into other complex plastids of red algal origin (see also chapter "Translocation of Proteins into Complex Plastids with Three Envelope Membranes"). Within the *Alveolata*, various other plastid containing groups are found, which however, are mostly non-photosynthetic [32]. Interestingly, the non-photosynthetic plastids of the parasitic *Apicomplexa* are surrounded by four envelope membranes [79] (with some contradictory findings published in [80]), and the protein targeting signals and mechanisms are similar to other four membrane-bound plastids of red algal origin [81, 82]. This also holds true for the four membrane-bound plastids of the closest known photosynthetic relatives to *Apicomplexa*, *Chromera velia,* and *Vitrella brassicaformis* [83, 84]. These differences compared to dinoflagellate plastids (together with other differences, in pigment content or plastid genomes), prompt hypotheses about independent plastid origins within *Alveolata* [84–86].

4 Protein Targeting into Plastids

4.1 Primary Plastids

The massive transfer of about 4500 genes [87] from the endosymbiont/plastid genome to the nuclear genome, although within evolutionary time scales, created a tremendous problem for the cells. Even though some of the transferred genes might have been redundant, thus obsolete and disappearing over time, a larger number of the transferred genes encode proteins that are essential for plastid processes like photosynthesis, plastid metabolism, and maintenance of the plastid genetic apparatus [1]. This becomes very obvious when considering that plastid genomes of photosynthetic algae and plants have been found to contain between 100–200 genes only [88]. Therefore, the mechanisms that initially had evolved for control of the symbiont by the host became increasingly important to transport the gene products of nucleus-encoded plastid genes after transcription/translation across the envelope membranes into the organelles. Intense research on the structure and function of these import systems revealed that primary plastids and mitochondria in principle have developed similar strategies to target proteins to the organelles [89–91]. In both cases, nuclear-encoded genes are transcribed in the nucleus, the respective mRNA is translated in the cytosol, and the proteins are imported mostly post-translationally, which is after translation by ribosomes, with the help of two membrane complexes named Tic and Toc in chloroplasts (translocon of the inner/outer chloroplast membrane) and Tim/Tom in mitochondria (translocon of the inner/outer mitochondrial membrane) [92–94] (see also chapter "Translocation of Proteins into Primary Plastids"). Protein import into primary plastids and mitochondria is dependent on an N-terminal pre-sequence [95] that is cleaved by an organellar signal peptidase once the pre-protein has entered the plastid stroma [96]. Different pathways and protein components exist to either integrate proteins into one of the two envelope membranes, or for targeting proteins to the intermembrane space [97]. Those proteins that need further targeting from the stroma to the thylakoid membranes may possess additional targeting sequences (for details see below).

4.2 Complex Plastids

For complex plastids with more than two envelope membranes, protein targeting into the plastids is more complicated, due to additional membranes and compartments that have to be crossed by the nuclear-encoded pre-proteins. As most of the work on protein targeting into complex plastids has been done in diatoms and *Cryptophyta*, we will focus here mostly on the results obtained in these organisms.

The Vesicular Transport Model for Protein Import

There has been a debate on the import mode of plastid pre-proteins into complex four membrane plastids like there are in diatoms. Based on evidence from a variety of electron microscopic data showing that vesicular or reticular structures can be found in the periplastidic space of heterokont algae [57], Sarah Gibbs was the first to postulate that a vesicular shuttle system might be involved in targeting of nucleus-encoded plastid proteins. She performed experiments with inhibitors of plastidic or cytosolic protein synthesis, resulting in an increasing or decreasing amount of such vesicular structures, respectively [98]. From this, Smith-Johannsen and Gibbs drew the conclusion that the vesicular structures are important for plastid protein targeting. Several authors then proposed scenarios for a vesicular protein import [56, 99, 100]. All these hypotheses postulate that the first step of protein transport to the complex plastid is a signal-peptide dependent, co-translational import of the cargo protein into the lumen of the cER. Furthermore, the vesicular transport hypotheses assumed that the periplastidic space, according to the endosymbiotic theory, essentially represents the cytosolic compartment of the endosymbiont cell. As the ancestral endosymbiotic alga in secondary endosymbiosis is of eukaryotic nature, this compartment originally contained all components required for intracellular vesicular targeting, including Golgi and ER components. A re-dedication of these components indeed could have led to vesicular transport from the second to the third envelope membrane. However, such a vesicular exchange between the cER and the inter-envelope space would also imply that proteins could shuttle between these compartments passively, which does not seem to be the case [101]. Furthermore, because diatoms and related algae do not possess a nucleomorph, the "Secretory Transport Hypothesis" [100] would require finding proteins that (i) are known to be involved in vesicular transport processes, (ii) are duplicated in the genome, and (iii) that include isoforms with and without a targeting sequence for the periplastidic space [102]. As, up to now, no genes for such proteins have been identified in diatoms [103, 104], investigations on protein targeting into complex plastids subsequently focused on an alternative hypothesis: protein import into complex plastids via individual membrane-bound translocators. This translocator model also postulates co-translational transport of the preproteins into the chloroplast ER (cER) lumen, and the subsequent involvement of specific translocators in each of the residual three membranes [100].

Pre-sequences and Targeting Signals

For a long time, in was unclear how nucleus-encoded proteins are imported into diatom plastids. First evidence came from a report by Bhaya and Grossman [105], who studied antenna proteins (FCP) in diatoms by sequencing genomic DNA fragments [106]. They observed that these plastid proteins possessed an N-terminal signal peptide for co-translational ER import. Signal peptides typically have a positive N-terminus, followed by a hydrophobic domain [107]. Signal peptides meanwhile can be easily predicted including their potential cleavage site [108]. The whole N-terminal extensions of the FCP pre-proteins rather short (less than 50 amino acids), which is why at that time it was not recognized that the pre-sequences actually are of bipartite nature. The second available gene sequence at that time, encoding the gamma subunit of the chloroplast ATPase (atpC), then revealed that, besides an ER sequence, the proteins also possess a second pre-sequence domain that is similar to transit peptides from plant plastid pre-proteins [109]. Such transit peptides usually possess a central part with a large proportion of hydroxylated amino acids like serine and threonine [110]. More and more bipartite pre-sequences consisting of two cleavable subsequent pre-sequences were identified, and also experimentally tested to be functional in transport essays and mutational experiments [27, 111–117], and meanwhile have been found to be the standard for *Stramenopiles* proteins imported into the plastid stroma (Fig. 3). Moreover, those proteins that need to be further translocated to the thylakoid lumen, like a subunit of the oxygen evolving complex, may even have tripartite pre-sequences, composed of a signal, a transit, and a second signal peptide [114, 122].

The large number of protein sequences as well as published import experiments, mostly via EGFP fusion proteins, allowed to compare sequences and to detect a conserved motif between the signal and the transit peptide that consists of variations of the amino acids ASA-FAP (or residues with similar properties), which is cleaved by a signal peptidase between amino acids A and F [114]. Functional experiments demonstrated that the phenylalanine (F) is the most crucial amino acid in this motif, which can only be replaced by other aromatic residues (Y or W), or by the bulky hydrophobic residue (L) [112, 114]. Further features of the transit peptide domain are a positive net charge [116] and the presence of ubiquitination sites [123], both of which are essential for successful plastid protein import [116, 123].

Pre-proteins possessing a signal and a transit peptide, but lacking the "ASAFAP"-motif, in most cases accumulate in what has been dubbed "blob"-like structures (BLS) [114]). Based on the observation of natural pre-sequences leading to accumulation in BLS [124, 125], and especially on the discovery that a protein that is

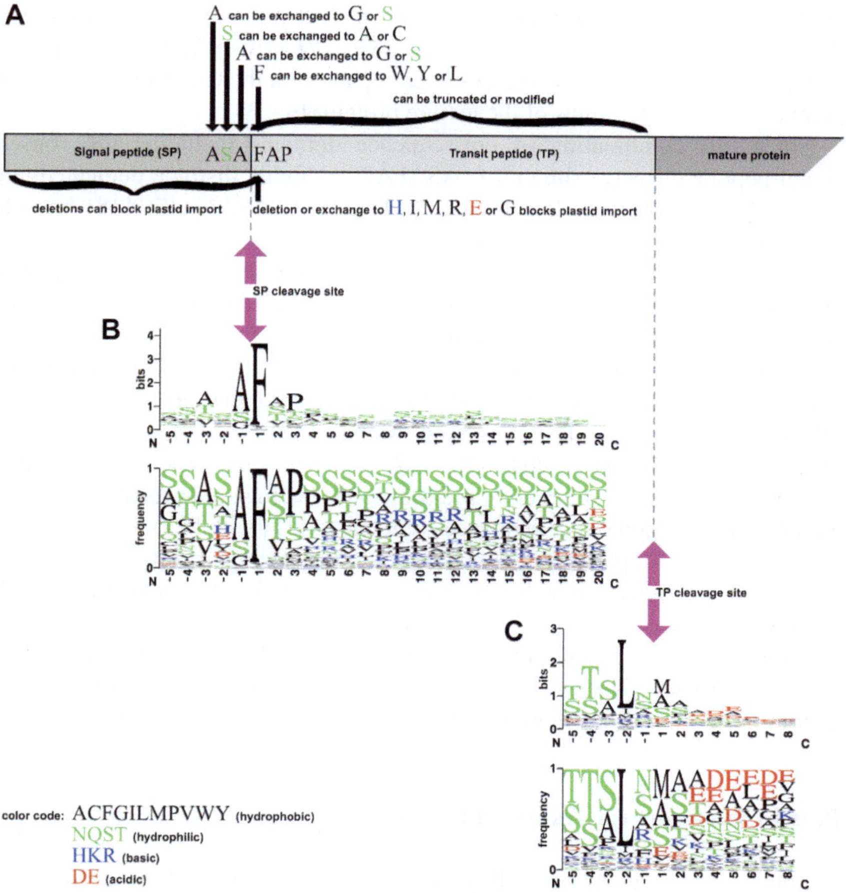

Fig. 3 Targeting signals of nucleus encoded plastid proteins in cells with complex plastids of red algal origin. **A** Results of mutational experiments (redrawn from [112]). **B** Sequence motif at the signal peptide cleavage site (based on 83 *Thalassiosira pseudonana* plastid proteins in Table S1 from [118]). **C** Sequence motif at the transit peptide cleavage site (based on 63 native N-termini of *T. pseudonana* plastid proteins identified by Huesgen et al. [119] position correction for sequences with methionine at the start of the mature protein applied, see Table S7 from [119]). The panels are arranged according to the position of the motifs in the targeting pre-sequences. Sequence logos [120] were created with the help of WebLogo [121]

most likely located in the periplastidic space of *Cryptophyta* [126] causes the same phenotype in *P. tricornutum*, it is thought that these structures correspond to the periplastidic space of diatoms [127].

In vitro experiments with microsomal vesicles from canine pancreas, and with isolated pea plastids showed that the two parts of the bipartite targeting signals are functional on their own, and that they are cleaved after import of the precursor protein [115]. Cleavage of the pre-sequences has also been demonstrated by western-blot analyses of fusion constructs with the green fluorescent protein [114, 126], and by mass-spectrometric identification of native N-termini in the plastid proteome of the diatom *Cyclotella nana* (commonly known as *Thalassiosira pseudonana*) [119], which also revealed a conserved cleavage site motif at the C-terminus of the transit peptide and the N-terminus of the mature protein (Fig. 3).

Since this combination of pre-sequence domains including the conserved signal-peptide cleavage site motif "ASAFAP" is highly distinct, dedicated targeting prediction programs like "Hectar" [128] and "ASAFind" [118] have been developed, which allow for targeting predictions with high statistical performance [25, 129–131]. Despite the fact that these bipartite plastid targeting signals are quite different from mitochondrial transit peptides, also dual targeting to plastids and mitochondria has been shown in diatoms [132].

It is interesting to see that the targeting signals described here, are highly similar within the *Stramenopiles* [128], and also with *Cryptophyta*, *Haptophyta,* or even *Dinoflagellata* [117]. The targeting signals are cross-functional between *Stramenopiles* and *Cryptophyta* [102, 126], and in vitro experiments conducted with cryptophyte plastid targeting pre-sequences led to similar results compared to diatom plastid targeting pre-sequences (compare [115] with [133]).

The similarity of plastid targeting pre-sequences among organisms with complex plastids of red algal origin has also proven useful for the annotation of whole genome data, for example in the cryptophyte *Guilardia theta* [63] or in the alveolate *Chromera velia* [83], however there are examples where experimental results of targeting essays between *C. velia*, the apicomplexan *Toxoplasma gondii*, and the diatom *Phaeodactylum tricornutum* need to be interpreted differently [134].

Protein Transport via Transport Complexes

Diatom plastids are surrounded by an ER membrane. Accordingly, based on the finding that diatom pre-proteins possess an N-terminal signal peptide, the hypothesis that a sec-based protein import system could be the first step of protein targeting was a likely assumption (Fig. 4 and Table 1). SecYEG systems are one of the most ancient membrane translocator system that can already be found in bacteria [146]. A related Sec61 protein has been found to serve as a translocator in eukaryotic cells, allowing co-translational import into the ER lumen [147, 148]. In both cases the sec systems consist of multiple subunit transmembrane complexes which enable ATP-dependent protein transport [149]. A functional relation of signal peptides from diatom plastid pre-proteins and ER targeted proteins has been demonstrated experimentally to be essential for translocation both types of pre-proteins into microsomal ER vesicles [115]. Mutational analyses of pre-sequences fused to a reporter protein have also shown that the signal peptides of ER proteins are exchangeable with the signal peptides of plastid proteins [111, 114]. Further support for the routing of plastid pre-proteins through the ER comes from the presence of glycoproteins in complex plastids, and from the glycosylation of reporter proteins during the transport to these plastids [143].

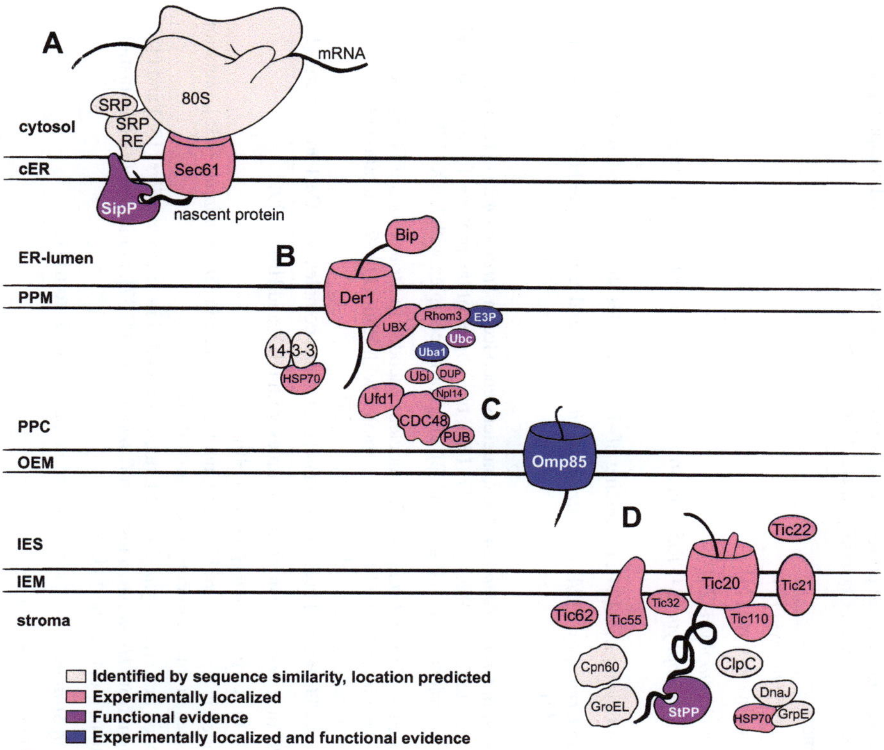

Fig. 4 Protein complexes potentially involved in the targeting of proteins into secondary plastids across the four envelope membranes. For proteins names and experimental data see Table 1. The figure is based on Gould et al. [24], with modifications by the authors based on integration of newer data according to Table 1, drawing by Houda Ouns Maaroufi. **A** Sec translocon, **B** SELMA system, **C** Toc translocon, **D** Tic translocon, see text for details. Abbreviations: cER chloroplast endoplasmic reticulum; ER endoplasmic reticulum, IEM inner envelope membrane, IES inter-envelope space, mRNA messenger RNA, OEM outer envelope membrane, PPC periplastidial compartment, PPM periplastidial membrane, Sec secretory pathway, SELMA symbiont-specific ERAD-like machinery, Tic translocator of the inner chloroplast membrane, Toc translocator of the outer chloroplast membrane

The Selma Complex

Studying the nucleomorph of *Guillardia theta*, Sommer et al. [135] detected genes for components of an ERAD (ER associated degradation system), together with ERAD genes in the nucleus, concluding that the nucleomorph-encoded proteins may have a plastid-related function. In general, ERAD complexes are ER membrane-based parts of a quality control system involved in exporting mal-folded proteins from the ER back into the cytosol, where they can be degraded by the proteasome. As the major function of ERAD is the export from the ER, it was concluded that the complex could be involved in transporting proteins across the second plastid membrane, thus from the ER lumen to the periplastidic space [135]. Similar plastidal

Table 1 Putative components of the protein transport complexes for nucleus encoded plastid proteins in cells with four membrane-bound complex plastids of red algal origin, and levels of evidence for their involvement (all data from *Stramenopiles*, *Cryptophyta*, and *Haptophyta*)

Name	Function	Comments/References
80S	Ribosome assembly	Nucleus-encoded in *Cryptophyta* [24]
SRP	Signal recognition particle	Nucleus-encoded in *Cryptophyta* [24]
SRP Re	SRP Receptor	Nucleus-encoded in *Cryptophyta* [24]
Sec61	Channel translocon	Nucleus-encoded in *Cryptophyta* [24, 135] and *Stramenopiles* [135], ER targeted in *Phaeodactylum tricornutum* [136]
SiPP	Signal peptide peptidase	Nucleus-encoded in *Cryptophyta* [24]; processing of signal peptides shown in western-blots [114]
Bip	Binding immunoglobulin protein (chaperone)	Nucleus-encoded in *Cryptophyta* [24, 135] *Stramenopiles* [135] and *Haptophyta* [137]; located in the cER [111, 113, 137]
Der1	Derlin-1, Cdc48-receptor and/or part of the retrotranslocon	Nucleomorp-encoded in *Cryptophyta* [24, 135], nucleus encoded in *Stramenopiles* [135] and *Haptophyta* [137]; located in the PPC [135, 137, 138], two copies of the proteins form homo- and hetero-oligomeric complexes (Der1-1 and Der1-2), also forms complex with UBX [139]
E3P (Hrd1)	RING-finger ubiquitin ligase (E3)	Nucleomorp-encoded in *Cryptophyta* [24, 135], nucleus encoded in *Stramenopiles* [135], located in the PPM in *P. tricornutum* (ptE3P) [140], forms complex with Rhom3 [139], mutation of ubiquitination sites in cargo proteins blocks import [123]
UBX	Ubiquitin-regulatory protein	Nucleus encoded and PPC targeted in *P. tricornutum* [141], forms complex with Der1 and Rhom3 [139]
Rhom3	Rhomboid protease	Located in the PPC in *P. tricornutum*, forms complex with UBX and Hrd1 [139]
DUP	De-ubiquitinating enzyme	Located in the PPC in *P. tricornutum* (ptDUP) [140]
Ubc	Polyubiquitin-C (E2)	Nucleomorp-encoded in *Cryptophyta* [24, 135], nucleus encoded in *Stramenopiles* [135], mutation of ubiquitination sites in cargo proteins blocks import [123]
Uba1	Ubiquitin-activating enzyme (E1)	Nucleomorp-encoded in *Cryptophyta* [24, 135], nucleus encoded in *Stramenopiles* [135] and [137], PPC located (protein from *Haptophyta* tested in *P. tricornutum*) [137], mutation of ubiquitination sites in cargo proteins blocks import [123]
Ubi	Ubiquitin	Nucleus encoded in *Stramenopiles* [135], located in the PPC in diatoms [135]
14-3-3	Regulatory protein	Nucleus-encoded in *Cryptophyta* [24]

Hsp70	70 kD heat shock protein (chaperone)	Two copies in *Cryptophyta*, one nucleomorp-encoded [24, 135], the other plastid-encoded [24], targeted to the PPC in *P. tricornutum* [126, 136, 138, 139, 142]
Cdc48	Protein unfolding, AAA-type ATPase	Nucleomorp-encoded in *Cryptophyta* [24, 135], nucleus encoded in *Stramenopiles* [135] and *Haptophyta* [137], located in the PPC [135, 137], two copies in *P. tricornutum* [104, 139]
Npl14	Cdc48 cofactor	Nucleus encoded and PPC targeted in *P. tricornutum* [141]
PUB	Unclear	Forms complex with Cdc48 [139]
Ufd1	Ubiquitin fusion degradation protein 1, Activator of Cdc48	Nucleomorp-encoded in *Cryptophyta* [24, 135], nucleus encoded in *Stramenopiles* [135] and *Haptophyta* [137], targeted to the PPC in *P. tricornutum* [143]
Toc75	Protein translocon	Nucleus-encoded in *Stramenopiles*; located in the OEM; functional as membrane pore [144], plastid targeted in *P. tricornutum* [136]
Tic22	Chaperone	Two copies, one nucleus-encoded, the other nucleomorp-encoded in *Cryptophyta* [24], located in the IES in *P. tricornutum* [145]
Tic20	Component of translocation channel	Nucleus-encoded in *Cryptophyta* [24], integral membrane protein located in the IEM in *P. tricornutum* [145]
Tic21	Channel component, iron permease	Located in the IEM in *P. tricornutum* [145]
Tic 32	Redox regulator	Peripheral membrane protein in plastids of *P. tricornutum* [145]
Tic55	Redox regulator	Integral membrane protein in plastids of *P. tricornutum* [145]
Tic62	Redox sensor	Nucleus-encoded in *Cryptophyta* [24], soluble protein in plastids of *P. tricornutum* [145]
Tic110	Component of translocation channel	Nucleomorp-encoded in *Cryptophyta* [24], integral membrane protein in plastids of *P. tricornutum* [145]
Cpn60	Chaperonin	Nucleomorp-encoded in *Cryptophyta* [24]
GroEL	Chaperonin	Plastid-encoded in *Cryptophyta* [24]
ClpC	Protease	Plastid-encoded in *Cryptophyta* [24]
StPP	Stromal processing peptidase	Nucleus-encoded in *Cryptophyta* [24]; processed native N-termini of potential substrates identified [119]
DnaJ	Chaperone	Nucleus-encoded in *Cryptophyta* [24]
GrpE	Co-chaperone	Nucleus-encoded in *Cryptophyta* [24]

Abbreviations: *cER*: chloroplast endoplasmic reticulum, *IEM*: inner envelope membrane, *IES*: inter envelope space, *OEM*: outer envelope membrane, *PPC*: periplastidial compartment, *PPM*: periplastidial membrane, please note that in many of the cited publications, protein names are preceded by a lower case “s”, for symbiont, to contrast the SELMA components from the host ERAD system.

ERAD components have also been detected in *Stramenopiles* [135] that do not have a nucleomorph [39]. In these cases, the respective ERAD components have bipartite plastid targeting sequences [135, 137]. To differentiate the two ERAD systems in the algae, the plastid ERAD complex was termed SELMA (Fig. 4 and Table 1), standing for "symbiont-specific ERAD-like machinery" [138, 150]. SELMA apparently contains many of the core components found in the "regular" ERAD complexes [103, 123, 139], including several membrane proteins located in the periplastidic membrane, such as the Derlins (the putative channel proteins). Also, an E3-ligase has been identified, which could be involved in ubiquitination of malfolded proteins, tagging them for subsequent degradation in the periplastidic space [138, 141]. Furthermore, potential soluble ERAD components located in the periplastidic space have been identified [130, 135, 137, 138, 141], like UBX domain containing proteins, potentially regulating the activity of the Cdc48 ATPase complex, of which two symbiotic copies of Cdc48 could be identified in *P. tricornutum* [141]. In SELMA mediated transport, ubiquitination of the transport substrate is thought to be decoupled from proteasomal degradation, and it has been suggested that in the PPC of organisms with complex plastids of red algal origin, the proteasomal degradation in turn does not depend on ubiquitinated substrates [141]. And indeed, lysine residues in the N-terminal portion of pre-proteins destined to the plastid, which can act as ubiquitination sites, are required for the successful completion of the import [123].

Variants of the SELMA complex are found in all groups of algae with complex plastids of red algal origin (but not in those with plastids of green algal origin), and this is a remarkable commonality that points to a shared history of plastid acquisition. Cavalier-Smith concluded, from the phylogeny of Derlin proteins, that the ones involved in SELMA (as opposed to the ones in the regular ERAD system) are not related to the nucleomorph-encoded derlin genes in *Cryptophyta*, and that therefore all algal groups with SELMA systems must trace back to a single endosymbiotic origin [151]. However, it should be noted that, in a higher order endosymbiosis, also nuclear-encoded genes of an endosymbiont can become available to horizontal acquisition by the host organism.

Toc/Omp85

The further transport across the inner two envelope membranes depends on the first amino acid of the transit peptide after cleavage of the signal peptide. The inner two membranes are thought to represent the plastid envelope of the eukaryotic symbiont, corresponding to the cyanobacterial double cell membrane. Accordingly, one can assume that nucleus-encoded plastid proteins utilize Tic/Toc complexes similar to those from higher plants to enter the stroma. However, analyses of the genomes of the diatoms *C. nana* and *P. tricornutum* did not yield putative components of the Toc apparatus. It was proposed that a component derived from the cyanobacterial outer membrane protein Omp85 [144], which is related to the Toc75 protein in higher plants, could act as a phenylalanine specific receptor and membrane channel across the third outermost membrane (Fig. 4 and Table 1). The conserved "F" based cleavage site motif in the plastid targeting pre-sequences, which we described above, has also been linked to the recognition of the protein cargo by Omp85/Toc75 [152–154], and it has been suggested that the lack of this motif in green algae and higher plants was enabled by a mutation in Toc75, which in green algae and higher plants, itself contains the phenylalanine [155].

Tic

Seven putative subunits of the translocons at the inner envelope membrane of chloroplasts (Tic) are encoded in the genome of the diatoms *P. tricornutum* and *C. nana* (Fig. 4 and Table 1). Six Tic components from the two diatoms *P. tricornutum* (Pt) and *C. nana* (Tp, for the synonym *T. pseudonana*) have been identified and studied by [145]: PtTic20, PtTic21, PtTic55, TpTic55, TpTic62, and TpTic110. All of them possess a bipartite plastid targeting pre-sequence. GFP fusion proteins of these TIC components are all localized to the plastid envelope membranes when expressed in *P. tricornutum*, although the microscopical images did not allow distinguishing between the envelope membranes.

Transport into Thylakoids

In land plant plastids, thylakoid proteins can be transported either directly into the thylakoid lumen or being inserted into the thylakoid membrane [156, 157]. Cleavage of the transit peptide for stromal targeting may expose a secondary targeting sequence for further transport to/into the thylakoids. There are two systems for entering the thylakoid lumen: the twin-arginine translocase (TAT) for folded proteins [158], or the SEC translocase for unfolded proteins [159]. Furthermore, the signal recognition particle (SRP) pathway allows insertion of integral thylakoid membrane proteins [160]. These systems are also present in *Cryptophyta* [161].

Experimentally, rather little is known about thylakoid targeting in complex plastids. First analyses of the pre-sequence of a subunit, the oxygen evolving complex (OEE3) from *P. tricornutum* revealed a tripartite pre-sequence that is functional in importing the protein into the plastids in vivo [114]. The diatom FCP pre-proteins instead have a bipartite pre-sequence and lack a third domain for thylakoid targeting [122]. Similar to chlorophyll a/b-binding proteins, there is apparently no pre-sequence present for targeting to the thylakoid membrane. An in vitro import assay for diatom thylakoids demonstrated that integration of the diatom FCP protein into the thylakoid membrane depends on the presence of stromal factors and GTP [122]. This indicates that a pathway involving signal recognition particles (SRP) is involved in membrane integration just as shown for LHCs in higher plants [162]. The diatom FCP can also integrate into thylakoids of higher plants and vice versa, also SRP-dependent targeting of LHCs from pea and *Arabidopsis* into diatom thylakoids. One component required for integration is the ALBINO3 (ALB3) protein [163]. A loss of ALB3b in *P. tricornutum* leads to a color change of the cells from brownish to green due to a strong decrease of FCP proteins in the thylakoids. In plants and green algae, ALB3 proteins interact with members of the chloroplast signal recognition particle (SRP) pathway through a Lys-rich C-terminal domain [164]. As in land plant plastids, a chloroplast signal recognition particle of 54 kDa (CpSRP54) has been found to be active in *P. tricornutum* [163]. CpSRP54 loss mutations in *P. tricornutum* results in reduced amounts of subunits of the photosynthetic complexes, indicating a role in the co-translational part of the CpSRP pathway. However, in contrast to plants, CpSRP54 apparently is not involved in the integration of light-harvesting antenna complexes.

5 Conclusions

Protein transport is a crucial process in the evolutionary establishment of organelles, and especially in the case of complex plastids of red algal origin points to common ancestry of these plastids. It is less clear how the former host cells of these organisms are related, and how often such plastids have been established across the eukaryotic diversity. Gene transfer to the nucleus has re-arranged the intracellular distribution of the proteome, and this has led to unusual physiology of organisms with complex plastids compared to organisms with primary plastids. The presence of targeting signals can be predicted from genomic data, and thus gives physiological, as well as evolutionary insights. Transport into four membrane-bound plastids of red algal origin is accomplished via the evolutionary recruitment of pre-existing protein transport complexes, the Sec system, the SELMA translocon, and the Toc and Tic systems.

Acknowledgments We would like to thank Ayush Sharma and Dr. Miroslav Oborník for helpful discussions, and Dr. Houda Ouns Maaroufi for the preparation of Fig. 4. AG is thankful for support by the Institute of Parasitology of the Biology Centre of the Czech Academy of Sciences, by the Czech Science Foundation (21-26115S), and by the ELIXIR CZ Research Infrastructure (ID LM2023055, MEYS CR). PGK is thankful for support by the German Research Foundation (DFG, SFB969) and by the University of Konstanz.

References

1. Archibald JM (2015) Endosymbiosis and eukaryotic cell evolution. Curr Biol 25(19):R911–R921. https://doi.org/10.1016/j.cub.2015.07.055
2. Gruber A, Oborník M (2024) Evolution of plastids and mitochondria in diatoms. In: Goessling JW, Serôdio J, Lavaud J (eds) Diatom photosynthesis: from primary production to high value. Scrivener Publishing LLC, Berverly, pp 79–110
3. Archibald JM (2015) Genomic perspectives on the birth and spread of plastids. Proc Natl Acad Sci U S A 112:10147. https://doi.org/10.1073/pnas.1421374112
4. Gruber A (2019) What's in a name? How organelles of endosymbiotic origin can be distinguished from endosymbionts. Microbial Cell 6(2):123–133. https://doi.org/10.15698/2F mic2019.02.668

5. Oborník M (2019) In the beginning was the word: how terminology drives our understanding of endosymbiotic organelles. Microbial Cell 6(2):134–141. https://doi.org/10.15698/mic2019.02.669
6. Roger AJ, Muñoz-Gómez SA, Kamikawa R (2017) The origin and diversification of Mitochondria. Curr Biol 27(21):R1177–r1192. https://doi.org/10.1016/j.cub.2017.09.015
7. Gruber A, Medlin LK (2023) Complex plastids and the evolution of the marine Phytoplankton. J Marine Sci Eng 11(10):1903. https://doi.org/10.3390/jmse11101903
8. McFadden GI (2014) Origin and evolution of plastids and photosynthesis in eukaryotes. Cold Spring Harb Perspect Biol 6(4):a016105. https://doi.org/10.1101/2Fcshperspect.a016105
9. Schreiber M, Rensing SA, Gould SB (2022) The greening ashore. Trends Plant Sci 27(9):847–857. https://doi.org/10.1016/j.tplants.2022.05.005
10. Spanner C, Darienko T, Filker S, Sonntag B, Pröschold T (2022) Morphological diversity and molecular phylogeny of five *Paramecium bursaria* (Alveolata, Ciliophora, Oligohymenophorea) syngens and the identification of their green algal endosymbionts. Sci Rep 12(1):18089. https://doi.org/10.1038/s41598-022-22284-z
11. Haferkamp I, Schmitz-Esser S, Linka N, Urbany C, Collingro A, Wagner M, Horn M, Neuhaus HE (2004) A candidate NAD+ transporter in an intracellular bacterial symbiont related to Chlamydiae. Nature 432(7017):622–625. https://doi.org/10.1038/nature03131
12. Rumpho ME, Summer EJ, Manhart JR (2000) Solar-powered sea slugs. Mollusc/algal chloroplast symbiosis. Plant Physiol 123(1):29–38. https://doi.org/10.1104/pp.123.1.29
13. Yamada N, Bolton JJ, Trobajo R, Mann DG, Dąbek P, Witkowski A, Onuma R, Horiguchi T, Kroth PG (2019) Discovery of a kleptoplastic 'dinotom' dinoflagellate and the unique nuclear dynamics of converting kleptoplastids to permanent plastids. Sci Rep 9(1):10474. https://doi.org/10.1038/s41598-019-46852-y
14. Keeling PJ, Archibald JM (2008) Organelle evolution: what's in a name? Curr Biol 18(8):R345–R347. https://doi.org/10.1016/j.cub.2008.02.065
15. Nowack ECM, Weber APM (2018) Genomics-informed insights into endosymbiotic organelle evolution in photosynthetic eukaryotes. Annu Rev Plant Biol 69(1):51–84. https://doi.org/10.1146/annurev-arplant-042817-040209
16. Tyler H, Valentina C, Loconte KA, Bieke T-K, Esther VWK, Shunyan M, Axel C, Jian-Hua E, Kyoko C, Yoshihito H, Tomohiro T, Masao N, Mark A, Carolyn LG, Larabell JP, Zehr (2024) Nitrogen-fixing organelle in a marine alga. Science 384(6692):217–222. https://doi.org/10.1126/science.adk1075
17. Adl SM, Bass D, Lane CE, Lukeš J, Schoch CL, Smirnov A, Agatha S, Berney C, Brown MW, Burki F et al (2019) Revisions to the classification, nomenclature, and diversity of eukaryotes. J Eukaryot Microbiol 66(1):4–119. https://doi.org/10.1111/jeu.12691
18. Gabr A, Grossman AR, Bhattacharya D (2020) *Paulinella*, a model for understanding plastid primary endosymbiosis. J Phycol 56(4):837–843. https://doi.org/10.1111/jpy.13003
19. Nowack EC, Melkonian M, Glockner G (2008) Chromatophore genome sequence of *Paulinella* sheds light on acquisition of photosynthesis by eukaryotes. Curr Biol 18(6):410–418. https://doi.org/10.1016/j.cub.2008.02.051
20. Irisarri I, Strassert JFH, Burki F (2021) Phylogenomic insights into the origin of primary plastids. Syst Biol 71(1):105–120. https://doi.org/10.1093/sysbio/syab036
21. Yoon HS, Nakayama T, Reyes-Prieto A, Andersen RA, Boo SM, Ishida K, Bhattacharya D (2009) A single origin of the photosynthetic organelle in different *Paulinella* lineages. BMC Evol Biol 9:98. https://doi.org/10.1186/1471-2148-9-98
22. Worden AZ, Follows MJ, Giovannoni SJ, Wilken S, Zimmerman AE, Keeling PJ (2015) Rethinking the marine carbon cycle: factoring in the multifarious lifestyles of microbes. Science 347(6223):1257594. https://doi.org/10.1126/science.1257594
23. Sanchez-Puerta MV, Delwiche CF (2008) A hypothesis for plastid evolution in chromalveolates. J Phycol 44:1097–1107. https://doi.org/10.1111/j.1529-8817.2008.00559.x
24. Gould SB, Waller RF, McFadden GI (2008) Plastid evolution. Annu Rev Plant Biol 59:491–517. https://doi.org/10.1146/annurev.arplant.59.032607.092915

25. Gruber A, Kroth PG (2017) Intracellular metabolic pathway distribution in diatoms and tools for genome-enabled experimental diatom research. Philos Trans R Soc B-Biol Sci 372(1728):20160402. https://doi.org/10.1098/rstb.2016.0402
26. Bowler C, Allen AE, Badger JH, Grimwood J, Jabbari K, Kuo A, Maheswari U, Martens C, Maumus F, Otillar RP et al (2008) The Phaeodactylum genome reveals the evolutionary history of diatom genomes. Nature 456(7219):239–244. https://doi.org/10.1038/nature07410
27. Kroth PG (2002) Protein transport into secondary plastids and the evolution of primary and secondary plastids. Int Rev Cytol 221:191–255. https://doi.org/10.1016/s0074-7696(02)21013-x
28. Martin W (2003) Gene transfer from organelles to the nucleus: frequent and in big chunks. Proc Natl Acad Sci U S A 100(15):8612–8614. https://doi.org/10.1073/pnas.1633606100
29. Martin W, Herrmann RG (1998) Gene transfer from organelles to the nucleus: how much, what happens, and Why? Plant Physiol 118(1):9–17. https://doi.org/10.1104/pp.118.1.9
30. Stegemann S, Hartmann S, Ruf S, Bock R (2003) High-frequency gene transfer from the chloroplast genome to the nucleus. Proc Natl Acad Sci U S A 100(15):8828–8833. https://doi.org/10.1073/pnas.1430924100
31. Doolittle WF (1998) You are what you eat: a gene transfer ratchet could account for bacterial genes in eukaryotic nuclear genomes. Trends Genet 14(8):307–311. https://doi.org/10.1016/s0168-9525(98)01494-2
32. Oborník M (2019) Endosymbiotic evolution of algae, secondary heterotrophy and parasitism. Biomolecules 9(7):266. https://doi.org/10.3390/biom9070266
33. Allen JF (2003) The function of genomes in bioenergetic organelles. Philos Trans R Soc Lond B Biol Sci 358(1429):19–38. https://doi.org/10.1098/2Frstb.2002.1191
34. Allen JF (2017) The CoRR hypothesis for genes in organelles. J Theor Biol 434:50–57. https://doi.org/10.1016/j.jtbi.2017.04.008
35. Gibbs SP (1978) The chloroplasts of *Euglena* may have evolved from symbiotic green algae. Can J Bot 56(22):2883–2889
36. Gibbs SP (2006) Looking at life: from binoculars to the electron microscope. Annu Rev Plant Biol 57(1):1–17. https://doi.org/10.1146/annurev.arplant.57.032905.105413
37. Melkonian M (1996) Phylogeny of photosynthetic protists and their plastids. Verh Dtsch Zool Ges 89:71–96
38. Melkonian M (2001) Systematics and evolution of the algae. I. Genomics meets phylogeny. Prog Botan 62:340–382
39. Gould SB, Maier UG, Martin WF (2015) Protein Import and the origin of red complex plastids. Curr Biol 25(12):R515–R521. https://doi.org/10.1016/j.cub.2015.04.033
40. Cavalier-Smith T (2000) Membrane heredity and early chloroplast evolution. Trends Plant Sci 5(4):174–182. https://doi.org/10.1016/s1360-1385(00)01598-3
41. Nakatogawa H (2020) Autophagic degradation of the endoplasmic reticulum. Proc Jpn Acad Ser B Phys Biol Sci 96(1):1–9. https://doi.org/10.2183/2Fpjab.96.001
42. Vacula R, Slàvikovà S, Schwartzbach SD (2007) Protein trafficking to the complex chloroplasts of euglena. In: van der Giezen M (ed) Protein targeting protocols. Humana Press, Totowa, pp 219–237. https://doi.org/10.1007/978-1-59745-466-7_15
43. Tomečková L, Tomčala A, Oborník M, Hampl V (2020) The lipid composition of *Euglena gracilis* middle plastid membrane resembles that of primary plastid envelopes. Plant Physiol 184(4):2052–2063. https://doi.org/10.1104/pp.20.00505
44. Patron NJ, Waller RF, Archibald JM, Keeling PJ (2005) Complex protein targeting to dinoflagellate plastids. J Mol Biol 348(4):1015–1024. https://doi.org/10.1016/j.jmb.2005.03.030
45. Baurain D, Brinkmann H, Petersen J, Rodríguez-Ezpeleta N, Stechmann A, Demoulin V, Roger AJ, Burger G, Lang BF, Philippe H (2010) Phylogenomic evidence for separate acquisition of plastids in cryptophytes, haptophytes, and stramenopiles. Mol Biol Evol 27(7):1698–1709. https://doi.org/10.1093/molbev/msq059
46. Martin R, Walther P, Tomaschko K-H (2015) Variable retention of kleptoplast membranes in cells of sacoglossan sea slugs: plastids with extended, shortened and non-retained durations. Zoomorphology 134(4):523–529. https://doi.org/10.1007/s00435-015-0278-3

47. Hehenberger E, Burki F, Kolisko M, Keeling PJ (2016) Functional relationship between a dinoflagellate host and Its diatom endosymbiont. Mol Biol Evol 33(9):2376–2390. https://doi.org/10.1093/molbev/msw109
48. Sun J, Rutherford ST, Silhavy TJ, Huang KC (2022) Physical properties of the bacterial outer membrane. Nat Rev Microbiol 20(4):236–248. https://doi.org/10.1038/2Fs41579-021-00638-0
49. Höhr AIC, Lindau C, Wirth C, Qiu J, Stroud DA, Kutik S, Guiard B, Hunte C, Becker T, Pfanner N et al (2018) Membrane protein insertion through a mitochondrial β-barrel gate. Science 359(6373):eaah6834. https://doi.org/10.1126/science.aah6834
50. Botte CY, Marechal E (2014) Plastids with or without galactoglycerolipids. Trends Plant Sci 19(2):71–78. https://doi.org/10.1016/j.tplants.2013.10.004
51. Sato N, Awai K (2016) Diversity in biosynthetic pathways of galactolipids in the light of endosymbiotic origin of chloroplasts. Front Plant Sci 7:117. https://doi.org/10.3389/fpls.2016.00117
52. Botte CY, Yamaryo-Botte Y, Janouskovec J, Rupasinghe T, Keeling PJ, Crellin P, Coppel RL, Marechal E, McConville MJ, McFadden GI (2011) Identification of plant-like galactolipids in *Chromera velia*, a photosynthetic relative of malaria parasites. J Biol Chem 286(34):29893–29903. https://doi.org/10.1074/jbc.m111.254979
53. Chen MH, Huang LF, Li HM, Chen YR, Yu SM (2004) Signal peptide-dependent targeting of a rice alpha-amylase and cargo proteins to plastids and extracellular compartments of plant cells. Plant Physiol 135(3):1367–1377. https://doi.org/10.1104/2Fpp.104.042184
54. Villarejo A, Buren S, Larsson S, Dejardin A, Monne M, Rudhe C, Karlsson J, Jansson S, Lerouge P, Rolland N et al (2005) Evidence for a protein transported through the secretory pathway en route to the higher plant chloroplast. Nat Cell Biol 7(12):1224–1231. https://doi.org/10.1038/ncb1330
55. Gruber A, Haferkamp I (2019) Nucleotide transport and metabolism in diatoms. Biomolecules 9(12):761. https://doi.org/10.3390/2Fbiom9120761
56. Gibbs SP (1981) The chloroplast endoplasmic reticulum: structure, function, and evolutionary significance. Int Rev Cytol 72:49–99. https://doi.org/10.1016/S0074-7696(08)61194-8
57. Gibbs SP (1979) The route of entry of cytoplasmically synthesized proteins into chloroplasts of algae possessing chloroplast ER. J Cell Sci 35:253–266. https://doi.org/10.1242/jcs.35.1.253
58. Flori S, Jouneau P-H, Finazzi G, Maréchal E, Falconet D (2016) Ultrastructure of the periplastidial compartment of the Diatom Phaeodactylum tricornutum. Protist 167(3):254–267. https://doi.org/10.1016/j.protis.2016.04.001
59. Archibald JM (2007) Nucleomorph genomes: structure, function, origin and evolution. Bioessays 29(4):392–402. https://doi.org/10.1002/bies.20551
60. Douglas S, Zauner S, Fraunholz M, Beaton M, Penny S, Deng LT, Wu XN, Reith M, Cavalier-Smith T, Maier UG (2001) The highly reduced genome of an enslaved algal nucleus. Nature 410(6832):1091–1096. https://doi.org/10.1038/35074092
61. Moore CE, Archibald JM (2009) Nucleomorph genomes. Annu Rev Genet 43:251–264. https://doi.org/10.1146/annurev-genet-102108-134809
62. Maier UG (1992) The four genomes of the alga *Pyrenomonas salina* (Cryptophyta). Biosystems 28(1–3):69–73. https://doi.org/10.1016/0303-2647(92)90009-n
63. Curtis BA, Tanifuji G, Burki F, Gruber A, Irimia M, Maruyama S, Arias MC, Ball SG, Gile GH, Hirakawa Y et al (2012) Algal genomes reveal evolutionary mosaicism and the fate of nucleomorphs. Nature 492(7427):59–65
64. Sato T, Nagasato C, Hara Y, Motomura T (2014) Cell cycle and nucleomorph division in *Pyrenomonas helgolandii* (Cryptophyta). Protist 165(2):113–122. https://doi.org/10.1016/j.protis.2014.01.003
65. Lewitus AJ, Caron DA, Miller KR (1991) Effects of light and glycerol on the organization of the photosynthetic apparatus in the facultative Heterotroph *Pyrenomonas Salina* (Cryptophyceae)1. J Phycol 27(5):578–587. https://doi.org/10.1111/j.0022-3646.1991.00578.x

66. Fu G, Nagasato C, Oka S, Cock JM, Motomura T (2014) Proteomics analysis of heterogeneous Flagella in Brown Algae (Stramenopiles). Protist 165(5):662–675. https://doi.org/10.1016/j.protis.2014.07.007
67. Uwizeye C, Decelle J, Jouneau P-H, Flori S, Gallet B, Keck J-B, Bo DD, Moriscot C, Seydoux C, Chevalier F et al (2021) Morphological bases of phytoplankton energy management and physiological responses unveiled by 3D subcellular imaging. Nat Commun 12(1):1049. https://doi.org/10.1038/s41467-021-21314-0
68. Bedoshvili YD, Popkova TP, Likhoshway YV (2009) Chloroplast structure of diatoms of different classes. Cell Tissue Biol 3(3):297–310
69. Flori S, Jouneau P-H, Bailleul B, Gallet B, Estrozi LF, Moriscot C, Bastien O, Eicke S, Schober A, Bártulos CR et al (2017) Plastid thylakoid architecture optimizes photosynthesis in diatoms. Nat Commun 8(1):15885. https://doi.org/10.1038/ncomms15885
70. Shimakawa G, Demulder M, Flori S, Kawamoto A, Tsuji Y, Nawaly H, Tanaka A, Tohda R, Ota T, Matsui H et al Diatom pyrenoids are encased in a protein shell that enables efficient CO_2 fixation. bioRxiv 2023:2023.2010.2025.564039. https://doi.org/10.1101/2023.10.25.564039
71. Horn S, Ehlers K, Fritzsch G, Gil-Rodríguez MC, Wilhelm C, Schnetter R (2007) *Synchroma grande* spec. nov. (Synchromophyceae class. nov. Heterokontophyta): an amoeboid marine alga with unique plastid complexes. Protist 158(3):277–293. https://doi.org/10.1016/j.protis.2007.02.004
72. Wetherbee R, Jackson CJ, Repetti SI, Clementson LA, Costa JF, van de Meene A, Crawford S, Verbruggen H (2019) The golden paradox – a new heterokont lineage with chloroplasts surrounded by two membranes. J Phycol 55(2):257–278. https://doi.org/10.1111/jpy.12822
73. Bhattacharya D, Ehlting J (1995) Actin coding regions: gene family evolution and use as a phylogenetic marker. Archiv für Protistenkunde 145(3):155–164. https://doi.org/10.1016/s0003-9365(11)80312-x
74. Lechtreck K-F (2004) An immunofluorescence study of the haptonema of *Chrysochromulina parva* (Prymnesiophyceae). Phycologia 43(5):635–640. https://doi.org/10.2216/i0031-8884-43-5-635.1
75. Kawachi M, Inouye I, Maeda O, Chihara M (1991) The haptonema as a food-capturing device: observations on *Chrysochromulina hirta* (Prymnesiophyceae). Phycologia 30(6):563–573. https://doi.org/10.2216/i0031-8884-30-6-563.1
76. Beattie A, Hirst EL, Percival E (1961) Studies on the metabolism of the Chrysophyceae. Comparative structural investigations on leucosin (chrysolaminarin) separated from diatoms and laminarin from the brown algae. Biochem J 79(3):531–537. https://doi.org/10.1042/bj0790531
77. Gagat P, Bodyl A, Mackiewicz P, Stiller JW (2014) Tertiary plastid endosymbioses in Dinoflagellates. In: Endosymbiosis. Springer. https://doi.org/10.1007/978-3-7091-1303-5_13
78. Howe C, Nisbet RE, Barbrook A (2008) The remarkable chloroplast genome of dinoflagellates. J Exp Bot 59:1035–1045. https://doi.org/10.1093/jxb/erm292
79. Waller RF, McFadden GI (2005) The Apicoplast: a review of the derived plastid of apicomplexan parasites. Curr Issues Mol Biol 7(1):57–80. https://doi.org/10.21775/cimb.007.057
80. Kohler S (2005) Multi-membrane-bound structures of Apicomplexa: I. the architecture of the Toxoplasma gondii apicoplast. Parasitol Res 96(4):258–272. https://doi.org/10.1007/s00436-005-1338-2
81. Tonkin CJ, Kalanon M, McFadden GI (2008) Protein targeting to the malaria parasite plastid. Traffic 9(2):166–175. https://doi.org/10.1111/j.1600-0854.2007.00660.x
82. Tonkin CJ, Struck NS, Mullin KA, Stimmler LM, McFadden GI (2006) Evidence for Golgi-independent transport from the early secretory pathway to the plastid in malaria parasites. Mol Microbiol 61(3):614–630. https://doi.org/10.1111/j.1365-2958.2006.05244.x
83. Füssy Z, Faitová T, Oborník M (2019) Subcellular compartments interplay for carbon and nitrogen allocation in *Chromera velia* and *Vitrella brassicaformis*. Genome Biol Evol 11(7):1765–1779. https://doi.org/10.1093/2Fgbe/2Fevz123
84. Moore RB, Obornik M, Janouskovec J, Chrudimsky T, Vancova M, Green DH, Wright SW, Davies NW, Bolch CJ, Heimann K et al (2008) A photosynthetic alveolate closely related to apicomplexan parasites. Nature 451(7181):959–963. https://doi.org/10.1038/nature06635

85. Füssy Z, Oborník M (2017) Chapter Six – Chromerids and their plastids. In: Hirakawa Y (ed) Advances in botanical research, vol 84. Academic Press, pp 187–218. https://doi.org/10.1016/bs.abr.2017.07.001
86. Ševčíková T, Horák A, Klimeš V, Zbránková V, Demir-Hilton E, Sudek S, Jenkins J, Schmutz J, Přibyl P, Fousek J et al (2015) Updating algal evolutionary relationships through plastid genome sequencing: did alveolate plastids emerge through endosymbiosis of an ochrophyte? Sci Rep 5:10134. https://doi.org/10.1038/srep10134
87. Martin W, Rujan T, Richly E, Hansen A, Ansorge W, Cornelsen S, Lins T, Leister D, Stoebe B, Hasegawa M et al (2002) Evolutionary analysis of *Arabidopsis*, cyanobacterial, and chloroplast genomes reveals plastid phylogeny and thousands of cyanobacterial genes in the nucleus. Proc Natl Acad Sci U S A 99(19):12246–12251. https://doi.org/10.1073/pnas.182432999
88. Sibbald SJ, Archibald JM (2020) Genomic insights into plastid evolution. Genome Biol Evol 12(7):978–990. https://doi.org/10.1093/gbe/evaa096
89. Soll J, Schleiff E (2004) Protein import into chloroplasts. Nat Rev Mol Cell Biol 5(3):198–208. https://doi.org/10.1038/nrm1333
90. Carrie C, Giraud E, Whelan J (2009) Protein transport in organelles: dual targeting of proteins to mitochondria and chloroplasts. FEBS J 276(5):1187–1195. https://doi.org/10.1111/j.1742-4658.2009.06876.x
91. Herrmann JM, Bykov Y (2023) Protein translocation in mitochondria: sorting out the Toms, Tims, Pams, Sams and Mia. FEBS Lett 597(12):1553–1554. https://doi.org/10.1002/1873-3468.14614
92. Schwenkert S, Dittmer S, Soll J (2018) Structural components involved in plastid protein import. Essays Biochem 62(1):65–75. https://doi.org/10.1042/ebc20170093
93. Kessler F, Schnell DJ (2006) The function and diversity of plastid protein import pathways: a multilane GTPase highway into plastids. Traffic 7(3):248–257. https://doi.org/10.1111/j.1600-0854.2005.00382.x
94. Hansen KG, Herrmann JM (2019) Transport of proteins into mitochondria. Protein J 38(3):330–342. https://doi.org/10.1007/s10930-019-09819-6
95. Lee DW, Hwang I (2021) Understanding the evolution of endosymbiotic organelles based on the targeting sequences of organellar proteins. New Phytol 230(3):924–930. https://doi.org/10.1111/nph.17167
96. Inoue K, Baldwin AJ, Shipman RL, Matsui K, Theg SM, Ohme-Takagi M (2005) Complete maturation of the plastid protein translocation channel requires a type I signal peptidase. J Cell Biol 171(3):425–430. https://doi.org/10.1083/2Fjcb.200506171
97. Ballabani G, Forough M, Kessler F, Shanmugabalaji V (2023) The journey of preproteins across the chloroplast membrane systems. Front Physiol 14:1213866. https://doi.org/10.3389/fphys.2023.1213866
98. Smith-Johannsen H, Gibbs SP (1972) Effects of chloramphenicol on chloroplast and mitochondrial ultrastructure in *Ochromonas danica*. J Cell Biol 52(3):598–614. https://doi.org/10.1083/jcb.52.3.598
99. Bodyl A, Mackiewicz P, Stiller JW (2009) Early steps in plastid evolution: current ideas and controversies. Bioessays 31(11):1219–1232. https://doi.org/10.1002/bies.200900073
100. Kilian O, Kroth PG (2003) Evolution of protein targeting into "complex" plastids: the "secretory transport hypothesis". Plant Biol 5:350–358. https://doi.org/10.1055/s-2003-42710
101. Vugrinec S, Gruber A, Kroth PG (2011) Protein targeting into complex plastids – support for the translocator model. J Endocytob Cell Res 21:59–63.
102. Gould SB, Sommer MS, Hadfi K, Zauner S, Kroth PG, Maier UG (2006) Protein targeting into the complex plastid of cryptophytes. J Mol Evol 62(6):674–681. https://doi.org/10.1007/s00239-005-0099-y
103. Maier UG, Zauner S, Hempel F (2015) Protein import into complex plastids: cellular organization of higher complexity. Eur J Cell Biol 94(7–9):340–348. https://doi.org/10.1016/j.ejcb.2015.05.008

104. Moog D, Stork S, Zauner S, Maier UG (2011) In silico and in vivo investigations of proteins of a minimized eukaryotic cytoplasm. Genome Biol Evol 3:375–382. https://doi.org/10.1093/gbe/evr031
105. Bhaya D, Grossman A (1991) Targeting proteins to diatom plastids involves transport through an endoplasmic reticulum. Mol Gen Genet 229:400–404. https://doi.org/10.1007/bf00267462
106. Bhaya D, Grossman AR (1993) Characterization of gene clusters encoding the fucoxanthin chlorophyll proteins of the diatom *Phaeodactylum tricornutum*. Nucleic Acids Res 21(19):4458–4466. https://doi.org/10.1093/2Fnar/2F21.19.4458
107. Emanuelsson O, Brunak S, von HG, Nielsen H (2007) Locating proteins in the cell using TargetP, SignalP and related tools. Nat Protoc 2(4):953–971. https://doi.org/10.1038/nprot.2007.131
108. Bendtsen JD, Nielsen H, von Heijne G, Brunak S (2004) Improved prediction of signal peptides: SignalP 3.0. J Mol Biol 340(4):783–795. https://doi.org/10.1016/j.jmb.2004.05.028
109. Pancic PG, Strotmann H (1993) Structure of the nuclear encoded gamma subunit of CF_0 CF_1 of the diatom *Odontella sinensis* including its presequence. FEBS Lett 320(1):61–66. https://doi.org/10.1016/0014-5793(93)81658-M
110. Emanuelsson O, Nielsen H, Brunak S, von Heijne G (2000) Predicting subcellular localization of proteins based on their N-terminal amino acid sequence. J Mol Biol 300(4):1005–1016. https://doi.org/10.1006/jmbi.2000.3903
111. Apt KE (2002) In vivo characterization of diatom multipartite plastid targeting signals. J Cell Sci 115(21):4061–4069. https://doi.org/10.1242/jcs.00092
112. Gruber A, Vugrinec S, Hempel F, Gould SB, Maier UG, Kroth PG (2007) Protein targeting into complex diatom plastids: functional characterisation of a specific targeting motif. Plant Mol Biol 64(5):519–530. https://doi.org/10.1007/s11103-007-9171-x
113. Kilian O, Kroth PG (2004) Presequence acquisition during secondary endocytobiosis and the possible role of introns. J Mol Evol 58(6):712–721. https://doi.org/10.1007/s00239-004-2593-z
114. Kilian O, Kroth PG (2005) Identification and characterization of a new conserved motif within the presequence of proteins targeted into complex diatom plastids. Plant J 41(2):175–183. https://doi.org/10.1111/j.1365-313x.2004.02294.x
115. Lang M, Apt KE, Kroth PG (1998) Protein transport into "complex" diatom plastids utilizes two different targeting signals. J Biol Chem 273(47):30973–30978. https://doi.org/10.1074/jbc.273.47.30973
116. Felsner G, Sommer MS, Maier UG (2010) The physical and functional borders of transit peptide-like sequences in secondary endosymbionts. BMC Plant Biol 10(1):223. https://doi.org/10.1186/1471-2229-10-223
117. Patron NJ, Waller RF (2007) Transit peptide diversity and divergence: a global analysis of plastid targeting signals. Bioessays 29(10):1048–1058. https://doi.org/10.1002/bies.20638
118. Gruber A, Rocap G, Kroth PG, Armbrust EV, Mock T (2015) Plastid proteome prediction for diatoms and other algae with secondary plastids of the red lineage. Plant J 81(3):519–528. https://doi.org/10.1111/tpj.12734
119. Huesgen PF, Alami M, Lange PF, Foster LJ, Schroder WP, Overall CM, Green BR (2013) Proteomic amino-termini profiling reveals targeting information for protein import into complex plastids. PloS One 8(9):e74483. https://doi.org/10.1371/journal.pone.0074483
120. Schneider TD, Stephens RM (1990) Sequence logos: a new way to display consensus sequences. Nucleic Acids Res 18(20):6097–6100. https://doi.org/10.1093/nar/18.20.6097
121. Crooks GE, Hon G, Chandonia J-M, Brenner SE (2004) WebLogo: a sequence logo generator: figure 1. Genome Res 14(6):1188–1190. https://doi.org/10.1101/gr.849004
122. Lang M, Kroth PG (2001) Diatom fucoxanthin chlorophyll a/c-binding protein (FCP) and land plant light-harvesting proteins use a similar pathway for thylakoid membrane Insertion. J Biol Chem 276(11):7985–7991. https://doi.org/10.1074/jbc.m006417200
123. Lau JB, Stork S, Moog D, Sommer MS, Maier UG (2015) N-terminal lysines are essential for protein translocation via a modified ERAD-system in complex plastids. Mol Microbiol 96:609. https://doi.org/10.1111/mmi.12959

124. Gruber A, Weber T, Río Bártulos C, Vugrinec S, Kroth PG (2009) Intracellular distribution of the reductive and oxidative pentose phosphate pathways in two diatoms. J Basic Microbiol 49(1):58–72. https://doi.org/10.1002/jobm.200800339
125. Weber T, Gruber A, Kroth PG (2009) The presence and localization of thioredoxins in diatoms, unicellular algae of secondary endosymbiotic origin. Mol Plant 2(3):468–477. https://doi.org/10.1093/mp/ssp010
126. Gould SB, Sommer MS, Kroth PG, Gile GH, Keeling PJ, Maier UG (2006) Nucleus-to-nucleus gene transfer and protein retargeting into a remnant cytoplasm of cryptophytes and diatoms. Mol Biol Evol 23(12):2413–2422. https://doi.org/10.1093/molbev/msl113
127. Gruber A, McKay C, Oborník M, Rocap G (2023) Multi class intracellular protein targeting predictions in diatoms and other algae with complex plastids: ASAFind 2.0. In: vol. arXiv:2303.02488: arXiv. https://doi.org/10.48550/arXiv.2303.02488
128. Gschloessl B, Guermeur Y, Cock JM (2008) HECTAR: a method to predict subcellular targeting in heterokonts. BMC Bioinform 9(1):393. https://doi.org/10.1186/1471-2105-9-393
129. Gruber A, Mckay C, Rocap G, Obornik M (2020) Comparison of different versions of SignalP and TargetP for diatom plastid protein predictions with ASAFind. Matters 6(6):e202005000001
130. Moog D (2018) In silico tools for the prediction of protein import into secondary plastids. Methods Mol Biol 1829:381–394. https://doi.org/10.1007/978-1-4939-8654-5_25
131. Gruber A, Kroth PG (2014) Deducing intracellular distributions of metabolic pathways from genomic data. Methods Mol Biol (Clifton, NJ) 1083:187–211. https://doi.org/10.1007/978-1-62703-661-0_12
132. Gile GH, Moog D, Slamovits CH, Maier U-G, Archibald JM (2015) Dual organellar targeting of aminoacyl-tRNA synthetases in diatoms and cryptophytes. Genome Biol Evol 7(6):1728–1742. https://doi.org/10.1093/gbe/evv095
133. Wastl J, Maier UG (2000) Transport of proteins into cryptomonads complex plastids. J Biol Chem 275(30):23194–23198. https://doi.org/10.1074/jbc.m003125200
134. Richtová J, Sheiner L, Gruber A, Yang S-M, Kořený L, Striepen B, Oborník M (2021) Using Diatom and Apicomplexan models to study the heme pathway of *Chromera velia*. Int J Mol Sci 22(12):6495. https://doi.org/10.3390/ijms22126495
135. Sommer MS, Gould SB, Lehmann P, Gruber A, Przyborski JM, Maier UG (2007) Der1-mediated preprotein import into the periplastid compartment of chromalveolates? Mol Biol Evol 24(4):918–928. https://doi.org/10.1093/molbev/msm008
136. Liu X, Hempel F, Stork S, Bolte K, Moog D, Heimerl T, Maier UG, Zauner S (2016) Addressing various compartments of the diatom model organism Phaeodactylum tricornutum via sub-cellular marker proteins. Algal Res 20:249–257. https://doi.org/10.1016/j.algal.2016.10.018
137. Felsner G, Sommer MS, Gruenheit N, Hempel F, Moog D, Zauner S, Martin W, Maier UG (2011) ERAD components in organisms with complex red plastids suggest recruitment of a preexisting protein transport pathway for the periplastid membrane. Genome Biol Evol 3:140–150. https://doi.org/10.1093/gbe/evq074
138. Hempel F, Bullmann L, Lau J, Zauner S, Maier UG (2009) ERAD-derived pre-protein transport across the 2nd outermost plastid membrane of diatoms. Mol Biol Evol 26:1781–1790. https://doi.org/10.1093/molbev/msp079
139. Lau JB, Stork S, Moog D, Schulz J, Maier UG (2016) Protein–protein interactions indicate composition of a 480 kDa SELMA complex in the second outermost membrane of diatom complex plastids. Mol Microbiol 100(1):76–89. https://doi.org/10.1111/mmi.13302
140. Hempel F, Felsner G, Maier UG (2010) New mechanistic insights into pre-protein transport across the second outermost plastid membrane of diatoms. Mol Microbiol 76(3):793–801. https://doi.org/10.1111/j.1365-2958.2010.07142.x
141. Stork S, Moog D, Przyborski JM, Wilhelmi I, Zauner S, Maier UG (2012) Distribution of the SELMA translocon in secondary plastids of red algal origin and predicted uncoupling of ubiquitin-dependent translocation from degradation. Eukaryot Cell 11(12):1472–1481. https://doi.org/10.1128/2FEC.00183-12

142. Marter P, Schmidt S, Kiontke S, Moog D (2020) Optimized mRuby3 is a suitable fluorescent protein for in vivo colocalization studies with GFP in the diatom *Phaeodactylum tricornutum*. Protist 171(1):125715. https://doi.org/10.1016/j.protis.2020.125715
143. Peschke M, Moog D, Klingl A, Maier UG, Hempel F (2013) Evidence for glycoprotein transport into complex plastids. Proc Natl Acad Sci U S A 110(26):10860–10865. https://doi.org/10.1073/pnas.1301945110
144. Bullmann L, Haarmann R, Mirus O, Bredemeier R, Hempel F, Maier UG, Schleiff E (2009) Filling the gap: the evolutionary conserved Omp85 in plastids of chromalveolates. J Biol Chem 285:6848–6856. https://doi.org/10.1074/jbc.m109.074807
145. Vugrinec S (2011) Protein targeting into Diatom Plastids. Dissertation, University of Konstanz, Konstanz. http://nbn-resolving.de/urn:nbn:de:bsz:352-138759
146. Allen WJ, Collinson I (2023) A unifying mechanism for protein transport through the core bacterial Sec machinery. Open Biol 13(8):230166. https://doi.org/10.1098/rsob.230166
147. Müller SB, Rensing SA, Martin WF (1994) Maier UG: cDNA cloning of a Sec61 homologue from the cryptomonad alga *Pyrenomonas salina*. Curr Genet 26:410–414. https://doi.org/10.1007/bf00309927
148. Gamerdinger M, Kobayashi K, Wallisch A, Kreft SG, Sailer C, Schlömer R, Sachs N, Jomaa A, Stengel F, Ban N et al (2019) Early scanning of nascent polypeptides inside the ribosomal tunnel by NAC. Mol Cell 75(5):996–1006.e1008. https://doi.org/10.1016/j.molcel.2019.06.030
149. Denks K, Vogt A, Sachelaru I, Petriman N-A, Kudva R, Koch H-G (2014) The Sec translocon mediated protein transport in prokaryotes and eukaryotes. Mol Membr Biol 31(2–3):58–84. https://doi.org/10.3109/09687688.2014.907455
150. Bolte K, Gruenheit N, Felsner G, Sommer MS, Maier UG, Hempel F (2011) Making new out of old: recycling and modification of an ancient protein translocation system during eukaryotic evolution. Mechanistic comparison and phylogenetic analysis of ERAD, SELMA and the peroxisomal importomer. Bioessays 33(5):368–376. https://doi.org/10.1002/bies.201100007
151. Cavalier-Smith T (2018) Kingdom Chromista and its eight phyla: a new synthesis emphasising periplastid protein targeting, cytoskeletal and periplastid evolution, and ancient divergences. Protoplasma 255(1):297–357. https://doi.org/10.1007/s00709-017-1147-3
152. Steiner JM, Berghofer J, Yusa F, Pompe JA, Klösgen RB, Löffelhardt W (2005) Conservative sorting in a primitive plastid. The cyanelle of *Cyanophora paradoxa*. FEBS J 272(4):987–998. https://doi.org/10.1111/j.1742-4658.2004.04533.x
153. Steiner JM, Löffelhardt W (2005) Protein translocation into and within cyanelles. Mol Membr Biol 22(1–2):123–132. https://doi.org/10.1080/09687860500041411
154. Steiner JM, Yusa F, Pompe JA, Loffelhardt W (2005) Homologous protein import machineries in chloroplasts and cyanelles. Plant J 44(4):646–652. https://doi.org/10.1111/j.1365-313x.2005.02559.x
155. Wunder T, Martin R, Löffelhardt W, Schleiff E, Steiner JM (2007) The invariant phenylalanine of precursor proteins discloses the importance of Omp85 for protein translocation into cyanelles. BMC Evol Biol 7:236. https://doi.org/10.1186/1471-2148-7-236
156. Cline K, Henry R, Li C, Yuan J (1993) Multiple pathways for protein transport into or across the thylakoid membrane. EMBO J 12(11):4105–4114. https://doi.org/10.1002/2Fj.1460-2075.1993.tb06094.x
157. Zhu D, Xiong H, Wu J, Zheng C, Lu D, Zhang L, Xu X (2022) Protein targeting into the thylakoid membrane through different pathways. Front Physiol 12:802057. https://doi.org/10.3389/2Ffphys.2021.802057
158. New CP, Ma Q, Dabney-Smith C (2018) Routing of thylakoid lumen proteins by the chloroplast twin arginine transport pathway. Photosynth Res 138(3):289–301. https://doi.org/10.1007/s11120-018-0567-z
159. Fernandez DE (2018) Two paths diverged in the stroma: targeting to dual SEC translocase systems in chloroplasts. Photosynth Res 138(3):277–287. https://doi.org/10.1007/s11120-018-0541-9

160. Schünemann D (2007) Mechanisms of protein import into thylakoids of chloroplasts. Biol Chem 388(9):907–915. https://doi.org/10.1515/bc.2007.111
161. Gould SB (2008) Ariadne's thread: guiding a protein across five membranes in cryptophytes 1. J Phycol 44(1):23–26. https://doi.org/10.1111/j.1529-8817.2007.00437.x
162. Schuenemann D, Gupta S, Persello-Cartieaux F, Klimyuk VI, Jones JDG, Nussaume L, Hoffman NE (1998) A novel signal recognition particle targets light-harvesting proteins to the thylakoid membranes. Proc Natl Acad Sci U S A 95(17):10312–10316. https://doi.org/10.1073/pnas.95.17.10312
163. Nymark M, Grønbech Hafskjold MC, Volpe C, Fonseca DDM, Sharma A, Tsirvouli E, Serif M, Winge P, Finazzi G, Bones AM (2021) Functional studies of CpSRP54 in diatoms show that the mechanism of thylakoid protein insertion differs from that in plants and green algae. Plant J 106(1):113–132. https://doi.org/10.1111/tpj.15149
164. Ziehe D, Dünschede B, Schünemann D (2018) Molecular mechanism of SRP-dependent light-harvesting protein transport to the thylakoid membrane in plants. Photosynth Res 138(3):303–313. https://doi.org/10.1007/s11120-018-0544-6

Translocation of Proteins into the Relict Plastid of Apicomplexan Parasites

Syrian G. Sanchez, Eléa A. Renaud, and Sébastien Besteiro

Abstract Parasites of the Apicomplexa phylum include species like *Plasmodium falciparum* and *Toxoplasma gondii* that are responsible for pathologies affecting humans and other animals. Most Apicomplexa contain a plastid-like organelle called the apicoplast that they inherited by secondary endosymbiosis. Because of its evolutionary history, this plastid is surrounded by four membranes across which the vast majority of apicoplast proteins, which are mostly encoded by the nucleus, must be imported in order to maintain proper organelle function. Translocation of proteins across these physical barriers is a complex journey involving sophisticated targeting mechanisms. Apicoplast proteins transit through the secretory pathway and usually carry a bipartite targeting signal composed of a signal peptide followed by a transit peptide important for specific steps of the import process. This combines classical translocons of the inner and outer chloroplast membranes with an additional endoplasmic reticulum-derived translocon machinery and its associated ubiquitin-like modifiers. Basic understanding of the molecular players that drive protein import into the apicoplast is of paramount importance, as these proteins are involved in several metabolic pathways hosted by the organelle that are vital for several apicomplexan developmental stages and are largely absent from their mammalian hosts, which makes them good potential drug targets to combat Apicomplexa-caused diseases.

Keywords Apicomplexa · Apicoplast · *Toxoplasma* · Toxoplasmosis · *Plasmodium* · Malaria · Metabolism · Protein import · Complex plastid · Protein translocation

S. G. Sanchez · E. A. Renaud · S. Besteiro (✉)
LPHI, Univ Montpellier, CNRS, INSERM, Montpellier, France
e-mail: elea.renaud@umontpellier.fr; sebastien.besteiro@inserm.fr

S. D. Schwartzbach et al. (eds.), *Endosymbiotic Organelle Acquisition*,
https://doi.org/10.1007/978-3-031-57446-7_16

1 Introduction

The phylum Apicomplexa encompasses a large number of obligate intracellular protozoan parasites, including the causative agents of important human diseases like malaria (*Plasmodium*) [1], cryptosporidiosis (*Cryptosporidium*) [2], and toxoplasmosis (*Toxoplasma*) [3]. Numerous other apicomplexan parasites, like those of the genera *Babesia, Eimeria, Theileria, Neospora,* and *Sarcocystis*, cause infectious diseases in wild and domesticated animals [4]. As many Apicomplexa species are of medical and economic importance, they have received considerable attention, although the most studied are *P. falciparum* and *T. gondii.*

P. falciparum is the main causative agent of human malaria, which is responsible for more than half a million deaths each year, mostly affecting children living in sub-Saharan *Africa* [5]. Transmission to humans occurs through the bite of an infected female mosquito, and after an initial developmental stage in the liver, the parasite undergoes repeated cycles of asexual replication in the blood that is responsible for the clinical symptoms of the disease. In some areas of Africa, there are very high mortality rates for malaria in children under the age of 5 years, but even in adults (where immunity against the parasite can develop more quickly), the infection can occasionally lead to a fatal episode of severe malaria [6]. *T. gondii* has a much lower burden on human health, but it is a very ubiquitous parasite that is able to infect virtually any warm blooded vertebrate, and is present in up to a third of the human population [7]. In healthy individuals, the infection is generally asymptomatic, but in the case of immunocompromised patients or in developing fetuses, it can lead to severe ocular or neurological outcomes that are caused by the spread of a highly multiplicative and invasive stage called tachyzoite [8]. This developmental stage, which is responsible for the acute form of toxoplasmosis is the most studied and the best characterized. Apicomplexan parasites have complex life cycles and not all species or developmental stages have benefited from extensive research efforts, although recent methodological advances offer the possibility of using functional genetics approaches in more cellular models [9]. Fortunately, not all groundbreaking discoveries rely on the use of cutting-edge molecular deciphering techniques, and the discovery of the vestigial plastid of apicomplexan parasites is an example of how observational biology coupled to the power of molecular biology approaches led to the identification of this unusual compartment [10].

2 Evolutionary History of the Apicoplast

Early electron microscopy studies dating back to the 1960s revealed the presence of a single organelle surrounded by multiple membranes in several apicomplexan parasites, but it was initially given different names depending on the organism (see [11] for an overview). In the 1970s, a circular extrachromosomal DNA molecule was observed in *Plasmodium* which was found to display a cross-shaped configuration

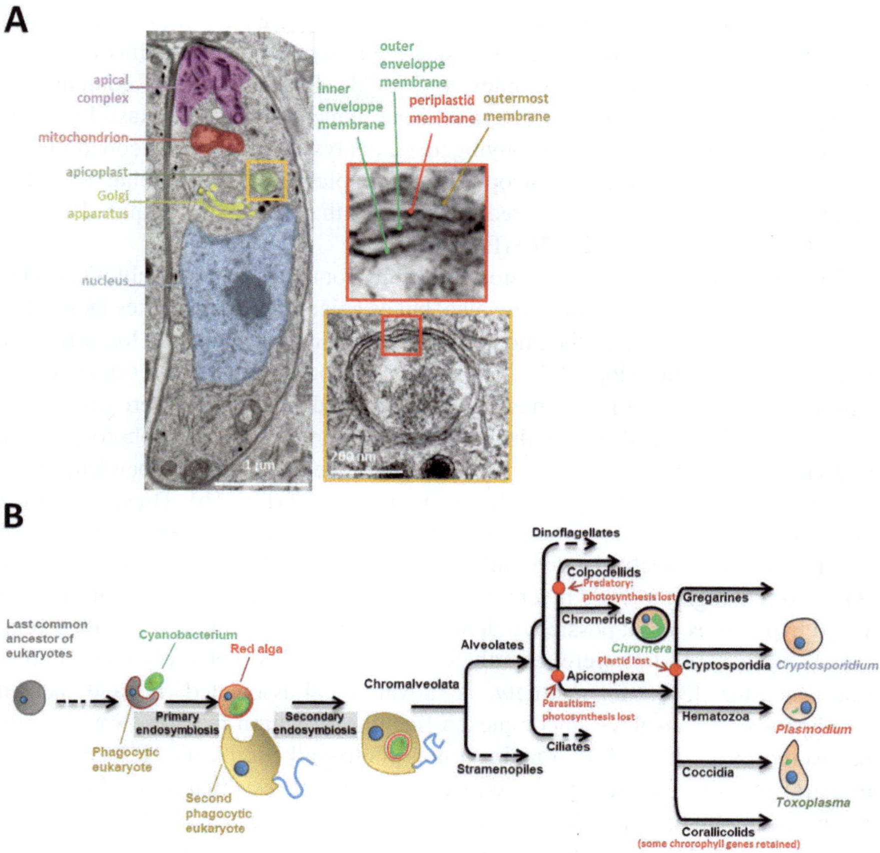

Fig. 1 Ultrastructure and evolutionary origin of the apicoplast. (**a**) Electron micrograph showing the ultrastructure of a *T. gondii* tachyzoite. The apical complex (comprising organelles used for host cell invasion), the mitochondrion, Golgi apparatus, nucleus, and apicoplast are highlighted with false colors. The insets show magnified views of the apicoplast and its four surrounding membranes (electron micrographs courtesy of Laurence Berry). (**b**) Schematic representation of the evolutionary history of the apicoplast

[12], hinting for the presence of inverted sequence repeats, which is one of the hallmarks of plastid DNA [13]. In situ hybridization studies in *Plasmodium* and *Toxoplasma* later revealed that the extrachromosomal genome resides in the multimembrane compartment that was observed previously, and sequencing of this genome confirmed its plastidic nature: the organelle was thus called "apicoplast" (for *apico*mplexan *plast*id) [14–16]. Detailed ultrastructural observations of the apicoplast revealed that it is surrounded by four membranes in several species [16–20], hinting at a complex evolutionary origin (Fig. 1a).

It is commonly accepted that plastids arose by the engulfment of a free-living ancestral cyanobacterium by an early eukaryote, which gave rise to the supergroup Archaeplastida possessing plastids with two envelope membranes [21]. This important event in the evolutionary history of eukaryotes gave rise to red and green algae, as well as land plants. Then, secondary endosymbiotic events involving two eukaryotes occurred, adding complexity [22, 23]. Plastid genomes comparisons have proven difficult because of divergent gene sequence or overall plastid genome organization [24], but molecular phylogenetic analyses of nuclear-encoded plastid genes have confirmed that the apicoplast, like the plastid of closely related dinoflagellates, is a complex plastid of red algal origin that was likely acquired from a secondary endosymbiosis [25, 26] (Fig. 1b).

The proto-apicomplexan ancestor resulting from these endosymbiotic events was likely a free-living phototrophic alga, but as apicomplexan parasites lack genes involved in photosynthesis, the photosynthetic function was probably lost when the parasitic lifestyle was adopted [27, 28]. In fact, apicomplexan parasites have photosynthetic relatives like the chromerids *Chromera* and *Vitrella* that were discovered recently [29, 30], but they are also close phylogenetically to non-photosynthetic colpodellid predators [31], suggesting the occurrence of several independent losses of photosynthesis in early apicomplexan evolution [32] (Fig. 1b). Thus, a common ancestor of Apicomplexa, chromerids, colpodellids, and also the above mentioned dinoflagellates (collectively denominated as Myzozoa, [33]) contained a plastid [34], and although the wide diversity of the plastidic features in different lineages fueled hypotheses about possible independent gains in Apicomplexa and their relatives [35], their plastids were nevertheless likely inherited vertically from a common ancestor [36]. Interestingly, a novel coral-associated lineage named corallicolids, that displays apicomplexan-like ultrastructural features, was recently discovered [37]; although it contains a non-photosynthetic plastid, it has retained ancestral genes for chlorophyll biosynthesis, which suggests that it may be an evolutionary intermediate. Finally, it should be noted that some apicomplexan parasites like those from the genus *Cryptosporidium*, as well as several members of the related group of gregarines (which are invertebrates-infecting Apicomplexa), seem to be lacking an apicoplast [38, 39].

After acquisition of the red algal endosymbiont by the proto-apicomplexan ancestor, there was a series of evolutionary reductions, leaving only a small genome and a stroma bound by four membranes (Fig. 1a). From the innermost to the outermost, these membranes are: the inner and outer envelope membranes, which are the two membranes surrounding the original red algal plastid, the periplastid membrane (which was the plasma membrane of the endosymbiont), and lastly a membrane originating from the host's own endomembrane system (which could be phagosome and/or endoplasmic reticulum (ER)-derived) [40, 41]. Translocation across organellar membranes involves considerable energetic costs, which perhaps played a role in the reductive evolution of endosymbiotic organelles [42, 43]. After the primary phagocytic event, the original phagosomal membrane was either lost or two of the membranes were merged during the transition of the endosymbiont into an organelle. Moreover, the original cyanobacterium-derived thylakoid membranes disappeared as photosynthesis was lost by the proto-apicomplexan ancestor. In other complex plastids, like the archetypical plastid of the related dinoflagellates lineage

Table 1 Features of selected apicoplast genomes

	Apicoplast genome size (kb)	Encoded proteins	tRNA	rRNA	A + T %	Reference	Specific comments
Plasmodium falciparum	35	31	25	2	86.9	[15]	
Toxoplasma gondii	35	29	24	2	78.6	[155]	rRNA genes duplicated
Babesia bovis	33	32	25	2	78.2	[156]	Transcription from only one strand
Babesia microti	29	31	24	2	86	[157]	Transcription from only one strand
Eimeria tenella	35	29	24	2	79.4	[158]	
Theileria parva	39.5	44	24	2	80.5	[159]	Transcription from only one strand
Cyclospora cayetanensis	35	29	33	2	78.1	[160]	rRNA genes duplicated
Leucocytozoon caulleryi	35	30	25	2	85.1	[161]	

for instance, one additional membrane has been lost (most likely the third from the inside, that corresponds to the plasma membrane of the algal endosymbiont) [44].

While the average chloroplast genome size in land plants is around 150 kb (ranging from 130 to 170 kb in most species), the apicoplast genome is much smaller (about 30–40 kb in length depending on the apicomplexan species, Table 1), hinting that several deletions and rearrangements occurred in the organellar genome during evolution. This streamlined, but high coding density, plastid genome is usually present in variable copy numbers depending on the Apicomplexa, but has a comparable size and is highly A/T-rich (Table 1). Apicoplast genomes typically encode proteins involved in housekeeping functions like transcription and translation, including ribosomal ribonucleic acids (rRNAs) associated to small and large ribosome subunits, about 24 transfer RNA species, 16 ribosomal proteins, the EF-Tu translation elongation factor and 3 subunits of a bacterial-type RNA polymerase.

The housekeeping proteins encoded by the apicoplast genome have prokaryotic features, thus molecules inhibiting translation in the apicoplast have been efficiently used as anti-apicomplexan drugs [45]. Only a few proteins encoded by the apicoplast genome are not involved in the expression of apicoplast genes. They include the ClpC protease, which is part of a key complex regulating proteostasis in the organelle [46], and SufB, which is part of the vital iron–sulfur cluster synthesis pathway hosted by the organelle [47, 48]. There are also a handful of hypothetical proteins whose function is currently unknown, but that seem to be conserved across several apicoplast genomes.

The predicted apicoplast proteome is clearly much more complex and diverse than the few proteins encoded by the organelle's genome: it supposedly contains hundreds of proteins [49, 50]. This implies that most of the apicoplast-resident proteins became encoded by the nucleus during plastid evolution. It is commonly assumed that this endosymbiotic gene transfer allowed the host to gain control over the expression of the plastid genes [51], and created a dependency over time to secure the definitive establishment of the endosymbiont as an organelle. One consequence of transfer to the host nuclear genome is that the products have to be imported back into the organelle to exert their function, through a specific machinery that is the main subject of this chapter.

During the evolution, division of the apicoplast became dependent on, and coordinated with division of the host [52], and is driven by association with the centrosomes in *Toxoplasma* [53]. As one particular example of the tight coordination needed between apicoplast and cellular division, during the course of replication of *Plasmodium* asexual stages, the organelle becomes highly branched but then rapidly segregates into newly forming daughter cells upon completion of nuclear division [54]. Interestingly, apicomplexan parasites have not retained any of the proteins associated with plastid division in plants and red algae, but instead they use an alternative dynamin for organelle fission [55] and seem to rely, in a non-canonical way, on part of the autophagy-related machinery for proper organelle inheritance [56, 57].

3 The Apicoplast Is an Important Metabolic Hub

In spite of having lost its photosynthetic activity, the plastid was retained by most apicomplexan parasites, hinting that key functions associated with the organelle are vital for these pathogens. Bioinformatics approaches utilizing the genomic data available for Apicomplexa like *Plasmodium* and *Toxoplasma*, with known metabolic maps of plant chloroplasts and bacteria as a scaffold, allowed predicting the metabolic pathways potentially harbored by the organelle [58, 59]. The four main apicoplast-localized biosynthetic pathways are for the biosynthesis of: heme (together with the mitochondrion), fatty acids (FAs, via a prokaryotic FASII pathway), isoprenoid precursors (through the so-called non-mevalonate or 1-deoxy-D-xylulose 5-phosphate -DOXP- pathway), and iron–sulfur (Fe–S) clusters (Fig. 2).

All these pathways have been mostly functionally investigated in *Plasmodium* and *Toxoplasma*, and they were generally found to be important for parasite fitness, albeit with varying degrees of essentiality depending on the developmental stages considered. For instance, de novo heme biosynthesis is dispensable in the blood stages of *P. falciparum*, but it is required for its liver stages [60–62], and it is also important for *Toxoplasma* tachyzoites [63]. Similarly, FASII seems to be only essential for the development of *Plasmodium* liver stages [64], and in *Toxoplasma* while FASII was shown previously to be critical for tachyzoite fitness [65], it seems

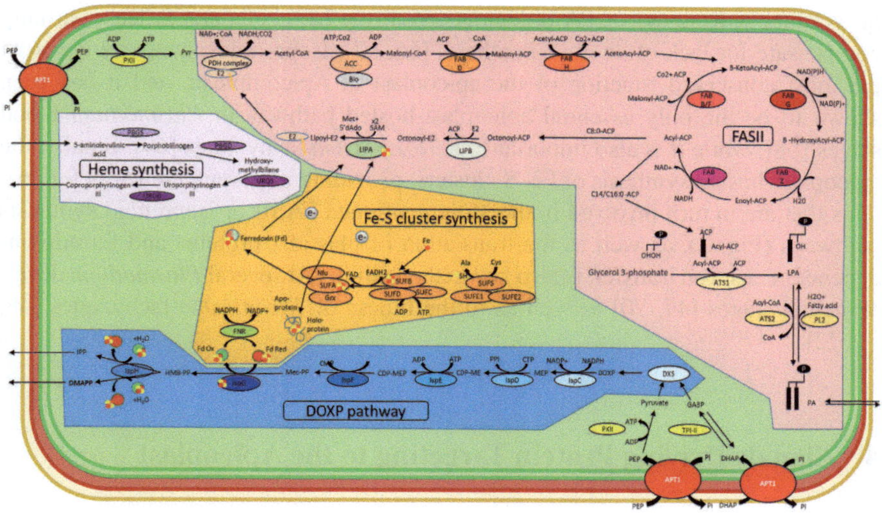

Fig. 2 Schematic representation of the main metabolic pathways hosted by the apicoplast. These are the Fatty Acid Synthesis (FAS) II pathway, the DOXP pathway (generating precursors of isoprenoids), the Iron–sulfur (Fe–S) cluster synthesis pathway, and part of the heme synthesis pathway. Abbreviations: APT1 Apicoplast triosephosphate translocator 1, PEP Phosphoenolpyruvate, DHAP Dihydroxyacetone phosphate, Pi Inorganic phosphate, PKII Pyruvate kinase II, ADP Adenosine diphosphate, ATP Adenosine triphosphate, Pyr Pyruvate, PDH complex Pyruvate dehydrogenase complex, E2 E2 Subunit of the PDH complex (shown with a yellow tail when lipoylated), NAD(P)+/NAD(P)H Nicotinamide adenine dinucleotide (phosphate) oxidized or reduced, CoA Coenzyme A, CO2 Carbon dioxide, Acetyl-CoA Acetyl-Coenzyme A, ACC Acetyl-CoA carboxylase, Bio Biotin, FABD Malonyl CoA-acyl carrier protein transacylase, FABH β-ketoacyl-[ACP] synthases III, FABG Ketoacyl-acyl carrier protein reductases, FABZ β-hydroxyacyl-Fdoxacyl carrier protein dehydratase, FABI enoyl-acyl carrier reductase, LIPB lipoyl(octanoyl)-ACP protein transferase, LIPA lipoic acid synthase, SAM S-Adenosylmethionine, Met Methionine, 5′ dAdo 5′ desoxyadenosine, ATS1 Glycerol 3-phosphate acyltransferase, LPA Lysophosphatidic acid, ATS2 Acyl-glycerol-3-phosphate acyltransferases, PL2 Patatin-like phospholipase, P.A Phosphatidic Acid, TPI-II Triose phosphate isomerase II, GA3P Glyceraldehyde-3-phosphate, DXS 1-deoxy-D-xylulose-5-phosphate synthase, DOXP 1-deoxy-D-xylulose-5-phosphate, ISPC 1-deoxy-D-xylulose 5-phosphate reductoisomerase, MEP 2-C-methyl-D-erythritol 4-phosphate, ISPD 4-Diphosphocitidyl-2C-methyl-D-erythritol synthase, CTP Cytidine 5′-triphosphate, PPi inorganic diphosphate, CDP-ME 4-Diphosphocytidyl-2-C-methyl-D-erythritol, ISPE 4-Diphosphocytidyl-2-C-methylerythritol kinase, CDP-MEP 4-Diphosphocytidyl-2-C-methyl-D-erythritol 2-phosphate, ISPF 2-C-methyl-D-erythritol 2,4-cyclodiphosphate synthase, CMP Cytidine 5-monophosphate, Mec-PP Methylerythritol cyclodiphosphate, ISPG 4-hydroxy-3-methylbut-2-en-1-yl diphosphate synthase, HMB-PP 4-hydroxy-3-methyl-but-2-enyl-pyrophosphate, ISPH 4-hydroxy-3-methylbut-2-enyl diphosphate reductase, IPP Isopentenyl pyrophosphate, DMAPP Dimethylallyl pyrophosphate, Ala Alanine, Cys Cysteine, SUFS Cysteine desulfurase, SUFE1/E2 Cysteine desulfurase partners, SUFB, SUFC, and SUFD scaffolding complex, SUFA, Grx and Nfu Fe-S cluster transfer proteins, FNR Ferredoxin NADP+ oxidoreductase, e^- electrons, PBGS Porphobilinogen synthase, PBGD Porphobilinogen deaminase, UROS Uroporphyrinogen Synthase, UROD Uroporphyrinogen decarboxylase

in fact that scavenging of FA precursors from the host cells can at least partially compensate for lack of de novo synthesis [66, 67]. Isoprenoids synthesis is one of the most conserved function of the apicoplast in Apicomplexa, and it has been shown to be the only essential apicoplast-hosted pathway in *Plasmodium* blood stages [68], while it is also important for *Toxoplasma* tachyzoites [69]. Finally, the apicoplast-hosted synthesis of Fe–S clusters, providing important cofactors for proteins that are in turn involved in the aforementioned FASII or isoprenoid synthesis pathways (Fig. 2), as well as the translation of plastidic proteins and ferredoxin-dependent electron transfer, is also key to the viability of several *Plasmodium* developmental stages [47, 70] as well as to the fitness of *Toxoplasma* tachyzoites [48, 71, 72].

4 Signals Driving Protein Targeting to the Apicoplast

The import of nuclear-encoded proteins into organelles originating from primary endosymbiosis, like the mitochondrion or primary plastids, typically involves outer and inner membrane translocons recognizing specific targeting presequences present in cytosolic-synthesized precursors. The overall origin of the mitochondrial targeting sequences and plastid-targeting transit peptides (TP), whose sequences are usually poorly conserved, is currently unknown: they might have arisen from convergent evolution by de novo acquisition, or might even have emerged from host-derived antimicrobial peptides [73].

For import of stromal proteins into secondary plastids, which in the case of the apicoplast contains two additional membranes, an additional step is needed. A bipartite N-terminal pre-sequence consisting of an ER targeting signal peptide that drives entry into the secretory pathway for access to the outermost membranes, is then followed by a more classical TP used for subsequent import by translocons through the innermost membranes of the apicoplast [74, 75] (Fig. 3).

Although apicoplast TPs show similarities to plant TPs, they do not contain any conserved specific motif [76]. They vary in length (up to 200 amino acids) and amino acid composition [77, 78], but the presence of positively-charged residues seems to be a conserved feature [49, 79]. Interestingly, in spite of differences in overall amino acid usage (sometimes driven by variations in the A/T content of the genome [76]), several examples in the literature have shown that TPs from other apicomplexan species could successfully drive targeting to the organelle (i.e., using a *Toxoplasma* TP for expression in *Plasmodium*), implying a good degree of functional conservation [74, 80].

As mentioned above, mitochondria are, like plastids, organelles of endosymbiotic origin for which a large proportion of the proteome is nuclear-encoded. Thus, many proteins of the single mitochondrion present in apicomplexan parasites are directly imported post-translationally into the organelle through the use of a mitochondrial targeting sequence [81]. Sequences for targeting proteins to the mitochondrion or to simple plastids have particular features driving specificity for organellar

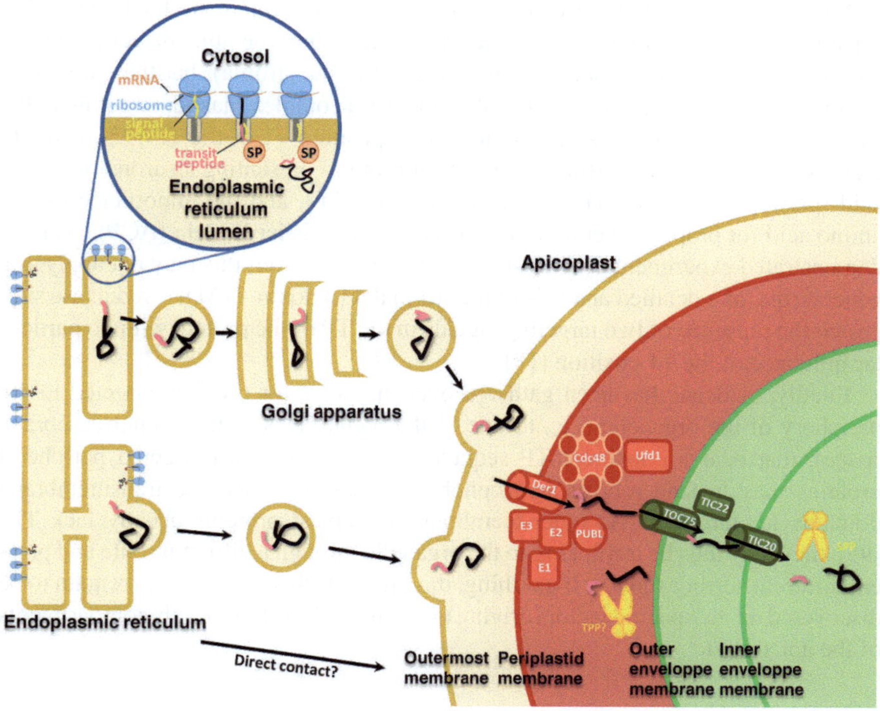

Fig. 3 Schematic representation of protein trafficking to the apicoplast. The polypeptide to be trafficked to the stroma of the apicoplast is represented in black, the transit peptide is in pink and the signal peptide in yellow. SP Signal peptide peptidase, SPP Stromal processing peptidase, TP Transit peptide, TPP Transit peptide peptidase, DER1 degradation, Cdc48 cell division cycle 48, PUBL plastid ubiquitin-like protein, Ufd1 ubiquitin fusion protein degradation 1, E1 ubiquitin-activating enzyme, E2 ubiquitin-conjugating enzyme, E3 ubiquitin ligase, TOC translocon of the outer apicoplast membrane, TIC translocon of the inner apicoplast membrane. See corresponding text for additional details

targeting [82], but some degree of compositional similarities can sometimes make them interchangeable and lead to dual-targeting to the plastid and the mitochondrion. These 'ambiguous' targeting signals have been predominantly documented in plants [83–85]. There are also several examples of a single presequence allowing transport to both the mitochondrion and the apicoplast in Apicomplexa. Perhaps the suboptimal nature of some signal peptide sequences may also contribute to the targeting to both organelles. Dually-targeted apicomplexan proteins include antioxidant enzymes superoxide dismutase and thioredoxin-dependent peroxidase [86–88], Fe–S protein aconitase [86], a lipoate protein ligase [89], a pyruvate kinase [90], and a serine hydroxymethyltransferase [91]. As mitochondria evolved before plastids [92], dual targeting of some specific enzymes to the apicoplast may have been subsequently acquired to perhaps provide the parasites with higher metabolic flexibility.

Specific signals in the presequence or in the mature protein are likely needed to drive the final localization of the targeted protein within the apicoplast (associated with one of the inner membranes, or addressed to the stroma), but they are so far largely unknown. In diatoms, which also possess a complex plastid, targeting to the stroma depends on an aromatic amino acid at position +1 of the TP [93]. Overall, there is a significant proportion of apicoplast proteins presenting an aromatic amino acid (particularly phenylalanine) at this position [76], and the importance of this amino acid for proper targeting of the stromal Acyl Carrier Protein (ACP) has been demonstrated experimentally in *Toxoplasma* [94]. However, this may not apply as a general rule, as a detailed analysis of the stromal ferredoxin–NADP+ reductase suggested the presence of two targeting signals in the TP of the protein with no particular influence of the +1 position [78].

Finally, evidence has been gathered over the years for proteins targeted to the periphery of the organelle (i.e., to one of the membranes or to an intermembrane space), that bear no apparent TP sequence. These apicoplast-targeted peripheral proteins may instead rely on hydrophobic sequences or specific transmembrane domains for targeting. As outer-membrane apicoplast proteins usually lack TPs [95], they may thus be trafficking to the organelle through a different route and present different sorting signals. If anything, this suggests that many things remain to be discovered on the specific factors driving targeting to the different sub-compartments of the apicoplast.

5 Transport to the Outer Membrane

The exact nature of the outermost membrane that surrounds the apicoplast is not known [41]. It may be of phagosomal origin, thus bearing similarities to the phagocytotic vacuole membrane of the host that was used during the secondary endosymbiosis. Yet, in several eukaryotic lineages like diatoms and cryptomonads that also contain complex plastids, the outermost membrane is a continuity of the host's ER and is studded with ribosomes, suggesting that the first step of protein import is a cotranslational translocation across the outer membrane of these plastids [96], potentially through the Sec61 translocon [97]. In apicomplexan parasites, there is no evidence of a continuity between the ER and the outermost apicoplast membrane, although they can be in close contact and direct protein transfer between these two compartments through membrane contact sites may not be ruled out [20, 98].

Entry into the ER seems, nevertheless, to be a common feature of proteins targeted to secondary plastids. The signal peptide is required for apicoplast targeting, since its deletion in apicoplast proteins by genetic engineering leads to cytosolic or mitochondrial localization of the proteins [74, 75, 78]. On the other hand, the signal peptide of an apicoplast protein can be exchanged with signal peptides from other secretory proteins and the protein will still be targeted correctly [99]. Nuclear-encoded apicoplast proteins are translated as preproteins on membrane bound ribosomes, and after co-translational translocation of the protein into the ER (presumably across the Sec61 complex), the signal peptide is cleaved off by a signal peptidase to eventually expose the TP (Fig. 3).

The proteins then traffic from the ER to the apicoplast, likely involving vesicular trafficking rather than direct "membrane to membrane" contact, but whether or not it involves export through the Golgi apparatus is unclear. Transfer from the ER to the Golgi apparatus is usually the default secretory pathway. From the Golgi apparatus proteins are subsequently sorted to diverse cellular destinations (including Apicomplexa-specific secretory organelles like rhoptries and micronemes), through the use of dedicated targeting sequences [100].

As proteins with a typical ER retention sequence are retrieved from the cis-Golgi compartment but are recycled back to the ER, one would expect that the addition of a C-terminal ER retention sequence (i.e., HDEL or SDEL) would block their transport to the apicoplast if a transport through the Golgi apparatus is required. The addition of such a C-terminal retention sequence, as well as the use of brefeldin A (a drug that blocks ER-to-Golgi protein trafficking), was found not to have any impact on the targeting of the apicoplast stroma-localized ACP [99, 101]. Targeting of ACP is also resistant to low temperature, which supposedly disrupts ER to Golgi trafficking [101]. On the other hand, fusion of ACP to conditional aggregation domains (like those of FK506 binding proteins) caused accumulation of the protein in the ER and blocked further trafficking to the apicoplast [101]. This led to the assumption that the trafficking of apicoplast luminal proteins might bypass the Golgi apparatus. However, other investigations using similar approaches suggested that, in fact, there could be different routes depending on the nature of the apicoplast proteins, and that they could either involve trafficking through the Golgi apparatus or direct from the ER to the apicoplast [87, 102, 103]. In any case, the nature and composition of the trafficking vesicles, whether they go through the Golgi system or shuttle directly from the ER, is unknown.

Both genetic and drug-based approaches have been used in *Toxoplasma* and in *Plasmodium* to induce the loss of the plastid, and although this is ultimately lethal for the parasites, the moderately slow onset of action allowed the observation of apicoplast proteins that accumulate in what looks like vesicular trafficking intermediates [68, 104]. Several peripheral apicoplast proteins were also observed by immuno-electron microscopy within vesicles in the vicinity of the apicoplast during normal parasite development, especially as the plastid elongated in preparation for division: this was observed for integral membrane proteins like the AAA protease FtsH1 and the phosphate transporter APT1, but also for the Atrx1 thioredoxin, a soluble or peripheral membrane protein residing in the apicoplast intermembrane spaces [105–107]. The formation of these vesicles does not seem to require an intact Golgi apparatus (but it does not rule out that their cargo is not usually Golgi-derived, as in the absence of a functional organelle they may accumulate as ER-derived vesicles instead). However, these vesicles do not seem to contain stromal protein content, which again suggests that peripheral and stromal apicoplast proteins go through different trafficking routes [104]. Interestingly, Golgi apparatus and apicoplast-localized proteins belonging to the SNARE family, which are key molecules involved in vesicular trafficking, were recently shown to be important for the targeting of apicoplast proteins in *Toxoplasma* [108, 109]. Because of the different models, experimental approaches, and protein markers that have been used over the years, overall, the current literature does not allow the establishment of a clear model for a differential trafficking of stromal and peripheral apicoplast proteins through specific ER- or Golgi apparatus-based routes. Some conflicting findings by different research groups clearly highlight the need for further investigation [102, 104, 110, 111].

6 Use of an ER-Derived Machinery for Import Through the Periplastid Membrane

The periplastid compartment and membrane represent a remnant of the original endosymbiont [112]. Transport across the periplastid membrane of complex plastids from chromalveolates like diatoms and Apicomplexa has been shown to unexpectedly involve a specific version of the ER-associated degradation (ERAD) complex [113, 114]. The ERAD is a quality control process of the secretory pathway that depends on the ubiquitin/proteasome, implying that misfolded proteins have to be transported from the ER back into the cytoplasm by a translocon [115]. Polyubiquitination is essential for the transport of substrates, and is mediated by ubiquitin-activating (E1), ubiquitin-conjugating (E2), and ubiquitin ligase (E3) enzymes. The addition of ubiquitin to target proteins involves a series of reactions with these E1/E2/E3 enzymes. The transport itself relies on transmembrane protein Der1 and it involves recognition by the Cdc48 ubiquitin-binding factor. Ubiquitination is thus not only a key mark of classical ERAD for targeting proteins to the proteasome for degradation, but it is also potentially a requirement for protein transport across the translocon [116].

SELMA (symbiont specific ERAD-like machinery) is a form of the ERAD transport machinery repurposed for a transport task at the periplastid membrane, with a main difference that it is not coupled to the proteasome system [117]. It is found in the second-outermost membrane of complex plastids and involves Der1-like proteins, a membrane-bound E3 ligase, and a Cdc48-like complex (Cdc48-Ufd1) to move polypeptides across this membrane [118]. This SELMA machinery has been shown to localize at the apicoplast in *Plasmodium* [119, 120], but also in *Toxoplasma* [121], in which most functional studies have been performed. Mutating the apicoplast-localized *Toxoplasma* homologues of Der1, the candidate membrane pore component of the translocon [122], and of Cdc48, the AAA ATPase that, together with its ubiquitin-binding cofactor Ufd1, provides the mechanical force allowing protein passage across the membrane [123] (Fig. 3), led to a loss of apicoplast protein import [121, 124].

The ubiquitination system is also present at the apicoplast in *Plasmodium* and *Toxoplasma*, and is important for the SELMA function in protein import [125].

The primarily unstructured nature of the apicoplast TP may favor recognition by the SELMA system [126], which classically recognizes misfolded proteins, but the features that would drive the specificity for SELMA versus ERAD have not yet been identified. Interestingly, a plastid-specific ubiquitin-like protein (PUBL) has been discovered in *Toxoplasma* and found to be essential for transport of proteins across the periplastid membrane [124]. However, while conserved in many Apicomplexa, PUBL seems to be absent from *Plasmodium*, which might rely on a different apicoplast-specific ubiquitin-like molecule [120]. In spite of considerable sequence differences between PUBL and canonical *Toxoplasma* ubiquitin, when the latter was addressed at the apicoplast, it was able to complement PUBL function, suggesting that this organelle-specific version is primarily involved in a general role for ubiquitination rather than a diversified and specialized function [124]. By analogy with the ERAD system, the translocation of a precursor protein through the SELMA translocon would be initiated by Der1, and when reaching the periplastidal compartment, the precursor protein would get ubiquitinated, followed by its complete extraction by the Cdc48 ATPase and its cofactor Ufd1 (Fig. 3). After translocation, the precursor protein likely undergoes deubiquitination and is either folded in the periplastid space or passed on to the translocon in the subsequent membrane.

7 Getting There: Crossing of the Two Inner Membranes

The two innermost membranes of the apicoplast correspond to those of the primary plastid originating from the first endosymbiosis event. Import into primary plastids has been extensively studied [127]: the protein transport machinery involves membrane transport components originating from the cyanobacterial endosymbiont and additional components adapted from the host [128]. The importing translocons are typically made up of two different protein complexes allowing transport at the outer and inner envelope membranes: the so-called TOC and TIC machineries (for Translocon at the Outer/Inner envelope membrane of Chloroplast, respectively). Given the conservation of function of these translocon systems in primary plastids, it is not surprising that apicomplexan parasites also use TOC and TIC-like proteins for protein transfer across the two inner membranes of the apicoplast (Fig. 3).

The composition of the TOC and, more particularly, of the TIC complexes is still debated [127]. In fact, these complexes are structurally and functionally distinct depending on the type of organism, plastid, or even tissue (in land plants for instance) considered. Beyond the core machinery, during evolution, additional components from the host were likely integrated to adapt for the specificity of transport or its regulation [129, 130]. The major component of the protein import channel at the outer membrane is Toc75 [131], which likely derives from a cyanobacterial outer membrane β-barrel protein called Omp85 [132]. Toc75 binds TPs and also chaperones like Tic22, which are thought to facilitate the transit of the preprotein across the intermembrane space [133]. The exact nature of the TIC channel and the composition of the associated protein complex have been investigated and debated for years, but evidence suggests that Tic20 plays a key role and is probably the central component of this import channel [134, 135]. In fact, Tic20 is the only TIC component shared by almost all plastid-bearing lineages, whether they are photosynthetic or not [136]. Upon import, the N-terminal TP is cleaved by the stromal processing peptidase [137], yielding mature functional proteins within the organelle.

Only a minimal set of TOC/TIC proteins has been identified in apicomplexan parasites so far, which correspond to the core components described above (Fig. 3). A Toc75 homologue has been localized to the apicoplast in *Plasmodium* and *Toxoplasma*, where it was shown to be specifically involved in the import of stromal proteins, but not directly of peripheral proteins, which is consistent with a role in transport through the innermost membranes of the organelle [94]. An apicoplast Tic22 homologue was identified in *Plasmodium* and *Toxoplasma* [119, 138] and also shown to be essential for protein import into the organelle [138]. Similarly, a *Toxoplasma* Tic20 homologue, which was demonstrated to localize to the innermost apicoplast membrane by an elegant split-green fluorescent protein assay, was shown to be crucial for stromal protein import [139]. Apicomplexan parasites also possess several putative homologues of the stromal processing peptidase, one of which was characterized experimentally in *Plasmodium* [140]. However, even upon depletion of Toc75 in *Toxoplasma,* the processing of peripheral apicoplast proteins is unaffected, suggesting the presence of a TP processing peptidase in the outer compartments upstream of the TOC machinery [94] (Fig. 3). Finally, correct folding and assembly of imported proteins likely also depends on stroma-localized chaperones/cochaperones, and some of these like Cpn60 or Cpn20 have been localized to the apicoplast in *Toxoplasma* and *Plasmodium* [121, 141].

Overall, although only a small set of TOC/TIC proteins has been identified in apicomplexan parasites, they constitute the core of the import machinery, with the most important and most conserved proteins described so far. By comparison with primary plastids, this streamlined machinery for the apicoplast may be explained by the fact that the TOC/TIC machinery of secondary plastids potentially interacts with a more reduced repertoire of proteins. As these were already filtered by the early steps of translocation across previous membranes, perhaps less receptors and adapters are required for recognizing proteins to be imported through the innermost membranes [142]. It is also possible that some TOC/TIC complex proteins are Apicomplexa-specific and cannot be identified through homology searches. Experimental evidence, like for example in a *Toxoplasma* mutant for which in spite of complete depletion of the Tic20 protein import still occurs to some extent [139], suggests that yet unidentified additional members of the translocon may also be involved.

8 Targeting the Targeting: Opportunities for Therapeutic Perspectives?

Overall, as biosynthetic functions of the apicoplast are essential at some point or another of the life cycle of many Apicomplexa parasites, and because of its prokaryotic origin, it is a particularly attractive organelle to look for potential drug targets [143]. It is in fact already targeted for pharmaceutical prophylaxis or for treatment of Apicomplexa-caused diseases: for instance, drugs that inhibit translation in prokaryotic-like systems like clindamycin or doxycycline inhibit parasite development by acting primarily on the apicoplast [144, 145].

As most proteins hosted by the apicoplast are not being synthesized by the organelle itself, understanding how they are imported has also gathered considerable interest in recent years. This is not only true for stromal proteins and enzymes but also for peripheral proteins such as transporters, which, although largely uncharacterized at the moment, could be interesting targets to prioritize as they are likely essential to the metabolic function of the organelle [146]. Hence, the functions of the organelle depend largely on the efficient transport of nuclear-encoded proteins, which is why targeting specifically protein trafficking to the organelle could be an efficient way of inhibiting parasite growth [147]. Ensuring a stable proteome in the apicoplast through proper protein import [95], but also efficient folding and proteostasis [46], is of critical importance. In fact, the functional characterization of components of the transport machineries described in this chapter has already highlighted the importance of these proteins for apicoplast maintenance and overall parasite viability.

However, it should be noted that a number of drug-based strategies specifically targeting the apicoplast lead to the so-called "delayed death," during which parasite growth is seemingly unaffected after the first lytic cycle of inhibitor treatment but is strongly impacted following the second lytic cycle (even after drug removal) [144]. This is particularly true for drugs targeting housekeeping functions, while those impacting apicoplast anabolic pathways usually lead to a more rapid death [145]. This slow onset-of-action should be taken into consideration as it may limit the clinical use of these drugs.

There are only a few examples so far of specific inhibitors of protein import to the apicoplast. 15-Deoxyspergualin (a synthetic derivative of bacterial antibiotic spergualin) is supposedly inhibiting protein transport to the organelle by interfering with chaperone-TP interactions and efficiently inhibits *Plasmodium* growth in vitro, yet with a delayed death effect [148, 149]. More recently, a study has shown that expressing and targeting to the apicoplast of *Plasmodium* a murine dihydrofolate reductase that can be conditionally stabilized by a small molecule, leads to a rapid loss of the organelle and, in this case, quick demise of the parasites [150]. This could be due to the protein blocking one of the translocons for import, or alternatively the overstabilized protein may be disrupting proteostasis in the organelle stroma and cause toxicity. Nevertheless, it suggests that imbalance in apicoplast protein import likely has deleterious effects, and that blocking translocons could be a promising strategy to interfere with parasite viability that should definitely be investigated further in the future.

9 Future Directions

As illustrated in this chapter, although there has been considerable progress made in the last 20 years, our understanding of the trafficking of nuclear-encoded proteins to the complex plastid of apicomplexan parasites is still quite patchy and many questions remain unanswered.

Current evidence suggests that several trafficking routes (i.e., passing through the Golgi apparatus or not) exist, and the nature and composition of the vesicular intermediates is still largely elusive. Peripheral and stromal proteins (and perhaps also proteins with dual mitochondrial/plastidic targeting) likely have different targeting signals and trafficking mechanisms. However, several of the features determining the sorting of these proteins in the ER to the different trafficking pathways remain largely unknown.

As for other complex plastids, the chimeric nature of the different translocon complexes, involving components originating from both host and symbiont and forming an intricate patchwork of proteins repurposed from other machineries, renders the complete identification of their core components difficult [41]. However, comparison with other models like diatoms can provide invaluable information on the apicomplexan machinery. The recent identification of chromerids, which are the photosynthetic relatives of Apicomplexa [29, 30], also offers the prospect of discovering novel evolutionary adaptations in plastid transport, although these algae are not yet amenable to genetic modification and functional investigations are thus currently limited. In any case, it should be kept in mind that some lineage-specific evolutionary adaptations possibly evolved for protein transport to the apicoplast, and these would not be easily identified through comparison with other complex plastids.

In this regard, knowing the exact nature of the apicoplast proteome would not only help better understand the motifs and sorting signals of apicoplast-targeted proteins, but also help identify other translocon components and provide a more exhaustive view of the metabolic pathways hosted by the organelle. Yet, while a large number of apicoplast proteins have been predicted by analogy with plastid-hosted pathways and the presence of a putative TP [49, 50, 77, 151], relatively few have been experimentally verified, especially for peripheral proteins. Classical mass spectrometry-based identification of proteins has not yet been performed due to difficulties in isolating and purifying the organelle [152]. However, interesting alternative approaches have been used to identify novel candidates in *Toxoplasma*, like using similarity in transcriptomic expression profiles and phyletic distribution [153], or through a spatial proteomics method called hyperLOPIT (Hyperplexed Localization of Organelle Proteins by Isotope Tagging) [154]. Hopefully, some of these approaches could also be transposed to *Plasmodium*, as for this particular model, only a few proteomics-based identification studies have been performed [50]. Dataset comparison between these different apicomplexan parasites should thus be very informative.

Overall, further identification and functional investigation of the machinery for trafficking to the apicoplast is an important priority. A better understanding of how proteins reach this crucial organelle may not only provide a more complete view of the evolution of complex plastids, but also help in identifying new potential drug targets and thus provide interesting perspectives for combating the diseases caused by apicomplexan parasites.

Acknowledgement We thank Laurence Berry for providing the electron microscopy images used on Fig. 1a. We also thank curators and data providers for the www.VEupathDB.org and www.ToxoDB.org databases.

References

1. Ashley EA, Pyae Phyo A, Woodrow CJ (2018) Malaria. Lancet 391:1608–1621
2. Davies AP, Chalmers RM (2009) Cryptosporidiosis. BMJ 339:b4168–b4168
3. Montoya JG, Liesenfeld O (2004) Toxoplasmosis. Lancet 363:1965–1976
4. Florin-Christensen M, Schnittger L (eds) (2018) Parasitic protozoa of farm animals and pets. Springer International Publishing. Available from: http://link.springer.com/10.1007/978-3-319-70132-5
5. World Health Organization (2020) World malaria report 2020: 20 years of global progress and challenges. World Health Organization, Geneva. Available from: https://apps.who.int/iris/handle/10665/337660
6. Ghazanfari N, Mueller SN, Heath WR (2018) Cerebral malaria in mouse and man. Front Immunol 9:2016. https://doi.org/10.3389/fimmu.2018.02016
7. Robert-Gangneux F, Dardé ML (2012) Epidemiology of and diagnostic strategies for toxoplasmosis. Clin Microbiol Rev 25:264–296. https://doi.org/10.1128/cmr.05013-11
8. Sanchez SG, Besteiro S (2021) The pathogenicity and virulence of *Toxoplasma gondii*. Virulence 12:3095–3114. https://doi.org/10.1080/21505594.2021.2012346
9. Seeber F, Steinfelder S (2016) Recent advances in understanding apicomplexan parasites. F1000Res 5:F1000 Faculty Rev-1369. https://doi.org/10.12688/2Ff1000research.7924.1
10. McFadden GI, Yeh E (2017) The apicoplast: now you see it, now you don't. Int J Parasitol 47:137–144. https://doi.org/10.1016/j.ijpara.2016.08.005
11. Siddall ME (1992) Hohlzylinders. Parasitol Today 8:90–91. https://doi.org/10.1016/0169-4758(92)90244-V
12. Kilejian A (1975) Circular mitochondrial DNA from the avian malarial parasite *Plasmodium lophurae*. Biochim Biophys Acta 390:276–284
13. Glöckner G, Rosenthal A, Valentin K (2000) The structure and gene repertoire of an ancient red algal plastid genome. J Mol Evol 51:382–390. https://doi.org/10.1007/s002390010101
14. McFadden GI, Reith ME, Munholland J, Lang-Unnasch N (1996) Plastid in human parasites. Nature 381:482–482
15. Wilson RJ, Denny PW, Preiser PR, Rangachari K, Roberts K, Roy A et al (1996) Complete gene map of the plastid-like DNA of the malaria parasite *Plasmodium falciparum*. J Mol Biol 261:155–172. https://doi.org/10.1006/jmbi.1996.0449
16. Köhler S, Delwiche CF, Denny PW, Tilney LG, Webster P, Wilson RJM et al (1997) A plastid of probable green algal origin in apicomplexan parasites. Science 275:1485–1489. https://doi.org/10.1006/jmbi.1996.0449
17. Diniz JAP, Silva EO, Lainson R, de Souza W (2000) The fine structure of *Garnia gonadati* and its association with the host cell. Parasitol Res 86:971–977. https://doi.org/10.1007/pl00008528

18. Tomova C, Geerts WJC, Müller-Reichert T, Entzeroth R, Humbel BM (2006) New comprehension of the apicoplast of Sarcocystis by transmission electron tomography. Biol Cell 98:535–545. https://doi.org/10.1042/bc20060028
19. Caballero MC, Pedroni MJ, Palmer GH, Suarez CE, Davitt C, Lau AOT (2012) Characterization of acyl carrier protein and LytB in *Babesia bovis* apicoplast. Mol Biochem Parasitol 181:125–133. https://doi.org/10.1016/2Fj.molbiopara.2011.10.009
20. Lemgruber L, Kudryashev M, Dekiwadia C, Riglar DT, Baum J, Stahlberg H et al (2013) Cryo-electron tomography reveals four-membrane architecture of the *Plasmodium apicoplast*. Malar J 12:25. https://doi.org/10.1186/1475-2875-12-25
21. Sibbald SJ, Archibald JM (2020) Genomic insights into plastid evolution. Genome Biol Evol 12:978–990. https://doi.org/10.1093/gbe/evaa096
22. Burki F (2017) The convoluted evolution of eukaryotes with complex plastids. In: Advances in botanical research. Elsevier, p 1–30. Available from: https://linkinghub.elsevier.com/retrieve/pii/S0065229617300484
23. Gentil J, Hempel F, Moog D, Zauner S, Maier UG (2017) Review: origin of complex algae by secondary endosymbiosis: a journey through time. Protoplasma 254:1835–1843. https://doi.org/10.1007/s00709-017-1098-8
24. Zhang Z, Green BR, Cavalier-Smith T (2000) Phylogeny of ultra-rapidly evolving dinoflagellate chloroplast genes: a possible common origin for sporozoan and dinoflagellate plastids. J Mol Evol 51:26–40. https://doi.org/10.1007/s002390010064
25. Fast NM, Kissinger JC, Roos DS, Keeling PJ (2001) Nuclear-encoded, plastid-targeted genes suggest a single common origin for apicomplexan and dinoflagellate plastids. Mol Biol Evol 18:418–426. https://doi.org/10.1093/oxfordjournals.molbev.a003818
26. Janouškovec J, Horák A, Oborník M, Lukeš J, Keeling PJ (2010) A common red algal origin of the apicomplexan, dinoflagellate, and heterokont plastids. Proc Natl Acad Sci 107:10949–10954. https://doi.org/10.1073/pnas.1003335107
27. van Dooren GG, Striepen B (2013) The algal past and parasite present of the apicoplast. Annu Rev Microbiol 67:271–289. https://doi.org/10.1146/annurev-micro-092412-155741
28. Oborník M (2020) Photoparasitism as an intermediate state in the evolution of apicomplexan parasites. Trends Parasitol 36:727–734. https://doi.org/10.1016/j.pt.2020.06.002
29. Moore RB, Oborník M, Janouškovec J, Chrudimský T, Vancová M, Green DH et al (2008) A photosynthetic alveolate closely related to apicomplexan parasites. Nature 451:959–963. https://doi.org/10.1038/nature06635
30. Woo YH, Ansari H, Otto TD, Klinger CM, Kolisko M, Michálek J et al (2015) Chromerid genomes reveal the evolutionary path from photosynthetic algae to obligate intracellular parasites. Elife 4:e06974. https://doi.org/10.7554/elife.06974
31. Simpson AGB, Patterson DJ (1996) Ultrastructure and identification of the predatory flagellate *Colpodella pugnax* Cienkowski (Apicomplexa) with a description of Colpodella turpis n. sp. and a review of the genus. Syst Parasitol 33:187–198. https://doi.org/10.1007/BF01531200
32. Janouškovec J, Tikhonenkov DV, Burki F, Howe AT, Kolísko M, Mylnikov AP et al (2015) Factors mediating plastid dependency and the origins of parasitism in apicomplexans and their close relatives. Proc Natl Acad Sci U S A 112:10200–10207. https://doi.org/10.1073/pnas.1423790112
33. Cavalier-Smith T, Chao EE (2004) Protalveolate phylogeny and systematics and the origins of Sporozoa and dinoflagellates (phylum Myzozoa nom. nov.). Eur J Protistol 40:185–212. https://doi.org/10.1016/j.ejop.2004.01.002
34. Muñoz-Gómez SA, Slamovits CH (2018) Plastid genomes in the Myzozoa. In: Advances in botanical research. Elsevier, p 55–94. Available from: https://linkinghub.elsevier.com/retrieve/pii/S0065229617300873
35. Bodył A, Stiller JW, Mackiewicz P (2009) Chromalveolate plastids: direct descent or multiple endosymbioses? Trends Ecol Evol 24:119–121. https://doi.org/10.1016/j.tree.2008.11.003
36. Janouškovec J, Paskerova GG, Miroliubova TS, Mikhailov KV, Birley T, Aleoshin VV et al (2019) Apicomplexan-like parasites are polyphyletic and widely but selectively dependent on cryptic plastid organelles. Elife 8:e49662. https://doi.org/10.7554/elife.49662

37. Kwong WK, del Campo J, Mathur V, Vermeij MJA, Keeling PJ (2019) A widespread coral-infecting apicomplexan with chlorophyll biosynthesis genes. Nature 568:103–107. https://doi.org/10.1038/s41586-019-1072-z
38. Zhu G, Marchewka MJ, Keithly JS (2000) *Cryptosporidium parvum* appears to lack a plastid genome. Microbiology 146:315–321. https://doi.org/10.1099/00221287-146-2-315
39. Toso MA, Omoto CK (2007) Gregarina niphandrodes may lack both a plastid genome and organelle. J Eukaryot Microbiol 54:66–72. https://doi.org/10.1111/j.1550-7408.2006.00229.x
40. Cavalier-Smith T (2000) Membrane heredity and early chloroplast evolution. Trends Plant Sci 5:174–182. https://doi.org/10.1016/s1360-1385(00)01598-3
41. Gould SB, Maier UG, Martin WF (2015) Protein import and the origin of red complex plastids. Curr Biol 25:R515–R521. https://doi.org/10.1016/j.cub.2015.04.033
42. Shi LX, Theg SM (2013) Energetic cost of protein import across the envelope membranes of chloroplasts. Proc Natl Acad Sci U S A 110:930–935. https://doi.org/10.1073/pnas.1115886110
43. Kelly S (2021) The economics of organellar gene loss and endosymbiotic gene transfer. Genome Biol 22:345. https://doi.org/10.1186/s13059-021-02567-w
44. Nassoury N, Cappadocia M, Morse D (2003) Plastid ultrastructure defines the protein import pathway in dinoflagellates. J Cell Sci 116:2867–2874. https://doi.org/10.1242/jcs.00517
45. Goodman CD, Pasaje CFA, Kennedy K, McFadden GI, Ralph SA (2016) Targeting protein translation in organelles of the Apicomplexa. Trends Parasitol 32:953–965. https://doi.org/10.1016/j.pt.2016.09.011
46. Florentin A, Stephens DR, Brooks CF, Baptista RP, Muralidharan V (2020) Plastid biogenesis in malaria parasites requires the interactions and catalytic activity of the Clp proteolytic system. Proc Natl Acad Sci U S A 117:13719–13729. https://doi.org/10.1073/pnas.1919501117
47. Gisselberg JE, Dellibovi-Ragheb TA, Matthews KA, Bosch G, Prigge ST (2013) The Suf iron-sulfur cluster synthesis pathway is required for apicoplast maintenance in malaria parasites. PLoS Pathog 9:e1003655. https://doi.org/10.1371/2Fjournal.ppat.1003655
48. Pamukcu S, Cerutti A, Bordat Y, Hem S, Rofidal V, Besteiro S (2021) Differential contribution of two organelles of endosymbiotic origin to iron-sulfur cluster synthesis and overall fitness in *Toxoplasma*. PLoS Pathog 17:e1010096. https://doi.org/10.1371/journal.ppat.1010096
49. Foth BJ, Ralph SA, Tonkin CJ, Struck NS, Fraunholz M, Roos DS et al (2003) Dissecting apicoplast targeting in the malaria parasite *Plasmodium falciparum*. Science 299:705–708. https://doi.org/10.1126/science.1078599
50. Boucher MJ, Ghosh S, Zhang L, Lal A, Jang SW, Ju A et al (2018) Integrative proteomics and bioinformatic prediction enable a high-confidence apicoplast proteome in malaria parasites. PLoS Biol 16:e2005895. https://doi.org/10.1371/journal.pbio.2005895
51. Ponce-Toledo RI, López-García P, Moreira D (2019) Horizontal and endosymbiotic gene transfer in early plastid evolution. New Phytol 224:618–624. https://doi.org/10.1111/nph.15965
52. Vaishnava S, Striepen B (2006) The cell biology of secondary endosymbiosis ? how parasites build, divide and segregate the apicoplast. Mol Microbiol 61:1380–1387. https://doi.org/10.1111/j.1365-2958.2006.05343.x
53. Striepen B, Crawford MJ, Shaw MK, Tilney LG, Seeber F, Roos DS (2000) The plastid of *Toxoplasma gondii* is divided by association with the centrosomes. J Cell Biol 151:1423–1434. https://doi.org/10.1083/2Fjcb.151.7.1423
54. Stanway RR, Witt T, Zobiak B, Aepfelbacher M, Heussler VT (2009) GFP-targeting allows visualization of the apicoplast throughout the life cycle of live malaria parasites. Biol Cell 101:415–435. https://doi.org/10.1042/bc20080202
55. van Dooren GG, Reiff SB, Tomova C, Meissner M, Humbel BM, Striepen B (2009) A novel dynamin-related protein has been recruited for apicoplast fission in *Toxoplasma gondii*. Curr Biol 19:267–276. https://doi.org/10.1016/2Fj.cub.2008.12.048
56. Lévêque MF, Berry L, Cipriano MJ, Nguyen HM, Striepen B, Besteiro S (2015) Autophagy-related protein ATG8 Has a noncanonical function for apicoplast inheritance in *Toxoplasma gondii*. MBio 6:e01446–e01415. https://doi.org/10.1128/mbio.01446-15

57. Walczak M, Ganesan SM, Niles JC, Yeh E (2018) ATG8 Is Essential Specifically for an Autophagy-Independent Function in Apicoplast Biogenesis in Blood-Stage Malaria Parasites. MBio 9:e02021. https://doi.org/10.1128/mbio.02021-17
58. Ralph SA, van Dooren GG, Waller RF, Crawford MJ, Fraunholz MJ, Foth BJ et al (2004) Metabolic maps and functions of the *Plasmodium falciparum* apicoplast. Nat Rev Microbiol 2:203–216. https://doi.org/10.1038/nrmicro843
59. Seeber F, Soldati-Favre D (2010) Metabolic pathways in the apicoplast of apicomplexa. Int Rev Cell Mol Biol 281:161–228. https://doi.org/10.1016/S1937-6448(10)81005-6
60. Nagaraj VA, Sundaram B, Varadarajan NM, Subramani PA, Kalappa DM, Ghosh SK et al (2013) Malaria parasite-synthesized heme is essential in the mosquito and liver stages and complements host heme in the blood stages of infection. PLoS Pathog 9:e1003522. https://doi.org/10.1371/journal.ppat.1003522
61. Ke H, Sigala PA, Miura K, Morrisey JM, Mather MW, Crowley JR et al (2014) The heme biosynthesis pathway is essential for *Plasmodium falciparum* development in mosquito stage but not in blood stages. J Biol Chem 289:34827–34837. https://doi.org/10.1074/jbc.m114.615831
62. Sigala PA, Crowley JR, Henderson JP, Goldberg DE (2015) Deconvoluting heme biosynthesis to target blood-stage malaria parasites. Elife 4:e09143. https://doi.org/10.7554/elife.09143
63. Bergmann A, Floyd K, Key M, Dameron C, Rees KC, Thornton LB et al (2020) *Toxoplasma gondii* requires its plant-like heme biosynthesis pathway for infection. PLoS Pathog 16:e1008499. https://doi.org/10.1371/2Fjournal.ppat.1008499
64. Vaughan AM, O'Neill MT, Tarun AS, Camargo N, Phuong TM, Aly ASI et al (2009) Type II fatty acid synthesis is essential only for malaria parasite late liver stage development. Cell Microbiol 11:506–520. https://doi.org/10.1111/2Fj.1462-5822.2008.01270.x
65. Mazumdar J, Wilson H, E, Masek K, A Hunter C, Striepen B. (2006) Apicoplast fatty acid synthesis is essential for organelle biogenesis and parasite survival in *Toxoplasma gondii.* Proc Natl Acad Sci U S A 103:13192–13197. https://doi.org/10.1073/pnas.0603391103
66. Amiar S, Katris NJ, Berry L, Dass S, Duley S, Arnold CS et al (2020) Division and adaptation to host environment of apicomplexan parasites depend on apicoplast lipid metabolic plasticity and host organelle remodeling. Cell Rep 30:3778–3792.e9. https://doi.org/10.1016/j.celrep.2020.02.072
67. Liang X, Cui J, Yang X, Xia N, Li Y, Zhao J et al (2020) Acquisition of exogenous fatty acids renders apicoplast-based biosynthesis dispensable in tachyzoites of Toxoplasma. J Biol Chem 295:7743–7752. https://doi.org/10.1074/jbc.ra120.013004
68. Yeh E, DeRisi JL (2011) Chemical rescue of malaria parasites lacking an apicoplast defines organelle function in blood-stage *Plasmodium falciparum.* PLoS Biol 9:e1001138. https://doi.org/10.1371/journal.pbio.1001138
69. Nair SC, Brooks CF, Goodman CD, Sturm A, Strurm A, McFadden GI et al (2011) Apicoplast isoprenoid precursor synthesis and the molecular basis of fosmidomycin resistance in *Toxoplasma gondii.* J Exp Med 208:1547–1559. https://doi.org/10.1084/jem.20110039
70. Charan M, Singh N, Kumar B, Srivastava K, Siddiqi MI, Habib S (2014) Sulfur mobilization for Fe-S cluster assembly by the essential SUF pathway in the *Plasmodium falciparum* apicoplast and its Inhibition. Antimicrob Agents Chemother 58:3389–3398. https://doi.org/10.1128/2FAAC.02711-13
71. Henkel S, Frohnecke N, Maus D, McConville MJ, Laue M, Blume M et al (2022) *Toxoplasma gondii* apicoplast-resident ferredoxin is an essential electron transfer protein for the MEP isoprenoid-biosynthetic pathway. J Biol Chem 298:101468. https://doi.org/10.1016/j.jbc.2021.101468
72. Renaud EA, Pamukcu S, Cerutti A, Berry L, Lemaire-Vieille C, Yamaryo-Botté Y et al (2022) Disrupting the plastidic iron-sulfur cluster biogenesis pathway in *Toxoplasma gondii* has pleiotropic effects irreversibly impacting parasite viability. J Biol Chem 298:102243. https://doi.org/10.1016/j.jbc.2022.102243
73. Wollman FA (2016) An antimicrobial origin of transit peptides accounts for early endosymbiotic events. Traffic 17:1322–1328. https://doi.org/10.1111/tra.12446

74. Waller RF (2000) Protein trafficking to the plastid of *Plasmodium falciparum* is via the secretory pathway. EMBO J 19:1794–1802. https://doi.org/10.1093/2Femboj/2F19.8.1794
75. DeRocher A, Hagen CB, Froehlich JE, Feagin JE, Parsons M (2000) Analysis of targeting sequences demonstrates that trafficking to the *Toxoplasma gondii* plastid branches off the secretory system. J Cell Sci 113(Pt 22):3969–3977. https://doi.org/10.1242/jcs.113.22.3969
76. Ralph SA, Foth BJ, Hall N, McFadden GI (2004) Evolutionary pressures on apicoplast transit peptides. Mol Biol Evol 21:2183–2194. https://doi.org/10.1093/molbev/msh233
77. Zuegge J, Ralph S, Schmuker M, McFadden GI, Schneider G (2001) Deciphering apicoplast targeting signals--feature extraction from nuclear-encoded precursors of *Plasmodium falciparum* apicoplast proteins. Gene 280:19–26. https://doi.org/10.1016/s0378-1119(01)00776-4
78. Harb OS, Chatterjee B, Fraunholz MJ, Crawford MJ, Nishi M, Roos DS (2004) Multiple functionally redundant signals mediate targeting to the apicoplast in the apicomplexan parasite *Toxoplasma gondii*. Eukaryot Cell 3:663–674. https://doi.org/10.1128/2FEC.3.3.663-674.2004
79. Tonkin CJ, Roos DS, McFadden GI (2006) N-terminal positively charged amino acids, but not their exact position, are important for apicoplast transit peptide fidelity in *Toxoplasma gondii*. Mol Biochem Parasitol 150:192–200. https://doi.org/10.1016/j.molbiopara.2006.08.001
80. Jomaa H, Wiesner J, Sanderbrand S, Altincicek B, Weidemeyer C, Hintz M et al (1999) Inhibitors of the nonmevalonate pathway of isoprenoid biosynthesis as antimalarial drugs. Science 285:1573–1576. https://doi.org/10.1126/science.285.5433.1573
81. Mallo N, Fellows J, Johnson C, Sheiner L (2018) Protein Import into the endosymbiotic organelles of apicomplexan parasites. Genes 9:412. https://doi.org/10.3390/2Fgenes9080412
82. McKinnon L, Theg SM (2019) Determinants of the specificity of protein targeting to chloroplasts or mitochondria. Mol Plant 12:893–895. https://doi.org/10.1016/j.molp.2019.05.004
83. Silva-Filho MC (2003) One ticket for multiple destinations: dual targeting of proteins to distinct subcellular locations. Curr Opin Plant Biol 6:589–595. https://doi.org/10.1016/j.pbi.2003.09.008
84. Baudisch B, Langner U, Garz I, Klösgen RB (2014) The exception proves the rule? Dual targeting of nuclear-encoded proteins into endosymbiotic organelles. New Phytol 201:80–90. https://doi.org/10.1111/nph.12482
85. Kunze M, Berger J (2015) The similarity between N-terminal targeting signals for protein import into different organelles and its evolutionary relevance. Front Physiol [Internet] 2015 [cité 2023 mars 10];6. Available from: http://journal.frontiersin.org/Article/10.3389/fphys.2015.00259/abstract
86. Pino P, Foth BJ, Kwok LY, Sheiner L, Schepers R, Soldati T et al (2007) Dual targeting of antioxidant and metabolic enzymes to the mitochondrion and the apicoplast of Toxoplasma gondii. PLoS Pathog 3:e115. https://doi.org/10.1371/journal.ppat.0030115
87. Chaudhari R, Narayan A, Patankar S (2012) A novel trafficking pathway in *Plasmodium falciparum* for the organellar localization of glutathione peroxidase-like thioredoxin peroxidase. FEBS J 279:3872–3888. https://doi.org/10.1111/j.1742-4658.2012.08746.x
88. Mastud P, Patankar S (2019) An ambiguous N-terminus drives the dual targeting of an antioxidant protein Thioredoxin peroxidase (TgTPx1/2) to endosymbiotic organelles in *Toxoplasma gondii*. PeerJ 7:e7215. https://doi.org/10.7717/peerj.7215
89. Günther S, Wallace L, Patzewitz EM, McMillan PJ, Storm J, Wrenger C et al (2007) Apicoplast lipoic acid protein ligase B is not essential for *Plasmodium falciparum*. PLoS Pathog 3:e189. https://doi.org/10.1371/2Fjournal.ppat.0030189
90. Saito T, Nishi M, Lim MI, Wu B, Maeda T, Hashimoto H et al (2008) A novel GDP-dependent pyruvate kinase isozyme from *Toxoplasma gondii* localizes to both the apicoplast and the mitochondrion. J Biol Chem 283:14041–14052. https://doi.org/10.1371/2Fjournal.ppat.0030189
91. Read M, Müller IB, Mitchell SL, Sims PFG, Hyde JE (2010) Dynamic subcellular localization of isoforms of the folate pathway enzyme serine hydroxymethyltransferase (SHMT) through the erythrocytic cycle of Plasmodium falciparum. Malar J 9:351. https://doi.org/10.1186/1475-2875-9-351

92. Kutschera U, Niklas KJ (2005) Endosymbiosis, cell evolution, and speciation. Theory Biosci 124:1–24. https://doi.org/10.1016/j.thbio.2005.04.001
93. Kilian O, Kroth PG (2005) Identification and characterization of a new conserved motif within the presequence of proteins targeted into complex diatom plastids. Plant J 41:175–183. https://doi.org/10.1111/j.1365-313x.2004.02294.x
94. Sheiner L, Fellows JD, Ovciarikova J, Brooks CF, Agrawal S, Holmes ZC et al (2015) *Toxoplasma gondii* Toc75 functions in import of stromal but not peripheral apicoplast proteins. Traffic 16:1254–1269. https://doi.org/10.1111/tra.12333
95. Parsons M, Karnataki A, Derocher AE (2009) Evolving insights into protein trafficking to the multiple compartments of the apicomplexan plastid. J Eukaryot Microbiol 56:214–220. https://doi.org/10.1111/j.1550-7408.2009.00405.x
96. Bhaya D, Grossman A (1991) Targeting proteins to diatom plastids involves transport through an endoplasmic reticulum. Mol Gen Genet 229:400–404. https://doi.org/10.1007/bf00267462
97. Zimmermann R, Eyrisch S, Ahmad M, Helms V (2011) Protein translocation across the ER membrane. Biochimica et Biophysica Acta (BBA) – Biomembranes 1808:912–924. https://doi.org/10.1016/j.bbamem.2010.06.015
98. Tomova C, Humbel BM, Geerts WJC, Entzeroth R, Holthuis JCM, Verkleij AJ (2009) Membrane contact sites between apicoplast and ER in *Toxoplasma gondii* revealed by electron tomography. Traffic 10:1471–1480. https://doi.org/10.1111/j.1600-0854.2009.00954.x
99. Tonkin CJ, Struck NS, Mullin KA, Stimmler LM, McFadden GI (2006) Evidence for Golgi-independent transport from the early secretory pathway to the plastid in malaria parasites. Mol Microbiol 61:614–630. https://doi.org/10.1111/j.1365-2958.2006.05244.x
100. Besteiro S, Klinger CM, Meissner M, Carruthers VB (2020) Endomembrane trafficking pathways in *Toxoplasma*. In: Toxoplasma gondii. Academic Press, p 705–41. https://doi.org/10.1016/B978-0-12-815041-2.00015-3
101. DeRocher A, Gilbert B, Feagin JE, Parsons M (2005) Dissection of brefeldin A-sensitive and -insensitive steps in apicoplast protein targeting. J Cell Sci 118:565–574. https://doi.org/10.1128/2FmBio.01380-21
102. Heiny SR, Pautz S, Recker M, Przyborski JM (2014) Protein traffic to the *Plasmodium falciparum* apicoplast: evidence for a sorting branch point at the Golgi. Traffic 15:1290–1304. https://doi.org/10.1111/tra.12226
103. Prasad A, Mastud P, Patankar S (2021) Dually localised proteins found in both the apicoplast and mitochondrion utilize the Golgi-dependent pathway for apicoplast targeting in *Toxoplasma gondii*. Biol Cell 113:58–78. https://doi.org/10.1111/boc.202000050
104. Bouchut A, Geiger JA, DeRocher AE, Parsons M (2014) Vesicles bearing Toxoplasma apicoplast membrane proteins persist following loss of the relict plastid or Golgi body disruption. PloS One 9:e112096. https://doi.org/10.1371/journal.pone.0112096
105. Karnataki A, DeRocher A, Coppens I, Nash C, Feagin JE, Parsons M (2007) Cell cycle-regulated vesicular trafficking of Toxoplasma APT1, a protein localized to multiple apicoplast membranes. Mol Microbiol 63:1653–1668. https://doi.org/10.1111/j.1365-2958.2007.05619.x
106. Karnataki A, DeRocher AE, Coppens I, Feagin JE, Parsons M (2007) A membrane protease is targeted to the relict plastid of *Toxoplasma* via an internal signal sequence. Traffic 8:1543–1553. https://doi.org/10.1111/j.1600-0854.2007.00637.x
107. DeRocher AE, Coppens I, Karnataki A, Gilbert LA, Rome ME, Feagin JE et al (2008) A thioredoxin family protein of the apicoplast periphery identifies abundant candidate transport vesicles in *Toxoplasma gondii*. Eukaryot Cell 7:1518–1529. https://doi.org/10.1128/ec.00081-08
108. Cao S, Yang J, Fu J, Chen H, Jia H (2021) The dissection of SNAREs reveals key factors for vesicular trafficking to the endosome-like compartment and apicoplast via the secretory system in *Toxoplasma gondii*. MBio 12:e0138021. https://doi.org/10.1128/mbio.01380-21
109. Fu J, Zhao L, Pang Y, Chen H, Yamamoto H, Chen Y et al (2022) Apicoplast biogenesis mediated by ATG8 requires the ATG12–ATG5-ATG16L and SNAP29 complexes in Toxoplasma gondii. Autophagy 19:1–19. https://doi.org/10.1080/15548627.2022.2123639

110. Jia H, Cao S, Yang J, Fu J (2021) Reply to Prasad and Patankar, "Recognition of two distinct pathways for trafficking of proteins to the apicoplast". MBio 12:e03107–e03121. https://doi.org/10.1128/mBio.03107-21
111. Prasad A, Patankar S (2021) Recognition of two distinct pathways for trafficking of proteins to the apicoplast. MBio 12:e02634–e02621. https://doi.org/10.1111/boc.202000050
112. Grosche C, Hempel F, Bolte K, Zauner S, Maier UG (2014) The periplastidal compartment: a naturally minimized eukaryotic cytoplasm. Curr Opin Microbiol 22:88–93. https://doi.org/10.1016/j.mib.2014.09.017
113. Sommer MS, Gould SB, Lehmann P, Gruber A, Przyborski JM, Maier UG (2007) Der1-mediated preprotein import into the periplastid compartment of Chromalveolates? Mol Biol Evol 24:918–928. https://doi.org/10.1093/molbev/msm008
114. Hempel F, Bullmann L, Lau J, Zauner S, Maier UG (2009) ERAD-derived preprotein transport across the second outermost plastid membrane of diatoms. Mol Biol Evol 26:1781–1790. https://doi.org/10.1093/molbev/msp079
115. Römisch K (2005) Endoplasmic reticulum-associated degradation. Annu Rev Cell Dev Biol 21:435–456. https://doi.org/10.1146/annurev.cellbio.21.012704.133250
116. Ye Y, Meyer HH, Rapoport TA (2003) Function of the p97-Ufd1-Npl4 complex in retrotranslocation from the ER to the cytosol: dual recognition of nonubiquitinated polypeptide segments and polyubiquitin chains. J Cell Biol 162:71–84. https://doi.org/10.1083/jcb.200302169
117. Broad W, Ling Q, Jarvis P (2016) New insights into roles of ubiquitin modification in regulating plastids and other endosymbiotic organelles. Int Rev Cell Mol Biol 325:1–33. https://doi.org/10.1016/bs.ircmb.2016.02.007
118. Stork S, Moog D, Przyborski JM, Wilhelmi I, Zauner S, Maier UG (2012) Distribution of the SELMA translocon in secondary plastids of red algal origin and predicted uncoupling of ubiquitin-dependent translocation from degradation. Eukaryot Cell 11:1472–1481. https://doi.org/10.1128/2FEC.00183-12
119. Kalanon M, Tonkin CJ, McFadden GI (2009) Characterization of two putative protein translocation components in the apicoplast of *Plasmodium falciparum*. Eukaryot Cell 8:1146–1154. https://doi.org/10.1128/2FEC.00061-09
120. Spork S, Hiss JA, Mandel K, Sommer M, Kooij TWA, Chu T et al (2009) An unusual ERAD-like complex is targeted to the apicoplast of Plasmodium falciparum. Eukaryot Cell 8:1134–1145. https://doi.org/10.1128/2FEC.00083-09
121. Agrawal S, van Dooren GG, Beatty WL, Striepen B (2009) Genetic evidence that an endosymbiont-derived endoplasmic reticulum-associated protein degradation (ERAD) system functions in import of apicoplast proteins. J Biol Chem 284:33683–33691. https://doi.org/10.1074/2Fjbc.M109.044024
122. Ye Y, Shibata Y, Yun C, Ron D, Rapoport TA (2004) A membrane protein complex mediates retro-translocation from the ER lumen into the cytosol. Nature 429:841–847. https://doi.org/10.1038/nature02656
123. Ye Y, Meyer HH, Rapoport TA (2001) The AAA ATPase Cdc48/p97 and its partners transport proteins from the ER into the cytosol. Nature 414:652–656. https://doi.org/10.1038/414652a
124. Fellows JD, Cipriano MJ, Agrawal S, Striepen B (2017) A plastid protein that evolved from ubiquitin and is required for apicoplast protein import in *Toxoplasma gondii*. MBio 8:e00950–e00917. https://doi.org/10.1038/414652a
125. Agrawal S, Chung DWD, Ponts N, van Dooren GG, Prudhomme J, Brooks CF et al (2013) An apicoplast localized ubiquitylation system is required for the import of nuclear-encoded plastid proteins. PLoS Pathog 9:e1003426. https://doi.org/10.1371/journal.ppat.1003426
126. Gallagher JR, Matthews KA, Prigge ST (2011) Plasmodium falciparum apicoplast transit peptides are unstructured in vitro and during apicoplast import. Traffic 12:1124–1138. https://doi.org/10.1111/j.1600-0854.2011.01232.x
127. Nakai M (2018) New perspectives on chloroplast protein import. Plant Cell Physiol 59:1111–1119. https://doi.org/10.1093/pcp/pcy083

128. Day PM, Theg SM (2018) Evolution of protein transport to the chloroplast envelope membranes. Photosynth Res 138:315–326. https://doi.org/10.1007/s11120-018-0540-x
129. Gross J, Bhattacharya D (2009) Revaluating the evolution of the Toc and Tic protein translocons. Trends Plant Sci 14:13–20. https://doi.org/10.1016/j.tplants.2008.10.003
130. Richardson LGL, Schnell DJ (2020) Origins, function, and regulation of the TOC–TIC general protein import machinery of plastids. J Exp Bot 71:1226–1238. https://doi.org/10.1093/jxb/erz517
131. Ganesan I, Theg SM (2019) Structural considerations of folded protein import through the chloroplast TOC/TIC translocons. FEBS Lett 593:565–572. https://doi.org/10.1002/1873-3468.13342
132. Day PM, Potter D, Inoue K (2014) Evolution and targeting of Omp85 homologs in the chloroplast outer envelope membrane. Front Plant Sci 5:535. https://doi.org/10.3389/2Ffpls.2014.00535
133. Paila YD, Richardson LG, Inoue H, Parks ES, McMahon J, Inoue K et al (2016) Multifunctional roles for the polypeptide transport associated domains of Toc75 in chloroplast protein import. Elife 5:e12631. https://doi.org/10.7554/2FeLife.12631
134. Kikuchi S, Oishi M, Hirabayashi Y, Lee DW, Hwang I, Nakai M (2009) A 1-megadalton translocation complex containing Tic20 and Tic21 mediates chloroplast protein import at the inner envelope membrane. Plant Cell 21:1781–1797. https://doi.org/10.1105/2Ftpc.108.063552
135. Kovács-Bogdán E, Benz JP, Soll J, Bölter B (2011) Tic20 forms a channel independent of Tic110 in chloroplasts. BMC Plant Biol 11:133. https://doi.org/10.1186/1471-2229-11-133
136. Kikuchi S, Bédard J, Hirano M, Hirabayashi Y, Oishi M, Imai M et al (2013) Uncovering the protein translocon at the chloroplast inner envelope membrane. Science 339:571–574. https://doi.org/10.1126/science.1229262
137. Richter S, Lamppa GK (1998) A chloroplast processing enzyme functions as the general stromal processing peptidase. Proc Natl Acad Sci U S A 95:7463–7468. https://doi.org/10.1126/science.1229262
138. Glaser S, van Dooren GG, Agrawal S, Brooks CF, McFadden GI, Striepen B et al (2012) Tic22 is an essential chaperone required for protein import into the apicoplast. J Biol Chem 287:39505–39512. https://doi.org/10.1074/2Fjbc.M112.405100
139. van Dooren GG, Tomova C, Agrawal S, Humbel BM, Striepen B (2008) *Toxoplasma gondii* Tic20 is essential for apicoplast protein import. Proc Natl Acad Sci U S A 105:13574–13579
140. van Dooren GG, Su V, D'Ombrain MC, McFadden GI (2002) Processing of an apicoplast leader sequence in *Plasmodium falciparum* and the identification of a putative leader cleavage enzyme. J Biol Chem 277:23612–23619. https://doi.org/10.1073/pnas.0803862105
141. Sato S, Wilson RJM, (Iain). (2005) Organelle-specific cochaperonins in apicomplexan parasites. Mol Biochem Parasitol 141:133–143. https://doi.org/10.1016/j.molbiopara.2005.01.010
142. Hirakawa Y, Burki F, Keeling PJ (2012) Genome-based reconstruction of the protein import machinery in the secondary plastid of a chlorarachniophyte alga. Eukaryot Cell 11:324–333. https://doi.org/10.1128/ec.05264-11
143. Biddau M, Sheiner L (2019) Targeting the apicoplast in malaria. Biochem Soc Trans 47:973–983. https://doi.org/10.1042/bst20170563
144. Kennedy K, Crisafulli EM, Ralph SA (2019) Delayed death by plastid inhibition in apicomplexan parasites. Trends Parasitol 35:747–759. https://doi.org/10.1016/j.pt.2019.07.010
145. Uddin T, McFadden GI, Goodman CD (2018) Validation of putative apicoplast-targeting drugs using a chemical supplementation assay in cultured human malaria parasites. Antimicrob Agents Chemother 62:e01161–e01117. https://doi.org/10.1128/aac.01161-17
146. Kloehn J, Lacour CE, Soldati-Favre D (2021) The metabolic pathways and transporters of the plastid organelle in Apicomplexa. Curr Opin Microbiol 63:250–258. https://doi.org/10.1016/j.mib.2021.07.016
147. Lindner J, Meissner KA, Schettert I, Wrenger C (2013) Trafficked Proteins—Druggable in *Plasmodium falciparum* ? Int J Cell Biol 2013:1–13. https://doi.org/10.1016/j.mib.2021.07.016

148. Ramya TNC, Karmodiya K, Surolia A, Surolia N (2007) 15-deoxyspergualin primarily targets the trafficking of apicoplast proteins in *Plasmodium falciparum*. J Biol Chem 282:6388–6397. https://doi.org/10.1074/jbc.m610251200
149. Banerjee T, Singh RR, Gupta S, Surolia A, Surolia N (2012) 15-Deoxyspergualin hinders physical interaction between basic residues of transit peptide in PfENR and Hsp70-1. IUBMB Life 64:99–107. https://doi.org/10.1002/iub.583
150. Boucher MJ, Yeh E (2019) Disruption of apicoplast biogenesis by chemical stabilization of an imported protein evades the delayed-death phenotype in malaria parasites. mSphere 4:e00710–e00718. https://doi.org/10.1002/iub.583
151. Cilingir G, Lau AOT, Broschat SL (2013) ApicoAP: the first computational model for identifying apicoplast-targeted transmembrane proteins in Apicomplexa. J Microbiol Methods 95:313–319. https://doi.org/10.1371/2Fjournal.pone.0036598
152. Botté CY, McFadden GI, Yamaryo-Botté Y (2018) Isolating the *Plasmodium falciparum* apicoplast using magnetic beads. Methods Mol Biol 1829:205–212. https://doi.org/10.1007/978-1-4939-8654-5_14
153. Sheiner L, Demerly JL, Poulsen N, Beatty WL, Lucas O, Behnke MS et al (2011) A systematic screen to discover and analyze Apicoplast proteins identifies a conserved and essential protein import factor. PLoS Pathog. 2011 ;7. Available from: http://www.ncbi.nlm.nih.gov/pmc/articles/PMC3228799/
154. Barylyuk K, Koreny L, Ke H, Butterworth S, Crook OM, Lassadi I et al (2020) A comprehensive subcellular atlas of the *Toxoplasma proteome* via hyperLOPIT provides spatial context for protein functions. Cell Host Microbe 28:752. https://doi.org/10.1016/j.chom.2020.09.011
155. Denny P, Preiser P, Williamson D, Wilson I (1998) Evidence for a single origin of the 35 kb plastid DNA in Apicomplexans. Protist 149:51–59. https://doi.org/10.1016/s1434-4610(98)70009-4
156. Brayton KA, Lau AOT, Herndon DR, Hannick L, Kappmeyer LS, Berens SJ et al (2007) Genome sequence of *Babesia bovis* and comparative analysis of apicomplexan hemoprotozoa. PLoS Pathog 3:1401–1413. https://doi.org/10.1371/journal.ppat.0030148
157. Garg A, Stein A, Zhao W, Dwivedi A, Frutos R, Cornillot E et al (2014) Sequence and annotation of the apicoplast genome of the human pathogen *Babesia microti*. PloS One 9:e107939. https://doi.org/10.1371/journal.pone.0107939
158. Cai X, Fuller AL, McDougald LR, Zhu G (2003) Apicoplast genome of the coccidian *Eimeria tenella*. Gene 321:39–46. https://doi.org/10.1016/j.gene.2003.08.008
159. Gardner MJ, Bishop R, Shah T, de Villiers EP, Carlton JM, Hall N et al (2005) Genome sequence of *Theileria parva,* a bovine pathogen that transforms lymphocytes. Science 309:134–137. https://doi.org/10.1016/j.gene.2003.08.008
160. Cinar HN, Qvarnstrom Y, Wei-Pridgeon Y, Li W, Nascimento FS, Arrowood MJ et al (2016) Comparative sequence analysis of *Cyclospora cayetanensis* apicoplast genomes originating from diverse geographical regions. Parasit Vectors 9:611. https://doi.org/10.1186/s13071-016-1896-4
161. Imura T, Sato S, Sato Y, Sakamoto D, Isobe T, Murata K et al (2014) The apicoplast genome of *Leucocytozoon caulleryi*, a pathogenic apicomplexan parasite of the chicken. Parasitol Res 113:823–828. https://doi.org/10.1007/s00436-013-3712-9

Index

S. D. Schwartzbach et al. (eds.), *Endosymbiotic Organelle Acquisition*,
https://doi.org/10.1007/978-3-031-57446-7

MIX
Papier aus verantwortungsvollen Quellen
Paper from responsible sources
FSC® C105338

If you have any concerns about our products,
you can contact us on
ProductSafety@springernature.com

In case Publisher is established outside the EU,
the EU authorized representative is:
Springer Nature Customer Service Center GmbH
Europaplatz 3, 69115 Heidelberg, Germany

Printed by Libri Plureos GmbH
in Hamburg, Germany